Calculus I
First Edition

Tunc Geveci

www.cognella.com 800.200.3908

Contents

CONTENTS

Preface

This is the first volume of my calculus series, **Calculus I, Calculus II** and **Calculus III**. This series is designed for the usual three semester calculus sequence that the majority of science and engineering majors in the United States are required to take. Some majors may be required to take only the first two parts of the sequence.

Calculus I covers the usual topics of the first semester: **Limits, continuity, the derivative, the integral and special functions such exponential functions, logarithms, and inverse trigonometric functions.** **Calculus II** covers the material of the second semester: **Further techniques and applications of the integral, improper integrals, linear and separable first-order differential equations, infinite series, parametrized curves and polar coordinates.** **Calculus III** covers topics in **multivariable calculus: Vectors, vector-valued functions, directional derivatives, local linear approximations, multiple integrals, line integrals, surface integrals, and the theorems of Green, Gauss and Stokes.**

An important feature of my book is its **focus on the fundamental concepts, essential functions and formulas of calculus.** Students should not lose sight of the basic concepts and tools of calculus by being bombarded with functions and differentiation or antidifferentiation formulas that are not significant. I have written the examples and designed the exercises accordingly. I believe that "less is more". That approach enables one to demonstrate to the students the beauty and utility of calculus, without cluttering it with ugly expressions. Another important feature of my book is **the use of visualization as an integral part of the exposition.** I believe that the most significant contribution of technology to the teaching of a basic course such as calculus has been the effortless production of graphics of good quality. Numerical experiments are also helpful in explaining the basic ideas of calculus, and I have included such data.

Appendix A and the first two sections of Chapter 1 provide a review of the essential precalculus material that the student should know in order to meet the challenges of calculus. The student should be comfortable with the language and notation that are necessary in order to refer to functions unambiguously. That is why I have included such material in the beginning of the first chapter. The main goal of that chapter is to introduce the concepts of limits and continuity, and to provide the student with the necessary tools that are helpful in the calculation of limits. I find it practical to introduce the concepts of limits and continuity simultaneously in terms of the understanding of the concepts and the evaluation of limits. Thus, my treatment differs from the usual calculus text. I also treat infinite limits and limits at infinity with more care than the usual calculus text, and provide the student with more tools for the evaluation of such limits. The limit of a sequence is treated in this chapter since the language of sequences is convenient in the discussion of Newton's method and the convergence of certain Riemann sums to the integral. The usual calculus text postpones the discussion of sequences to the chapter on infinite series.

Chapter 2 introduces the derivative. I deviate from the usual calculus text by discussing local linear approximations and the differential at an early stage. Indeed, the idea of derivative is intimately linked to local linear approximations with a certain form of the error, and that is how the concept is generalized to functions of several variables. At this point, the early discussion of local linear approximations is helpful in justifying the identification of the derivative of a function with its rate of change and the slope of its graph. Such discussion is also helpful in providing plausibility arguments for the product rule and the chain rule.

Chapter 3 discusses the link between the sign of the derivative, the increasing/decreasing behavior of a function, and its local and global extrema. Unlike the usual text, I do not start by the stating the theorem on the existence of the absolute extrema of a continuous function on a closed and bounded interval and the Mean Value Theorem. Thus, my approach is more practical, and does not give the impression to the student that the only time you can talk about absolute extrema is when you have a continuous function on a closed and bounded interval. I also discuss the link between the second derivative, the increasing/decreasing behavior of the first derivative, and the concavity of a graph.

Chapter 4 introduces special functions such as exponential and logarithmic functions and inverse trigonometric functions. The introduction of inverse functions in the usual text is confusing. My introduction is more practical and more careful at the same time. I postpone the introduction of the exponential and logarithmic functions to this chapter since I find it impossible to motivate the significance of the natural exponential function before introducing the derivative. Besides, powers, sine, cosine, and their combinations are adequate for the illustration of the derivative and its applications prior to this chapter. I discuss the different orders of magnitude of powers, exponential and logarithms independently of L'Hôpital's rule. I find this approach much more illuminating than a mechanical applications of L'Hôpital's rule (that is covered in the last section of the chapter).

Chapter 5 introduces the integral. I introduce the part of the Fundamental Theorem of Calculus which states that

$$\int_a^b F'(x)\,dx = F(b) - F(a)$$

first, since that enables the student to compute many integrals before the introduction of the idea of a function that is defined via an integral. Thus, the student has a better chance of understanding the meaning of the part of the Fundamental Theorem which says that

$$\frac{d}{dx}\int_a^x f(t)dt = f(x)$$

(provided that f is continuous). Many texts introduce both parts of the Theorem suddenly, and do not present them in a way that establishes the link between the derivative and the integral clearly. I find it amusing, but not helpful, when I see a title such as "total change theorem", as if something other than the Fundamental Theorem is involved.

My own preference is to cover special functions before the integral since that makes it possible to provide a richer collection of examples and problems in the following chapter. On the other hand, some people prefer the elegance of the "late transcendentals" approach whereby the natural logarithm is introduced as an integral. The pdf files for the versions of chapter 4 and chapters 5 that introduce the integral before logarithms, exponentials and inverse trigonometric functions will be provided upon request.

Remarks on some icons: I have indicated the end of a proof by ■, the end of an example by □ and the end of a remark by ◊.

Supplements: An **instructors' solution manual** that contains the solutions of all the problems is available as a PDF file that can be sent to an instructor who has adopted the book. The student who purchases the book can access the **students' solutions manual** that contains the solutions of odd numbered problems via **www.cognella.com**.

Acknowledgments: ScientificWorkPlace enabled me to type the text and the mathematical formulas easily in a seamless manner. **Adobe Acrobat Pro** has enabled me to convert the LaTeX files to pdf files. **Mathematica** has enabled me to import high quality graphics to my documents. I am grateful to the producers and marketers of such software without which I would not have had he patience to write and rewrite the material in this volume.

I would also like to acknowledge my gratitude to two wonderful mathematicians who have influenced me most by demonstrating the beauty of Mathematics and teaching me to write clearly and precisely: **Errett Bishop and Stefan Warschawski**.

Last, but not the least, I am grateful to **Simla** for her encouragement and patience while I spent hours in front a computer screen.

Tunc Geveci (tgeveci@math.sdsu.edu)
San Diego, March 2010

Chapter 1

Functions, Limits and Continuity

The first two sections of this chapter review some basic facts about **functions** defined by **rational powers of x, polynomials, rational functions** and **trigonometric functions**. Appendix A contains additional precalculus review material. We will discuss exponential, logarithmic and inverse trigonometric functions in Chapter 4.

The main body of the chapter is devoted to the discussion of the fundamental concepts of continuity and limits. Roughly speaking, a function f is said to be **continuous at a point a** if $f(x)$ approximates $f(a)$ when x is close to a. It may happen that $f(x)$ approximates a specific number L if x is close to a but $x \neq a$, even if f is not defined at a, or irrespective of the value of f at a. The relevant concept is **the limit of f at a**. We will also discuss **infinite limits, limits at infinity** and **the limits of sequences**.

1.1 Powers of x, Sine and Cosine

We will deal with a variety of functions in calculus. We will begin by reviewing the relevant terminology and notation. Then we will review some basic facts about $\sin(x)$, $\cos(x)$ and rational powers of x. These are the building blocks for a rich collection of functions.

Terminology and Notation

Definition 1 Let D be a subset of the set of real numbers $\mathbb{R}$. A real-valued **function** of a real variable with **domain** D is a rule that assigns to each element of D a unique real number.

We may refer to a function by a letter such as f or g. Some functions are special since they occur frequently. Such functions have names such as sine or cosine, and their names have specific abbreviations, such as sin or cos. We will denote an arbitrary element of the domain of a function by a letter such as x or t. If we choose the letter x, and the function in question is f, then x is **the independent variable of** f. The unique real number that is assigned by f to x is **the value of f at x** and is denoted by $f(x)$ (read "f of x"). The value of a special function will be denoted by using the abbreviation reserved for that function. For example, the value of the sine function at x will be denoted by $\sin(x)$. If we set $y = f(x)$, then y is **the dependent variable of** f. We can refer to a function f by the letter that denotes the dependent variable and set $y = y(x) \, (= f(x))$.

Example 1 Let

$$f(x) = \frac{1}{x} \text{ where } x \neq 0.$$

We can express the domain of f as $(-\infty, 0) \cup (0, +\infty)$, the union of the interval $(-\infty, 0)$ and $(0, +\infty)$ (Section A3 of Appendix A contains a review of the number line and intervals). The independent variable is x. If we set $y = 1/x$, the dependent variable of f is y. We can replace x by any nonzero real number to obtain the corresponding value of the function. For example, the value of f at $\sqrt{2}$ is

$$f(\sqrt{2}) = \frac{1}{\sqrt{2}} \cong 0.707107,$$

rounded to 6 significant digits (we count the number of significant digits of a decimal starting with the first nonzero digit, as discussed in Section A4 of Appendix A). Thus, the value of the dependent variable y that corresponds to the value $\sqrt{2}$ of the independent variable is $1/\sqrt{2}$. $\square$

Example 2 Assume that a car is traveling at a constant speed of 60 miles per hour. If we denote time by t (in hours), the distance s covered by the car in t hours is $60t$ miles. Let us set $s = f(t) = 60t$. The letter t denotes the independent variable and can be assigned any nonnegative real number. Thus, the domain of f is the set of all nonnegative real numbers and can be expressed as the interval $[0, \infty)$. The letter s denotes a dependent variable. $\square$

Example 3 The surface area A of a sphere of radius r is $4\pi r^2$. Let us set $A = g(r) = 4\pi r^2$ for any $r > 0$. The domain of the function g is the interval $(0, +\infty)$. The independent variable is r and the dependent variable is A. We may choose to refer to the function by the letter that denotes the dependent variable, and set $A(r) = 4\pi r^2$. $\square$

Example 4 Let

$$f(x) = \begin{cases} x & \text{if} \quad x < 1, \\ x+1 & \text{if} \quad x \geq 1. \end{cases}$$

We have defined a function whose domain is the set of real numbers $\mathbb{R}$, since the above rule assigns a unique value to any $x \in \mathbb{R}$, even though the expression for $f(x)$ is not the same for each $x \in \mathbb{R}$. For example, $f(-1) = -1$ since $-1 < 1$, $f(1) = 1 + 1 = 2$, and $f(2) = 2 + 1 = 3$ since $2 > 1$. We will refer to such a function as a **piecewise defined function**. $\square$

Eventually, we will consider relationships between variable entities that need not be real numbers. For example, the variable in question can be a point whose coordinates in the Cartesian coordinate plane are real-valued functions of a real variable. **We will speak of real-valued functions of a real variable simply as functions**, until we consider more general relationships.

Assume that $f(x)$ is a single expression for each x. **The natural domain** of f consists of all $x \in \mathbb{R}$ such that the expression $f(x)$ is a real number. We may refer to "the function $f(x)$". In this case, it should be understood that the domain of the function f is its natural domain. For example, if $f(x) = 1/x$, the natural domain of f consists of all nonzero real numbers. We may refer to f as "the function $1/x$".

The graph of a function is very helpful in visualizing the relationship between the dependent and independent variable. If x denotes the independent variable of the function f and y denotes the dependent variable, the graph of f is the subset of the xy-plane that consists of the points (x, y) where x is in the domain of f and $y = f(x)$:

$$\{(x, f(x) : x \text{ is in the domain of } f\}.$$

Example 5 Let $f(x) = 1/x$, as in Example 1. In the xy-plane, the graph of f consists of all points (x, y) such that $x \neq 0$ and $y = 1/x$. We can express the graph of f as

$$\left\{ \left(x, \frac{1}{x}\right) : x \neq 0 \right\}.$$

Since $1/x$ attains values of arbitrarily large magnitude if x is near 0, a graphing utility shows us only part of the graph of f corresponding to an interval that contains 0. We will say that **the viewing window is** $[a, b] \times [c, d]$ if x and y are restricted to the intervals $[a, b]$ and $[c, d]$, respectively. Figure 1 shows the part of the graph of f in the viewing window $[-3, 3] \times [-5, 5]$. We will usually refer to such a picture simply as the graph of the relevant function. $\square$

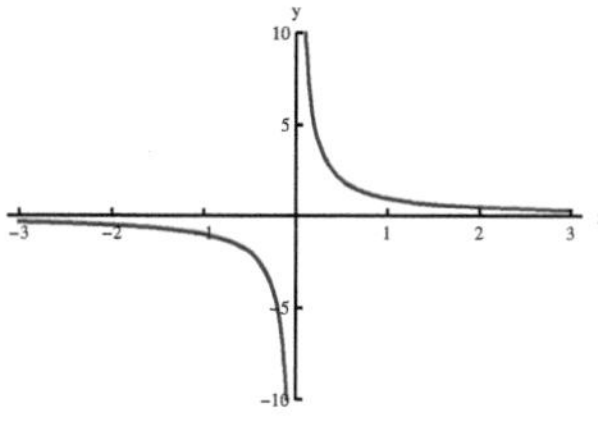

Figure 1

You are already familiar with the graphs of certain types of functions from precalculus courses. You should be able to provide rough sketches of functions that are not too complicated. In fact, calculus will provide you with certain tools that will enable you to come up with sketches that reflect the behavior of many functions correctly. A **graphing calculator** or a software package such as **Maple, Mathematica, Matlab** or **MuPAD** is helpful in obtaining the graph a function. We will refer to such a device as a **graphing utility** if we wish to emphasize its graphing capabilities, and as a **computational utility** if we wish to emphasize its computational capabilities. If a particular example or exercise requires both computational and graphical capabilities, we may simply refer generically to **"a calculator"**. Maple, Mathematica and MuPAD are **computer algebra systems**, i.e., they can work with symbols such as x and y, as well as with numbers (Matlab has a "symbolic toolbox" that is powered by Maple). We will use the abbreviation "**CAS**" when we refer to computer algebra systems.

Figure 1 was generated with the help of a graphing utility. Note that the scale on the vertical axis is not the same as the scale on the horizontal axis. That will be the case in many of the graphs that will be displayed in this book, since a graphing utility adjusts the scales on the coordinate axes automatically for better viewing. We will impose equal scales on the axes if that serves a definite purpose. Also note that we have labeled both axes in Figure 1. We may not label the vertical axis in some cases where we do not assign a letter to the dependent variable of the relevant function.

Example 6 Let

$$f(x) = \begin{cases} x & \text{if } x < 1, \\ x + 1 & \text{if } x \geq 1, \end{cases}$$

as in Example 4. Figure 2 shows the graph of f. In Figure 2, the point $(1, 2)$ is shown as a black dot in order to indicate that it is on the graph of f (we have $f(1) = 2$). $\square$

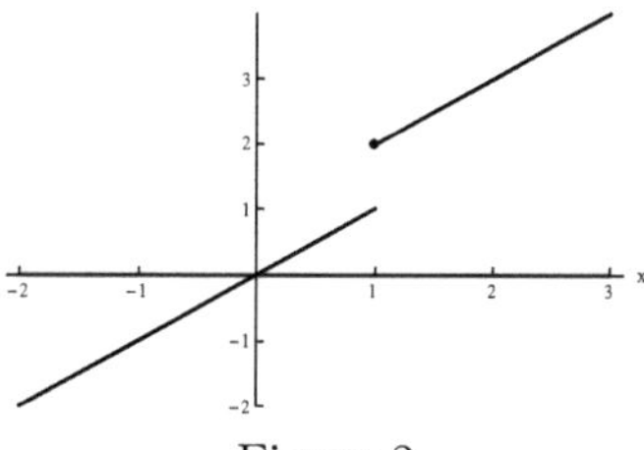

Figure 2

In Section A5 of Appendix A we review the graphs of some equations in the xy-plane. If $f(x)$ is defined by the same equation for each x in its domain, the graph of f in the xy-plane is the same as the graph of the equation $y = f(x)$. On the other hand, it is not true that the graph of an equation is always the graph of a function. **The vertical line test** is helpful in determining whether a graph is the graph of a function:

Assume that f is a function. Given a vertical line $x = a$, either the graph of f has no point on the line, or there is a single point where the graph of f and the line $x = a$ intersect.

Indeed, if a is not in the domain of f then f does not assign a value to a, so that the graph of f does not have any point on the line $x = a$. If a is in the domain of f then f assigns the *unique* value $f(a)$ to a, so that the only point on the line $x = a$ that belongs to the graph of f is $(a, f(a))$.

Example 7 The graph of the equation $x = y^2$ in the xy-plane is a parabola, as in Figure 3. For each $a > 0$, the equation $a = y^2$ has two distinct solutions, $y = \sqrt{a}$ and $y = -\sqrt{a}$. Therefore, the vertical line $x = a$ intersects the parabola $x = y^2$ at the distinct points $(a, \sqrt{a})$ and $(a, -\sqrt{a})$. Thus, the graph of the equation $x = y^2$ fails the vertical line test, and cannot be the graph of a function of x. $\square$

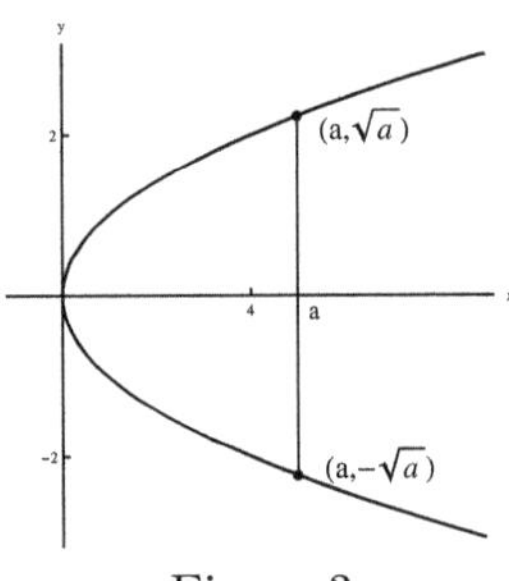

Figure 3

Remark 1 (Warning: Spurious line segments) By the vertical line test, the graph of a function cannot contain vertical line segments. On the other hand, a graphing utility may produce pictures that contain line segments which appear to be vertical.

Let

$$f(x) = \begin{cases} x & \text{if} \quad x < 1, \\ x + 1 & \text{if} \quad x \geq 1, \end{cases}$$

as in Example 6. Figure 4 shows another computer generated graph of f. $\Diamond$

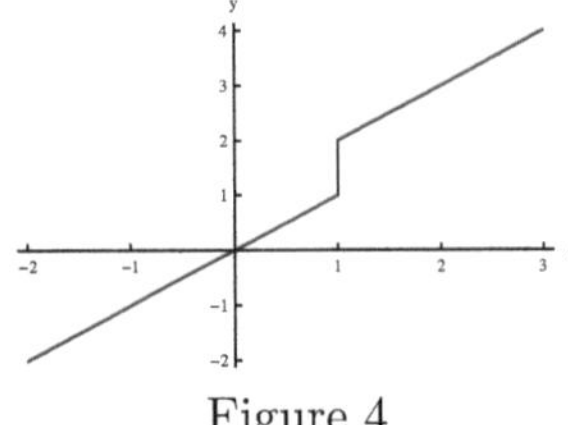

Figure 4

In Figure 4 it appears as if the vertical line segment that connects the points $(1,1)$ and $(1,2)$ is part of the graph of f, and that is not the case. We will refer to such a line as a **spurious line segment**. The only point on that line segment that belongs to the graph of f is the point $(1,2)$. We can explain the picture as follows: The graphing utility that produced Figure 4 sampled a value of x slightly to the left of 1, another value of x slightly to the right of 1, and connected the corresponding points by a line segment. One point is very close to $(1,1)$ and the other point is very close to $(1,2)$. The spurious line segment is the line segment that joins these points. Figure 2 was produced by a graphing utility that makes special provision for the sudden jump in the value of the function at $x = 1$. Some pictures in this book may contain spurious line segments, just as the pictures that are produced by your graphing calculator. As long as we interpret the pictures correctly, there should be no misunderstanding. $\Diamond$

We should be clear about the meaning of **the equality of functions**: We say that **the function f is equal to the function g and write $f = g$ if the domain of f and the domain of g are the same, and $f(x) = g(x)$ for each x in the common domain.**

Example 8 Determine whether $f = g$ if

a) $f(r) = 4\pi r^2$ for any $r \in \mathbb{R}$, and $g(x) = 4\pi x^2$ for any $x \in \mathbb{R}$,

b) $f(x) = 4\pi x^2$ for any $x \in \mathbb{R}$ and $g(x) = 4\pi x^2$ for any $x \geq 0$.

Solution

a) The domain of g is the same as the domain of f and consists of all real numbers. Since

$$g(x) = 4\pi x^2 = f(x)$$

for each $x \in \mathbb{R}$, we have $g = f$, even though we have used different letters to denote the independent variables.

b) Even though $g(x) = f(x)$ for each $x \geq 0$, the function g is not equal to the function f, since the domain of g is the set of nonnegative real numbers $[0, +\infty)$, whereas the domain of f is the set of all real numbers $\mathbb{R}$. The graph of g is part of the graph of f, as shown in Figure 5. $\square$

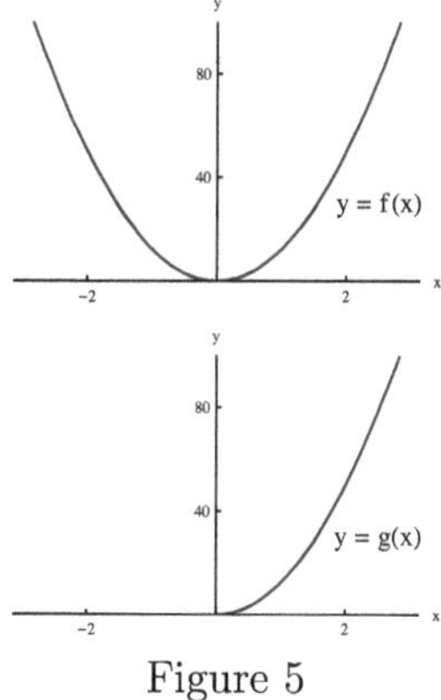

Figure 5

Example 8 illustrates the **restriction** of a function: The function g is the **restriction of the function f to the set S** if the domain of g is S, the set S is contained in the domain of f, and $g(x) = f(x)$ for each $x \in S$. In this case, f is said to be an **extension** of g. Thus, if f and g are as in part b) of Example 8, then g is the restriction of f to $[0, +\infty)$. The function f is an extension of g.

We have many uses for the absolute value, as reviewed in Section A3 of Appendix A. Let us look at the relevant function:

Example 9 (The absolute-value function) Let

$$f(x) = |x| = \begin{cases} x & \text{if } x \geq 0, \\ -x & \text{if } x < 0. \end{cases}$$

The graph of the absolute-value function f is shown in Figure 6. Since $f(x) = x$ for each $x \geq 0$, the graph of f on the interval $[0, +\infty)$ coincides with the line $y = x$. Since $f(x) = -x$ for each $x < 0$, the graph of f on the interval $(-\infty, 0)$ coincides with the line $y = -x$. $\square$

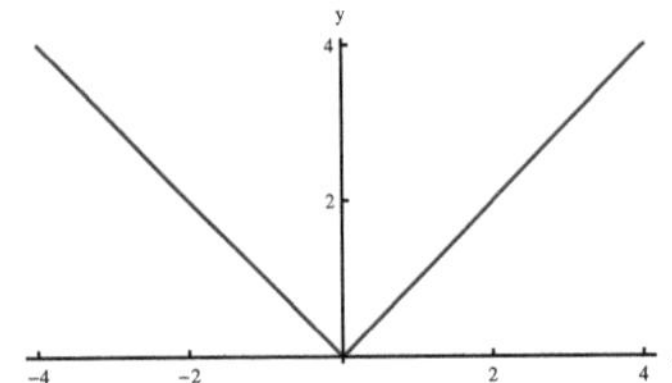

Figure 6: The absolute-value function

Rational Powers of x

Functions defined by rational powers of x are building blocks for a rich family of functions, as we will see throughout calculus. Let's review some basic facts about such functions and look at some samples.

Let $f(x) = x^r$, where the **exponent** r is a rational number. **The rules for exponents** should be familiar. If a and b are rational numbers, we have

$$\begin{aligned} x^a x^b &= x^{a+b}, \\ x^0 &= 1, \\ x^{-a} &= \frac{1}{x^a}, \\ (x^a)^b &= x^{ab}, \end{aligned}$$

provided that the expressions are defined.

If $f(x) = x^n$, where n is a positive integer, $f(x)$ is defined for each $x \in \mathbb{R}$, so that we can identify the natural domain of f with the entire number line.

If $f(x) = x$, then f is a linear function. The graph of f is a line with slope 1 that passes through the origin. Since the value of f at x is the same as x, f can be referred to as **the identity function**.

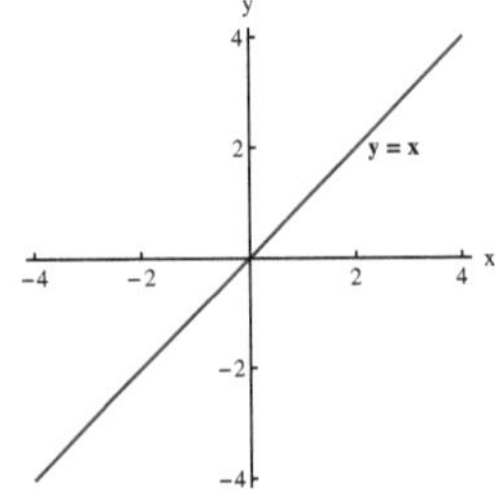

Figure 7: The identity function

If $f(x) = x^2$, then f is a quadratic function, and the graph of f is a parabola. The function is a prototype of the functions defined by x^n, where n is an even positive integer. Figure 8 shows the graphs of $y = x^2$ and $y = x^4$.

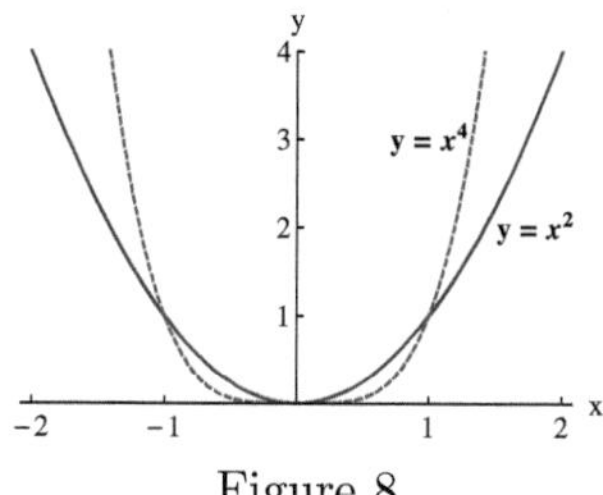

Figure 8

The function defined by x^3 is a prototype of functions defined by x^n, where n is an odd integer and $n \geq 3$. Figure 9 shows the graphs of $y = x^3$ and $y = x^5$.

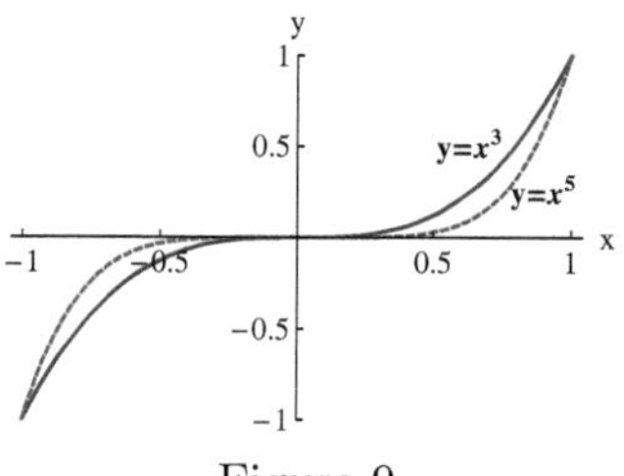

Figure 9

A function f is an **even function** if $f(x) = f(-x)$ for each x in the domain of f. A function f is an **odd function** if $f(-x) = -f(x)$ for each x in the domain of f. **The graph of an even function is symmetric with respect to the vertical axis, and the graph of an odd function is symmetric with respect to the origin.** Indeed, if f is even and (x, y) is on the graph of f, we have $y = f(x) = f(-x)$. Therefore, $(-x, y)$ is also on the graph of f, and the points (x, y) and $(-x, y)$ are symmetric with respect to the vertical axis. If f is odd and (x, y) is on the graph of f, then $y = f(x) = -f(-x)$, so that $-y = f(-x)$. Therefore, $(-x, -y)$ is also on the graph of f, and the points (x, y) and $(-x, -y)$ are symmetric with respect to the origin.

If $f(x) = x^n$, where n is an even positive integer, then $f(-x) = (-x)^n = x^n = f(x)$, so that f is an even function. Therefore, the graph of f is symmetric with respect to the vertical axis. Figure 8 that shows the graphs of $y = x^2$ and $y = x^4$ is consistent with that fact.

If $f(x) = x^n$, where n is an odd positive integer, then $f(-x) = (-x)^n = -x^n = -f(x)$, so that f is an odd function. Therefore, the graph of f is symmetric with respect to the origin. Figure 9 that shows the graphs of $y = x^3$ and $y = x^5$ is consistent with that fact.

If n is a positive integer, and

$$f(x) = x^{-n} = \frac{1}{x^n},$$

the natural domain of f consists of all real numbers x such that $x \neq 0$. We can identify the domain of f with the union of the open intervals $(-\infty, 0)$ and $(0, +\infty)$. Such a function is odd if n is odd and even if n is even.

Figure 10 shows the graph of f, where $f(x) = x^{-1} = 1/x$.

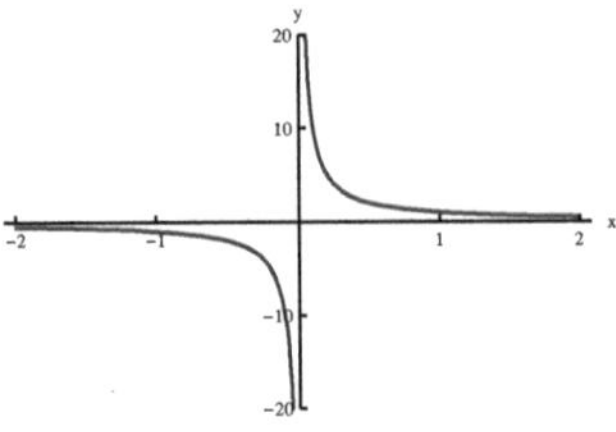

Figure 10

Note that the graph of f is symmetric with respect to the origin, consistent with the fact that f is an odd function. Also note that $|1/x|$ becomes arbitrarily large if we imagine that x takes on values that are closer and closer to 0. The function f is a prototype of functions defined by x^{-n}, where n is an odd positive integer.

Figure 11 shows the graph of g, where $g(x) = x^{-2} = 1/x^2$.

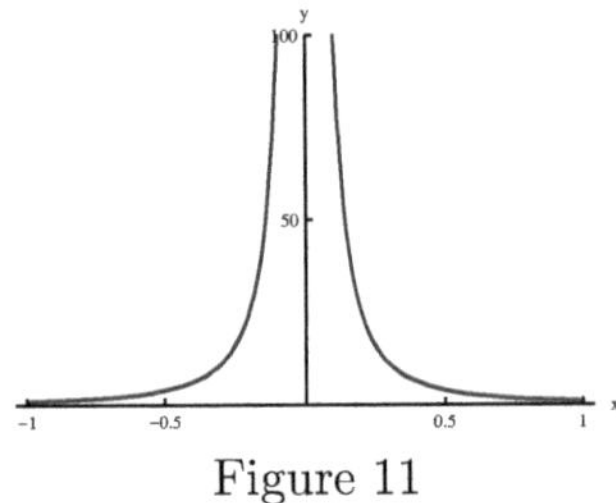

Figure 11

Note that the graph of g is symmetric with respect to the vertical axis, consistent with the fact that g is an even function. Also note that $1/x^2$ becomes arbitrarily large if we imagine that x takes on values that are closer and closer to 0. The function g is a prototype of functions defined by x^{-n}, where n is an even positive integer.

If n is an even positive integer and $x \geq 0$, we set

$$y = x^{1/n} = \sqrt[n]{x} \text{ if } x = y^n \text{ and } y \geq 0.$$

Thus, the natural domain of a function defined by $x^{1/n}$ is the interval $[0, +\infty)$ if n is an even positive integer.

The **square-root function** defined by $x^{1/2}$ is a prototype of functions defined by $x^{1/n}$, where n is an even positive integer:

$$y = x^{1/2} = \sqrt{x} \text{ if } x = y^2 \text{ and } y \geq 0.$$

Figure 12 shows the graphs of $y = x^{1/2}$ and $y = x^{1/4}$.

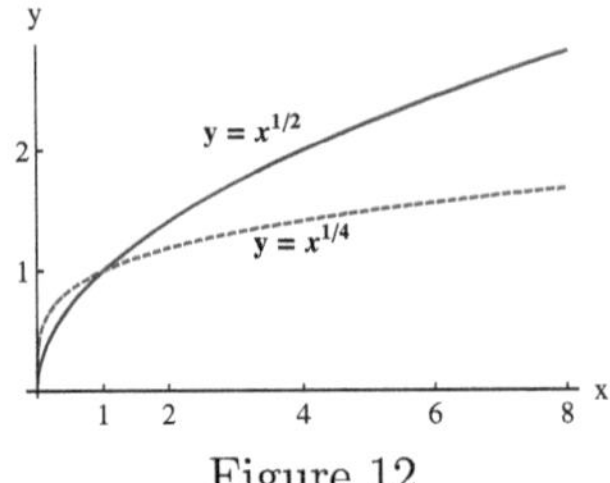

Figure 12

If n is an odd positive integer, we set

$$y = x^{1/n} = \sqrt[n]{x} \text{ if } x = y^n.$$

Here, x can be an arbitrary real number. Thus, the natural domain of a function defined by $x^{1/n}$, where n is an odd positive integer, is the entire number line.

The **cube-root function** defined by $x^{1/3}$ is a prototype of functions defined by $x^{1/n}$, where n is an odd positive integer:

$$y = x^{1/3} = \sqrt[3]{x} \text{ if } x = y^3.$$

Figure 13 shows the graphs of $y = x^{1/3}$ and $y = x^{1/5}$. Note that the graphs are symmetric with respect to the origin, since the underlying functions are odd (confirm).

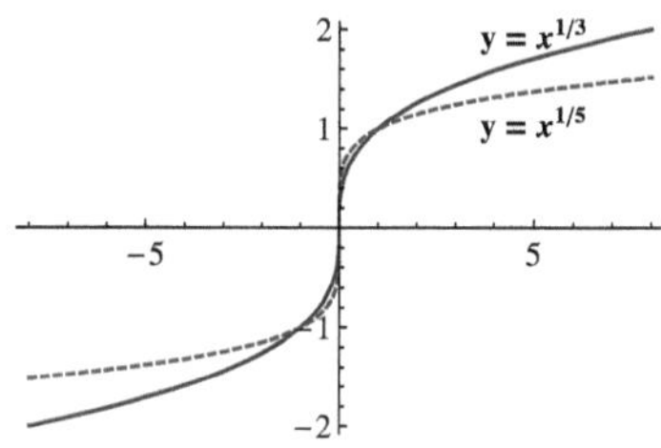

Figure 13

If n is a positive integer, and

$$f(x) = x^{-1/n} = \frac{1}{x^{1/n}},$$

then $f(x)$ is defined for $x > 0$ if n is even, and for each $x \neq 0$ if n is odd. Therefore, if n is even the natural domain of f is $(0, +\infty)$, and if n is odd the natural domain of f is $(-\infty, 0) \cup (0, +\infty)$. Figure 14 shows the graph of $y = x^{-1/2}$.

In general, graphing utilities do not picture the part of the graph of f near the vertical axis very well. The picture may give the impression that part of the graph coincides with the vertical axis. This is not the case, of course. The function is not defined at 0. In fact, $f(x)$ takes on arbitrarily large values if x is close to 0.

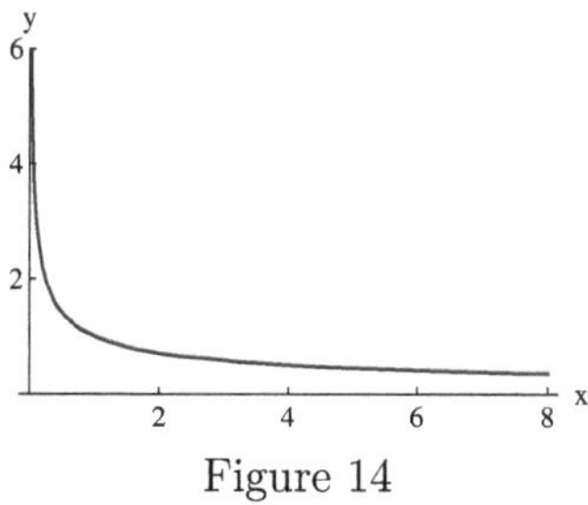

Figure 14

If r is an arbitrary rational number, we can express r as m/n, where m is an integer, n is a positive integer, and $|m|$ and n do not have a common factor other than 1. Thus,

$$x^r = x^{m/n} = \left(x^{1/n}\right)^m.$$

For example, if we set

$$f(x) = x^{2/3} = \left(x^{1/3}\right)^2,$$

then $f(x)$ is defined for each $x \in \mathbb{R}$. The function is even since

$$f(-x) = (-x)^{2/3} = \left((-x)^{1/3}\right)^2 = \left(-x^{1/3}\right)^2 = \left(x^{1/3}\right)^2 = x^{2/3} = f(x).$$

Figure 15 shows the graph of f.

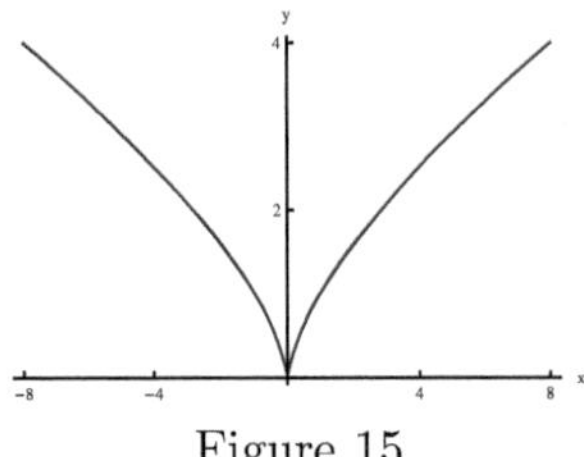

Figure 15

Sine and Cosine

Let's begin by reviewing the meaning of **radian measure.** Consider the circle that is centered at the origin and has radius 1. We will refer to this circle as **the unit circle.** The unit circle is the graph of the equation $u^2 + v^2 = 1$ in the uv-plane. We associate a unique point on the unit circle with each real number x as follows: If $x = 0$, the associated point is $A = (1, 0)$. Imagine a particle that travels on the unit circle, starting at the point A. If $x > 0$, we associate with x the point P that the particle reaches by traveling the distance x along the unit circle in the counterclockwise direction. With reference to Figure 16, x is the radian measure of the angle AOP. Thus, the radian measure of the angle AOP is the length of the arc AP.

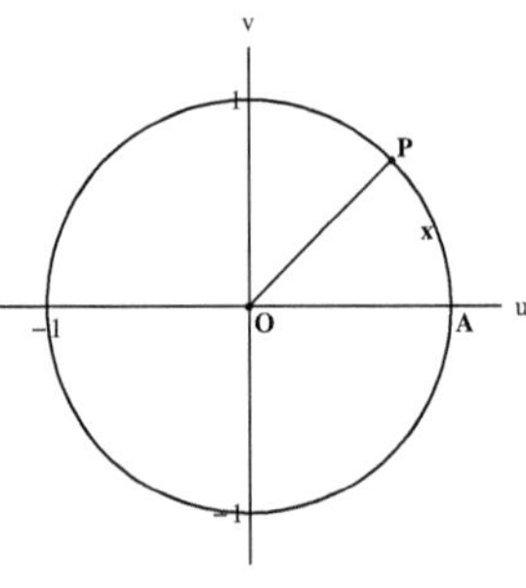

Figure 16

If $x < 0$, imagine that the particle travels the distance $|x| = -x$ along the unit circle in the clockwise direction and reaches the point P. We associate the point P with the number x. With reference to Figure 17, x is the radian measure of the angle AOP.

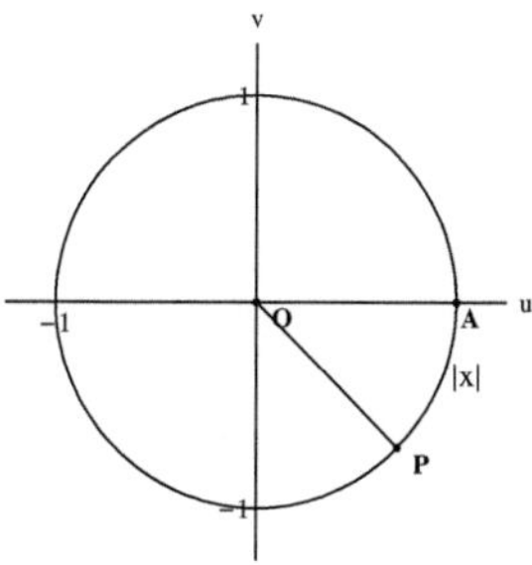

Figure 17

The length of the unit circle is 2π. Thus, the point that is associated with $x + 2n\pi$, where $n = 0, \pm 1, \pm 2, \ldots$, is the same as the point that is associated with x.

In everyday usage, we refer to angles in degrees. Since one revolution around the unit circle corresponds to $360°$, if the radian measure of an angle is x and the degree measure of an angle is $\theta°$, we have the relationship

$$\frac{\theta°}{x} = \frac{360}{2\pi} = \frac{180}{\pi}.$$

In particular, $180°$ corresponds to π radians, $90°$ corresponds to $\pi/2$ radians, $60°$ corresponds to $\pi/3$ radians, $45°$ corresponds to $\pi/4$ radians, and $30°$ corresponds to $\pi/6$ radians. **In this book, "the default measure" for angles will be radians, so that if we refer to "the angle x", it should be understood that x is in radians, unless stated otherwise.** There will be ample opportunity for you to appreciate the fact that the radian measure is the natural measure for angles in calculus.

The calculation of the coordinates of the point on the unit circle that is associated with a real number x leads to the trigonometric functions sine and cosine:

Definition 2 If $P = (u, v)$ is the point on the unit circle that is associated with the real number x, we define the horizontal coordinate u to be the value of the function **cosine** at x, and the vertical coordinate v as the value of function **sine** at x. We abbreviate sine as sin and cosine as cos, so that $P = (\cos(x), \sin(x))$.

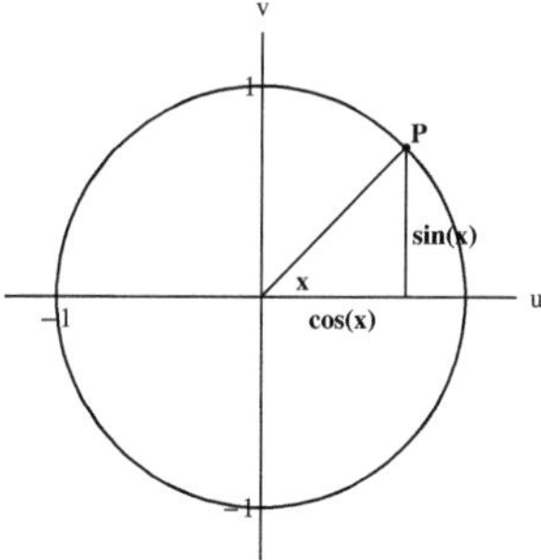

Figure 18: The definition of $\sin(x)$ and $\cos(x)$

The notations $\cos(x)$ and $\sin(x)$ are consistent with notation $f(x)$ that denotes the value of a function f at x. Traditionally, $\cos x$ and $\sin x$ denote $\cos(x)$ and $\sin(x)$, respectively. The traditional imprecise notation may lead to confusion in describing functions such as $\sin(x^2)$. Besides, a calculator will demand precise syntax. Therefore, the notations $\cos(x)$ and $\sin(x)$ will be used in this book, and you are encouraged to do the same.

We have

$$-1 \leq \sin\left(x\right) \leq 1 \text{ and } -1 \leq \cos\left(x\right) \leq 1$$

for each real number x. This follows from the geometric definitions of the functions: If $P = \left(\cos\left(x\right), \sin\left(x\right)\right)$, the horizontal and vertical coordinates of the point P are between -1 and 1, since P is on the unit circle.

The functions sine and cosine are **periodic functions**. First, let us make sure that we agree on the meaning of periodicity:

Definition 3 A function f is said to be **periodic with period** p $(p \neq 0)$ if $f(x+p) = f(x)$ for each real number x.

If f is periodic with period p, we have $f(x \pm np) = f(x+p)$ for $n = 1, 2, 3, \ldots$, so that $\pm np$ is also a period of f. The smallest positive period of f is referred to as its **fundamental period**.

The functions sine and cosine are periodic with period 2π. Indeed, let $P = \left(\cos(x), \sin(x)\right)$ be the point on the unit circle that is associated with the real number x. The point P is associated with $x + 2\pi$, as well, since the particle that travels on the unit circle comes back to the same point after a full revolution around the unit circle. Therefore,

$$\left(\cos\left(x + 2\pi\right), \sin\left(x + 2\pi\right)\right) = \left(\cos\left(x\right), \sin\left(x\right)\right),$$

so that $\cos\left(x + 2\pi\right) = \cos\left(x\right)$ and $\sin\left(x + 2\pi\right) = \sin\left(x\right)$.

Since sine and cosine are periodic with period 2π, the shape of the graph of each function on an interval of the form $\left[-4\pi, -2\pi\right]$, $\left[-2\pi, 0\right]$ or $\left[2\pi, 4\pi\right]$ is the same as the shape of the graph of the function on $\left[0, 2\pi\right]$. We may choose any interval of length 2π to discuss an issue pertaining to sine or cosine. The most popular choices are the intervals $\left[0, 2\pi\right]$ and $\left[-\pi, \pi\right]$. Figure 19 shows the graphs of sine and cosine on the interval $\left[-2\pi, 2\pi\right]$.

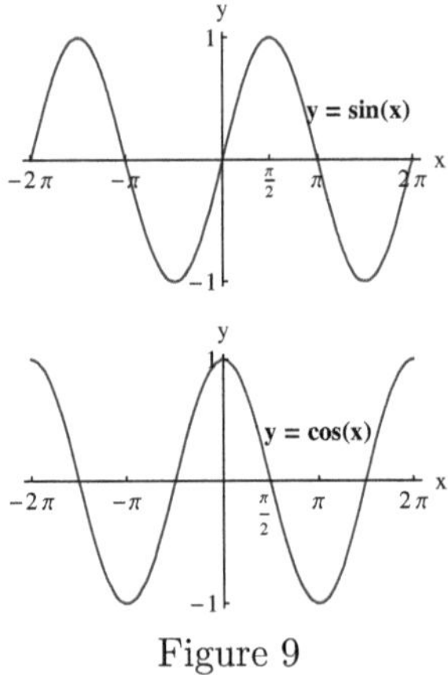

Figure 9

Figure 19 indicates that the graph of sine is symmetric with respect to the origin, and the graph of cosine is symmetric with respect to the vertical axis. These graphical observations reflect "the symmetry" of each function:

Sine is an odd function and cosine is an even function, i.e.,

$$\sin(-x) = -\sin(x) \text{ and } \cos(-x) = \cos(x)$$

for each real number x.

Figure 20 illustrates these facts for x between 0 and $\pi/2$.

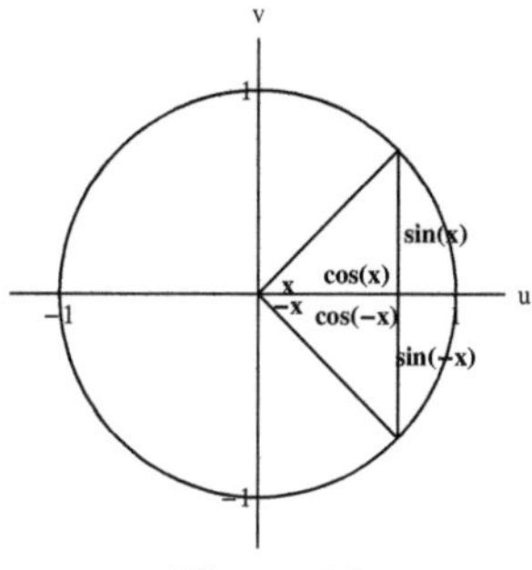

Figure 20

The functions sine and cosine are "built-in" functions in any calculator that you may be using in conjunction with calculus, in the sense that the calculator is equipped with the capability to compute $\sin(x)$ and $\cos(x)$ accurately and fast. A computer algebra system can compute the exact values of $\sin(x)$ and $\cos(x)$ if x is a special angle such as $\pi/3$ or $\pi/6$. In general, a calculator will provide you with approximations to $\sin(x)$ and $\cos(x)$. Your own accuracy requirements, and the capability of your calculator may vary. As discussed in Section A4 of Appendix A, the relevant numbers in this book have been calculated with the help of a calculator that bases its calculations on rounding decimals to 14 significant digits.

Example 10 Determine the points on the unit circle that are associated with $\pm\pi/6$. Sketch the unit circle and indicate the location of each point on the unit circle.

Solution

As discussed in Section A6 of Appendix A,

$$\cos\left(\frac{\pi}{6}\right) = \frac{\sqrt{3}}{2} \text{ and } \sin\left(\frac{\pi}{6}\right) = \frac{1}{2}.$$

Therefore, the point P that is associated with $\pi/6$ is

$$\left(\cos\left(\frac{\pi}{6}\right), \sin\left(\frac{\pi}{6}\right)\right) = \left(\frac{\sqrt{3}}{2}, \frac{1}{2}\right).$$

Since cosine is an even function and sine is an odd function, we have

$$\cos\left(-\frac{\pi}{6}\right) = \cos\left(\frac{\pi}{6}\right) = \frac{\sqrt{3}}{2} \text{ and } \sin\left(-\frac{\pi}{6}\right) = -\sin\left(\frac{\pi}{6}\right) = -\frac{1}{2}.$$

Therefore, point Q that is associated with $-\pi/6$ is

$$\left(\cos\left(-\frac{\pi}{6}\right), \sin\left(-\frac{\pi}{6}\right)\right) = \left(\frac{\sqrt{3}}{2}, -\frac{1}{2}\right).$$

The points P and Q are shown in Figure 21. $\square$

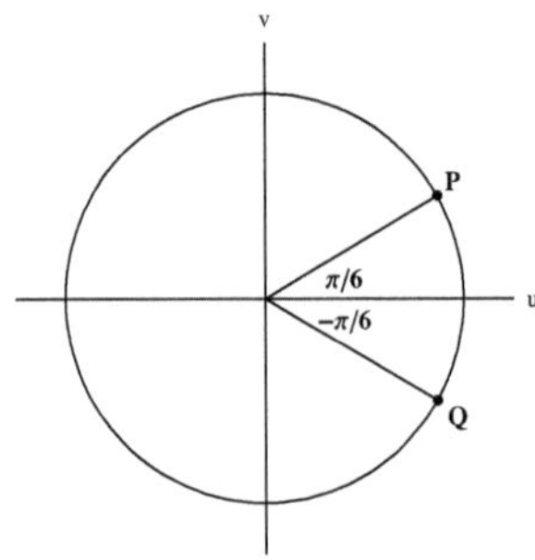

Figure 21

You can find more examples and a review of the laws of sine and cosine in Section A6 of Appendix A.

In the next section we will see examples of functions that are constructed from $\cos(x)$, $\sin(x)$ and rational powers of x by arithmetic operations and compositions.

Problems

1. Let V be the volume of a spherical balloon of radius r. Express V as a function of r. What is the volume of such a balloon of radius 40 cm.?

2. Let V be the volume of a right circular cylinder of height 20 inches and base radius r inches. Express V as a function r. What is the volume of such a cylinder of base radius 5 inches?.

3. Let A be the area of the lateral surface of a right circular cylinder of base radius 50 cm. and height h cm. Express A as a function of h. What is the lateral area of such a cylinder of height 40 cm?

4. Let V be the volume of a cone with a circular base of radius 2 meters and height h meters. Express V as a function of h. What is the volume of such a cone of height 5 meters?

In problems 5 and 6, determine whether the given recipe defines a function. Justify your response.

5.
$$f(x) = \begin{cases} x^2 & \text{if } x \le 2, \\ 2x + 1 & \text{if } x \ge 2. \end{cases}$$

6.
$$f(x) = \begin{cases} -2x & \text{if } x \le 1, \\ x - 3 & \text{if } x \ge 1. \end{cases}$$

In problems 7 and 8, sketch the graph of the given piecewise defined function.

7.
$$f(x) = \begin{cases} 3x + 1 & \text{if } x < 1, \\ x - 1 & \text{if } x \ge 1. \end{cases}$$

8.
$$f(x) = \begin{cases} x^2 + 4 & \text{if } x < 0, \\ -x^2 + 1 & \text{if } x \ge 0. \end{cases}$$

In problems 9 - 18, determine **the natural domain** D of the given function. Express the natural domain as a union of intervals.

9.
$$f(x) = 1 + x + \frac{1}{2}x^2 + \frac{1}{6}x^3 + \frac{1}{24}x^4$$

10.
$$f(x) = \frac{4 + x^2}{9 - x^2}$$

11.
$$f(x) = \frac{1 + x + x^2 + x^3}{(x^2 + 1)^2 \, (x^2 - 1)}$$

12.
$$f(x) = \sqrt{x - 3}$$

13.
$$f(x) = \sqrt{4 - x^2}$$

14.
$$f(x) = \frac{1}{x^{1/3}}$$

15.
$$f(x) = \frac{1}{x^{3/4}}$$

16.
$$f(x) = (4 - x^2)^{2/3}$$

17.
$$f(x) = (x^2 - 9)^{3/4}$$

18.
$$f(x) = \sin(x) + \frac{1}{3}\sin(3x) + \frac{1}{5}\sin(5x)$$

In problems 19 - 22, show that the graph of the given equation is not the graph of a function of x. Sketch the graphs.

19.
$$x = y^2 - 4y + 7$$

20.
$$x^2 - 4x + y^2 - 2y + 1 = 0$$

21.
$$\frac{x^2}{9} + \frac{y^2}{4} = 1$$

22.
$$\frac{x^2}{9} - \frac{y^2}{4} = 1$$

In problems 23 and 24, make use of your graphing utility to plot the graph of the given function. Does the picture include spurious line segments that appear to be vertical?

23.
$$f(x) = \begin{cases} 4x & \text{if} \quad x < 3, \\ 2x & \text{if} \quad x > 3. \end{cases}$$

24.
$$f(x) = \frac{1 - x^2}{x^2 + x - 6}$$

25. Let $f(x) = x^2 - 4x + 7$ for each $x \in \mathbb{R}$ and let g be the restriction of f to the interval $[2, +\infty)$.
a) Sketch the graphs of f and g.
b) Show that g is an increasing function. Is your sketch consistent with this fact?

26. Let $f(x) = x^2 + 6x + 13$ for each $x \in \mathbb{R}$ and let g be the restriction of f to the interval $(-\infty, -3]$..
a) Sketch the graphs of f and g.
b) Show that g is a decreasing function. Is your sketch consistent with this fact?

27 [C] Make use of your graphing utility to plot the graphs of the following functions for $x \in [-1, 1]$. Are the pictures consistent with the fact that the functions are odd?
a) x^5

b) $\dfrac{1}{x^3}$

28. [C] Make use of your graphing utility to plot the graphs of the following functions for $x \in [-1, 1]$. Are the pictures consistent with the fact that the functions are even?
a) x^6

b) $\dfrac{1}{x^4}$

29. Let g be the restriction of the sine function to the interval $[-\pi/2, \pi/2]$.
a) Sketch the graph of g. Is your sketch consistent with the fact that g is an increasing function?
b) Is g an odd function or an even function? Is your sketch consistent with your response?

30. Let g be the restriction of the cosine function to the interval $[0, \pi]$.
a) Sketch the graph of g. Is your sketch consistent with the fact that g is a decreasing function?
b) Does it make sense to enquire whether g is an even or odd function?

31. Sketch the point on the unit circle that correspond to x (radians). Determine $\cos(x)$ and $\sin(x)$. Is the picture consistent with the fact that sine is an odd function and cosine is an even function?

a) $x = \pi/3$ and $x = -\pi/3$

b) $x = 5\pi/6$ and $x = -5\pi/6$

32. [C] Make use of your computational utility to calculate $\sin(x)$ and $\cos(x)$ (x is in the default mode of radians). Round decimals to 6 significant digits, as discussed in Appendix A4. Are the numbers consistent with the fact that sine is an odd function and cosine is an even function?

a) $x = 1$ and $x = -1$

b) $x = 2.5$ and $x = -2.5$

33. [C] Make use of your computational utility to calculate $\sin(x)$ and $\cos(x)$ (x is in the default mode of radians). Round decimals to 6 significant digits. Are the numbers consistent with the fact that sine and cosine are periodic functions with period 2π?

a) $x = 1$ and $x = 1 + 4\pi$

b) $x = 2.5$ and $x = 2.5 - 6\pi$.

34. [C] Make use of your computational utility to plot the graphs f and g. Are the pictures consistent with the fact that sine and cosine are periodic functions with period 2π?

a) $f(x) = \sin(x)$, $x \in [0, 2\pi]$, $g(x) = \sin(x)$, $x \in [4\pi, 6\pi]$

b) $f(x) = \cos(x)$, $x \in [-\pi, \pi]$, $g(x) = \cos(x)$, $x \in [\pi, 3\pi]$

1.2 Combinations of Functions

The combinations of basic functions such as those defined by rational powers of x, $\sin(x)$ and $\cos(x)$ lead to a rich collection of functions that enables us to express the operations of calculus and apply them to diverse phenomena. The arithmetic operations of addition, multiplication and division lead to the natural definitions of **sums, products and quotients of functions**. In this section we will examine such **arithmetic combinations of functions** and the **composition of functions**.

Arithmetic Combinations of Functions

Let f and g denote arbitrary functions. Recall that the **sum**, **product** and **quotient** of f and g are defined via the corresponding operations on their values: For each x that belongs to the domain of f and the domain of g,

$$(f + g)(x) = f(x) + g(x),$$

$$(fg)(x) = f(x)\, g(x),$$

and

$$\left(\frac{f}{g}\right)(x) = \frac{f(x)}{g(x)}$$

provided that $g(x) \neq 0$.

Example 1 Let $f(x) = x$ and $g(x) = \cos(x)$. Then

$$(f + g)(x) = f(x) + g(x) = x + \cos(x)$$

for each $x \in \mathbb{R}$.

Figure 1 shows the graph of $f + g$ on the interval $[-2\pi, 2\pi]$. The dashed line in Figure 1 is the line $y = f(x) = x$. Note that the point $(x + \cos(x), x)$ on the graph of $f + g$ can be obtained by moving the point (x, x) on the graph of f vertically by $g(x) = \cos(x)$ (up or down, depending on the sign of $\cos(x)$). $\square$

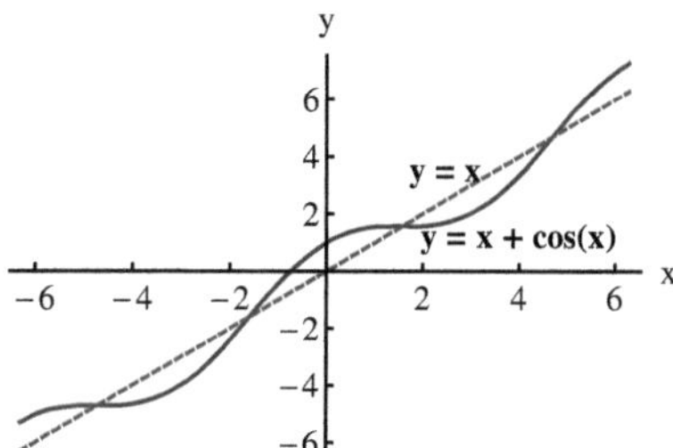

Figure 1: $(f + g)(x) = x + \cos(x)$

Example 2 Let $f(x) = \sqrt{x}$ and $g(x) = \sin(x)$. Then

$$(fg)(x) = f(x)g(x) = \sqrt{x}\sin(x)$$

for each such that $\sqrt{x}$ and $\sin(x)$ are defined, i.e., for each $x \geq 0$. Thus, the domain of the product fg is the interval $[0, +\infty)$. Since $-1 \leq \sin(x) \leq 1$, we have

$$-\sqrt{x} \leq \sqrt{x}\sin(x) \leq \sqrt{x} \text{ for each } x \in \mathbb{R}.$$

Therefore, the graph of fg lies between the graphs of $y = \sqrt{x}$ and $y = \sqrt{x}$. Figure 2 shows the graph of fg. $\square$

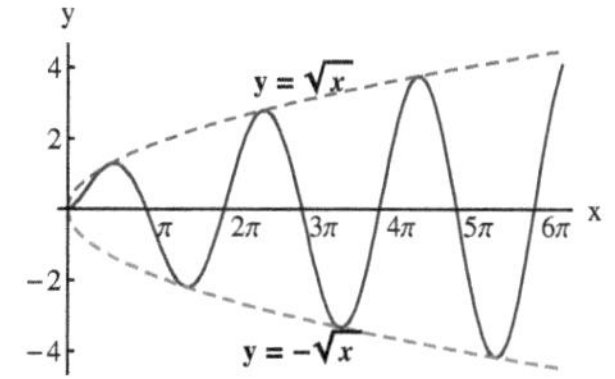

Figure 2: $\sqrt{x} \leq \sqrt{x}\sin(x) \leq \sqrt{x}$

Example 3 Let $f(x) = \sqrt{x}$ and $g(x) = \sqrt{x} + 2$. Then $g(x) = f(x) + 2$. Figure 3 shows the graphs of f and g. If we imagine that the graph of f is shifted vertically upward by by 4 units, the resulting curve will coincide with the graph of g.

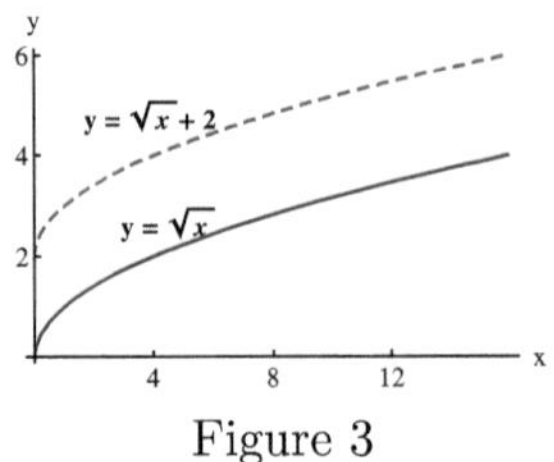

Figure 3

If $G(x) = \sqrt{x} - 2$, then $G(x) = f(x) - 2$. Figure 4 shows the graphs of f and G. If we imagine that the graph of f is shifted vertically downward by 4 units, the resulting curve will coincide with the graph of G. $\square$

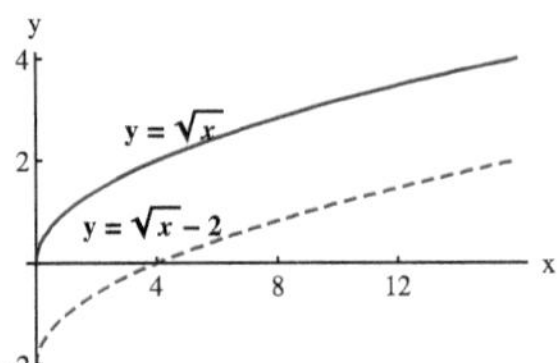

Figure 4

Example 3 illustrates the addition of a constant to a function. The graph of the resulting function can be sketched by shifting the graph of the original function vertically upward or downward, depending on the sign of the constant.

We will use special terminology to refer to the products of constant functions with arbitrary functions: If c is a constant, **the constant multiple** cf of the function f is defined so that

$$(cf)(x) = cf(x)$$

for each x in the domain of f.

Example 4 Let $f(x) = x^2$. Then

$$(2f)(x) = 2f(x) = 2x^2,$$
$$\left(\frac{1}{2}f\right)(x) = \frac{1}{2}f(x) = \frac{1}{2}x^2,$$
$$\left(-\frac{1}{2}f\right)(x) = -\frac{1}{2}f(x) = -x^2,$$
$$(-2f)(x) = -2f(x) = -2x^2$$

for each $x \in \mathbb{R}$.

Figure 5 displays the graphs f, $2f$, $-2f$, $(1/2)f$ and $(-1/2)f$. The graph of $2f$ can be sketched by stretching the graph of f vertically by a factor of 2. The graph of $f/2$ can be sketched by shrinking the graph of f vertically by a factor of 2. The graph of $-2f$ can be sketched by reflecting the graph of $2f$ with respect to the horizontal axis. The graph of $-f/2$ can sketched by reflecting the graph of $f/2$ with respect to the horizontal axis. $\square$

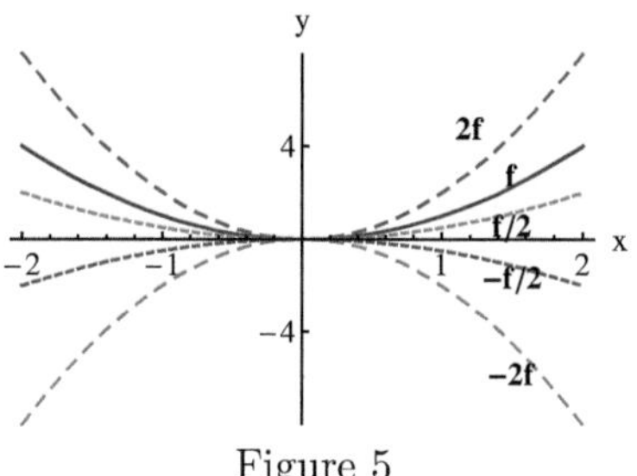

Figure 5

Example 4 illustrates the multiplication of a function by a constant. The graph of the resulting function can be obtained from stretching or shrinking the graph of the original function vertically if the constant is positive. If the constant is negative, the new graph can be obtained from the original graph by a reflection with respect to the horizontal axis, followed by vertical stretching or shrinking.

Many new functions are formed by adding constant multiples of basic functions:

Definition 1 A linear combination of the functions f and g is a function of the form $c_1 f + c_2 g$, where c_1 and c_2 are constants.

Thus,

$$(c_1 f + c_2 g)(x) = c_1 f(x) + c_2 g(x)$$

for each x such that both $f(x)$ and $g(x)$ are defined.

We can form linear combinations of more than two functions in the obvious manner. For example, if c_1, c_2 and c_3 are constants, and f_1, f_2, f_3 are given functions,

$$(c_1 f_1 + c_2 f_2 + c_3 f_3)(x) = c_1 f_1(x) + c_2 f_2(x) + c_3 f_3(x)$$

for each x such that $f_1(x)$, $f_2(x)$ and $f_3(x)$ are defined.

A word of caution: You must distinguish between the meaning of "linear" as in a "linear function", and "linear" as in "a linear combination". For example, the quadratic function $1 + x + x^2$ is a linear combination of 1, x and x^2, but the function is not linear.

Note that a linear function $f(x) = mx + b$ is a linear combination of the constant function 1 and x. A quadratic function $ax^2 + bx + c$ is a linear combination of 1, x and x^2. More generally, a polynomial is a linear combination of 1 and positive integer powers of x:

Definition 2 Given constants $a_0, a_1, \ldots, a_n$, where $a_n \neq 0$, the expression

$$P(x) = a_0 + a_1 x + a_2 x^2 + \cdots + a_n x^n$$

is a **polynomial of degree n. The constant term is** a_0, and a_k is the **coefficient of** x^k for $k = 0, 1, \ldots.n$. We will refer to a polynomial of degree less than or equal to n as a polynomial of **order n.**

We may refer to the function defined by the polynomial $P(x)$ simply as "the polynomial $P(x)$".

A polynomial of order 1 is **a linear function**, and a polynomial of degree 2 is a **quadratic function**. The basic facts about linear and quadratic functions are reviewed in Section A5 of Appendix A. Polynomials of degree higher than 2 can be complicated. In fact, the tools that we will develop in calculus will help us understand the behavior of such functions.

Example 5 Let

$$P(x) = 1 - \frac{1}{2}x^2 + \frac{1}{24}x^4.$$

Thus, $P(x)$ is a polynomial of degree 4. Figure 6 displays the graph of P. $\square$

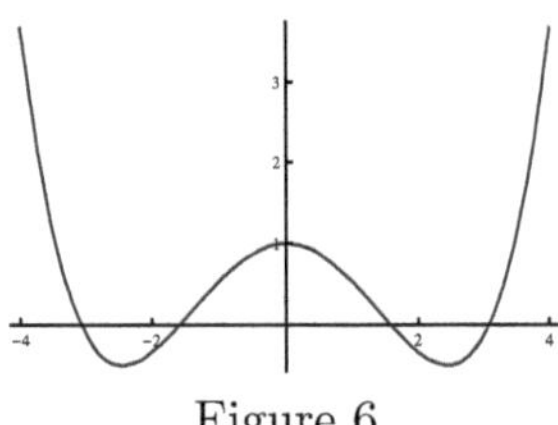

Figure 6

Definition 3 A **rational function** is a quotient of polynomials. Thus, f is a rational function if

$$f(x) = \frac{p(x)}{q(x)},$$

where $p(x)$ and $q(x)$ are polynomials.

Note that the natural domain of a rational function $f = p/q$ consists of all $x \in R$ such that $q(x) \neq 0$. Also note that negative-integer powers of x define rational functions, since

$$x^{-n} = \frac{1}{x^n}.$$

for each positive integer n.

Example 6 Let

$$f(x) = \frac{x^2 + 1}{x^2 - 1},$$

so that f is a rational function, as a quotient of quadratic functions (i.e., polynomials of degree 2).

a) Express the domain of f as a union of intervals.
b) Plot the graph of f with the help of your graphing utility.

Solution

a) We have

$$x^2 - 1 = 0 \Leftrightarrow x = 1 \text{ or } x = -1.$$

Thus, the denominator of the expression for $f(x)$ vanishes iff and only if $x = 1$ or $x = -1$. Therefore, the domain of f is

$$(-\infty, -1) \cup (-1, 1) \cup (1, +\infty).$$

b) Figure 7 displays the graph of f, as produced by a graphing utility. In Figure 6 the viewing window is $[-2, 2] \times [-10, 10]$. Since $|F(x)|$ is very large if x is near 1 or -1 (you should sample some values), the choice of an appropriate viewing window may require some experimentation. We will discuss the behavior of a function such as f near points such as ± 1 in some detail in Section 1.6. $\square$

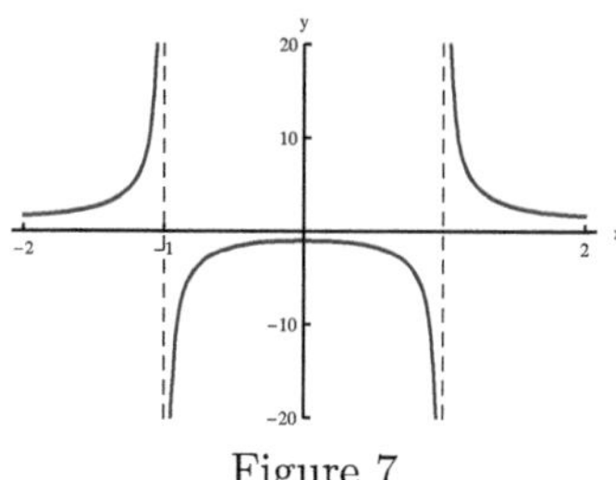

Figure 7

The Functions Tangent and Secant

The trigonometric function **tangent** is the quotient of sine and cosine:

$$\tan(x) = \frac{\sin(x)}{\cos(x)}$$

at each x such that $\cos(x) \neq 0$. Since $\cos(x) = 0$ if and only if x is an odd multiple of $\pm\pi/2$, i.e.,

$$x = \pm(2n+1)\frac{\pi}{2}$$

where $n = 0, 1, 2, \ldots$, a real number is in the natural domain of the tangent function if and only if it is not an odd multiple of $\pm\pi/2$.

The tangent function is periodic with period π. Indeed,

$$\begin{aligned}
\tan(x+\pi) &= \frac{\sin(x+\pi)}{\cos(x+\pi)} = \frac{\sin(x)\cos(\pi) + \cos(x)\sin(\pi)}{\cos(x)\cos(\pi) - \sin(x)\sin(\pi)} \\
&= \frac{\sin(x)(-1) + \cos(x)(0)}{\cos(x)(-1) - \sin(x)(0)} \\
&= \frac{\sin(x)}{\cos(x)} = \tan(x)
\end{aligned}$$

(we have made use of the addition formulas for sine and cosine, as indicated). Since tangent is periodic with period π, the graph of tangent on the interval $(-\pi/2, \pi/2)$ gives an idea about the tangent function on the entire number line. Figure 8 displays the graph of tangent on the interval $(-3\pi/2, 3\pi/2)$.

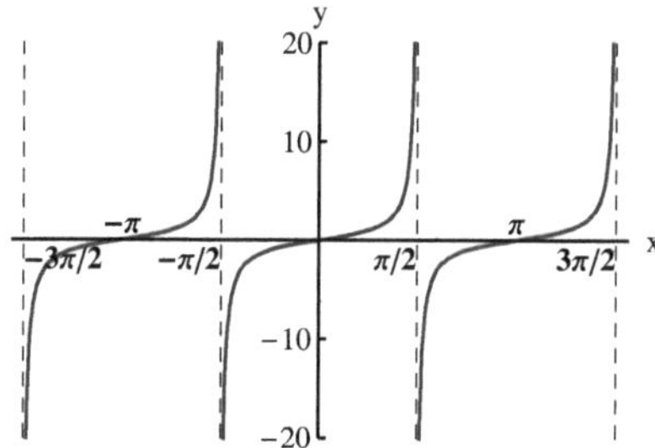

Figure 8: $y = \tan(x)$

The picture indicates that the graph of tangent is symmetric with respect to the origin. Indeed, tangent is an odd function:

$$\tan(-x) = \frac{\sin(-x)}{\cos(-x)} = \frac{-\sin(x)}{\cos(x)} = -\tan(x),$$

since sine is odd and cosine is even.

You can find additional review material about tangent in Section A6 of Appendix A.

Another special trigonometric function is **secant** that is the reciprocal of cosine. Thus,

$$\sec(x) = \frac{1}{\cos(x)}$$

at each x such that $\cos(x) \neq 0$. As in the case of tangent, a real number x is in the natural domain of secant if x is not an odd multiple of $\pm\pi/2$.

Since cosine is periodic with period 2π, **secant has period** 2π as well. Figure 9 displays the graph of secant on the interval $(-3\pi/2, 3\pi/2)$. The picture indicates that the graph of secant is symmetric with respect to the vertical axis. Indeed, secant is an even function, as the reciprocal of the even function cosine:

$$\sec(-x) = \frac{1}{\cos(-x)} = \frac{1}{\cos(x)} = \sec(x).$$

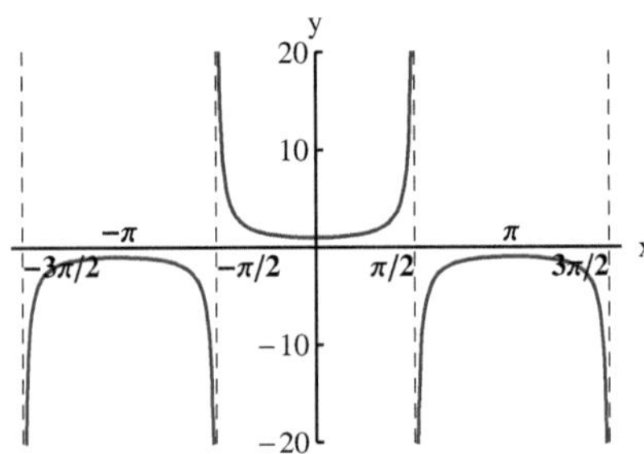

Figure 9: $y = \sec(x)$

The function **cotangent** is the reciprocal of tangent, and is abbreviated as **cot**:

$$\cot(x) = \frac{1}{\tan(x)} = \frac{\cos(x)}{\sin(x)}.$$

The function **cosecant** is the reciprocal of sine, and is abbreviated usually as **csc**:

$$\csc(x) = \frac{1}{\sin(x)}.$$

The trigonometric functions sine, cosine, tangent, and to a lesser extent secant, are important special functions of calculus. The functions cotangent and cosecant will not play a prominent role.

Composite Functions

Let's begin with a specific case.

Example 7 Let $F(x) = \sin(x^2)$. If we set $g(x) = x^2$ and $f(x) = \sin(x)$, we have $F(x) = f(g(x))$. We can also view a function as an input-output mechanism, and consider the evaluation of $F(x)$ as a two-stage operation: Given the input x, the function g produces the output x^2. This serves as the input of the sine function f and the result is the output $\sin(x^2)$. We can display this view of F schematically as follows:

$$x \xrightarrow{g} x^2 \xrightarrow{f} \sin(x^2)$$

$\square$

Example 7 illustrates the operation of composing functions:

Definition 4 Given functions f and g, the function $f \circ g$ (read "f **composed with** g") is defined so that

$$(f \circ g)(x) = f(g(x))$$

for each x such that $g(x)$ and $f(g(x))$ is defined.

Thus, the domain of $f \circ g$ consists of all x in the domain of g such that $g(x)$ is in the domain of f. The symbol "$\circ$" is a little circle, and $f \circ g$ should not be confused with the product of f and g. A function that can be expressed as $f \circ g$ will be referred to as **the composition of f and g.** Such a function is **a composite function.** We can describe the composite function $f \circ g$ schematically:

$$x \xrightarrow{g} g(x) \xrightarrow{f} f(g(x))$$

Example 8 Let $f(x) = \sin(x)$ and $g(x) = x^2$, as in Example 7. Then

$$(f \circ g)(x) = f(g(x)) = f(x^2) = \sin(x^2)$$

for each $x \in \mathbb{R}$. $\square$

The order in which we compose functions matters, as illustrated by the following example:

Example 9 Let $f(x) = \sin(x)$ and $g(x) = x^2$, as in Example 8 Show that $g \circ f \neq f \circ g$.

Solution

In Example 8 we determined that $(f \circ g)(x) = \sin(x^2)$. On the other hand,

$$(g \circ f)(x) = g(f(x)) = g(\sin(x)) = (\sin(x))^2 = \sin^2(x).$$

Therefore, in order to show that $f \circ g \neq g \circ f$, we need to find x such that $\sin(x^2) \neq (\sin(x))^2$. For example, if we set $x = \sqrt{\pi}$, we have

$$\sin((\sqrt{\pi})^2) = \sin(\pi) = 0,$$

whereas,

$$(\sin(\sqrt{\pi}))^2 \cong .959\,882 \neq 0.$$

Thus, $f \circ g \neq g \circ f$.

Figure 10 shows the graph of $(f \circ g)(x) = \sin(x^2)$ on the interval $[-2\pi, 2\pi]$, and Figure 11 shows the graph of $(g \circ f)(x) = \sin^2(x)$ on the same interval . Since $(f \circ g)(x)$ oscillates between -1 and 1 on smaller and smaller intervals as the distance of x from the origin increases, a graphing utility may produce a picture that does not reflect the behavior of the function well. In any case, the graphs are quite different, so that they reinforce the observation that $f \circ g \neq g \circ f$. Figure 11 indicates that $g \circ f$ is periodic with period π (confirm). $\square$

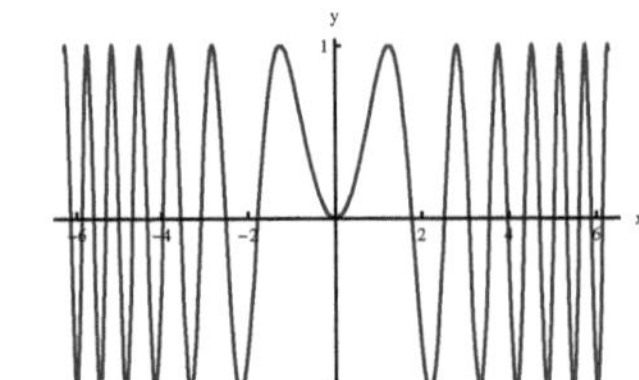

Figure 10: $y = (f \circ g)(x) = \sin\left(x^2\right)$

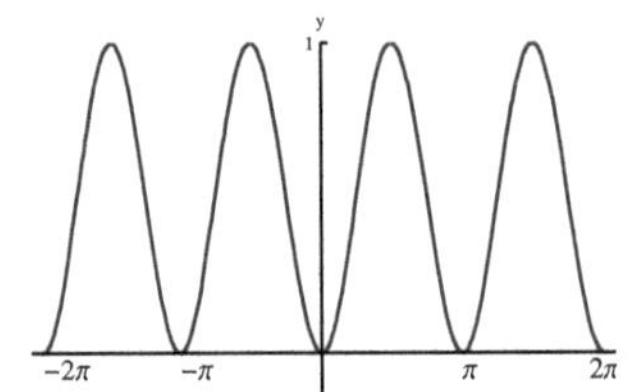

Figure 11: $y = (g \circ f)(x) = (\sin(x))^2$

In discussing the composition of the functions f and g, it is convenient to use the same letter for the independent variable of "the outer function" f and the dependent variable of "the inner function" g.

Example 10 Let $F(x) = \left(x^2 + 1\right)^4$. Express F as a composite function $f \circ g$.

Solution

Let us set $u = g(x) = x^2 + 1$ and $f(u) = u^4$. Then $F = f \circ g$. Indeed,

$$(f \circ g)(x) = f(g(x)) = f\left(x^2 + 1\right) = \left(x^2 + 1\right)^4 = F(x).$$

We can also view the evaluation of $F(x)$ as a two-stage process:

$$x \xrightarrow{g} \left(x^2 + 1\right) \xrightarrow{f} \left(x^2 + 1\right)^4.$$

$\square$

We can consider the composition of any number of functions. For example, given functions f, g and h, we define the composite function $f \circ g \circ h$ by setting

$$(f \circ g \circ h)(x) = f(g(h(x)))$$

for each x such that the expression $f(g(h(x)))$ is defined. Thus, x must be in the domain of h, $h(x)$ must be in the domain of g, and $g(h(x))$ must be in the domain of f. It is not easy to keep track of the parentheses. But the input-output picture helps:

$$x \xrightarrow{h} h(x) \xrightarrow{g} g(h(x)) \xrightarrow{f} f(g(h(x))).$$

We can group the operations of the functions involved in a composite function in different ways. For example,

$$f_1 \circ f_2 \circ f_3 = (f_1 \circ f_2) \circ f_3 = f_1 \circ (f_2 \circ f_3).$$

This should be clear from the input-output picture of composition.

Example 11 Let $F(x) = \cos^2(4x) = (\cos(4x))^2$. Express F as the composition of three functions.

Solution

We can describe the evaluation of $F(x)$ as a three-stage operation:

$$x \rightarrow 4x \rightarrow \cos(4x) \rightarrow (\cos(4x))^2.$$

Each stage corresponds to the action of a function. For bookkeeping purposes, it is useful to denote the independent and dependent variables of the functions systematically, starting with "the innermost function". Let us set $v = h(x) = 4x$, $u = g(v) = \cos(v)$, and $f(u) = u^2$. Then,

$$f(g(h(x))) = f(g(4x)) = f(\cos(4x)) = (\cos(4x))^2 = F(x).$$

Therefore, $F = f \circ g \circ h$. We can indicate this fact schematically:

$$x \xrightarrow{h} 4x \xrightarrow{g} \cos(4x) \xrightarrow{f} (\cos(4x))^2.$$

$\square$

Given a function f and a positive constant a, we can express a function that is defined by an expression of the form $f(x \pm a)$ as the composite function $f \circ h$, where $h(x) = x \pm a$. **If $g(x) = f(x - a)$, the graph of g can be sketched by shifting the graph of f horizontally to the right by a units. If $g(x) = f(x + a)$ the graph of g can be sketched by shifting the graph of f horizontally to the left by a units.**
Indeed, if $g(x) = f(x - a)$,

$$y = f(x) = f((x + a) - a) = g(x + a),$$

so that (x, y) is on the graph of f if and only if $(x + a, y)$ is on the graph of g.
Similarly, if $g(x) = f(x + a)$,

$$y = f(x) = f((x - a) + a) = g(x - a),$$

so that (x, y) is on the graph of f if and only if $(x - a, y)$ is on the graph of g.

Example 12 Let $f(x) = x^2$ and $g(x) = (x - 2)^2$. Then, $g(x) = f(x - 2)$. Figure 12 shows the graphs of f and g. If we imagine that the graph of f is shifted to the right by 2 units, the resulting curve will coincide with the graph of g.

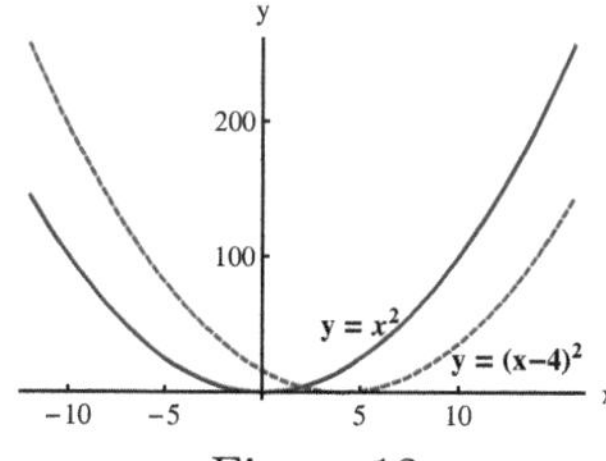

Figure 12

If $f(x) = x^2$ and $G(x) = (x + 2)^2$, then $G(x) = f(x + 4)$. Figure 13 shows the graphs of f and g. If we imagine that the graph of f is shifted to the left by 2 units, the resulting curve will coincide with the graph of G. $\square$

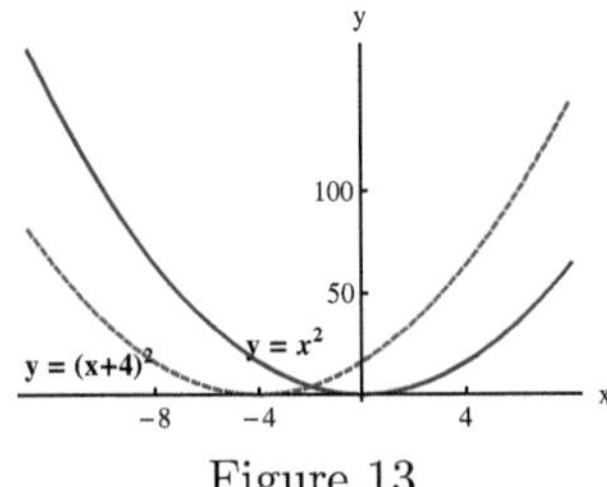

Figure 13

Example 13 Let $f(x) = x^2 - 4x + 7$. The completion of the square leads to the expression

$$f(x) = (x-2)^2 + 3.$$

Thus, if $g(x) = x^2$, then

$$f(x) = g(x-2) + 3.$$

Therefore, the graph of f can be obtained by shifting the graph of g to the right by 2 units and lifting the resulting graph vertically upward by 3 units. Note that the graph of f is a parabola with vertex $(2,3)$ (Figure 14). $\square$

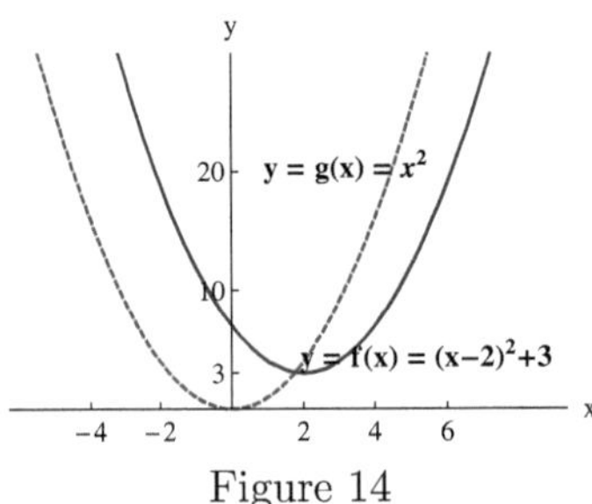

Figure 14

Let's look at the graphs of sine and cosine again. Figure 15 indicates that we can obtain the graph of sine if we shift the graph of cosine to the right by $\pi/2$.

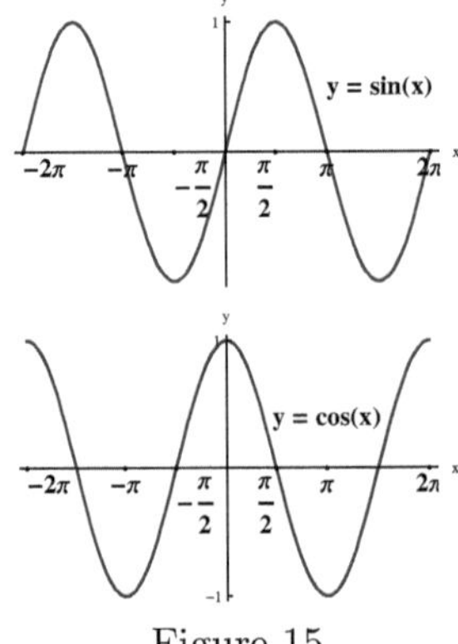

Figure 15

Therefore, we should have

$$\cos(x - \frac{\pi}{2}) = \sin(x).$$

This is indeed the case: We make use of the difference formula for cosine, as reviewed in Section A6 of Appendix A:

$$\cos\left(x - \frac{\pi}{2}\right) = \cos(x)\cos(\frac{\pi}{2}) + \sin(x)\sin(\frac{\pi}{2}) = \cos(x)(0) + \sin(x)(1) = \sin(x).$$

Note that

$$\cos(\frac{\pi}{2} - x) = \cos(x - \frac{\pi}{2}) = \sin(x)$$

since cosine is an even function. Figure 16 illustrates the above relationship for an angle x between 0 and $\pi/2$.

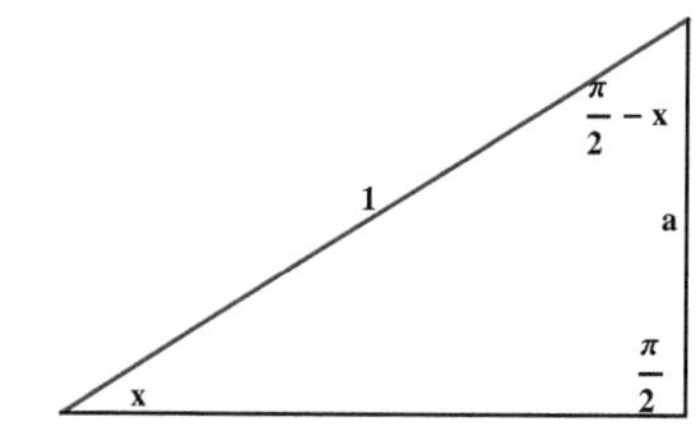

Figure 16: $\cos\left(\frac{\pi}{2} - x\right) = a = \sin(x)$

Functions of the form $\sin(\omega x)$ and $\cos(\omega x)$ where ω is some constant are the building blocks of a rich and useful collection of functions, just as x^n, where $n = 1, 2, 3, \dots.$ These functions are periodic. If $\omega = 0$, they are merely constant functions: $\sin(0) = 0$ and $\cos(0) = 1$. If $\omega \neq 0$ then such a function has period $2\pi/\omega$. Indeed,

$$\sin(\omega(x + \frac{2\pi}{\omega})) = \sin(\omega x + 2\pi) = \sin(\omega x),$$

since sine has period 2π. The case of $\cos(\omega x)$ is similar.

Example 14

a) Let $f(x) = 2\sin(4x)$. Determine the fundamental period p of f. Sketch the graph of f on the interval $[-p/2, p/2]$. Indicate the points at which the graph of f intersects the x-axis, and the points at which f has value 1 or -1.

b) Let

$$g(x) = \frac{1}{2}\sin\left(\frac{x}{4}\right).$$

Perform the tasks that were prescribed in part a) for the function f on the function g.

Solution

a) The fundamental period of f is

$$\frac{2\pi}{4} = \frac{\pi}{2}.$$

Figure 17 displays the graph of f on the interval $[-\pi/4, \pi/4]$. Notice that the shape of the graph of $2\sin(4x)$ is similar to the shape of the graph of $\sin(x)$ on the interval $[-\pi, \pi]$. You can imagine that the graph of sine has shrunk horizontally by a factor of 4 and has been stretched vertically by a factor of 2.

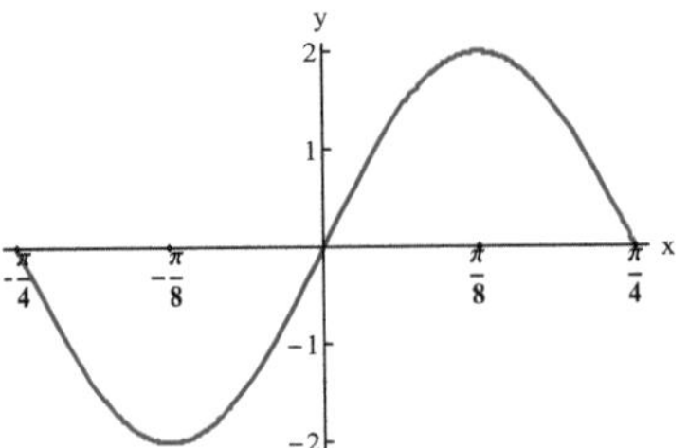

Figure 17: $y = 2\sin(4x)$

b) The fundamental period of g is

$$\frac{2\pi}{1/4} = 8\pi.$$

Figure 18 displays the graph of g on the interval $[-4\pi, 4\pi]$. Notice that the shape of the graph of g on $[-4\pi, 4\pi]$ is similar to the shape of the graph of $\sin(x)$ on $[-\pi, \pi]$. You can imagine that the graph of sine has been stretched horizontally by a factor of 4 and has shrunk vertically by a factor of 2. $\square$

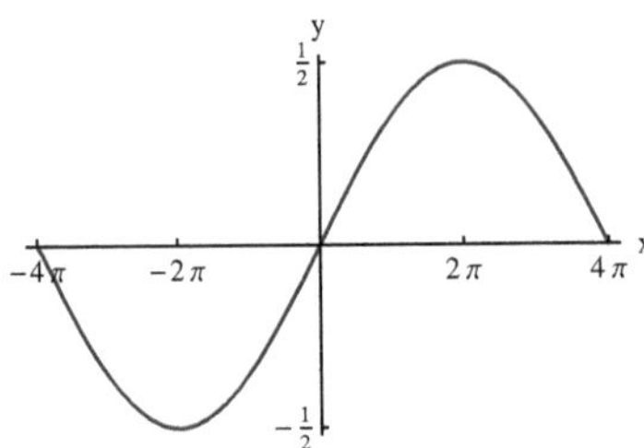

Figure 18: $y = \frac{1}{2}\sin\left(\frac{x}{4}\right)$

Trigonometric Polynomials

Linear combinations of $\sin(nx)$ and $\cos(nx)$ play a prominent role in calculus and many applications:

Definition 5 A **trigonometric polynomial** is a linear combination of the constant function 1, and functions of the form $\sin(nx)$ and $\cos(nx)$, where n is an integer.

A trigonometric polynomial is periodic with period 2π, and can be expressed as a linear combination of products of powers of $\sin(x)$ and $\cos(x)$. As we will discuss briefly in Chapter 9, trigonometric polynomials provide useful approximations to arbitrary periodic functions and appear in many applications such as signal processing.

Example 15 Let

$$f(x) = \sin(x) + \frac{1}{3}\sin(3x),$$

so that f is a trigonometric polynomial, as a linear combination of $\sin(x)$ and $\sin(3x)$.

a) Confirm that f is periodic with period 2π.
b) Express $f(x)$ as a linear combination of products of powers of $\sin(x)$ and $\cos(x)$.
c) Plot the graph of f on the interval $[-2\pi, 2\pi]$ with the help of your calculator.

Solution

a) We have

$$f(x + 2\pi) = \sin\left(x + 2\pi\right) + \frac{1}{3}\sin(3\left(x + 2\pi\right)) = \sin(x) + \frac{1}{3}\sin(3x + 6\pi)$$

$$= \sin(x) + \frac{1}{3}\sin(3x) = f(x),$$

since the sine function has period 2π.

b) We will make use of some trigonometric identities, as reviewed in Section A6 of Appendix A
. We have

$$\sin(2x) = 2\sin(x)\cos(x).$$

We also have

$$\sin(3x) = \sin(2x + x)$$
$$= \sin(2x)\cos(x) + \cos(2x)\sin(x)$$
$$= (2\sin(x)\cos(x))\cos(x) + \left(\cos^2\left(x\right) - \sin^2\left(x\right)\right)\sin(x)$$
$$= 2\sin(x)\cos^2\left(x\right) + \cos^2\left(x\right)\sin(x) - \sin^3\left(x\right)$$
$$= 3\sin\left(x\right)\cos^2\left(x\right) - \sin^3\left(x\right).$$

Therefore,

$$f(x) = \sin\left(x\right) + \frac{1}{3}\sin\left(3x\right)$$
$$= \sin(x) + \frac{1}{3}\left(3\sin\left(x\right)\cos^2\left(x\right) - \sin^3\left(x\right)\right)$$
$$= \sin(x) + \sin\left(x\right)\cos^2\left(x\right) - \frac{1}{3}\sin^3\left(x\right).$$

Thus, we have expressed $f\left(x\right)$ as a linear combination of $\sin(x)$, $\sin(x)\cos^2\left(x\right)$ and $\sin^3\left(x\right)$.
Even though such an alternative expression may or may not offer an advantage for the discussion
of the function, it does explain why such a function is referred to as a trigonometric polynomial.

c) Figure 19 shows the graph of f on the interval $[-2\pi, 2\pi]$. $\square$

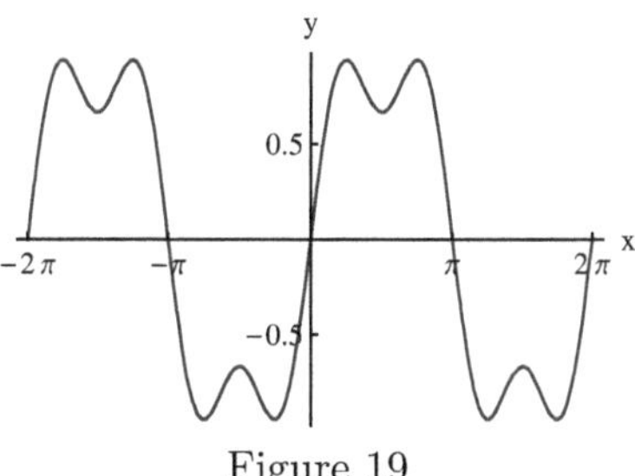

Figure 19

Problems

In problems 1 - 4, determine $f + g$, $3f$, fg, $1/f$, and f/g and $2f - 3g$ (specify the domain of
each function).

1. $f(x) = x^2$, $g(x) = x^3$

2. $f(x) = \sqrt{x}$, $g(x) = x^2$

3. $f(x) = \dfrac{1}{x}$, $g(x) = x^2$

4. $f(x) = \dfrac{1}{\sqrt{x}}$, $g(x) = \dfrac{1}{x}$

In problems 5 and 6,
a) Determine the linear function f such that its graph passes through the point P_1 and P_2 (obtain the point-slope form of the equation with basepoint P_1 and the slope-intercept form of the equation),
b) Sketch the graph of f. Indicate the points at which the graph of f intersects the coordinate axes.

5. $P_1 = (4,3)$, $P_2 = (6,1)$

6. $P_1 = (-2,4)$, $P_2 = (1,6)$

In problems 7 and 8,
a) Determine the maximum or minimum value of the quadratic function f by completing the square, provided that such a value exists,
b) Determine the solutions of $f(x) = 0$,
c) Sketch the parabola that is the graph of f. Indicate the vertex of the parabola and and the points at which it intersects the coordinate axes.

7. $f(x) = 3x^2 - 12x + 17$

8. $f(x) = -2x^2 + 12x - 14$

In problems 9 - 12, determine $f \circ g$, $g \circ f$, $f \circ f$ and $g \circ g$ (specify the domain of each function).

9. $f(x) = \dfrac{1}{x}$, $g(x) = x^2$

10. $f(x) = \sqrt{x}$, $g(x) = x^2 - 9$

11. $f(x) = \sqrt{x}$, $g(x) = 4 - x^2$

12. $f(x) = x^{1/3}$, $g(x) = \dfrac{1}{x^2}$

In problems 13-16, determine $f \circ g$ and $g \circ f$ (you need to specify the domains of the functions). Show that $f \circ g \neq g \circ f$.

13. $f(x) = \sqrt{x}$ and $g(x) = |x|$

14. $f(x) = x^3$ and $g(x) = \cos(x)$.

15. $f(u) = u^{1/4}$ and $g(x) = \sin(x)$.

16. $f(u) = |u|$ and $g(x) = \cos(x)$

In problems 17 - 24,
a) Determine the natural domain of f,
b) Express f as $F \circ G$, where $G(x)$ is a polynomial or a rational function.

17. $f(x) = \sqrt{16 - x^2}$.

18. $f(x) = \left(x^2 - 9\right)^{3/4}$

19. $f(x) = \sin\left(x^2\right)$

20. $f(x) = \cos\left(\dfrac{1}{x}\right)$

21. $f(x) = \sin(4x)$

22. $f(x) = \cos\left(\dfrac{x}{4}\right)$

23. $f(x) = \tan(2x)$

24. $f(x) = \sec\left(\dfrac{x}{2}\right)$

In problems 25 and 26,
a) Determine the part of the natural domain of f that is contained in $[-\pi, \pi]$,
b) Express $f(x)$ as $(F \circ G)(x)$, where $F(u)$ is a fractional power of u.

25. $f(x) = \sin^{3/4}(x)$

26. $f(x) = \cos^{1/4}(x)$

In problems 27 - 34,

a) Determine the fundamental period p of f,
b) Determine the part of the natural domain of f in the interval $[-p/2, p/2]$ and whether f is an odd or even function,
c) Sketch the graph of f on the interval $[-p/2, p/2]$.

27. $f(x) = \sin(6x)$

28. $f(x) = \cos(x/3)$

29. $f(x) = \tan(\pi x)$

30. $f(x) = \sec(x/4)$.

31. $f(x) = \sqrt{\sin(x)}$

32. $f(x) = \sqrt{\cos(x)}$

33. $f(x) = \sqrt{\tan(x)}$

34. $f(x) = \sqrt{\sec(x)}$

1.3 Limits and Continuity: The Concepts

In this section we will discuss the concept of the limit of a function at a point. This concept provides a general framework for problems such as the determination of the slope of a tangent line to the graph of a function or the velocity of an object in motion. We will also discuss the related concept of continuity.

The Slope of a Tangent Line

How should we determine the slope of the graph of a function? Let's begin with a case where we know the answer. Let f be **a linear function** so that $f(x) = mx + b$, where m and b are constants. The graph of f is a line with **slope** m. If a is an arbitrary point on the number line and $x \neq a$, then

$$\frac{f(x) - f(a)}{x - a} = \frac{(mx + b) - (ma + b)}{x - a} = \frac{m(x - a)}{x - a} = m.$$

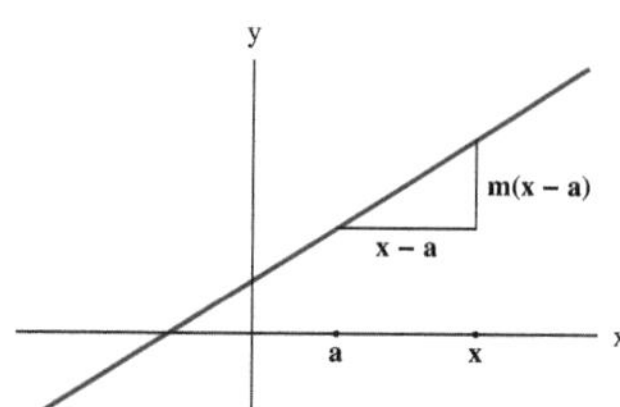

Figure 1: The line $y = mx + b$ has slope m

The case of a **nonlinear function** is not that straightforward. Let's consider a specific case.

Example 1 Let $F(x) = x^2$. The graph of F is a familiar parabola, as shown in Figure 2.

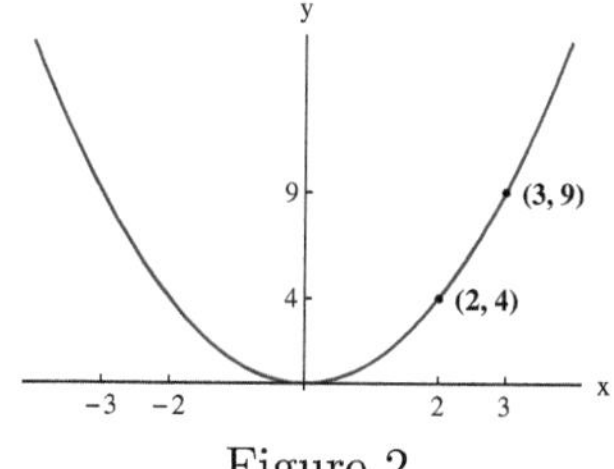

Figure 2

A reasonable notion of the slope of the graph of F cannot yield a single number, unlike the case of a linear function. For example, if we compare the behavior of the graph of F near $(3, F(3)) = (3, 9)$ and $(2, F(2)) = (2, 4)$, the graph appears to rise more steeply near $(3, 9)$.

Let's focus our attention on the behavior of the function near the point 3. If $x \neq 3$, we will refer to the line that passes through the points $(3, F(3))$ and $(x, F(x))$ as **a secant line**. The slope of such a secant line is

$$\frac{F(x) - F(3)}{x - 3} = \frac{x^2 - 9}{x - 3}.$$

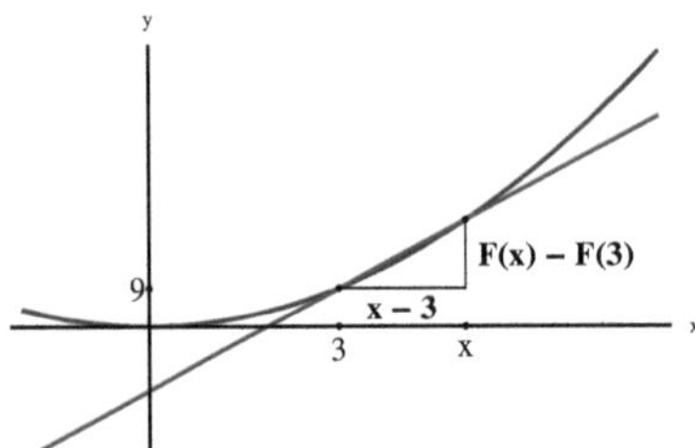

Figure 3: A secant line

Since the secant line is almost **"tangential"** to the graph of F at $(3, F(3))$ if x is close to 3, its slope should approximate the slope of the tangent line to the graph of F at $(3, F(3))$. Table 1 displays the slope of the secant line that passes through the points $(3, F(3))$ and $(x, F(x))$ for $x = 3 + 10^{-n}$, where $n = 1, 2, 3, 4, 5$. The numbers indicate that the slope of the secant line approximates 6 if x is close to 3.

x	$\frac{F(x) - F(3)}{x - 3}$
$3 + 10^{-1}$	6.1
$3 + 10^{-2}$	6.01
$3 + 10^{-3}$	6.001
$3 + 10^{-4}$	6.0001
$3 + 10^{-5}$	6.00001

Table 1

A little algebra clarifies the situation. We cannot merely replace x by 3 in the expression for the slope of a secant line. This leads to the undefined or **indeterminate expression** $0/0$. On the other hand, we can simplify the expression for any $x \neq 3$:

$$\frac{F(x) - F(3)}{x - 3} = \frac{x^2 - 9}{x - 3} = \frac{(x - 3)(x + 3)}{x - 3} = x + 3.$$

If x is close to 3, then $x + 3 \cong 3 + 3 = 6$. It seems reasonable to declare that **the slope of the tangent line to the graph of** F **at** $(3, F(3))$ is 6. Since the tangent line should pass through $(3, F(3))$ and has slope 6, it is the graph of the equation

$$y = F(3) + 6(x - 3) = 9 + 6(x - 3)$$

in the xy-plane (this is the point-slope form of the equation of the tangent line with basepoint 3). Figure 4 shows the graph of F and the tangent line to the graph of F at $(3, F(3))$. The picture is consistent with our intuitive notion of a tangent line. We will identify **the slope of**

the graph of F at $(3, F(3))$ with the slope of the tangent line to the graph of F at that point.
$\square$

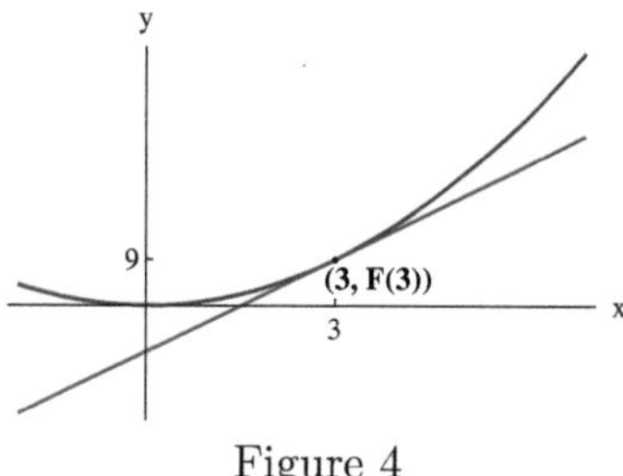

Figure 4

The Informal Definitions of Limits and Continuity

Let $F(x) = x^2$, as in Example 1 The slope of the secant line that passes through $(3, F(3))$ and $(x, F(x))$ is a function of x in its own right. Let's set

$$f(x) = \frac{F(x) - F(3)}{x - 3} = \frac{x^2 - 9}{x - 3},$$

so that $f(x)$ is defined if $x \neq 3$. We have $f(x) \cong 6$ if $x \neq 3$ and x is close to 3. Given an arbitrary function f and a point a on the number line, it will be useful to discuss whether $f(x)$ approximates a certain number L if $x \neq a$ and x is close to a, even if $f(x)$ is not necessarily the slope of a secant line. This leads to the concept of the limit of a function at a point.

Definition 1 (THE LIMIT OF A FUNCTION AT A POINT) Assume that $f(x)$ is defined for each x in some open interval that contains the point a, with the possible exception of a itself. **The limit of $f(x)$ at a is L** if $f(x)$ is as close to L as desired provided that $\boldsymbol{x \neq a}$ and $\boldsymbol{x}$ is sufficiently close to a. In this case we write

$$\lim_{x \to a} f(x) = L$$

(read "**the limit of $\boldsymbol{f(x)}$ as $\boldsymbol{x}$ approaches $\boldsymbol{a}$ is $\boldsymbol{L}$**").

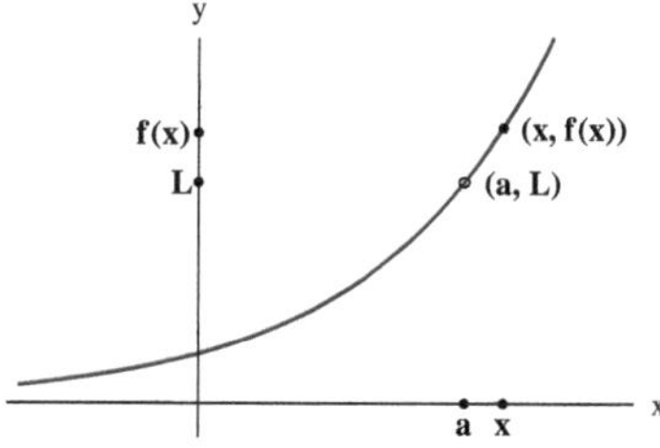

Figure 5: $f(x)$ is close to the limit L at a if $x \neq a$ and x is close to a

You can imagine that $f(x)$ gets closer and closer to L as x approaches a, with the restriction that $x \neq a$. We will refer to Definition 1 has been dubbed as "the informal definition of the limit", since the phrases "as close to L as desired " and "sufficiently close" have not been quantified. For almost all our purposes in calculus, an intuitive understanding of the concept of the limit will be adequate. The precise definition of the limit of a function at a point will be discussed in the next section.

Example 2 Let

$$f\left(x\right) = \frac{x^2 - 9}{x - 3}.$$

Then $f\left(x\right)$ is defined if $x \neq 3$. As in Example 1,

$$f\left(x\right) = \frac{x^2 - 9}{x - 3} = \frac{\left(x - 3\right)\left(x + 3\right)}{x - 3} = x + 3,$$

so that $f\left(x\right) \cong 6$ if $x \cong 3$ and $x \neq 3$. We have

$$\left|f(x) - 6\right| = \left|(x + 3) - 6\right| = \left|x - 3\right|,$$

so that $f\left(x\right)$ is as close to 6 as desired if x is sufficiently close to 3. Therefore, the limit of f at 3 is 6:

$$\lim_{x \to 3} f\left(x\right) = \lim_{x \to 3} \frac{x^2 - 9}{x - 3} = 6.$$

$\square$

We can set $x = a + h$ so that h represents the deviation of x from a. We have $h \neq 0$ corresponding to $x \neq a$ and h approaches 0 if and only if x approaches a. Thus,

$$\lim_{x \to a} f\left(x\right) = \lim_{h \to 0} f\left(a + h\right).$$

Example 3 Let

$$f\left(x\right) = \frac{\sqrt{x} - 2}{x - 4} \text{ if } x \neq 4.$$

a) Calculate $f\left(4 + h\right)$ for $h = \pm 10^{-n}$, $n = 1, 2, 3, 4$. Do the numbers lead to a conjecture about $\lim_{x \to 4} f\left(x\right)$?
b) Plot the graph of with a graphing device. Does the picture support your conjecture?

Solution

a) We have

$$f\left(4 + h\right) = \frac{\sqrt{4 + h} - 2}{h}.$$

Table 2 displays $f\left(4 + h\right)$ for $h = \pm 10^{-n}$, $n = 1, 2, 3, 4$. The numbers indicate that $\left|f\left(4 + h\right) - 0.25\right|$ becomes smaller and smaller as $\left|h\right|$ becomes small. Thus, we can expect that $\lim_{x \to 4} f\left(x\right) = 1/4$ ($\left|f\left(4 + h\right) - 0.25\right|$ is rounded to 2 significant digits).

h	$f(4 + h)$	$\left|f\left(4 + h\right) - 0.25\right|$
10^{-1}	$0.248\,457$	1.5×10^{-3}
-10^{-1}	$0.251\,582$	1.6×10^{-3}
10^{-2}	$0.249\,844$	1.6×10^{-4}
-10^{-2}	$0.250\,156$	1.6×10^{-4}
10^{-3}	$0.249\,984$	1.6×10^{-5}
-10^{-3}	$0.250\,016$	1.6×10^{-5}
10^{-4}	$0.249\,998$	1.6×10^{-6}
-10^{-4}	$0.250\,002$	1.6×10^{-6}

Table 2

b) Figure 6 supports the conjecture that $\lim_{x \to 4} f(x) = 0.25$. In fact, the program that generated the picture seems to be unaware of the fact that f is not defined at 4. This is not surprising, since such a device samples some values of x, calculates the corresponding values of the function, and joins the resulting points by line segments. It is immaterial that f is not defined at 4, as long as $f(x) \cong 4$ when x is near 0. $\square$

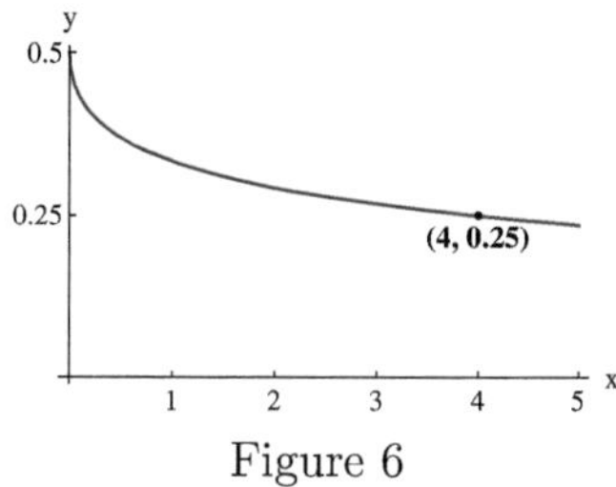

Figure 6

A function may fail to have a limit at a point:

Example 4 Let

$$f(x) = \begin{cases} x & \text{if} \quad x < 1, \\ x+2 & \text{if} \quad x \geq 1. \end{cases}$$

a) Sketch the graph of f.
b) Show that f does not have a limit at 1.

Solution

a) Figure 7 shows the graph of f.

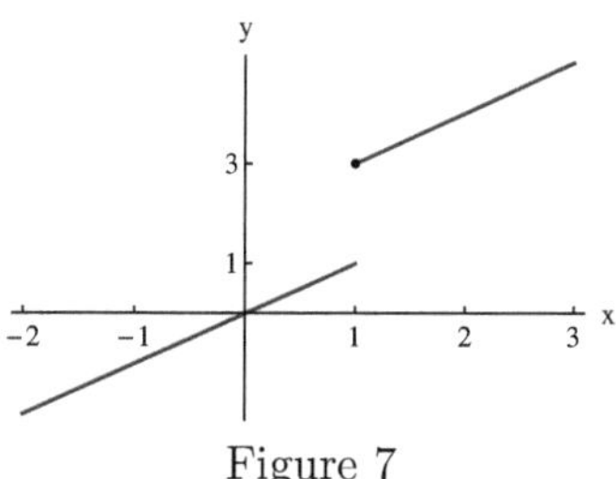

Figure 7

b) If $x < 1$ and x is close to 1, then $f(x) = x \cong 1$. On the other hand, if $x > 1$ and x is near 1, $f(x) = x+2 \cong 3$. Thus, $f(x)$ does not approximate a definite number if $x \neq 1$ and x is near 1. Therefore f does not have a limit at 1. $\square$

The limit of a function at a point need not be the same as the value of the function at that point, as in the following example.

Example 5 Let

$$f(x) = \begin{cases} x & \text{if} \quad x \neq 1 \\ 2 & \text{if} \quad x = 1. \end{cases}$$

a) Sketch the graph of f.
b) Show that $\lim_{x \to 1} f(x) \neq f(1)$.

Solution

a) Figure 8 shows the graph of f. The fact that $f(1) \neq 2$ is indicated by a little hollow circle. The point $(1, 2)$ belongs to the graph of f since $f(1) = 2$.

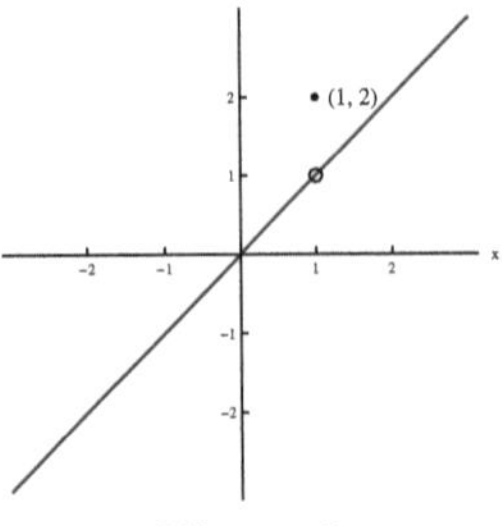

Figure 8

b) Since $f(x) = x$ for each $x \neq 1$,

$$\lim_{x \to 1} f(x) = \lim_{x \to 1} x = 1.$$

On the other hand, $f(1) = 2$. Therefore, $\lim_{x \to 1} f(x) \neq f(1)$. $\square$

We use special terminology that applies to a case where the limit of a function at a point is the value of the function at that point:

Definition 2 (CONTINUITY) A function f is said to be **continuous at a point** a if $f(x)$ is defined in some open interval that contains a and $\lim_{x \to a} f(x) = f(a)$.

Example 6 Let $g(x) = x + 3$. Then g is continuous at 3. Indeed,

$$\lim_{x \to 3} g(x) = \lim_{x \to 3} (x + 3) = 6 = g(3).$$

$\square$

Since $\lim_{x \to a} f(x) = \lim_{h \to 0} f(a + h)$, a function f is continuous at a point a if and only if

$$\lim_{h \to 0} f(a + h) = f(a).$$

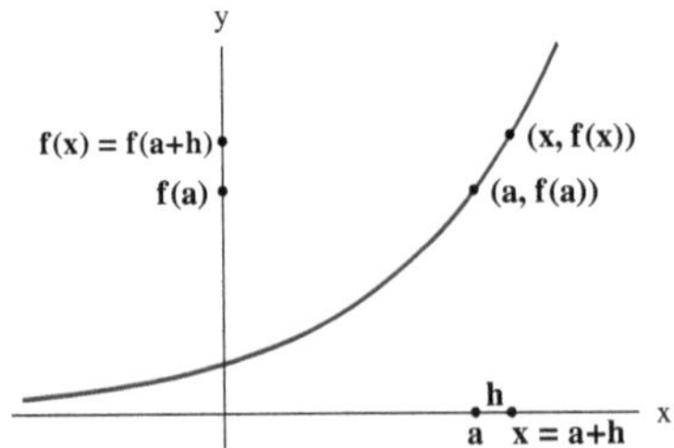

Figure 9: $f(x)$ is close to $f(a)$ if x is close to a

Example 7 Calculate $\sin(\pi/6 + 10^{-n})$ for $n = 2, 3, 4, 5$. Do the numbers suggest that the sine function is continuous at $\pi/6$?

Solution

We have $\sin(\pi/6) = 0.5$. The numbers in Table 3 indicate that $|\sin(\pi/6 + h) - \sin(\pi/6)|$ should be as small as desired provided that $|h|$ is small enough, and suggest that sine is continuous at $\pi/6$. $\square$

x	$\sin(x)$	$|\sin(x) - 0.5|$
$\frac{\pi}{6} + 10^{-2}$	$0.508\,635$	8.6×10^{-3}
$\frac{\pi}{6} + 10^{-3}$	$0.500\,866$	8.7×10^{-4}
$\frac{\pi}{6} + 10^{-4}$	$0.500\,087$	8.7×10^{-5}
$\frac{\pi}{6} + 10^{-5}$	$0.500\,009$	8.7×10^{-6}

Table 3

We will say that a function f is **discontinuous** at a if f is not continuous at a.

Example 8 Let

$$f(x) = \frac{x^2 - 9}{x - 3} \text{ if } x \neq 3,$$

as in Example 2. The function f is discontinuous at 3 since f is not defined at 3, even though $\lim_{x \to 3} f(x)$ exists. $\square$

Example 9 Let

$$f(x) = \begin{cases} x & \text{if } x < 1, \\ x + 2 & \text{if } x \geq 1, \end{cases}$$

as in Example 4. The function f is discontinuous at 1 since $\lim_{x \to 1} f(x)$ does not exist. $\square$

Example 10 Let

$$f(x) = \begin{cases} x & \text{if } x \neq 1 \\ 2 & \text{if } x = 1, \end{cases}$$

as in Example 5. The function f is discontinuous at 1 since $\lim_{x \to 1} f(x) = \lim_{x \to 1} x = 1 \neq f(1)$. $\square$

Let's take another look at the function of Example 4:

$$f(x) = \begin{cases} x & \text{if } x < 1, \\ x + 2 & \text{if } x \geq 1. \end{cases}$$

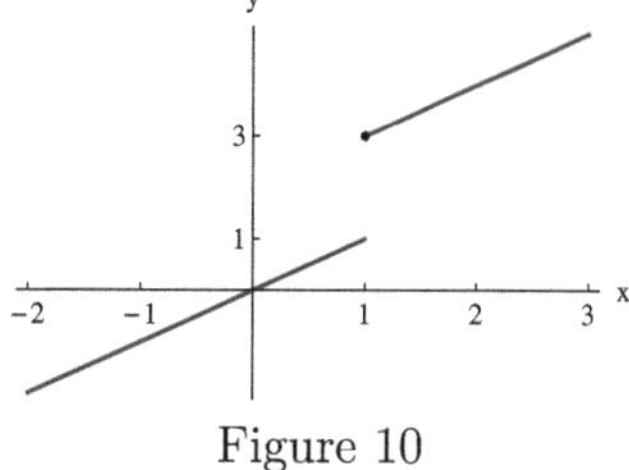

Figure 10

We observed that f does not have a limit at 1, since $f(x)$ approaches 1 if x approaches 1 from the left, and $f(x)$ approaches 3 if x approaches 1 from the right. These are examples of "one-sided limits":

Definition 3

a) **The right-limit of f at a is L_+** if $f(x)$ is as close to L_+ as desired provided that $x > a$ and x is sufficiently close to a. In this case we write

$$\lim_{x \to a+} f(x) = L_+$$

(read "the limit of $f(x)$ as x approaches a from the right is L_+").
b) **The left-limit of f at a is L_-** if $f(x)$ is as close to L_- as desired provided that $x < a$ and x is sufficiently close to a. In this case we write

$$\lim_{x \to a-} f(x) = L_-$$

(read "the limit of $f(x)$ as x approaches a from the left is L_-").

The language is suggestive: You can imagine that $f(x)$ approaches L_+ as x approaches a from the right on the number line, and that $f(x)$ approaches L_- as x approaches a from the left.

If f is the function of Example 4, then

$$\lim_{x \to 1+} f(x) = \lim_{x \to 1+} (x + 2) = 3,$$

and

$$\lim_{x \to 1-} f(x) = \lim_{x \to 1-} x = 1.$$

Definition 4 We say that f has a **jump discontinuity** at a if $\lim_{x \to a+} f(x)$ and $\lim_{x \to a-} f(x)$ exist but are not equal.

Thus, the function of Example 9 has a jump discontinuity at 1.

Clearly, $\lim_{x \to a} f(x)$ exists if and only if both one-sided limits of f at a exist and $\lim_{x \to a+} f(x) = \lim_{x \to a-} f(x)$. If this is the case,

$$\lim_{x \to a} f(x) = \lim_{x \to a-} f(x) = \lim_{x \to a+} f(x).$$

The notion of one-sided continuity is related to one-sided limits:

Definition 5 A function f is said to be **continuous at a from the right** if f is defined at a and $\lim_{x \to a+} f(x) = f(a)$. Similarly, f is **continuous at a from the left** if $\lim_{x \to a-} f(x) = f(a)$.

The function f of Example 4 is continuous at 1 from the right, since

$$\lim_{x \to 1+} f(x) = 3 = f(2).$$

The function is discontinuous at 1 from the left, since

$$\lim_{x \to 1-} f(x) = 1 \neq f(1).$$

By the definition of continuity from the right and from the left, a function f is continuous at a point a if and only if f is continuous at a from the right and from the left.

As discussed in Section A3 of Appendix A, a point is in **the interior of an interval** if it belongs to the interval but it is not an endpoint of the interval.

Definition 6 A function f **is continuous on the interval** J if f is continuous at each point in the interior of J, and the appropriate one-sided continuity is valid at any endpoint of J that is in J.

Note that there is a break in the graph of the function of Example 9, corresponding to the discontinuity at 1. If f is the function of Example 10, then f is also discontinuous at 1. The point $(1,2)$ which is on the graph of f seems to have left a hole at the point $(1,1)$. Such breaks or holes in the graph of a function indicate discontinuities. **If f is continuous at each point of an interval, the graph of f on that interval is a "continuous curve" without any breaks or holes.** We will refer to such a portion of the graph of a function simply as a **continuous curve.**

A word of caution: Figure 7 that displays the graph of the function f of Example 4 was generated by a program that takes into account the discontinuities of a function. Figure 11 shows another computer generated graph for the same function. The picture shows a continuous curve as the graph of the function, even though f is discontinuous at 3. The picture includes a line segment that appears to be vertical and seems to connect the points $(1,1)$ and $(1,3)$. That is **a spurious line segment** and is not part of the graph of f (the graph of a function cannot contain a vertical line segment!). The graphing utility that produced Figure 11 sampled values of x immediately to the left of 1 and immediately to the right of 1, and connected the corresponding points on the graph of f with a line segment.

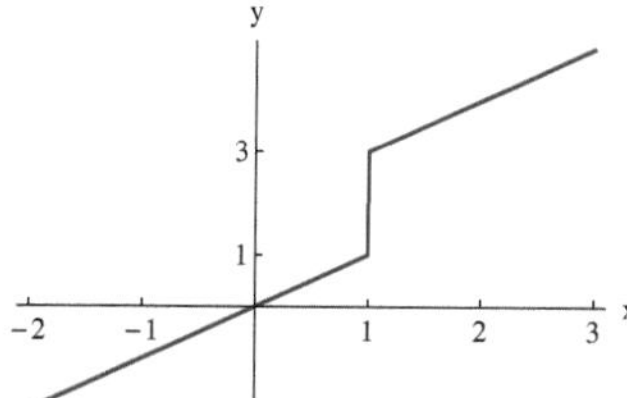

Figure 11: Virtual continuity

Example 11 Let $f(x) = 1/x$.

Since $f(x)$ attains arbitrarily large positive values as x approaches 0 from the right, the function does not have a right-limit at 0. The function does not have a left-limit at 0 either, since $f(x)$ attains negative values of arbitrarily large magnitude as x approaches 0 from the left. For example,

$$f\left(\frac{1}{10^n}\right) = 10^n \text{ and } f\left(-\frac{1}{10^n}\right) = -10^n,$$

where n is arbitrarily large. $\square$

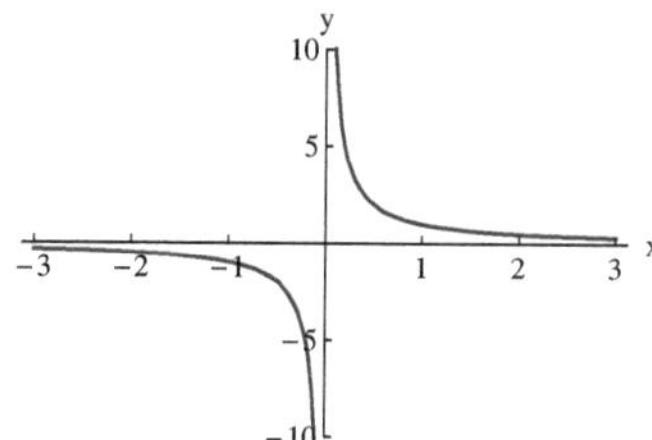

Figure 12: Unbounded discontinuity at 0

Definition 7 We will say that f has an **unbounded discontinuity** at a if $f(x)$ attains values of arbitrarily large magnitude as x approaches a from the right or from the left.

Thus, the function f of Example 11 has an unbounded discontinuity at 0.

A function can have a discontinuity other than a jump discontinuity or an unbounded discontinuity:

Example 12 Let $f(x) = \sin(1/x)$. Figure 13 shows a computer generated graph of f.

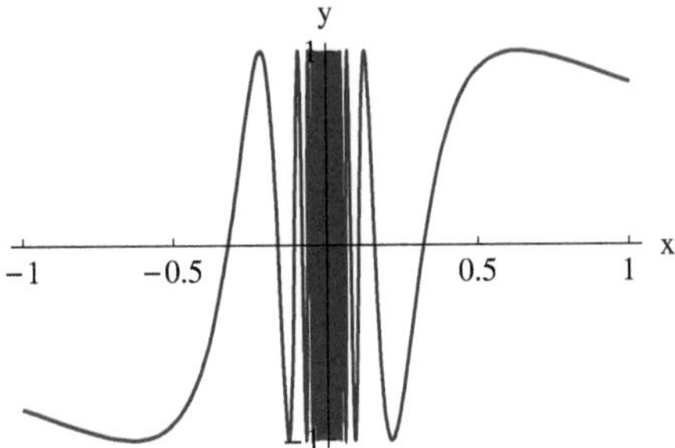

Figure 13: Discontinuity at 0 due to oscillations.

There appears to be a dark blob around the interval $[-1, 1]$ on the vertical axis. In particular, the picture does not indicate the existence of a definite number to which $f(x)$ approaches as x approaches 0. Thus, the picture suggests that the function does not have a limit at 0. Indeed, there are points that are arbitrarily close to 0 at which the function has values 1 or -1 (determine such points as an exercise). You can imagine that the graph of f oscillates between 1 and -1 "infinitely often" near 0. $\square$

Problems

In problems 1 - 8 the graph of a function f is displayed. Determine $\lim_{x \to a+} f(x)$, $\lim_{x \to a-} f(x)$ and $\lim_{x \to a} f(x)$, as indicated by the picture, provided that such values exists. Based on your response, is f continuous at a?

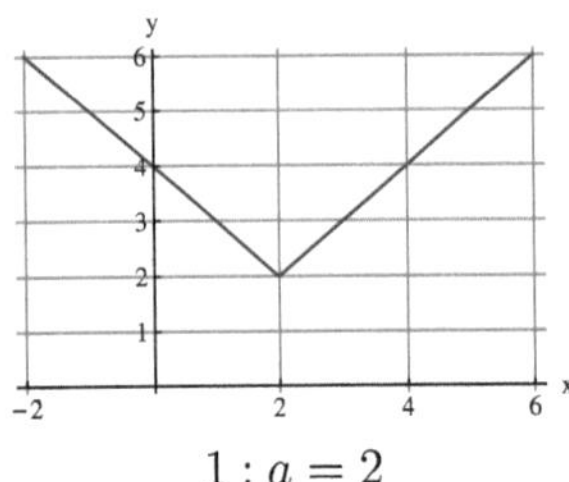

$1 : a = 2$

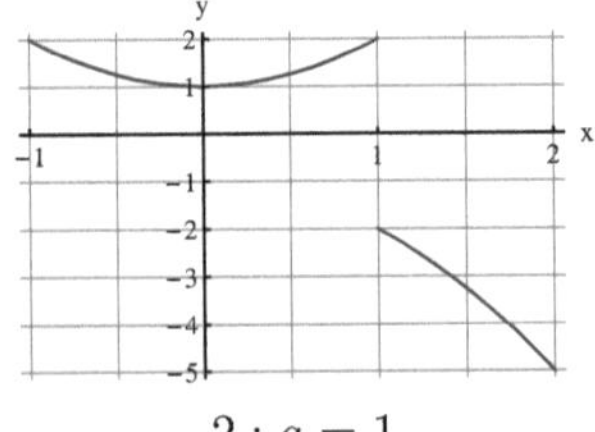

$2 : a = 1$

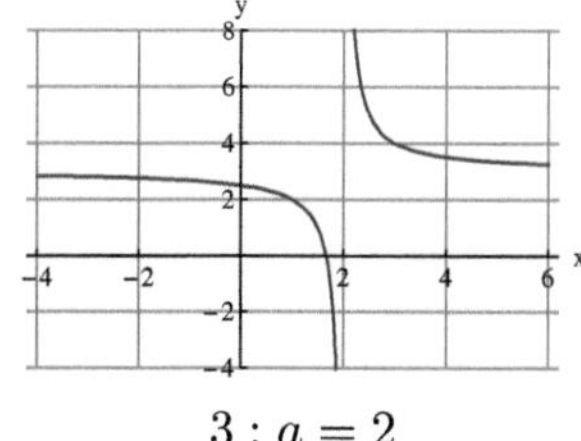

$3: a = 2$

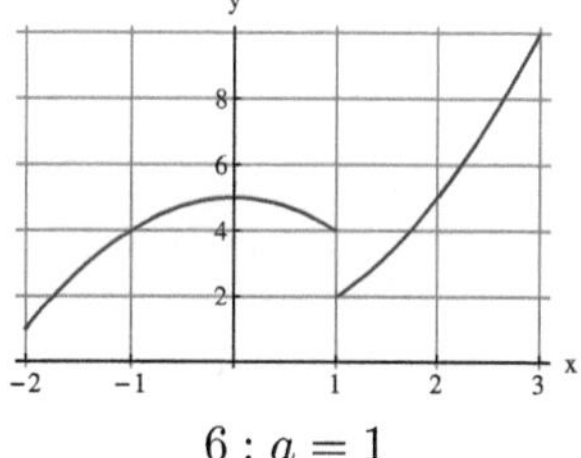

$6: a = 1$

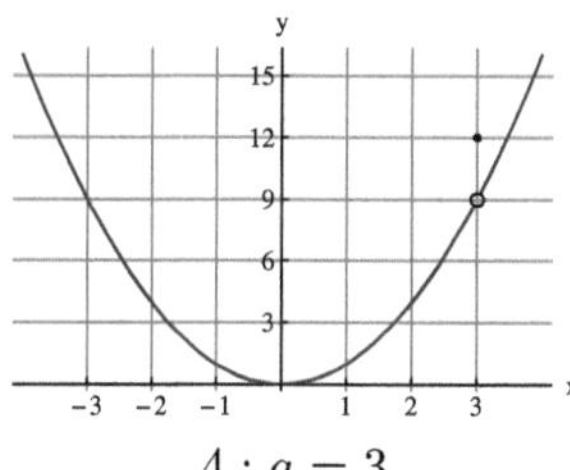

$4: a = 3$

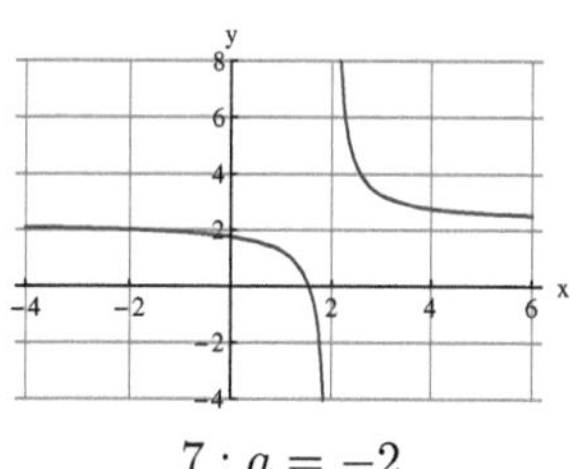

$7: a = -2$

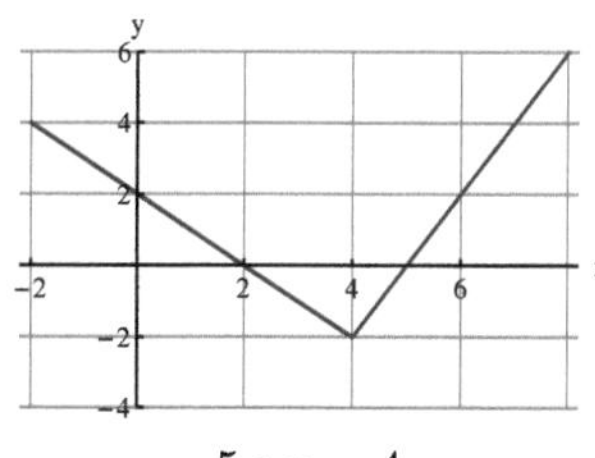

$5: a = 4$

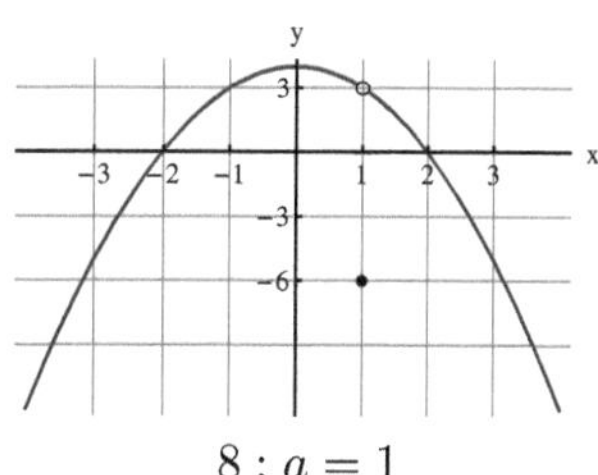

$8: a = 1$

In problems 9 - 12, calculate $f(a+h)$ for $h = \pm 10^{-n}$, $n = 2, 3, 4, 5$. Do the numbers approach a definite number that should be the limit of f at a?. If that is the case, is f continuous at a?

9.
$$a = 4 \text{ and } f(x) = \begin{cases} -2x + 3 & \text{if } x < 4, \\ 3x - 7 & \text{if } x \geq 4. \end{cases}$$

10.
$$a = 2 \text{ and } f(x) = \begin{cases} x^2 + 4 & \text{if } x < 2, \\ x^3 & \text{if } x \geq 2. \end{cases}$$

11.
$$a = 3 \text{ and } f(x) = \begin{cases} \dfrac{1}{x-3} & \text{if } x < 3, \\ x & \text{if } x \geq 3. \end{cases}$$

12.
$$a = \pi \text{ and } f(x) = \sin\left(\frac{x}{4}\right)$$

1.4 The Precise Definitions (Optional)

We said that the limit of the function f at the point a is L if $f(x)$ is as close to L as desired provided that $x \neq a$ and x is sufficiently close to a. The phrases "as close to L as desired " and "sufficiently close" need to be quantified in order to make the definition precise. Let $\varepsilon > 0$ represent an "error tolerance" that can be arbitrarily small. We should be able to ensure that $|f(x) - L| < \varepsilon$ by having $x \neq a$ and $|x - a| < \delta$, where δ is a sufficiently small positive number.

Definition 1 (The precise definition of the limit of a function at a point) Assume that $f(x)$ is defined for each x in some open interval that contains a, with the possible exception of a itself. **The limit of f at a is L** if, given any $\varepsilon > 0$ there exists $\delta > 0$ such that $|f(x) - L| < \varepsilon$ provided that $x \neq a$ and $|x - a| < \delta$.

Note that $|x - a| < \delta$ if and only if $a - \delta < x < a + \delta$, and $|f(x) - L| < \varepsilon$ if and only if $L - \varepsilon < f(x) < L + \varepsilon$. Therefore, the limit of f at a is L if we can ensure that $f(x)$ is in the open interval $(L - \varepsilon, L + \varepsilon)$, where $\varepsilon > 0$ is as small as desired, by restricting x to be in an interval of the form $(a - \delta, a + \delta)$, where $\delta > 0$ is small enough, and $x \neq a$.

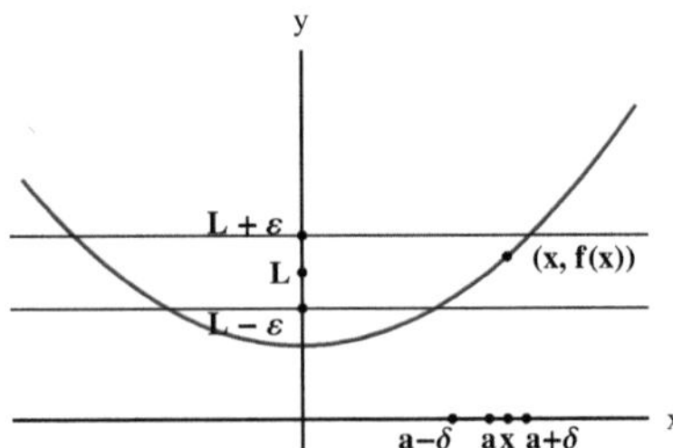

Figure 1: The $\varepsilon - \delta$ definition of the limit

Example 1 Let
$$f(x) = \frac{4x^2 - 36}{x - 3}.$$

a) Determine $L = \lim_{x \to 3} f(x)$.
b) Show that L is the limit of of f at 3 in accordance with the precise definition of the limit.

Solution

a) We have
$$\lim_{x \to 3} f(x) = \lim_{x \to 3} \frac{4x^2 - 36}{x - 3} = \lim_{x \to 3} \frac{4\left(x^2 - 9\right)}{x - 3} = \lim_{x \to 3} 4(x + 3) = 4(6) = 24.$$

b) If $x \neq 3$,
$$|f(x) - 24| = |4(x + 3) - 24| = |4x - 12| = |4(x - 3)| = 4|x - 3|.$$

Let $\varepsilon > 0$ be the given error tolerance. In order to ensure that $|f(x) - 24| < \varepsilon$, it is sufficient to have $4|x - 3| < \varepsilon$, i.e., $|x - 3| < \varepsilon/4$. With reference to Definition 1, we can set $\delta = \varepsilon/4$. $\square$

A function f is continuous at a point a if and only if $\lim_{x \to a} f(x) = f(a)$. By the precise definition of the limit, given $\varepsilon > 0$ there exists $\delta > 0$ such that $|f(x) - f(a)| < \varepsilon$ if $x \neq a$ and $|x - a| < \delta$. If $x = a$ then $|f(x) - f(a)| = |f(a) - f(a)| = 0$, and 0 is less than any $\varepsilon > 0$. Thus, we can state the precise definition of continuity as follows:

Definition 2 (The precise definition of continuity) Assume that $f(x)$ is defined for each x in some open interval that contains a. The function f **is continuous at** a if, given $\varepsilon > 0$ there exists $\delta > 0$ such that $|f(x) - f(a)| < \varepsilon$ provided that $|x - a| < \delta$.

Thus, f is continuous at a if, given $\varepsilon > 0$ there exists $\delta > 0$ such that $f(a) - \varepsilon < f(x) < f(a) + \varepsilon$, provided that $a - \delta < x < a + \delta$.

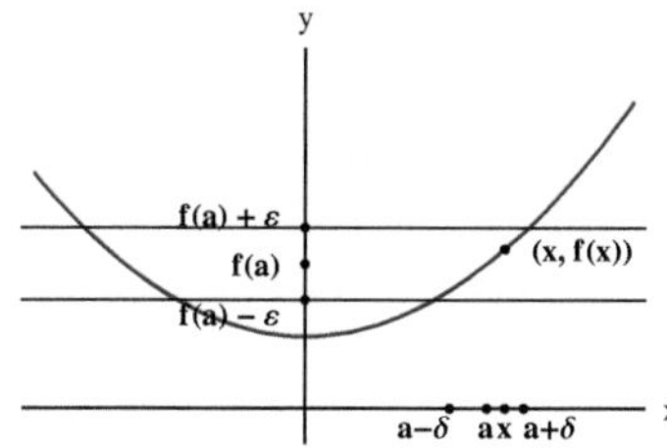

Figure 2: The $\varepsilon - \delta$ definition of continuity

Example 2 Let $f(x) = x^2$. Prove that f is continuous at 4 in accordance with the precise definition of continuity.

Solution

We have
$$|f(x) - f(4)| = \left|x^2 - 16\right| = |(x+4)(x-4)| = |x+4|\,|x-4|.$$

By the triangle inequality,
$$|f(x) - f(4)| \le (|x| + 4)\,|x - 4|.$$

Since we are entitled to have $|x - 4|$ as small as necessary in order to meet a requirement of the form $|f(x) - f(4)| < \varepsilon$, let us restrict x so that $|x - 4| < 1$. In this case,
$$|x| = |(x-4) + 4| \le |x - 4| + 4 < 1 + 4 = 5$$

(with the help of the triangle inequality). Therefore,
$$|f(x) - f(4)| \le (|x| + 4)\,|x - 4| < (5 + 4)\,|x - 4| = 9\,|x - 4|,$$

provided that $|x - 4| < 1$. If we are required to have $|f(x) - f(4)| < \varepsilon$, where $\varepsilon > 0$ is an arbitrary error tolerance, it is sufficient to have $|x - 4| < 1$ and $9\,|x - 4| < \varepsilon$. With reference to the precise definition of continuity, we can set
$$\delta = \min\left(1, \frac{\varepsilon}{9}\right).$$

$\square$

Example 3 Let $f(x) = 1/x$. Prove that f is continuous at $1/2$.

Solution

We have
$$f(x) - f(1/2) = \frac{1}{x} - 2 = \frac{1 - 2x}{x}.$$

Therefore,
$$|f(x) - f(1/2)| = \frac{|1 - 2x|}{|x|} = \frac{|2x - 1|}{|x|} = \frac{2\,|x - 1/2|}{|x|}.$$

If $|x - 1/2| < 1/4$, then $1/4 < x < 3/4$. Therefore

$$\frac{1}{|x|} = \frac{1}{x} < 4.$$

Thus,

$$|f(x) - f(1/2)| = \frac{2\,|x - 1/2|}{|x|} < 8\,|x - 1/2|$$

if $|x - 1/2| < 1/4$. With reference to the precise definition of continuity, given $\varepsilon > 0$, we can set

$$\delta = \min\left(1/4, \varepsilon/8\right).$$

$\square$

If we set $x = a + h$, then x approaches a as h approaches 0. Therefore, we have

$$\lim_{x \to a} f(x) = \lim_{h \to 0} f(a + h).$$

Thus, we can express the precise definitions of the limit of a function and continuity as follows:

Definition 3 (The alternative definitions of a limit and continuity) The **limit of a function** f **at a point** a **is** L if, given $\varepsilon > 0$ there exists $\delta > 0$ such that $|f(a + h) - L| < \varepsilon$ provided that $h \neq 0$ and $|h| < \delta$. The function f **is continuous at the point** a, if given $\varepsilon > 0$ there exists $\delta > 0$ such that $|f(a + h) - f(a)| < \varepsilon$ provided that $|x - a| < \delta$.

Thus, the limit of f at a is L if, given any $\varepsilon > 0$ there exists $\delta > 0$ such that $L - \varepsilon < f(a + h) < L + \varepsilon$, provided that $h \neq 0$ and $-\delta < h < \delta$. The function f is continuous at a if, given any $\varepsilon > 0$ there exists $\delta > 0$ such that $f(a) - \varepsilon < f(a + h) < f(a) + \varepsilon$, provided that $-\delta < h < \delta$.

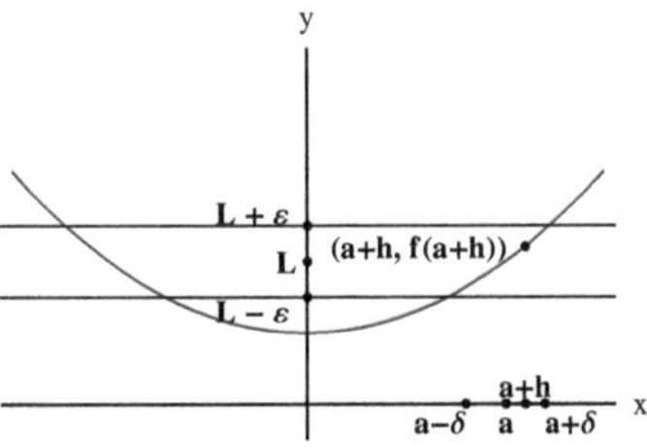

Figure 3: $f(a + h)$ is close to L if $|h|$ is small

Example 4   Prove that

$$\lim_{x \to 2} \frac{x^3 - 8}{x - 2} = 12$$

with reference to the alternative definition of the limit.

Solution

We set

$$f(x) = \frac{x^3 - 8}{x - 2}.$$

If $h \neq 0$,

$$f(2 + h) = \frac{(2 + h)^3 - 8}{h} = \frac{8 + 12h + 6h^2 + h^3 - 8}{h} = \frac{h\left(12 + 6h + h^2\right)}{h} = 12 + 6h + h^2.$$

Therefore,

$$|f(2+h) - 12| = \left|(12 + 6h + h^2) - 12\right| = \left|6h + h^2\right| \le 6\,|h| + h^2 = |h|\,(6 + |h|)$$

(with the help of the triangle inequality). If $h \ne 0$ and $|h| < 1$, we have

$$|f(2+h) - 12| \le |h|\,(6 + |h|) < |h|\,(6 + 1) = 7\,|h|.$$

Given $\varepsilon > 0$, in order to have $|f(2+h) - 12| < \varepsilon$ it is sufficient to have $h \ne 0$, $|h| < 1$ and $7\,|h| < \varepsilon$, i.e., $|h| < \varepsilon/7$. With reference to the alternative definition of the limit, we can set

$$\delta = \min\left(1, \frac{\varepsilon}{7}\right).$$

$\square$

Example 5 Let $f(x) = x^3$. Prove that f is continuous at an arbitrary point $a \in \mathbb{R}$, with reference to the alternative definition of continuity.

Solution

We have

$$f(a+h) - f(a) = (a+h)^3 - a^3 = a^3 + 3a^2h + 3ah^2 + h^3 - a^3 = 3a^2h + 3ah^2 + h^3.$$

Therefore,

$$\begin{aligned}
|f(a+h) - f(a)| = \left|3a^2h + 3ah^2 + h^3\right| &\le 3a^2\,|h| + 3\,|a|\,|h| + |h|^3 \\
&= \left(3a^2 + 3\,|a|\,|h| + h^2\right)|h|
\end{aligned}$$

(with the help of the triangle inequality). If $|h| < 1$,

$$|f(a+h) - f(a)| \le \left(3a^2 + 3\,|a|\,|h| + h^2\right)|h| < \left(3a^2 + 3\,|a| + 1\right)|h|.$$

Therefore, given $\varepsilon > 0$, it is sufficient to have $\left(3a^2 + 3\,|a| + 1\right)|h| < \varepsilon$ and $|h| < 1$ in order to ensure that $|f(a+h) - f(a)| < \varepsilon$. With reference to the alternative definition of continuity, we can set

$$\delta = \min\left(1, \frac{\varepsilon}{3a^2 + 3\,|a| + 1}\right).$$

$\square$

Here are the precise definitions of **one-sided limits and continuity:**

Definition 4 (One-sided limits and continuity)

The right limit of f at a is L_+ if, given any $\varepsilon > 0$ there exists $\delta > 0$ such that $|f(x) - L_+| < \varepsilon$ provided that $a < x < a + \delta$.
The left limit of f at a is L_- if, given any $\varepsilon > 0$ there exists $\delta > 0$ such that $|f(x) - L_-| < \varepsilon$ provided that $a - \delta < x < a$.
The function f is continuous at a from the right if, given any $\varepsilon > 0$ there exists $\delta > 0$ such that $|f(x) - f(a)| < \varepsilon$ provided that $a \le x < a + \delta$.
The function f is continuous at a from the left if, given any $\varepsilon > 0$ there exists $\delta > 0$ such that $|f(x) - f(a)| < \varepsilon$ provided that $a - \delta < x \le a$.

Example 6 Let

$$f(x) = \begin{cases} 3x & \text{if} \quad x < 2, \\ 2x + 4 & \text{if} \quad x \ge 2. \end{cases}$$

a) Prove that $\lim_{x \to 2-} f(x) = 6$ in accordance with the precise definition of a left limit.
b) Prove that f is continuous at 2 from the right in accordance with the precise definition.

Solution

a) If $x < 2$,
$$|f(x) - 6| = |3x - 6| = |3(x - 2)| = 3|x - 2| = 3(2 - x)$$

Therefore, in order to have $|f(x) - 6| < \varepsilon$, where ε is an arbitrary positive number, it is sufficient to have $x < 2$ and $3(2 - x) < \varepsilon$, i.e., $2 - x < \varepsilon/3$. This is the case if
$$2 - \frac{\varepsilon}{3} < x < 2.$$

With reference to Definition 4, we can set $\delta = \varepsilon/3$. Therefore, $\lim_{x \to 2-} f(x) = 6$.
b) If $x \geq 2$,
$$|f(x) - f(2)| = |f(x) - 8| = |(2x + 4) - 8| = |2x - 4|$$
$$= |2(x - 2)| = 2|x - 2| = 2(x - 2).$$

Therefore, in order to have $|f(x) - 8| < \varepsilon$, where ε is an arbitrary positive number, it is sufficient to have $x > 2$ and $2(x - 2) < \varepsilon$, i.e., $x - 2 < \varepsilon/2$. This is the case if
$$2 < x < 2 + \frac{\varepsilon}{2}.$$

With reference to Definition 4, we can set $\delta = \varepsilon/2$. Therefore, f is continuous at 2 from the right. $\square$

Problems

In problems 1 - 4, justify the indicated limit in accordance with the precise definition (you may refer to the alternative definition).

1.
$$\lim_{x \to 1} \frac{2x^2 - 2}{x - 1} = 4$$

3.
$$\lim_{x \to 2} \frac{x + 2}{x^2 + 3x + 2} = \frac{1}{3}$$

2.
$$\lim_{x \to 3} \frac{x^2 + x - 12}{x - 3} = 7$$

4.
$$\lim_{x \to -1} \frac{x^3 + x^2 + 4x + 4}{x + 1} = 5$$

In problems 5 - 8, justify the continuity of f at a in accordance with the precise definition (you may refer to the alternative definition).

5.
$$f(x) = 3x^2 + x - 1, \ a = 1$$

7.
$$f(x) = \frac{1}{x - 2}, \ x = 4$$

6.
$$f(x) = x^3 + 4x, \ x = -1$$

8.
$$f(x) = \frac{1}{x^2}, \ x = 2$$

In problems 9 and 10, justify the indicated one-sided limit of f in accordance with the precise definition.

9.
$$f(x) = \begin{cases} \dfrac{x^2 + x - 12}{x - 3} & \text{if } x < 3, \\ x^4 + x^2 + 1 & \text{if } x \geq 3. \end{cases} \quad , \quad \lim_{x \to 3-} f(x) = 7$$

10.
$$f(x) = \begin{cases} \sin(x) & \text{if } x < 1, \\[2mm] \dfrac{x-1}{x^2-1} & \text{if } x > 1. \end{cases} \qquad , \qquad \lim_{x\to 1+} f(x) = \frac{1}{2}$$

In problems 11 and 12, justify the indicated one-sided continuity of f in accordance with the precise definition.

11. If
$$f(x) = \begin{cases} \dfrac{1}{x+2} & \text{if } x < -2, \\[2mm] x^2+4 & \text{if } x \geq -2, \end{cases}$$

then f is continuous at -2 from the right.

12. If
$$f(x) = \begin{cases} x^3 + 2x^2 & \text{if } x \leq 1, \\[2mm] \dfrac{1}{x} & \text{if } x > 1, \end{cases}$$

then f is continuous at 1 from the left.

1.5   The Calculation of Limits

In this section we will provide guidelines for the determination of limits.

A Portfolio of Continuous Functions

Since the limit of a function f at a point a is simply the value of f at a, let's begin by taking stock of a rich collection of continuous functions:

Polynomials, rational functions, sine, cosine, tangent and secant are continuous on their respective natural domains.

You can find the justification of these facts in Appendix B.

In particular, a polynomial is continuous on the entire number line.

Example 1 Let
$$f(x) = 1 - \frac{1}{2}x^2 + \frac{1}{24}x^4.$$

Evaluate $\lim_{x \to \sqrt{2}} f(x)$.

Solution

Figure 1 shows the graph of f. The graph is a curve without any breaks or holes, consistent with the continuity of f on the number line.

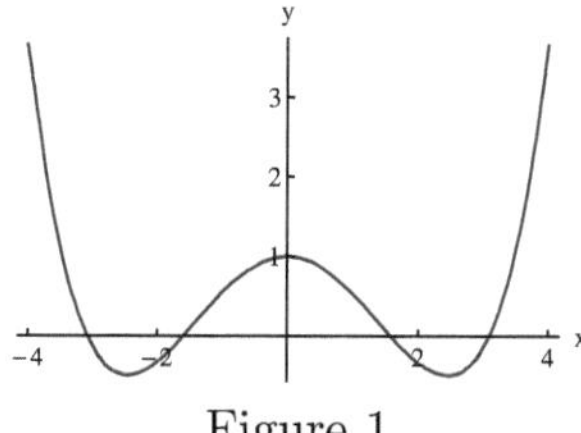

Figure 1

By the continuity of f at $\sqrt{2}$,

$$\lim_{x \to \sqrt{2}} f(x) = f\left(\sqrt{2}\right) = 1 - \frac{1}{2}\left(\sqrt{2}\right)^2 + \frac{1}{24}\left(\sqrt{2}\right)^4 = \frac{1}{6}.$$

$\square$

Example 2 Let

$$f(x) = \frac{x^2 + 1}{x^2 - 1}.$$

a) Determine the points at which f is discontinuous.

b) Evaluate $\lim_{x \to 2} f(x)$.

Solution

a) The rational function f is continuous at any point of its natural domain, i.e., at any point where the denominator does not vanish, Since

$$x^2 - 1 = 0 \Leftrightarrow x = \pm 1,$$

f is continuous at a if $a \neq 1$ and $a \neq -1$.

b) By the continuity of f at 2,

$$\lim_{x \to 2} f(x) = f(2) = \frac{5}{3}.$$

Figure 2 shows the graph of f. Since f is continuous at each point of the intervals $(-\infty, -1)$, $(-1, 1)$ and $(1, +\infty)$, the corresponding parts of the graph of f are continuous curves. On the other hand, f is not defined at 1 or -1, so that f is discontinuous at these points, and the graph has breaks at $x = 1$ and $x = -1$. $\square$

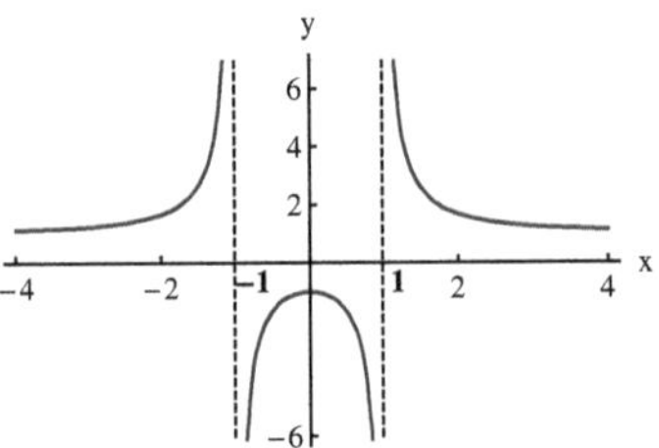

Figure 2: The function has discontinuities at ± 1

The trigonometric functions sine and cosine are periodic functions that are defined on the entire number line. Figure 3 shows their graphs on the interval $[-2\pi, 2\pi]$.

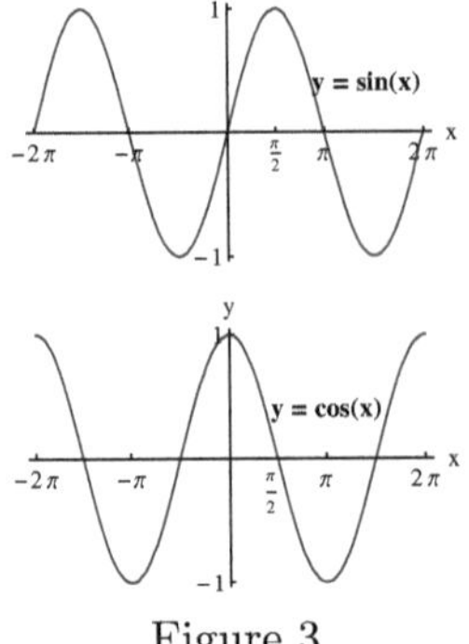

Figure 3

The graphs are "continuous waves", consistent with the fact that the functions are periodic functions that are continuous on the number line.

Example 3 Determine

$$\lim_{x \to \pi/6} \sin(x) \quad \text{and} \quad \lim_{x \to \pi/6} \cos(x).$$

Solution

By the continuity of sine at $\pi/6$

$$\lim_{x \to \pi/6} \sin(x) = \sin\left(\frac{\pi}{6}\right) = \frac{1}{2}.$$

By the continuity of cosine at $\pi/6$,

$$\lim_{x \to \pi/6} \cos(x) = \cos\left(\frac{\pi}{6}\right) = \frac{\sqrt{3}}{2}.$$

$\square$

The natural domain of

$$\tan(x) = \frac{\sin(x)}{\cos(x)}$$

consists of all x such that $\cos(x) \neq 0$. Thus, tangent is continuous at each x that is not an odd multiple of $\pm\pi/2$. Figure 4 shows the graph of $y = \tan(x)$ on $(-3\pi/2, 3\pi/2)$. The parts of the graph on the intervals $(-3\pi/2, -\pi/2)$, $(-\pi/2, \pi/2)$ and $(\pi/2, 3\pi/2)$ are continuous curves, consistent with the continuity of the function at each point of such an interval.

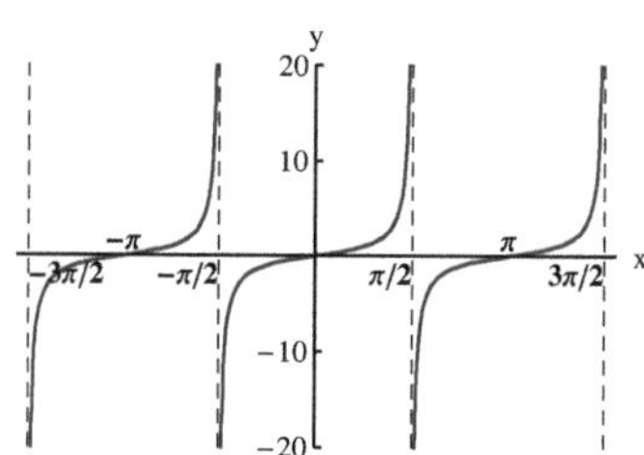

Figure 4: $y = \tan(x)$

The natural domain of

$$\sec(x) = \frac{1}{\cos(x)}$$

also consists of all x such that $\cos(x) \neq 0$. Thus, secant is continuous at each x that is not an odd multiple of $\pm\pi/2$. Figure 5 shows the graph of secant on the interval $(-3\pi/2, 3\pi/2)$. The parts of the graph of secant on the intervals $(-3\pi/2, -\pi/2)$, $(-\pi/2, \pi/2)$ and $(\pi/2, 3\pi/2)$ are continuous curves, consistent with the continuity of secant on these intervals.

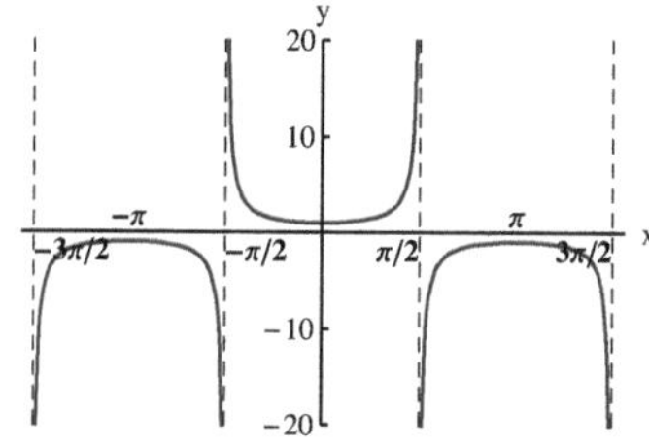

Figure 5: $y = \sec(x)$

Example 4 Evaluate

$$\lim_{x \to \pi/4} \tan(x) \quad \text{and} \quad \lim_{x \to \pi/4} \sec(x).$$

Solution

By the continuity of tangent at $\pi/4$,

$$\lim_{x \to \pi/4} \tan(x) = \tan\left(\frac{\pi}{4}\right) = 1.$$

By the continuity of secant at $\pi/4$,

$$\lim_{x \to \pi/4} \sec(x) = \sec(\pi/4) = \frac{1}{\cos(\pi/4)} = \frac{1}{\frac{1}{\sqrt{2}}} = \sqrt{2}.$$

$\square$

A rational power of x defines a function that is continuous at each point of its natural domain:

If r is a rational number the function defined by x^r is continuous at each point of is natural domain, with the understanding that continuity is from the right or from the left only, if appropriate.

You can find the proof of this fact in Appendix B.

Example 5 Let $f(x) = \sqrt{x} = x^{1/2}$. **The square-root function f** is continuous on the interval $[0, +\infty)$. The continuity of f at 0 is only from the right. The graph of f on any interval that is contained in $[0, +\infty)$ is a continuous curve, consistent with the continuity of f on any such interval. $\square$

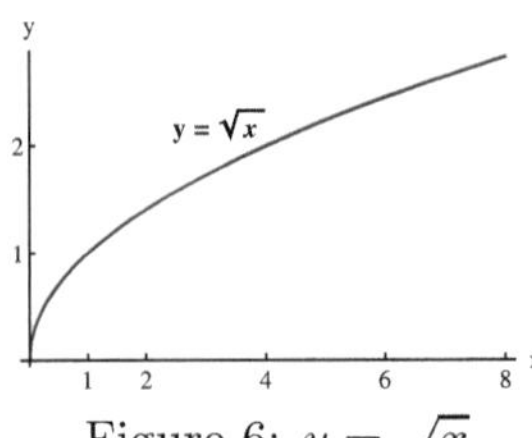

Figure 6: $y = \sqrt{x}$

The **cube-root function** is a prototype of functions defined by $x^{1/n}$, where n is an odd positive integer:

Example 6 Let $f(x) = x^{1/3}$. Then, f is continuous on the entire number line. The graph of the cube-root function on any interval is a continuous curve. $\square$

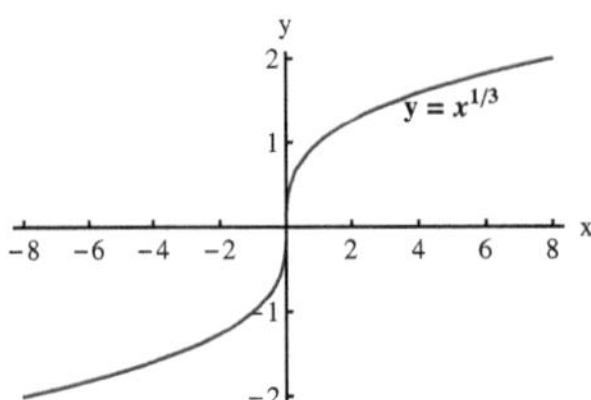

Fiugre 7: $y = x^{1/3}$

Example 7 Let $f(x) = x^{3/4} = \left(x^{1/4}\right)^3$. Then, f is continuous on $[0, +\infty)$. The continuity of f at 0 is only from the right. $\square$

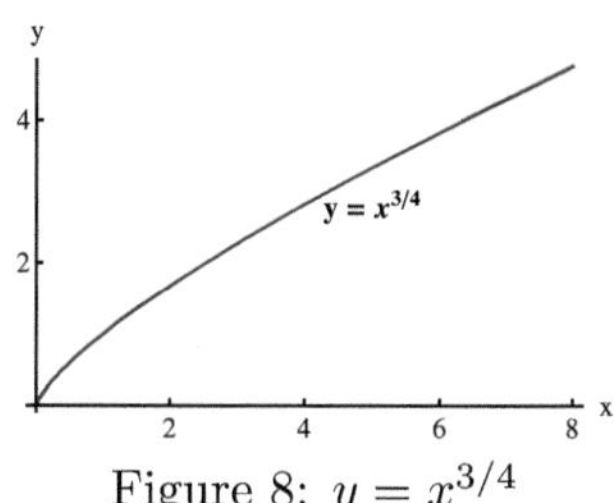

Figure 8: $y = x^{3/4}$

Example 8 Let $f(x) = x^{2/3} = \left(x^{1/3}\right)^2$. Then, f is continuous on the entire number line. $\square$

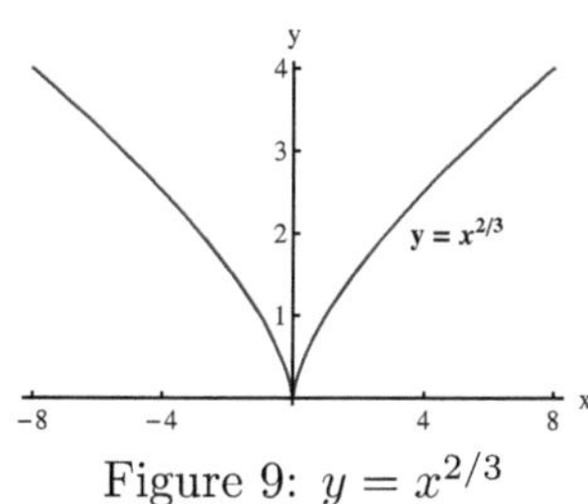

Figure 9: $y = x^{2/3}$

Limits and Removable Discontinuities

The following observation enables us to compute a limit by making use of our knowledge about continuous functions:

Assume that $f(x) = g(x)$ for each x in an open interval J that contains the point a, with the possible exception of a itself, and that g is continuous at a. Then,

$$\lim_{x \to a} f(x) = g(a).$$

Proof

Since g is continuous at a, we have $\lim_{x \to a} g(x) = g(a)$. Since $f(x) = g(x)$ for each x in J such that $x \neq a$, and the definition of the limit of f at a does not involve a,

$$\lim_{x \to a} f(x) = \lim_{x \to a} g(x) = g(a).$$

■

Example 9 Let

$$f(x) = \frac{4x - 16}{x^2 - 16}.$$

Determine $\lim_{x \to 4} f(x)$.

Solution

Since

$$x^2 - 16 = 0 \Leftrightarrow x = 4 \text{ or } x = -4,$$

the rational function f is not defined at 4. We are led to the indeterminate form $0/0$ if we try to replace x in the expression for $f(x)$ by 4. Let us simplify the expression for $f(x)$:

$$f(x) = \frac{4x - 16}{x^2 - 16} = \frac{4(x - 4)}{(x - 4)(x + 4)} = \frac{4}{x + 4}$$

if $x \neq 4$. Thus, if we set

$$g(x) = \frac{4}{x + 4},$$

then $f(x) = g(x)$ for each x such that $x \neq 4$ and $x \neq -4$. The rational function g *is* defined and continuous at $x = 4$. Since $f(x) = g(x)$ for each x near 4 such that $x \neq 4$. Therefore

$$\lim_{x \to 4} f(x) = \lim_{x \to 4} g(x) = g(4) = \frac{4}{8} = \frac{1}{2}.$$

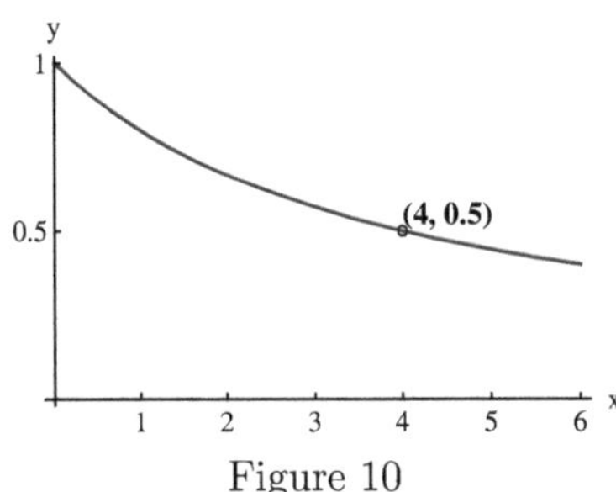

Figure 10

Figure 10 shows a computer generated graph of f on the interval $[0, 6]$, as generated by a graphing utility. The computer does not seem to be aware of the fact that f is not defined at 4, and has produced a continuous curve. It would have produced the same picture if it had been asked to plot the graph of the function g which is continuous at each point of $[0, 6]$, including 4, since $g(x) = f(x)$ for each $x \neq 4$ in that interval. The picture is consistent with the fact that $\lim_{x \to 4} f(x) = 0.5$. It appears that we can "remove " the discontinuity of f at 4 by declaring that its value at 4 is 0.5. $\square$

Definition 1 Assume that a function f is defined in an open interval J that contains the point a, with the possible exception of a itself, and that f is discontinuous at a. We say that f **has a removable discontinuity at** a if $\lim_{x \to a} f(x)$ exists.

The terminology is appropriate since g is continuous at a if

$$g(x) = \begin{cases} f(x) & \text{if } \quad x \neq a \text{ and } x \in J, \\ \lim_{x \to a} f(x) & \text{if } \qquad x = a. \end{cases}$$

We can say that the discontinuity of f at a is removed by defining or redefining its value at a properly.

Note that the function f of Example 9 has a removable discontinuity at 4.

Limits and Continuity of Combinations of Functions

The following rules are relevant to the calculation of the limits of sums and products of functions:

LIMITS OF ARITHMETIC COMBINATIONS OF FUNCTIONS

Assume that $\lim_{x\to a} f(x)$ and $\lim_{x\to a} g(x)$ exist and that c is a constant.

1. The constant multiple rule for limits:

$$\lim_{x\to a} cf(x) = c \lim_{x\to a} f(x)$$

2. The sum rule for limits:

$$\lim_{x\to a} (f(x) + g(x)) = \lim_{x\to a} f(x) + \lim_{x\to a} g(x)$$

(the limit of a sum is the sum of the limits).

3. The product rule for limits:

$$\lim_{x\to a} f(x)g(x) = \left(\lim_{x\to a} f(x)\right)\left(\lim_{x\to a} g(x)\right)$$

(the limit of a product is the product of the limits).

4. The quotient rule for limits: *If $\lim_{x\to a} g(x) \neq 0$,*

$$\lim_{x\to a} \frac{f(x)}{g(x)} = \frac{\lim_{x\to a} f(x)}{\lim_{x\to a} g(x)}$$

(the limit of a quotient is the quotient of the limits).

The above rules are plausible: If $\lim_{x\to a} f(x) = L_1$ and $\lim_{x\to a} g(x) = L_2$, we have $f(x) \cong L_1$ and $g(x) \cong L_2$ if $x \neq a$ and $x \cong a$. Therefore,

$$cf(x) \cong cL_1, \ f(x) + g(x) \cong L_1 + L_2, \ f(x)g(x) \cong L_1 L_2,$$

and

$$\frac{f(x)}{g(x)} \cong \frac{L_1}{L_2},$$

if $L_2 \neq 0$. You can find the proofs of the above statements in Appendix B.

Since a function f is continuous at a point a if $\lim_{x\to a} f(x) = f(a)$, the rules for the limits of arithmetic combinations of functions lead to the continuity of arithmetic combinations of continuous functions:

Assume that f and g are continuous at a and c is a constant. Then the constant multiple cf, the sum $f + g$ and the product fg are continuous at a. If $g(a) \neq 0$, the quotient f/g is also continuous at a.

Example 10 Evaluate
$$\lim_{x\to \pi/4} \sqrt{x}\cos(x).$$

Solution

Since $\sqrt{x}$ defines a continuous function on $[0, \infty)$ and cosine is continuous on the entire number line, the product $\sqrt{x}\cos(x)$ defines a function that is continuous at $\pi/4$. Therefore,

$$\lim_{x\to \pi/4} \sqrt{x}\cos(x) = \sqrt{\frac{\pi}{4}}\cos\left(\frac{\pi}{4}\right) = \frac{\sqrt{\pi}}{2}\left(\frac{\sqrt{2}}{2}\right) = \frac{\sqrt{2\pi}}{4}.$$

$\square$

Example 11 Determine $\lim_{h \to 0} f(h)$ if

$$f(h) = \frac{\sqrt{4 + h} - 2}{h}.$$

Solution

The function f is not defined at 0. The attempt to replace h by 0 leads to the indeterminate form $0/0$. We will obtain another expression for $f(h)$ by rationalizing the numerator. If $h \neq 0$,

$$\frac{\sqrt{4 + h} - 2}{h} = \left(\frac{\sqrt{4 + h} - 2}{h} \right) \left(\frac{\sqrt{4 + h} + 2}{\sqrt{4 + h} + 2} \right)$$

$$= \frac{\left(\sqrt{4 + h}\right)^2 - 2^2}{h\left(\sqrt{4 + h} + 2\right)} = \frac{(4 + h) - 4}{h\left(\sqrt{4 + h} + 2\right)} = \frac{h}{h\left(\sqrt{4 + h} + 2\right)} = \frac{1}{\sqrt{4 + h} + 2}$$

If we set

$$g(h) = \frac{1}{\sqrt{4 + h} + 2},$$

the function g *is* defined at 0. In fact, g is continuous at 0 since the square-root function is continuous at 4 and the denominator is nonzero at $h = 0$. Since $f(h) = g(h)$ if $h \neq 0$ and $|h|$ is small enough,

$$\lim_{h \to 0} f(h) = \lim_{h \to 0} g(h) = g(0) = \frac{1}{\sqrt{4} + 2} = \frac{1}{4}.$$

Figure 11 shows the graph of f on the interval $[-2, 2]$, as plotted by a graphing utility. The graphing utility would have produced the same picture if it had been asked to plot the graph of g. The picture is consistent with the fact that $\lim_{h \to 0} f(h) = 1/4$. $\square$

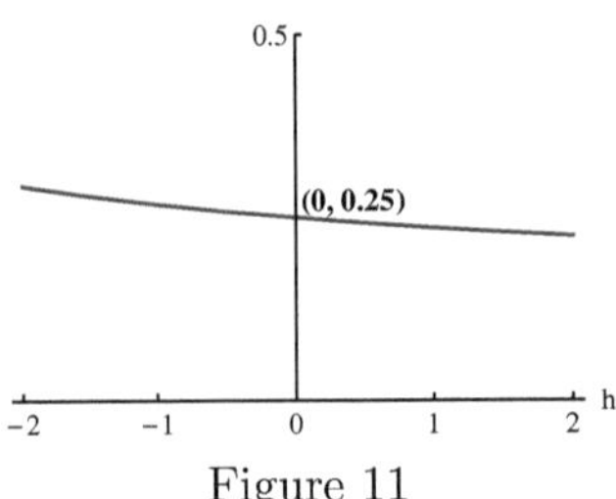

Figure 11

Many functions are formed by composing simpler functions. Therefore, we should be able to calculate limits involving composite functions:

THE LIMIT OF A COMPOSITE FUNCTION

Assume that $\lim_{x \to a} g(x) = L$ and f is continuous at L. Then,

$$\lim_{x \to a} (f \circ g)(x) = \lim_{x \to a} f(g(x)) = f(L).$$

It is easy to remember this fact in the following form:

$$\lim_{x \to a} f(g(x)) = f\left(\lim_{x \to a} g(x)\right).$$

You can find the proof of the above statement in Appendix B. The statement is plausible: As x approaches a, $g(x)$ approaches L. Therefore, $f(g(x))$ approaches $f(L)$ by the continuity of f at L.

If g is continuous at a, we have $\lim_{x \to a} g(x) = g(a)$. Therefore, the above fact about the limits of composite functions leads to the continuity of compositions of continuous functions:

Assume that g is continuous at a and f is continuous at $g(a)$. Then, $f \circ g$ is continuous at a. We have

$$\lim_{x \to a} f(g(x)) = f(g(a)).$$

Example 12 Evaluate

$$\lim_{x \to 1} \cos\left(\frac{\pi(x^2 - 1)}{6(x - 1)}\right)$$

Solution

We have

$$\lim_{x \to 1} \frac{\pi(x^2 - 1)}{6(x - 1)} = \lim_{x \to 1} \frac{\pi(x - 1)(x + 1)}{6(x - 1)} = \lim_{x \to 1} \frac{\pi(x + 1)}{6} = \frac{2\pi}{6} = \frac{\pi}{3}.$$

Since cosine is continuous at $\pi/3$,

$$\lim_{x \to 1} \cos\left(\frac{\pi(x^2 - 1)}{6(x - 1)}\right) = \cos\left(\lim_{x \to 1} \frac{\pi(x^2 - 1)}{6(x - 1)}\right) = \cos\left(\frac{\pi}{3}\right) = \frac{1}{2}.$$

$\square$

Example 13 Let

$$F(x) = \tan\left(\frac{3\pi}{4(x^2 - 1)}\right).$$

Justify the continuity of F at 2. Determine $\lim_{x \to 2} F(x)$.

Solution

If we set

$$u = g(x) = \frac{3\pi}{4(x^2 - 1)} \quad \text{and} \quad f(u) = \tan(u),$$

then $F(x) = f(g(x))$ so that $F = f \circ g$. The rational function g is continuous at 2 and $g(2) = \pi/4$. The tangent function f is continuous at $\pi/4$. Therefore $F = f \circ g$ is continuous at 2. Thus,

$$\lim_{x \to 2} F(x) = F(2) = f(g(2)) = \tan\left(\frac{\pi}{4}\right) = 1.$$

$\square$

We will often encounter functions of the form $\sin(\omega x)$ and $\cos(\omega x)$, where ω is a constant. Such a function is continuous at any point on the number line, since it can be expressed as $f \circ g$, where $g(x) = \omega x$ and $f(u) = \sin(u)$ or $f(u) = \cos(u)$, and both f and g are continuous at any point. Recall that **a trigonometric polynomial** is a linear combination of functions of the form $\sin(nx)$ and $\cos(nx)$, where n is an integer (as we saw in Section 1.2). Since such functions are continuous at each point on the number line, a trigonometric polynomial is continuous at each $x \in \mathbb{R}$.

Example 14 Evaluate

$$\lim_{x \to \pi/2} \left(\sin(x) + \frac{1}{3}\sin(3x)\right)$$

Solution

By the continuity of a trigonometric polynomial,

$$\lim_{x \to \pi/2} \left(\sin(x) + \frac{1}{3}\sin(3x) \right) = \sin\left(\frac{\pi}{2}\right) + \frac{1}{3}\sin\left(\frac{3\pi}{2}\right) = 1 + \frac{1}{3}(-1) = \frac{2}{3}.$$

Figure 12 shows the graph of

$$f(x) = \sin(x) + \frac{1}{3}\sin(3x)$$

on the interval $[-2\pi, 2\pi]$. The graph is a continuous curve, consistent with the continuity of f on $[-2\pi, 2\pi]$. $\square$

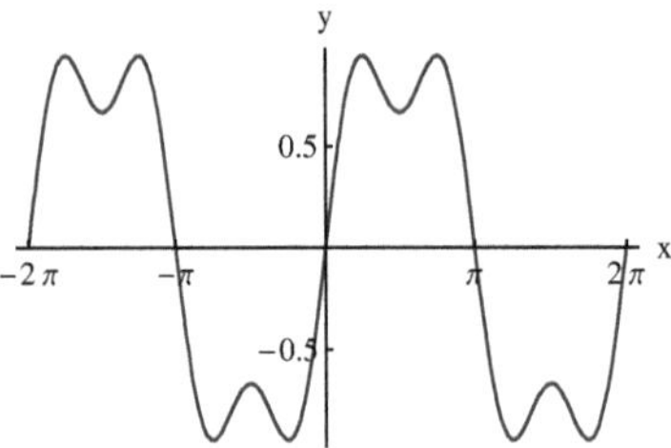

Figure 12 : A trigonometric polynomial is continuous

In some cases, the following theorem is helpful to determine the limit of a function:

THE SQUEEZE THEOREM   **Assume that $h(x) \le f(x) \le g(x)$ for all $x \ne a$ in an open interval containing a, and**

$$\lim_{x \to a} h(x) = \lim_{x \to a} g(x) = L.$$

Then $\lim_{x \to a} f(x) = L$ as well.

The squeeze theorem is intuitively plausible: If the values of f are squeezed between the corresponding values of h and g, and both $h(x)$ and $g(x)$ approach the same limit L as x approaches a, we should have $\lim_{x \to a} f(x) = L$. You can find the proof of the squeeze theorem in Appendix B.

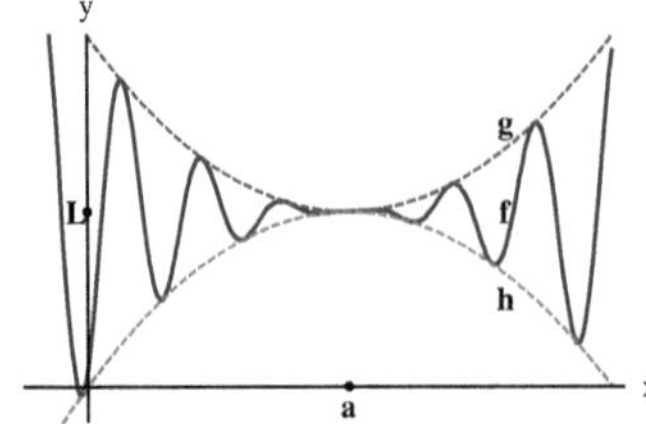

Figure 13: The illustration of the squeeze theorem

Example 15   Determine

$$\lim_{x\to 0} x^2 \sin\left(\frac{1}{x}\right).$$

Solution

Since $-1 \le \sin(1/x) \le 1$, we have

$$-x^2 \le x^2 \sin(1/x) \le x^2$$

for $x \ne 0$. We have

$$\lim_{x\to 0}\left(-x^2\right) = \lim_{x\to 0}\left(x^2\right) = 0.$$

Therefore,

$$\lim_{x\to 0} x^2 \sin\left(\frac{1}{x}\right) = 0,$$

by the Squeeze Theorem. $\square$

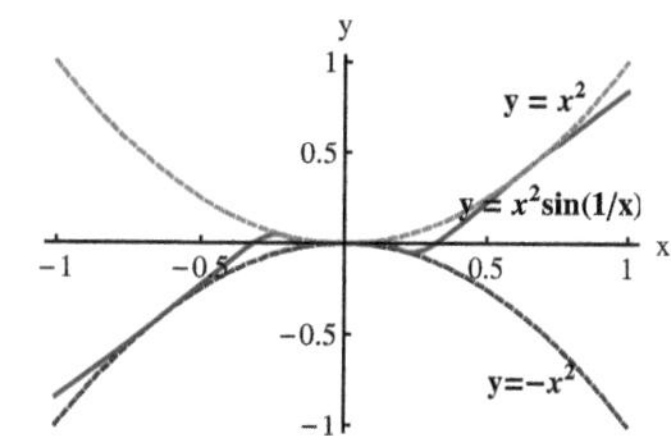

Figure 14: $-x^2 \le x^2 \sin(1/x) \le x^2$

Problems

In problems 1 - 12,
a) Determine whether f is continuous at a. Justify your response,
b) Determine $\lim_{x\to a} f(x)$.

1.
$$f(x) = \frac{\cos(x)}{\sin^2(x)}, \quad a = \pi/6.$$

2.
$$f(x) = \frac{\cos(x)}{\sin^2(x)}, \quad a = \pi.$$

3.
$$f(x) = \frac{x^4 - 16}{x - 2}, \quad a = 3.$$

4.
$$f(x) = \frac{x^3 + x^2 + 1}{x + 3}, \quad a = 2.$$

5.
$$f(x) = \sqrt{x^2 - 9}, \quad a = 2.$$

6.
$$f(x) = \sqrt{4 - x^2}, \quad a = 3.$$

7.
$$f(x) = \sqrt{x^2 - 9}, \quad a = 4.$$

8.
$$f(x) = \left(x^2 - 1\right)^{3/4}, \quad a = 5.$$

9.
$$f(x) = \begin{cases} x^2 - 4 & \text{if } x \le 2, \\ 4 - x^2 & \text{if } x > 2. \end{cases}, \quad a = 2.$$

10.
$$f(x) = \begin{cases} x^2 - 1 & \text{if } x \le 2, \\ 4 - x^2 & \text{if } x > 2. \end{cases}, \quad a = 2.$$

11.
$$f(x) = \begin{cases} 4x - 3 & \text{if } x \ne 2, \\ -3 & \text{if } x = 2. \end{cases}, \quad a = 2.$$

12.
$$f(x) = \begin{cases} x^2 + 1 & \text{if } x \ne 3, \\ 8 & \text{if } x = 3. \end{cases}, \quad a = 3.$$

In problems 13 -20,
a) Determine the function g that is continuous at a such that $g(x) = f(x)$ if $x \neq a$ and x is near a.
b) Evaluate the limit of f at a.

13.
$$f(x) = \frac{x^2 - 4}{x^3 - 8}, \quad a = 2$$

14.
$$f(x) = \frac{2x^2 + 5x - 3}{x^2 + x - 6}, \quad a = -3$$

15.
$$f(x) = \tan(x)\cos(x), \quad a = \pi/2$$

16.
$$f(x) = \frac{\sec(x)}{\tan(x)}, \quad a = 3\pi/2$$

17.
$$f(x) = \frac{x^3 - 27}{x - 3}, \quad a = 3$$

18.
$$f(x) = \frac{\frac{1}{x^2} - \frac{1}{9}}{x - 3}, \quad a = 3$$

19.
$$f(x) = \frac{\sqrt{x} - 4}{x - 16}, \quad a = 16$$

20.
$$f(x) = \frac{x^{1/3} - 2}{x - 8}, \quad a = 8$$

In problems 21-26, evaluate the indicated limit. Indicate the steps that lead to the final result:

21.
$$\lim_{x \to \pi/4} \cos^2(x)$$

22.
$$\lim_{x \to \pi/3} \frac{\sin(x)}{\cos^2(x)}$$

23.
$$\lim_{x \to 4} \frac{x^2 - x - 12}{2x^2 - 12x + 16}$$

24.
$$\lim_{h \to 0} \frac{\sqrt{9 + h} - 3}{h}$$

25.
$$\lim_{x \to -3+} \frac{x^2 - 4x + 3}{\sqrt{x - 3}}$$

26.
$$\lim_{h \to 0} \frac{(4 + h)^3 - 64}{h}$$

In problems 27 - 30, express F as $f \circ g$ and evaluate $\lim_{x \to a} F(x)$.

27.
$$F(x) = \sqrt{\frac{x^2 - 9}{x - 3}}, \quad a = 3$$

28.
$$F(x) = \left(\frac{x^2 - 16}{x - 4}\right)^{1/3}, \quad a = 4$$

29.
$$F(x) = \sqrt{\sin(x)}, \quad a = \pi/6$$

30.
$$F(x) = \cos\left(\frac{\pi x^2 - 4\pi}{3x - 6}\right), \quad a = 2$$

1.6 Infinite Limits

A function may attain arbitrarily large values near a point. In this section we will discuss such cases.

The Definitions

Let's begin with a specific case:

Example 1 Let
$$f(x) = \frac{1}{x - 1}.$$

The function is not defined at 1. If $x > 1$ and x is close to 1, $f(x)$ is large. For example, if n is an arbitrary positive integer,

$$f\left(1+\frac{1}{n}\right) = \frac{1}{\left(1+\frac{1}{n}\right)-1} = \frac{1}{\frac{1}{n}} = n,$$

and n can be as large as we please. If $x < 1$ and x is close to 1, $f(x)$ is a negative number of large magnitude. For example,

$$f\left(1-\frac{1}{n}\right) = \frac{1}{\left(1-\frac{1}{n}\right)-1} = -\frac{1}{\frac{1}{n}} = -n.$$

Figure 1 shows the graph of f.

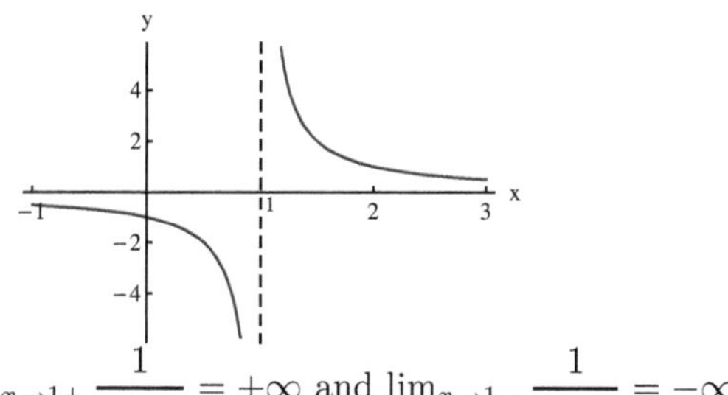

$$\text{Figure 1: } \lim_{x\to 1+}\frac{1}{x-1} = +\infty \text{ and } \lim_{x\to 1-}\frac{1}{x-1} = -\infty$$

We express the fact that $1/(x-1)$ attains arbitrarily large values as x comes closer and closer to 1 from the right by saying that $1/(x-1)$ **tends to** $+\infty$ as x approaches 1 from the right, and write

$$\lim_{x\to 1+}\frac{1}{x-1} = +\infty.$$

We see that a point $(x, f(x))$ on the graph of f approaches the vertical line $x = 1$ as x approaches 1 from the right. We say that the line $x = 1$ is a **vertical asymptote** for the graph of f.
We express the fact that $1/(x-1)$ attains negative values of arbitrarily large magnitude as x comes closer and closer to 1 from the left by saying that $1/(x-1)$ **tends to** $-\infty$ as x approaches 1 from the left, and write

$$\lim_{x\to 1-}\frac{1}{x-1} = -\infty.$$

$\square$

Here are the relevant definitions:

Definition 1 **The right-limit of f at a is** $+\infty$ if $f(x)$ exceeds any number, however large, provided that $x > a$ and x is sufficiently close to a. In this case we write

$$\lim_{x\to a+} f(x) = +\infty$$

(read "the limit of $f(x)$ as x approaches a from the right is plus infinity").
The right-limit of f at a is $-\infty$ if, $f(x) < 0$ and $|f(x)| = -f(x)$ exceeds any number, however large, provided that $x > a$ and x is sufficiently close to a. In this case we write

$$\lim_{x\to a+} f(x) = -\infty$$

(read "the limit of $f(x)$ as x approaches a from the right is minus infinity").

Note that the statement $\lim_{x \to a+} f(x) = -\infty$ is equivalent to the statement

$$\lim_{x \to a+} (-f(x)) = +\infty.$$

The definitions of **infinite left-limits** are similar. We consider $x < a$ instead of $x > a$. We write

$$\lim_{x \to a-} f(x) = \pm\infty$$

to express the fact that the left-limit of f at a is $\pm\infty$ (read "the limit of $f(x)$ as x approaches a from the left is plus (minus) infinity").

Definition 2 The line $x = a$ is a **vertical asymptote for the graph of f** if

$$\lim_{x \to a+} f(x) = \pm\infty \text{ or } \lim_{x \to a-} f(x) = \pm\infty$$

(the "or" in the definition is "inclusive or", i.e., the limits from the right and from the left can be both infinite).

A function may be defined only on one side of a vertical asymptote, as in the following example:

Example 2 Let

$$f(x) = \frac{1}{\sqrt{x}}.$$

We have $\lim_{x \to 0+} f(x) = +\infty$. Therefore, the vertical axis is a vertical asymptote for the graph of f. $\square$

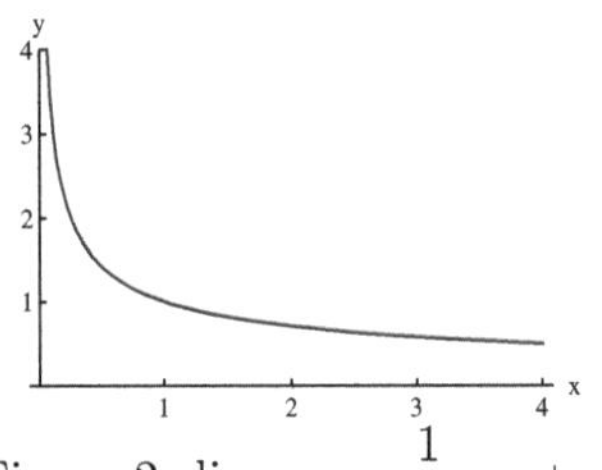

Figure 2: $\lim_{x \to 0+} \dfrac{1}{\sqrt{x}} = +\infty$

A word of caution: In any of the cases covered by Definition 1, the relevant one-sided limit of f does not exist. For example, if $\lim_{x \to a+} f(x) = L$, then $f(x)$ approximates L if x is near a and $x > a$, so that $f(x)$ must be between certain bounds for such values of x. Here, we have an example of "mathematical doublespeak": We are using the same word "limit", and the same symbol "lim" in connection with "finite limits" and "infinite limits". The doublespeak is traditional and convenient, and we will use it. The particular context should clarify which usage of the word "limit" we have in mind. Nevertheless, if there is any possibility of confusion, we may stress that we are talking about a **"finite limit"**, or an **"infinite limit"** in the sense of Definition 1.

If a function has infinite right and left limits of the same sign at a point, we can speak of an infinite limit at that point:

Definition 3 The limit of f at a is $+\infty$ if, given any $M > 0$, however large, $f(x) > M$ provided that $x \neq a$ and x is sufficiently close to a. In this case we write

$$\lim_{x \to a} f(x) = +\infty$$

(read "the limit of $f(x)$ as x approaches a is $+\infty$").

The limit of f at a is $-\infty$ if, given any $M > 0$, however large, $f(x) < -M$ provided that $x \neq a$ and x is sufficiently close to a. In this case we write

$$\lim_{x \to a} f(x) = -\infty$$

(read "the limit of $f(x)$ as x approaches a is $-\infty$).

Clearly,

$$\lim_{x \to a} f(x) = +\infty \Leftrightarrow \lim_{x \to a+} f(x) = \lim_{x \to a-} f(x) = +\infty,$$

and

$$\lim_{x \to a} f(x) = -\infty \Leftrightarrow \lim_{x \to a+} f(x) = \lim_{x \to a-} f(x) = -\infty.$$

Example 3 Let $f(x) = 1/x^2$, $x \neq 0$.

Since $1/x^2$ exceeds any given number if x is sufficiently close to 0, irrespective of the sign of x, we have

$$\lim_{x \to 0+} \frac{1}{x^2} = \lim_{x \to 0-} \frac{1}{x^2} = +\infty,$$

so that

$$\lim_{x \to 0} \frac{1}{x^2} = +\infty.$$

Figure 3 shows the graph of f. The picture is consistent with the fact that $\lim_{x \to 0} f(x) = +\infty$. The vertical axis is a vertical asymptote for the graph of f. $\square$

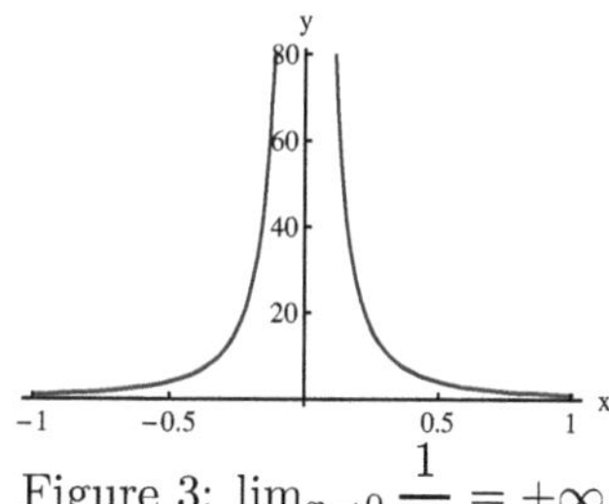

Figure 3: $\lim_{x \to 0} \dfrac{1}{x^2} = +\infty$

The Determination of Infinite Limits

We may guess that a function has an infinite limit at a point, and support our claim with graphical and numerical data. Nevertheless, it will be useful to provide some guidelines that are helpful in the determination of infinite limits.

If $f(x)$ is positive and small, its reciprocal $1/f(x)$ is large. For example, if $f(x) = 10^{-6}$ then $1/f(x) = 10^6$. On the other hand, if $f(x)$ is negative and has small magnitude then $1/f(x)$ is a negative number of large magnitude. For example, if $f(x) = -10^{-6}$ then $1/f(x) = -10^6$. Therefore, the following guideline should be plausible:

If $f(x) > 0$ when x is close to a, $x \neq a$, and $\lim_{x \to a} f(x) = 0$ then

$$\lim_{x \to a} \frac{1}{f(x)} = +\infty.$$

If $f(x) < 0$ when x is close to a, $x \neq a$, and $\lim_{x \to a} f(x) = 0$ then

$$\lim_{x \to a} \frac{1}{f(x)} = -\infty.$$

Similar statements are valid for one-sided limits.

You can find the proof of the above guideline in Appendix B.

Example 4 Determine $\lim_{x \to \pi/2\pm} \sec(x)$.

Solution

We have

$$\sec(x) = \frac{1}{\cos(x)}.$$

If $0 < x < \pi/2$ then $\cos(x) > 0$, and

$$\lim_{x \to \pi/2-} \cos(x) = \lim_{x \to \pi/2} \cos(x) = \cos\left(\frac{\pi}{2}\right) = 0.$$

Therefore,

$$\lim_{x \to \pi/2-} \sec(x) = \lim_{x \to \pi/2-} \frac{1}{\cos(x)} = +\infty.$$

If $\pi/2 < x < 3\pi/2$ then $\cos(x) < 0$, and $\lim_{x \to \pi/2} \cos(x) = 0$. Therefore,

$$\lim_{x \to \pi/2+} \sec(x) = \lim_{x \to \pi/2+} \frac{1}{\cos(x)} = -\infty.$$

Figure 4 shows the graph of secant on the interval $[\pi, \pi]$. The picture is consistent with the infinite limits that we calculated. The line $x = \pi/2$ is a vertical asymptote for the graph of secant. Since secant is an even function, the graph is symmetric with respect to the vertical axis, and the line $x = -\pi/2$ is also a vertical asymptote. $\square$

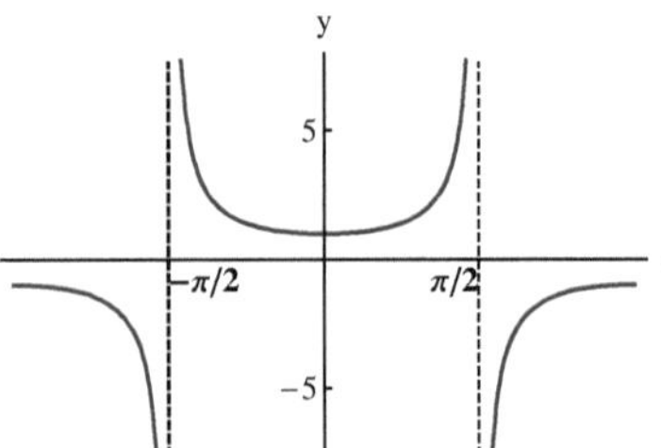

Figure4: $\lim_{x \to \pi/2-} \sec(x) = +\infty$ and $\lim_{x \to \pi/2+} \sec(x) = -\infty$

If $f(x) > c > 0$ and $g(x)$ exceeds any $M > 0$ when x is sufficiently close to a, then $f(x)g(x) > cM$, so that the product also becomes arbitrarily large as x approaches a. Therefore the following guideline should be plausible:

If $\lim_{x \to a} f(x) > 0$ or $\lim_{x \to a} f(x) = +\infty$, and $\lim_{x \to a} g(x) = +\infty$ then

$$\lim_{x \to a} f(x)g(x) = +\infty.$$

Similar statements are valid for one-sided limits.

You can find the proof of the above guideline in Appendix B. Changes in sign lead to similar statements. For example, if $\lim_{x \to a} f(x) < 0$ and $\lim_{x \to a} g(x) = +\infty$ then

$$\lim_{x \to a} f(x)g(x) = -\infty.$$

Example 5

a) Determine the vertical asymptotes for the graph of $y = \tan(x)$ on the interval $[-\pi, \pi]$ and the relevant limits.
b) Plot the graph of $y = \tan(x)$ on the interval $[-\pi, \pi]$ with the help of your graphing utility. Does the graph support your response to part a)?
c) Compute $\tan\left(\pi/2 + 10^{-k}\right)$ for $k = 3, 4, 5, 6$. Do the numbers support your response to part a)?

Solution

a) The only discontinuities of

$$\tan(x) = \frac{\sin(x)}{\cos(x)}$$

in the interval $[-\pi, \pi]$ are the points at which $\cos(x) = 0$, i.e., $\pi/2$ and $-\pi/2$. Let's check whether the lines $x = \pi/2$ and $x = -\pi/2$ are vertical asymptotes for the graph of tangent. We have

$$\lim_{x \to \pi/2} \sin(x) = \sin\left(\frac{\pi}{2}\right) = 1 > 0.$$

As in Example 4,

$$\lim_{x \to \pi/2-} \frac{1}{\cos(x)} = +\infty.$$

Therefore,

$$\lim_{x \to \pi/2-} \tan(x) = \lim_{x \to \pi/2-} \sin(x)\left(\frac{1}{\cos(x)}\right) = +\infty.$$

As in Example 4,

$$\lim_{x \to \pi/2+} \frac{1}{\cos(x)} = -\infty.$$

Therefore,

$$\lim_{x \to \pi/2+} \tan(x) = \lim_{x \to \pi/2+} \sin(x)\left(\frac{1}{\cos(x)}\right) = -\infty.$$

The line $x = \pi/2$ is a vertical asymptote for the graph of tangent.

Similarly,

$$\lim_{x \to -\pi/2-} \tan(x) = +\infty \quad \text{and} \quad \lim_{x \to -\pi/2+} \tan(x) = -\infty.$$

The line $x = -\pi/2$ is also a vertical asymptote for the graph of tangent.

b) Figure 5 shows the graph of $y = \tan(x)$ on the interval $[-\pi, \pi]$. Figure 5 is consistent with the infinite limits that we calculated.

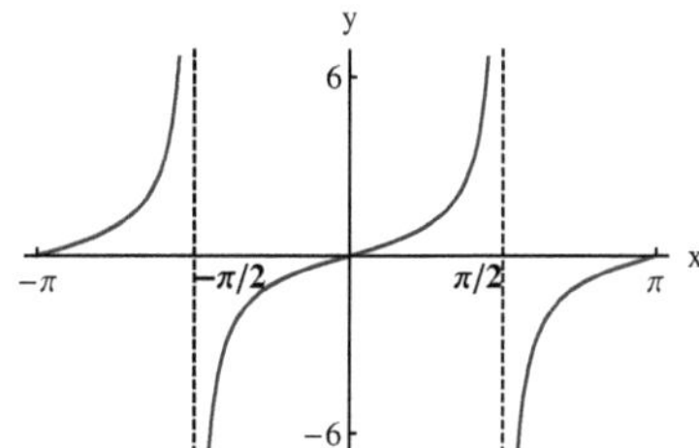

Figure 5: $\lim_{x \to \pi/2-} \tan(x) = +\infty$ and $\lim_{x \to \pi/2+} \tan(x) = -\infty$

c) Table 2 displays $\tan(x)$ (in scientific notation and rounded to 2 significant digits), where $x = \pi/2 - 10^{-k}$, $k = 3, 4, 5, 6$, The numbers support the claim that $\lim_{x \to \pi/2-} \tan(x) = +\infty$.

$\square$

x	$\tan(x)$
$\pi/2 - 10^{-3}$	1.0×10^3
$\pi/2 - 10^{-4}$	1.0×10^4
$\pi/2 - 10^{-5}$	1.0×10^5
$\pi/2 - 10^{-6}$	1.0×10^6

Table 2

A word of caution: We cannot make a general statement about $\lim_{x \to a+} f(x) g(x)$ if $\lim_{x \to a+} f(x) = 0$ and $\lim_{x \to a+} g(x) = +\infty$. **The expression $0 \cdot \infty$ is indeterminate.**

For example, we have

$$\lim_{x \to 1} \left(x^2 - 1\right) = 0, \quad \lim_{x \to 1+} \frac{1}{x - 1} = +\infty,$$

and

$$\lim_{x \to 1+} \left(x^2 - 1\right) \left(\frac{1}{x - 1}\right) = \lim_{x \to 1+} (x + 1) = 2.$$

In this case the indeterminate expression $\infty \cdot 0$ seems to hide the number 2.

Similarly,

$$\lim_{x \to 1} (x - 1) = 0, \quad \lim_{x \to 1+} \frac{1}{x - 1} = +\infty,$$

and

$$\lim_{x \to 1+} (x - 1) \left(\frac{1}{x - 1}\right) = \lim_{x \to 1+} 1 = 1.$$

In this case, $\infty \cdot 0$ seems to hide the number 1. $\Diamond$

Example 6 Let

$$f(x) = \frac{10x^2 + 20x}{x^2 - x - 2}.$$

a) Determine the vertical asymptotes for the graph of f and the corresponding infinite limits.
b) Plot the graph of f with the help of your graphing utility. Does the graph support your response to part b)?

Solution

a) We have

$$f(x) = \frac{10x^2 + 20x}{x^2 - x - 2} = \frac{10x(x + 2)}{(x + 1)(x - 2)}.$$

Since the denominator vanishes if $x = -1$ or $x = 2$, the rational function f is continuos at any x such that $x \neq -1$ and $x \neq 2$. Therefore, the only candidates for a vertical asymptote are the lines $x = -1$ and $x = 2$. We need to calculate the corresponding one-sided limits.

In order to investigate the behavior of $f(x)$ when x is near 2, we can express $f(x)$ as

$$f(x) = \left(\frac{10x(x + 2)}{x + 1}\right) \left(\frac{1}{x - 2}\right),$$

and observe that

$$\lim_{x \to 2} \frac{10x\,(x+2)}{x+1} = \frac{80}{3} > 0.$$

Since $x - 2 > 0$ if $x > 2$ and $\lim_{x \to 2}(x-2) = 0$ we have

$$\lim_{x \to 2+} \frac{1}{x-2} = +\infty.$$

Therefore,

$$\lim_{x \to 2+} f(x) = \lim_{x \to 2+}\left(\frac{10x\,(x+2)}{x+1}\right)\left(\frac{1}{x-2}\right) = +\infty.$$

Since $x - 2 < 0$ if $x < 2$ and $\lim_{x \to 2}(x-2) = 0$ we have

$$\lim_{x \to 2-} \frac{1}{x-2} = -\infty.$$

Therefore,

$$\lim_{x \to 2-} f(x) = \lim_{x \to 2-}\left(\frac{10x\,(x+2)}{x+1}\right)\left(\frac{1}{x-2}\right) = -\infty.$$

The line $x = 2$ is indeed a vertical asymptote for the graph of f.

Similarly,

$$\lim_{x \to -1+} f(x) = +\infty \quad \text{and} \quad \lim_{x \to --1} f(x) = -\infty.$$

The line $x = -1$ is a vertical asymptote for the graph of f.

b) Figure 6 displays the graph of f. The picture is consistent with our response to part a). $\square$

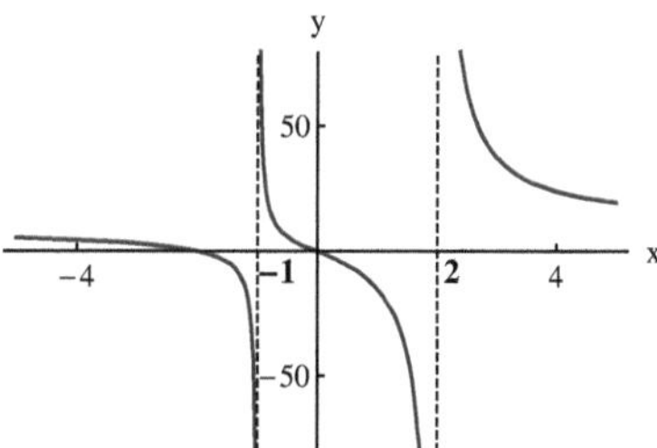

Figure 6: Vertical asymptotes at $x = -1$ and $x = 2$

Another word of caution: As in the above example, when we see that the denominator of a quotient approaches 0, we may expect to encounter infinite limits and hence a vertical asymptote. That is not always the case. The numerator may also approach 0. **The expression 0/0 is indeterminate,** and does not have to lead to an infinite limit. For example, let

$$F(x) = \frac{2x^2 - 18}{x - 3}.$$

We have

$$\lim_{x \to 3}\left(2x^2 - 18\right) = 0 \quad \text{and} \quad \lim_{x \to 3}(x - 3) = 0.$$

We also have

$$\lim_{x \to 3} F(x) = \lim_{x \to 3} \frac{2x^2 - 18}{x - 3} = \lim_{x \to 3} \frac{2\,(x-3)\,(x+3)}{x-3} = \lim_{x \to 3} 2\,(x+3) = 12.$$

Since F has a finite limit at 3, the line $x = 3$ is not a vertical asymptote for the graph of F. In fact, a computer generated graph of F coincides with the graph of the linear function defined by $2(x + 3)$, as in Figure 7. The function F has a removable discontinuity at 3. In this case, the indeterminate expression hides the (finite) number 12. ◊

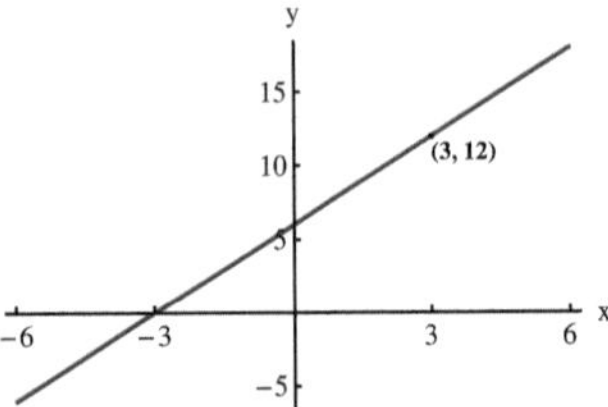

Figure 7: The function has a removable discontinuity at 3

If we add two large numbers we get another large number. Therefore, the following guideline should be plausible:

If $\lim_{x \to a} f(x) = L$, where L is finite, or $\lim_{x \to a} f(x) = +\infty$, and $\lim_{x \to a} g(x) = +\infty$ then

$$\lim_{x \to a} (f(x) + g(x)) = +\infty.$$

Similar statements are valid for one-sided limits and with sign changes. You can find the proof of the above guideline in Appendix B.

Example 7 Let

$$F(x) = 4 + x^2 + \frac{1}{x}.$$

Determine $\lim_{x \to 0+} F(x)$ and $\lim_{x \to 0-} F(x)$.

Solution

We have

$$\lim_{x \to 0} (4 + x^2) = 4, \quad \lim_{x \to 0+} \frac{1}{x} = +\infty \text{ and } \lim_{x \to 0-} \frac{1}{x} = -\infty.$$

Therefore,

$$\lim_{x \to 0+} F(x) = \lim_{x \to 0+} \left(4 + x^2 + \frac{1}{x}\right) = +\infty \text{ and } \lim_{x \to 0-} F(x) = -\infty.$$

Figure 8 shows the graph of F. The picture is consistent with our conclusions. □

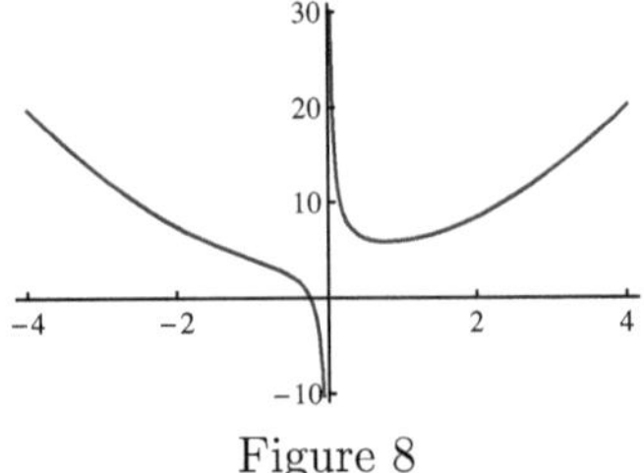

Figure 8

Still another word of caution: We cannot make a general statement about $\lim_{x \to a+} (f(x) + g(x))$ if $\lim_{x \to a+} f(x) = +\infty$ and $\lim_{x \to a+} g(x) = -\infty$ (or vice versa). **The expression $\infty - \infty$ is indeterminate.** In particular, **you cannot declare that $\infty - \infty$ is 0.**

For example,

$$\lim_{x \to 0+} \frac{1}{x} = +\infty, \quad \lim_{x \to 0+} \left(\frac{1}{x^2} \right) = +\infty.$$

We have

$$\lim_{x \to 0+} \left(\frac{1}{x} - \frac{1}{x^2} \right) = \lim_{x \to 0+} \frac{x - 1}{x^2} = -\infty$$

since

$$\lim_{x \to 0} (x - 1) = -1 < 0, \ x^2 > 0 \text{ if } x \neq 0 \text{ and } \lim_{x \to 0} \left(x^2 \right) = 0.$$

The Precise Definition of an Infinite Limit (Optional)

Definition 1 is made precise by quantifying the expression "sufficiently close":

Definition 4

The right-limit of f at a is $+\infty$ if, given any $M > 0$, there exists $\delta > 0$ such that $f(x) > M$ provided that $a < x < a + \delta$.
The right-limit of f at a is $-\infty$ if, given any $M > 0$, there exists $\delta > 0$ such that $f(x) < -M$ provided that $a - \delta < x < a$.

Example 8 Let

$$f(x) = \frac{1}{(x + 3)(x - 2)}.$$

Prove that $\lim_{x \to 2+} f(x) = +\infty$, with reference to the precise definition.

Solution

We will restrict x so that $2 < x < 3$. Thus $5 < x + 3 < 6$, so that

$$\frac{1}{x + 3} > \frac{1}{6}.$$

Therefore,

$$\frac{1}{(x + 3)(x - 2)} > \frac{1}{6(x - 2)}.$$

Thus, given $M > 0$, in order to ensure that

$$\frac{1}{(x + 3)(x - 2)} > M,$$

it is sufficient to have

$$\frac{1}{6(x - 2)} > M.$$

This is the case if

$$x - 2 < \frac{1}{6M}, \text{ i.e., } x < 2 + \frac{1}{6M},$$

with the restriction that $2 < x < 3$. With reference to the precise definition, we can set

$$\delta = \min \left(1, \frac{1}{6M} \right).$$

We have $f(x) > M$ if $2 < x < 2 + \delta$. $\square$

Problems

In problems 1 - 14, determine the infinite limits $\lim_{x \to a-} f(x)$, $\lim_{x \to a+} f(x)$ and if applicable, $\lim_{x \to a} f(x)$. Indicate how you reach your conclusions by making use of the guidelines that are provided in Section 1.6.

1.
$$f(x) = \frac{x-2}{x-4}, \ a = 4$$

2.
$$f(x) = \frac{x+1}{x+2}, \ a = -2$$

3.
$$f(x) = \frac{2x}{x^2 - 1}, \ a = 1$$

4.
$$f(x) = \frac{x+1}{(x+2)(x-3)}, \ a = -2$$

5.
$$f(x) = \frac{2x - x^2 - 6}{(1+x^2)(x+3)^3}, \ a = -3$$

6.
$$f(x) = \frac{x-6}{x^2 - x - 12}, \ a = 4$$

7.
$$f(x) = \sec(x), \ a = \pi/2$$

8.
$$f(x) = \frac{\cos(x)}{\sin^2(x)}, \ a = \pi$$

9.
$$f(x) = \frac{1}{\sin(x)}, \ a = \pi$$

10.
$$f(x) = \frac{\cos(x)}{\sin(x)}, \ a = 0$$

11.
$$f(x) = x^2 - \frac{2}{(x-3)^2}, \ a = 3$$

12.
$$f(x) = 2x + \frac{3}{(x-5)^3} - \frac{1}{x-3}, \ a = 5$$

13.
$$f(x) = 4x^2 + 1 + \frac{1}{x-4} - \frac{3}{x-2}, \ a = 2$$

14.
$$f(x) = 4x^2 + 1 + \frac{1}{x-4} - \frac{3}{x-2}, \ a = 4$$

1.7   Limits at Infinity

In this section we will examine the behavior of a function as the independent variable tends to $\pm\infty$. Such information is valuable in many applications.

Finite Limits at Infinity and Horizontal Asymptotes

Let's begin with a specific case:

Example 1 Let
$$f(x) = 5 + \frac{1}{x}.$$

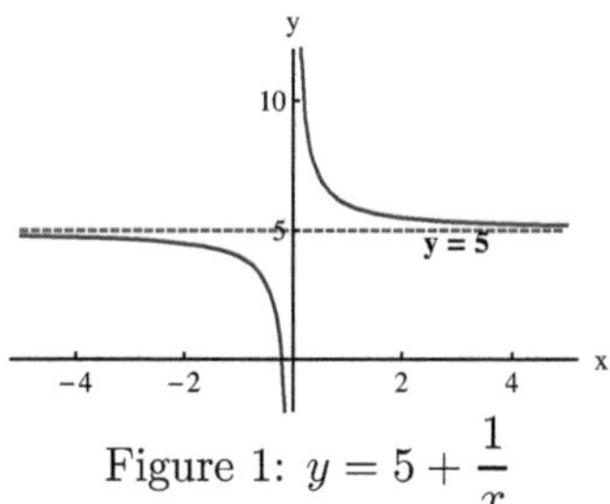

Figure 1: $y = 5 + \dfrac{1}{x}$

Figure 1 shows the graph of f. The picture indicates that the graph of f approaches the line $y = 5$ as x attains larger and larger values. Indeed,

$$f\left(x\right) - 5 = \left(5 + \frac{1}{x}\right) - 5 = \frac{1}{x}.$$

Since $1/x$ becomes small as x becomes large, $f\left(x\right)$ approaches 5. We express this by saying that the limit of $f\left(x\right)$ as x tends to ∞ is 5, and write

$$\lim_{x \to +\infty} \left(5 + \frac{1}{x}\right) = 5.$$

We say that the line $y = 5$ is a **horizontal asymptote for the graph of f at $+\infty$**.

Similarly, $f\left(x\right)$ approaches 5 as x attains negative values of large magnitude. We express this by saying that the limit of $f\left(x\right)$ as x tends to $-\infty$ is 5, and write

$$\lim_{x \to -\infty} \left(5 + \frac{1}{x}\right) = 5.$$

We say that the line $y = 5$ is a **horizontal asymptote for the graph of f at $-\infty$**. $\square$

Here are the relevant definitions:

Definition 1 The limit of f at $+\infty$ is L if $f\left(x\right)$ is as close to L as desired provided that x is sufficiently large. In this case we write

$$\lim_{x \to +\infty} f\left(x\right) = L$$

(read "the limit of $f(x)$ as x approaches $+\infty$ is L"). The line $y = L$ is **a horizontal asymptote for the graph of f at $+\infty$**.

The limit of f at $-\infty$ is L if $f\left(x\right)$ is as close to L as desired provided that $x < 0$ and $|x| = -x$ is sufficiently large. In this case we write

$$\lim_{x \to -\infty} f(x) = L$$

(read "the limit of $f\left(x\right)$ as x approaches $-\infty$ is L"). The line $y = L$ **a horizontal asymptote for the graph of f at $-\infty$**.

The rules for the limits of arithmetic combinations of functions are valid for limits at $\pm\infty$.

Example 2 Let

$$f\left(x\right) = 2 - \frac{2}{x+1} + \frac{1}{x-3}.$$

Determine $\lim_{x \to \pm\infty} f\left(x\right)$ and the horizontal asymptotes for the graph of f.

Solution

We have

$$\lim_{x \to \pm\infty} f(x) = \lim_{x \to \pm\infty} \left(2 - \frac{2}{x+1} + \frac{1}{x-3} \right)$$

$$= \lim_{x \to \pm\infty} 2 - 2 \lim_{x \to \pm\infty} \frac{1}{x+1} + \lim_{x \to \pm\infty} \frac{1}{x-3}.$$

Clearly,

$$\lim_{x \to \pm\infty} \frac{1}{x+1} = 0 \ \text{ and } \ \lim_{x \to \pm\infty} \frac{1}{x-3} = 0.$$

Therefore, $\lim_{x \to \pm\infty} f(x) = 2$. Thus, the line $y = 2$ is a horizontal asymptote for the graph of f at $+\infty$ and $-\infty$. Figure 2 shows the graph of f. The picture is consistent with our analysis. The picture indicates that the lines $x = -1$ and $x = 3$ are vertical asymptotes (confirm). $\square$

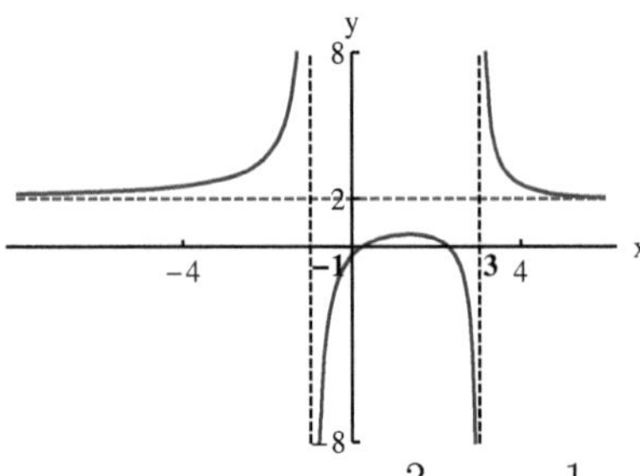

$$\text{Figure 2: } y = 2 - \frac{2}{x+1} + \frac{1}{x-3}$$

The rule for **the limit of a composite function** is valid for limits at infinity:

Assume that $\lim_{x \to +\infty} g(x) = L$ and f is continuous at L. Then,

$$\lim_{x \to \pm\infty} f(g(x)) = f\left(\lim_{x \to \pm\infty} g(x) \right).$$

Example 3 Determine

$$\lim_{x \to +\infty} \cos\left(\frac{\pi}{3} + \frac{1}{x^2} \right).$$

Solution

We have

$$\lim_{x \to +\infty} \left(\frac{\pi}{3} + \frac{1}{x^2} \right) = \frac{\pi}{3} + \lim_{x \to +\infty} \frac{1}{x^2} = \frac{\pi}{3}.$$

Since cosine is continuous everywhere,

$$\lim_{x \to +\infty} \cos\left(\frac{\pi}{3} + \frac{1}{x^2} \right) = \cos\left(\lim_{x \to +\infty} \left(\frac{\pi}{3} + \frac{1}{x^2} \right) \right) = \cos\left(\frac{\pi}{3} \right) = \frac{1}{2}.$$

$\square$

There is a version of **the Squeeze Theorem** :

Assume that $h(x) \le f(x) \le g(x)$ if x is sufficiently large, and

$$\lim_{x \to +\infty} h(x) = \lim_{x \to +\infty} g(x) = L.$$

Then $\lim_{x \to +\infty} f(x) = L$ as well.

There is an obvious counterpart for limits at $-\infty$.

Example 4 Determine

$$\lim_{x\to+\infty} \frac{\sin(x)}{x}.$$

Solution

Since $-1 \le \sin(x) \le 1$ for each $x \in \mathbb{R}$, we have

$$-\frac{1}{x} \le \frac{\sin(x)}{x} \le \frac{1}{x}$$

if $x > 0$. We also have

$$\lim_{x\to+\infty} \left(-\frac{1}{x}\right) = \lim_{x\to+\infty} \frac{1}{x} = 0.$$

Therefore,

$$\lim_{x\to+\infty} \frac{\sin(x)}{x} = 0,$$

as well. Figure 3 supports this fact. $\square$

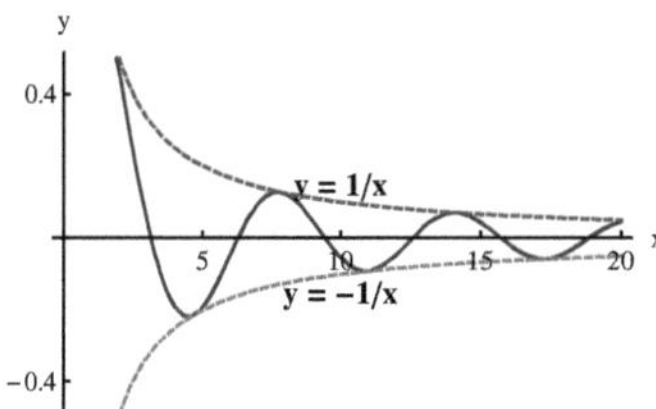

Figure 3: $\sin(x)/x$ is squeezed between $-1/x$ and $1/x$

Infinite Limits at Infinity

In Section 1.6 we discussed the infinite limits of a function at a point. The values $f(x)$ may become arbitrarily large in magnitude as x tends to $\pm\infty$. Thus, the infinite limits of a function at $\pm\infty$ are also of interest:

Definition 2 The limit of f at $+\infty$ is $+\infty$ if, given any $M > 0$, however large, $f(x) > M$ provided that x is sufficiently large. In this case we write

$$\lim_{x\to+\infty} f(x) = +\infty$$

(read "the limit of $f(x)$ as x approaches $+\infty$ is $+\infty$").

The limit of f at $+\infty$ is $-\infty$ *if, given any $M > 0$, we have $f(x) < -M$ provided that x is sufficiently large. In this case we write*

$$\lim_{x\to+\infty} f(x) = -\infty$$

(read "the limit of $f(x)$ as x approaches $+\infty$ is $-\infty$").

The definitions of

$$\lim_{x\to-\infty} f(x) = \pm\infty$$

are similar: We consider $x < 0$ such that $|x|$ is large, instead of large positive x.

Example 5 Let $f(x) = x^3$. since x^3 exceeds any given number if x is sufficiently large, $\lim_{x \to +\infty} f(x) = +\infty$. If $x < 0$, then $x^3 < 0$ and $\left| x^3 \right| = |x|^3$ exceeds any given number if $|x|$ is large enough. Therefore, $\lim_{x \to -\infty} x^3 = -\infty$. $\square$

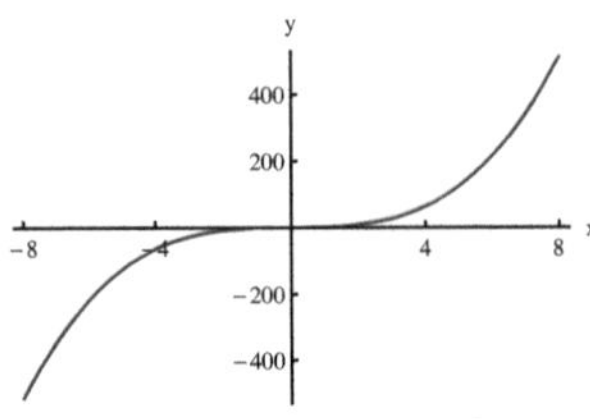

Figure 4: $y = x^3$

In Section 1.6 we provided some guidelines that were helpful in the determination of infinite limits. Those guidelines have useful counterparts for infinite limits at infinity:

a) Assume that $f(x) > 0$ if x is sufficiently large and $\lim_{x \to +\infty} f(x) = 0.$. Then

$$\lim_{x \to +\infty} \frac{1}{f(x)} = +\infty$$

b) If $\lim_{x \to +\infty} f(x) > 0$ or $\lim_{x \to +\infty} f(x) = +\infty$, and $\lim_{x \to +\infty} g(x) = +\infty$ then

$$\lim_{x \to +\infty} f(x)g(x) = +\infty.$$

c) If $\lim_{x \to +\infty} f(x) = L$, where L is finite, or $\lim_{x \to +\infty} f(x) = +\infty$, and $\lim_{x \to +\infty} g(x) = +\infty$, then

$$\lim_{x \to +\infty} (f(x) + g(x)) = +\infty.$$

There are obvious counterparts of the above facts when the signs change.

When we discuss the limits of a polynomial at infinity, the "dominant term" is the term that has the highest power of x, in the sense that the magnitude of that term is much larger than the magnitudes of the other terms when $|x|$ is large. We can determine the required limit easily by factoring the highest power of x, as in the following example.

Example 6 Let

$$f(x) = x - \frac{1}{6}x^3.$$

Determine $\lim_{x \to \pm\infty} f(x)$.

Solution

We cannot reach any conclusion about the required limits by stating that

$$\lim_{x \to \pm\infty} x = \pm\infty \quad \text{and} \quad \lim_{x \to \pm\infty} x^3 = \pm\infty,$$

since the expression $\infty - \infty$ is indeterminate. Thus, we have to follow another route. The dominant term in the expression for $f(x)$ involves x^3. Let's factor x^3:

$$f(x) = x - \frac{1}{6}x^3 = x^3 \left(\frac{1}{x^2} - \frac{1}{6} \right)$$

Since

$$\lim_{x \to +\infty} \left(\frac{1}{x^2} - \frac{1}{6} \right) = -\frac{1}{6} < 0 \text{ and } \lim_{x \to +\infty} x^3 = +\infty,$$

we have $\lim_{x \to +\infty} f(x) = -\infty$. Similarly,

$$\lim_{x \to -\infty} \left(\frac{1}{x^2} - \frac{1}{6} \right) = -\frac{1}{6} < 0 \text{ and } \lim_{x \to -\infty} x^3 = -\infty,$$

so that $\lim_{x \to -\infty} f(x) = +\infty$.

Figure 5 shows the graph of f. The picture supports our conclusions $\square$

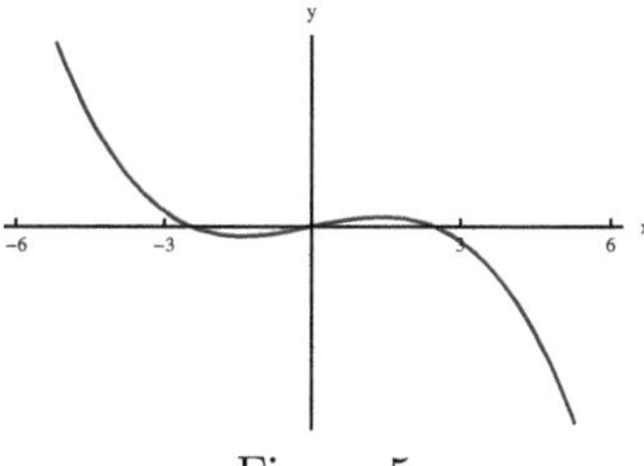

Figure 5

The Behavior of Rational Functions at $\pm\infty$

A practical strategy for determining the limits of a rational function at $\pm\infty$ is to factor the terms with the highest powers of x in the numerator and the denominator, and proceed after possible cancellations, as in the following example.

Example 7 Let

$$f(x) = \frac{2x^2 - 5x + 1}{x^2 - 2x - 3}$$

Determine $\lim_{x \to \pm\infty} f(x)$.

Solution

The highest power of x in the numerator and the denominator is 2. We factor x^2:

$$f(x) = \frac{2x^2 - 5x + 1}{x^2 - 2x - 3} = \frac{x^2 \left(2 - \dfrac{5}{x} + \dfrac{1}{x^2} \right)}{x^2 \left(1 - \dfrac{2}{x} - \dfrac{3}{x^2} \right)} = \frac{2 - \dfrac{5}{x} + \dfrac{1}{x^2}}{1 - \dfrac{2}{x} - \dfrac{3}{x^2}}.$$

Therefore,

$$\lim_{x \to \pm\infty} f(x) = \lim_{x \to \pm\infty} \frac{2 - \dfrac{5}{x} + \dfrac{1}{x^2}}{1 - \dfrac{2}{x} - \dfrac{3}{x^2}} = 2.$$

Thus, the line $y = 2$ is a horizontal asymptote for the graph of f at $\pm\infty$. Figure 6 shows the graph of f. The picture is consistent with our analysis. The picture also indicates that the lines $x = -1$ and $x = 3$ are vertical asymptotes (confirm). $\square$

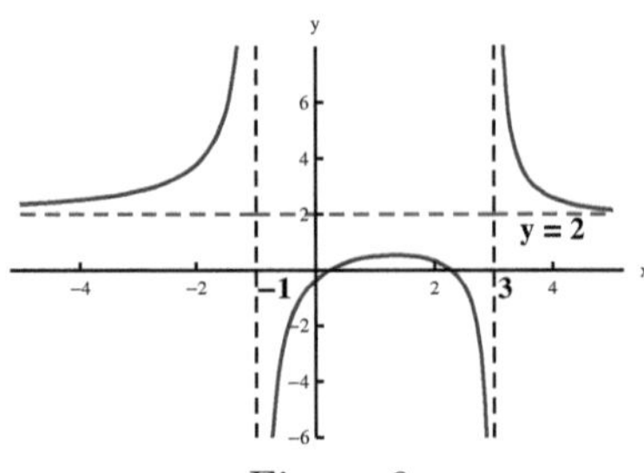

Figure 6

Remark 1 With reference to the function f of Example 7, an attempt to evaluate $\lim_{x \to +\infty} f(x)$ by setting

$$\lim_{x \to +\infty} \frac{2x^2 - 5x + 1}{x^2 - 2x - 3} = \frac{\lim_{x \to +\infty} \left(2x^2 - 5x + 1\right)}{\lim_{x \to +\infty} \left(x^2 - 2x - 3\right)}$$

would have resulted in **the indeterminate form** ∞/∞. $\Diamond$

The graph of a rational function can get closer and closer to a line that is not horizontal as the distance from the origin increases:

Definition 3 The line $y = mx + b$, where $m \neq 0$, is an **oblique asymptote for the graph of** the rational function f at $\pm\infty$ if

$$\lim_{x \to \pm\infty} \left(f(x) - (mx + b)\right) = 0.$$

Example 8 Let

$$f(x) = \frac{2x^2 - 6x + 1}{x - 3}$$

a) Determine $\lim_{x \to \pm\infty} f(x)$.

b) Show that there is an oblique asymptote for the graph of f.

Solution

a) Note that an attempt to evaluate the limit as a quotient of limits leads to the indeterminate expression ∞/∞. As in Example 7, we will factor the highest powers of x in the denominator and numerator:

$$\frac{2x^2 - 6x + 1}{x - 3} = \frac{x^2 \left(2 - \dfrac{6}{x} + \dfrac{1}{x^2}\right)}{x \left(1 - \dfrac{3}{x}\right)} = x \left(\frac{2 - \dfrac{6}{x} + \dfrac{1}{x^2}}{1 - \dfrac{3}{x}}\right)$$

Since

$$\lim_{x \to \pm} \frac{2 - \dfrac{6}{x} + \dfrac{1}{x^2}}{1 - \dfrac{3}{x}} = 2 > 0,$$

$\lim_{x \to +\infty} x = +\infty$ and $\lim_{x \to -\infty} x = -\infty$, we have

$$\lim_{x \to +\infty} f(x) = +\infty \quad \text{and} \quad \lim_{x \to -\infty} f(x) = -\infty.$$

b) Since

$$f\left(x\right) = x\left(\frac{2 - \dfrac{6}{x} + \dfrac{1}{x^2}}{1 - \dfrac{3}{x}}\right) \cong 2x$$

if $|x|$ is large, the line $y = 2x$ is expected to be an oblique asymptote for the graph of f. Indeed,

$$f\left(x\right) - 2x = \frac{2x^2 - 6x + 1}{x - 3} - 2x = \frac{2x^2 - 6x + 1 - 2x^2 + 6x}{x - 3} = \frac{1}{x - 3}.$$

Therefore,

$$\lim_{x \to \pm\infty}\left(f\left(x\right) - 2x\right) = \lim_{x \to \pm\infty}\frac{1}{x - 3} = 0.$$

Figure 7 shows the graph of f. We see that the graph of f is close to the line $y = 2x$ if $|x|$ is large. This is consistent with our analysis. $\square$

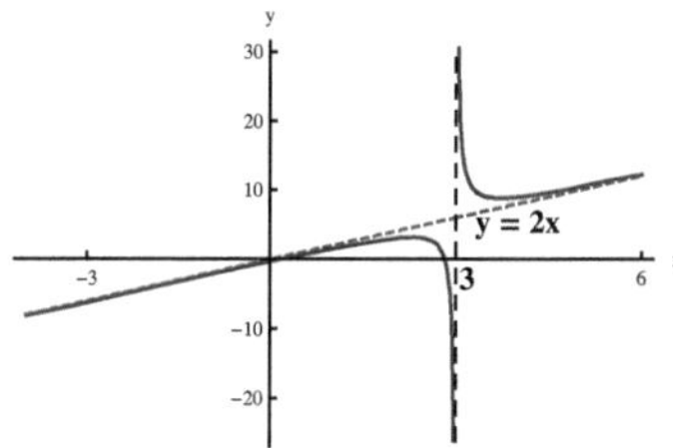

Figure 7: An oblique asymptote for the graph of a rational function

The graph of a rational function may approach the graph of a nonlinear polynomial as the distance from the origin increases, as in the following example.

Example 9 Let

$$f(x) = \frac{x^3 - x^2 + 1}{x - 1}.$$

a) Determine $\lim_{x \to \pm\infty} f(x)$.
b) Determine the polynomial p such that $\lim_{x \to \pm\infty}\left(f\left(x\right) - p\left(x\right)\right) = 0$.

Solution

a) We have

$$f\left(x\right) = \frac{x^3 - x^2 + 1}{x - 1} = \frac{x^3\left(1 - \dfrac{1}{x} + \dfrac{1}{x^3}\right)}{x\left(1 - \dfrac{1}{x}\right)} = x^2\left(\frac{1 - \dfrac{1}{x} + \dfrac{1}{x^3}}{1 - \dfrac{1}{x}}\right)$$

Since $\lim_{x \to \pm\infty} x^2 = +\infty$ and

$$\lim_{x \to \pm\infty}\frac{1 - \dfrac{1}{x} + \dfrac{1}{x^3}}{1 - \dfrac{1}{x}} = 1 > 0,$$

$\lim_{x \to \pm\infty} f\left(x\right) = +\infty$.

b) Since

$$f\left(x\right) = x^2\left(\frac{1 - \dfrac{1}{x} + \dfrac{1}{x^3}}{1 - \dfrac{1}{x}}\right) \cong x^2$$

if $|x|$ is large, we expect that $\lim_{x \to \pm\infty}\left(f\left(x\right) - x^2\right) = 0$. Indeed,

$$f\left(x\right) - x^2 = \frac{x^3 - x^2 + 1}{x - 1} - x^2 = \frac{x^3 - x^2 + 1 - x^3 + x^2}{x - 1} = \frac{1}{x - 1}.$$

Therefore,

$$\lim_{x \to \pm\infty}\left(f(x) - x^2\right) = \lim_{x \to \pm\infty}\frac{1}{x - 1} = 0.$$

Figure 8 displays the graphs of f and the parabola $y = x^2$. We see that the graph of f gets closer to the parabola $y = x^2$ as $|x|$ becomes larger. This is consistent with the fact that $\lim_{x \to \pm\infty}\left(f(x) - x^2\right) = 0$. It may be appropriate to refer to the curve $y = x^2$ as **"a curved asymptote for the graph of f"**. $\square$

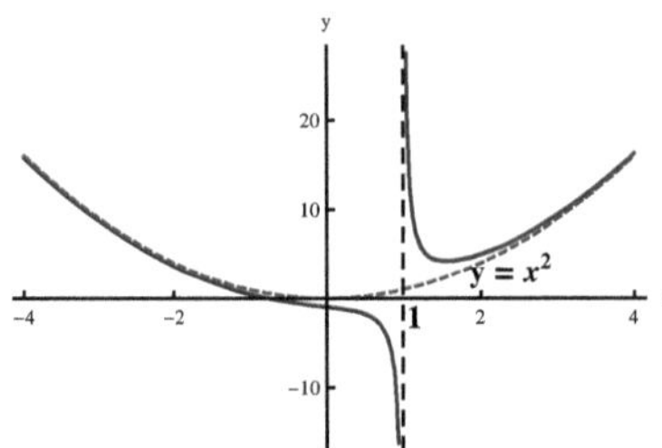

Figure 8: The graph of f is close to the parabola $y = x^2$ if $|x|$ is large

The Precise Definitions (Optional)

The precise definitions of finite limits at $\pm\infty$ are as follows:

Definition 4 The limit of f at $+\infty$ is L if, given any $\varepsilon > 0$ there exists $A > 0$ such that $|f\left(x\right) - L| < \varepsilon$ for each $x > A$. **The limit of f at $-\infty$ is L** if, given any $\varepsilon > 0$ there exists $A > 0$ such that $|f\left(x\right) - L_+| < \varepsilon$ for each $x < -A$.

Example 10 Let

$$f\left(x\right) = \frac{2x}{x + 3}.$$

Show that $\lim_{x \to +\infty} f\left(x\right) = 2$ in accordance with Definition 4.

Solution

We have

$$|f(x) - 2| = \left|\frac{2x}{x + 3} - 2\right| = \left|\frac{2x - 2\left(x + 3\right)}{x + 3}\right| = \frac{6}{|x + 3|}.$$

Therefore, if $x > -3$,

$$|f\left(x\right) - 2| = \frac{6}{x + 3}.$$

Let $\varepsilon > 0$ be given, and assume that $x > -3$. Then,

$$\frac{6}{x+3} < \varepsilon \Leftrightarrow \frac{x+3}{6} > \frac{1}{\varepsilon} \Leftrightarrow x+3 > \frac{6}{\varepsilon} \Leftrightarrow x > \frac{6}{\varepsilon} - 3.$$

It is certainly sufficient to have $x > 6/\varepsilon$. With reference to Definition 4, we can set $A = 6/\varepsilon$. By the above calculations, if $x > A$ we have

$$|f(x) - 2| < \varepsilon.$$

Therefore, $\lim_{x \to +\infty} f(x) = 2$.
Figure 9 illustrates the determination of A for a given $\varepsilon > 0$. $\square$

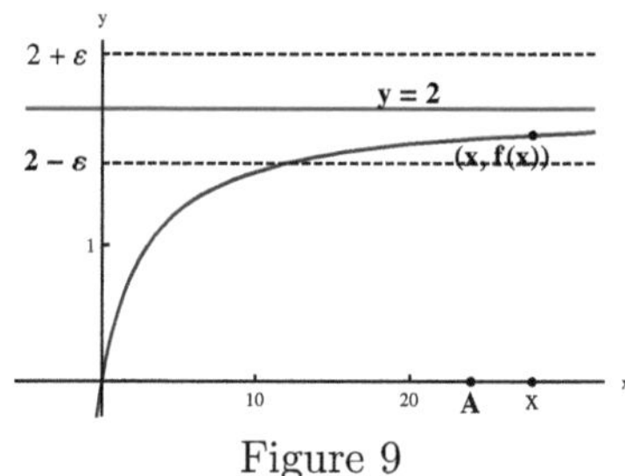

Figure 9

The precise definitions of infinite limits at $+\infty$ are as follows (the modifications for limits at $-\infty$ are straightforward):

Definition 5 The limit of f at $+\infty$ is $+\infty$ if, given any $M > 0$, there exists $A > 0$ such that $f(x) > M$ for each $x > A$. **The limit of f at $+\infty$ is $-\infty$** if, given any $M > 0$, there exists $A > 0$ such that $f(x) < -M$ for each $x > A$.

Example 11 Let $f(x) = x^2 - x$. Show that $\lim_{x \to +\infty} f(x) = +\infty$ in accordance with Definition 5.

Solution

We have

$$f(x) = x^2 - x = x^2 \left(1 - \frac{1}{x}\right).$$

If $x > 2$, then

$$\frac{1}{x} < \frac{1}{2} \Rightarrow -\frac{1}{x} > -\frac{1}{2}.$$

Therefore,

$$f(x) = x^2 \left(1 - \frac{1}{x}\right) > x^2 \left(1 - \frac{1}{2}\right) = \frac{x^2}{2}.$$

Let $M > 0$ be given. By the above inequality, in order to have $f(x) > M$ it is sufficient to have $x > 2$ and

$$\frac{x^2}{2} > M.$$

This is the case if $x > 2$ and $x > \sqrt{2M}$. With reference to Definition 5, we can set A to be the maximum of 2 and $\sqrt{2M}$. Thus, $f(x) > M$ if $x > A$. Therefore, $\lim_{x \to +\infty} f(x) = +\infty$.
Figure 10 illustrates the choice of A for a given M. $\square$

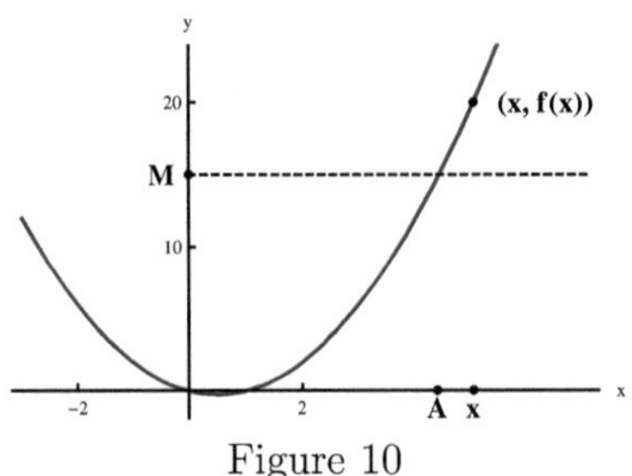

Figure 10

Problems

In problems 1 - 6, the graph of a function f is displayed. Determine the finite or infinite limit of f at $\pm\infty$, and the corresponding horizontal asymptote for the graph of f, if applicable.

1.

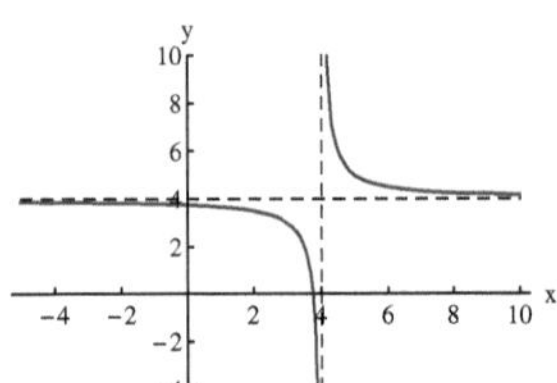

4.

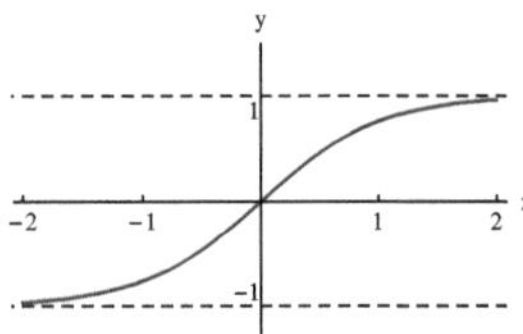

2.

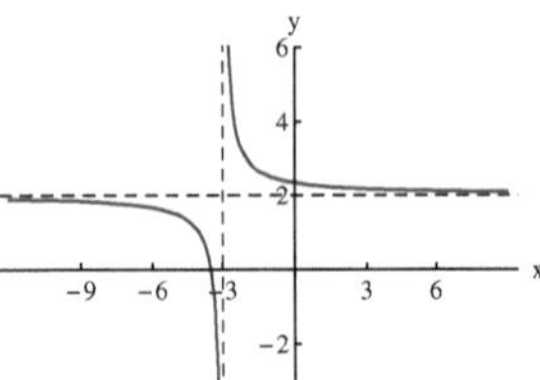

5.

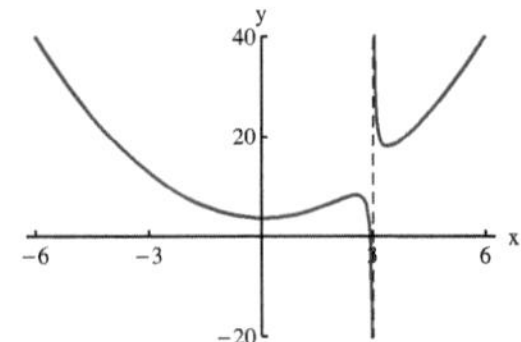

3.

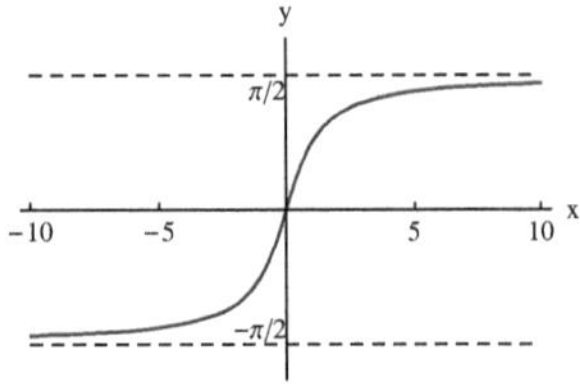

6.

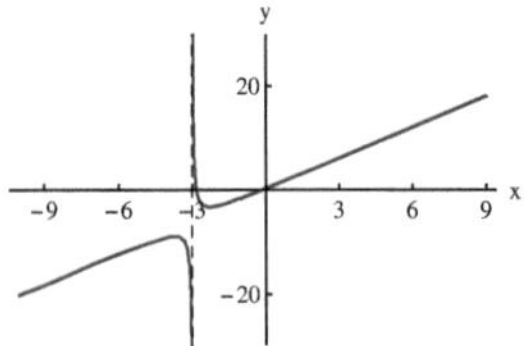

In problems 7 - 12,
a) Determine the required finite or infinite limit at $\pm\infty$, and the corresponding horizontal asymptote for the graph of f, if applicable,

b) [C] Plot the graph of the relevant function with the help of your graphing utility. Is the picture consistent with your response to part a)?

7.
$$\lim_{x \to \pm\infty} \frac{3x^2 - 2x - 26}{x^2 - x - 12}$$

10.
$$\lim_{x \to +\infty} \frac{\sqrt{x^3 - 8}}{3x + 4}$$

8.
$$\lim_{x \to \pm\infty} \frac{4x^3 + x + 1}{3x^3 + x^2 + 4}$$

11.
$$\lim_{x \to \pm\infty} \frac{x^4 + x^2 + 1}{2x^3 + 9}$$

9.
$$\lim_{x \to \pm\infty} \frac{\sqrt{x^2 + 2x}}{4x + 5}$$

12.
$$\lim_{x \to \pm\infty} \frac{4x^4 + 1}{2x^2 + 5}$$

In problems 13 - 16,
a) Determine the line $y = l_1(x)$ that is an asymptote to the graph of f at $+\infty$, and the line $y = l_2(x)$ that is asymptotic to the graph of f at $-\infty$. You need to justify your assertions by showing that
$$\lim_{x \to +\infty} (f(x) - l_1(x)) = 0 \text{ and } \lim_{x \to -\infty} (f(x) - l_2(x)) = 0.$$

b) [C] Plot the graph of the relevant function with the help of your graphing utility. Is the picture consistent with your response to part a)?

13.
$$f(x) = -4x + 3 + \frac{2x - 1}{x^2 + x}$$

Hint: Divide
15.
$$f(x) = x + \sqrt{x^2 + 1}$$

14.
$$f(x) = \frac{2x^2 - 3x - 8}{x - 3}$$

16.
$$f(x) = \frac{1}{x + \sqrt{x^2 + 1}}$$

In problems 17 and 18.
a) Determine the quadratic function q_1 whose graph is an asymptote to the graph of f at $+\infty$, and the the quadratic function q_2 whose graph is an asymptote to the graph of f at $-\infty$. You need to justify your assertions by showing that
$$\lim_{x \to +\infty} (f(x) - q_1(x)) = 0 \text{ and } \lim_{x \to -\infty} (f(x) - q_2(x)) = 0.$$

b) [C] Plot the graph of the relevant function with the help of your graphing utility. Is the picture consistent with your response to part a)?

17.
$$f(x) = 2x^2 + 3 + \frac{1}{x - 2}$$

18.
$$f(x) = \frac{-3x^4 + 16x^2 - 16 + x}{x^2 - 4}$$

Hint: Divide.

1.8 The Limit of a Sequence

A **sequence** is a list of numbers $a_1, a_2, a_3, \ldots, a_n, \ldots$, such as

$$1, \frac{1}{2}, \frac{1}{3}, \frac{1}{4}, \ldots, \frac{1}{n}, \ldots$$

It is usually of interest to determine whether a_n approaches a definite number as n becomes larger and larger. The relevant mathematical concept is the limit of a sequence.

Terminology and Notation

We may denote a **sequence** $a_1, a_2, a_3, \ldots, a_n, \ldots$ as $\{a_n\}_{n=1}^{\infty}$, or simply as $\{a_n\}$ if we don't feel the need to specify the starting value of the **index** n. Thus, the sequence

$$1, \frac{1}{2}, \frac{1}{3}, \frac{1}{4}, \ldots, \frac{1}{n}, \ldots$$

can be denoted as

$$\left\{\frac{1}{n}\right\}_{n=1}^{\infty} \quad \text{or} \quad \left\{\frac{1}{n}\right\}.$$

The index n is a **"dummy index"**, and can be replaced by any other letter. Thus,

$$\left\{\frac{1}{n}\right\}_{n=1}^{\infty} \quad \text{and} \quad \left\{\frac{1}{k}\right\}_{k=1}^{\infty}$$

denote the same sequence.

The starting value of the index can be an integer other than 1. For example, if we consider the sequence

$$\frac{5}{1}, \frac{6}{2}, \frac{7}{3}, \ldots, \frac{n}{n-4}, \ldots,$$

the starting value of the index is 5. We can denote the sequence as

$$\left\{\frac{n}{n-4}\right\}_{n=5}^{\infty}.$$

We may even refer to a sequence simply as "the sequence a_n". In this case, it should be understood that the starting value of the index is its smallest value such that the expression a_n is defined. For example, if we refer to "the sequence $n/(n-4)$", it is understood that the starting value of n is 5. The number a_n is referred to as **the nth term** of the sequence $a_1, a_2, a_3, \ldots, a_n, \ldots$. Thus, $1/n$ is the nth term of the sequence $\{1/n\}_{n=1}^{\infty}$. In the sequence

$$\frac{5}{1}, \frac{6}{2}, \frac{7}{3}, \ldots, \frac{n}{n-4}, \ldots$$

the nth term is not $n/(n-4)$. In such a case we will refer to a_n as **the term corresponding to n.**

Example 1 Recall that $n!$ (n factorial) is the product of the first n positive integers:

$$n! = (1)(2)(3) \cdots (n-1)(n).$$

We can form the sequence $\{n!\}_{n=1}^{\infty}$ whose nth term is $n!$. The first few terms of the sequence are

$$1! = 1, \ 2! = 2, \ 3! = 6, \ 4! = 24, \ 5! = 120.$$

We may simply refer to "the sequence $n!$". $\square$

Remark 1 We can view a sequence $\{a_n\}_{n=N}^{\infty}$, where the starting value of the index n is the integer N, as a function whose domain is the set of integers $\{N, N+1, N+2, N+3, \ldots\}$. If we refer to this function as f, then $f(n) = a_n$ for $n = N, N+1, N+2, \ldots$. $\Diamond$

The definitions of the graph and the range of a sequence are consistent with the view of a sequence as a function:

Definition 1 The graph of the sequence $\{a_n\}_{n=N}^{\infty}$ is the set of points of the form (n, a_n) in the Cartesian coordinate plane, where $n = N, N+1, N+2, \ldots$. **The range of the sequence** $\{a_n\}_{n=N}^{\infty}$ is the range of the function f such that $f(n) = a_n$, $n \geq N$.

Just as in the case of a function that is defined on an interval, the graph of a sequence helps us visualize the sequence. The graph of a sequence consists of isolated points in the plane, unlike the graph of a function that is defined at all points of an interval. We may also visualize a sequence simply by sketching its range on the number line.

Example 2 Let

$$a_n = \frac{n}{n-4}, \quad n = 5, 6, 7, \ldots$$

The graph of the sequence $\{a_n\}_{n=5}^{\infty}$ is the set of points in the Cartesian coordinate plane in the form

$$\left(n, \frac{n}{n-4}\right),$$

where $n = 5, 6, 7, \ldots$. Figure 1 shows the points in the graph of the sequence corresponding to $n = 5, 6, 7, \ldots, 20$. $\square$

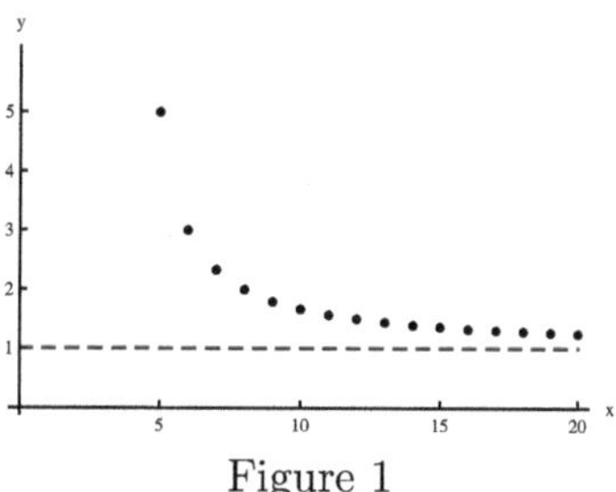

Figure 1

In the above examples, the terms of a sequence were defined by an expression such as

$$\frac{1}{n}, \ n! \ \text{or} \ \frac{n}{n-4}.$$

We will also come across sequences $\{a_n\}$ where we don't have a single expression that defines a_n for all n, but the terms can be evaluated successively. Here is an example:

Example 3 Let $a_0 = 2$ and

$$a_n = \frac{1}{2}a_{n-1} + \frac{1}{a_{n-1}}$$

for $n = 1, 2, 3, \ldots$.

We have defined a sequence, since this recipe enables us to determine the terms of the sequence, one after the other. For example,

$$a_1 = \frac{1}{2}a_0 + \frac{1}{a_0} = \frac{1}{2}(2) + \frac{1}{2} = 1.5,$$

$$a_2 = \frac{1}{2}a_1 + \frac{1}{a_1} = \left(\frac{1}{2}\right)(1.5) + \frac{1}{1.5} \cong 1.416\,67.$$

This is the sort of computation where we definitely need a computational utility. It is easy to give the instruction for the generation of the terms of such a sequence. If we set

$$g(x) = \frac{1}{2}x + \frac{1}{x},$$

we have $a_n = g(a_{n-1})$. Table 1 displays a_n (rounded to 8 significant digits), where $n = 0, 1, 2, 4, 5$. $\square$

n	a_n
0	2
1	1.5
2	1.4166667
3	1.4142157
4	1.4142136
5	1.4142136

Table 1

If $a_n = g(a_{n-1})$, as in Example 3, we say that the sequence $\{a_n\}$ is generated **recursively**, or **iteratively**.

The Limit of a Sequence

Figure 1 indicates that the terms of the sequence in Example 2 get closer and closer to 1 as n increases. Intuitively, a sequence $\{a_n\}_{n=N}^{\infty}$ is said to converge to the limit L if a_n approaches L as n becomes larger and larger.

Definition 2 The number L is **the limit of the sequence** $\{a_n\}_{n=N}^{\infty}$ if $|a_n - L|$ is as small as desired provided that n is sufficiently large. In this case we say that **the sequence** $\{a_n\}_{n=N}^{\infty}$ **converges to** L, and write

$$\lim_{n \to \infty} a_n = L$$

(read "**the limit of** a_n **as** n **approaches infinity is equal to** L".

You can find the precise definition of the limit of a sequence at the end of this section.

Remark 2 Note the similarity between $\lim_{n \to \infty} a_n$ and $\lim_{x \to +\infty} f(x)$, where f is a function that is defined on an interval of the form $[A, +\infty)$. If $a_n = f(n)$ and $\lim_{x \to +\infty} f(x) = L$, then $\lim_{n \to \infty} a_n = \lim_{x \to +\infty} f(x)$. Therefore, we can determine the limit of such as sequence by making use of the knowledge that we acquired in Section 1.7. $\Diamond$

Example 4 Let

$$a_n = \frac{n}{n-4}, \quad n = 5, 6, 7, \ldots$$

as in Example 2. Determine $\lim_{n \to \infty} a_n$.

Solution

If we set

$$f(x) = \frac{x}{x-4},$$

then $a_n = f(n)$. Therefore, $\lim_{n \to +\infty} a_n = \lim_{x \to +\infty} f(x)$. We know how to determine the limit of a rational function at $+\infty$: Since

$$f(x) = \frac{x}{x\left(1 - \dfrac{4}{x}\right)} = \frac{1}{1 - \dfrac{4}{x}},$$

we have

$$\lim_{x \to +\infty} f(x) = \lim_{x \to +\infty} \frac{1}{1 - \dfrac{4}{x}} = 1.$$

Thus,

$$\lim_{n \to \infty} \frac{n}{n - 4} = 1,$$

as well. $\square$

Example 5 Let

$$a_n = \frac{2n^2 + 1}{n^2 + 1}, \quad n = 1, 2, 3, \ldots.$$

a) Determine $\lim_{n \to \infty} a_n$.
b) Plot the points in the graph of the sequence a_n corresponding to $n = 1, 2, \ldots, 10$ with the help of your graphing utility. Is the picture consistent with the result of part a)?

Solution

a) Note that $a_n = f(n)$, where

$$f(x) = \frac{2x^2 + 1}{x^2 + 1}, \quad x \in \mathbb{R}.$$

Therefore, $\lim_{n \to \infty} a_n = \lim_{x \to +\infty} f(x)$, and f is a rational function. Let's apply the technique that we used in Section 1.7 in order to evaluate the limit of a rational function at infinity:

$$f(x) = \frac{2x^2 + 1}{x^2 + 1} = \frac{x^2\left(2 + \dfrac{1}{x^2}\right)}{x^2\left(1 + \dfrac{1}{x^2}\right)} = \frac{2 + \dfrac{1}{x^2}}{1 + \dfrac{1}{x^2}}.$$

Therefore,

$$\lim_{n \to \infty} a_n = \lim_{x \to +\infty} f(x) = \lim_{x \to +\infty} \frac{2 + \dfrac{1}{x^2}}{1 + \dfrac{1}{x^2}} = 2.$$

b) Figure 2 displays the points in the graph of the sequence corresponding to $n = 1, 2, \ldots, 10$, superimposed on the graph of f. We see that the points (n, a_n) approach the line $y = 2$ as n gets larger. This is consistent with the fact that $\lim_{n \to \infty} a_n = 2$. Note that $y = 2$ is a horizontal asymptote for the graph of f. $\square$

The determination of the limit of a sequence that is defined recursively is not as straightforward as the above examples:

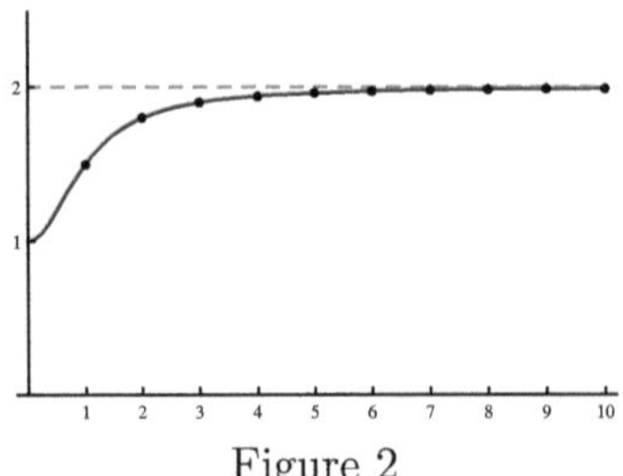

Figure 2

Example 6 Let $a_0 = 2$ and

$$a_n = \frac{1}{2}a_{n-1} + \frac{1}{a_{n-1}}$$

for $n = 1, 2, 3, \ldots$, as in Example 3.

The numbers in Table 1 indicate that the sequence a_n converges to a number that is approximately 1.4142136. Indeed, $\lim_{n\to\infty} a_n = \sqrt{2}$ ($\sqrt{2} \cong 1.4142136$, rounded to 8 significant digits). We will not prove this fact. You will find more about the background of this particular sequence in Section 2.10. $\square$

A sequence need not have a limit:

Example 7 Let $a_n = (-1)^n$, $n = 1, 2, 3, \ldots$, so that the sequence is $-1, 1, -1, 1, \ldots$.

The only candidates for the limit of the sequence are 1 and -1, and the sequence does not converge to either number. $\square$

If c is a constant, we will refer to the sequence $c, c, c, \ldots$ as a **constant sequence**. Obviously, **the limit of a constant sequence is the same constant:**

$$\lim_{n\to\infty} c = c.$$

The rules for the limits of arithmetic combinations of sequences are parallel to the rules for the limits of arithmetic combinations of functions:

Assume that $\lim_{n\to\infty} a_n$ and $\lim_{n\to\infty} b_n$ exist, and that c is a constant. Then,

$$\lim_{n\to\infty} ca_n = c \lim_{n\to\infty} a_n,$$
$$\lim_{n\to\infty} (a_n + b_n) = \lim_{n\to\infty} a_n + \lim_{n\to\infty} b_n,$$
$$\lim_{n\to\infty} a_n b_n = \left(\lim_{n\to\infty} a_n \right) \left(\lim_{n\to\infty} b_n \right).$$

If $\lim_{n\to\infty} b_n \neq 0$,

$$\lim_{n\to\infty} \frac{a_n}{b_n} = \frac{\lim_{n\to\infty} a_n}{\lim_{n\to\infty} b_n}.$$

There is a useful connection between the convergence of sequences and the continuity of a function at a point:

Theorem 1 If a function f is continuous at a point a, and $\lim_{n\to\infty} a_n = a$ then

$$\lim_{n\to\infty} f(a_n) = f(a).$$

We can express the conclusion of Theorem 1 as follows:

$$\lim_{n \to \infty} f\left(a_n\right) = f\left(\lim_{n \to \infty} a_n\right).$$

You can find the proof of Theorem 1 in Appendix B. The statement of the theorem is plausible: If $\lim_{n \to \infty} a_n = a$ then $a_n \cong a$ if n is large. In that case, $f(a_n) \cong f(a)$ since f is continuous at a. Thus, we should have $\lim_{n \to \infty} f\left(a_n\right) = f\left(a\right)$.

Example 8 Determine

$$\lim_{n \to \infty} \sqrt{\frac{4n}{n+1}}.$$

Solution

Set $f\left(x\right) = \sqrt{x}$, so that

$$\sqrt{\frac{4n}{n+1}} = f\left(\frac{4n}{n+1}\right).$$

We have

$$\lim_{n \to \infty} \frac{4n}{n+1} = \lim_{n \to \infty} \frac{4n}{n\left(1+\dfrac{1}{n}\right)} = \lim_{n \to \infty} \frac{4}{1+\dfrac{1}{n}} = 4.$$

Since the square-root function f is continuous at 4,

$$\lim_{n \to \infty} \sqrt{\frac{4n}{n+1}} = \lim_{n \to \infty} f\left(\frac{4n}{n+1}\right) = f\left(\lim_{n \to \infty} \frac{4n}{n+1}\right) = f\left(4\right) = \sqrt{4} = 2.$$

$\square$

Remark 3 There is a version of **the squeeze theorem for sequences**: Assume that $b_n \leq a_n \leq c_n$ for $n = 1, 2, 3, \ldots$, and

$$\lim_{n \to \infty} b_n = \lim_{n \to \infty} c_n.$$

Then, $\lim_{n \to \infty} a_n$ exists and

$$\lim_{n \to \infty} a_n = \lim_{n \to \infty} b_n = \lim_{n \to \infty} c_n.$$

Intuitively, if both b_n and c_n approach the same number L as n becomes large, then a_n must also approach L as n tends to infinity, since a_n is squeezed between b_n and c_n. $\Diamond$

Example 9 Determine

$$\lim_{n \to \infty} \frac{\sin\left(n\right)}{\sqrt{n}}.$$

Solution

Since $-1 \leq \sin\left(n\right) \leq 1$, we have

$$-\frac{1}{\sqrt{n}} \leq \frac{\sin\left(n\right)}{\sqrt{n}} \leq \frac{1}{\sqrt{n}}.$$

We also have

$$\lim_{n \to \infty} \left(-\frac{1}{\sqrt{n}}\right) = \lim_{n \to \infty} \frac{1}{\sqrt{n}} = 0.$$

Therefore,

$$\lim_{n \to \infty} \frac{\sin(n)}{\sqrt{n}} = 0,$$

by the squeeze theorem for sequences. Figure 3 illustrates the convergence of the sequence. The dashed curves are the graphs of $y = 1/\sqrt{x}$ and $y = -1/\sqrt{x}$. $\square$

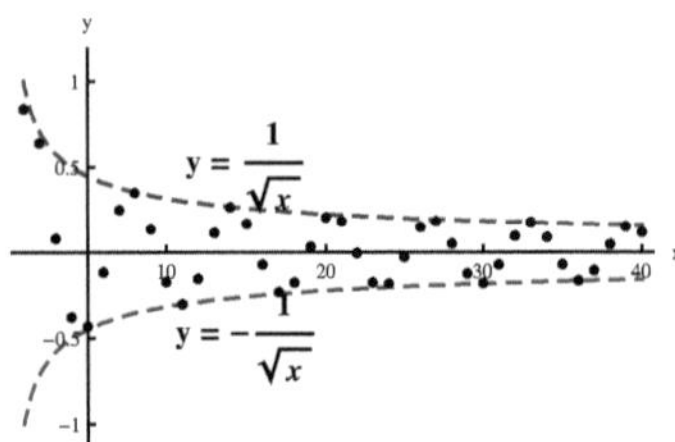

Figure 3

The following observation is useful:

Proposition 1 We have $\lim_{n \to \infty} a_n = 0$ if and only if $\lim_{n \to \infty} |a_n| = 0$.

Indeed, $\lim_{n \to \infty} a_n = 0$ means that $|a_n - 0| = |a_n|$ is as small as desired provided that n is large enough. This says that $\lim_{n \to \infty} |a_n| = 0$.

Example 10 Show that

$$\lim_{n \to \infty} (-1)^n \frac{1}{n} = 0.$$

Solution

We have

$$\left| (-1)^n \frac{1}{n} \right| = \frac{1}{n},$$

and

$$\lim_{n \to \infty} \frac{1}{n} = 0.$$

Therefore,

$$\lim_{n \to \infty} (-1)^n \frac{1}{n} = 0.$$

$\square$

Here is another useful observation:

Proposition 2 Let $|a| < 1$. Then $\lim_{n \to \infty} a^n = 0$.

For example,

$$\lim_{n \to \infty} \frac{1}{2^n} = 0 \text{ and } \lim_{n \to \infty} \left(-\frac{1}{2} \right)^n = 0.$$

The Proof of Proposition 2

By Proposition 1, we need to show that $\lim_{n \to \infty} |a^n| = \lim_{n \to \infty} |a|^n = 0$. Therefore, we can assume that $0 \leq a < 1$. The case $a = 0$ is trivial, since the corresponding sequence is the constant sequence $0, 0, 0, \ldots$. If $0 < a < 1$,

$$a = \frac{1}{1+h},$$

where $h > 0$. Therefore,

$$a^n = \frac{1}{(1+h)^n}.$$

By the Binomial Theorem,

$$(1+h)^n = 1 + nh + \frac{n(n-1)}{2}h^2 + \cdots h^n > 1 + nh,$$

since $h > 0$. Therefore,

$$0 < a^n = \frac{1}{(1+h)^n} < \frac{1}{1+nh} < \frac{1}{nh},$$

and

$$\lim_{n \to \infty} \frac{1}{nh} = \frac{1}{h} \lim_{n \to \infty} \frac{1}{n} = 0.$$

Thus, the sequence a^n is squeezed between the constant sequence $0, 0, 0, \ldots$ and another sequence whose limit is 0. Therefore $\lim_{n \to \infty} a^n = 0$ as well. ∎

Infinite Limits of Sequences

The concept of an infinite limit of a sequence is similar to the concept of an infinite limit of a function at infinity as we discussed in Section 1.7:

Definition 3 The sequence $\{a_n\}$ **diverges to** $+\infty$ if, given any number $M > 0$, however large, $a_n > M$ provided that n is large enough. In this case we write

$$\lim_{n \to \infty} a_n = +\infty$$

(read **"the limit of a_n as n tends to infinity is** $+\infty$"). The sequence a_n **diverges to** $-\infty$ if, given any number $M > 0$, $a_n < -M$ provided that n is large enough. In this case we write

$$\lim_{n \to \infty} a_n = -\infty$$

(read **"the limit of a_n as n tends to infinity is** $-\infty$").

Remark 4 As in the case of $\lim_{x \to +\infty} f(x) = \pm\infty$, where f is defined on an interval of the form $[a, +\infty)$, we are using the same symbol "lim" in two different contexts:

When we write $\lim_{n \to \infty} a_n = L$, where L is a number, the limit of the sequence $\{a_n\}$ is L. On the other hand, when we write $\lim_{n \to \infty} a_n = \pm\infty$, the symbol "lim" has to be interpreted in the sense of Definition 3. As in the case of finite limits, if $a_n = f(n)$, where f is defined on an interval of the form $[a, +\infty)$, and $\lim_{x \to +\infty} f(x) = \pm\infty$, then $\lim_{n \to \infty} a_n = \pm\infty$. ◊

Example 11 We have

$$\lim_{n \to \infty} n^2 = +\infty \text{ and } \lim_{n \to \infty} \left(-n^2\right) = -\infty.$$

Indeed, given any $M > 0$, however large, $n^2 > M$ and $-n^2 < -M$ if $n > \sqrt{M}$. □

The following guidelines for the determination of infinite limits of sequences are parallel to the guidelines that were provided in Section 1.7 for the determination of infinite limits of functions at infinity:

Theorem 2

a) If $a_n > 0$ provided that n is sufficiently large and $\lim_{n\to\infty} a_n = 0$, then

$$\lim_{n\to\infty} \frac{1}{a_n} = +\infty.$$

b) If $\lim_{n\to\infty} a_n > 0$ or $\lim_{n\to\infty} a_n = +\infty$, and $\lim_{n\to\infty} b_n = +\infty$, then $\lim_{n\to\infty} a_n b_n = +\infty$.

c) If $\lim_{n\to\infty} a_n = L$, where L is finite, or $\lim_{n\to\infty} a_n = +\infty$, and $\lim_{x\to+\infty} b_n = +\infty$, then

$$\lim_{n\to\infty} (a_n + b_n) = +\infty.$$

Example 12 Discuss the convergence or divergence of the sequence $\{a_n\}_{n=1}^{\infty}$, where

$$a_n = \frac{n^2 + 3n + 9}{2n + 1}.$$

Solution

Note that $a_n = f(n)$ where

$$f(x) = \frac{x^2 + 3x + 9}{2x + 1}.$$

Therefore, $\lim_{n\to\infty} a_n = \lim_{x\to+\infty} f(x)$. We will use the same strategy that is used in the evaluation of $\lim_{x\to+\infty} f(x)$, but express the steps in terms of the sequence a_n. We have

$$a_n = \frac{n^2 + 3n + 9}{2n + 1} = \frac{n^2\left(1 + \dfrac{3}{n} + \dfrac{9}{n^2}\right)}{2n\left(1 + \dfrac{1}{2n}\right)} = \left(\frac{n}{2}\right)\left(\frac{1 + \dfrac{3}{n} + \dfrac{9}{n^2}}{1 + \dfrac{1}{2n}}\right).$$

Since

$$\lim_{n\to\infty} \frac{1 + \dfrac{3}{n} + \dfrac{9}{n^2}}{1 + \dfrac{1}{2n}} = 1 > 0,$$

and $\lim_{n\to\infty} n/2 = +\infty$, we have $\lim_{n\to\infty} a_n = +\infty$, by Theorem 2. $\square$

The Precise Definitions

The following is the precise version of Definition 2:

Definition 4 The limit of the sequence $\{a_n\}$ is L if, given any $\varepsilon > 0$, there exists a positive integer N such that $|a_n - L| < \varepsilon$ if $n > N$.

Note that $|a_n - L| < \varepsilon$ if $L - \varepsilon < a_n < L + \varepsilon$.

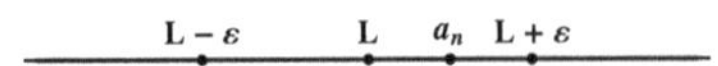

Figure 4: $L - \varepsilon < a_n < L + \varepsilon$ if $n > N$

Example 13 Let

$$a_n = \frac{n}{n+1}, \quad n = 5, 6, 7, \ldots.$$

Show that $\lim_{n\to\infty} a_n = 1$, in accordance with the precise definition of the limit of a sequence.

Solution

We have

$$a_n - 1 = \frac{n}{n+1} - 1 = \frac{n - (n+1)}{n+1} = \frac{-1}{n+1}.$$

Therefore,

$$|a_n - 1| = \frac{1}{n+1} < \frac{1}{n}.$$

Let $\varepsilon > 0$ be given. By the above inequality, in order to have $|a_n - 1| < \varepsilon$, it is sufficient to have

$$\frac{1}{n} < \varepsilon \Leftrightarrow n > \frac{1}{\varepsilon}.$$

Therefore, let's set N to be an integer such that $N > 1/\varepsilon$. If $n > N$, we have

$$|a_n - 1| < \frac{1}{n} < \frac{1}{N} < \varepsilon.$$

Thus, $\lim_{n\to\infty} a_n = 1$. $\square$

The following is the precise version of Definition 3:

Definition 5 We have $\lim_{n\to\infty} a_n = +\infty$ if, given any $M > 0$ there exists a positive integer N such that $a_n > M$ if $n > N$. We have $\lim_{n\to\infty} a_n = -\infty$ if, given any $M > 0$ there exists a positive integer N such that $a_n < -M$ if $n > N$.

Example 14 Show that

$$\lim_{n\to\infty} \sqrt{n} = +\infty \text{ and } \lim_{n\to\infty} \left(-\sqrt{n}\right) = -\infty$$

in accordance with the precise definitions.

Solution

Given $M > 0$, we have $\sqrt{n} > M$ if $n > M^2$. With reference to Definition 5, we can set N to be a positive integer such that $N > M^2$. If $n > N$, then $\sqrt{n} > \sqrt{N} > M$. Therefore, $\lim_{n\to\infty} \sqrt{n} = +\infty$.

Similarly, if the positive integer N is chosen so that $N > M^2$, then $-\sqrt{n} < -\sqrt{N} < -M$ if $n > N$. Therefore, $\lim_{n\to\infty} \left(-\sqrt{n}\right) = -\infty$. $\square$

Problems

In problems 1 - 6, list the first 4 terms of the sequence $\{a_n\}$.

1.

$$a_n = 2n - 1, \quad n = 1, 2, 3, \ldots$$

2.

$$a_n = \frac{n}{n^2 - 1}, \quad n = 2, 3, 4, \ldots$$

3.

$$a_n = (-1)^n \frac{1}{3^n}, \quad n = 0, 1, 2, \ldots$$

4.

$$a_n = \frac{n^2}{\sin\left((2n+1)\frac{\pi}{2}\right)}, \quad n = 1, 2, 3, \ldots$$

5.

$$a_n = \frac{2^n}{n!}, \ n = 1, 2, 3, \ldots$$

6.

$$a_n = 1 + \frac{1}{2} + \frac{1}{2^2} + \cdots + \frac{1}{2^{n-1}}, \ n = 1, 2, 3, \ldots$$

[C] In problems 7 - 10, determine the first 4 terms of the recursively-generated sequence.

7.

$$a_{n+1} = 2a_n - 1, \ n = 0, 1, 2, \ldots, \ a_0 = 2.$$

8.

$$a_{n+1} = a_n - \frac{1}{a_n}, \ n = 0, 1, 2, \ldots, \ a_0 = 4$$

9.

$$a_{n+1} = g\left(a_n\right), n = 0, 1, 2, \ldots, \ \text{where } g\left(x\right) = x^2 - 3 \text{ and } a_0 = 3$$

10.

$$a_{n+1} = g\left(a_n\right), n = 0, 1, 2, \ldots, \ \text{where } g\left(x\right) = 2 - \frac{\cos\left(x\right)}{\sin\left(x\right)} \text{ and } a_0 = 3.$$

In problems11 - 14.
a) Determine the limit of the given sequence,
b) [C] Make use of your graphing utility to plot (n, a_n) for the indicated values of n. Are the pictures consistent with your response to part a)?

11.

$$a_n = \frac{3n+2}{n+5}, \ n = 1, 2, \ldots, 40$$

12.

$$a_n = \frac{1}{\sqrt{n}}, \ n = 1, 2, \ldots, 20.$$

13.

$$a_n = \frac{\sin\left(n\right)}{n}, \ n = 1, 2, \ldots, 20$$

14.

$$a_n = \frac{4n^2 + 3n}{n^2 - n}, \ n = 2, 3, 4, \ldots, 10$$

In problems 15 - 28 , determine the finite or infinite limit, if such a limit exists. You need to display the steps that lead to your response, and provide an explanation if you claim that the limit does not exist.

15.

$$\lim_{n \to \infty} \frac{5n^2 + 9}{2n^2 + 1}$$

16.

$$\lim_{n \to \infty} \frac{n^2 + 10}{4n^3 + 1}$$

17.

$$\lim_{n \to \infty} \frac{n^4 + 4}{n^2 - 3}$$

18.

$$\lim_{n \to \infty} \frac{n^3 - 1}{2n^3 + 100}$$

19.

$$\lim_{n \to \infty} \sqrt{\frac{3n^2}{9n^2 - 2}}$$

20.

$$\lim_{n \to \infty} \left(\frac{16n^2 + 3}{2n^2 + 9n + 1}\right)^{1/3}$$

21.

$$\lim_{n \to \infty} \cos\left(\pi n\right)$$

22.

$$\lim_{n \to \infty} \sin\left(\frac{\pi n}{6n + 2}\right)$$

23.

$$\lim_{n \to \infty} \frac{(-1)^n n}{n + 4}$$

24.

$$\lim_{n \to \infty} \frac{(-1)^n n}{n^2 + 4}$$

25.

$$\lim_{n \to \infty} \left(-\frac{3}{4}\right)^n$$

26.

$$\lim_{n \to \infty} \left(\frac{5}{3}\right)^n$$

27.

$$\lim_{n \to \infty} \left(-\frac{3}{2} \right)^n$$

28.

$$\lim_{n \to \infty} \frac{\cos(n)}{\sqrt{n}}$$

Chapter 2

The Derivative

In this chapter we will introduce the fundamental concept of the **derivative**. The derivative of a function f at a point a can be interpreted as the **rate of change** of f at a or the **slope** of the graph of f at $(a, f(a))$. The **instantaneous velocity** at the instant t of an object in one-dimensional motion is the derivative of the relevant position function at t. The graph of the linear function that has the same value and the same derivative as f at a is tangent to the graph of f at $(a, f(a))$. That linear function is "**the best local linear approximation**" to f near a, in a sense that will be explained in this chapter. You will also learn how to compute the derivatives of the basic functions and their combinations.

2.1 The Concept of the Derivative

In this section we will introduce the concept of **the derivative**. The derivative of a function f at a point a can be interpreted as **the slope of the tangent line** to the graph of f at $(a, f(a))$. The tangent line is the graph of a linear function that is the best linear approximation to f near a in a sense that will be explained in this section.

The Derivative of a Function at a Point

In Section 1.3 we saw that the determination of the slope of a tangent line leads to the idea of the limit. Let's look at another example:

Example 1 Let $f(x) = x^2 - 2x + 4$. Let's determine the slope of the tangent line to the graph of f at $(3, f(3)) = (3, 7)$.

Figure 1 shows a secant line that passes through the point $(3, f(3)) = (3, 7)$.

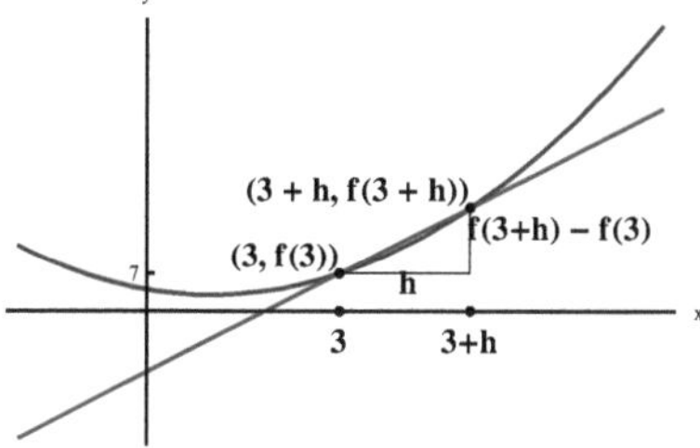

Figure 1

The slope of the secant line passing through $(3, f(3))$ and $(3 + h, f(3 + h))$ is

$$\frac{f(3+h) - f(3)}{h} = \frac{\left((3+h)^2 - 2(3+h) + 4\right) - 7}{h}$$

$$= \frac{\left(7 + 4h + h^2\right) - 7}{h} = \frac{h(4+h)}{h} = 4 + h.$$

Since we expect such a secant line to be almost "tangential" to the graph of f at $(3, f(3))$ if $|h|$ is small, it is reasonable to calculate the slope of the tangent line to the graph of f at $(3, f(3))$ as the limit of the slope of the secant line as h approaches 0:

$$\lim_{h \to 0} \frac{f(3+h) - f(3)}{h} = \lim_{h \to 0} (4 + h) = 4.$$

Since the tangent line has slope 4 and passes through $(3, f(3))$, it is the graph of the equation

$$y = f(3) + 4(x - 3) = 7 + 4(x - 3).$$

Figure 2 shows the graph of f and the tangent line at $(3, f(3))$. The picture is consistent with our intuitive notion of a tangent line. $\square$

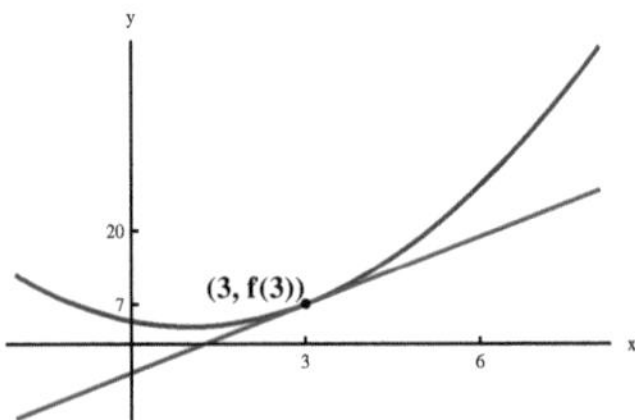

Figure 2: The tangent line to the graph of f at $(3, f(3))$

In the general case, assume that $f(x)$ is defined for each x in an open interval that contains the point a. If $h \neq 0$ and $|h|$ is small enough so that $f(a + h)$ is defined, **the slope of the secant line** that passes through the points $(a, f(a))$ and $(a + h, f(a + h))$ is

$$\frac{f(a+h) - f(a)}{h}.$$

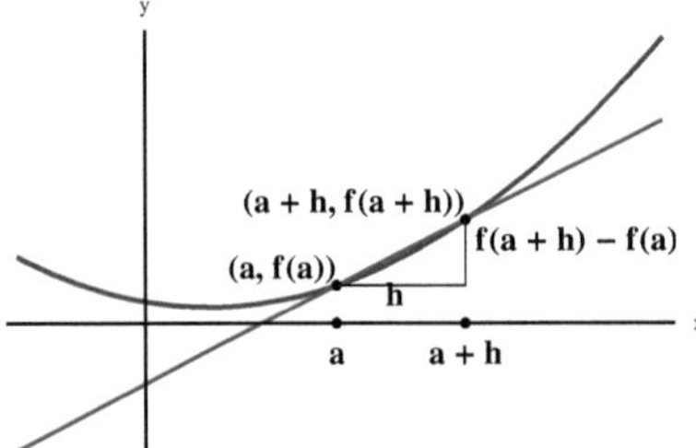

Figure 3: A secant line

Since the secant line that passes through the points $(a, f(a))$ and $(a + h, f(a + h))$ is almost "tangential" to the graph of f at $(a, f(a))$ if $|h|$ is small, we will define **the slope of the tangent line to the graph of f at $(a, f(a))$** as

$$\lim_{h \to 0} \frac{f(a+h) - f(a)}{h}.$$

The slope of a tangent line to the graph of a function can be treated within the framework of the concept of the derivative:

Definition 1 Assume that $f(x)$ is defined for each x in an open interval that contains the point a. **The derivative of f at a is**

$$\lim_{h \to 0} \frac{f(a+h) - f(a)}{h}$$

provided that the limit exists.

We denote the derivative of f at a as $f'(a)$ (read "f **prime at** a"), so that

$$f'(a) = \lim_{h \to 0} \frac{f(a+h) - f(a)}{h}.$$

Thus, $f'(a)$ **can be interpreted as the slope of the tangent line to the graph of f at $(a, f(a))$. The tangent line to the graph of** f at $(a, f(a))$ is the graph of the equation

$$y = f(a) + f'(a)(x - a)$$

(the point-slope form of the equation of the tangent line).
We will refer to the ratio

$$\frac{f(a+h) - f(a)}{h}$$

as a **difference quotient**, since $f(a+h) - f(a)$ is the difference between the values of f at $a+h$ and a, and h is the difference between $a+h$ and a. Graphically, a difference quotient can be interpreted as the slope of a secant line.

If we set $x = a + h$, then x approaches a as h approaches 0. Therefore,

$$f'(a) = \lim_{h \to 0} \frac{f(a+h) - f(a)}{h} = \lim_{x \to a} \frac{f(x) - f(a)}{x - a}.$$

We will favor the expression in terms of h.

Example 2 Assume that f is a linear function, so that $f(x) = mx + b$, where m and b are given constants. The graph of f is a line with slope m. Therefore, we should have $f'(a) = m$ at each point a. Indeed,

$$\frac{f(a+h) - f(a)}{h} = \frac{[m(a+h) + b] - [ma + b]}{h} = \frac{ma + mh + b - ma - b}{b}$$

$$= \frac{mh}{h} = m.$$

Therefore,

$$f'(a) = \lim_{h \to 0} \frac{f(a+h) - f(a)}{h} = \lim_{h \to 0} m = m.$$

$\square$

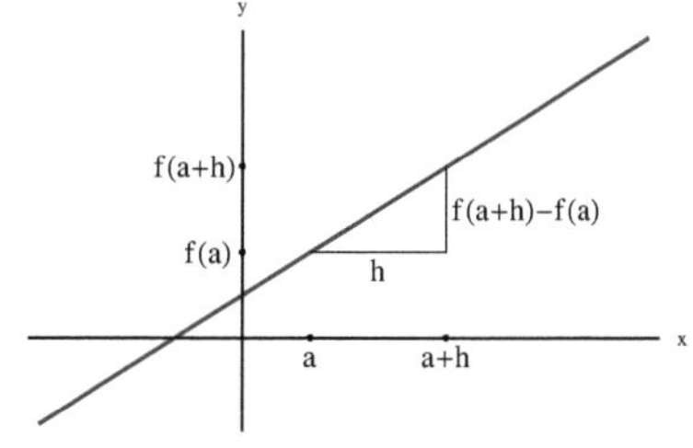

Figure 4

Example 3 Let $f(x) = x^3$.

a) Determine $f'(2)$.

b) Determine the tangent line to the graph of f at $(2, f(2))$.

Solution

a) The relevant difference quotient is

$$
\begin{aligned}
\frac{f(2+h) - f(2)}{h} &= \frac{(2+h)^3 - 2^3}{h} \\
&= \frac{2^3 + 3\left(2^2\right)h + 3\left(2\right)\left(h^2\right) + h^3 - 2^3}{h} \\
&= \frac{12h + 6h^2 + h^3}{h} \\
&= \frac{h\left(12 + 6h + h^2\right)}{h} \\
&= 12 + 6h + h^2.
\end{aligned}
$$

Therefore,

$$
f'(2) = \lim_{h \to 0} \frac{(2+h)^3 - 2^3}{h} = \lim_{h \to 0} \left(12 + 6h + h^2\right) = 12.
$$

b) Since the tangent line to the graph of f at $(2.f(2)) = (2, 8)$ has slope $f'(2) = 12$ and passes through $(2, f(2)) = (2, 8)$, it is the graph of the equation

$$
y = f(2) + f'(2)(x - 2) = 8 + 12(x - 2).
$$

Figure 5 displays the graph f and the tangent line at $(2, f(2))$.

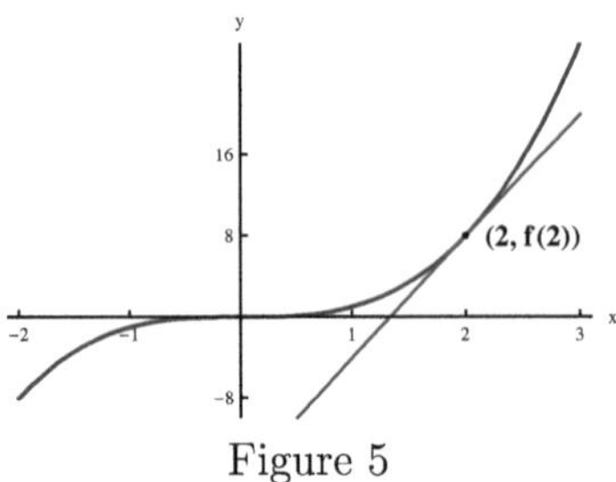

Figure 5

Figure 6 illustrates the effect of zooming in towards the point of contact $(2, f(2)) = (2, 8)$ (the dashed line is the tangent line). $\square$

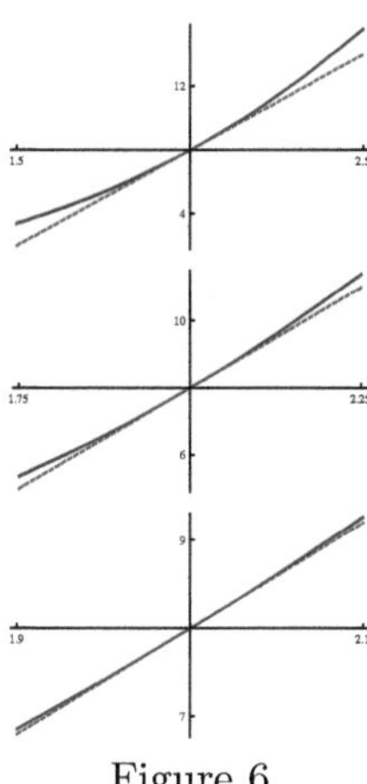

Figure 6

As in the above example, if a function f is differentiable at a the graph of f and the tangent line to the graph of f at $(a, f(a))$ are hardly distinguishable from each other near the point of contact $(a, f(a))$. **We will identify the slope of the graph of f at $(a, f(a))$ with the slope of the tangent line at $(a, f(a))$, i.e., with $f'(a)$.**

We may refer to the difference quotient

$$\frac{f(a+h) - f(a)}{h} = \frac{\text{change in } f(x)}{\text{change in } x}$$

as **the average rate of change of** $f(x)$ corresponding to the change in x from a to $a + h$. Since

$$f'(a) = \lim_{h \to 0} \frac{f(a+h) - f(a)}{h}$$

the derivative of f at a is the limit of the average rate of change of $f(x)$ as the x-increment approaches 0. **Therefore, we will identify the rate of change of f at a with the derivative of f at a.**

Example 4 Let $f(x) = x^2 - 2x + 4$, as in Example 1. We showed that $f'(3) = 4$, so that the rate of change of f at 3 is 4.

Table 1 displays the average rate of change and

$$\left| \frac{f(3+h) - f(3)}{h} - f'(3) \right|$$

for $h = -10^{-n}$, $n = 2, 3, 4$. The numbers are consistent with the fact that average rate of change approaches the rate of change at 3 as h approaches 0. $\square$

h	$\dfrac{f(3+h) - f(3)}{h}$	$\left\| \dfrac{f(3+h) - f(3)}{h} - f'(3) \right\|$	
-10^{-2}	3.99	10^{-2}	
-10^{-3}	3.999	10^{-3}	
-10^{-4}	3.9999	10^{-4}	

Table 1

Definition 2 We say that a function f is **differentiable** at a point a if the derivative of f at a exists.

A function need not be differentiable at a point even if it is continuous at that point, as in the following example.

Example 5 Let f be **the absolute-value function** so that $f(x) = |x|$. Show that f is not differentiable at 0.

Solution

Note that f is continuous at 0, since $\lim_{h\to 0} f(h) = \lim_{h\to 0} |h| = 0 = f(0)$. Figure 4 shows the graph of f.

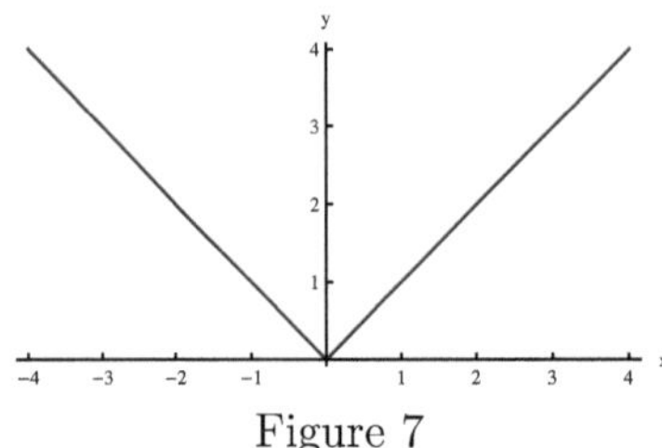

Figure 7

If $h > 0$, the slope of the secant line that passes through $(0, f(0)) = (0,0)$ and $(h, f(h))$ is

$$\frac{f(h) - f(0)}{h} = \frac{|h|}{h} = \frac{h}{h} = 1.$$

Therefore,

$$\lim_{h\to 0+} \frac{f(h) - f(0)}{h} = \lim_{h\to 0} 1 = 1.$$

If $h < 0$, the slope of the secant line that passes through $(0, f(0))$ and $(h, f(h))$ is

$$\frac{f(h) - f(0)}{h} = \frac{|h|}{h} = \frac{-h}{h} = -1.$$

Therefore,

$$\lim_{h\to 0-} \frac{f(h) - f(0)}{h} = \lim_{h\to 0} (-1) = -1.$$

Since

$$\lim_{h\to 0+} \frac{f(h) - f(0)}{h} \neq \lim_{h\to 0-} \frac{f(h) - f(0)}{h},$$

$\lim_{h\to 0} (f(h) - f(0))/h$ does not exist. Therefore, f is not differentiable at 0. $\square$

Even though the absolute-value function is not differentiable at 0, we saw that the one-sided limits of the relevant difference quotient exist. These are examples of one-sided derivatives:

Definition 3 We say that f **is differentiable at a from the right** if

$$\lim_{h\to 0+} \frac{f(a + h) - f(a)}{h}$$

exists. In this case, we define **the right-derivative of f at a** as that limit, and denote it by $f'_+(a)$. Thus,

$$f'_+(a) = \lim_{h\to 0+} \frac{f(a + h) - f(a)}{h}.$$

Similarly, we define $f'_-(a)$, **the left-derivative of f at a** as

$$f'_-(a) = \lim_{h \to 0-} \frac{f(a+h) - f(a)}{h},$$

provided that the limit exists.

Clearly, **a function f is differentiable at a point a if and only if $f'_+(a)$ and $f'_+(a)$ exist and $f'_+(a) = f'_+(a)$.** In the case of equality, the common value of the one-sided derivatives is $f'(a)$.

Example 6 Let $f(x) = |x|$, as in Example 5. We have

$$f'_+(0) = \lim_{h \to 0+} \frac{f(h) - f(0)}{h} = 1 \text{ and } f'_-(0) = \lim_{h \to 0-} \frac{f(h) - f(0)}{h} = -1.$$

$\square$

Here is another example of continuity without differentiability:

Example 7 Let $f(x) = x^{2/3}$. Show that f is not differentiable at 0.

Solution

The relevant difference quotient is

$$\frac{f(h) - f(0)}{h} = \frac{f(h)}{h} = \frac{h^{2/3}}{h} = \frac{1}{h^{1/3}}.$$

Therefore,

$$\lim_{h \to 0+} \frac{f(h) - f(0)}{h} = \lim_{h \to 0+} \frac{1}{h^{1/3}} = +\infty.$$

Since the difference quotient does not have a finite limit as h approaches from the right, $f'_+(0)$ does not exist. Thus, f is not differentiable at 0. Graphically, the secant line that passes through $(0,0)$ and $(h, f(h)) = (h, h^{2/3})$ becomes steeper and steeper as h approaches 0 from the right, as illustrated in Figure 8. $\square$

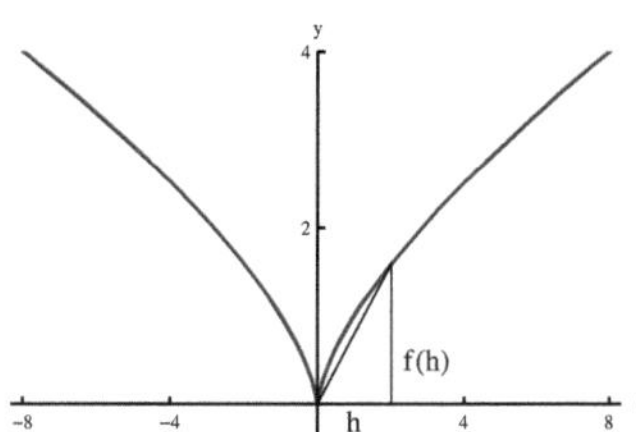

Figure 8: The slope of the secant line becomes steep as h approaches 0

Even though continuity does not imply differentiability, differentiability implies continuity:

Proposition 1 Assume that f is differentiable at a. Then, f is continuous at a.

Proof

We have

$$f(a+h) - f(a) = \left(\frac{f(a+h) - f(a)}{h} \right) h,$$

so that

$$f(a + h) = f(a) + \left(\frac{f(a + h) - f(a)}{h} \right) h.$$

Therefore,

$$\lim_{h \to 0} f(a + h) = f(a) + \lim_{h \to 0} \left(\left(\frac{f(a + h) - f(a)}{h} \right) h \right)$$

$$= f(a) + \lim_{h \to 0} \left(\frac{f(a + h) - f(a)}{h} \right) \lim_{h \to 0} h$$

$$= f(a) + f'(a)(0) = f(a).$$

Since $\lim_{h \to 0} f(a + h) = f(a)$ the function f is continuous at a. ∎

The Derivative as a Function

If x denotes the independent variable of f, it is natural to use the same letter to denote the variable basepoint at which the derivative is evaluated. Thus,

$$f'(x) = \lim_{h \to 0} \frac{f(x + h) - f(x)}{h}.$$

We treat x as being fixed in the evaluation of the limit. You can think of h as "the dynamic variable".

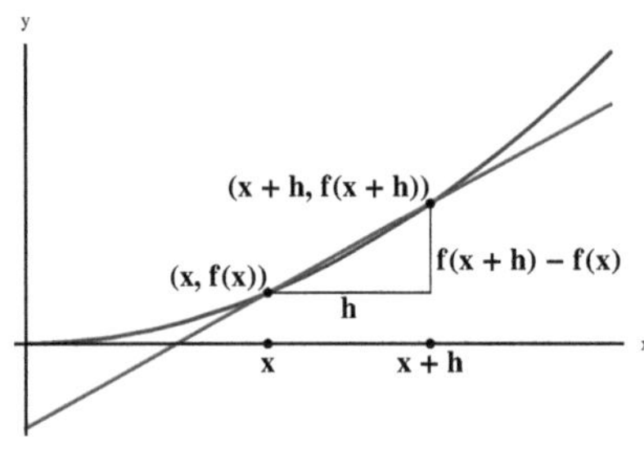

Figure 9

Definition 4 The domain of **the derivative function** corresponding to the function f consists of all x such that f is differentiable at x. **The value of the derivative function at such an x is $f'(x)$.**

We will denote the derivative function corresponding to f as f', so that you may read $f'(x)$ as "f prime of x", as well as "f prime at x". Graphically, the value of the derivative function f' at x is the slope of the tangent line to the graph of f at $(x, f(x))$, alias, the slope of the graph of f at $(x, f(x))$. Thus, the derivative function f' enables us to keep track of the way the slope of the graph of f changes as the basepoint varies. Usually, we will simply refer to "**the derivative of f**", instead of "the derivative function corresponding to f".

Example 8 Let f be **a linear function,** so that $f(x) = mx + b$, where m and b are constants. In Example we showed that $f'(a) = m$ at each $a \in \mathbb{R}$. If we replace a by the variable x, we have $f'(x) = m$ for each $x \in \mathbb{R}$. Thus, the derivative of a linear function is a constant function whose value is the slope of the line that is the graph of the function. As a special case, if f **is a constant function, then $f'(x) = 0$ for each $x \in \mathbb{R}$.** □

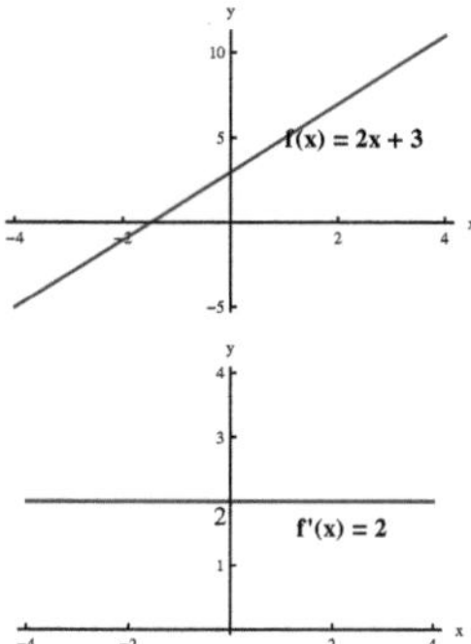

Figure 10: The derivative of a linear function is the slope of its graph

Example 9 Let $f(x) = x^2$. Determine the derivative function f'.

Solution

If x is an arbitrary point on the number line and $h \neq 0$,

$$\frac{f(x+h) - f(x)}{h} = \frac{(x+h)^2 - x^2}{h} = \frac{x^2 + 2xh + h^2 - x^2}{h}$$
$$= \frac{h(2x + h)}{h} = 2x + h.$$

Therefore,

$$f'(x) = \lim_{h \to 0} \frac{f(x+h) - f(x)}{h} = \lim_{h \to 0} (2x + h) = 2x.$$

Thus, $f'(x) = 2x$ for each $x \in \mathbb{R}$. We see that the derivative function that corresponds to the quadratic function f is a linear function. Figure 11 shows the graphs of f and f'.Note that the slope of the graph of f at $(x, f(x))$ is negative if $x < 0$, positive if $x > 0$ and 0 if $x = 0$. $\square$

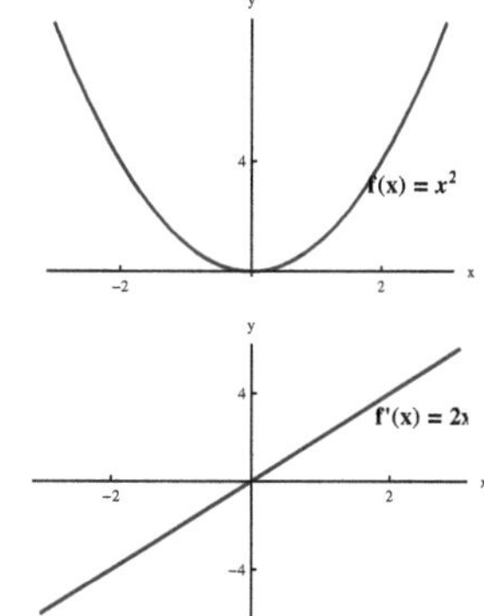

Figure 11: $f(x) = x^2$ and $f'(x) = 2x$

Example 10 Let $f(x) = x^3$. Determine f'.

Solution

If x is an arbitrary point on the number line and $h \neq 0$,

$$\frac{f(x+h) - f(x)}{h} = \frac{(x+h)^3 - x^3}{h} = \frac{x^3 + 3x^2h + 3xh^2 + h^3 - x^3}{h}$$

$$= \frac{h\left(3x^2 + 3xh + h^2\right)}{h}$$

$$= 3x^2 + 3xh + h^2.$$

Therefore,

$$f'(x) = \lim_{h \to 0} \frac{f(x+h) - f(x)}{h} = \lim_{h \to 0} \left(3x^2 + 3xh + h^2\right) = 3x^2.$$

Thus, the derivative function that corresponds to f is the quadratic function defined by $3x^2$. Figure 11 shows the graphs of f and f'. Note that the slope of the graph of f at $(x, f(x))$ is positive if $x \neq 0$ and 0 if $x = 0$. $\square$

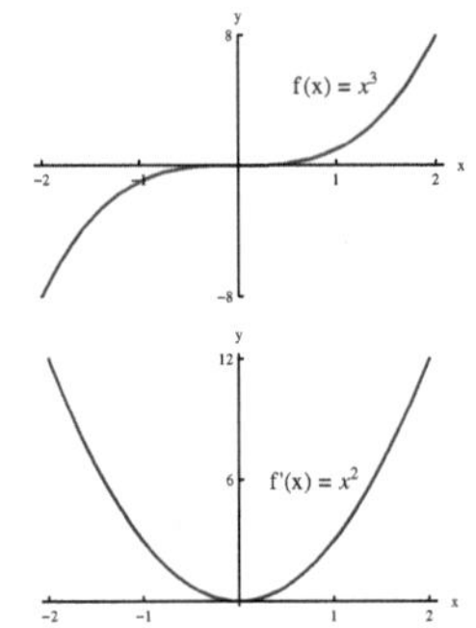

Figure 12: $f(x) = x^3$ and $f'(x) = 3x^2$

The computation of the derivative of a function f at a point x will be referred to as **the differentiation of f at x**. The determination of the derivative function f' corresponding to f will be referred to as **the differentiation of f**. Thus, differentiation is an operation that assigns a function to a given function, as in the above examples.

Example 11 Let f be the absolute-value function, so that $f(x) = |x|$ for each $x \in \mathbb{R}$. Determine f' (you must specify the domain of f').

Solution

In Example 5 we showed that f is not differentiable at 0. Let $x > 0$. Then $x + h$ is also positive if $|h|$ is small enough. Therefore,

$$f'(x) = \lim_{h \to 0} \frac{f(x+h) - f(x)}{h} = \lim_{h \to 0} \frac{|x+h| - |x|}{h}$$

$$= \lim_{h \to 0} \frac{(x+h) - x}{h} = \lim_{h \to 0} \frac{h}{h} = \lim_{h \to 0} (1) = 1.$$

If $x < 0$, we also have $x + h < 0$ if $|h|$ is small enough. Therefore,

$$f'(x) = \lim_{h \to 0} \frac{f(x+h) - f(x)}{h} = \lim_{h \to 0} \frac{|x+h| - |x|}{h}$$

$$= \lim_{h \to 0} \frac{-(x+h) - (-x)}{h} = \lim_{h \to 0} \frac{-h}{h} = \lim_{h \to 0} (-1) = -1.$$

Thus,

$$f'(x) = \begin{cases} 1 & \text{if } x > 0, \\ -1 & \text{if } x < 0. \end{cases}$$

Figure 12 shows the graphs of f and f'. The slope of the graph at $(x, f(x))$ is 1 if $x > 0$, and the slope of the graph of f at $(x, f(x))$ is -1 if $x < 0$. $\square$

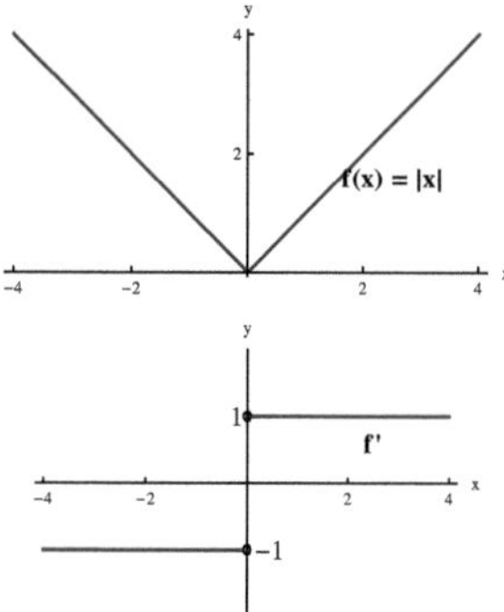

Figure 13: The absolute-value function and its derivative

The Leibniz Notation

Leibniz and **Newton** are recognized as the cofounders of calculus. Newton used the notation $\dot{f}$ for the derivative of f. You may come across Newton's notation in older books on mechanics. The notation that was devised by Leibniz has been more popular and did not lose its popularity over the centuries, since it is practical to use, as you will see in the following sections. We will continue using "the prime notation" as well.

We have

$$f'(x) = \lim_{\Delta x \to 0} \frac{f(x + \Delta x) - f(x)}{\Delta x}.$$

We can replace $f(x + \Delta x) - f(x)$ by Δf, as illustrated in Figure 13. Thus,

$$f'(x) = \lim_{\Delta x \to 0} \frac{\Delta f}{\Delta x}.$$

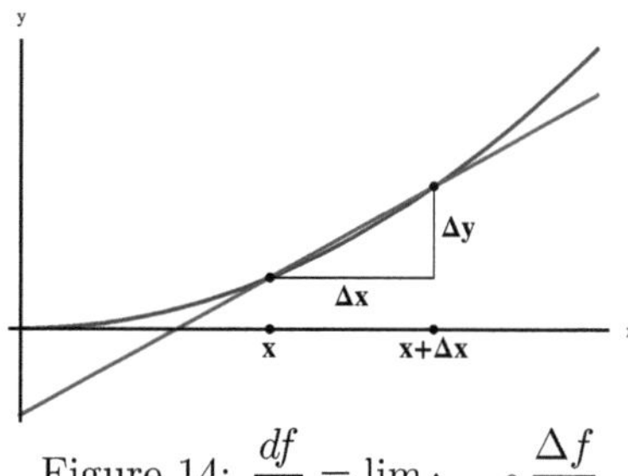

Figure 14: $\dfrac{df}{dx} = \lim_{\Delta x \to 0} \dfrac{\Delta f}{\Delta x}$

The Leibniz notation for the derivative of f at x is

$$\frac{df}{dx}(x).$$

Thus,

$$\frac{df}{dx}(x) = \lim_{\Delta x \to 0} \frac{\Delta f}{\Delta x}.$$

Note that we have replaced Δ in the expression for the difference quotient by the letter d. We may write simply

$$\frac{df}{dx}.$$

The symbol

$$\frac{df}{dx}$$

is not a genuine fraction, i.e., it is not the ratio of some quantity df and some quantity dx. We may refer to it as a " **symbolic fraction**". As long as we are aware of the fact that we are not dealing with an ordinary fraction, an initial advantage of the Leibniz notation is that it reminds us of the definition of the derivative: The derivative is obtained as the limit of a genuine fraction, namely, $\Delta f/\Delta x$, as Δx approaches 0.

We may type the derivative of f in the Leibniz notation as

$$\frac{df}{dx}(x), \quad \frac{df}{dx}, \quad \frac{df(x)}{dx} \quad \text{or} \quad \frac{d}{dx}f(x).$$

The Leibniz notation is convenient in expressing differentiation rules. Let us display the results of some of the examples of this section by using the Leibniz notation:

$$\frac{d}{dx}(mx + b) = m, \text{ where } m \text{ and } b \text{ are constants,}$$

$$\frac{d}{dx}\left(x^2\right) = 2x,$$

$$\frac{d}{dx}\left(x^3\right) = 3x^2.$$

We may refer to a function by the name of the dependent variable. Assume that x is the independent variable and y is the dependent variable of a function. We may speak of "the function $y = y(x)$". In such a case, we will denote the difference quotient as

$$\frac{y(x + \Delta x) - y(x)}{\Delta x} = \frac{\Delta y}{\Delta x},$$

so that Δy denotes the increment of the dependent variable corresponding to the increment Δx of the independent variable, as illustrated in Figure 14. This leads to the Leibniz notation dy/dx for the derivative of y as a function of x:

$$\frac{dy}{dx} = \lim_{\Delta x \to 0} \frac{\Delta y}{\Delta x}.$$

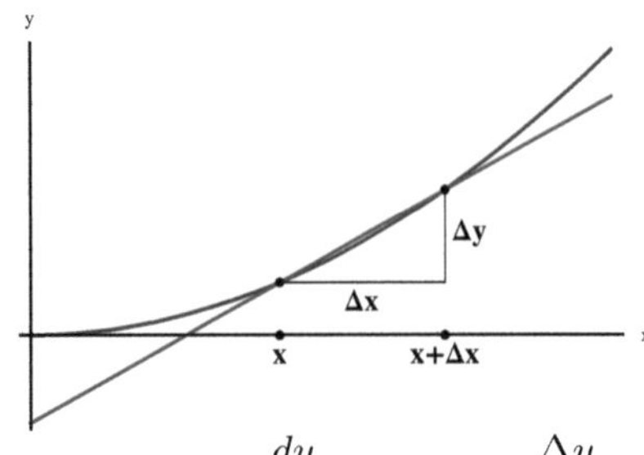

Figure 15: $\dfrac{dy}{dx} = \lim_{\Delta x \to 0} \dfrac{\Delta y}{\Delta x}$

For example, if $y = x^2$,

$$\frac{dy}{dx} = \lim_{\Delta x \to 0} \frac{\Delta y}{\Delta x} = \lim_{\Delta x \to 0} \frac{(x + \Delta x)^2 - x^2}{\Delta x} = 2x.$$

If we use the Leibniz notation for the derivative and we wish to indicate that the derivative of f is to be evaluated at a specific point a, we may use the notations

$$\left.\frac{df(x)}{dx}\right|_{x=a}.$$

For example, if $f(x) = x^3$, then

$$\frac{df}{dx}(x) = 3x^2 \Rightarrow \frac{df}{dx}(2) = 12.$$

We may also express this fact as follows:

$$\left.\frac{d(x^3)}{dx}\right|_{x=2} = \left. 3x^2 \right|_{x=2} = 12.$$

Problems

In problems 1 - 4,
a) Determine the slope of the tangent line to the graph of f at $(a, f(a))$ as a limit of the slopes of secant lines, and the point-slope form of the equation of the tangent line to the graph of f at $(a, f(a))$ with basepoint a.
b) [C] Make use of your graphing utility to plot the graph of f, the tangent line to the graph of f at $(a, f(a))$, and the secant line that passes through the points $(a, f(a))$ and $\left(a + h, f\left(a + 10^{-3}\right)\right)$. Does the picture indicate that the secant line is almost "tangential" to the graph of f at $(a, f(a))$ when $|h|$ is small?
c) [C] Compute the slope of the secant line secant line that passes through the points $(a, f(a))$ and $(a + h, f(a + h))$.for $h = 10^{-n}$, $n = 1, 2, 3, 4$. Do the numbers approach the the slope of the tangent line to the graph of f at $(a, f(a))$?

1. $f(x) = x^2 - 6x + 11$, $a = 4$ 3. $f(x) = x^3 - 4x$, $a = 2$

2. $f(x) = -x^2 - 4x - 1$, $a = -1$ 4. $f(x) = \sqrt{x + 6}$, $x = -2$

In problems 5 - 8, a function f and a number a are given. In each case. Determine $f'(a)$, the derivative of f at a directly from the definition of $f'(a)$ as a limit.

5. $f(x) = 3x^2$, $a = 2$ 7. $f(x) = x^3 + x$, $a = 1$

6. $f(x) = x^2 + 2x$, $a = -3$ 8. $f(x) = \dfrac{1}{x^2}$, $a = \dfrac{1}{2}$

In problems 9 - 12, the limit is the derivative of a function f at a point a. Determine f and a.

9. $\displaystyle\lim_{h \to 0} \frac{\sqrt{16 + h} - 4}{h}$ 11. $\displaystyle\lim_{x \to 2} \frac{x^3 - 8}{x - 2}$

10. $\displaystyle\lim_{h \to 0} \frac{\dfrac{1}{(2 + h)^2} - \dfrac{1}{4}}{h}$ 12. $\displaystyle\lim_{x \to \pi} \frac{\cos(x) + 1}{x - \pi}$

In problems 13 and 14,
a) Determine the average rate of change of the function $f(x)$ as x changes from a to $a + h$.
b) Determine the rate of change of f at a.

13. $f(x) = x^2 - 3x$, $a = 1$

14. $f(x) = \dfrac{4}{x - 5}$, $a = 10$

15. Let $f(x) = |x - 2|$.
a) Determine $f'_+(2)$ and $f'_-(2)$.
b) Is f differentiable at 2?. Justify your response.

16. Let $f(x) = |x^2 - 9|$.
a) Determine the right and left derivatives of f at 3.
b) Is f is differentiable at 3?. Justify your response.

17. Let

$$f(x) = \begin{cases} x^2 & \text{if} \quad x \geq 0, \\ x & \text{if} \quad x < 0. \end{cases}$$

a) Show that f is continuous at 0.
b) Is f differentiable at 0? Justify your response. Determine $f'(0)$ if you claim that f is differentiable at 0.

18. Let

$$f(x) = \begin{cases} x - 1 & \text{if} \quad x \geq 4, \\ 3 & \text{if} \quad x < 4. \end{cases}$$

a) Show that f is continuous at 4.
b) Is f differentiable at 4? Justify your response. Determine $f'(4)$ if you claim that f is differentiable at 4.

19. Let $f(x) = (x - 2)^{1/4}$ if $x \geq 2$. Does f have a right-derivative at 2? Justify your response. Determine $f'_+(2)$ if you claim that $f'_+(2)$ exists.

20. Let $f(x) = (x - 3)^{4/5}$. Is f differentiable at 3? Justify your response. Determine $f'(3)$ if you claim that f is differentiable at 3

21. Let $f(x) = (x - 3)^{6/5}$. Is f differentiable at 3? Justify your response. Determine $f'(3)$ if you claim that f is differentiable at 3

In problems 22 - 24,
a) Determine $f'(a)$, the derivative of f at a directly from the definition of $f'(a)$.
b) Determine the tangent line to the graph of f at $(a, f(a))$ (the point-slope form of the equation with basepoint a will do).

22.
$$f(x) = \sqrt{x}, \; a = 16$$

23.
$$f(x) = 9x^2$$

24.
$$f(x) = \dfrac{1}{x - 1}, \; a = 4$$

In problems 25 - 29, determine the derivative (function) f' directly from the definition of the derivative as a limit.

25. $f(x) = 3x^2$

26. $f(x) = x^2 + 2x$

27. $f(x) = x^3 + x$

28. $f(x) = \dfrac{1}{x - 1}$

29. $f(x) = \dfrac{1}{x^2}$

In problems 30 and 31 determine df/dx as the limit

$$\lim_{\Delta x \to 0} \frac{f(x + \Delta x) - f(x)}{\Delta x}.$$

30. $f(x) = x^2 - 3x$

31. $f(x) = \dfrac{4}{x - 5}$

In problems 32 -35, the limit is the derivative of a function f at x Find f.

32. $\displaystyle\lim_{\Delta x \to 0} \frac{(x + \Delta x)^3 - x^3}{\Delta x}$

34. $\displaystyle\lim_{x \to -1} \frac{(x + \Delta x)^4 - x^4}{\Delta x}$

33. $\displaystyle\lim_{h \to 0} \frac{\dfrac{1}{(x + \Delta x)^4} - \dfrac{1}{x^4}}{\Delta x}$

35. $\displaystyle\lim_{x \to \pi} \frac{\sin(2x + 2\Delta x) + \sin(2x)}{\Delta x}$

In problems 36 and 37,
a) Determine the derivative (function) f' directly from the definition (you need to specify the domain of f'),
b) Sketch the graphs of f and f'.

36. $f(x) = |x - 2|$

37. $f(x) = |x^2 - 9|$.

2.2   The Derivatives of Powers and Linear Combinations

The Derivatives of Rational Powers of x

Rational powers of x are basic building blocks for a rich collection of functions. The rule for the differentiation of powers of x is easy to remember:

THE POWER RULE   **If r is a nonzero rational number, then**

$$\frac{d}{dx}x^r = rx^{r-1}$$

provided that x^r and x^{r-1} are defined.

We will prove the power rule if the exponent r is a positive or negative integer. You can find the proof for an arbitrary rational number at the end of this section.

Let's begin with the case of a positive integer.n. By **the Binomial Theorem** (as reviewed in Section A2 of Appendix A),

$$(x + h)^n = x^n + nx^{n-1}h + \frac{n(n-1)}{2}x^{n-2}h^2 + \cdots + \binom{n}{k}x^{n-k}h^k + \cdots + h^n.$$

For any $x \in \mathbb{R}$ and $h \neq 0$,

$$\frac{f(x+h) - f(x)}{h} = \frac{(x+h)^n - x^n}{h}$$

$$= \frac{\left(x^n + nx^{n-1}h + \dfrac{n(n-1)}{2}x^{n-2}h^2 + \cdots + h^n \right) - x^n}{h}$$

$$= \frac{h \left(nx^{n-1} + \dfrac{n(n-1)}{2}x^{n-2}h + \cdots + h^{n-1} \right)}{h}$$

$$= nx^{n-1} + \frac{n(n-1)}{2}x^{n-2}h + \cdots + h^{n-1}.$$

Therefore,

$$f'(x) = \lim_{h \to 0} \left(nx^{n-1} + \frac{n(n-1)}{2}x^{n-2}h + \cdots + h^{n-1} \right) = nx^{n-1}.$$

The case of a negative integer exponent follows from the first case. Let $r = -n$, where n is a positive integer. If $f(x) = x^r = x^{-n}$, we need to show that

$$f'(x) = rx^{r-1} = -nx^{-n-1}$$

for any $x \neq 0$ so that $f(x)$ is defined. If $h \neq 0$ and $|h|$ is small enough $f(x+h)$ is also defined. The relevant difference quotient is

$$\frac{f(x+h) - f(x)}{h} = \frac{(x+h)^{-n} - x^{-n}}{h} = \frac{1}{h}\left(\frac{1}{(x+h)^n} - \frac{1}{x^n} \right)$$

$$= \frac{1}{h}\left(\frac{x^n - (x+h)^n}{(x+h)^n x^n} \right)$$

$$= \left(-\frac{(x+h)^n - x^n}{h} \right)\left(\frac{1}{(x+h)^n x^n} \right).$$

Therefore,

$$f'(x) = \lim_{h \to 0} \frac{f(x+h) - f(x)}{h} = \lim_{h \to 0} \left(-\frac{(x+h)^n - x^n}{h} \right)\left(\frac{1}{(x+h)^n x^n} \right)$$

$$= \left(-\lim_{h \to 0} \frac{(x+h)^n - x^n}{h} \right)\left(\lim_{h \to 0} \frac{1}{(x+h)^n x^n} \right).$$

By the first case,

$$-\lim_{h \to 0} \frac{(x+h)^n - x^n}{h} = -\frac{d}{dx}x^n = -nx^{n-1},$$

and $\lim_{h \to 0} (x+h)^n = x^n$. Therefore,

$$f'(x) = \left(-nx^{n-1}\right)\left(\frac{1}{x^n x^n} \right) = -\frac{nx^{n-1}}{x^{2n}} = -nx^{-n-1}.$$

■

Example 1   Let $f(x) = x^4$. Then $f'(x) = 4x^3$. Figure 1 shows the graphs of f and f'. $\square$

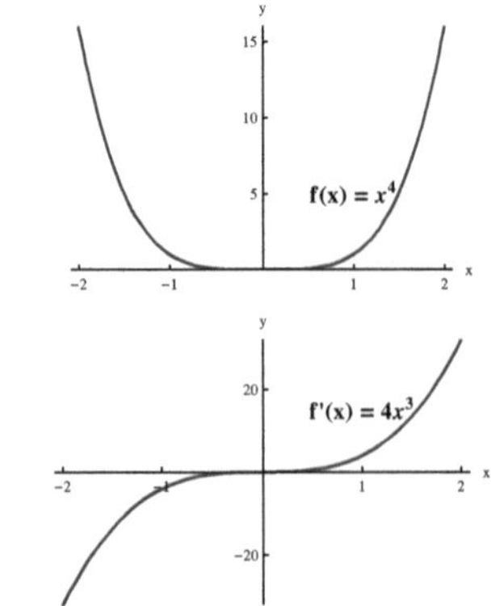

Figure 1: $f(x) = x^4$, $f'(x) = 4x^3$

Example 2 Let

$$f(x) = \frac{1}{x^2}.$$

a) Determine f'.

b) Determine the points at which f is not differentiable and the relevant limits of f'. Interpret the results graphically.

Solution

a) By the power rule,

$$f'(x) = \frac{d}{dx}\left(x^{-2}\right) = -2x^{-3} = -\frac{2}{x^3}.$$

The above expression is valid for each $x \neq 0$.

b) The function f is differentiable at $x \neq 0$. Since $f(x)$ is not defined at $x = 0$, f is certainly not differentiable at 0. On the other hand, we can discuss the limits of $f'(x)$ as x approaches 0 from the right and from the left.

We have

$$f'(x) = -\frac{2}{x^3} = (-2)\left(\frac{1}{x^3}\right).$$

Since $x^3 > 0$ if $x > 0$ and $\lim_{x \to 0} x^3 = 0$,

$$\lim_{x \to 0+} \frac{1}{x^3} = +\infty.$$

Since $\lim_{x \to 0}(-2) = -2 < 0$,

$$\lim_{x \to 0+} f'(x) = \lim_{x \to 0+} (-2)\left(\frac{1}{x^3}\right) = -\infty.$$

On the other hand, $x^3 > 0$ if $x > 0$ so that

$$\lim_{x \to 0-} \frac{1}{x^3} = -\infty$$

Therefore,

$$\lim_{x \to 0-} f'(x) = \lim_{x \to 0+} (-2)\left(\frac{1}{x^3}\right) = +\infty.$$

Therefore, the vertical axis is a vertical asymptote for the graphs of f' (and f). Figure 2 shows the graphs of f and f'. The picture is consistent with our analysis. The tangent line to the graph of f at $(x, f(x))$ becomes steeper and steeper as x approaches 0 from either side. The slope is negative to the right of 0, and positive to the left of 0. This example is a prototype for $1/x^n$ where n is an even positive integer. $\square$

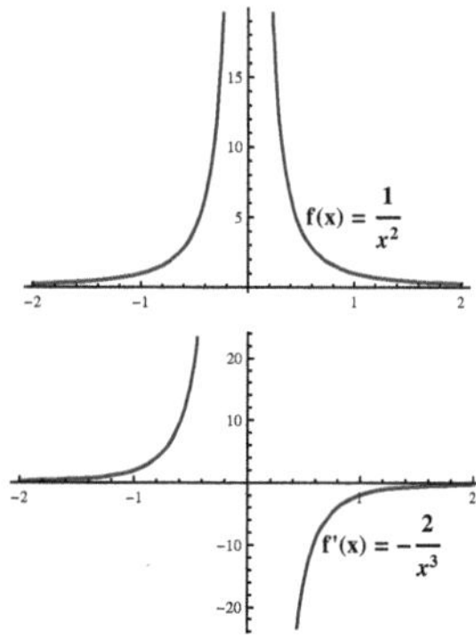

Figure 2

Example 3 Let $f(x) = \sqrt{x} = x^{1/2}$.

a) Determine f' directly from the definition of the derivative.
b) Show that f is not differentiable at 0 from the right.
c) Determine $\lim_{x \to 0+} f'(x)$. Interpret the result graphically and in terms of the rate of change of f.
d) Determine $\lim_{x \to +\infty} f'(x)$. Interpret the result as in part c)

Solution

a) We will use a time-honored trick to express the relevant difference quotient in a way that will lead to the derivative of f:

$$
\begin{aligned}
\frac{f(x+h) - f(x)}{h} &= \frac{\sqrt{x+h} - \sqrt{x}}{h} \\
&= \left(\frac{\sqrt{x+h} - \sqrt{x}}{h} \right) \left(\frac{\sqrt{x+h} + \sqrt{x}}{\sqrt{x+h} + \sqrt{x}} \right) \\
&= \frac{(x+h) - x}{h \left(\sqrt{x+h} + \sqrt{x} \right)} \\
&= \frac{h}{h \left(\sqrt{x+h} + \sqrt{x} \right)} \\
&= \frac{1}{\sqrt{x+h} + \sqrt{x}}.
\end{aligned}
$$

Therefore,

$$
f'(x) = \lim_{h \to 0} \frac{f(x+h) - f(x)}{h} = \lim_{h \to 0} \frac{1}{\sqrt{x+h} + \sqrt{x}} = \frac{1}{2\sqrt{x}}
$$

for any $x > 0$.

b) We have

$$
\lim_{h \to 0+} \frac{f(0+h) - f(0)}{h} = \lim_{h \to 0+} \frac{\sqrt{h}}{h} = \lim_{h \to 0+} \frac{1}{\sqrt{h}} = +\infty.
$$

Therefore, f is not differentiable at 0 from the right even though it is continuous at 0 from the right.

c) We also have

$$
\lim_{x \to 0+} f'(x) = \lim_{x \to 0+} \frac{1}{2\sqrt{x}} = +\infty.
$$

The tangent line to the graph of f at $(x, f(x))$ becomes steeper and steeper as x approaches 0 from the right. Since $f'(x)$ is the rate of change of f at x, the rate of change of the square-root function f increases beyond all bounds as x approaches 0 from the right.

d) We have

$$
\lim_{x \to +\infty} f'(x) = \lim_{x \to +\infty} \frac{1}{2\sqrt{x}} = 0.
$$

Thus, the slope of the tangent line to the graph of the square-root function at $(x, \sqrt{x})$ approaches 0 as x becomes larger and larger. Since $f'(x)$ is the rate of change of f at x, the rate of change of the square root function decreases towards to 0 as x becomes large.

Figure 3 shows the graphs of f and f'. The picture is consistent with our analysis. Since $\lim_{x\to 0+} f'(x) = +\infty$, the vertical axis is a vertical asymptote for the graph of f'. The square-root function is a prototype of functions defined by $x^{1/n}$, where n is an even positive integer. $\square$

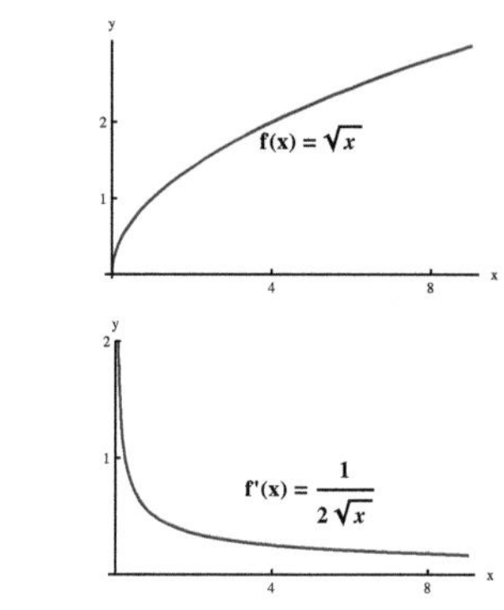

$$\text{Figure 3: } f(x) = \sqrt{x}, \quad f'(x) = \frac{1}{2\sqrt{x}}$$

Example 4 Let $f(x) = x^{1/3}$.

a) Determine f'.
b) Show that f is not differentiable at 0.
c) Determine $\lim_{x\to 0} f'(x)$. Interpret the result graphically.

Solution

a) By the power rule,

$$f'(x) = \frac{d}{dx}\left(x^{1/3}\right) = \frac{1}{3}x^{1/3-1} = \frac{1}{3}x^{-2/3} = \frac{1}{3x^{2/3}}.$$

The above expression is valid if $x \neq 0$.

b) We have

$$\lim_{h\to 0} \frac{f(0+h) - f(0)}{h} = \lim_{h\to 0} \frac{h^{1/3}}{h} = \lim_{h\to 0} \frac{1}{h^{2/3}} = +\infty,$$

since $h^{2/3} = \left(h^{1/3}\right)^2 > 0$ for each $h \neq 0$ and $\lim_{h\to 0} h^{2/3} = 0$. Thus, the function is not differentiable at 0, even though it is continuous at 0.

c) We also have

$$\lim_{x\to 0} f'(x) = \lim_{x\to 0} \frac{1}{3x^{2/3}} = +\infty.$$

Therefore, the tangent line to the graph of f at $(x, f(x))$ becomes steeper and steeper as x approaches 0. The slope is positive on either side of 0. Since $\lim_{x\to 0} f'(x) = +\infty$, the vertical axis is a vertical asymptote for the graph of f. Figure 4 shows the graphs of f and f'. The picture is consistent with our analysis. The cube-root function is a prototype of functions defined by $x^{1/n}$, where n is an odd positive integer. $\square$

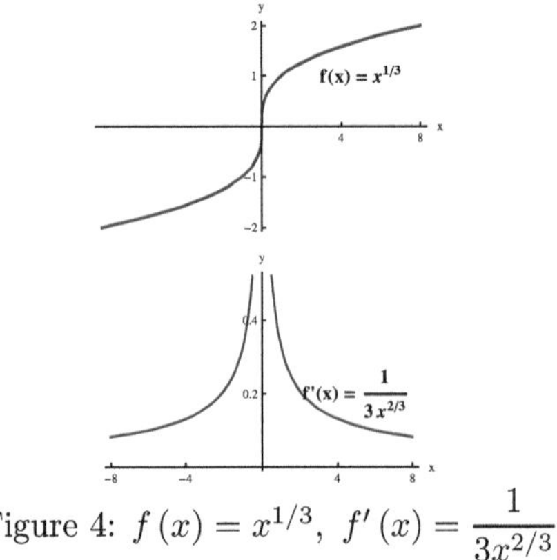

Figure 4: $f\left(x\right) = x^{1/3}$, $f'\left(x\right) = \dfrac{1}{3x^{2/3}}$

There is terminology that describes the behavior of the derivative of a function that is similar to the behavior of the derivatives of the square-root function and the cube-root function near 0:

Definition 1 The graph of f has a **vertical tangent** at $(a, f\left(a\right))$ if

$$f \text{ is continuous at } a \text{ and } \lim_{x \to a} f'\left(x\right) = \pm\infty,$$

or

$$f\left(x\right) \text{ is defined only if } x \geq a, \ f \text{ is continuous at } a \text{ from the right, and } \lim_{x \to a+} f'(x) = \pm\infty,$$

or

$$f(x) \text{ is defined only if } x \leq a, \ f \text{ is continuous at } a \text{ from the left, and } \lim_{x \to a-} f'(x) = \pm\infty.$$

Example 5 Show that the graph of f has a vertical tangent at 0 if

a) $f\left(x\right) = \sqrt{x}$, *b)* $f\left(x\right) = x^{1/3}$.

Solution

a) Let $f\left(x\right) = \sqrt{x}$. Then $f\left(x\right)$ is defined only for $x \geq 0$ and f is continuous at 0 from the right. In Example 3 we showed that $\lim_{x \to 0+} f'\left(x\right) = +\infty$. Therefore, the graph of f has a vertical tangent at 0. Indeed, in Figure 3 the vertical axis appears to be tangential to the graph of f at $(0,0) = (0, f\left(0\right))$.

b) Let $f\left(x\right) = x^{1/3}$. Then $f\left(x\right)$ is defined for each $x \in \mathbb{R}$ and f is continuous at 0. In Example 4 we showed that $\lim_{x \to 0} f\left(x\right) = +\infty$. Therefore, the graph of f has a vertical tangent at 0. Indeed, in Figure 4 the vertical axis appears to be tangential to the graph of f at $(0,0) = (0, f\left(0\right))$. $\square$

Example 6 Let $f(x) = x^{2/3}$.

a) Determine f'.
b) Show that f is not differentiable at 0.
c) Determine $\lim_{x \to 0+} f'\left(x\right)$ and $\lim_{x \to 0-} f'\left(x\right)$. Interpret the result graphically.

Solution

a) By the power rule,

$$f'(x) = \frac{d}{dx} x^{2/3} = \frac{2}{3} x^{-1/3} = \frac{2}{3x^{1/3}}.$$

The above expression is valid if $x \neq 0$.

b) We have

$$\lim_{h \to 0+} \frac{f(0+h) - f(0)}{h} = \lim_{h \to 0+} \frac{h^{2/3}}{h} = \lim_{h \to 0+} \frac{1}{h^{1/3}} = +\infty,$$

since $h^{1/3} > 0$ if $h > 0$ and $\lim_{h \to 0} h^{1/3} = 0$. Therefore, the function is not differentiable at 0.

c) Similarly,

$$\lim_{x \to 0+} f'(x) = \lim_{x \to 0+} \frac{2}{3x^{1/3}} = +\infty.$$

We have

$$\lim_{x \to 0-} f'(x) = \lim_{x \to 0-} \frac{2}{3x^{1/3}} = -\infty,$$

since $x^{1/3} < 0$ if $x < 0$ and $\lim_{x \to 0} x^{1/3} = 0$.

Therefore, the tangent line to the graph of f at $(x, f(x))$ becomes steeper and steeper as x approaches 0. The sign of the slope is positive to the right of the origin and negative to the left of the origin.

Figure 5 shows the graphs of f and f'. Since $\lim_{x \to 0+} f'(x) = +\infty$ and $\lim_{x \to 0-} f'(x) = -\infty$, the vertical axis is a vertical asymptote for the graph of f'. $\square$

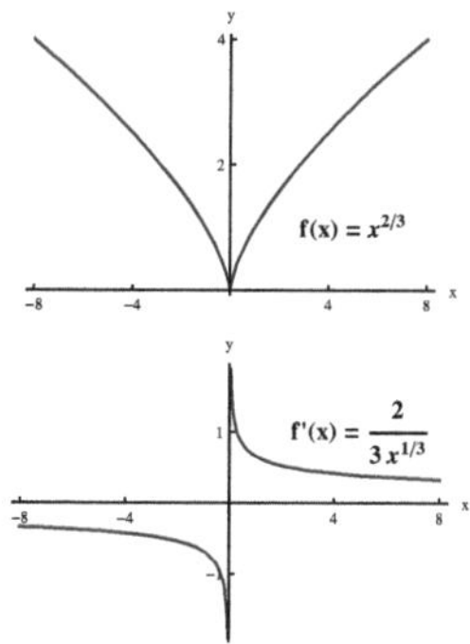

$$\text{Figure 5: } f(x) = x^{2/3}, \ f'(x) = \frac{2}{3x^{1/3}}$$

There is terminology that describes the behavior of the derivative of a function that is similar to the behavior of the derivative of the function in Example 6:

Definition 2 The graph of f has a **cusp** at $(a, f(a))$ if f is continuous at a, and

$$\lim_{x \to a+} f'(x) = +\infty, \quad \lim_{x \to a-} f'(x) = -\infty,$$

or

$$\lim_{x \to a+} f'(x) = -\infty \text{ and } \lim_{x \to a-} f'(x) = +\infty.$$

Example 7 Let $f(x) = x^{2/3}$, as in Example 6. Since $\lim_{x \to 0+} f'(x) = +\infty$ and $\lim_{x \to 0-} f'(x) = -\infty$, the graph of f has a cusp at $(0, 0) = (0, f(0))$. Figure 5 is typical when the graph of a function has a cusp. If a function f has a cusp at $(a, f(a))$, the graph of f seems to have a "sharp beak" at $(a, f(a))$. $\square$

In the expression

$$\frac{d}{dx}x^r = x^{r-1},$$

the derivative should be interpreted as a right derivative at $x = 0$ if x^r and x^{r-1} are defined if and only if $x \geq 0$. For example, if $f(x) = x^{5/4} = \left(x^{1/4}\right)^5$, then the domain of f is the interval $[0, +\infty)$. By the power rule,

$$f'(x) = \frac{d}{dx}\left(x^{5/4}\right) = \frac{5}{4}x^{1/4}.$$

The above expression defines $f'(x)$ if $x > 0$. Both expression $x^{5/4}$ and $x^{1/4}$ are defined at $x = 0$ and have the value 0. We have $f'_+(0) = 0$. Figure 6 shows the graphs of f and f'.

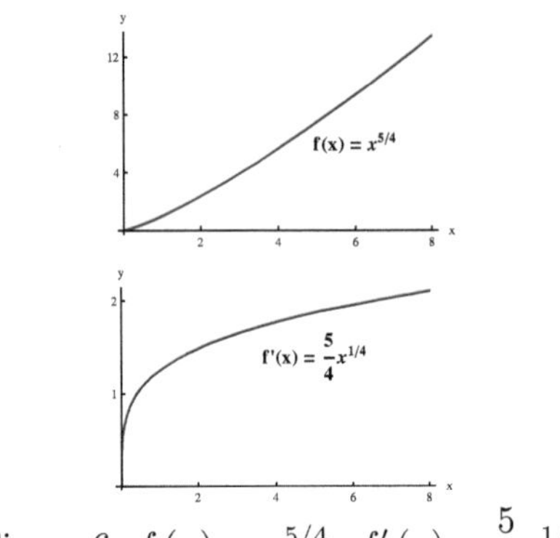

$$\text{Figure 6: } f(x) = x^{5/4}, \ f'(x) = \frac{5}{4}x^{1/4}$$

The Derivatives of Linear Combinations

THE CONSTANT MULTIPLE RULE FOR DIFFERENTIATION　　Assume that f is differentiable at x, and that c is a constant. Then, cf is also differentiable at x, and we have

$$(cf)'(x) = cf'(x).$$

In the Leibniz notation,

$$\frac{d}{dx}(cf(x)) = c\frac{d}{dx}f(x).$$

Proof

The difference quotient corresponding to cf, x and $h \neq 0$ is

$$\frac{(cf)(x+h) - (cf)(x)}{h} = \frac{cf(x+h) - cf(x)}{h} = c\left(\frac{f(x+h) - f(x)}{h}\right).$$

By the constant multiple rule for limits,

$$(cf)'(x) = \lim_{h \to 0}\left(c\left(\frac{f(x+h) - f(x)}{h}\right)\right) = c\lim_{h \to 0}\frac{f(x+h) - f(x)}{h} = cf'(x).$$

∎

Example 8 Let $f(x) = x^3$. Then,

$$\frac{d}{dx}\left(2f\left(x\right)\right) = \frac{d}{dx}\left(2x^3\right) = 2\left(\frac{d}{dx}\left(x^3\right)\right) = 2\left(3x^2\right) = 6x.$$

$\square$

The derivative of a sum of functions is the sum of their derivatives:

THE SUM RULE FOR DIFFERENTIATION   **Assume that f and g are differentiable at x. Then, the sum $f + g$ is also differentiable at x, and we have**

$$(f+g)'(x) = f'(x) + g'(x).$$

In the Leibniz notation,

$$\frac{d}{dx}\left(f\left(x\right) + g(x)\right) = \frac{d}{dx}f\left(x\right) + \frac{d}{dx}g(x).$$

Proof

The difference quotient corresponding to $f + g$, x and $h \neq 0$ is

$$\begin{aligned}
\frac{(f+g)\left(x+h\right) - (f+g)\left(x\right)}{h} &= \frac{f(x+h) + g(x+h) - (f(x) + g(x))}{h} \\
&= \frac{f(x+h) + g(x+h) - f(x) - g(x)}{h} \\
&= \frac{f(x+h) - f(x)}{h} + \frac{g(x+h) - g(x)}{h}.
\end{aligned}$$

By the sum rule for limits,

$$\begin{aligned}
(f+g)'\left(x\right) &= \lim_{h \to 0}\left(\frac{f(x+h) - f(x)}{h} + \frac{g(x+h) - g(x)}{h}\right) \\
&= \lim_{h \to 0}\frac{f(x+h) - f(x)}{h} + \lim_{h \to 0}\frac{g(x+h) - g(x)}{h} \\
&= f'(x) + g'(x).
\end{aligned}$$

$\blacksquare$

Example 9

$$\frac{d}{dx}\left(x + x^2\right) = \frac{d}{dx}\left(x\right) + \frac{d}{dx}\left(x^2\right) = 1 + 2x,$$

by the sum rule and the power rule. $\square$

The sum rule extends to the sum of an arbitrary number of functions. Thus,

$$\frac{d}{dx}\left(f_1(x) + f_2\left(x\right) + \cdots + f_n\left(x\right)\right) = \frac{d}{dx}f_1\left(x\right) + \frac{d}{dx}f_2\left(x\right) + \cdots + \frac{d}{dx}f_n\left(x\right),$$

where n is an arbitrary positive integer.

Recall that a **linear combination** of the functions f and g is a function of the form $c_1 f + c_1 g$, where c_1 and c_2 are constants. **The derivative of a linear combination of functions is the linear combination of the corresponding derivatives**, with the same coefficients:

DIFFERENTIATION IS A LINEAR OPERATION   Assume that f and g are differentiable at x. If c_1 and c_2 are constants, then the linear combination $c_1 f + c_2 g$ is also differentiable at x, and we have

$$(c_1 f + c_2 g)'(x) = c_1 f'(x) + c_2 g'(x).$$

In the Leibniz notation,

$$\frac{d}{dx}(c_1 f(x) + c_2 g(x)) = c_1 \frac{d}{dx} f(x) + c_2 \frac{d}{dx} g(x).$$

Proof

We apply the sum rule and the constant multiple rule for differentiation:

$$\frac{d}{dx}(c_1 f(x) + c_2 g(x)) = \frac{d}{dx}(c_1 f(x)) + \frac{d}{dx}(c_2 g(x)) = c_1 \frac{d}{dx} f(x) + c_2 \frac{d}{dx} g(x).$$

∎

The above rule extends to linear combinations of any number of functions:

$$\frac{d}{dx}(c_1 f_1 x) + c_2 f_2(x) + \cdots + c_n f_n(x)) = c_1 \frac{d}{dx} f_1(x) + c_2 \frac{d}{dx} f_2(x) + \cdots + c_n \frac{d}{dx} f_n(x),$$

where $c_1, c_2, \ldots, c_n$ are constants.

Example 10 Let

$$f(x) = 1 - \frac{1}{2}x^2 + \frac{1}{24}x^4.$$

a) Determine $f'(x)$.

b) Determine the tangent line to the graph of f at $(1, f(1))$.

Solution

a) By the linearity of differentiation and the power rule,

$$\frac{d}{dx} f(x) = \frac{d}{dx}(1) - \frac{1}{2}\frac{d}{dx}(x^2) + \frac{1}{24}\frac{d}{dx}(x^4)$$

$$= 0 - \frac{1}{2}(2x) + \frac{1}{24}(4x^3)$$

$$= -x + \frac{1}{6}x^3.$$

b) We have

$$f(1) = \frac{13}{24} \text{ and } f'(1) = -\frac{23}{24}$$

Therefore, the tangent line to the graph of f at $(1, f(1))$ is the graph of the equation

$$y = f(1) + f'(1)(x - 1) = \frac{13}{24} - \frac{23}{24}(x - 1).$$

Figure 7 shows the graph of f and the tangent line to the graph of f at $(1, f(1))$. □

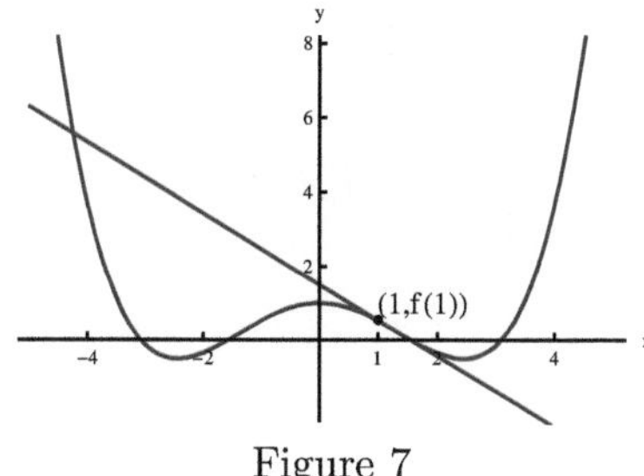

Figure 7

A **polynomial** is a linear combination of the constant 1 and positive integer powers of x. Therefore, we can differentiate any polynomial, as in Example 10, thanks to the linearity of differentiation: If

$$f(x) = c_0 + c_1 x + c_2 x^2 + \cdots + c_n x^n,$$

where $c_0, c_1, \ldots, c_n$ are given constants, then

$$f'(x) = c_1 + 2c_2 x + \cdots + n c_n x^{n-1}.$$

Note that $f'(x)$ is a polynomial of degree $\leq n - 1$.

Example 11 Let

$$f(x) = 3x^2 + \frac{1}{\sqrt{x}}$$

Determine $f'(x)$.

Solution

By the linearity of differentiation and the power rule,

$$\frac{df}{dx} = \frac{d}{dx}\left(3x^2 + x^{-1/2}\right) = 3\frac{d}{dx}\left(x^2\right) + \frac{d}{dx}\left(x^{-1/2}\right)$$

$$= 3(2x) - \frac{1}{2}x^{-3/2} = 6x - \frac{1}{2x^{3/2}}.$$

Note that the expression is valid if $x > 0$. $\square$

Higher-Order Derivatives

The second derivative of a function f is the derivative of f'. In the "prime notation", the second derivative is denoted as f''. Thus, $f''(x) = (f')'(x)$ if f' is differentiable at x. If we use the Leibniz notation to denote the derivative, we have

$$f''(x) = \frac{d}{dx}\left(\frac{d}{dx}f(x)\right).$$

This suggests the Leibniz notation

$$\frac{d^2 f}{dx^2}$$

for the second derivative of f. We may type this as

$$\frac{d^2 f}{dx^2}(x), \quad \frac{d^2 f(x)}{dx^2} \quad \text{or} \quad \frac{d^2}{dx^2}f(x).$$

Just as in the case of the derivative, the Leibniz notation for the second derivative is convenient to use, as long as you don't try to attach a meaning other than

$$\frac{d}{dx}\left(\frac{d}{dx}f(x)\right)$$

to the symbol

$$\frac{d^2 f}{dx^2}.$$

The above expression does *not* involve raising a quantity d/dx to the second power. The derivative of f'' is **the third derivative** f''' of f:

$$f'''(x) = (f'')'(x).$$

The Leibniz notation for the third derivative is

$$\frac{d^3 f}{dx^3}.$$

Thus,

$$\frac{d^3 f}{dx^3} = \frac{d}{dx}\left(\frac{d^2}{dx^2}f(x)\right).$$

This may be typed as

$$\frac{d^3 f(x)}{dx^3} \quad \text{or} \quad \frac{d^3}{dx^3}f(x).$$

The second derivative of f is also referred to as **the second-order derivative of f**, and the third derivative is **the third-order derivative of f**. More generally, we obtain **the nth order derivative of f** by differentiating the derivative of order $n-1$. As n increases, the prime notation becomes unwieldy. We may denote the nth order derivative of f by $f^{(n)}$. There is no difficulty to express higher-order derivatives by the Leibniz notation:

$$f^{(n)}(x) = \frac{d^n}{dx^n}f(x) = \frac{d}{dx}\left(\frac{d^{n-1}}{dx^{n-1}}f(x)\right).$$

Example 12 Let $f(x) = x^4$. Determine the second derivative of f.

Solution

By the power rule,

$$f'(x) = \frac{d}{dx}\left(x^4\right) = 4x^3,$$

$$f''(x) = \frac{d^2 f}{dx^2}(x) = \frac{d}{dx}\left(\frac{d}{dx}f(x)\right) = \frac{d}{dx}\left(4x^3\right) = 4\frac{d}{dx}\left(x^3\right) = 4\left(3x^2\right) = 12x^2,$$

$$f^{(3)}(x) = \frac{d^3 f}{dx^3}(x) = \frac{d}{dx}\left(\frac{d^2}{dx^2}f(x)\right) = \frac{d}{dx}\left(12x^2\right) = 12\frac{d}{dx}\left(x^2\right) = 12\left(2x\right) = 24x,$$

$$f^{(4)}(x) = \frac{d^4 f}{dx^4}(x) = \frac{d}{dx}\left(\frac{d^3}{dx^3}f(x)\right) = \frac{d}{dx}\left(24x\right) = 24,$$

and

$$f^{(n)}(x) = \frac{d^n f}{dx^n}(x) = 0$$

for $n = 5, 6, 7, \ldots$ $\square$

The Proof of the Power Rule for Arbitrary Rational Powers

We have already established the power rule for positive and negative integer powers of x. Let's begin by deriving the rule for the derivative of $x^{1/n}$ for any positive integer n. We will only consider $x > 0$. The proof is similar if $x < 0$ and $x^{1/n}$ is defined.

The difference quotient that is relevant to the calculation of

$$\frac{d}{dx} x^{1/n}$$

is

$$\frac{(x + \Delta x)^{1/n} - x^{1/n}}{\Delta x},$$

where $\Delta x \neq 0$.

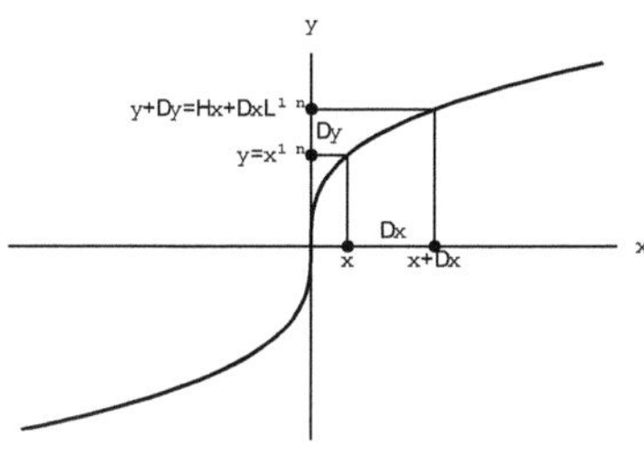

Figure 8

With reference to Figure 8, let's set

$$y = x^{1/n} \text{ and } y + \Delta y = (x + \Delta x)^{1/n},$$

so that

$$x = y^n, \ x + \Delta x = (y + \Delta y)^n \text{ .and } \Delta x = (y + \Delta y)^n - y^n.$$

Thus, $\Delta x = (y + \Delta y)^n - y^n$. Note that $\Delta y \neq 0$ since $\Delta x \neq 0$. Therefore,

$$\frac{(x + \Delta x)^{1/n} - x^{1/n}}{\Delta x} = \frac{(y + \Delta y) - y}{(y + \Delta y)^n - y^n} = \frac{\Delta y}{(y + \Delta y)^n - y^n} = \frac{1}{\dfrac{(y + \Delta y)^n - y^n}{\Delta y}}$$

By the continuity of the function defined by $x^{1/n}$,

$$\lim_{\Delta x \to 0} \Delta y = \lim_{\Delta x \to 0} \left((x + \Delta x)^{1/n} - x^{1/n} \right) = 0.$$

Therefore,

$$\lim_{\Delta x \to 0} \frac{(x + \Delta x)^{1/n} - x^{1/n}}{\Delta x} = \lim_{\Delta x \to 0} \frac{1}{\dfrac{(y + \Delta y)^n - y^n}{\Delta y}}$$

$$= \frac{1}{\lim_{\Delta x \to 0} \dfrac{(y + \Delta y)^n - y^n}{\Delta y}} = \frac{1}{\lim_{\Delta y \to 0} \dfrac{(y + \Delta y)^n - y^n}{\Delta y}},$$

provided that

$$\lim_{\Delta y \to 0} \frac{(y + \Delta y)^n - y^n}{\Delta y} \neq 0.$$

By the power rule,

$$\lim_{\Delta y \to 0} \frac{(y + \Delta y)^n - y^n}{\Delta y} = \frac{d}{dy}(y^n) = ny^{n-1} > 0$$

since $y = x^{1/n} > 0$. Therefore,

$$\frac{d}{dx}\left(x^{1/n}\right) = \lim_{\Delta x \to 0} \frac{(x + \Delta x)^{1/n} - x^{1/n}}{\Delta x} = \frac{1}{ny^{n-1}} = \frac{1}{n\left(x^{1/n}\right)^{n-1}} = \frac{1}{nx^{1-1/n}} = \frac{1}{n}x^{1/n-1}$$

if $x > 0$. $\blacksquare$

Now we are ready to establish the rule for an arbitrary rational number. Thus, let $r = m/n$ where n is a positive integer and m a positive or negative integer. The difference quotient that is relevant to the calculation of

$$\frac{d}{dx}x^{m/n}$$

is

$$\frac{(x + \Delta x)^{m/n} - x^{m/n}}{\Delta x},$$

where $\Delta x \neq 0$. Let's set $u = x^{1/n}$ and $(x + \Delta x)^{1/n} = u + \Delta u$.

$$\frac{(x + \Delta x)^{m/n} - x^{m/n}}{\Delta x} = \frac{(u + \Delta u)^m - u^m}{\Delta x}$$

$$= \left(\frac{(u + \Delta u)^m - u^m}{\Delta u}\right)\left(\frac{\Delta u}{\Delta x}\right).$$

Since $\Delta u = (x + \Delta x)^{1/n} - x^{1/n}$, and $x^{1/n}$ defines a continuous function,

$$\lim_{\Delta x \to 0} \Delta u = 0.$$

Therefore,

$$\lim_{\Delta x \to 0} \frac{(x + \Delta x)^{m/n} - x^{m/n}}{\Delta x} = \left(\lim_{\Delta x \to 0} \frac{(u + \Delta u)^m - u^m}{\Delta u}\right)\left(\lim_{\Delta x \to 0} \frac{\Delta u}{\Delta x}\right)$$

$$= \left(\lim_{\Delta u \to 0} \frac{(u + \Delta u)^m - u^m}{\Delta u}\right)\left(\frac{du}{dx}\right)$$

$$= \left(\frac{d}{du}u^m\right)\left(\frac{d}{dx}\left(x^{1/n}\right)\right)$$

$$= \left(mu^{m-1}\right)\left(\frac{1}{n}x^{1/n-1}\right)$$

$$= \frac{m}{n}\left(x^{1/n}\right)^{m-1}\left(x^{1/n-1}\right)$$

$$= \frac{m}{n}x^{m/n-1/n+1/n-1}$$

$$= \frac{m}{n}x^{m/n-1} = rx^{r-1}.$$

$\blacksquare$

Problems

In problems 1-6, determine the indicated derivative by using the power rule. You need to specify the domain of the derivative function.

1. $\dfrac{d}{dx}\left(x^5\right)$

2. $\dfrac{d}{dx}\left(\dfrac{1}{x^3}\right)$

3. $\dfrac{d}{dx}\left(x^{1/7}\right)$

4. $\dfrac{d}{dx}\left(x^{5/4}\right)$

5. $\dfrac{d}{dx}\left(x^{-3/5}\right)$

6. $\dfrac{d}{dx}\left(\dfrac{1}{x^{1/6}}\right)$

In problems 7-12,
a) Determine f'. You need to specify the domain of the derivative function.
b) Determine the points at which the graph of f has a vertical tangent or a cusp. You need to justify your assertions.
c) [C] Plot the graphs of f and f' with the help of your graphing utility. Are the pictures consistent with your assertions in parts a) and b)?

7. $f(x) = x^3 - 3$

8. $f(x) = \dfrac{1}{x} + 4$

9. $f(x) = x^{1/4} + 2$

10. $f(x) = x^{1/7} - 2$

11. $f(x) = x^{3/4}$

12. $f(x) = x^{4/5} + 1$

In problems 13 and 14,
a) Determine the tangent line to the graph of f at $(a, f(a))$ (the point-slope form with basepoint a will do).
b) [C] Make use of your graphing utility to plot the graph of f and the tangent line at $(a, f(a))$. Zoom in towards $(a, f(a))$ until you are unable to distinguish between the graph of f and the tangent line. Do the pictures reinforce the identification of the slope of the graph of f at $(a, f(a))$ with the slope of the tangent line at that point?

13. $f(x) = x^{2/3}$, $a = 8$

14. $f(x) = x^{1/4}$, $a = 16$

In problems 15 - 20, determine f'. You need not specify the domain of f'.

15. $f(x) = -4\sqrt{x}$

16. $f(x) = \dfrac{2}{x^{1/3}}$

17. $f(x) = 3x^2 - 4x + 8$

18. $f(x) = 5x^4 - 7x^2 + 3x - 10$

19. $f(x) = 2x - 3\sqrt{x} + 9x^{2/3}$

20. $f(x) = \dfrac{4}{x^{7/4}} - 10x^{-5} + 100$

2.3   The Derivatives of Sine and Cosine

In this section we will determine the derivatives of the basic trigonometric functions **sine** and **cosine**. Angles are measured in radians, unless stated otherwise.

The Derivatives of Sine and Cosine at 0

Let's begin by differentiating sine and cosine at 0. By the definition of the derivative,

$$\frac{d}{dx}\sin(x)\bigg|_{x=0} = \lim_{h\to 0}\frac{\sin(h) - \sin(0)}{h} = \lim_{h\to 0}\frac{\sin(h)}{h}.$$

Let's set

$$F(h) = \frac{\sin(h)}{h}.$$

Graphically, $F(h)$ is the slope of the secant line that passes through the points $(0,0)$ and $(h, \sin(h))$, as illustrated in Figure 1.

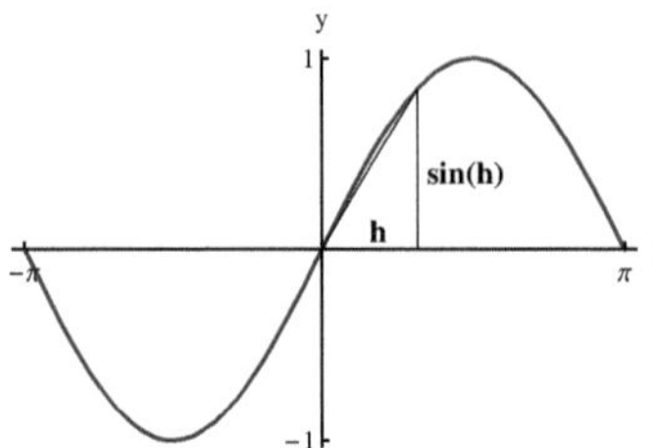

Figure 1: The secant line through $(0,0)$ and $(h, \sin(h))$

Figure 2 shows the graph of F, as produced by a graphing utility. The picture indicates that

$$\lim_{h \to 0} F(h) = \lim_{h \to 0} \frac{\sin(h)}{h} = 1.$$

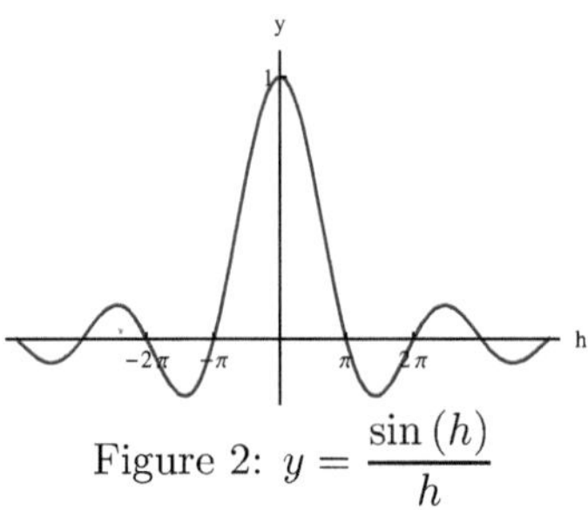

Figure 2: $y = \dfrac{\sin(h)}{h}$

Table 1 displays $\sin(h)/h$ (rounded to 10 significant digits) and $1 - \sin(h)/h$ (rounded to 2 significant digits) for $h = 10^{-k}$, where $k = 1, 2, 3, 4, 5$. The numbers in Table 1 definitely support the claim that $\lim_{h \to 0} \sin(h)/h = 1$ (the last row displays $F(10^{-5})$ as 1.0, rounded to 10 significant digits).

h	$\sin(h)/h$	$1 - \sin(h)/h$
10^{-1}	.9983341665	1.7×10^{-3}
10^{-2}	.9999833334	1.7×10^{-5}
10^{-3}	.9999998333	1.7×10^{-7}
10^{-4}	.9999999983	1.7×10^{-9}
10^{-5}	1.0	1.7×10^{-11}

Table 1

The above graphs and numbers indicate that

$$\left. \frac{d}{dx} \sin(x) \right|_{x=0} = 1.$$

By the definition of the derivative,

$$\left. \frac{d}{dx} \cos(x) \right|_{x=0} = \lim_{h \to 0} \frac{\cos(h) - \cos(0)}{h} = \lim_{h \to 0} \frac{\cos(h) - 1}{h}.$$

Let's set

$$G(h) = \frac{\cos(h) - 1}{h}.$$

Graphically, $G(h)$ is the slope of the secant line that passes through the points $(0, 1)$ and $(h, \cos(h))$, as illustrated in Figure 3.

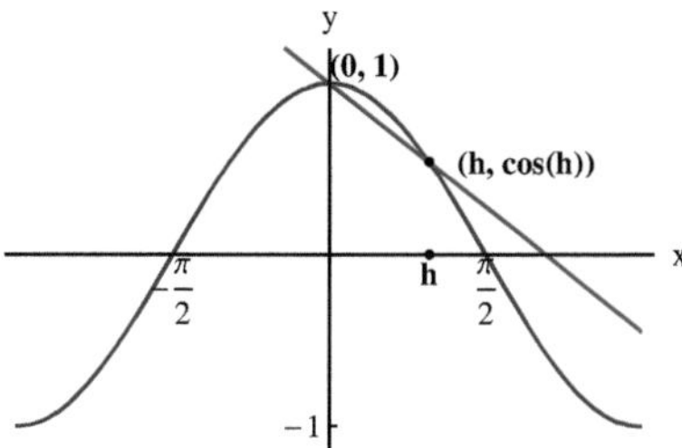

Figure 3: The secant line through $(0, 1)$ and $(h, \cos(h))$

Figure 4 displays the graph of G, as produced by a graphing utility. The picture indicates that

$$\lim_{h \to 0} G(h) = \lim_{h \to 0} \frac{\cos(h) - 1}{h} = 0.$$

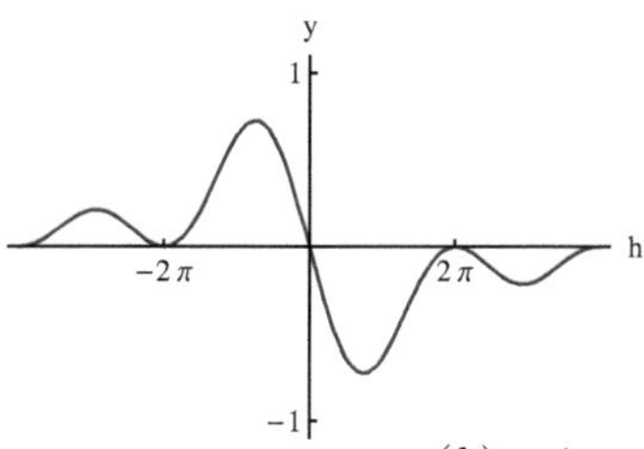

Figure 4: $y = \dfrac{\cos(h) - 1}{h}$

Table 2 displays $(\cos(h) - 1)/h$ (rounded to 6 significant digits) for $h = 10^{-k}$, where $k = 1, 2, 3, 4, 5$. The numbers support the claim that $\lim_{h \to 0}(\cos(h) - 1)/h = 0$.

h	$\dfrac{\cos(h) - 1}{h}$
10^{-1}	$-4.995\,83 \times 10^{-2}$
10^{-2}	$-4.999\,96 \times 10^{-3}$
10^{-3}	-5×10^{-4}
10^{-4}	-5×10^{-5}
10^{-5}	-5.0×10^{-6}

Table 2

The above graphs and numbers indicate that

$$\frac{d}{dx} \cos(x) \bigg|_{x=0} = 0.$$

Proposition 1 The derivative of sine at 0 is 1, and the derivative of cosine at 0 is 0:

$$\frac{d}{dx}\sin(x)\Big|_{x=0} = \lim_{h\to 0}\frac{\sin(h)}{h} = 1,$$

and

$$\frac{d}{dx}\cos(x)\Big|_{x=0} = \lim_{h\to 0}\frac{\cos(h)-1}{h} = 0$$

You can find the proof of Proposition 1 at the end of this section.

Example 1 Determine the tangent line to the graph of sine at $(0, \sin(0))$.

Solution

By Proposition 1,

$$\frac{d}{dx}\sin(x)\Big|_{x=0} = 1.$$

Therefore, the tangent line to the graph of sine at $(0, \sin(0))$ is the graph of the equation

$$y = \sin(0) + \left(\frac{d}{dx}\sin(x)\Big|_{x=0}\right)(x-0) = x.$$

Figure 5 shows the graph of sine and the line $y = x.\square$

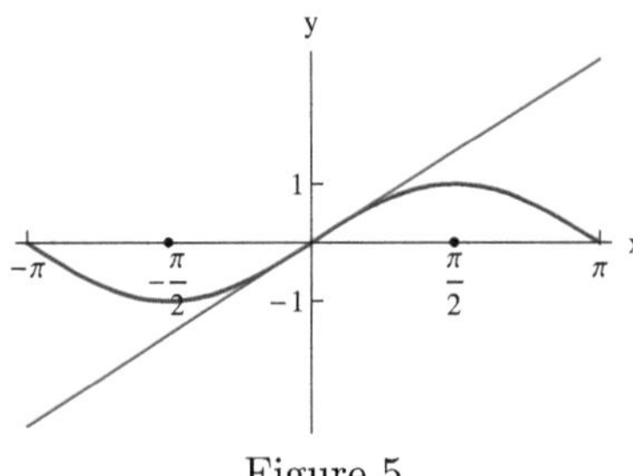

Figure 5

Example 2 Determine the tangent line to the graph of cosine at $(0, \cos(0))$.

Solution

By Proposition1,

$$\frac{d}{dx}\cos(x)\Big|_{x=0} = 0.$$

Therefore, the tangent line to the graph of cosine at $(0, \cos(0))$ is the graph of

$$y = \cos(0) + \left(\frac{d}{dx}\cos(x)\Big|_{x=0}\right)(x-0) = 1.$$

Thus, the horizontal line $y = 1$ is tangent to the graph of cosine at $(0, 1)$, as illustrated in Figure 6. the tangent line to the graph of sine at $(\pi/3, \sin(\pi/3))$. $\square$

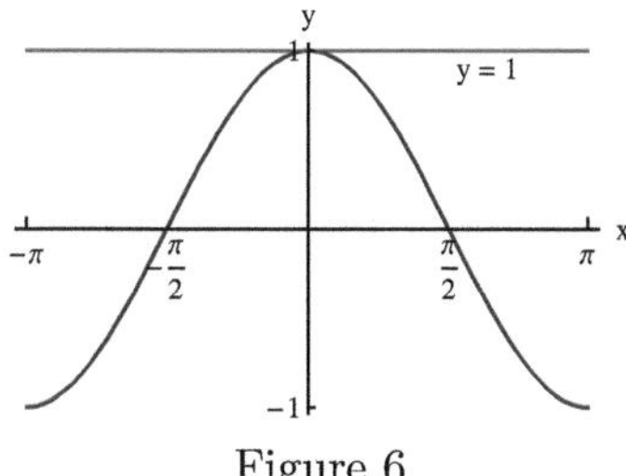

Figure 6

The Derivative Functions Corresponding to Sine and Cosine

The formulas of the derivatives of sine and cosine are elegant and easy to remember:

Theorem 1 We have

$$\frac{d}{dx}\sin(x) = \cos(x) \text{ and } \frac{d}{dx}\cos(x) = -\sin(x)$$

for each real number x.

Proof

The above expressions follow from Proposition 1 on the derivatives of sine and cosine at 0, with the help of the addition formulas for these functions.

Let's begin with sine. For any $x \in \mathbb{R}$ and increment $h \neq 0$,

$$\frac{\sin(x+h) - \sin(x)}{h} = \frac{\sin(x)\cos(h) + \cos(x)\sin(h) - \sin(x)}{h}$$

$$= \sin(x)\left(\frac{\cos(h) - 1}{h}\right) + \cos(x)\left(\frac{\sin(h)}{h}\right).$$

Therefore,

$$\frac{d}{dx}\sin(x) = \lim_{h \to 0} \frac{\sin(x+h) - \sin(x)}{h}$$

$$= \lim_{h \to 0}\left(\sin(x)\left(\frac{\cos(h) - 1}{h}\right) + \cos(x)\left(\frac{\sin(h)}{h}\right)\right)$$

$$= \sin(x)\lim_{h \to 0}\frac{\cos(h) - 1}{h} + \cos(x)\lim_{h \to 0}\frac{\sin(h)}{h}$$

$$= \sin(x)(0) + \cos(x)(1)$$

$$= \cos(x).$$

We have made use of the sum and constant multiple rules for limits (as far as the limit process is concerned, $\sin(x)$ and $\cos(x)$ are constants, since x is kept fixed), and Proposition 1.

The expression for the derivative of cosine is derived in a similar manner. We make use of the addition formula for cosine:

$$\frac{\cos(x+h) - \cos(x)}{h} = \frac{\cos(x)\cos(h) - \sin(x)\sin(h) - \cos(x)}{h}$$

$$= \cos(x)\left(\frac{\cos(h) - 1}{h}\right) - \sin(x)\left(\frac{\sin(h)}{h}\right).$$

Therefore,

$$\frac{d}{dx}\cos\left(x\right) = \lim_{h\to 0}\left(\cos(x)\left(\frac{\cos(h)-1}{h}\right) - \sin(x)\left(\frac{\sin(h)}{h}\right)\right)$$
$$= \cos(x)\lim_{h\to 0}\frac{\cos(h)-1}{h} - \sin(x)\lim_{h\to 0}\frac{\sin(h)}{h}$$
$$= \cos(x)\left(0\right) - \sin(x)\left(1\right)$$
$$= -\sin(x).$$

■

Example 3 Let $f\left(x\right) = \sin\left(x\right)$. Determine the tangent line to the graph of sine at $\left(\pi/3, \sin\left(\pi/3\right)\right)$.

Solution

We have

$$f\left(\frac{\pi}{3}\right) = \sin\left(\frac{\pi}{3}\right) = \frac{\sqrt{3}}{2},$$

and

$$f'\left(\frac{\pi}{3}\right) = \frac{d}{dx}\sin\left(x\right)\bigg|_{x=\pi/3} = \cos\left(x\right)|_{x=\pi/3} = \frac{1}{2}.$$

Therefore, the tangent line to the graph of sine at $\left(\pi/3, \sin\left(\pi/3\right)\right)$ is the graph of the equation

$$y = f\left(\frac{\pi}{3}\right) + f'\left(\frac{\pi}{3}\right)\left(x - \frac{\pi}{3}\right) = \frac{\sqrt{3}}{2} + \frac{1}{2}\left(x - \frac{\pi}{3}\right).$$

Figure 7 shows this tangent line and the graph of sine. □

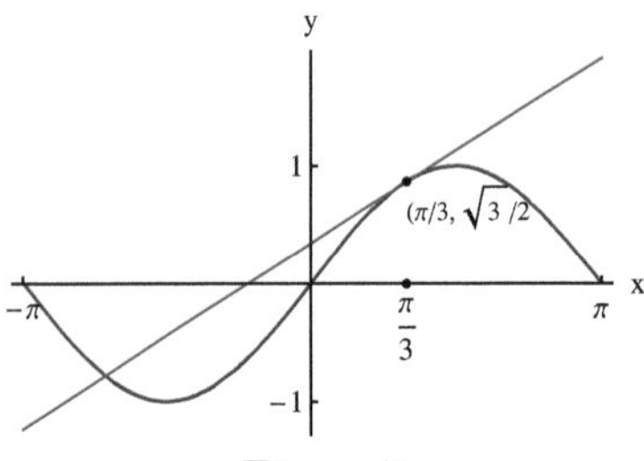

Figure 7

The Proof of Proposition 1

In order to prove Proposition 1 we will establish the following inequalities: If $h \neq 0$ and $-\pi/2 < h < \pi/2$,

a)

$$0 < \frac{\sin\left(h\right)}{h} < 1,$$

b)

$$0 < 1 - \cos\left(h\right) < \frac{h^2}{2},$$

c)

$$\frac{\sin\left(h\right)}{h} > \cos\left(h\right).$$

Assume that $0 < h < \pi/2$. With reference to Figure 9, the area of triangle AOP is less than the area of the circular sector determined by the points A, O and P. The area of the triangle AOP is

$$\frac{1}{2} \times \text{base} \times \text{height} = \frac{1}{2}\left(1\right)\left(\sin(h)\right) = \frac{1}{2}\sin\left(h\right).$$

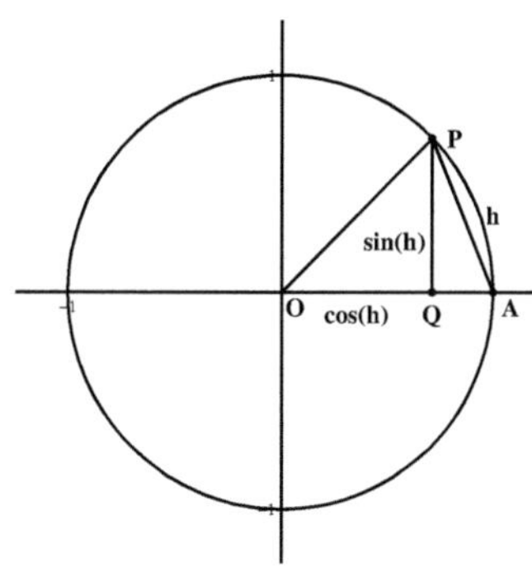

Figure 9

The area of a sector of a disk of radius r that corresponds to the angle h (in radians) is

$$\frac{1}{2}hr^2.$$

Therefore, the area of the circular sector determined by A, O and P is $h/2$. Thus,

$$0 < \frac{1}{2}\sin(h) < \frac{1}{2}h \Rightarrow 0 < \sin(h) < h \Rightarrow 0 < \frac{\sin\left(h\right)}{h} < 1$$

Now assume that $-\pi/2 < h < 0$. Since sine is an odd function,

$$\frac{\sin\left(h\right)}{h} = \frac{-\sin\left(-h\right)}{h} = \frac{\sin\left(-h\right)}{-h}.$$

Since $0 < -h < \pi/2$,

$$0 < \frac{\sin\left(-h\right)}{-h} < 1.$$

Therefore,

$$0 < \frac{\sin\left(h\right)}{h} < 1$$

if $h \neq 0$ and $-\pi/2 < h < \pi/2$. Thus, we have established inequality a).

In order to establish inequality b), we will use the identity

$$\cos\left(h\right) = 1 - 2\sin^2\left(\frac{h}{2}\right).$$

Indeed, by the addition formula for cosine,

$$\cos(h) = \cos\left(\frac{h}{2} + \frac{h}{2}\right) = \cos^2\left(\frac{h}{2}\right) - \sin^2\left(\frac{h}{2}\right)$$

$$= \left(1 - \sin^2\left(\frac{h}{2}\right)\right) - \sin^2\left(\frac{h}{2}\right) = 1 - 2\sin^2\left(\frac{h}{2}\right).$$

Assume that $h \neq 0$ and $-\pi/2 < h < \pi/2$. Since

$$0 < \frac{\sin(h/2)}{h/2} < 1,$$

we have

$$\frac{\sin^2(h/2)}{h^2/4} < 1,$$

so that

$$\sin^2(h/2) < \frac{h^2}{4}.$$

Therefore,

$$\cos(h) = 1 - 2\sin^2\left(\frac{h}{2}\right) > 1 - 2\left(\frac{h^2}{4}\right) = 1 - \frac{h^2}{2}.$$

Thus,

$$\frac{h^2}{2} > 1 - \cos(h).$$

If $h \neq 0$ and $-\pi/2 < h < \pi/2$, then $\cos(h) < 1$. Therefore, $1 - \cos(h) > 0$. Thus,

$$0 < 1 - \cos(h) < \frac{h^2}{2}.$$

We have established inequality b).

Now we will establish inequality c). Assume that $0 < h < \pi/2$. With reference to Figure 10, the line AT is tangential to the unit circle at A.

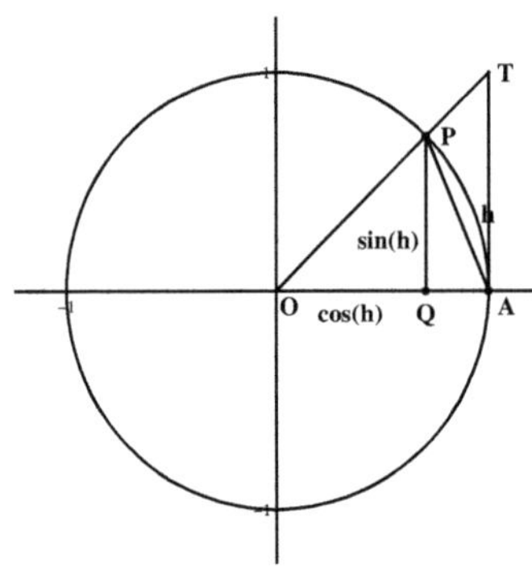

Figure 10

Since

$$\frac{\text{length of } AT}{\text{length of } OA} = \frac{\text{length of } AT}{1} = \tan(h),$$

the area of the triangle AOT is

$$\frac{1}{2} \times \text{base} \times \text{height} = \frac{1}{2}(1)(\tan(h)).$$

As in part a), the area of the circular sector AOP is $h/2$. The area of the circular sector AOP is less than the area of the triangle AOT. Therefore,

$$\frac{1}{2}h < \frac{1}{2}\tan(h).$$

Thus,

$$\tan(h) > h \text{ if } 0 < h < \frac{\pi}{2},$$

i.e.,

$$\frac{\sin(h)}{\cos(h)} > h \text{ if } 0 < h < \frac{\pi}{2}.$$

Therefore,

$$\frac{\sin(h)}{h} > \cos(h) \text{ if } 0 < h < \frac{\pi}{2}.$$

If $-\pi/2 < h < 0$, we have $0 < (-h) < \pi/2$, so that

$$\frac{\sin(-h)}{(-h)} > \cos(-h).$$

Since cosine is an even function and sine is an odd function, we obtain

$$-\frac{\sin(h)}{-h} > \cos(h) \Rightarrow \frac{\sin(h)}{h} > \cos(h).$$

Therefore,

$$\frac{\sin(h)}{h} > \cos(h) \text{ if } -\frac{\pi}{2} < h < \frac{\pi}{2} \text{ and } h \neq 0.$$

Thus, we have established inequality c).

Now we will show that Proposition 1 follows from inequalities a), b) and c). We restrict h so that $h \neq 0$ and $-\pi/2 < h < \pi/2$.

By inequalities a) and c),

$$\cos(h) < \frac{\sin(h)}{h} < 1$$

Figure 11 illustrates the above inequality.

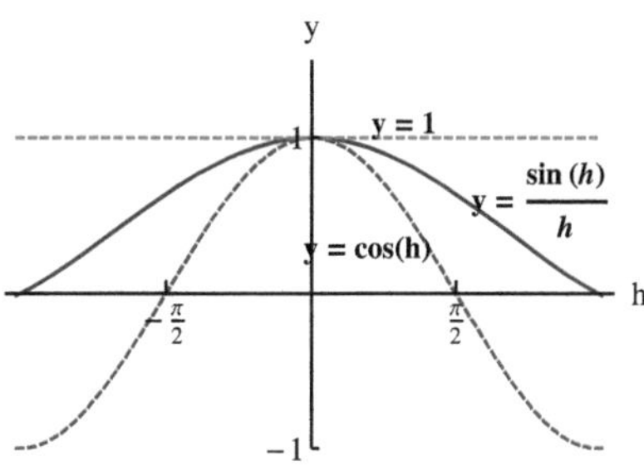

Figure 11

We see that the graph of $y = \sin(h)/h$ is squeezed between the graphs of $y = \cos(h)$ and $y = 1$. Since cosine is continuous at 0, $\lim_{h \to 0} \cos(h) = \cos(0) = 1$. Of course, $\lim_{h \to 0} 1 = 1$. Therefore,

$$\lim_{h \to 0} \frac{\sin(h)}{h} = 1$$

as well, by the Squeeze Theorem.

Now we will show that

$$\lim_{h \to 0} \frac{\cos(h) - 1}{h} = 0.$$

By inequality b),

$$0 < 1 - \cos(h) < \frac{h^2}{2}$$

Since $h^2 = |h|^2$,

$$\frac{1 - \cos(h)}{|h|} < \frac{|h|}{2}.$$

Since $1 - \cos(h) > 0$, $|\cos(h) - 1| = 1 - \cos(h)$. Therefore,

$$\left| \frac{\cos(h) - 1}{h} \right| < \frac{|h|}{2}$$

Since

$$\left| \frac{\cos(h) - 1}{h} \right| < \frac{|h|}{2} \Leftrightarrow -\frac{|h|}{2} < \frac{\cos(h) - 1}{h} < \frac{|h|}{2},$$

the graph of

$$y = \frac{\cos(h) - 1}{h}$$

on the interval $[-\pi/2, \pi/2]$ is squeezed between the lines $y = -|h|/2$ and $y = |h|/2$, as illustrated in Figure 12.

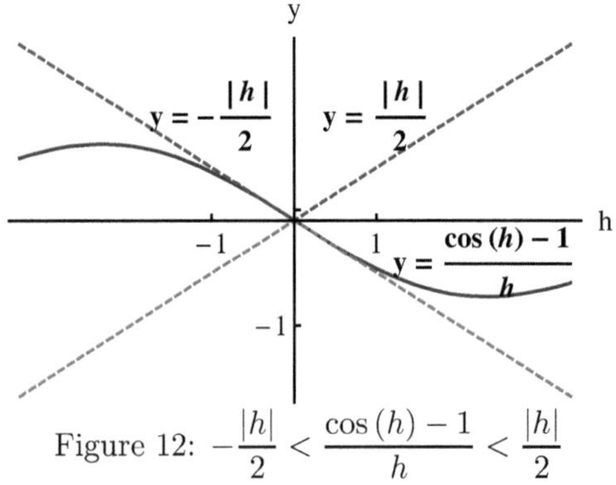

Figure 12: $-\dfrac{|h|}{2} < \dfrac{\cos(h) - 1}{h} < \dfrac{|h|}{2}$

Since

$$\lim_{h \to 0} \left(-\frac{|h|}{2} \right) = \lim_{h \to 0} \frac{|h|}{2} = 0,$$

we have

$$\lim_{h \to 0} \frac{\cos(h) - 1}{h} = 0,$$

as well, again by the Squeeze Theorem. ∎

Problems

In problems 1 - 4, determine the limit.

1. $\lim\limits_{x \to 0} \dfrac{\sin(4x)}{x}$

2. $\lim\limits_{x \to 0} \dfrac{\sin(6x)}{\sin(3x)}$

3. $\lim_{x \to 0} \dfrac{\cos(2x) - 1}{4x}$

4. $\lim\limits_{x \to 0} \dfrac{\cos(x) - 1}{3\sin(x)}$

In problems 5 - 12,

a) Determine the expression for $f'(x)$,

b) Evaluate $f'(a)$.

5. $f(x) = 2\sin(x)$, $a = \dfrac{\pi}{3}$

6. $f(x) = 4\cos(x)$, $a = \dfrac{\pi}{6}$

7. $f(x) = 3\sin(x) - 4\cos(x)$, $a = \dfrac{\pi}{4}$

8. $f(x) = 4\sqrt{x} + 3\sin(x)$, $a = \dfrac{\pi}{3}$

9. $f(x) = 4\cos(x) - \dfrac{2}{x}$, $a = \pi$

10. $f(x) = x^2 - 3x + 7 + \cos(x)$, $a = 1$

11. $f(x) = 2\sin(x) - 3\cos(x)$, $a = \dfrac{\pi}{2}$

12. $f(x) = 4\sin(x) + 2\cos(x) - 3x^4$, $a = 0$

In problems 13 and 14,

a) Determine the tangent line to the graph of f at $(a, f(a))$ (the point-slope form with basepoint a will do),

b) [C] Make use of your graphing utility to plot the graph of f and the tangent line at $(a, f(a))$. Zoom in towards $(a, f(a))$ until you are unable to distinguish between the graph of f and the tangent line. Do the pictures reinforce the identification of the slope of the graph of f at $(a, f(a))$ with the slope of the tangent line at that point?

13.

$$f(x) = \sin(x), \ a = \pi/6.$$

14.

$$f(x) = \cos(x), \ a = \pi/4.$$

2.4 Velocity and Acceleration in One-Dimensional Motion

In this section we will discuss the velocity and acceleration of an object in **one-dimensional motion**. **Instantaneous velocity** is the rate of change of the position of the object with respect to time. Therefore we can identify instantaneous velocity with a derivative. **Instantaneous acceleration is** the rate of change of velocity with respect to time and can be computed as the derivative of the velocity function.

We will consider the simplest kind of motion that involves an object which moves along a line. The object can be a car travelling along a straight stretch of a highway, a ball falling down from the Tower of Pisa, or a weight that is attached to a vibrating spring. We model the object as a point on the number line. We place the number line so that the positive direction coincides with the direction of motion that is selected as the positive direction. Let $f(t)$ denote the position of the object at time t. Time is measured in units such as hours or seconds, and distance is measured in units such as miles or centimeters (we may not bother to indicate specific units in every example or problem). We will refer to this kind of mathematical model as **one-dimensional motion,** and to f as **the position function** of the object in one dimensional motion.

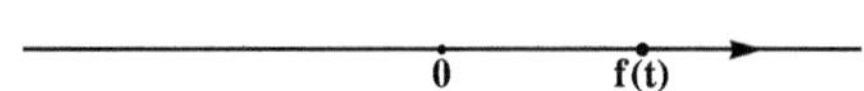

Figure 1: One-dimensional motion

Let's look at some examples.

Example 1 Let $f(t) = 60t$ be the position at time t of a car travelling along a straight stretch of a highway. Distances are measured in miles.

We have $f(0) = 0$, so that the origin of the number line corresponds to the position of the car at the time we start to monitor its motion. Figure 2 illustrates the motion of the car in **space-time**, i.e., the graph of f in the ty-plane. The graph of f is the part of the line $y = 60t$ that corresponds to $t \geq 0$. $\square$

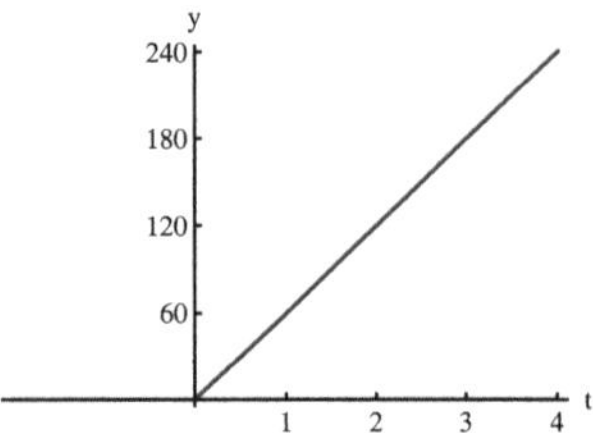

Figure 2: The motion of the car as illustrated in space-time

Example 2 Assume that a ball is dropped from a high tower (perhaps by Galileo from the top of the leaning tower of Pisa).

Let's model the ball as a point on the y-axis that points downward and let the origin coincide with the point at which the ball is released, as illustrated in Figure 3.

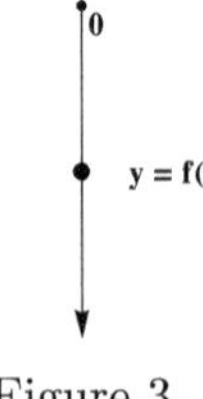

Figure 3

Assume that $f(t) = 4.9t^2$ is the position of the ball (in meters), if t is such that the ball has not hit the ground yet. Figure 4 illustrates the motion in space-time. The graph of $y = t$ in the ty-plane is part of the parabola $y = 4.9t^2$. $\square$

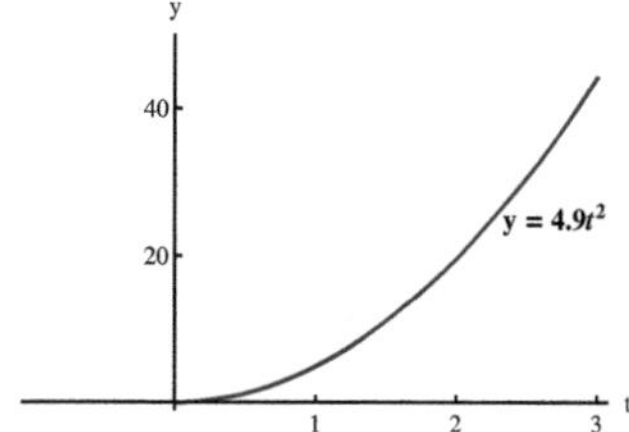

Figure 4: The motion of a falling object as illustrated in space-time

Example 3 Assume that a projectile is launched vertically from the ground level, rises up to a certain height, and falls back to the ground.

We assume that the motion is along a vertical line during the relevant time interval. Let

$$y = f(t) = 196t - 4.9t^2$$

be the position of the projectile above the ground at time t. We measure distances in meters and time in seconds. The motion is along the y-axis. The positive direction coincides with the upward movement of the projectile.

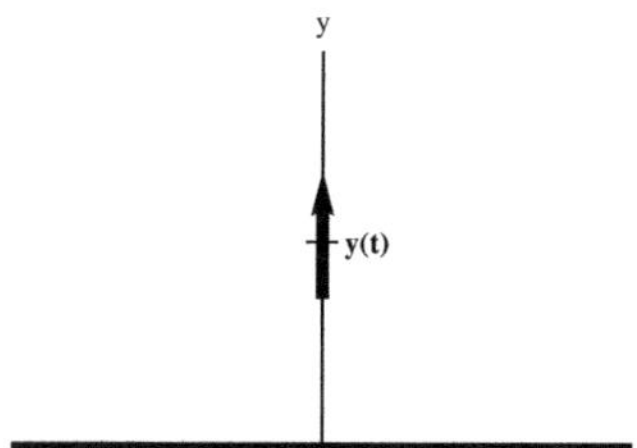

Figure 5: A projectile that is climbing vertically

We have
$$f(t) = 0 \Leftrightarrow 196t - 4.9t^2 = t = 0 \text{ or } t = \frac{196}{4.9} = 40.$$

Thus, the projectile is fired from the ground level at $t = 0$, and hits the ground at $t = T = 40$ (seconds).

We can determine the maximum height of the projectile by completing the square: Since

$$f(t) = -4.9t^2 + 196t = -4.9\left(t^2 - 40t\right) = -4.9\left(t - 20\right)^2 + 1960,$$

the graph of f is part of a parabola with vertex at $(20, 1960)$. Thus, the rocket reaches a height of 1960 meters, then falls back, and hits the ground 40 seconds after it has been launched. Figure 6 shows the graph of f. $\square$

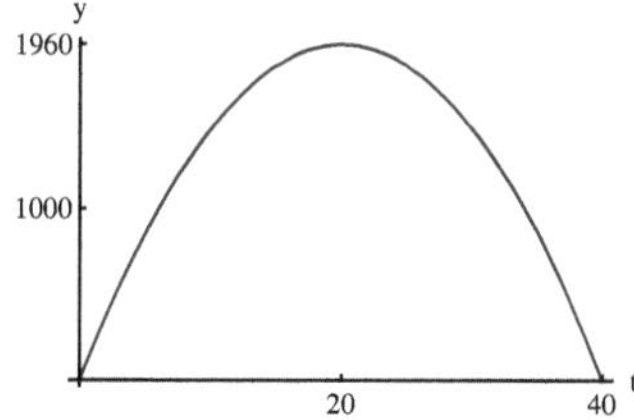

Figure 6: The path of a projectile in space-time

How should we determine the velocity of an object in one-dimensional motion? If the position function is linear as in Example 1, this is straightforward: if $f(t) = 60t$ and $\Delta t > 0$ represents a time increment, **the average velocity** of the car in the time interval $[t, \Delta t]$ is

$$\frac{\text{change in position}}{\text{elapsed time}} = \frac{f(t + \Delta t) - f(t)}{\Delta t} = \frac{60(t + \Delta t) - 60t}{\Delta t} = \frac{60\Delta t}{\Delta t} = 60$$

(miles per hour.) This quantity is independent of t and Δt. We can say that the velocity at any instant t is 60 miles per hour. Note that 60 is also the slope of the line that is the graph of the linear function f. Similarly, if $f(t) = mt + b$, then

$$\frac{f(t + \Delta t) - f(t)}{\Delta t} = \frac{[m(t + \Delta t) + b] - [mt + b]}{\Delta t}$$
$$= \frac{mt + m\Delta t + b - mt - b}{\Delta t} = \frac{m\Delta t}{\Delta t} = m$$

for any t and Δt, so that the velocity of the object has the constant value m at any instant t. This number is also the slope of the graph of f.

Let's now assume that the position function f is nonlinear, as in Example 2 and Example 3. Let's consider a specific instant t (you can imagine that time is frozen at t) and let Δt be an arbitrary positive time increment. As in the linear case, we can determine **the average velocity** over the time interval $[t, t + \Delta t]$ as

$$\frac{\text{change in position}}{\text{elapsed time}} = \frac{f(t + \Delta t) - f(t)}{\Delta t}.$$

In general, average velocity depends on t and Δt. It seems reasonable to define the instantaneous velocity at the instant t as the limit of the average velocity as the time increment approaches 0:

Definition 1 The instantaneous velocity $v(t)$ at the instant t of an object in one-dimensional motion is the derivative of the position function f at t:

$$v(t) = \frac{df}{dt}(t) = \lim_{\Delta t \to 0} \frac{f(t + \Delta t) - f(t)}{\Delta t}.$$

We may refer to $v(t)$ simply as the velocity at the instant t. Since we have identified the rate of change of a function at a point as its derivative at that point, $v(t)$ **is the rate of change of the position function at the instant** t.

Note that we have not restricted Δt to be positive in the above definition. Indeed, if $\Delta t < 0$ then

$$\frac{f(t + \Delta t) - f(t)}{\Delta t} = \frac{f(t) - f(t + \Delta t)}{(-\Delta t)},$$

so that the difference quotient may be interpreted as the average velocity of the object on the time interval $[t + \Delta t, t]$.

Graphically, the average velocity over the time interval $[t, t + \Delta t]$ is the slope of the secant line that passes through the points $(t, f(t))$ and $(t + \Delta t, f(t + \Delta t))$ on the graph of the position function, and the instantaneous velocity at the instant t is the slope of the tangent line to the graph of f at $(t, f(t))$.

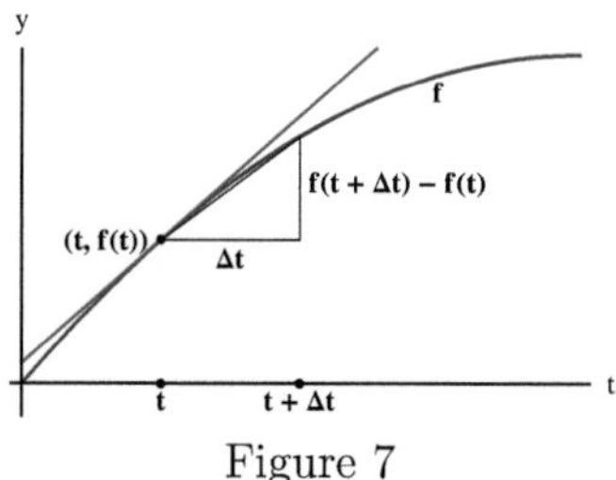

Figure 7

Example 4 Let $f(t) = 4.9t^2$, as in Example 2.

The average velocity of the ball over the time interval determined by t and $t + \Delta t$ is

$$\frac{f(t + \Delta t) - f(t)}{\Delta t} = \frac{4.9(t + \Delta t)^2 - 4.9t^2}{\Delta t} = 4.9\left(\frac{t^2 + 2t\Delta t + (\Delta t)^2 - t^2}{\Delta t}\right) = 4.9(2t + \Delta t).$$

The instantaneous velocity of the ball at the instant t is the limit of the average velocity as the time increment Δt approaches 0:

$$v(t) = \frac{df}{dt}(t) = \lim_{\Delta t \to 0} \frac{f(t + \Delta t) - f(t)}{\Delta t} = \lim_{\Delta t \to 0} 4.9(2t + \Delta t) = 9.8t$$

(meters per second).. Thus, velocity is not constant over time, unlike the case of a linear position function. Figure 8 displays the graph of the velocity function. $\square$

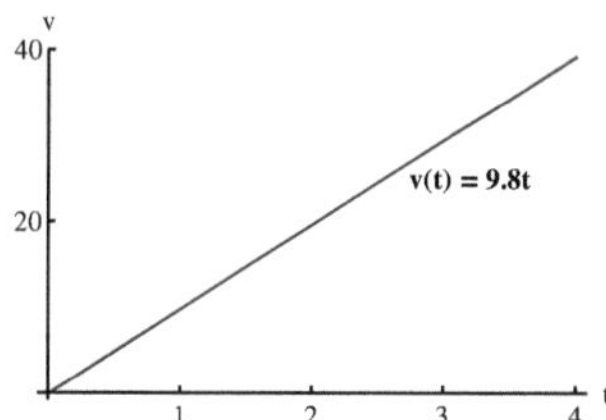

Figure 8: The velocity function of Example 4

If an object is moving in the direction that has been assigned as the positive direction, and $\Delta t > 0$ then

$$\frac{f(t + \Delta t) - f(t)}{\Delta t} \geq 0.$$

Therefore,

$$v(t) = \lim_{\Delta t \to 0} \frac{f(t + \Delta t) - f(t)}{\Delta t} \geq 0$$

Similarly, $v(t) \leq 0$ if the object is moving in the opposite direction. **The speed** of the object at time t is defined as the magnitude of velocity at time t, i.e., $|v(t)|$.

Example 5 Let

$$f(t) = 196t - 4.9t^2,$$

as in Example 3. Determine the velocity and the speed of the projectile as functions of t. Interpret the sign of the velocity function with reference to the direction of motion of the projectile.

Solution

a) As we discussed in Example 3, the relevant time interval is $[0, 40]$. The velocity at the instant t is

$$v\left(t\right) = \frac{d}{dt}\left(196t - 4.9t^2\right) = 196 - 4.9\left(2t\right) = 196 - 9.8t.$$

Therefore,

$$v\left(t\right) = 0 \Leftrightarrow t = \frac{196}{9.8} = 20$$

(seconds). We have

$$v\left(t\right) > 0 \text{ if } 0 < t < 20 \text{ and } v(t) < 0 \text{ if } 0 < t < 40.$$

Since the positive direction is upward in this example, the projectile climbs up in the time interval $[0, 20]$ and falls back towards the ground in the time interval $[20, 40]$. The maximum height of the projectile is

$$f\left(20\right) = 196t - 4.9t^2\big|_{t=20} = 1960 \text{ meters.}$$

The instantaneous velocity at $t = 20$ is 0. You can imagine that the projectile is momentarily stationary at that very instant, before it starts to fall back to earth. The velocity at the time of impact is $v\left(40\right) = -196$ meters/second. At that instant, speed is $|-196| = 196$ meters/second. Figure 9 shows the graph of the velocity function. $\square$

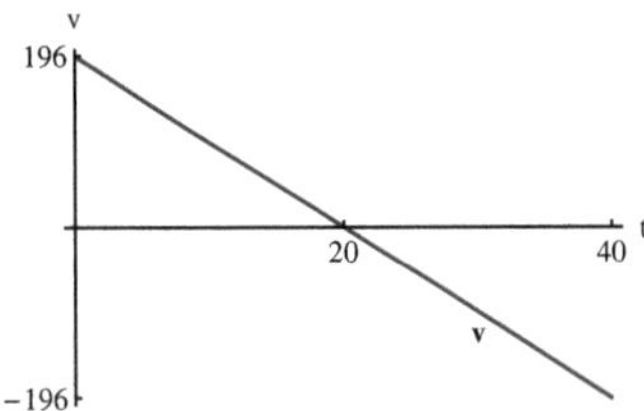

Figure 9: The velocity function of Example 5

Intuitively, acceleration is the rate of change of velocity. We translate "rate of change" to "derivative":

Definition 2 Let v be the velocity function of an object in one-dimensional motion. The **(instantaneous) acceleration** $a\left(t\right)$ of the object at the instant t is the derivative of the velocity at t:

$$a\left(t\right) = \frac{d}{dt}v\left(t\right).$$

Thus, **acceleration is the second derivative of the position function** $f\left(t\right)$:

$$a\left(t\right) = \frac{d}{dt}v\left(t\right) = \frac{d}{dt}\left(\frac{d}{dt}f\left(t\right)\right) = \frac{d^2}{dt^2}f\left(t\right).$$

We may refer to $a\left(t\right)$ simply as the acceleration at the instant t. Since the unit of velocity is (unit of distance)/(unit of time), the unit of acceleration is (unit of distance)/(unit if time)/(unit of time) = (unit of distance)/(unit of time)^2. For example, if distance is measured in meters and time is measured in seconds, acceleration is measured in meters/second/second = meters/second^2.

Example 6 Let $f(t) = 60t$, as in Example 1. We have $v(t) = 60$ (miles/hour). The acceleration of the car is

$$a(t) = \frac{d}{dt}v(t) = \frac{d}{dt}(60) = 0.$$

Indeed, the velocity is a constant, so that its rate of change of is 0. $\square$

Example 7 Let $f(t) = 4.9t^2$, as in Example 4. We have $v(t) = 9.8t$ (meters/second). Therefore, the acceleration of the falling ball is

$$a(t) = \frac{d}{dt}v(t) = \frac{d}{dt}(9.8t) = 9.8 \text{ (meters/second/second)}.$$

The above expression for acceleration is consistent with the assumptions that the only force acting on the object is due to gravitational acceleration of 9.8 meters/second^2, and that the opposing force due to air resistance can be neglected . Indeed, by **Newton's second law of motion,**

$$\text{Force} = \text{mass} \times \text{acceleration.}$$

Therefore, if the object has mass m (kilograms), the force that is acting on the object due to gravitational acceleration g is the weight mg of the object. This force is $9.8m$, if it is assumed that $g = 9.8$ meters/second^2. Thus, we have the equation

$$ma(t) = mg \Rightarrow a(t) = 9.8$$

In Chapter 5 we will see that this expression for acceleration leads to $v(t) = 9.8t$ and $f(t) = 4.9t^2$. $\square$

Example 8 Let $f(t) = 196t - 4.9t^2$, as in Example 5. We calculated the velocity as $v(t) = 196 - 9.8t$ Therefore,

$$a(t) = \frac{dv}{dt} = \frac{d}{dt}(196 - 9.8t) = -9.8$$

The sign is $(-)$ since the positive direction is upward in this example. $\square$

Example 9 Assume that $f(t) = 2\cos(t)$ is the position of an object at time t. Thus, the object oscillates about the origin. Determine the velocity and acceleration functions.

Solution
The velocity at time t is

$$v(t) = \frac{d}{dt}f(t) = \frac{d}{dt}(2\cos(t)) = -2\sin(t).$$

The acceleration at time t is

$$a(t) = \frac{d}{dt}v(t) = \frac{d}{dt}(-2\sin(t)) = -2\cos(t).$$

Note that

$$a(t) = -f(t).$$

The motion is periodic with period 2π. Figure 10 displays the graphs of position, velocity and acceleration functions on the interval $[0, 2\pi]$. $\square$

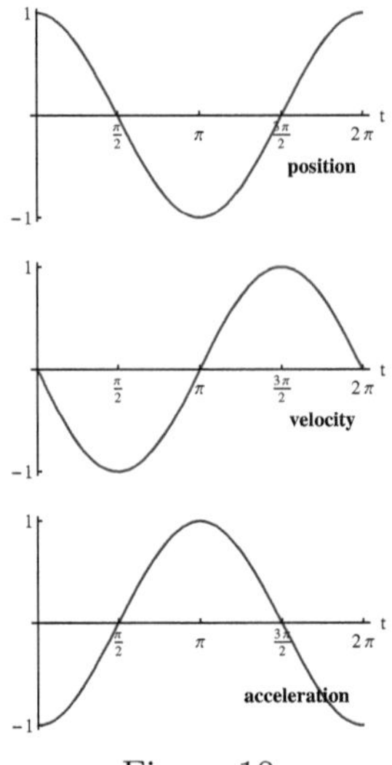

Figure 10

Problems

In problems 1 - 4, $f(t)$ is the position at time t of an object in one-dimensional motion.
a) Determine $v(t)$, the velocity of the object at time t, and $a(t)$, the acceleration of the object at time t.
b) Calculate $v(t_0)$ and $a(t_0)$.

1.
$$f(t) = 200t - 5t^2, \ t_0 = 1$$

2.
$$f(t) = 5t^2 + 100; \ t_0 = 4$$

3.
$$f(t) = 10\sin(t), \ t_0 = \pi/6$$

4.
$$f(t) = 3\sin(t) + 8\cos(t), \ t_0 = \pi/2$$

2.5   Local Linear Approximations and the Differential

The derivative of a function f at a point a can be interpreted as **the slope of the tangent line** to the graph of f at $(a, f(a))$. The tangent line is the graph of a linear function that is the best linear approximation to f near a in a sense that will be explained in this section.

Local Linear Approximations

Given a function f that is differentiable at the point a, the tangent line to the graph of f at $(a, f(a))$ is the graph of the equation

$$y = f(a) + f'(a)(x - a).$$

We will give a name to the underlying linear function:

Definition 1 The linear approximation to f based at a is

$$L_a(x) = f(a) + f'(a)(x - a).$$

We refer to a as **the basepoint.**

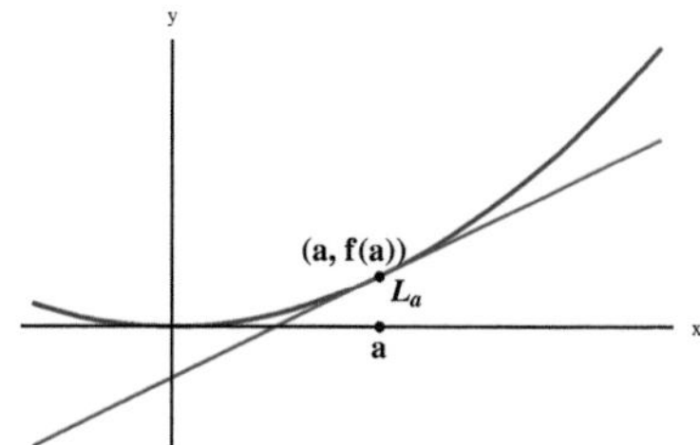

Figure 1: The graph of L_a is a tangent line

Example 1 Let $f(x) = x^2 - 2x + 4$, as in Example 1 of Section 2.1. We showed that $f'(3) = 4$ and the tangent line to to graph of f at $(3, f(3))$ is the graph of the equation

$$y = f(3) + f'(3)(x - 3) = 7 + 4(x - 3).$$

Thus, the linear approximation to f based at 3 is

$$L_3(x) = 7 + 4(x - 3).$$

Figure 2 illustrates the effect of zooming in towards the point $(3, f(3)) = (3, 7)$. Note that we can hardly distinguish between the graphs of f and L_3 in the third frame. This indicates that $L_3(x)$ approximates $f(x)$ very well if x is close to the basepoint 3. On the other hand, we do not expect $L_3(x)$ to approximate $f(x)$ when x is far from 3. The linear function L_3 is a "local approximation" to f. $\square$

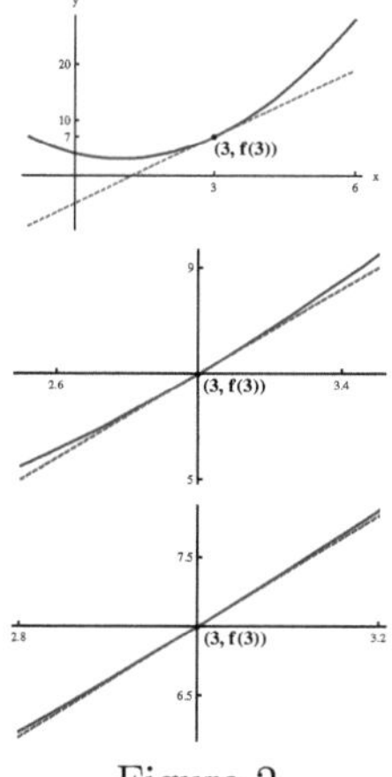

Figure 2

Let's assess the error in the approximation of $f(x)$ by $L_3(x)$ algebraically. Since $L_3(x)$ is expected to be a good approximation to f when x is near 3, it is convenient to set $x = 3 + h$, so that $h \ (= x - 3)$ represents the deviation of x from the basepoint 3. We have

$$L_3(3 + h) = 7 + 4(x - 3)|_{x=3+h} = 7 + 4h.$$

Therefore,

$$f(3 + h) - L_3(3 + h) = (3 + h)^2 - 2(3 + h) + 4 - (7 + 4h)$$
$$= 9 + 6h + h^2 - 6 - 2h + 4 - 7 - 4h$$
$$= h^2$$

Thus, the absolute error is

$$|f(3+h) - L_3(3+h)| = h^2.$$

Note that h^2 is much smaller than $|h|$ if $|h|$ is small. For example,

$$\left(10^{-2}\right)^2 = 10^{-4} \text{ and } \left(10^{-3}\right)^2 = 10^{-6}.$$

Thus, the absolute error in the approximation of $f(x)$ by $L_3(x)$ is much smaller than the distance of x from the basepoint 3 if x is close to 3. This numerical fact is consistent with our graphical observation. $\square$

Example 2 Let

$$f(x) = \frac{1}{x}.$$

a) Determine L_2, the linear approximation to f based at 2.
b) Calculate $f(2+h)$ and $L_2(2+h)$ for $h = -10^{-n}$, $n = 1, 2, 3$. Compare $|f(2+h) - L_2(2+h)|$ with $|h|$.

Solution

a) By the power rule,

$$f'(x) = \frac{d}{dx}\left(\frac{1}{x}\right) = \frac{d}{dx}\left(x^{-1}\right) = -x^{-2} = -\frac{1}{x^2}.$$

Therefore, $f'(2) = -1/4$ and

$$L_2(x) = f(2) + f'(2)(x-2) = \frac{1}{2} - \frac{1}{4}(x-2).$$

Thus,

$$L_2(2+h) = \frac{1}{2} - \frac{1}{4}h.$$

Figure 3 shows the graphs of f and L_2 (the dashed line) in a small viewing window that is centered at $(2, f(2)) = (2, 0.5)$.

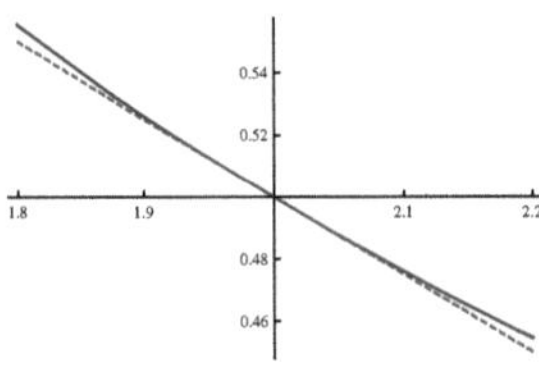

Figure 3

b) Table 1 displays the required data. We see that $|f(2+h) - L_2(2+h)|$ is much smaller than $|h|$ for the values of h that are considered. Indeed, $f(2 - 10^{-3})$ and $L_2(2 - 10^{-3})$ are represented by the same decimal, rounded to 6 significant digits. Therefore, the numbers support our analysis of the error in linear approximations. $\square$

h	$f(2+h)$	$L_2(2+h)$	$\lvert f(2+h) - L_2(2+h)\rvert$
-10^{-1}	0.526316	0.525	1.3×10^{-3}
-10^{-2}	0.502513	0.5025	1.3×10^{-5}
-10^{-3}	0.50025	0.50025	1.3×10^{-7}

Table 1

Remark We have identified **the rate of change of a function** f at a point a with $f'(a)$, and $f'(a)$ is the rate of the linear function L_a. The fact that $L_a(x)$ approximates $f(x)$ very well if x is close to a justifies this identification. After all, there is no question that the rate of change of the linear function

$$L_a(x) = f(a) + f'(a)(x - a)$$

is $f'(a)$ at any point. $\Diamond$

In particular, if $f'(a) = 0$ we declare that the rate of change of f at a is 0. This does not mean that we have $f(x) = f(a)$ for each x in some interval centered at a. On the other hand,

$$L_a(x) = f(a) + f'(a)(x - a) = f(a),$$

and the rate of change of the constant function L_a *is* 0. Since

$$f(x) \cong L_a(x) = f(a),$$

and the magnitude of the error can be expected to be much smaller than $|x - a|$ if $|x - a|$ is small, the restriction of f to a small interval centered at a is almost a constant function. Therefore, it is reasonable to declare that the rate of change of f at a is 0.

Example 3 As in Example 2 of Section 2.3, where we determined the tangent line to the graph of cosine at $(0, 0)$, the linear approximation to cosine based at 0 is

$$L_0(x) = \cos(0) + \left(\frac{d}{dx} \cos(x) \Big|_{x=0} \right) x = 1.$$

Thus, L_0 is a constant function and its graph, i.e., the tangent line to the graph of cosine at $(1, 0)$, is a horizontal line, as shown in Figure 4.

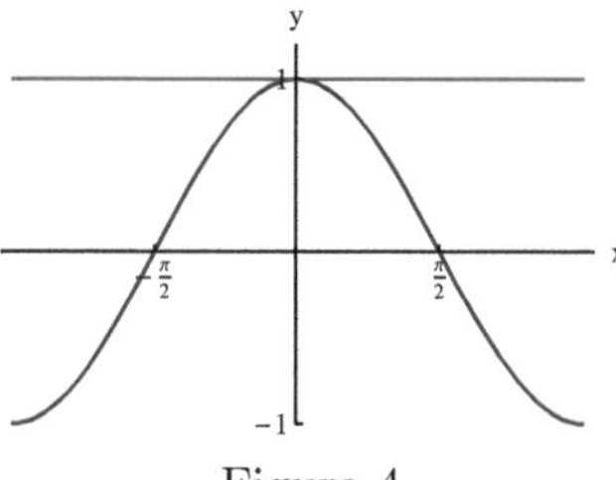

Figure 4

Obviously, the rate of change of L_0 is 0. We declare that the rate of change of cosine at 0 is also 0, even though $\cos(x) \neq 0$ if x deviates from 0 slightly. This is justified in view of the fact that $\cos(x) \cong 1$ if $x \cong 0$, and the absolute error in the approximation is much smaller than $|x|$ is $|x|$ is small. For example, $\cos(0.01) \cong 0.99995$, $|\cos(0.01) - 1| \cong 5.0 \times 10^{-5}$, and 5.0×10^{-5} is much smaller than 10^{-2}. $\square$

The Differential

It is useful to consider all the local linear approximations to a given function at once by considering the basepoint to be a variable. In this case it is convenient to work with differences and a change in the notation seems to be in order. We will denote an increment along the x-axis by Δx. Thus,

$$f'(x) = \lim_{h \to 0} \frac{f(x+h) - f(x)}{h} = \lim_{\Delta x \to 0} \frac{f(x + \Delta x) - f(x)}{\Delta x}.$$

Therefore,

$$\frac{f(x + \Delta x) - f(x)}{\Delta x} \cong f'(x)$$

if $|\Delta x|$ is small, so that

$$f(x + \Delta x) - f(x) \cong f'(x)\,\Delta x.$$

We will give the expression $f'(x)\,\Delta x$ a special name:

Definition 2 The differential of the function f is

$$df = f'(x)\Delta x.$$

Thus, **df is a function of two independent variables**, the basepoint x and the increment Δx. We can indicate this explicitly by writing

$$df(x, \Delta x) = f'(x)\Delta x.$$

We have

$$f(x + \Delta x) - f(x) \cong df(x, \Delta x)$$

if $|\Delta x|$ is small. The idea behind the differential is the same as the idea of local linear approximations. **The differential merely keeps track of local linear approximations to a function as the basepoint varies.** Note that $df(x, \Delta x)$ **is the change corresponding to the increment Δx** along the tangent line to the graph of f at $(x, f(x))$, as illustrated in Figure 5.

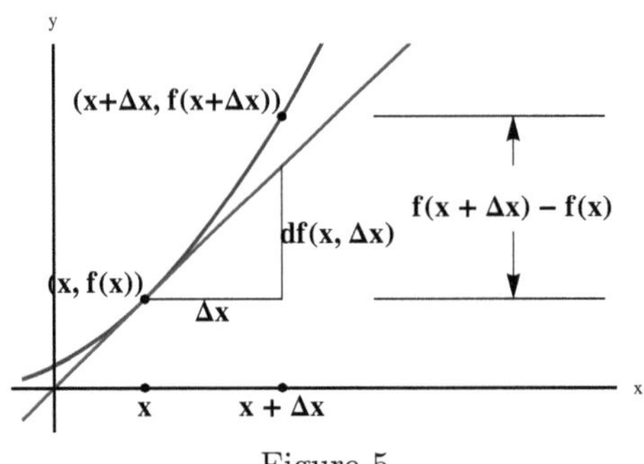

Figure 5

Example 4 Let $f(x) = \sqrt{x}$. Approximate $\sqrt{4.1}$ via the differential of f.

Solution

Since

$$f'(x) = \frac{d}{dx}\sqrt{x} = \frac{1}{2\sqrt{x}},$$

The differential of f is

$$df(x, \Delta x) = f'(x)\,\Delta x = \frac{1}{2\sqrt{x}}(\Delta x) = \frac{\Delta x}{2\sqrt{x}}.$$

It is natural to set $x = 4$ and $\Delta x = 0.1$ for the approximation of $\sqrt{4.1} = f(4.1)$ since $f(4) = \sqrt{4} = 2$. Thus,

$$\sqrt{4.1} - 2 = f(4.1) - f(4) \cong df(4, 0.1) = \frac{0.1}{2\sqrt{4}} = \frac{0.1}{4} = 0.025.$$

Therefore,

$$\sqrt{4.1} = 2 + \left(\sqrt{4.1} - 2\right) \cong 2 + 0.025 = 2.025.$$

We have

$$\sqrt{4.1} \cong 2.024\,85,$$

rounded to 6 significant digits, and

$$\left|\sqrt{4.1} - 2.025\right| \cong 1.5 \times 10^{-4}.$$

Note that the absolute error in the approximation of $\sqrt{4.1}$ via the differential is much smaller than $\Delta x = 0.1$. $\square$

Remark 1 As we saw in Section 2.4, the rate of change of the position $f(t)$ of an object in one-dimensional motion at time t is the instantaneous velocity $v(t)$. If the time increment is $\Delta t > 0$ is small then

$$f(t + \Delta t) - f(t) \cong df(t, \Delta t) = f'(t)\,\Delta t = v(t)\,\Delta t.$$

Thus, the displacement over the time time interval $[t, t + \Delta t]$ is approximately $v(t)\,\Delta t$ if Δt is small.
For example, if $f(t) = \cos(t)$ then $v(t) = -\sin(t)$ so that

$$f(t + \Delta t) - f(t) \cong -\sin(t)\,\Delta t.$$

In particular,

$$f\left(\frac{\pi}{6} + 0.1\right) - f\left(\frac{\pi}{6}\right) \cong -\sin\left(\frac{\pi}{6}\right)(0.1) = -\frac{0.1}{2} = -0.05.$$

The $(-)$ sign indicates that the motion is in the negative direction. $\Diamond$

The Traditional Notation for the Differential

We wrote

$$df(x, \Delta x) = f'(x)\Delta x.$$

Traditionally, the increment Δx is denoted by dx within the context of differentials. Thus,

$$df(x, dx) = f'(x)dx.$$

If we use the Leibniz notation for $f'(x)$, we have

$$df(x, dx) = \frac{df}{dx}(x)\,dx.$$

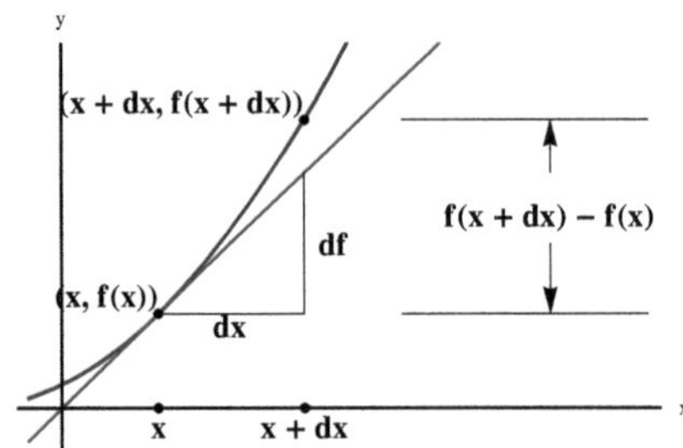

Figure 6: The geometric meaning of the differential

We usually do not bother to indicate that the differential depends on x and dx, and write

$$df = \frac{df}{dx}dx.$$

This is convenient and traditional notation, but you should keep in mind that the "fraction"

$$\frac{df}{dx}$$

is a symbolic fraction, and that the symbol dx that appears as the denominator does not have the same meaning as dx that stands for the increment in the value of the independent variable. The expression

$$df = \frac{df}{dx}dx$$

is analogous to the expression

$$\Delta f = \frac{\Delta f}{\Delta x}\Delta x,$$

where $\Delta x \neq 0$ and $\Delta f = f(x + \Delta x) - f(x)$.

If we refer to the function as $y = y(x)$, we can write

$$dy = \frac{dy}{dx}dx$$

The above expression is analogous to the expression

$$\Delta y = \frac{\Delta y}{\Delta x}\Delta x,$$

where $\Delta x \neq 0$ and $\Delta y = y(x + \Delta x) - y(x)$.

Example 5 Let $f(x) = x^{1/3}$

a) Determine the differential df.
b) Make use of the differential df to approximate $(8.01)^{1/3}$. Determine the absolute error in the approximation by treating the value that is obtained from your calculator as the exact value. Compare with the deviation from the basepoint that you have chosen.

Solution

a)

$$df = \frac{df}{dx}dx = \left(\frac{d}{dx}\left(x^{1/3}\right)\right)dx = \left(\frac{1}{3}x^{-2/3}\right)dx = \frac{1}{3x^{2/3}}dx.$$

b) Since 8.01 is close to 8, and $f(8) = 8^{1/3} = 2$, the natural choice for the basepoint is 8. Thus, $dx = 8.01 - 8 = 0.01$. The value of the differential corresponding to $x = 8$ and $dx = 0.01$ is

$$\left(\frac{1}{3\left(8^{2/3}\right)}\right)(0.01) = \frac{0.01}{3\,(4)} = \frac{0.01}{12}.$$

Therefore,

$$(8.01)^{1/3} - 2 = f(8.01) - f(8) \cong \frac{0.01}{12}$$

so that

$$(8.01)^{1/3} \cong 2 + \frac{0.01}{12} \cong 2.\,000\,83$$

A calculator tells us that $(8.01)^{1/3} \cong 2.\,000\,83$, rounded to 6 significant digits. Thus, the approximation via the differential gave us the same decimal, rounded to 6 significant digits. There is a nonzero of course. Indeed,

$$\left|\left(2 + \frac{0.01}{12}\right) - (8.01)^{1/3}\right| \cong 3.\,5 \times 10^{-7}.$$

Thus, the absolute error in the approximation is much smaller than 0.01, the deviation of 8.01 from the basepoint 8. $\square$

Example 6 The volume of a spherical ball of radius r is

$$V = \frac{4}{3}\pi r^3.$$

a) Determine the differential dV.
b) Use the differential to approximate the change in the volume of the ball if the ball is inflated and its radius increases from 20 centimeters to 20.1 centimeters.

Solution

a) We have

$$\frac{dV}{dr} = \frac{d}{dr}\left(\frac{4}{3}\pi r^3\right) = \frac{4}{3}\pi\frac{d}{dr}\left(r^3\right) = \frac{4}{3}\pi\left(3r^2\right) = 4\pi r^2.$$

Therefore,

$$dV = \frac{dV}{dr}dr = 4\pi r^2 dr.$$

Note that $4\pi r^2$ is the surface area of sphere of radius r. Therefore, the change in the volume of a spherical ball that corresponds to a small change in the radius can be approximated by the product of the area of its boundary and the increment of the radius.

b) In particular,

$$V(20.1) - V(20) \cong 4\pi\left(20^2\right)(0.1) \cong 502.655$$

(cm^3). The actual change in the volume is

$$V(20.1) - V(20) = \frac{4}{3}\pi\,(20.1)^3 - \frac{4}{3}\pi\,(20)^3 \cong 505.172$$

(cm^3). Therefore, the error in the approximation of the change in the volume via the differential is approximately $2.517\left(cm^3\right)$. This may not be considered to be a small number. On the other hand, the relative error is usually more appropriate in assessing error. Thus,

$$\frac{(V(20.1) - V(20)) - 4\pi\left(20^2\right)(0.1)}{V(20)} \cong \frac{2.517}{33510.3} \cong 7.5 \times 10^{-5},$$

and this number *is* small.

We can also approximate the relative change in the volume, i.e.,

$$\frac{V\,(20.1) - V\,(20)}{V\,(20)},$$

via the differential by calculating

$$\frac{dV\,(20, 0.1)}{V\,(20)} = \frac{4\pi\,(20^2)\,(0.1)}{V\,(20)} \cong 1.5 \times 10^{-2}.$$

This approximates

$$\frac{V\,(20.1) - V\,(20)}{V\,(20)} \cong 1.507\,51 \times 10^{-2}$$

with an error that is approximately 7×10^{-5}. $\square$

The Accuracy of Local Linear Approximations

Theorem 1 Assume that f is differentiable at a, and that L_a is the linear approximation to f *based at* a. We have

$$f\,(a + h) = L_a(a + h) + hq\,(h),$$

where

$$\lim_{h \to 0} q(h) = 0.$$

Thus, $hq\,(h)$ represents the error in the approximation of f by L_a at $x = a + h$. Since the error is the product of h and $q\,(h)$, and $q\,(h) \to 0$ as $h \to 0$, its magnitude is much smaller than $|h| = |x - a|$ if x is close to the basepoint a.

With reference to Example 1, $q\,(h) = h^2$.

Proof

As in Example 1, we will set $x = a + h$, so that $h = x - a$ represents the deviation of x from a and has small magnitude if x is near a. We have

$$L_a\,(a + h) = f\,(a) + f'\,(a)\,(x - a)\big|_{x - a = h} = f\,(a) + f'\,(a)\,h.$$

Therefore,

$$\begin{aligned}
f\,(a + h) - L_a\,(a + h) &= f\,(a + h) - (f\,(a) + f'\,(a)\,h) \\
&= (f\,(a + h) - f\,(a)) - f'\,(a)\,h \\
&= h\left(\frac{f\,(a + h) - f\,(a)}{h} - f'\,(a)\right)
\end{aligned}$$

Let's set

$$q\,(h) = \frac{f\,(a + h) - f\,(a)}{h} - f'\,(a),$$

so that $q\,(h)$ is the difference between the difference quotient and the derivative. We have

$$\lim_{h \to 0} q\,(h) = \lim_{h \to 0}\left(\frac{f\,(a + h) - f\,(a)}{h} - f'\,(a)\right) = 0,$$

since the difference quotient approaches the derivative as $h \to 0$.

Thus,

$$f(a+h) - L_a(a+h) = hq(h),$$

so that

$$f(a+h) = L_a(a+h) + hq(h),$$

where $\lim_{h \to 0} q(h) = 0$. $\blacksquare$

The analysis of the error in the approximation of differences via the differential is along similar lines:

Theorem 2 Assume that f is differentiable at x. Then,

$$f(x + \Delta x) - f(x) = df(x, \Delta x) + \Delta x\, q(\Delta x),$$

where

$$\lim_{\Delta x \to 0} q(\Delta x) = 0.$$

Proof

We have

$$
\begin{aligned}
f(x + \Delta x) - f(x) - df(x, \Delta x) &= f(x + \Delta x) - f(x) - f'(x)\,\Delta x \\
&= \Delta x \left(\frac{f(x + \Delta x) - f(x)}{\Delta x} - f'(x) \right).
\end{aligned}
$$

If we set

$$q(\Delta x) = \frac{f(x + \Delta x) - f(x)}{\Delta x} - f'(x),$$

then

$$f(x + \Delta x) - f(x) - df(x, \Delta x) = \Delta x q(\Delta x).$$

We have

$$\lim_{\Delta x \to 0} q(\Delta x) = \lim_{\Delta x \to 0} \left(\frac{f(x + \Delta x) - f(x)}{\Delta x} - f'(x) \right) = 0,$$

since

$$\lim_{\Delta x \to 0} \frac{f(x + \Delta x) - f(x)}{\Delta x} - .f'(x).$$

Thus,

$$f(x + \Delta x) - f(x) - df(x, \Delta x) = \Delta x q(\Delta x),$$

where $\lim_{\Delta x \to 0} q(\Delta x) = 0.$

Problems

In problems 1 - 6,
a) Determine L_a, the linear approximation to f based at a,
b) Make use of $L_a(b)$ to approximate $f(b)$ if such a point b is indicated,
c) [C] Calculate the absolute error in the approximation of $f(b)$ by $L_a(b)$ and compare with $|b - a|$.
d) [C] Plot the graphs of f and L_a in a sufficiently small viewing window centered at $(a, f(a))$ that demonstrates the accuracy of the linear approximation near a.

1.
$$f(x) = x^2 + x, \ a = 3, \ b = 3.01$$

2.
$$f(x) = x^4, \ a = 1, b = 0.98$$

3.
$$f(x) = \frac{1}{x^2}, \ a = 0.5, \ b = 0.502$$

4.
$$f(x) = x^{2/3}, \ a = 8, \ b = 7.8$$

5.
$$f(x) = x^{1/4}, \ a = 16, \ b = 16.2$$

6.
$$f(x) = \sin(x), \ a = \frac{\pi}{3}, \ b = \frac{\pi}{3} - 0.1$$

In problems 7 and 8,
a) Determine the differential of f,
b) Approximate $f(b)$ via the the differential of f'

7.
$$f(x) = \sqrt{x}, \ b = 24.9.$$

8.
$$f(x) = \frac{1}{x^2}, \ b = 2.2.$$

In problems 9 - 12 approximate the given number via the differential (You need to choose an appropriate function and basepoint).

9. $(26.5)^{1/3}$

10.. $\dfrac{1}{\sqrt{3.9}}$

11. $\sin\left(\dfrac{3\pi}{4} + 0.1\right)$

12. $\cos\left(-\dfrac{\pi}{6} - 0.2\right)$

13. Let $A(r)$ be the area of a disk of radius r.
a) Approximate the change in the area corresponding to a change in the radius from 10 meters to 10.1 meters via the differential.
b) Calculate the relative error in the approximation of part a).

14. Let $V(r)$ be the volume of a right circular cone of height 10 meter and base radius r meters.
a) Approximate the change in the volume corresponding to a change in the radius from 4 meters to 4.2 meters via the differential.
b) Calculate the relative error in the approximation of part a).

2.6   The Product Rule and the Quotient Rule

You know how to differentiate functions such as those defined by rational powers of x, $\sin(x)$, $\cos(x)$, and linear combinations of these functions, without going back to the definition of the derivative. In this section you will learn how to compute **the derivatives of products and quotients** of such functions.

The Product Rule

THE PRODUCT RULE   Assume that f and g are differentiable at x. The product fg is also differentiable at x and we have

$$(fg)'(x) = f'(x)g(x) + f(x)g'(x).$$

In the Leibniz notation,

$$\frac{d}{dx}(f(x)g(x)) = \left(\frac{df(x)}{dx}\right)g(x) + f(x)\left(\frac{dg(x)}{dx}\right).$$

Proof

The difference quotient that is relevant to the calculation of $(fg)'(x)$ is

$$\frac{f(x+\Delta x)\,g(x+\Delta x) - f(x)\,g(x)}{\Delta x}.$$

Let's set

$$u = f(x),\ \Delta u = f(x+\Delta x) - f(x),\ v = g(x)\ \text{and}\ \Delta v = g(x+\Delta x) - g(x),$$

so that

$$f(x+\Delta x) = u + \Delta u\ \text{and}\ g(x+\Delta x) = v + \Delta v.$$

Thus,

$$
\begin{aligned}
\frac{f(x+\Delta x)\,g(x+\Delta x) - f(x)\,g(x)}{\Delta x} &= \frac{(u+\Delta u)(v+\Delta v) - uv}{\Delta x} \\
&= \frac{uv + v(\Delta u) + u(\Delta v) + (\Delta u)(\Delta v) - uv}{\Delta x} \\
&= \frac{v(\Delta u) + u(\Delta v) + (\Delta u)(\Delta v)}{\Delta x} \\
&= v\left(\frac{\Delta u}{\Delta x}\right) + u\left(\frac{\Delta v}{\Delta x}\right) + \left(\frac{\Delta u}{\Delta x}\right)\Delta v.
\end{aligned}
$$

Note that

$$\lim_{\Delta x\to 0}\left(\frac{\Delta u}{\Delta x}\right)\Delta v = \left(\lim_{\Delta x\to 0}\frac{f(x+\Delta x) - f(x)}{\Delta x}\right)\left(\lim_{\Delta x\to 0}(g(x+\Delta x) - g(x))\right) = f'(x)(0) = 0,$$

since g is continuous at x. Therefore,

$$
\begin{aligned}
(fg)'(x) &= \lim_{\Delta x\to 0}\frac{f(x+\Delta x)\,g(x+\Delta x) - f(x)\,g(x)}{\Delta x} \\
&= v\lim_{\Delta x\to 0}\frac{\Delta u}{\Delta x} + u\lim_{\Delta x\to 0}\frac{\Delta v}{\Delta x} \\
&= g(x)\lim_{\Delta x\to 0}\frac{f(x+\Delta x) - f(x)}{\Delta x} + f(x)\lim_{\Delta x\to 0}\frac{g(x+\Delta x) - g(x)}{\Delta x} \\
&= g(x)f'(x) + f(x)g'(x).
\end{aligned}
$$

∎

Example 1 Let $F(x) = x^2\sin(x)$. Determine $F'(x)$.

Solution

If we set $f(x) = x^2$ and $g(x) = \sin(x)$, then $F = fg$. We know how to differentiate each factor:

$$f'(x) = \frac{d}{dx}\left(x^2\right) = 2x,\ \text{and}\ g'(x) = \frac{d}{dx}\sin(x) = \cos(x).$$

By the product rule,

$$F'(x) = f'(x)g(x) + f(x)g'(x) = (2x)\sin(x) + x^2(\cos(x)) = 2x\sin(x) + x^2\cos(x).$$

It is more practical to indicate the application of the product rule to such a case by using the Leibniz notation, as in the application of the other rules of differentiation:

$$\frac{d}{dx}\left(x^2\sin(x)\right) = \left(\frac{d}{dx}\left(x^2\right)\right)\sin(x) + x^2\left(\frac{d}{dx}\sin(x)\right) = 2x\sin(x) + x^2\cos(x).$$

□

A word of caution: The product rule does **not** say that the derivative of a product is the product of the derivatives. For example, if $f(x) = x$, we have $(f(x))^2 = x^2$, so that

$$\frac{d}{dx}\left((f(x))^2\right) = \frac{d}{dx}\left(x^2\right) = 2x.$$

On the other hand,

$$\left(\frac{d}{dx}f(x)\right)\left(\frac{d}{dx}f(x)\right) = \left(\frac{d}{dx}(x)\right)\left(\frac{d}{dx}(x)\right) = (1)(1) = 1 \neq 2x.$$

Example 2 Let $f(x) = \sqrt{x}\cos(x)$. Determine f'.

Solution

By the product rule,

$$f'(x) = \frac{d}{dx}\left(\sqrt{x}\cos(x)\right) = \left(\frac{d}{dx}\sqrt{x}\right)\cos(x) + \sqrt{x}\left(\frac{d}{dx}\cos(x)\right)$$

$$= \left(\frac{1}{2\sqrt{x}}\right)\cos(x) + \sqrt{x}\left(-\sin(x)\right)$$

$$= \frac{\cos(x)}{2\sqrt{x}} - \sqrt{x}\sin(x)$$

if $x > 0$. □

The Quotient Rule

Now we will discuss the rule for the differentiation of quotients of functions. Let us begin with a special case:

THE DERIVATIVE OF A RECIPROCAL   **Assume that g is differentiable at x and $g(x) \neq 0$. Then $1/g$ is also differentiable at x, and we have**

$$(\frac{1}{g})'(x) = -\frac{g'(x)}{g^2(x)}.$$

In the Leibniz notation,

$$\frac{d}{dx}(\frac{1}{g(x)}) = -\frac{\dfrac{dg(x)}{dx}}{g^2(x)}.$$

Proof

The relevant difference quotient is

$$\frac{\dfrac{1}{g(x+\Delta x)} - \dfrac{1}{g(x)}}{\Delta x} = \frac{1}{\Delta x}\left(\frac{1}{g(x+\Delta x)} - \frac{1}{g(x)}\right)$$

$$= \frac{1}{\Delta x}\left(\frac{g(x) - g(x+\Delta x)}{g(x+\Delta x)\,g(x)}\right)$$

$$= \left(-\frac{g(x+\Delta x) - g(x)}{\Delta x}\right)\left(\frac{1}{g(x+\Delta x)\,g(x)}\right).$$

Therefore,

$$\left(\frac{1}{g}\right)'(x) = \lim_{\Delta x \to 0} \frac{\dfrac{1}{g(x+\Delta x)} - \dfrac{1}{g(x)}}{\Delta x}$$

$$= \lim_{\Delta x \to 0}\left(\left(-\frac{g(x+\Delta x)-g(x)}{\Delta x}\right)\left(\frac{1}{g(x+\Delta x)\,g(x)}\right)\right)$$

$$= \lim_{\Delta x \to 0}\left(-\frac{g(x+\Delta x)-g(x)}{\Delta x}\right)\lim_{\Delta x \to 0}\left(\frac{1}{g(x+\Delta x)\,g(x)}\right)$$

$$= (-g'(x))\frac{1}{(\lim_{\Delta x \to 0} g(x+\Delta x))g(x)}.$$

Since g is differentiable at x, g is continuous at x. Therefore, $\lim_{\Delta x \to 0} g(x+\Delta x) = g(x)$. Thus,

$$\left(\frac{1}{g}\right)'(x) = (-g'(x))\frac{1}{(\lim_{\Delta x \to 0} g(x+\Delta x))g(x)} = (-g'(x))\frac{1}{g(x)g(x)} = -\frac{g'(x)}{g^2(x)}.$$

■

Example 3 Let

$$f(x) = \frac{1}{x^2 - 9}.$$

Determine f'.

Solution

By the rule on the derivative of a reciprocal,

$$f'(x) = \frac{d}{dx}\left(\frac{1}{x^2-9}\right) = -\frac{\dfrac{d}{dx}\left(x^2-9\right)}{\left(x^2-9\right)^2} = -\frac{2x}{\left(x^2-9\right)^2}$$

if $x^2 - 9 \neq 0$, i.e., if $x \neq 3$ and $x \neq -3$.

Note that f and f' are rational functions. Figure 1 displays the graphs of f and f'. Both graphs have vertical asymptotes at $x = \pm 3$. □

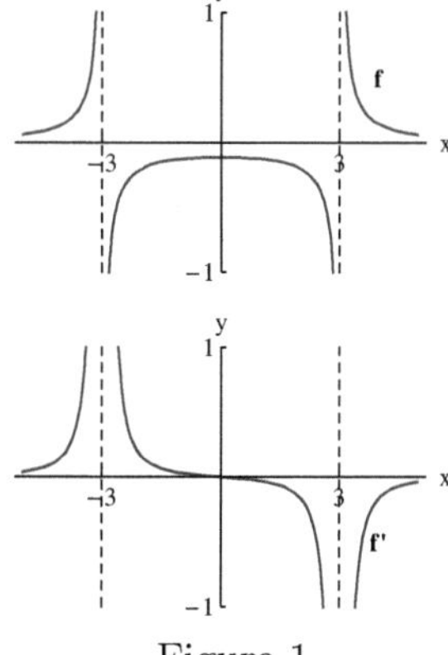

Figure 1

Example 4 Let

$$f(x) = \frac{x}{x^2 + 1}.$$

Determine f'.

Solution

We use the product rule and the rule for the differentiation of reciprocals:

$$\begin{aligned}
f'(x) &= \frac{d}{dx}\left(\frac{x}{x^2+1}\right) \\
&= \frac{d}{dx}\left(x\left(\frac{1}{x^2+1}\right)\right) \\
&= \left(\frac{d}{dx}x\right)\left(\frac{1}{x^2+1}\right) + x\left(\frac{d}{dx}\left(\frac{1}{x^2+1}\right)\right) \\
&= \frac{1}{x^2+1} + x\left(-\frac{\frac{d}{dx}(x^2+1)}{(x^2+1)^2}\right) \\
&= \frac{1}{x^2+1} + x\left(-\frac{2x}{(x^2+1)^2}\right) \\
&= \frac{1}{x^2+1} - \frac{2x^2}{(x^2+1)^2}.
\end{aligned}$$

There is no restriction on x since $x^2 + 1 \neq 0$ for any $x \in \mathbb{R}$. Figure 2 shows the graphs of f and f'. $\square$

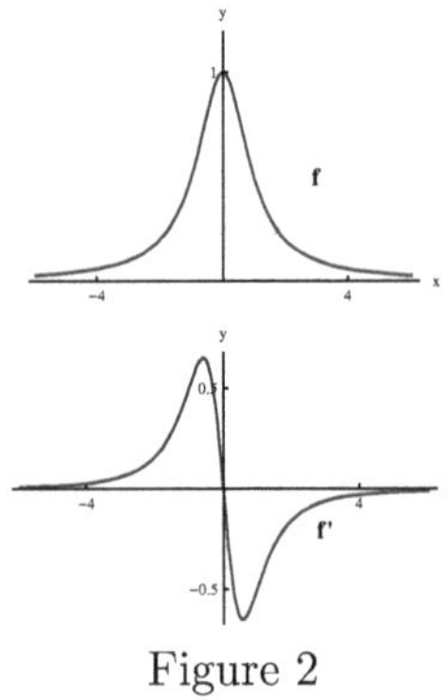

Figure 2

The procedure that was used in Example 4 leads to the general rule for the differentiation of quotients:

THE QUOTIENT RULE   Assume that f and g are differentiable at x, and $g^2(x) \neq 0$. **Then**

$$\left(\frac{f}{g}\right)'(x) = \frac{f'(x)g(x) - f(x)g'(x)}{g^2(x)}$$

In the Leibniz notation,

$$\frac{d}{dx}\left(\frac{f(x)}{g(x)}\right) = \frac{\left(\dfrac{df(x)}{dx}\right)g(x) - f(x)\left(\dfrac{dg(x)}{dx}\right)}{g^2(x)}.$$

Proof

We apply the product rule and the rule for the differentiation of reciprocals:

$$\frac{d}{dx}\left(\frac{f(x)}{g(x)}\right) = \frac{d}{dx}\left(f(x)\left(\frac{1}{g(x)}\right)\right)$$

$$= \frac{df}{dx}\left(\frac{1}{g(x)}\right) + f(x)\left(\frac{d}{dx}\left(\frac{1}{g(x)}\right)\right)$$

$$= \frac{df}{dx}\left(\frac{1}{g(x)}\right) + f(x)\left(-\frac{\frac{dg}{dx}}{g^2(x)}\right)$$

$$= \frac{\left(\frac{df}{dx}\right)g(x) - f(x)\left(\frac{dg}{dx}\right)}{g^2(x)}.$$

■

Example 5 Let

$$f(x) = \frac{x^3 - 2x}{x^2 - 4}.$$

Determine f'.

Solution

By the quotient rule,

$$\frac{df}{dx} = \frac{d}{dx}\left(\frac{x^3 - 2x}{x^2 - 4}\right)$$

$$= \frac{\left(\frac{d}{dx}(x^3 - 2x)\right)(x^2 - 4) - (x^3 - 2x)\left(\frac{d}{dx}(x^2 - 4)\right)}{(x^2 - 4)^2}$$

$$= \frac{(3x^2 - 2)(x^2 - 4) - (x^3 - 2x)(2x)}{(x^2 - 4)^2}$$

$$= \frac{x^4 - 10x^2 + 8}{(x^2 - 4)^2}.$$

The above expression is valid as long as $x^2 - 4 \neq 0$, i.e., if $x \neq 2$ and $x \neq -2$. Figure 3 shows the graphs of f and f'. Note that both graphs have vertical asymptotes at $x = \pm 2$. The graph of f' has the horizontal asymptote $y = 1$ at $\pm\infty$ (confirm by evaluating the relevant limits). □

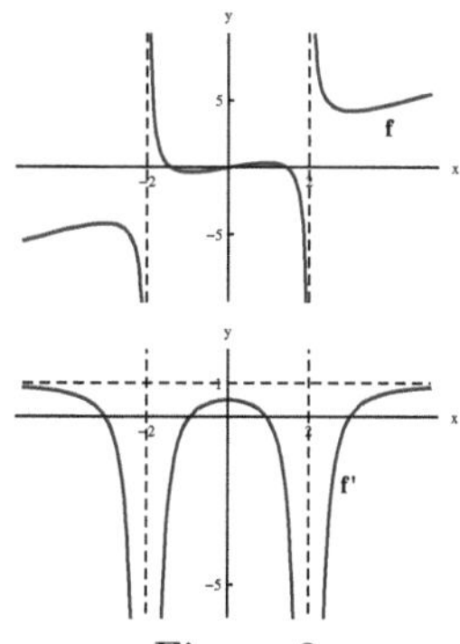

Figure 3

Now we are in a position to differentiate the trigonometric functions **tangent** and **secant**:

$$\frac{d}{dx}\tan(x) = \sec^2(x) \text{ and } \frac{d}{dx}\sec(x) = \sec(x)\tan(x)$$

if x is not an odd integer multiple of $\pm.\pi/2$.

Proof

Thanks to the quotient rule,

$$\frac{d}{dx}\tan(x) = \frac{d}{dx}\left(\frac{\sin(x)}{\cos(x)}\right) = \frac{\left(\frac{d}{dx}\sin(x)\right)\cos(x) - \sin(x)\left(\frac{d}{dx}\cos(x)\right)}{\cos^2(x)}$$

$$= \frac{(\cos(x))\cos(x) - \sin(x)(-\sin(x))}{\cos^2(x)} = \frac{\cos^2(x) + \sin^2(x)}{\cos^2(x)}.$$

Since $\cos^2(x) + \sin^2(x) = 1$,

$$\frac{d}{dx}\tan(x) = \frac{1}{\cos^2(x)} = \sec^2(x).$$

The expression is valid if $\cos(x) \neq 0$, i.e., if x is not an odd multiple of $\pm\pi/2$. ∎
Figure 4 shows the graphs of tangent and its derivative on the interval $[-3\pi/2, 3\pi/2]$.

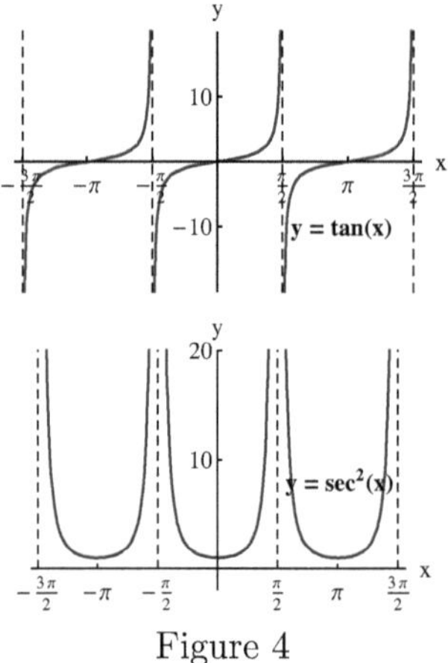

Figure 4

Since

$$\sec(x) = \frac{1}{\cos(x)},$$

we can apply the special case of the quotient rule for reciprocals:

$$\frac{d}{dx}\sec(x) = \frac{d}{dx}\left(\frac{1}{\cos(x)}\right) = -\frac{\frac{d}{dx}\cos(x)}{\cos^2(x)} = -\frac{(-\sin(x))}{\cos^2(x)} = \frac{\sin(x)}{\cos^2(x)}.$$

Therefore,

$$\frac{d}{dx}\sec(x) = \left(\frac{1}{\cos(x)}\right)\left(\frac{\sin(x)}{\cos(x)}\right) = \sec(x)\tan(x).$$

As in the case of tangent, the above expression is valid as long as $\cos(x) \neq 0$. ∎
Figure 5 show the graphs of secant and their derivative on the interval $[-3\pi/2, 3\pi/2]$.

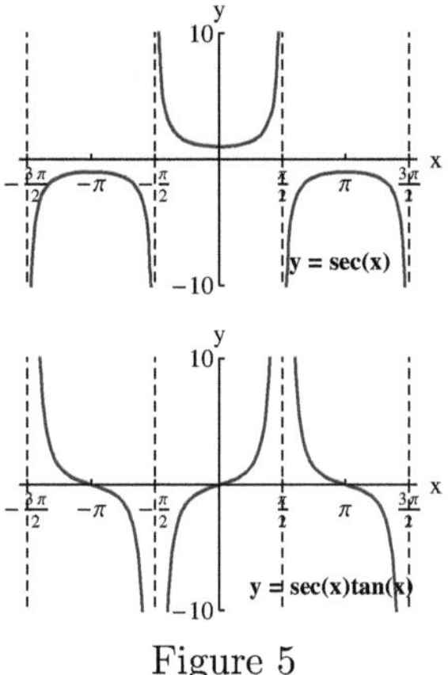

Figure 5

Example 6 Let
$$f(x) = \tan(x)$$

a) Determine the differential of f.
b) Use the differential of f to approximate $\tan(0.8)$. Compare the magnitude of the error with
the deviation from the basepoint that was chosen for the approximation.

Solution

a)
$$df(x, \Delta x) = \frac{df}{dx}\Delta x = \left(\frac{d}{dx}\tan(x)\right)\Delta x = \sec^2(x)\,\Delta x.$$

b) Since
$$\frac{\pi}{4} \cong 0.785\,398,$$

the point 0.8 is close to $\pi/4$, and we know that $\tan(\pi/4) = 1$. Therefore, $\pi/4$ is a good choice
as the basepoint. We have
$$df\left(\frac{\pi}{4}, \Delta x\right) = \sec^2\left(\frac{\pi}{4}\right)\Delta x = \frac{1}{\cos^2\left(\frac{\pi}{4}\right)}\Delta x = \frac{1}{\left(\frac{1}{\sqrt{2}}\right)^2}\Delta x = 2\Delta x.$$

Since we are interested in approximating $\tan(0.8)$, $\Delta x = (0.8) - \pi/4 \cong 1.\,460\,18 \times 10^{-2}$. Therefore,
$$\tan(0.8) - \tan\left(\frac{\pi}{4}\right) \cong df\left(\frac{\pi}{4}, \Delta x\right) = 2\Delta x \cong 2\left(1.\,460\,18 \times 10^{-2}\right) \cong 2.\,920\,36 \times 10^{-2}$$

Thus,
$$\tan(0.8) \cong \tan\left(\frac{\pi}{4}\right) + 2.\,920\,36 \times 10^{-2} \cong 1 + 2.\,920\,36 \times 10^{-2} \cong 1.\,029\,2.$$

A calculator will tell us that $\tan(0.8) \cong 1.\,029\,64$. Thus, the absolute error in the approximation
of $\tan(0.8)$ via the differential is approximately
$$|\tan(0.8) - 1.\,029\,2| \cong 4.4 \times 10^{-4}.$$

Note that 4.4×10^{-4} is much smaller than $\Delta x \cong 1.\,5 \times 10^{-2}$. □

Problems

In problems 1-5, determine f'.

1. $f(x) = \left(x^3 - 2x^2 + 9\right)\left(8x^2 - 7\right)$
2. $f(x) = x^{2/3}\sin(x)$
3. $f(x) = x^2\cos(x)$

4. $f(x) = \left(x^2 - 17x + 3\right)\sin(x)$

5. $f(x) = \left(8x^3 - 6x + 2\right)\cos(x)$

In problems 6-8, determine f' and f''.

6. $f(x) = x^4\sin(x)$

7. $f(x) = \dfrac{1}{x^2}\cos(x)$

8. $f(x) = \sqrt{x}\cos(x)$

In problems 9-14, determine $f'(x)$ and specify the domain of f':

9. $f(x) = \dfrac{1}{4x^2 + 1}$

10. $f(x) = \dfrac{1}{9x^2 - 4}$

11. $f(x) = \dfrac{2x + 1}{x^2 - 4}$

12. $f(x) = \dfrac{x^2 + 4}{x^2 - 9}$

13. $f(x) = \dfrac{3}{x^2 - 4} + 2x$

14. $f(x) = \dfrac{x^2 - 1}{x^3 - 2x}$

In problems 15-20, determine the indicated derivative.

15. $\dfrac{d}{dx}\left(\dfrac{x - 1}{x^2 - 9}\right)$

16. $\dfrac{d}{dx}\left(\dfrac{x - 5}{x^3 - x}\right)$

17. $\dfrac{d}{dx}\left(x^2\tan(x)\right)$

18. $\dfrac{d}{dx}\left(\dfrac{\tan(x)}{x^2 + 3}\right)$

19. $\dfrac{d}{dx}\left(\dfrac{x\sin(x)}{1 + x}\right)$

20. $\dfrac{d}{dx}\left(\dfrac{2\cos(x)}{x + \cos(x)}\right)$

21. Show that
$$\frac{d}{dx}\cot(x) = \frac{d}{dx}\frac{1}{\tan(x)} = -\csc^2(x).$$

22. Show that
$$\frac{d}{dx}\csc(x) = \frac{d}{dx}\frac{1}{\sin(x)} = -\csc(x)\cot(x).$$

In problems 23 and 24,
a) Determine L_a, the linear approximation to f based at a,
b) Make use of L_a to approximate $f(b)$. Determine the absolute error in the approximation, assuming the value that you obtain from your calculator to be the exact value of $f(b)$.

23. $f(x) = \dfrac{x - 4}{x + 4}$, $a = 8$, $b = 7.8$

24. $f(x) = \tan(x)$, $a = \pi/4$, $b = \pi/4 + 0.1$

2.7 The Chain Rule

In the previous sections of this chapter we discussed the rules for the differentiation of the sums, products and quotients of functions. In this section you will learn how to differentiate a function that can be expressed as a **composition** of functions with known derivatives. The relevant differentiation rule is **the chain rule.** For example, if $F(x) = \sin\left(x^2\right)$, the rules that you have learned until now do not lead to the derivative of F, at least not immediately. On the other hand, we can express F as $f \circ g$, where $f(u) = \sin(u)$ and $g(x) = x^2$, and we know how to differentiate both f and g. The chain rule will enable you to determine F' easily.

Introduction to the Chain Rule

THE CHAIN RULE   Assume that g is differentiable at x and f is differentiable at $g(x)$. Then $f \circ g$ is differentiable at x and we have

$$(f \circ g)'(x) = f'(g(x))g'(x).$$

Example 1 Let $F(x) = \sin\left(x^2\right)$. Determine $F'(x)$.

Solution

If we set $g(x) = x^2$ and $f(u) = \sin(u)$, then $f(g(x)) = f(x^2) = \sin(x^2)$. Therefore, $F = f \circ g$. We have

$$f'(u) = \frac{d}{du}\sin(u) = \cos(u), \; g'(x) = \frac{d}{dx}\left(x^2\right) = 2x.$$

By the chain rule,

$$F'(x) = (f \circ g)'(x) = f'(g(x))\, g'(x) = \cos\left(x^2\right)(2x) = 2x\cos\left(x^2\right).$$

$\square$

A Plausibility Argument for the Chain Rule

The difference quotient that is relevant to the differentiation of $f \circ g$ is

$$\frac{(f \circ g)(x + \Delta x) - (f \circ g)(x)}{\Delta x} = \frac{f(g(x + \Delta x)) - f(g(x))}{\Delta x}.$$

Let's set $g(x) = u$ and $g(x + \Delta x) = u + \Delta u$ so that

$$\Delta u = g(x + \Delta x) - g(x).$$

Thus,

$$\frac{f(g(x + \Delta x)) - f(g(x))}{\Delta x} = \frac{f(u + \Delta u) - f(u)}{\Delta x}$$

Assume that $|\Delta x|$ is small. Since g is differentiable at x it is continuous at x. Therefore $|\Delta u|$ is also small. As we have seen in Section 2.5,

$$f(u + \Delta u) - f(u) \cong df(u, \Delta u) = f'(u)\,\Delta u.$$

Thus,

$$\frac{f(g(x + \Delta x)) - f(g(x))}{\Delta x} = \frac{f(u + \Delta u) - f(u)}{\Delta x} \cong \frac{f'(u)\,\Delta u}{\Delta x}$$

Threfore we should have

$$
\begin{aligned}
(f \circ g)'(x) = \lim_{\Delta x \to 0} \frac{f(g(x+\Delta x)) - f(g(x))}{\Delta x} &= \lim_{\Delta x \to 0} \frac{f'(u)\,\Delta u}{\Delta x} \\
&= f'(u) \lim_{\Delta x \to 0} \frac{\Delta u}{\Delta x} \\
&= f'(g(x)) \lim_{\Delta x \to 0} \frac{g(x+\Delta x) - g(x)}{\Delta x} \\
&= f'(g(x))\, g'(x),
\end{aligned}
$$

as claimed.

You can find the proof of the chain rule at the end of this section. The proof is along the lines of the above plausibility argument.

Remark 1 (Caution) In order to determine the derivative of the composite function $f \circ g$ at x, we must evaluate g' at x and f' at $g(x)$. The chain rule does *not* say that

$$
(f \circ g)'(x) = f'(x)\, g'(x).
$$

For example, if $f(x) = g(x) = x^2$, then $(f \circ g)(x) = f(g(x)) = \left(x^2\right)^2 = x^4$, so that $(f \circ g)'(x) = 4x^3$, by the power rule. On the other hand, $f'(x)\, g'(x) = (2x)(2x) = 4x^2$. $\Diamond$

We can visualize the composite function $f \circ g$ schematically, where the functions are viewed as input-output mechanisms. The input for "the outer function" f is the output $g(x)$ of the "inner function" g:

$$
x \xrightarrow{g} g(x) \xrightarrow{f} f(g(x))
$$

Thus, it should be easy to remember to evaluate f' at $g(x)$ in the evaluation of $(f \circ g)'(x)$. $\Diamond$

Example 2 Let

$$
F(x) = \sqrt{x^2 + 1}.
$$

Determine F'.

Solution

If we set

$$
u = g(x) = x^2 + 1 \text{ and } f(u) = \sqrt{u},
$$

then $F(x) = f(g(x))$, so that $F = f \circ g$. We have

$$
f'(u) = \frac{d}{du} \sqrt{u} = \frac{1}{2\sqrt{u}},
$$

so that

$$
f'(g(x)) = f'(x^2 + 1) = \frac{1}{2\sqrt{x^2 + 1}}.
$$

We also have

$$
g'(x) = \frac{d}{dx}\left(x^2 + 1\right) = 2x.
$$

By the chain rule,

$$
F'(x) = f'(g(x))\, g'(x) = \left(\frac{1}{2\sqrt{x^2 + 1}}\right)(2x) = \frac{x}{\sqrt{x^2 + 1}}.
$$

The above expression is valid for each $x \in \mathbb{R}$ since $x^2 + 1 > 0$. Figure 1 shows the graphs of F and F'. Note that the graph of F' has the horizontal asymptote $y = -1$ at $-\infty$ and the horizontal asymptote $y = 1$ at $+\infty$ (confirm by evaluating the relevant limits). $\square$

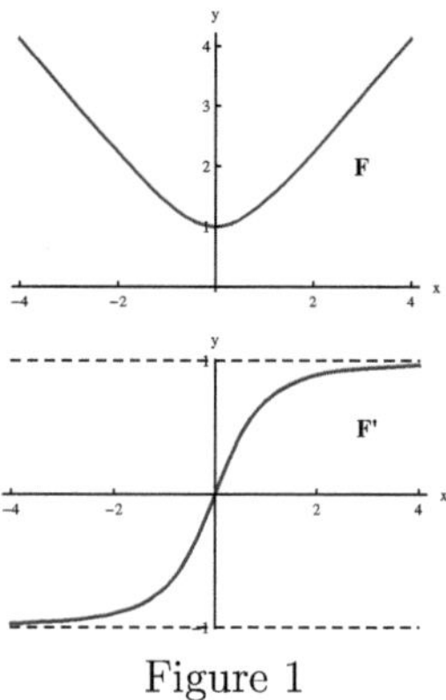

Figure 1

The Chain Rule in the Leibniz Notation

As in the implementation of the other rules for differentiation, it is usually more practical to use the Leibniz notation when we apply the chain rule. Assume that $F(x) = f(u(x))$. By the chain rule,

$$F'(x) = f'(u(x)) u'(x).$$

The above relationship can be expressed in the Leibniz notation as follows:

$$\frac{d}{dx} f(u(x)) = \left(\left. \frac{df}{du} \right|_{u=u(x)} \right) \frac{du}{dx} = \frac{df}{du}(u(x)) \frac{du}{dx}.$$

Example 3 Determine

$$\frac{d}{dx} \tan\left(x^3\right).$$

Solution

If we set $u(x) = x^3$ then $\tan\left(x^3\right) = \tan(u(x))$. Therefore,

$$\begin{aligned}
\frac{d}{dx} \tan\left(x^3\right) &= \left(\left. \frac{d}{du} \tan(u) \right|_{u=x^3} \right) \left(\frac{d}{dx}\left(x^3\right) \right) = \left(\left. \sec^2(u) \right|_{u=x^3} \right) \left(3x^2\right) \\
&= \left(\sec^2\left(x^3\right) \right) \left(3x^2\right) \\
&= 3x^2 \sec^2\left(x^3\right).
\end{aligned}$$

$\square$

The chain rule enables us to evaluate the derivative of **a translation of a function** easily: If c is a constant,

$$\frac{d}{dx} g(x - c) = \frac{dg}{du}(x - c).$$

Indeed, if we set $f(x) = g(x - c)$ and $u(x) = x - c$,

$$\frac{df}{dx}(x) = \frac{d}{dx}g(u(x)) = \left(\frac{dg}{du}(u(x))\right)\left(\frac{d}{dx}(x - c)\right)$$

$$= \left(\frac{dg}{du}(x - c)\right)(1)$$

$$= \frac{dg}{du}(x - c).$$

It is practical to implement the chain rule directly in a specific case, as in the following example.

Example 4 Determine

$$\frac{d}{dx}(x - 4)^{2/3}.$$

Solution

If we set $u(x) = x - 4$,

$$\frac{d}{dx}(x - 4)^{2/3} = \frac{d}{dx}(u(x))^{2/3} = \left(\left.\frac{d}{du}\left(u^{2/3}\right)\right|_{u=x-4}\right)\left(\frac{d}{dx}(x - 4)\right)$$

$$= \left(\left.\frac{2}{3}u^{-1/3}\right|_{u=x-4}\right)(1)$$

$$= \frac{2}{3(x - 4)^{1/3}}.$$

$\square$

We will come across many functions of the form $g(\omega x)$, where ω is a constant. If we set $u(x) = \omega x$,

$$\frac{d}{dx}g(\omega x) = \frac{d}{dx}g(u(x)) = \left(\left.\frac{dg}{du}\right|_{u=\omega x}\right)\left(\frac{d}{dx}(\omega x)\right)$$

$$= g'(\omega x)(\omega) = \omega g'(\omega x).$$

Again, it is practical to implement the chain rule directly in a specific case, as in the following example.

Example 5 Let ω be an arbitrary constant. then

$$\frac{d}{dx}\sin(\omega x) = \omega\cos(\omega x) \ \text{ and } \ \frac{d}{dx}\cos(\omega x) = -\omega\sin(\omega x).$$

We can derive these formulas with the help of the chain rule:

$$\frac{d}{dx}\sin(\omega x) = \left(\left.\frac{d}{du}\sin(u)\right|_{u=\omega x}\right)\left(\frac{d}{dx}(\omega x)\right) = (\cos(u)|_{u=\omega x})(\omega)$$

$$= \omega\cos(\omega x).$$

Similarly,

$$\frac{d}{dx}\cos(\omega x) = \left(\left.\frac{d}{du}\cos(u)\right|_{u=\omega x}\right)\left(\frac{d}{dx}(\omega x)\right) = (-\sin(u)|_{u=\omega x})(\omega)$$

$$= -\omega\sin(\omega x).$$

□

If y is the dependent variable of f, and we refer to $f(u)$ as $y(u)$, then the expression

$$\frac{d}{dx} f(u(x)) = \left(\frac{df}{du} \bigg|_{u=u(x)} \right) \frac{du}{dx}$$

reads

$$\frac{d}{dx} y(u(x)) = \left(\frac{dy}{du} \bigg|_{u=u(x)} \right) \frac{du}{dx}.$$

We can simply write

$$\frac{dy}{dx} = \frac{dy}{du} \frac{du}{dx},$$

with the understanding that the letter y on the left-hand side refers to $y(u(x))$, and dy/du is evaluated at $u(x)$. This somewhat imprecise expression for the chain rule is appealing due to its "symbolic correctness": If we pretend that we are dealing with genuine fractions, and not just symbolic fractions, the cancellation of du on the right-hand side of the expression yields dy/dx. Aside from its "symbolic correctness", an appealing feature of the above expression is its interpretation in terms of rates of change. Indeed, dy/dx is the rate of change of y with respect to x, dy/du is the rate of change of y with respect to u (at $u(x)$), and du/dx is the rate of change of u with respect to x. Therefore, we can read the chain rule as follows:

The rate of change of y with respect to x

$=$ (the rate of change of y with respect to u)

$\times$ (the rate of change of u with respect to x).

Remark 2 (Another Plausibility Argument for the Chain Rule)

Let's set $u = u(x)$, $\Delta u = u(x + \Delta x) - u(x)$, and $\Delta y = y(u(x + \Delta x)) - y(u(x))$ so that $\Delta y = y(u + \Delta u) - y(u)$. If we assume that $\Delta x \neq 0$ and $\Delta u \neq 0$,

$$\frac{\Delta y}{\Delta x} = \frac{\Delta y}{\Delta u} \frac{\Delta u}{\Delta x}.$$

We can read the above equality as follows:

The average rate of change of y with respect to x

$=$ (the average rate of change of y with respect to u)

$\times$ (the average rate of change of u with respect to x).

We have

$$\frac{dy}{dx} = \lim_{\Delta x \to 0} \frac{\Delta y}{\Delta x} = \lim_{\Delta x \to 0} \left(\frac{\Delta y}{\Delta u} \frac{\Delta u}{\Delta x} \right) = \left(\lim_{\Delta x \to 0} \frac{\Delta y}{\Delta u} \right) \left(\lim_{\Delta x \to 0} \frac{\Delta u}{\Delta x} \right),$$

assuming that $\Delta u \neq 0$ if $\Delta x \neq 0$. Since $\Delta u = u(x + \Delta x) - u(x)$ approaches 0 as Δx approaches 0 (differentiability implies continuity),

$$\frac{dy}{dx} = \left(\lim_{\Delta x \to 0} \frac{\Delta y}{\Delta u} \right) \left(\lim_{\Delta x \to 0} \frac{\Delta u}{\Delta x} \right) = \left(\lim_{\Delta u \to 0} \frac{\Delta y}{\Delta u} \right) \left(\lim_{\Delta x \to 0} \frac{\Delta u}{\Delta x} \right) = \frac{dy}{du} \frac{du}{dx}.$$

Thus, we can consider the chain rule to be the limiting case of an obvious fact about average rates of change. This plausibility argument does not lead to a rigorous proof, as in the case of the plausibility argument that relied on differentials, since we may have $\Delta u = u(x + \Delta x) - u(x) = 0$ even if $\Delta x \neq 0$. ◊

Example 6 Let $f(x) = \sin^{2/3}(x)$. Determine $f'(x)$.

Solution

We set $f(x) = y(x) = (\sin(x))^{2/3}$ and $u = \sin(x)$, so that $y(u) = u^{2/3}$. By the chain rule,

$$f'(x) = \frac{dy}{dx} = \frac{dy}{du}\frac{du}{dx} = \left(\frac{d}{du}u^{2/3}\right)\left(\frac{d}{dx}\sin(x)\right)$$

$$= \left(\frac{2}{3}u^{-1/3}\right)\cos(x) = \frac{2}{3}(\sin(x))^{-1/3}\cos(x) = \frac{2\cos(x)}{3\sin^{1/3}(x)}.$$

Therefore,

$$f'(x) = \frac{2\cos(x)}{3\sin^{1/3}(x)}$$

if $\sin(x) \neq 0$.

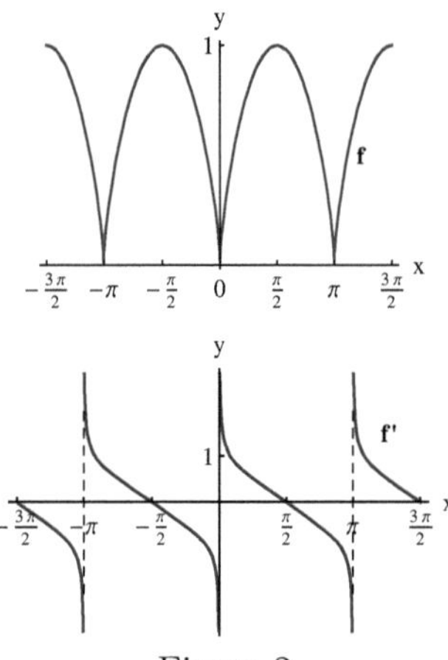

Figure 2

Figure 2 shows the graphs of f and f' on the interval $[-3\pi/2, 3\pi/2]$. Note that the graph of f has cusps at $-\pi, 0$ and π (Definition 2 of Section 2.2) and the graph of f' has vertical asymptotes at these points. For example,

$$\lim_{x \to 0-} \frac{2}{3}\cos(x) = \frac{2}{3} > 0,$$

and

$$\lim_{x \to 0-} \frac{1}{\sin^{1/3}(x)} = -\infty$$

since $\sin^{1/3}(x) < 0$ if $-\pi/2 < x < 0$ and

$$\lim_{x \to 0} \sin^{1/3}(x) = 0.$$

Therefore,

$$\lim_{x \to 0-} f'(x) = \lim_{x \to 0-}\left(\frac{2}{3}\cos(x)\right)\left(\frac{1}{\sin^{1/3}(x)}\right) = -\infty.$$

Similarly,

$$\lim_{x \to 0+} f'(x) = +\infty.$$

$\square$

Remark 3 As in Example 6, if a function is of the form $u^r(x)$, where r is a rational exponent, we can apply the chain rule to evaluate its derivative. Indeed, if we set $y = u^r(x) = (u(x))^r$ and $u = u(x)$, then $y = u^r$. By the chain rule and the power rule,

$$\frac{d}{dx}u^r(x) = \frac{dy}{dx} = \frac{dy}{du}\frac{du}{dx}$$
$$= \left(\frac{d}{du}u^r\right)\frac{du}{dx} = \left(ru^{r-1}\right)\frac{du}{dx} = ru^r(x)\frac{du}{dx}.$$

Thus,

$$\frac{d}{dx}u^r(x) = ru^{r-1}\frac{du}{dx}.$$

Since the above expression reduces to the power rule if $u(x) = x$, it may be referred to as **the function-power rule**. The implementation of the function-power rule is slightly faster than the direct implementation of the chain rule, and the rule is easy to remember due to the similarity with the ordinary power rule (don't neglect du/dx, though). $\Diamond$

Example 7 Determine

$$\frac{d}{dx}\cos^{10}(x).$$

Solution

By the function-power rule:

$$\frac{d}{dx}\cos^{10}(x) = 10\cos^9(x)\left(\frac{d}{dx}\cos(x)\right) = 10\cos^9(x)(-\sin(x)) = -10\cos^9(x)\sin(x).$$

The direct implementation of the chain rule is not much slower: Set $y(x) = (\cos(x))^{10}$ and $u = \cos(x)$ so that $y(u) = u^{10}$. By the chain rule and the power rule,

$$\frac{d}{dx}\cos^{10}(x) = \frac{dy}{dx} = \frac{dy}{du}\frac{du}{dx}$$
$$= \left(\frac{d}{dx}u^{10}\right)\left(\frac{d}{dx}\cos(x)\right) = (10u^9)(-\sin(x)) = -10\cos^9(x)\sin(x).$$

$\square$

The Chain Rule for more than two Functions

The chain rule can be extended to cover cases that involve the composition of more than two functions: For example, if $F = f \circ g \circ h$, then

$$F(x) = (f \circ g)(h(x)),$$

so that

$$F'(x) = (f \circ g)'(h(x))h'(x) = f'(g(h(x)))g'(h(x))h'(x).$$

The following schematic description of the composition should make it easier to remember where to evaluate the derivatives:

$$x \to h(x) \to g(h(x)) \to f(g(h(x)))$$

The expression of the chain rule in "the prime notation" is somewhat unwieldy when the composition of more than two functions is involved. We may refer to the functions with the symbols that denote their dependent variables, and use the Leibniz notation: If we set $y = y(u(v(x)))$, then

$$\frac{dy}{dx} = \frac{dy}{du}\frac{du}{dx} = \frac{dy}{du}\left(\frac{du}{dv}\frac{dv}{dx}\right),$$

so that

$$\frac{dy}{dx} = \frac{dy}{du}\frac{du}{dv}\frac{dv}{dx}.$$

The symbolic cancellations are helpful in checking that we are on the right track. Note that dy/du is evaluated at $u(v(x))$ and du/dv is evaluated at $v(x)$.

Example 8 Determine

$$\frac{d}{dx}\sqrt{\cos\left(\frac{1}{x}\right)}.$$

Solution

We set

$$y = \sqrt{\cos\left(\frac{1}{x}\right)}, \quad u = \cos\left(\frac{1}{x}\right) \text{ and } v = \frac{1}{x},$$

so that

$$y = \sqrt{u} \text{ and } u = \cos(v).$$

By the chain rule,

$$\frac{dy}{dx} = \frac{dy}{du}\frac{du}{dv}\frac{dv}{dx} = \left(\frac{d}{du}\sqrt{u}\right)\left(\frac{d}{dv}\cos(v)\right)\left(\frac{d}{dx}\left(x^{-1}\right)\right)$$

$$= \left(\frac{1}{2\sqrt{u}}\right)(-\sin(v))\left(-x^{-2}\right) = \frac{\sin\left(\frac{1}{x}\right)}{2x^2\sqrt{\cos\left(\frac{1}{x}\right)}}$$

The expression is valid if $x \neq 0$ and $\cos(1/x) > 0$. $\square$

The Proof of the Chain Rule

We set $u = g(x)$ and $\Delta u = g(x + \Delta x) - g(x)$, so that $g(x + \Delta x) = u + \Delta u$. Then,

$$(f \circ g)(x + \Delta x) - (f \circ g)(x) = f(g(x + \Delta x)) - f(g(x)) = f(u + \Delta u) - f(u).$$

As in Theorem 2 of Section 2.5,

$$f(u + \Delta u) - f(u) = f'(u)\Delta u + \Delta u q(\Delta u),$$

where

$$\lim_{\Delta u \to 0} q(\Delta u) = 0.$$

Therefore,

$$\frac{f(g(x + \Delta x)) - f(g(x))}{\Delta x} = \frac{f(u + \Delta u) - f(u)}{\Delta x}$$

$$= \frac{f'(u)\Delta u + \Delta u q(\Delta u)}{\Delta x}$$

$$= \frac{f'(g(x))\Delta u + \Delta u q(\Delta u)}{\Delta x}$$

$$= f'(g(x))\frac{\Delta u}{\Delta x} + \frac{\Delta u}{\Delta x}q(\Delta u),$$

where $\Delta x \neq 0$. We have

$$\lim_{\Delta x \to 0} \frac{\Delta u}{\Delta x} = \lim_{\Delta x \to 0} \frac{g\left(x + \Delta x\right) - g\left(x\right)}{\Delta x} = g'\left(x\right).$$

Since g is differentiable at x, it is continuous at x. Thus,

$$\lim_{\Delta x \to 0} \Delta u = \lim_{\Delta x \to 0} \left(g\left(x + \Delta x\right) - g\left(x\right)\right) = 0.$$

Therefore,

$$\lim_{\Delta x \to 0} q\left(\Delta u\right) = 0.$$

Thus,

$$\begin{aligned}
\left(f \circ g\right)'\left(x\right) &= \lim_{\Delta x \to 0} \frac{f\left(g\left(x + \Delta x\right)\right) - f\left(g\left(x\right)\right)}{\Delta x} \\
&= \lim_{\Delta x \to 0} \left(f'\left(g(x)\right)\frac{\Delta u}{\Delta x} + \frac{\Delta u}{\Delta x}q\left(\Delta u\right)\right) \\
&= f'\left(g\left(x\right)\right)\lim_{\Delta x \to 0}\frac{\Delta u}{\Delta x} + \left(\lim_{\Delta x \to 0}\frac{\Delta u}{\Delta x}\right)\left(\lim_{\Delta x \to 0}q\left(\Delta u\right)\right) \\
&= f'\left(g(x)\right)g'\left(x\right) + g'\left(x\right)\left(0\right) \\
&= f'\left(g\left(x\right)\right)g'\left(x\right).
\end{aligned}$$

∎

Problems

In problems 1-23, compute $f'(x)$ (It will be practical to use the Leibniz notation):

1.
$$f\left(x\right) = \sqrt{x^2 - 2x + 5}$$

2.
$$f(x) = x + \sqrt{x^2 + 4}$$

3.
$$f(x) = \left(x^2 - 16\right)^{2/3}$$

4.
$$f(x) = \frac{1}{\sqrt{x^4 + 9}}.$$

5.
$$f(x) = \sqrt{\frac{4 + x^2}{4 - x^2}}.$$

6.
$$f\left(x\right) = \left(x^2 - 4x + 8\right)^{2/3}$$

7.
$$f\left(x\right) = \sin\left(10x\right)$$

8.
$$f\left(x\right) = \cos\left(\frac{x}{4}\right)$$

9.
$$f\left(x\right) = \sin\left(\pi x\right)$$

10.
$$f\left(x\right) = \cos(x) + \frac{1}{9}\cos(3x) + \frac{1}{25}\cos(5x)$$

11.
$$f\left(x\right) = \sin(\pi x) - \frac{1}{2}\sin(2\pi x) + \frac{1}{3}\sin(3\pi x)$$

12.
$$f\left(x\right) = 10\cos\left(\frac{x}{4} - 1\right)$$

13.
$$f\left(x\right) = \frac{1}{4}\sin\left(\frac{1}{2}x + \frac{\pi}{6}\right)$$

14.
$$f(x) = \tan(2x)$$

15.
$$f(x) = \cos(x^2).$$

16.
$$f(x) = \cos(1/x)$$

17.
$$f(x) = \sin(\sqrt{x})$$

18.
$$f(x) = \sin^2(3x)$$

19.
$$f(x) = \sqrt{\sin(x/2)}$$

20.
$$f(x) = \cos\left(\sqrt{x^2 + 1}\right)$$

21.
$$f(x) = \sin^2\left(\frac{1}{x}\right)$$

22.
$$f(x) = \sqrt{4 - \cos^3(2x)}$$

23.
$$f(x) = \sqrt{\tan(x^2)}$$

In problems 24-26, compute $f'(x)$ and $f''(x)$:

24.
$$f(x) = \sin(4x)$$

25.
$$f(x) = \cos(1/x).$$

26.
$$f(x) = \sin^2(6x)$$

27. Let
$$f(x) = \frac{1}{\sqrt{x^2 + 16}}.$$

a) Determine L_3, the linear approximation to f based at 3,
b) Make use of L_3 to approximate $f(2.8)$.

28. Let
$$f(x) = \sin^2\left(x^2\right).$$

a) Determine the differential of f.
b) Make use of the differential of f in order to approximate $f\left(\sqrt{\pi/4} + 0.1\right)$

In problems 29 and 30,
a) Compute $f'(x)$, determine the fundamental period p of f, and specify the part of the domains of f and f' in the interval $[-p/2, p/2]$,
b) Determine whether the graph of f has vertical tangents or cusps on the interval $[-p/2, p/2]$,
c) [C] Make use of your graphing utility to plot the graphs of f and f' on $[-p/2, p/2]$ Are the pictures consistent with your response to part b)?

29.
$$f(x) = \cos^{2/3}(x)$$

30.
$$f(x) = \sqrt{\tan(x)}$$

The motion of an oscillating object such as a mass that is attached to a spring can be expresed by a position function of the form

$$y(t) = A\cos(\omega t - \alpha),$$

where t represents time, $A > 0$ and α are constants (friction forces are neglected). We say that the object is in **simple harmonic motion**. The motion has **period**

$$T = \frac{2\pi}{\omega}.$$

The **frequency** of the motion is the reciprocal of its period, i.e.,

$$\frac{1}{T} = \frac{\omega}{2\pi}.$$

Since $|A\cos(\omega t - \alpha)| = A|\cos(\omega t - \alpha)| \leq A$, the maximum distance of the object from the equilibrium position is A. We refer to A as **the amplitude** of the simple harmonic motion.

We can express $y(t)$ as

$$y(t) = A\cos(\omega t - \alpha) = A\cos\left(\omega\left(t - \frac{\alpha}{\omega}\right)\right).$$

Thus, the graph of $y(t)$ can be obtained by stretching or shrinking the graph of $A\cos(t)$ horizontally by a factor of ω, followed by a horizontal shift to the right or to the left by $|\alpha/\omega|$.

In problems 31 and 32, $y(t)$ is the position at time t of an object in simple harmonic motion.

a) Determine $v(t)$, the velocity of the object at time t, and $a(t)$, the acceleration of the object at time t,

b) Show that the acceleration function is a constant multiple of the position function.

c) Determine the fundamental period p and the amplitude of the simple harmonic motion.

d) [C] Make use of your graphing utility to plot the position, velocity and acceleration functions on the interval $[0, p]$.

31.
$$y(t) = 4\sin\left(6t - \frac{\pi}{4}\right)$$

32.
$$y(t) = 10\cos\left(\frac{1}{2}t - \frac{\pi}{6}\right)$$

2.8 Related Rate Problems

In this section we will look at some problems that involve the rates of change of certain quantities that are related to each other. The chain rule enables us to relate the rates of change of such quantities.

Example 1 Assume that a pebble that is dropped on the surface of a pond creates an expanding ripple that is always a perfect circle centered at the point where the pebble has been dropped. Assume that the radius of the circle is increasing at the rate of 20 centimeters per second. Determine the rate at which the area enclosed by the circular ripple is increasing at the instant the radius is 3 meters.

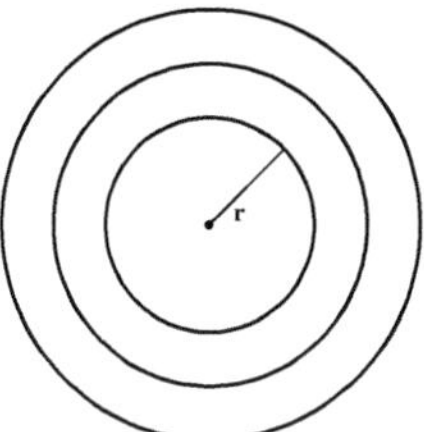

Figure 1: Expanding circular ripples

Solution

If we denote the radius of the circle by r and the area enclosed by the circle by A, we have $A = \pi r^2$. Thus, A is a function of r. We will measure r in meters (m), so that A is measured in m^2, and we will measure time t in seconds. The radius r is a function of t, so that $r = r(t)$, and $A = A(r(t))$. We are given that the rate of change of r with respect to t is 0.2 meters per second. Thus,

$$\frac{dr}{dt} = 0.2$$

(meters/second). The rate of change of A with respect to t is dA/dt. By the chain rule,

$$\frac{dA}{dt} = \frac{dA}{dr}\frac{dr}{dt} = \frac{dA}{dr}(0.2) = 0.2\frac{dA}{dr}.$$

We have

$$\frac{dA}{dr} = \frac{d}{dr}\left(\pi r^2\right) = 2\pi r$$

(note that this is the length of the circle of radius r). Therefore,

$$\frac{dA}{dt} = 0.2\,(2\pi r) = 0.4\pi r.$$

At the instant $r = 3$ meters,

$$\frac{dA}{dt} = 0.4\pi r\big|_{r=3} = 0.4\pi\,(3) = 1.2\pi \cong 3.77$$

$(m^2/\sec)$. $\square$

Example 2 Assume that helium is being pumped into a spherical balloon at a constant rate of 100 cubic centimeters per second. Also assume that the shape of the balloon is a perfect sphere as it is being inflated. Determine the rate at which the radius of the balloon is increasing at the instant its radius is 10 centimeters.

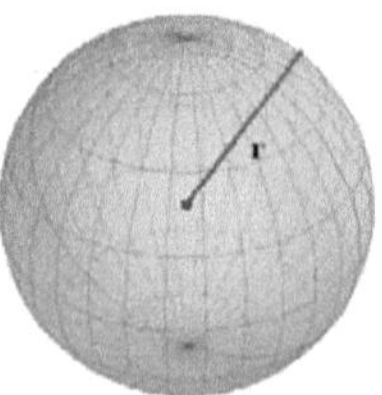

Figure 2

Solution

Let $r(t)$ denote the radius (in centimeters) and let $V(t)$ denote the volume (in cubic centimeters) of the balloon at time t (in seconds). Thus,

$$V(t) = \frac{4}{3}\pi r^3(t).$$

By the chain rule,

$$\frac{dV}{dt} = \frac{dV}{dr}\frac{dr}{dt},$$

i.e.,

the rate of change of V with respect to t

$=$ (the rate of change of V with respect to r)

$\times$ (rate of change of r with respect to t).

Thus,

$$\frac{dV}{dt} = \left(\frac{d}{dr} \left(\frac{4}{3} \pi r^3 \right) \right) \frac{dr}{dt} = 4\pi r^2 \frac{dr}{dt}.$$

(note that the rate of change of the volume inside a sphere with respect to the radius is the area of the sphere). We are given the information that gas is being pumped into the balloon at the rate of 100 cm^3/second. Thus, the rate of change of V with respect to time is 100, i.e., $dV/dt = 100$. Therefore,

$$100 = 4\pi r^2 \frac{dr}{dt}$$

Thus, the rate of change of the radius with respect to time can be expressed as

$$\frac{dr}{dt} = \frac{100}{4\pi r^2}.$$

This expression enables us to determine the rate of change of the radius with respect to time at the instant the radius is 10:

$$\left. \frac{dr}{dt} \right|_{r=10} = \frac{100}{4\pi \left(10\right)^2} = \frac{1}{4\pi} \cong 0.08 \ (\text{cm/second}).$$

$\square$

Example 3 An airplane is flying at an altitude of 2 miles with a speed of 200 miles/hour. It is being tracked by an observer on the ground with a searchlight. Find the rate at which the angle between the searchlight and the vertical direction changes at the instant the horizontal distance of the plane from the observer is 10 miles.

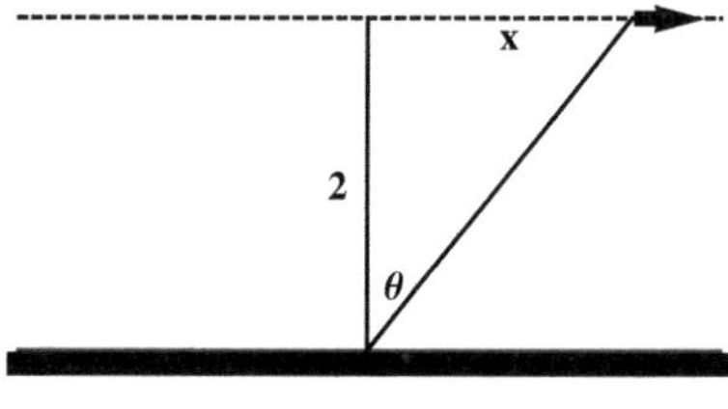

Figure 3

Solution

With reference Figure 3,

$$\tan\left(\theta\right) = \frac{x}{2}.$$

Both the "angle of elevation" θ (in radians), and the horizontal distance from the observer, x (in miles), are functions of time t (in hours). Since the above relationship is an identity, the derivatives of the functions represented by either side of the equality are the same:

$$\frac{d}{dt} \tan\left(\theta\right) = \frac{d}{dt} \left(\frac{x}{2} \right) = \frac{1}{2} \frac{dx}{dt}.$$

By the chain rule,

$$\frac{d}{dt} \tan\left(\theta\right) = \left(\frac{d}{d\theta} \tan\left(\theta\right) \right) \frac{d\theta}{dt} = \sec^2\left(\theta\right) \frac{d\theta}{dt} = \frac{1}{\cos^2\left(\theta\right)} \frac{d\theta}{dt}.$$

Therefore,

$$\frac{1}{\cos^2(\theta)}\frac{d\theta}{dt} = \frac{1}{2}\frac{dx}{dt}.$$

We have established a relationship between the rate of change of the angle of elevation, $\frac{d\theta}{dt}$, and the rate of change of the horizontal distance, $\frac{dx}{dt}$. Since we are given the information that the plane is traveling with a speed of 200 mi/hr, we have

$$\frac{dx}{dt} = 200.$$

Therefore,

$$\frac{d\theta}{dt} = \cos^2(\theta)\left(\frac{1}{2}\frac{dx}{dt}\right) = 100\cos^2(\theta).$$

We are asked to compute $\frac{d\theta}{dt}$ at the instant $x = 10$. At that instant,

$$\cos(\theta) = \frac{2}{\sqrt{2^2 + 10^2}} = \frac{2}{\sqrt{104}}.$$

Therefore, the rate of change of the angle of elevation at that instant is

$$\left.\frac{d\theta}{dt}\right|_{\cos(\theta)=2/\sqrt{104}} = 100\left(\frac{2}{\sqrt{104}}\right)^2 = \frac{400}{104} \cong 3.846 \text{ (radians/hr)}.$$

$\square$

Example 4 Assume that a ladder which is 10 feet long is leaning against a wall and its base is sliding away from the wall at the rate of 2 ft/sec. Determine the rate at which the top of the ladder is sliding down the wall at the instant the base of the ladder is 4 feet from the wall.

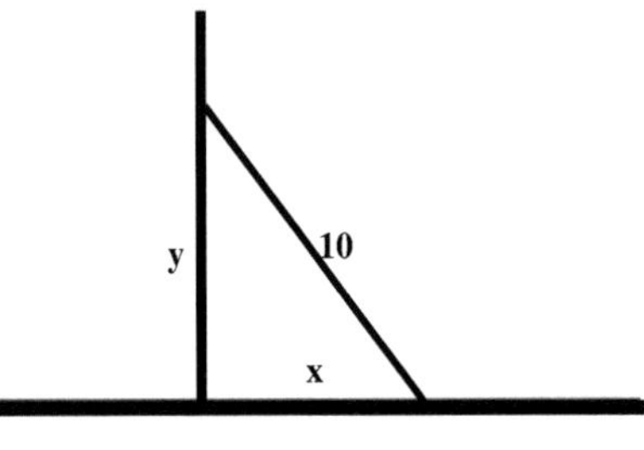

Figure 4

Solution

With reference to Figure 4, we have

$$x^2 + y^2 = 10^2 = 100,$$

by Pythagoras. Both the height of the top of the ladder, y (in feet) and the distance of its base from the wall, x (in feet), are functions of time t (in seconds). Since the above relationship is an identity, we have

$$\frac{d}{dt}(x^2 + y^2) = \frac{d}{dt}(100) = 0.$$

By the chain rule,

$$\frac{d}{dt}\left(x^2 + y^2\right) = 2x\frac{dx}{dt} + 2y\frac{dy}{dt} = 0.$$

We are given that the rate at which the bottom of the ladder is sliding away from the wall is 2 ft/sec. Thus, $dx/dt = 2$. Therefore,

$$4x + 2y\frac{dy}{dt} = 0.$$

We are asked to compute the rate at which the height of the top of the ladder is changing at the instant the bottom of the ladder is 4 feet from the wall, i.e., at the instant $x = 4$. At that instant, $4^2 + y^2 = 10^2$, so that $y = \sqrt{84}$. Therefore,

$$4(4) + 2\left(\sqrt{84}\right)\frac{dy}{dt} = 0.$$

Thus,

$$\frac{dy}{dt} = -\frac{16}{2\sqrt{84}} \cong -0.873 \ (\text{ft/sec}).$$

The $(-)$ sign corresponds to the fact that y is decreasing as the ladder is sliding down the wall. Thus, the top of the ladder is sliding down the wall at the rate of

$$\frac{16}{2\sqrt{84}} \ (\text{ft/sec})$$

at the instant the bottom of the ladder is 4 feet from the wall. $\square$

Example 5 Assume that an athlete who is running on a circular track of radius 100 meters runs at a constant speed and makes one revolution in 2 minutes. Assume that his trainer is at a point that is 50 meters from the center. With reference to Figure 5, how fast is the distance s from the trainer to the athlete changing at the instant θ is $\pi/3$ (radians)?

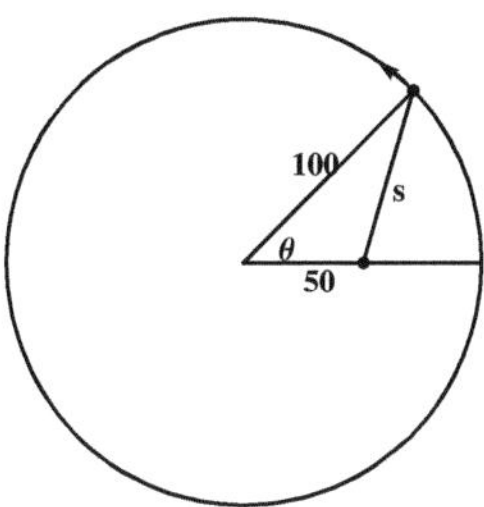

Figure 5

Solution

By the law cosines,

$$s^2 = 100^2 + 50^2 - 2\,(100)\,(50)\cos\left(\theta\right) = 10000 + 2500 - 10000\cos\left(\theta\right).$$

Therefore,

$$\frac{d}{dt}s^2 = -10000\frac{d}{dt}\cos\left(\theta\right).$$

By the chain rule,

$$2s\frac{ds}{dt} = 10000\sin\left(\theta\right)\frac{d\theta}{dt}.$$

Since the athlete makes one revolution in 2 minutes, θ changes at the constant rate of $2\pi/2 = \pi$ radians per minute. Thus, $d\theta/dt = \pi$. Therefore,

$$2s\frac{ds}{dt} = 10000\sin\left(\theta\right)\left(\pi\right) = 10000\pi\sin\left(\theta\right).$$

At the instant $\theta = \pi/3$,

$$2s\frac{ds}{dt} = 10000\pi\left(\frac{\sqrt{3}}{2}\right).$$

At that instant,

$$s^2 = 10000 + 2500 - 10000\cos\left(\frac{\pi}{3}\right) = 12500 - 10000\left(\frac{1}{2}\right) = 7500,$$

so that $s = \sqrt{7500}$. Therefore,

$$2s\frac{ds}{dt} = 10000\pi\left(\frac{\sqrt{3}}{2}\right) \Rightarrow 2\sqrt{7500}\frac{ds}{dt} = 5000\sqrt{3}\pi$$

$$\Rightarrow \frac{ds}{dt} = \frac{5000\sqrt{3}\pi}{2\sqrt{7500}} \cong 157 \text{ (meters per minute)}$$

at the instant $\theta = \pi/3$. $\square$

Problems

In problems 1-4, the variables x and y are functions of time t, so that $x = x\left(t\right)$ and $y = y\left(t\right)$. Use the given conditions to determine $\dfrac{dy}{dt}$.

1.

$$y = 4x^2, \ \frac{dx}{dt} = 3, \ x = 2$$

2.

$$x = y^3, \ \frac{dx}{dt} = -2, \ x = 8$$

3.

$$x = 10\sin\left(y\right), \ \frac{dx}{dt} = 4, \ y = \pi/6$$

4.

$$x^2 - y^2 = 1, \ \frac{dx}{dt} = 2, \ y = 3, \ x = \sqrt{10}.$$

5. Assume that an oil slick on the surface of the sea is expanding as a perfect disk with fixed center, and that the rate of change of its area is 100 m^2/hour. Determine the rate at which the radius of the oil slick is increasing at the instant its area is 1000 m^2.

6. Each side of a square is increasing at the rate of 10 cm/second. Determine the rate at which its area is increasing at the instant the length of each side is 100 cm.

7. Assume that the surface area of a sphere is decreasing at the rate of 10 cm^2/second. Determine the rate at which its radius is decreasing at the instant the radius is 5 cm.

8. A man who is 6 ft. tall walks away from a street lamp that is 20 ft. tall at a constant rate of 4 ft. per second. At what rate is the tip of his shadow moving away from the base of the street lamp?

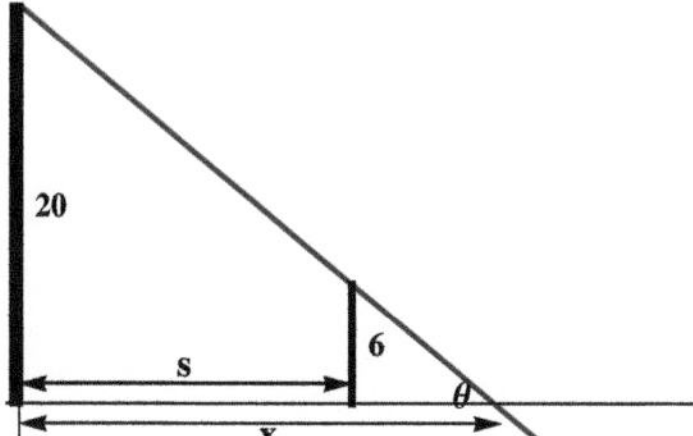

9. An oil tank in the shape of a right circular cylinder of radius 6 meters is being filled at a constant rate of 8 m^3/min. How fast is the level of the oil rising?

10. An east-west highway and a north-south highway intersect at a certain point. One car, traveling due north at 60 mph, crosses the intersection at 10 a.m. Another car, traveling due east at 70 mph, crosses the intersection at 11 a.m. Assuming that the cars maintain their respective directions and speeds, determine the rate at which the distance between the two cars is increasing at 3 p.m. that afternoon.

11. An airplane is flying at an altitude of 5 miles with a speed of 300 miles/hour. It is being tracked by an observer on the ground with a searchlight. Find the rate at which the distance between the observer and the airplane is increasing at the instant the horizontal distance of the plane from the observer is 20 miles.

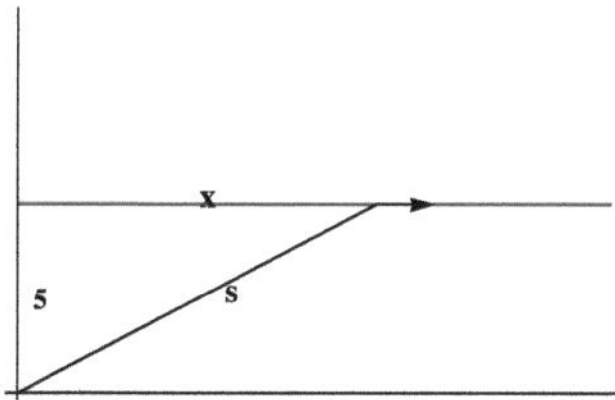

12. An observer is tracking the vertical lift-off of a rocket from a horizontal distance of 2 km. The rocket is climbing at a rate of 400 km/min. Determine the rate of change of the angle between the ground level and the observer's line of sight at the instant the rocket is at a height of 5 km.

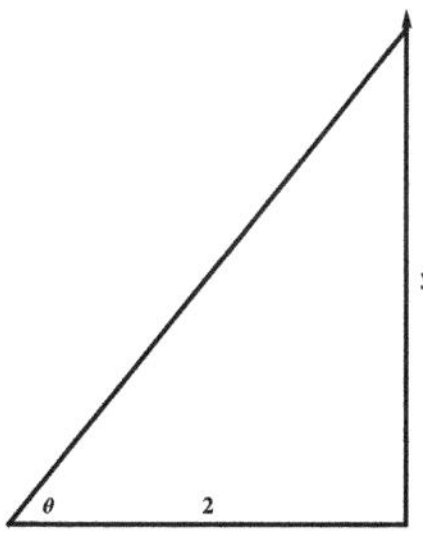

13 .Assume that a ladder which is 10 feet long is leaning against a wall and its base is sliding away from the wall at the rate of 2 ft/sec. Determine the rate at which the angle between the ground level and the ladder is changing at the instant the base of the ladder is 4 feet from the wall.

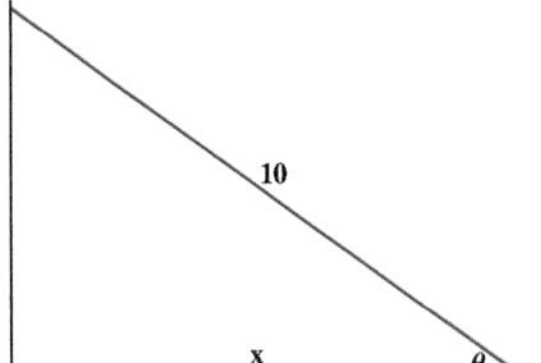

14. A conical tank has a depth of 10 m and a radius at the top of 3 m. If water is filling the tank at the rate of 8 m^3/ min, determine the rate at which the radius of the surface of the water is increasing at the instant when the depth of the water reaches 2 m. How fast is the water level rising at that instant?

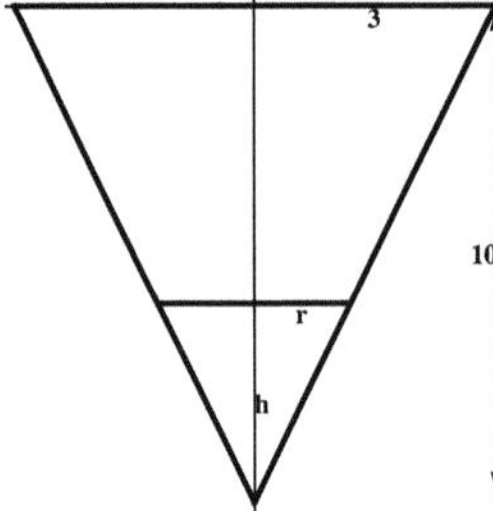

2.9   The Intermediate Value Theorem and Newton's Method

In some cases, we can compute the exact solutions of equations on our own, or with the help of a computer algebra system. In many cases, we can obtain only approximate solutions with the help of a computational utility. **The Intermediate Value Theorem** for continuous functions guarantees the existence of solutions in certain intervals. **Newton's method** is the basis of many professional equation solvers, such as the one on your calculator.

The Intermediate Value Theorem

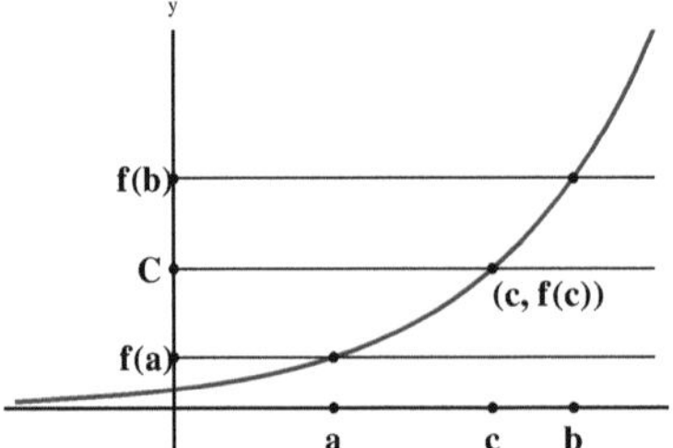

Figure 1: A continuous function f attains all values between $f(a)$ and $f(b)$

If the function f is continuous on the interval $[a, b]$, the graph of f on the interval $[a, b]$ is a "continuous curve" without any gaps. Therefore, if $f(a) \neq f(b)$ and C is a number between

$f(a)$ and $f(b)$, the line $y = C$ should intersect the graph of f at some point (c, C), where c is between a and b, as in Figure 1. That is the graphical counterpart of the following theorem:

Theorem 1 (The Intermediate Value Theorem) Assume that f is continuous on the interval $[a, b]$, $f(a) \neq f(b)$, and that C is a number (strictly) between $f(a)$ and $f(b)$. Then, there exists $c \in (a, b)$ such that $f(c) = C$.

The Intermediate Value Theorem predicts the existence of at least one solution of the equation $f(x) = C$ between a and b if C is an "intermediate value" between the values of f at the endpoints a and b, provided that f is continuous on $[a, b]$. In particular, if $f(a)$ and $f(b)$ have different signs, the equation $f(x) = 0$ must have a solution between a and b.

Corollary to the Intermediate Value Theorem   Assume that f is continuous in the interval $[a, b]$ and that $f(a)$ and $f(b)$ have opposite signs. Then, there exists $c \in (a, b)$ such that $f(c) = 0$.

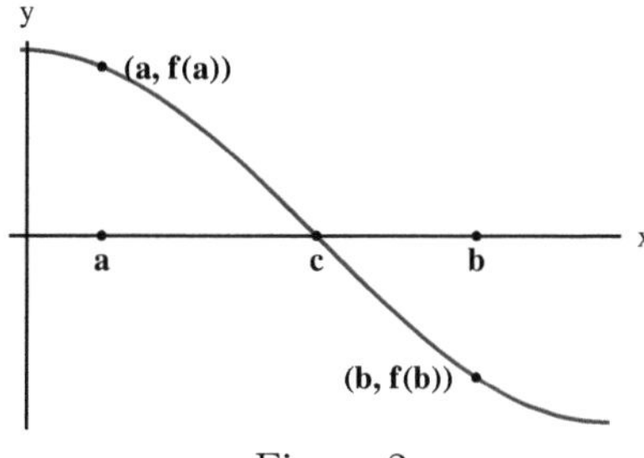

Figure 2

Figure 2 illustrates the graphical meaning of the corollary: The graph of f must intersect the x-axis at some point between a and b if $f(a)$ and $f(b)$ have different signs and f is continuous on $[a, b]$ (there may be several such points).

We leave the proof of the Intermediate Value Theorem to a course in advanced calculus.

Note that the equation $f(x) = 0$ need not have a solution if f has a discontinuity, as in the following case:

Let

$$f(x) = \begin{cases} x - 2 & \text{if} \quad x \leq 1, \\ x & \text{if} \quad x > 1. \end{cases}$$

The graph of f is displayed in Figure 3. We have $f(0) = -2 < 0$, and $f(2) = 2 > 0$, but there is no solution of the equation $f(x) = 0$ in the interval $(0, 2)$.

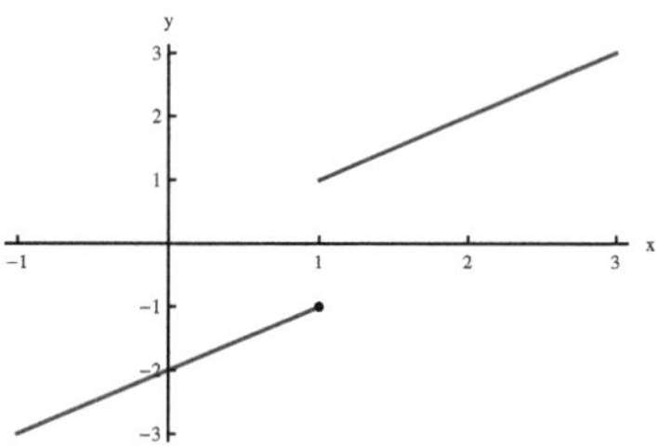

Figure 3

If $f(x)$ is a quadratic polynomial we have the quadratic formula for the solutions of the equation $f(x) = 0$. If $f(x)$ is a polynomial of degree 3 or 4, there are formulas for the solutions, even though they are not as user-friendly and popular as the quadratic formula. A computer algebra system can give you the exact values of the solutions based on such formulas. There are no general formulas for the solution of $f(x) = 0$, complicated or otherwise, if $f(x)$ is a polynomial of degree higher than 4. In general, when we are faced with a high degree polynomial, we will rely on the approximate equation solver of our computational utility, as in the following example, unless some of the solutions can be determined by inspection:

Example 1 Let

$$f(x) = x^5 + 4x^2 - 6x - 3.$$

a) Plot the graph of f with the help of your calculator. Does the picture indicate that the equation $f(x) = 0$ has a solution between 1 and 2?

b) Show that the equation $f(x) = 0$ has a solution r in the interval $(1, 2)$. Find an approximation to r with the help of your calculator.

Solution

a) Figure 4 indicates that the graph of f intersects the x-axis at a single point between 1 and 2. Therefore, there must be one solution of the equation $f(x) = 0$ in the interval $(1, 2)$.

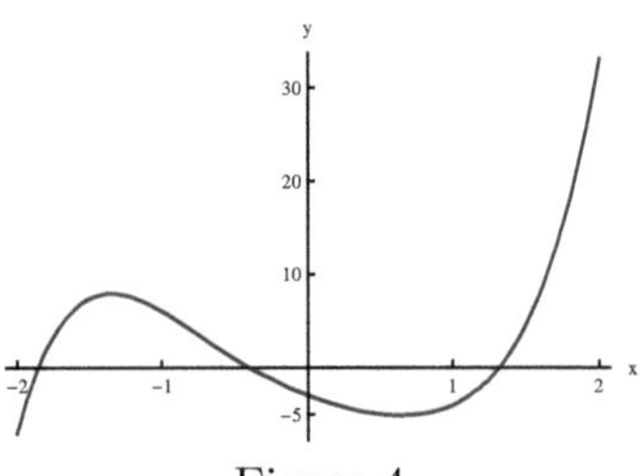

Figure 4

b) We have $f(1) = -4 < 0$ and $f(2) = 33 > 0$, and f is continuous on $[1, 2]$. By the Corollary to the Intermediate Value Theorem, there must be a solution of the equation $f(x) = 0$ in the interval $(1, 2)$. That solution is approximately 1.3171, rounded to 6 significant digits. $\square$

Example 2

a) Plot the graph of $y = \cos(x)$ and the line $y = 0.4$ on the interval $[0, 2\pi]$ with the help of your calculator. Show that the equation $\cos(x) = 0.4$ has two solutions in the interval $[0, 2\pi]$.

b) Compute approximations to the solutions of the equation $\cos(x) = 0.4$ that belong to the interval $[0, 2\pi]$, by making use of the approximate equation solver of your calculator.

Solution

a) Figure 5 displays the graph of cosine and the line $y = 0.4$ on the interval $[0, 2\pi]$.

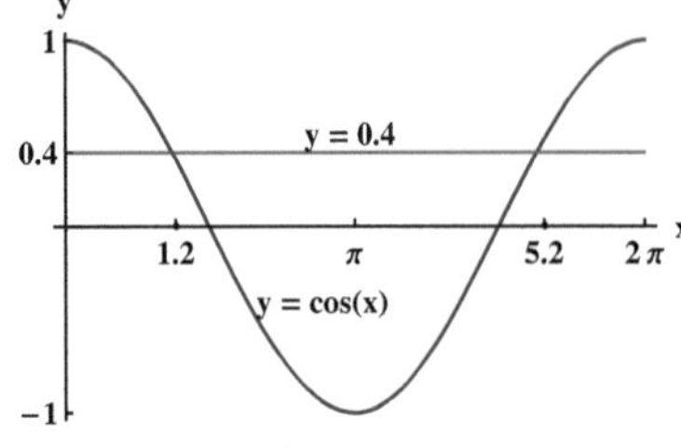

Figure 5

The picture indicates that the equation $\cos(x) = 0.4$ has solutions near 1.2 and 5.2 in the interval $[0, 2\pi]$. We can confirm the existence of the solutions with the help of the Intermediate Value Theorem. We have

$$\cos(1) \cong 0.540\,302 \text{ and } \cos(1.5) \cong 7.073\,72 \times 10^{-2},$$

so that $\cos(1) > 0.4 > \cos(1.5)$, and cosine is continuous on the interval $[1, 1.5]$. Therefore, there must exists c_1 between 1 and 1.5 such that $\cos(c_1) = 0.4$.

As for the existence of the other solution, we have

$$\cos(4) \cong -0.653\,644 \text{ and } \cos(5.4) \cong 0.634\,693,$$

so that $\cos(4) < 0.4 < \cos(5.4)$, and cosine is continuous on $[4, 5.4]$. Therefore, there must exist c_2 between 4 and 5.4 such that $\cos(c_2) = 0.4$.

b) We have $c_1 \cong 1.15928$ and $c_2 \cong 5.12391$, rounded to 6 significant digits. $\square$

Newton's Method

Assume that the function f is differentiable, so that f is continuous, and we suspect that there is a solution r of the equation $f(x) = 0$ in an interval $[b, c]$. An initial idea about the location of a solution of the equation can be gleaned from the graph of the function f, and numerically supported by observing that $f(x)$ changes sign. For example, if we notice that $f(b) < 0$ and $f(c) > 0$, we know that there must be a point r between b and c such that $f(r) = 0$, by the Corollary to the Intermediate Value Theorem. Let x_0 denote an "initial guess" for a solution of the equation $f(x) = 0$, and let

$$L_{x_0}(x) = f(x_0) + f'(x_0)(x - x_0)$$

be the linear approximation to f based at x_0. The graph of L_{x_0} is the tangent line to the graph of f at $(x_0, f(x_0))$. If $f'(x_0) \neq 0$, we can determine the solution of the equation $L_{x_0}(x) = 0$:

$$f(x_0) + f'(x_0)(x - x_0) = 0 \Leftrightarrow (x - x_0) = -\frac{f(x_0)}{f'(x_0)} \Leftrightarrow x = x_1 = x_0 - \frac{f(x_0)}{f'(x_0)}.$$

Thus, x_1 is the point at which the tangent line intersects the x-axis.

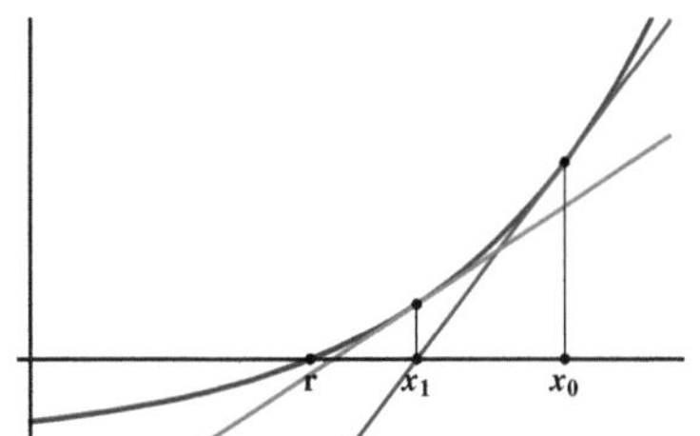

Figure 6: Newton's method follows the tangent lines

It is hoped that x_1 is a better approximation to the suspected solution of the equation $f(x) = 0$ than the initial guess x_0. The process is repeated: We determine x_2 as the point at which the tangent line to the graph of f at $(x_1, f(x_1))$ intersects the x-axis:

$$L_{x_1}(x) = 0 \Leftrightarrow f(x_1) + f'(x_1)(x - x_1) = 0 \Leftrightarrow x = x_2 = x_1 - \frac{f(x_1)}{f'(x_1)},$$

provided that $f'(x_1) \neq 0$. If we have already determined the points $x_0, x_1, x_2, \ldots, x_n$, we determine x_{n+1} as

$$x_{n+1} = x_n - \frac{f(x_n)}{f'(x_n)}.$$

This describes **Newton's Method** for the approximation of the solutions of the equation $f(x) = 0$. It is hoped that $\lim_{n \to \infty} x_n = r$.

Newton's method generates the sequence $x_0, x_1, x_2, \ldots, x_n, x_{n+1}, \ldots$ iteratively (or recursively). If we set

$$g(x) = x - \frac{f(x)}{f'(x)}$$

then

$$x_{n+1} = g(x_n), \quad n = 0, 1, 2, \ldots.$$

We will refer to g as **the iteration function**, and to x_n as the **nth iterate**.

Example 3 Let $f(x) = x^2 - 2$. The solutions of $f(x) = 0$ are $\sqrt{2}$ and $-\sqrt{2}$. Let's see how Newton's method works if we start with the initial guess $x_0 = 2$ for the positive solution of the equation $f(x) = 0$.

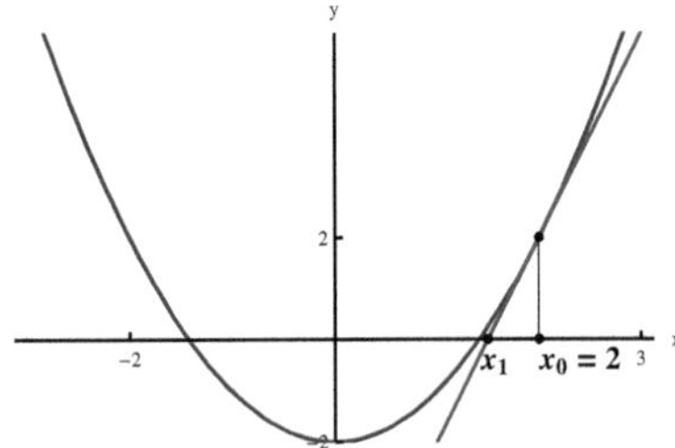

Figure 7: The first step of Newton's method for $x^2 - 2 = 0$

Newton's method generates the sequence $x_1, x_2, \ldots, x_n, x_{n+1}, \ldots$ recursively, according to the rule

$$x_{n+1} = x_n - \frac{f(x_n)}{f'(x_n)} = x_n - \frac{x^2 - 2}{2x_n} = \frac{2x_n^2 - x^2 + 2}{2x_n} = \frac{x_n^2 + 2}{2x_n}.$$

The function

$$g(x) = \frac{x^2 + 2}{2x}$$

is the iteration function: $x_{n+1} = g(x_n)$, $n = 0, 1, 2, \ldots$. For example,

$$x_1 = \frac{x_0^2 + 2}{2x_0} = \frac{2^2 + 2}{2(2)} = \frac{3}{2} = 1.5,$$

$$x_2 = \frac{x_1^2 + 2}{2x_1} = \frac{(1.5)^2 + 2}{2(1.5)} \cong 1.4166667.$$

Table 1 displays x_n, $n = 0, 1, 2, 3, 4$, rounded to 8 significant digits, and the absolute value of the error, i.e., $\left|\sqrt{2} - x_n\right|$ (rounded to 2 significant digits, as usual). We have $\sqrt{2} \cong 1.4142136$, rounded to 8 significant digits.

n	x_n	$\left|\sqrt{2} - x_n\right|$
0	2	0.59
1	1.5	8.6×10^{-2}
2	1.4166667	2.5×10^{-3}
3	1.4142157	2.1×10^{-6}
4	1.4142136	1.6×10^{-12}

Table 1

The numbers support the expectation that $\lim_{n\to\infty} x_n = \sqrt{2}$. Notice that the absolute value of the error decreases dramatically when we carry out the Newton iterations. In the numerical analysis jargon, Newton's method "converges rapidly". Indeed, $\left|\sqrt{2} - x_{n+1}\right|$ is approximately $\left(\sqrt{2} - x_n\right)^2$ for $n = 2$ and $n = 3$: The number 10^{-6} is small, but $\left(10^{-6}\right)^2 = 10^{-12}$ is even smaller! The rounding of the fourth iterate x_4 and the rounding of the decimal expansion of $\sqrt{2}$ to 8 significant digits results in the same decimal (even though $x_4 \neq \sqrt{2}$). $\square$

The fast convergence that we saw in Example 3 is quite typical. If $f'(r) \neq 0$, f'' is continuous and the initial guess x_0 is sufficiently close to r, the points $x_1, x_2, x_3, \ldots, x_n, \ldots$ generated by Newton's method **converge quadratically** to the solution r of the equation $f(x) = 0$, in the sense that

$$|x_{n+1} - r| \leq C(x_n - r)^2$$

where C is a constant that depends on the function f, the solution r of $f(x) = 0$, and the initial guess x_0. The analysis that leads to such an error estimate belongs to a post-calculus course. In practice, one needs a stopping criterion in the implementation of Newton's method. The simplest criterion is to stop when two successive iterates differ by a number which is less than a given "error tolerance". Thus, we may stop at x_n if $|x_{n+1} - x_n| < \varepsilon$, where ε is a positive number such as 5×10^{-7} and represents the error tolerance. Usually, this ensures that $|x_n - r|$ is approximately ε, where r is the relevant solution of the equation $f(x) = 0$. The approximation x_{n+1} is even better.

Example 4 Let $f(x) = \cos(x) + \cos(3x)$. Figure 8 displays the graph of f on $[0, \pi]$.

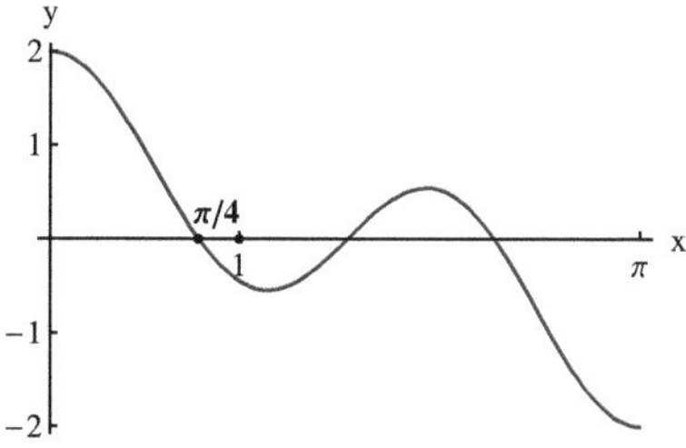

Figure 8

We have

$$f\left(\frac{\pi}{4}\right) = \cos\left(\frac{\pi}{4}\right) + \cos\left(\frac{3\pi}{4}\right) = \frac{\sqrt{2}}{2} - \frac{\sqrt{2}}{2} = 0,$$

and $\pi/4 \cong 0.785\,398$. Apply Newton's method to the equation $f(x) = 0$, with the initial guess $x_0 = 1$. Continue the iterations until $|x_{n+1} - x_n| \leq 10^{-3}$. Compare $|x_n - \pi/4|$ and $|x_{n+1} - \pi/4|$ with 10^{-3}. Do the numbers support the usual quadratic convergence of Newton's method?

Solution

The iteration function is

$$g(x) = x - \frac{f(x)}{f'(x)} = x - \frac{\cos(x) + \cos(3x)}{-\sin(x) - 3\sin(3x)} = x + \frac{\cos(x) + \cos(3x)}{\sin(x) + 3\sin(3x)},$$

so that $x_{n+1} = g(x_n)$, $n = 0, 1, 2, \ldots$.

Table 2 displays x_n and $|x_n - \pi/4|$ for $n = 0, 1, 2, 3, 4$, and $|x_{n+1} - x_n|$ for $n = 0, 1, 2, 3$. We see that $|x_{n+1} - x_n|$ is a good indication of the absolute error in the approximation of $\pi/4$ by x_n for $n = 1, 2, 3$. The numbers are consistent with the quadratic convergence of Newton's method: $|x_{n+1} - \pi/4| \cong (x_n - \pi/4)^2$ for $n = 2$ and $n = 3$. $\square$

n	x_n	$|x_{n+1} - x_n|$	$|x_n - \pi/4|$
0	1	3.6×10^{-1}	2.1×10^{-1}
1	.644466	1.3×10^{-1}	1.4×10^{-1}
2	.775045	10^{-2}	10^{-2}
3	.785296	10^{-4}	10^{-4}
4	.785 398		10^{-8}

Table 2

We can use Newton's method in order to approximate solutions of equations that are not given in the form $f(x) = 0$ initially:

Example 5 Determine the approximate values of x such that the graphs of $y = \sin(x)$ and $y = x/2$ intersect at the corresponding points, with the help of Newton's method. Continue with the iterations until the absolute value of the difference between successive iterates is at most 10^{-4}. Treat the approximate solutions that you obtain with the help of the equation solver of your computational utility as exact, and determine the absolute errors of the Newton iterations.

Solution

Figure 9 displays the graphs of $y = \sin(x)$ and $y = x/2$. The picture indicates the existence of r near 2 such that the graphs intersect at the corresponding point. Due to the symmetry with respect to the origin (confirm), there is another point of intersection corresponding to $-r$. We will approximate r.

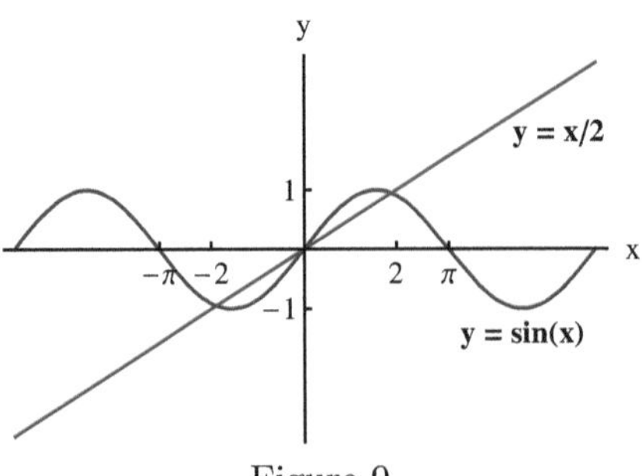

Figure 9

We must set the stage for the implementation of Newton's method. We have

$$\sin(x) = \frac{x}{2} \Leftrightarrow \frac{x}{2} - \sin(x) = 0.$$

We will set

$$f\left(x\right) = \frac{x}{2} - \sin\left(x\right),$$

and use Newton's method to approximate the positive solution of $f\left(x\right) = 0$. Figure 10 shows the graph of f. The picture indicates the existence of a unique solution r of the equation $f(x) = 0$ near 2 (you can confirm this with the help of the Intermediate Value Theorem, as in Example 1).

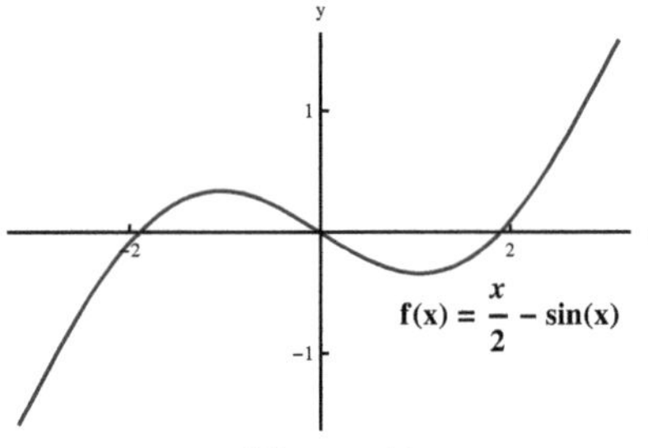

Figure 10

Let's pick x_0 as 2.5 to test Newton's method for the approximation of the solution of $f\left(x\right) = 0$ near 2 (if we choose 2 as the initial guess, the method converges so fast that we won't be able to display numbers that illustrate the convergence of Newton's method!). The iteration function is

$$g\left(x\right) = x - \frac{f\left(x\right)}{f'\left(x\right)} = x - \frac{\dfrac{x}{2} - \sin\left(x\right)}{\dfrac{1}{2} - \cos\left(x\right)},$$

so that $x_{n+1} = g\left(x_n\right)$ for $n = 0, 1, 2, \ldots$. We have $r \cong 1.895\,49$, rounded to 6 significant digits. Table 3 displays x_n and $|x_n - r|$ for $n = 0, 1, 2, 3, 4$, and $|x_{n+1} - x_n|$ for $n = 0, 1, 2, 3$ (as usual, differences are rounded to 2 significant digits). The numbers indicate that $|x_{n+1} - x_n|$ is a good measure of the accuracy with which x_n approximates the exact solution r. The numbers $|x_3 - r|$ and $|x_4 - r|$ support the quadratic convergence of Newton's method. $\square$

n	x_n	$|x_{n+1} - x_n|$	$|x_n - r|$
0	2.5	5×10^{-1}	6×10^{-1}
1	$1.999\,27$	9.8×10^{-2}	10^{-1}
2	$1.900\,92$	5.4×10^{-3}	5.4×10^{-3}
3	$1.895\,51$	1.7×10^{-5}	1.7×10^{-5}
4	$1.895\,49$		1.7×10^{-10}

Table 3

The above examples showed that Newton's method can be very effective for the approximation of solutions of equations. Nevertheless, let's take a look at some examples which show that we must be prepared for the occasional sub-optimal performance, even the failure of Newton's method.

The sequence generated by Newton's method may converge to a solution r of the equation $f\left(x\right) = 0$ even if $f'\left(r\right) = 0$, but the rate of convergence may not be as fast as in the case of r such that $f'\left(r\right) \neq 0$, as in the following example.

Example 6 Let $f(x) = \cos^2(x)$. A solution of the equation $f(x) = 0$ is $\pi/2$. The derivative of f is 0 at $\pi/2$ (confirm). Still, we can test the implementation of Newton's method for the approximation of the solution $\pi/2$ of the equation $f(x) = 0$.

Let us set $x_0 = 1$. The iteration function is

$$g(x) = x - \frac{f(x)}{f'(x)} = x - \frac{\cos^2(x)}{-2\cos(x)\sin(x)} = x + \frac{\cos(x)}{2\sin(x)}.$$

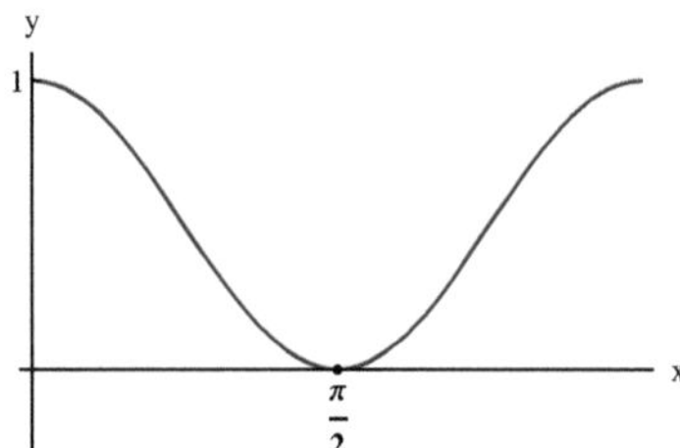

Figure 11: $y = \cos^2(x)$

Table 4 displays x_n and $|x_n - \pi/2|$ for $n = 1, 2, \ldots, 7$. The numbers are indicative of convergence ($\pi/2 \cong 1.5708$), but the evidence is not as overwhelming as in the previous examples. The numbers are not indicative of quadratic convergence: Even though $|x_{n+1} - \pi/2|$ is smaller than $|x_n - \pi/2|$, $|x_{n+1} - \pi/2|$ is not comparable to $(x_n - \pi/2)^2$. $\square$

n	x_n	$|x_n - \pi/2|$
1	1.32105	0.25
2	1.44858	0.12
3	1.51	6×10^{-2}
4	1.54043	3×10^{-2}
5	1.55562	1.5×10^{-2}
6	1.56321	7.6×10^{-3}
7	1.567	3.8×10^{-3}

Table 4

Example 7 An implementation of Newton's method may generate points that diverge to infinity:

Let

$$f(x) = \frac{x}{1 + x^2}$$

The only solution of the equation $f(x) = 0$ is 0.
Let's test Newton's method by taking $x_0 = 2$ as the starting point. We have

$$f'(x) = \frac{d}{dx}\left(\frac{x}{1+x^2}\right) = \frac{(1+x^2) - 2x^2}{(1+x^2)^2} = \frac{1 - x^2}{(1+x^2)^2}.$$

Therefore, the iteration function is

$$g(x) = x - \frac{f(x)}{f'(x)} = x - \frac{\dfrac{x}{1+x^2}}{\dfrac{1-x^2}{(1+x^2)^2}} = x - \frac{x(1+x^2)}{1-x^2}.$$

Table 5 displays x_n for $n = 2, 4, 6, 8, 10$. The numbers indicate that the sequence x_n diverges to infinity (you may calculate more points if you are skeptical).

n	x_n
2	11.0553
4	44.676
6	178.816
8	715.292
10	2861.18

Table 5

A picture such as Figure 12 provides further evidence of divergence. $\square$

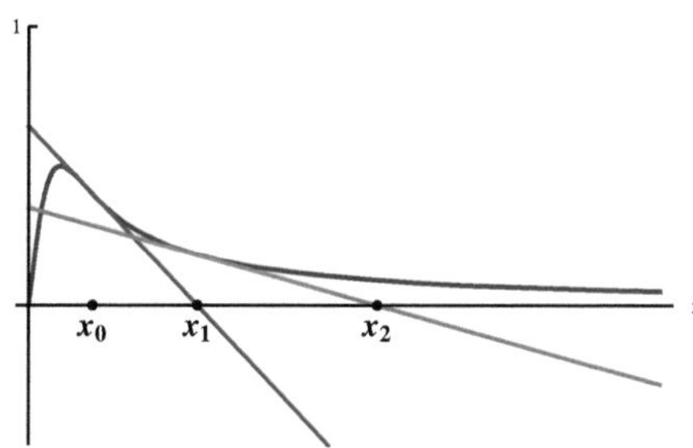

Figure 12: The Newton iterations may diverge to infinity

Problems

[C] In problems 1-4,
a) Make use of the corolary of the Intermediate Value Theorem to show that the equation $f(x) = 0$ has a solution in the interval J.
b) Make use of your graphing and computational utility to find approximations to all solutions of the equation $f(x) = 0$ in the interval J. Display 6 significant digits.

1.
$$f(x) = x - \frac{1}{x^2 + 4}, \quad J = [-1, 1].$$

2.
$$f(x) = 1 - \frac{1}{2}x^2 + \frac{1}{24}x^4, \quad J = [1, 2].$$

3.
$$f(x) = \sin(x) + \frac{1}{3}\cos(3x), \quad J = [2, 4].$$

4.
$$f(x) = \sin^2(x) + 2\cos^3(x), \quad J = [-3, -1]$$

[C] In problems 5-8,
a) Make use of your graphing utillity in order to plot the graph of f on the interval J and to determine the approximate locations of the solutions of the equation $f(x) = 0$ in J.
b) Make use of Newton's method to determine approximations to the solutions of the equation $f(x) = 0$ in the interval J. Stop the iterations when the absolute value of the difference between two iterations is less than 10^{-4}. Assuming that your computational utility provides exact solutions, calculate the absolute error in the approximations.

5.
$$f(x) = x^3 - 5x^2 - 8x + 40, \ J = [1, 6]$$

6.
$$f(x) = x^4 - 17x^2 + 50, \ J = [1, 4]$$

7.
$$f(x) = f, \ J = [0, 2]$$

8.
$$f(x) = \tan(x) + x - 2, \ \ J = [2, 4]$$

In problems 9 and 10,
a) Make use of your graphing utillity in order to plot the graphs of f and g on the interval J and determine the approximate locations of the solutions of the equation $f(x) = g(x)$ in J.
b) Make use of Newton's method to determine approximations to the solutions of the equation $f(x) = g(x)$ in the interval J. Stop the iterations when the absolute value of the difference between two iterations is less than 10^{-4}.

9.
$$f(x) = \frac{1}{x^2 + 1}, \ g(x) = -x^3, \ J = [-2, 2]$$

10.
$$f(x) = \frac{1}{1 - 2\cos(x)}, \ g(x) = \frac{1}{2 - \sin(x)}, \ J = [3, 5]$$

2.10   Implicit Differentiation

Assume that $F(x, y)$ is an expression that involves the variables x and y and that C is a constant. Let's consider the equation $F(x, y) = C$. Assume that this equation can be solved for y in terms of x, at least in principle, so that y can be expressed as a function of x. If we denote that function by f, and replace y by $f(x)$ in the equation $F(x, y) = C$ we have

$$F(x, f(x)) = C$$

for each x in some interval. In this case, we say that f **is defined implicitly as a function of x by the equation $F(x, y) = C$.** In some cases the expression $f(x)$ cannot be determined explicitly. In this section we will discuss the differentiation of functions which are defined implicitly.

Example 1 Consider the equation $x^2 + y^2 = 1$. The graph of the equation is the unit circle.

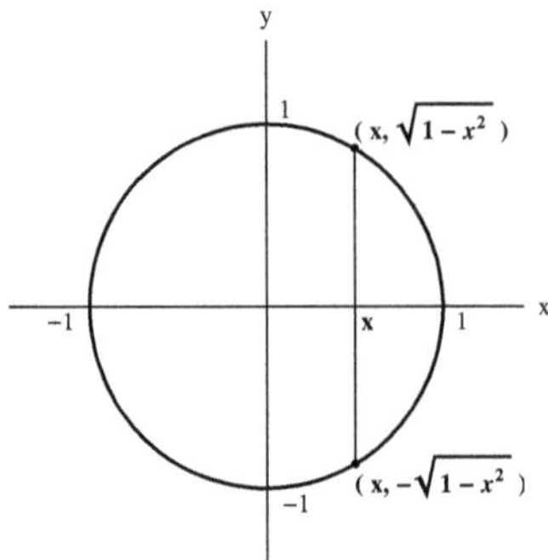

Figure 1: $x^2 + y^2 = 1$

We can solve the equation to express y in terms of x:

$$x^2 + y^2 = 1 \Leftrightarrow y^2 = 1 - x^2 \Leftrightarrow y = \pm\sqrt{1 - x^2}.$$

Since $\sqrt{1 - x^2} \neq -\sqrt{1 - x^2}$ if $-1 < x < 1$, the unit circle is not the graph of a function of x (it fails the vertical line test). On the other hand, if we set $f(x) = \sqrt{1 - x^2}$, we have defined a function whose domain is the interval $[-1, 1]$, and $x^2 + f^2(x) = 1$ for each $x \in [-1, 1]$. Therefore, f is defined implicitly by the equation $x^2 + y^2 = 1$. The graph of f is the upper half of the unit circle.

If we set $g(x) = -\sqrt{1 - x^2}$ for each $x \in [-1, 1]$, we also have $x^2 + g^2(x) = 1$ for each $x \in [-1, 1]$. Thus, the function g is also defined implicitly by the equation $x^2 + y^2 = 1$. The graph of g is the lower half of the unit circle. $\square$

In Example 1 we had explicit expressions for the functions that were defined implicitly by the given equation. That is not always possible, as in the following example.

Example 2 Consider the equation

$$y^5 + y^2 - y - x^2 + 1 = 0.$$

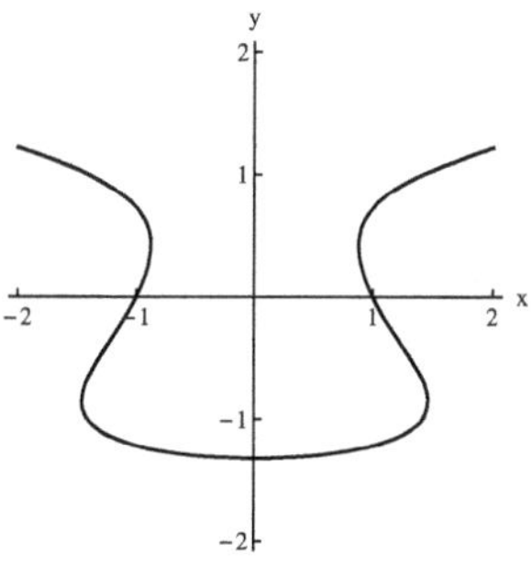

Figure 2

Figure 2 shows the graph of the equation. Even though the graph fails the vertical line test and cannot be the graph of a function of x, it does contain graphs of functions that are defined by the equation implicitly, as in Example **??**. We cannot obtain the expressions of such functions explicitly, though. The expression on the right-hand side of the equation is a polynomial of degree 5 in the variable y for each value of x. There is no formula for the solution of an equation that involves a polynomial of degree 5 that will enable us to express y as a function of x explicitly. $\square$

A procedure that is referred to as **implicit differentiation** enables us to determine the derivative of a function that is defined implicitly by an equation, even if the function cannot be expressed explicitly. Let's illustrate this procedure in the setting of Example 1.

Example 3 Consider the equation $x^2 + y^2 = 1$, as in Example 1.

Let $y(x)$ represent $f(x)$ or $g(x)$, where $f(x) = \sqrt{1 - x^2}$ and $g(x) = -\sqrt{1 - x^2}$. Thus,

$$x^2 + y^2(x) = 1$$

for each $x \in [-1, 1]$. Let's differentiate with respect to x:

$$\frac{d}{dx}\left(x^2 + y^2(x)\right) = \frac{d}{dx}(1) = 0$$

for each $x \in (-1, 1)$ (the endpoints require special consideration). By the linearity of differentiation and the chain rule (or the function-power rule that followed from the chain rule),

$$\frac{d}{dx}\left(x^2 + y^2\left(x\right)\right) = \frac{d}{dx}\left(x^2\right) + \frac{d}{dx}y^2\left(x\right) = 2x + 2y\left(x\right)\frac{dy}{dx}.$$

Therefore,

$$2x + 2y\left(x\right)\frac{dy}{dx} = 0 \Rightarrow \frac{dy}{dx} = -\frac{x}{y\left(x\right)},$$

provided that $y\left(x\right) \neq 0$. Note that $y\left(x\right) = \pm\sqrt{1 - x^2} \neq 0$ if $-1 < x < 1$. If $y\left(x\right) = f\left(x\right) = \sqrt{1 - x^2}$, then

$$\frac{dy}{dx} = -\frac{x}{f\left(x\right)} = -\frac{x}{\sqrt{1 - x^2}}.$$

If $y\left(x\right) = g\left(x\right) = -\sqrt{1 - x^2}$, then

$$\frac{dy}{dx} = -\frac{x}{g\left(x\right)} = \frac{x}{\sqrt{1 - x^2}}.$$

The expressions are valid if $-1 < x < 1$. As an exercise in the chain rule, you can confirm that

$$\frac{df}{dx} = -\frac{x}{\sqrt{1 - x^2}} \text{ and } \frac{dg}{dx} = \frac{x}{\sqrt{1 - x^2}}$$

by differentiating f and g "explicitly".

We have $y\left(\pm 1\right) = 0$, so that the expression

$$-\frac{x}{y\left(x\right)}$$

is not defined at ± 1. The functions f and g do not have even one-sided derivatives at these points anyway (confirm that the graphs of f and g have vertical tangents at ± 1).

In practice, we use more practical notation when we implement implicit differentiation. Starting with the equation $x^2 + y^2 = 1$, we treat y as a function of x, even though we don't bother to replace y by $y\left(x\right)$. Thus,

$$x^2 + y^2 = 1 \Rightarrow \frac{d}{dx}\left(x^2\right) + \frac{d}{dx}y^2 = 1 \Rightarrow 2x + 2y\frac{dy}{dx} = 0 \Rightarrow \frac{dy}{dx} = -\frac{x}{y}.$$

When we wish to make use of the above expression to calculate the derivative at a specific value of x, the corresponding value of y must be specified as well. For example, if $x = 1/2$, we have

$$y^2 = 1 - x^2 = 1 - \frac{1}{4} = \frac{3}{4},$$

so that $y = \pm\sqrt{3}/2$. If we specify that $y\left(1/2\right) = \sqrt{3}/2$, then

$$\left.\frac{dy}{dx}\right|_{x=1/2} = \left.-\frac{x}{y}\right|_{x=1/2,\, y=\sqrt{3}/2} = -\frac{\frac{1}{2}}{\frac{\sqrt{3}}{2}} = -\frac{1}{\sqrt{3}}.$$

Thus, the slope of the tangent line to unit circle $x^2 + y^2 = 1$ at $\left(1/2, \sqrt{3}/2\right)$ is $-1/\sqrt{3}$. The tangent line is the graph of the equation

$$y = \frac{\sqrt{3}}{2} - \frac{1}{\sqrt{3}}\left(x - \frac{1}{2}\right).$$

Figure 3 shows the unit circle and the tangent line that we determined.

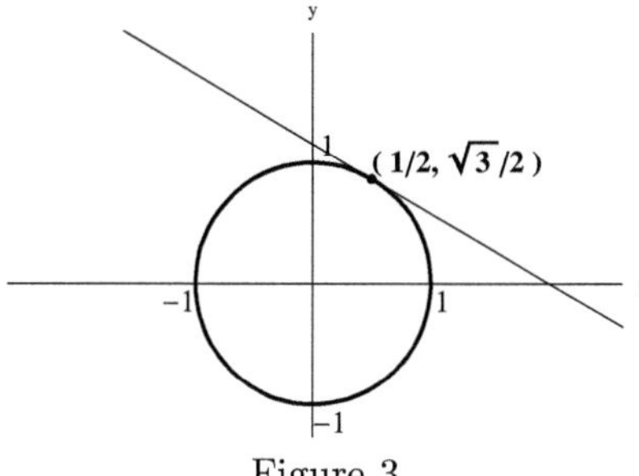

Figure 3

On the other hand, if $y\left(1/2\right) = -\sqrt{3}/2$, then

$$y'\left(\frac{1}{2}\right) = -\left.\frac{x}{y}\right|_{x=1/2 \text{ and } y=-\sqrt{3}/2} = -\frac{\frac{1}{2}}{-\frac{\sqrt{3}}{2}} = \frac{1}{\sqrt{3}}.$$

The line that is tangent to the unit circle $x^2 + y^2 = 1$ at the point $\left(1/2, -\sqrt{3}/2\right)$ is the graph of the equation

$$y = -\frac{\sqrt{3}}{2} + \frac{1}{\sqrt{3}}\left(x - \frac{1}{2}\right),$$

as illustrated in Figure 4. $\square$

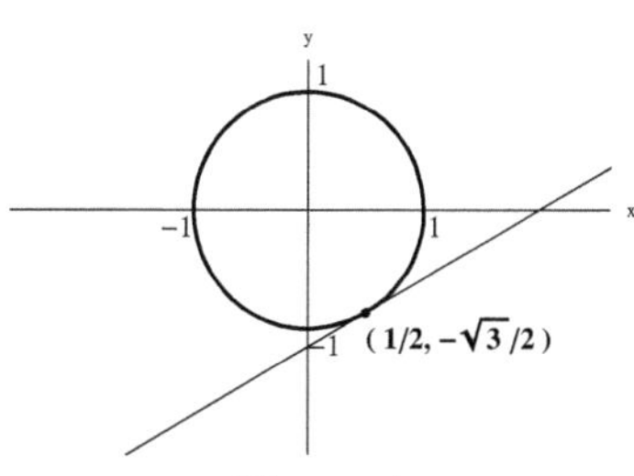

Figure 4

Example 4 Consider the equation

$$y^5 + y^2 - y - x^2 + 1 = 0,$$

as in Example 2.

a) Assume that $y\left(x\right)$ represents a function that is defined implicitly by the given equation. Determine $y'\left(x\right)$.

b) Evaluate $y'\left(1\right)$ if $y\left(1\right) = 0$. Determine the tangent line to the graph of the equation at $\left(1, 0\right)$.

Solution

a) We will implement implicit differentiation. We treat y as a function of x and apply the chain rule. Thus,

$$\frac{d}{dx}\left(y^5 + y^2 - y - x^2 + 1\right) = 0,$$

so that

$$5y^4\frac{dy}{dx} + 2y\frac{dy}{dx} - \frac{dy}{dx} - 2x = 0.$$

Therefore,

$$\left(5y^4 + 2y - 1\right)\frac{dy}{dx} = 2x.$$

Thus,

$$\frac{dy}{dx} = \frac{2x}{5y^4 + 2y - 1}..$$

The expression makes sense if $5y^4 + 2y - 1 \neq 0$.

b) If $y\left(1\right) = 0$, we have

$$\left.\frac{dy}{dx}\right|_{x=1} = \left.\frac{2x}{5y^4 + 2y - 1}\right|_{x=1 \text{ and } y=0} = \frac{2}{-1} = -2.$$

Therefore the line that is tangent to the graph of the equation

$$y^5 + y^2 - y - x^2 + 1 = 0$$

at $(1,0)$ is the graph of the equation

$$y = -2\left(x - 1\right).$$

Figure 5 shows the graph of the equation and the tangent line at $(1,0)$.$\square$

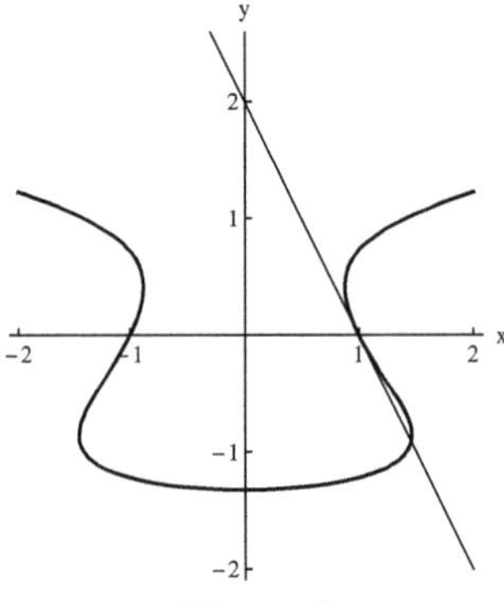

Figure 5

Remark   Assume that we are given an equation $F\left(x, y\right) = C$, where C is a constant, and that we obtain the expression

$$\frac{dy}{dx} = \frac{p(x, y)}{q\left(x, y\right)}$$

via implicit differentiation. If $F\left(x_0, y_0\right) = C$ and $q\left(x_0, y_0\right) \neq 0$ (and some smoothness conditions are satisfied), **the implicit function theorem** implies that there is a function f which is defined implicitly by the given equation such that $f\left(x_0\right) = y_0$ and

$$f'\left(x_0\right) = \frac{p(x_0, y_0)}{q\left(x_0, y_0\right)}$$

The proof of this fact belongs to an advanced calculus course. $\lozenge$

Example 5 Consider the equation

$$x^3 + y^3 = 9xy.$$

a) Assume that $y(x)$ represents a function that is defined implicitly by the given equation. Determine $y'(x)$.
b) Evaluate $y'(2)$ if $y(2) = 4$. Determine the tangent line to the graph of the equation at $(2, 4)$.

Solution

a) We differentiate both sides of the equation $x^3 + y^3 = 9xy$ with respect to x, treating y as a function of x. With the help of the product rule and the chain rule,

$$3x^2 + \frac{d}{dx}\left(y^3\right) = \frac{d}{dx}\left(9xy\right) \Rightarrow 3x^2 + 3y^2 \frac{dy}{dx} = 9y + 9xy.$$

Therefore,

$$\left(3y^2 - 9x\right)\frac{dy}{dx} = 9y - 3x^2,$$

so that

$$\frac{dy}{dx} = \frac{9y - 3x^2}{3y^2 - 9x} = \frac{3y - x^2}{y^2 - 3x}.$$

b) If $y(2) = 4$,

$$y'(2) = \left.\frac{3y - x^2}{y^2 - 3x}\right|_{x=2 \text{ and } y=4} = \frac{4}{5}.$$

Therefore, the tangent line to the graph of the equation at the point $(2, 4)$ is the graph of the equation

$$y = 4 + \frac{4}{5}\left(x - 2\right).$$

Figure 6 shows the graph of the equation and the tangent line at $(2, 4)$. $\square$

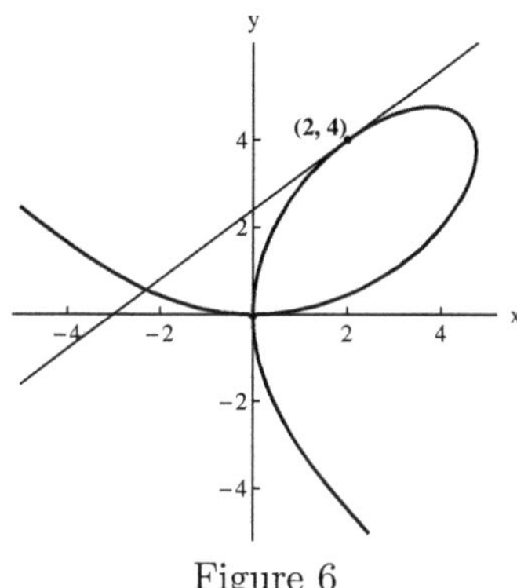

Figure 6

Example 6 Assume that $y(x)$ is defined implicitly by the equation $\sin(y) = x$.

a) Determine $y'(x)$.
b) If $y(1/2) = \pi/6$, evaluate $y'(1/2)$, and determine the tangent line to the graph of the given equation at $(1/2, \pi/6)$.

Solution

a) We have

$$\frac{d}{dx}\left(x\right) = \frac{d}{dx}\sin\left(y\right), \quad -1 < x < 1.$$

By the chain rule,

$$1 = \cos\left(y\right)\frac{dy\left(x\right)}{dx},$$

so that

$$\frac{dy}{dx} = \frac{1}{\cos\left(y\right)}.$$

b) We are given that $y(1/2) = \pi/6$. Therefore,

$$\left.\frac{dy}{dx}\right|_{x=1/2} = \frac{1}{\cos\left(\pi/6\right)} = \frac{1}{\frac{\sqrt{3}}{2}} = \frac{2}{\sqrt{3}}.$$

The tangent line to the graph of the equation at $(1/2, \pi/6)$ is the graph of the equation

$$y = \frac{\pi}{6} + \frac{2}{\sqrt{3}}\left(x - \frac{1}{2}\right).$$

Figure 7 displays the graph of the equation $x = \sin\left(y\right)$ in the window $[-2, 2] \times [-2\pi, 2\pi]$.

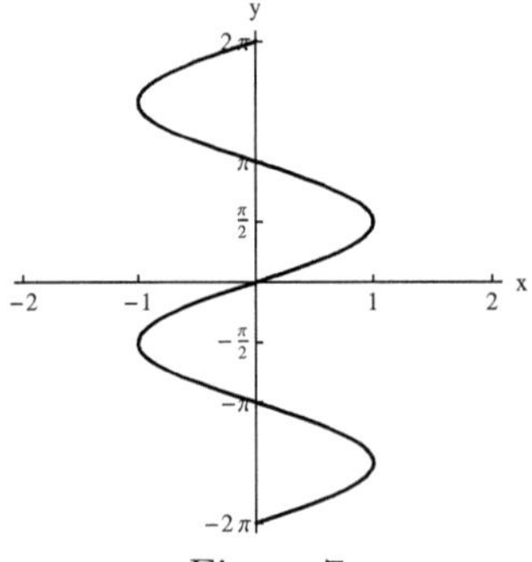

Figure 7

The part of the graph that is in the horizontal strip $-\pi/2 \le y \le \pi/2$ is the graph of a function, and is relevant to our problem. Figure 8 shows the graph of the equation $\sin\left(y\right) = x$ in the window $[-2, 2] \times [-\pi/2, \pi/2]$ and the tangent line at $(1/2, \pi/6)$. $\square$

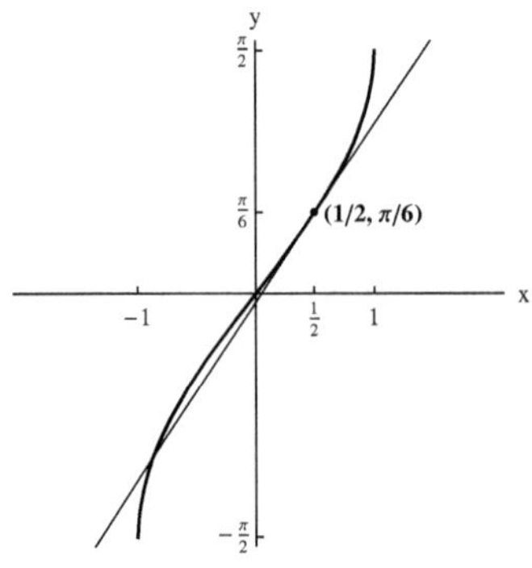

Figure 8

Problems

In problems 1-4, assume that $y(x)$ is implicitly defined by the given equation and $y(x_0) = y_0$.
a) Compute $y'(x_0)$ by implicit differentiation.
b) Determine $y(x)$ explicitly. Compute $y'(x_0)$ and check that the result is the same as that obtained in part (a).
c) [C] Make use of your graphing utility in order to plot the graph of the equation. Identify the graph of the equation and the graph of the function $y(x)$ that is defined implicitly by the given equation such that $y(x_0) = y_0$.

1.
$$y^2 - x^2 = 9, \ x_0 = 4, \ y_0 = -5.$$

2.
$$\frac{x^2}{9} + \frac{y^2}{16} = 1, \ x_0 = 2, \ y_0 = \frac{4\sqrt{5}}{3}.$$

3.
$$x = (y-2)^2 - 4, \ x_0 = 1, \ y_0 = 2 - \sqrt{5}$$

4.
$$(x-2)^2 + (y-3)^2 = 4, \ x_0 = 3, \ y_0 = 3 - \sqrt{3}$$

In problems 5 -10, assume that $y(x)$ is defined implicitly by the given equation Compute $y'(x)$ by implicit differentiation.

5.
$$x^2 - xy + 2y^2 = 3$$

6.
$$3x^2 - 4xy - 2y^2 = 9$$

7.
$$y^3 - 4xy = 8$$

8
$$x^3 + y^3 - y + 2x = 6$$

9.
$$\sin(x) + 2\cos(y) = 0$$

10.
$$x\cos(x) - y\sin(y) = 0$$

In problems 11-16, assume that $y(x)$ is implicitly defined by the given equation and $y(x_0) = y_0$.
a) Compute $y'(x)$ by implicit differentiation.
b) Determine the tangent line to the graph of the given equation at the point (x_0, y_0) (a point-slope form of the equation of the tangent line will do).
c) [C] Make use of your graphing utility in order to plot the graph of the equation and the tangent line at (x_0, y_0).

11.
$$y^3 - 9y - 3x = 0. \ x_0 = 0, \ y_0 = 3.$$

12.
$$x^{2/3} + y^{2/3} = 8, \ x_0 = 8, \ y_0 = 8.$$

13.
$$x^3 - 3xy^2 + y^3 = 1, \ x_0 = 2, \ y_0 = -1.$$

14.
$$x^3 - 2xy + 6y^2 = 24, \ x_0 = 2, \ y_0 = 2$$

15.
$$x = \tan(y), \ x_0 = 1, \ y_0 = \frac{\pi}{4}.$$

16.
$$x = \cos(y), \ x_0 = \frac{1}{2}, \ y_0 = \frac{\pi}{3}.$$

Chapter 3

Maxima and Minima

The sign of the derivative of a function provides us with valuable information about the **increasing/decreasing behavior** of the function and its **maximum** and **minimum** values. The sign of the second derivative of a function provides information about **the increasing or decreasing behavior of the derivative** of the function. We will make use of such information to determine the solutions of practical **optimization problems.**

3.1 Increasing/decreasing Behavior and Extrema

The sign of the derivative of a function provides information about the intervals on which the function is increasing or decreasing and the points at which it attains its maximum and minimum values.

Some Terminology

Let's begin by recalling some terminology. Let J denote an interval that can be a bounded open interval such as $(1, 3)$, a half-open interval such as $[1, 3)$, or an unbounded interval such as $(-\infty, 2]$. Recall that x is in **the interior** of an interval J if $x \in J$, but x is not an endpoint of J. Thus, the interior of $[1, 3)$ is the open interval $(1, 3)$, and the interior of $(-\infty, 2]$ is the open interval $(-\infty, 2)$. Also recall that a function f is said to be **increasing** on an interval J if, for any pair of points x_1 and x_2 in J such that $x_1 < x_2$, we have $f(x_1) < f(x_2)$. A function f is said to be **decreasing** on J if, for x_1 and x_2 in J such that $x_1 < x_2$, we have $f(x_1) < f(x_2)$. A function f is said to be **decreasing** on J if, for x_1 and x_2 in J such that $x_1 < x_2$, we have $f(x_1) > f(x_2)$. We will say that f is **monotone** on J if f is increasing, decreasing or constant on J.

A function f is said to have **a local maximum** or **a local minimum** at a point a if $f(a)$ is its maximum or minimum value, respectively, relative to some interval that contains a in its interior. **The absolute maximum** of f on a set D is *the maximum value* of f on D, **the absolute minimum** of f on D is *the minimum value* of f on D. Here are the precise expressions:

Definition 1 A function f has **a local maximum at** a if there exists an open interval J that contains a such that $f(a) \geq f(x)$ for each $x \in J$. The function f has **a local minimum at** a if there exists an open interval J that contains a such that $f(a) \leq f(x)$ for each $x \in J$. In either case, we say that f has **a local extremum at** a. **The absolute maximum** of f on a set D is M if there exists $c_M \in D$ such that $f(c_M) = M$ and $M \geq f(x)$ for each $x \in D$. **The absolute minimum** of f on a set D is m if there exists $c_m \in D$ such that $f(c_m) = m$ and $m \leq f(x)$ for each $x \in D$. We may refer to absolute maxima and minima as **absolute extrema.**

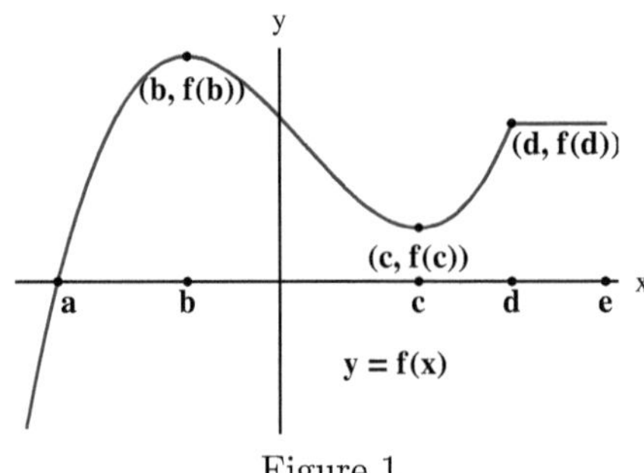

Figure 1

It appears that the function of Figure 1 has local maxima at b and d, and a local minimum at c. The picture indicates that $f(b)$ is the absolute maximum of f on $[a, e]$ and $f(a)(= 0)$ is the absolute minimum of f on $[a, e]$. Note that f is constant on the interval $[d, e]$.

In this section we will focus on finding the local extrema of a function. We will take up the issue of absolute extrema in the next section

Remark 1 According to our definition of a local extremum, if a function f has a local extremum at a point a, f must attain a maximum or minimum at a relative to the points in some *open* interval that contains a. Thus, $f(x) = \sqrt{x}$ does not have a local minimum at 0, even though $f(x) \geq f(0) = 0$ for each $x \in [0, +\infty)$. We have to single out the endpoints of an interval for special consideration when we investigate the absolute maximum and the absolute minimum of a function on the entire interval. $\Diamond$

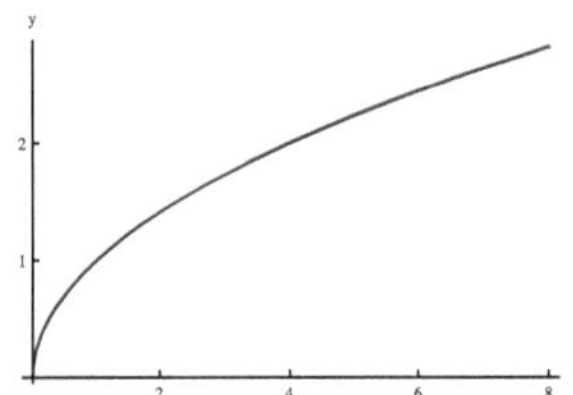

Figure 2: The square-root function does not have a local minimum at 0

The Derivative Test for Monotonicity and Extrema

Recall that the tangent line to the graph of f at $(a, f(a))$ is the graph of the linear function

$$L_a(x) = f(a) + f'(a)(x - a).$$

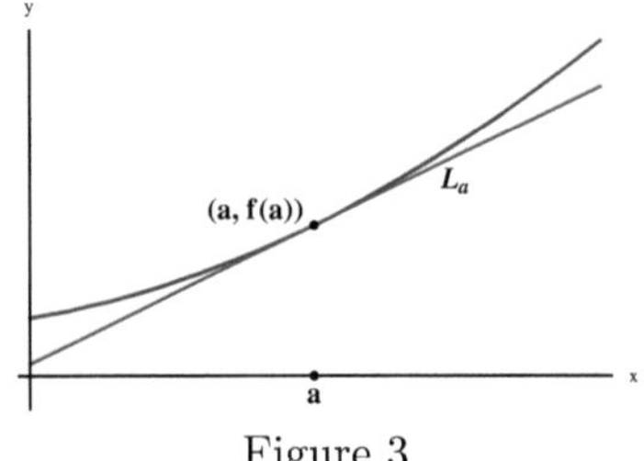

Figure 3

We have seen that $L_a(x)$ approximates $f(x)$ very well if x is close to a. If $f'(a) > 0$, the linear function L_a is an increasing function. Therefore it is reasonable to expect that f is also increasing in a small interval containing the point a. Similarly, if $f'(a) < 0$ then L_a is a decreasing function, and we would expect that f is decreasing in a sufficiently small interval containing a. Actually, the sign of the derivative of a function gives us information about the increasing/decreasing behavior of the function on any interval, irrespective of its size.

THE DERIVATIVE TEST FOR MONOTONICITY Assume that f is continuous on the interval J and that f is differentiable at each x in the interior of J. If $f'(x) > 0$ for each x in the interior of J, then f is increasing on J. If $f'(x) < 0$ for each x in the interior of J, then f is decreasing on J.

We will discuss the theoretical basis of the derivative test for monotonicity in the next section. An immediate corollary of the derivative test for monotonicity is that **a function has a local maximum or minimum at a point where its derivative changes sign:**

THE DERIVATIVE TEST FOR LOCAL EXTREMA Assume that f is continuous at a, and a is contained in open interval (c, d) such that $f'(x) > 0$ if $c < x < a$, and $f'(x) < 0$ if $a < x < d$. Then f has a local maximum at a. Similarly, if $f'(x) < 0$ if $c < x < a$, and $f'(x) > 0$ if $a < x < d$, then f has a local minimum at a.

Proof

a)

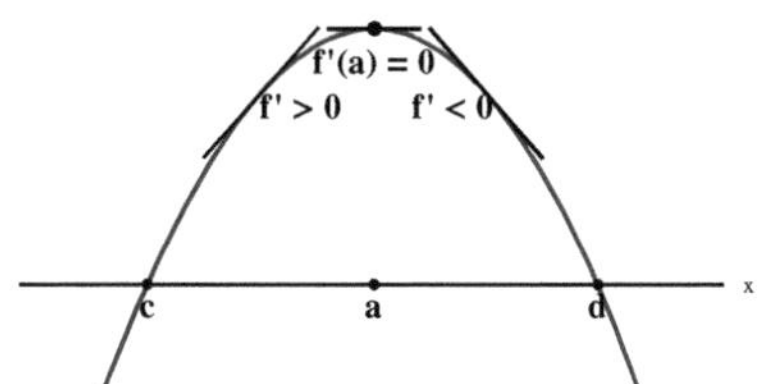

Figure 4: f has a local maximum at a

Since $f'(x) > 0$ if $x \in (c, a)$, f is increasing on $(c, a]$. Since $f'(x) < 0$ if $x \in (a, d)$, f is decreasing on $[a, d)$. Therefore, $f(a) \geq f(x)$ for each $x \in (c, d)$, so that f has a local maximum at a.

b)

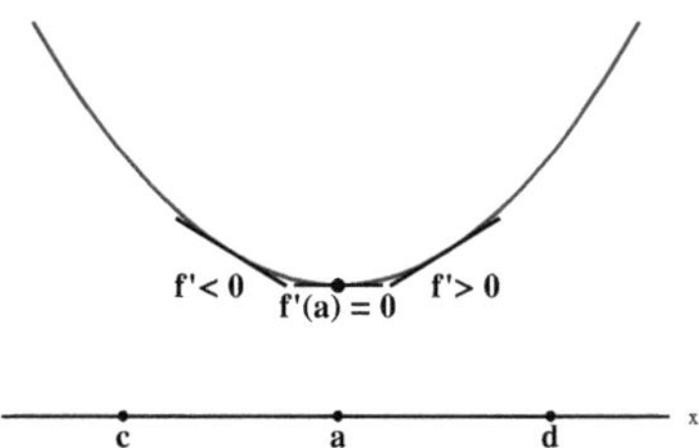

Figure 5: f has a local minimum at a

Since $f'(x) < 0$ if $x \in (c, a)$, f is decreasing on $(c, a]$. Since $f'(x) > 0$ if $x \in (a, d)$, f is increasing on $[a, d)$. Therefore, $f(a) \leq f(x)$ for each $x \in (c, d)$, so that f has a local minimum at a. ∎

Example 1 Let

$$f\left(x\right) = \frac{1}{3}x^3 - 9x.$$

a) *Determine the points at which f has a local maximum or minimum, and the corresponding values of f.*
b) Determine the absolute maximum and minimum values of f on the intervals $(-\infty, 0]$ and $[0, +\infty)$, provided that such values exist. Justify your responses if you claim that such values do not exist.

Solution

Figure 6 shows the graph of f. The picture indicates that f has a local maximum near -3 and a local,minimum near 3.

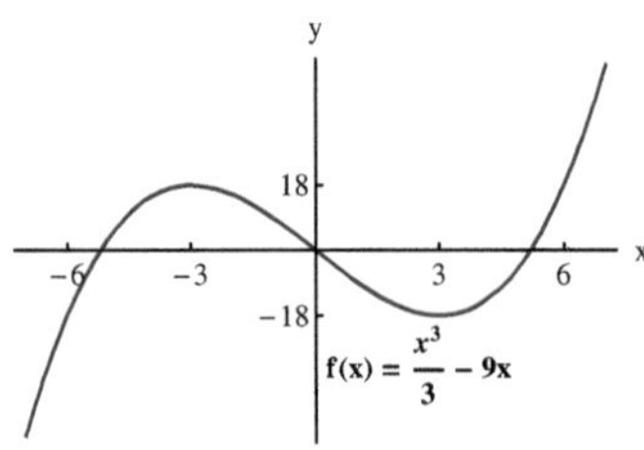

Figure 6

We will determine the exact values with the help of the first derivative test. We have

$$\begin{aligned}
f'\left(x\right) = \frac{d}{dx}\left(\frac{1}{3}x^3 - 9x\right) &= \frac{1}{3}\left(3x^2\right) - 9 \\
&= x^2 - 9 = \left(x+3\right)\left(x-3\right).
\end{aligned}$$

Figure 7 shows the graph of f'.

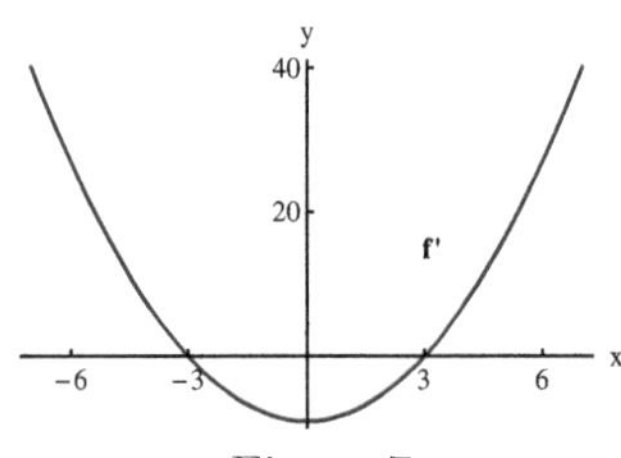

Figure 7

We have $f'\left(x\right) = 0$ if $x = \pm 3$, $f'\left(x\right) > 0$ if $x < -3$ or $x > 3$, and $f'\left(x\right) < 0$ if $-3 < x < 3$. By the derivative test for monotonicity, f is increasing on the intervals $(-\infty, -3]$ and $[3, +\infty)$, and decreasing on $[-3, 3]$. Since f is increasing on the interval $(-\infty, -3]$ and decreasing on $[-3, 3]$, $f\left(3\right) \geq f\left(x\right)$ for each $x \in (-\infty, 3)$. Therefore, f has a local maximum at -3. We have $f\left(-3\right) = 18$. Similarly, since f is decreasing on $[-3, 3]$ and increasing on $[3, +\infty)$, f has a local minimum at 3. We have $f\left(-3\right) = -18$. The pictures are consistent with our conclusions.

b) Since f is increasing on $(-\infty, -3]$ and decreasing on $[-3, 0]$, the function attains its absolute maximum on the interval $(-\infty, 0]$ at -3. We have $f\left(-3\right) = 18$. On the other hand, f does not have an absolute minimum on $(-\infty, 0]$ since f attains negative values of arbitrarily large magnitude. Indeed,

$$\lim_{x\to-\infty} f\left(x\right) = \lim_{x\to-\infty}\left(\frac{1}{3}x^3 - 9x\right) = \lim_{x\to-\infty} x^3\left(\frac{1}{3} - \frac{9}{x^2}\right) = -\infty$$

since

$$\lim_{x \to -\infty} x^3 = -\infty \text{ and } \lim_{x \to -\infty} \left(\frac{1}{3} - \frac{9}{x^2} \right) = \frac{1}{3} > 0.$$

$\square$

We saw that the derivative of the function of Example 1 vanishes at the points at which it has a local maximum or minimum. Thus, the tangent lines at the corresponding points on its graph are horizontal. This is not an accident. Assume that a function f has a local maximum or a local minimum at a point a, and that f' is continuous at a. If $f'(a) > 0$, then $f'(x) \cong f'(a) > 0$ if x is close to a, by the continuity of f' at a. Therefore, $f'(x) > 0$ for each x in some open interval that contains a. The derivative test for monotonicity implies that f is increasing on that interval. This contradicts the assumption that f has a local extremum at a. Similarly, if $f'(a) < 0$, then f is decreasing on an open interval containing a so that f cannot have a local extremum at a. Therefore, we must have $f'(a) = 0$. This is actually valid even if f' is not continuous at a:

FERMAT'S THEOREM   If f **has a local maximum or minimum at** a **and** f **is differentiable at** a **we have** $f'(a) = 0$.

You can find the proof of Fermat's Theorem at the end of this section.

Definition 2 We say that the point a is a **stationary point** of f if $f'(a) = 0$.

Thus, a is a stationary point of f if the tangent line to the graph of f at $(a, f(a))$ is horizontal. Fermat's Theorem says that a must be a stationary point of f if f is differentiable at a and has a local extremum at a. The term "stationary point" has the following meaning within the context of one-dimensional motion: Assume that $f(t)$ is the position at time t of an object in one-dimensional motion. The instantaneous velocity $v(t)$ at the instant t is the rate of change of f at t, i.e., $v(t) = f'(t)$. Thus, we have $v(t_0) = 0$ if t_0 is a stationary point of f. Since the instantaneous velocity of the object at t_0 is 0, we may imagine that the object is "instantaneously stationary" at the instant t_0.

A word of caution: You should not misread Fermat's Theorem. The condition, $f'(a) = 0$, is **necessary** for f to have a local maximum or minimum at a. The theorem does **not** say that the condition $f'(a) = 0$ is *sufficient* for f to have a local extremum at a.

Example 2 Let $f(x) = x^3$. Then $f'(x) = 3x^2 = 0$ if $x = 0$, but f does not have a local maximum or minimum at 0. The function is increasing on $(-\infty, +\infty)$. $\square$

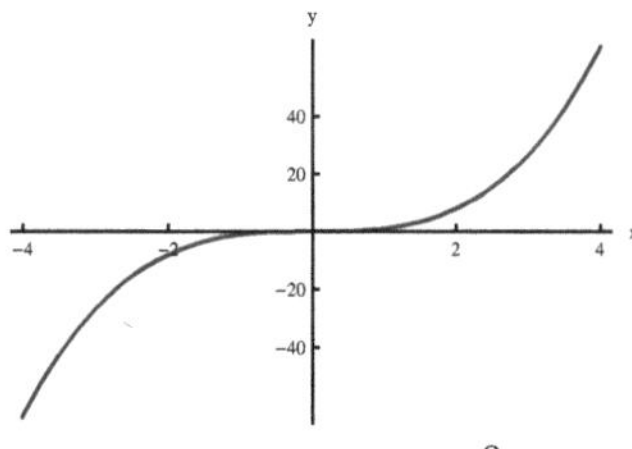

Figure 8: $y = x^3$

A function may have a local extremum at a point even though it is not differentiable at that point.

Example 3 Let $f(x) = x^{2/3}$.

The function is not differentiable at 0, as we discussed in Section 2.3. The graph of f has a cusp at $(0,0)$. The function has a local minimum at 0 ($f(0)$ is actually the absolute minimum of f on $\mathbb{R}$). $\square$

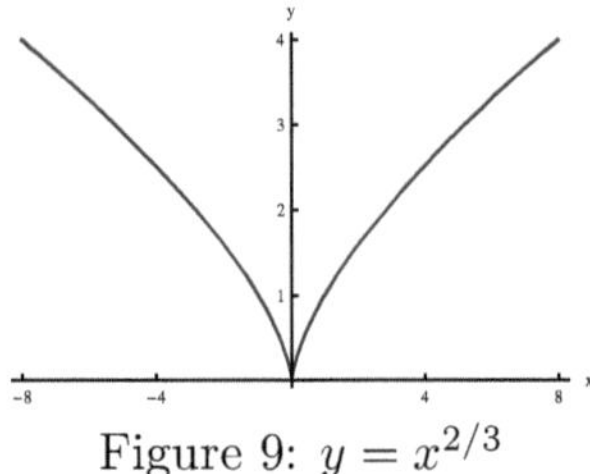

Figure 9: $y = x^{2/3}$

Definition 3 A point a is a **critical point** of the function f if a is a stationary point of f, i.e., $f'(a) = 0$, or f is defined at a but not differentiable at a.

Thus, 0 is the only critical point of the function of Example 3.

By Fermat's Theorem, if a function f is differentiable at a point a and f is differentiable at a, we must have $f'(a) = 0$. We just saw that a function can have a local extremum at a point where it is not differentiable. Therefore, we have the following **necessary condition for a local extremum:**

Assume that f has a local maximum or minimum at a. Then a is a critical point of f.

Remark 2 We have seen that a function does not have to have a local extremum at a point where its derivative is 0. Let's also note that a function need not have a local extremum at a critical point where it is not differentiable. For example, if $f(x) = x^{1/3}$, then f is not differentiable at 0, as we saw in Section 2.3, so that 0 is a critical point of f. Figure 10 shows the graph of f. The function does not have a local maximum or minimum at 0. Note that there is a vertical tangent to the graph of f at $(0,0)$. $\lozenge$

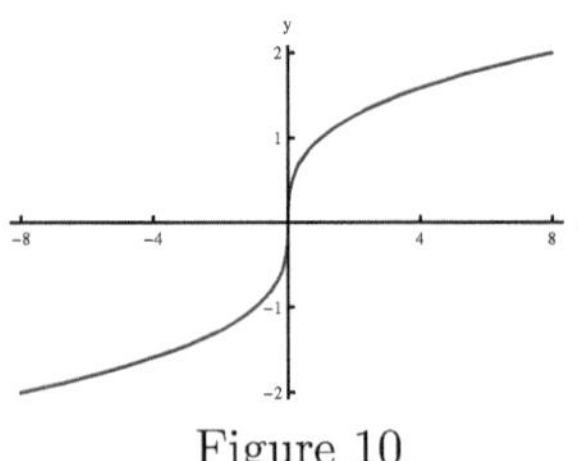

Figure 10

Our observations lead to the following strategy for the determination of the local extrema of a function:

1. **Determine the domains of f and f'.**

2. **Determine the critical points of f.**

3. **Make use of the derivative test for monotonicity to determine whether f is increasing or decreasing on the intervals that are separated from each other by the critical points of f or points at which f is not defined.**

4. **Determine the local extrema of f based on the results of item 3.**

Example 4 Let
$$f(x) = \frac{2}{3}x^3 + \frac{1}{4}x^4.$$

a) Determine the local maxima and minima of f.
b) Determine the absolute maximum and minimum values of f on the intervals $(-\infty, 0]$ and $[0, +\infty)$, provided that such values exist. Justify your responses if you claim that such values do not exist.

Solution

Figure 11 indicates that f has a local minimum near -2. The picture also inductees that the graph of f has a horizontal tangent line at the origin, even though it does not have a local extremum there. We can confirm all this with the help of the derivative test.

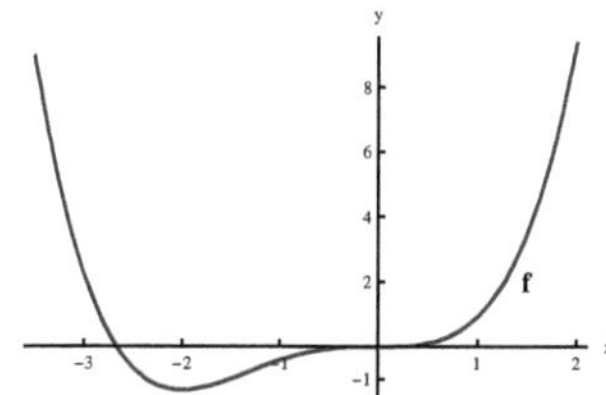

Figure 11: $f(x) = \frac{2}{3}x^3 + \frac{1}{4}x^4$

The polynomial f is differentiable on the entire number line. Therefore, the only critical points of f are its stationary points. We have

$$f'(x) = \frac{d}{dx}\left(\frac{2}{3}x^3 + \frac{1}{4}x^4\right) = \frac{2}{3}\left(3x^2\right) + \frac{1}{4}\left(4x^3\right)$$
$$= 2x^2 + x^3 = x^2\left(2 + x\right).$$

Therefore, $f'(x) = 0$ if $x = 0$ or $x = -2$. Thus, the stationary points of f are -2 and 0. Figure 12 displays the graph of f'.

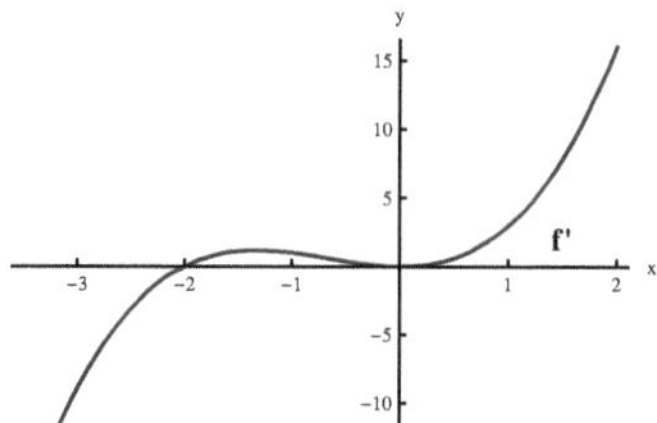

Figure 12: $f'(x) = x^2\left(2 + x\right)$

Since $x^2 > 0$ if $x \neq 0$, we see that the sign of $f'(x) = x^2\left(x + 2\right)$ is determined by the sign of the factor $x + 2$ if $x \neq 0$. Thus, $f'(x) < 0$ if $x < -2$, $f'(x) > 0$ if $-2 < x < 0$, and $f'(x) > 0$ if $x > 0$. Therefore, f is decreasing on the interval $(-\infty, -2]$ and increasing on the entire interval $[-2, +\infty)$. The conclusion is that f has a local minimum at -2. The function f does not have a local extremum at 0, even though 0 is a stationary point of f.

Table 1 summarizes the relationship between the sign of f' and the increasing/decreasing behavior of f, and indicates the local minimum of f..

x		-2		0	
$f'(x)$	$-$	0	$+$	0	$+$
f	decreasing	min	increasing		increasing

Table 1

b) Since f is decreasing on $(-\infty, -2]$ and increasing on $[-2, 0]$, the function attains its absolute minimum on $(-\infty, 0]$ at -2. We have $f(-2) = -4/3$.

Since f is increasing on $[0, +\infty)$, the function attains its absolute minimum on $[0, +\infty)$ at 0. We have $f(0) = 0$. The function does not attain an absolute maximum on $[0, +\infty)$ since it attains arbitrarily large values. Indeed,

$$\lim_{x \to +\infty} f(x) = \lim_{x \to +\infty} \left(\frac{2}{3}x^3 + \frac{1}{4}x^4\right) = \lim_{x \to +\infty} x^4 \left(\frac{2}{3x} + \frac{1}{4}\right) = +\infty$$

since

$$\lim_{x \to +\infty} x^4 = +\infty \text{ and } \lim_{x \to +\infty} \left(\frac{2}{3x} + \frac{1}{4}\right) = \frac{1}{4} > 0.$$

$\square$

A useful observation: In order to determine the sign of the derivative of a function f on an interval that does not contain a critical point or a point of discontinuity of f, it is sufficient to sample a single point in the interval and determine the sign of f' at that point, provided that f' is continuous on the interval. This is usually easier than working with inequalities. Indeed, assume that f' is continuous on the interval (a, b) and $f'(x) \neq 0$ for each $x \in (a, b)$. If x_1 and x_2 are in (a, b) and the sign of $f'(x_1)$ and the sign of $f'(x_2)$ are different, there must be a point c between x_1 and x_2 such that $f'(c) = 0$. This is a consequence of the Intermediate Value Theorem that was discussed in Section 2.10. Therefore, the sign of $f'(x)$ is the same for all $x \in (a, b)$.

Example 5 Let $f(x) = x(x-3)^{2/3}$.

a) Determine the critical points of f. Does the graph of f have vertical tangents or cusps at any of the critical points?
b) Determine the points at which f has a local maximum or minimum.

Solution

Figure 13 displays the graph of f. The picture suggests that the graph of f has a cusp at 3 where it has a local minimum The picture also indicates that f has a local maximum near 2.

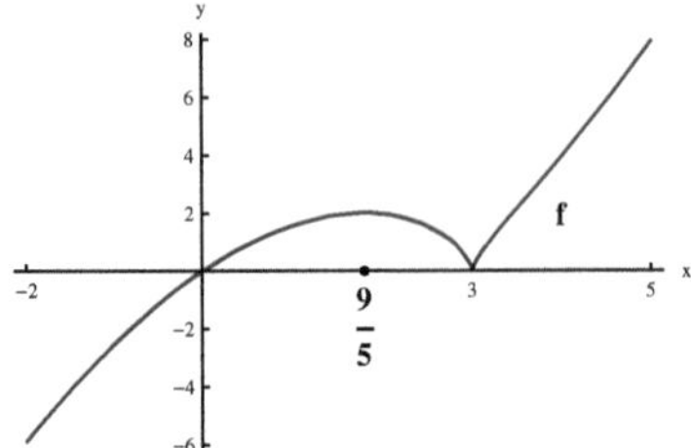

Figure 13: $f(x) = x(x-3)^{2/3}$

a) With the help of the product rule and the chain rule,

$$f'(x) = \frac{d}{dx}\left(x(x-3)^{2/3}\right) = \left(\frac{d}{dx}(x)\right)(x-3)^{2/3} + x\left(\frac{d}{dx}(x-3)^{2/3}\right)$$

$$= (1)(x-3)^{2/3} + x\left(\frac{2}{3}(x-3)^{-1/3}\right)$$

$$= (x-3)^{2/3} + \frac{2x}{3(x-3)^{1/3}}$$

$$= \frac{3(x-3)+2x}{3(x-3)^{1/3}} = \frac{5x-9}{3(x-3)^{1/3}}$$

if $x \neq 3$. Therefore,

$$f'(x) = 0 \Leftrightarrow 5x - 9 = 0 \Leftrightarrow x = \frac{9}{5}.$$

Therefore $9/5$ is the stationary point of f. The function is not differentiable at 3, as suggested by the expression for $f'(x)$ (you should be able to confirm this), although f is defined at 3. Therefore, the critical points of f are $9/5$ and 3. We have

$$\lim_{x \to 3-} f'(x) = \lim_{x \to 3-}\left(\frac{5x-9}{3(x-3)^{1/3}}\right) = -\infty,$$

and

$$\lim_{x \to 3+} f'(x) = \lim_{x \to 3+}\left(\frac{5x-9}{3(x-3)^{1/3}}\right) = +\infty$$

(confirm with the help of Proposition 1 and Proposition 2 of Section 1.6). Therefore, the graph of f has a cusp at $(3, f(3)) = (3, 0)$. Figure 14 shows the graph of f'. The picture is consistent with our conclusions.

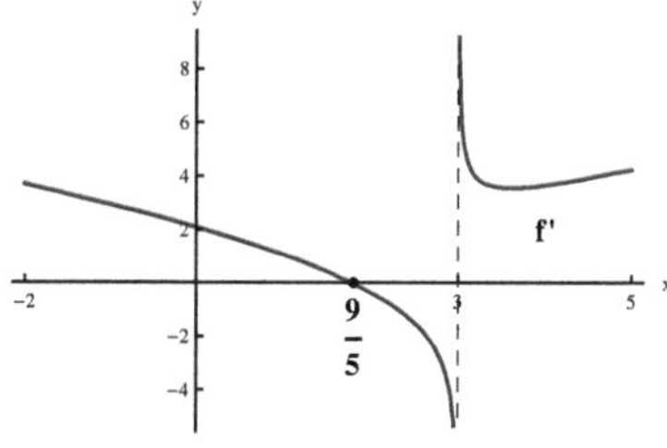

Figure 14

b) The function is defined everywhere. The intervals that are separated by the critical points of f are

$$(-\infty, \frac{9}{5}], [\frac{9}{5}, 3] \text{ and } [3, +\infty)$$

Since f' is continuous in the interior of each interval, and the only zero of f' is $9/5$, f' has a constant sign in the interior of each interval. It is practical to sample a point in the interior of each interval. Table 2 summarizes the relationship between the sign of f', the increasing/decreasing behavior of f, and the local extrema of f. By the derivative test for monotonicity, f is increasing on the interval $(-\infty, 9/5]$, decreasing on $[9/5, 3]$ and increasing on $[9/5, +\infty)$. Thus, f has a local maximum at $9/5$ and a local minimum at 3. $\square$

x		9/5		3	
$f'(x)$	+	0	−	undefined	+
$f(x)$	increasing	max	decreasing	min	increasing

Table 2

Example 6 Let

$$f(x) = x + \frac{1}{x-1}.$$

a) Determine the stationary points of f.
b) Determine whether f has a local maximum or minimum at each stationary point.
c) Determine the absolute maximum and minimum values of f on $(1, +\infty)$, provided that such values exist. Justify your responses if you claim that such values do not exist.

Solution

Figure 15 displays the graph of f. The picture suggests that the graph of f has a vertical asymptote at $x = 1$ and that f has a local maximum near 0 and a local minimum near 2.

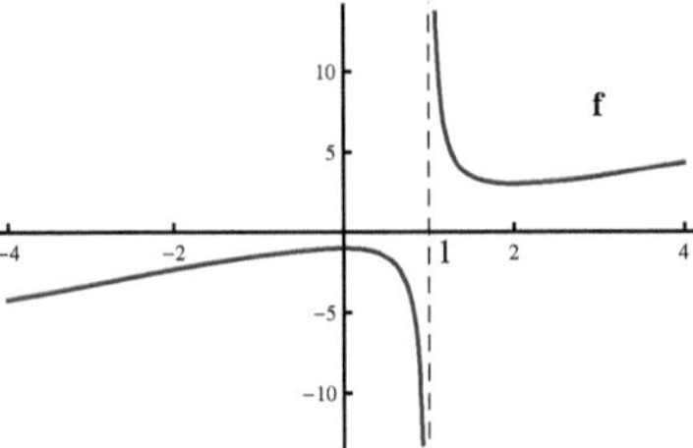

Figure 15: $f(x) = x + \dfrac{1}{x-1}$

a) We have

$$f'(x) = \frac{d}{dx}\left(x + \frac{1}{x-1}\right) = \frac{d}{dx}(x) + \frac{d}{dx}\left(\frac{1}{x-1}\right)$$

$$= 1 + \frac{-\dfrac{d}{dx}(x-1)}{(x-1)^2} = 1 - \frac{1}{(x-1)^2}$$

$$= \frac{(x-1)^2 - 1}{(x-1)^2} = \frac{x^2 - 2x}{(x-1)^2} = \frac{x(x-2)}{(x-1)^2}.$$

Therefore,

$$f'(x) = 0 \Leftrightarrow x = 0 \text{ or } x = 2.$$

Thus, the stationary points of f are 0 and 2.

b) We will apply the derivative test for monotonicity. We have to take into account the fact that f and f' are not defined at $x = 1$ (you can show that the line $x = 1$ is a vertical asymptote for the graphs of f and f'). The intervals that are separated from each other by the discontinuities or the stationary points of f are

$$(-\infty, 0], \ [0, 1), \ (1, 2] \text{ and } [2, +\infty).$$

Notice that we have excluded the point 1 from the relevant intervals, since f is not defined at 1. You can determine the sign of f' in the interior of each interval by making use of your knowledge about inequalities or by sampling appropriate values of f'. Figure 16 shows the graph of f'.

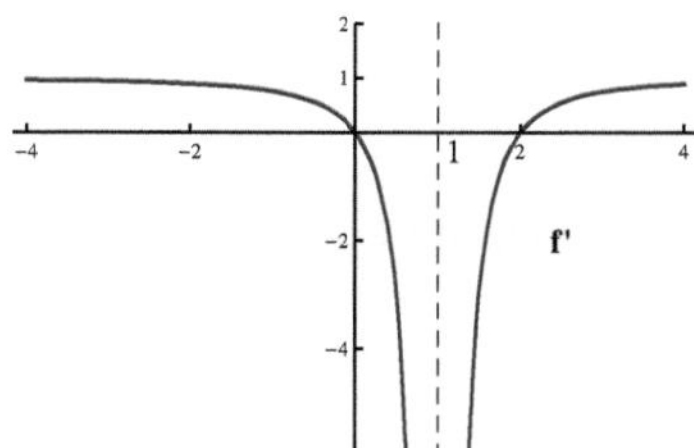

Table 3 summarizes the relationship between the sign of f', the increasing/decreasing behavior of f, and the local extrema of f. The function is increasing on $(-\infty, 0]$, decreasing on $[0, 1)$, decreasing on $(1, 2]$, and increasing on $[2, +\infty)$. Thus, f has a local maximum at 0 and a local minimum at 2. $\square$

x		0		1		2	
$f'(x)$	$+$	0	$-$	undefined	$-$	0	$+$
f	increasing	max	decreasing	undefined	decreasing	min	increasing

Table 3

c) Since f is decreasing on $(1, 2]$ and increasing on $[2, +\infty)$, it attains its absolute value on $(1, +\infty)$ at 2. We have $f(2) = 3$.

The function does not have an absolute maximum on $(1, +\infty)$ since it attains arbitrarily large values. We have

$$\lim_{x \to 1+} f(x) = \lim_{x \to +\infty} f(x) = +\infty$$

(confirm). $\square$

The Proof of Fermat's Theorem

Assume that f attains a local maximum at a. Then $f(a + h) \leq f(a)$, if $|h|$ is small enough. Therefore, $f(a + h) - f(a) \leq 0$ if $h > 0$ and h is sufficiently small. Thus.

$$\frac{f(a + h) - f(a)}{h} \leq 0$$

under these conditions. Therefore,

$$f'(a) = \lim_{h \to 0} \frac{f(a + h) - f(a)}{h} = \lim_{h \to 0+} \frac{f(a + h) - f(a)}{h} \leq 0.$$

Similarly, if we consider $h < 0$ such that $|h|$ is sufficiently small, $f(a + h) \leq f(a)$ so that $f(a + h) - f(a) \leq 0$. Since $h < 0$, we have

$$\frac{f(a + h) - f(a)}{h} \geq 0,$$

so that

$$f'(a) = \lim_{h \to 0} \frac{f(a + h) - f(a)}{h} = \lim_{h \to 0-} \frac{f(a + h) - f(a)}{h} \geq 0.$$

Since we deduced that $f'(a) \leq 0$ and $f'(a) \geq 0$, we must have $f'(a) = 0$, as claimed. The proof of the fact that $f'(a) = 0$ at a point a at which f attains a local minimum is similar.

∎

Remark 3 In the above proof we have used the fact that $\lim_{h \to 0+} g(h) \leq 0$ if $g(h) \leq 0$ when h is sufficiently small and positive (and a similar fact for $\lim_{h \to 0-} g(h)$). Indeed, if we assume that $\lim_{h \to 0+} g(h) = L > 0$, then $g(h)$ must also be positive if $h > 0$ and h is sufficiently small, since $g(h)$ is as close to L as desired if $h > 0$ and h is small enough. This contradicts the fact that $g(h) \leq 0$ when h is sufficiently small and positive. $\Diamond$

Problems

In problems 1 - 6, a function f is given and its graph is displayed. Find the critical points and the points at which f has a local maximum or minimum.

1.

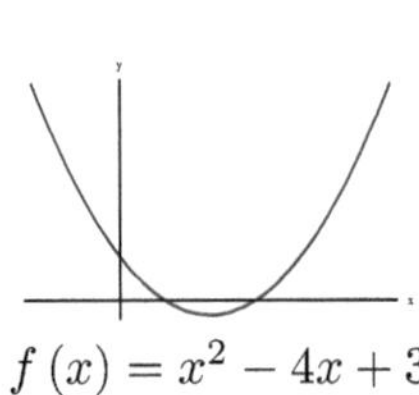

$$f(x) = x^2 - 4x + 3$$

2.

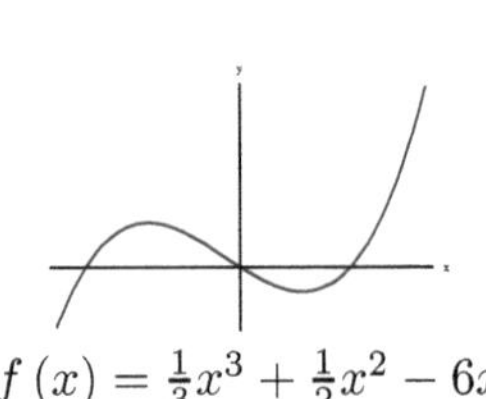

$$f(x) = \tfrac{1}{3}x^3 + \tfrac{1}{2}x^2 - 6x$$

3.

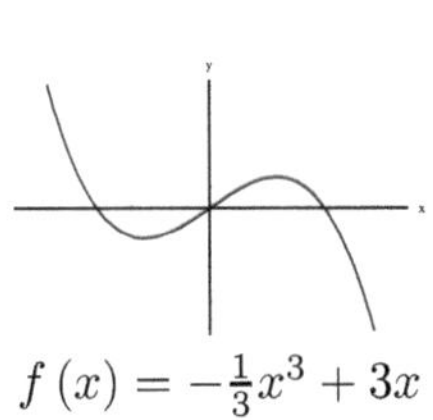

$$f(x) = -\tfrac{1}{3}x^3 + 3x$$

4. $f(x) = (x-2)^2 (x+3)^{2/3}$

5.

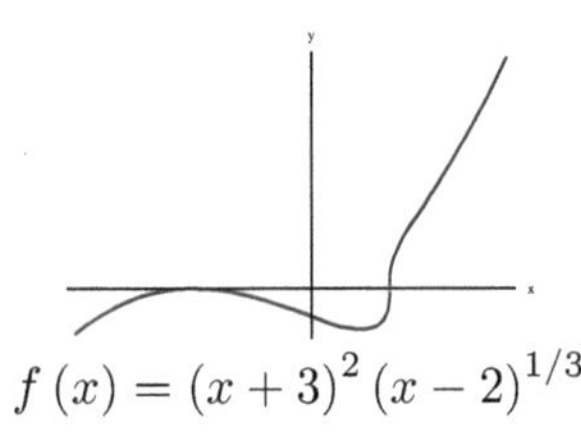

$$f(x) = (x+3)^2 (x-2)^{1/3}$$

6.

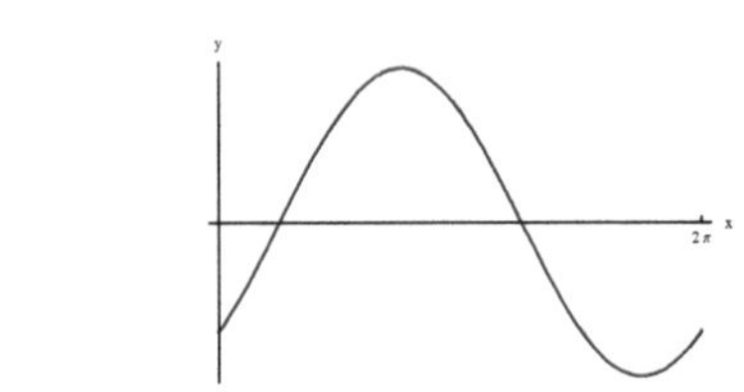

$$f(x) = \sin(x) - \cos(x) \text{ (only on the interval } [0, 2\pi])$$

In problems 7 - 18, given the function f,
a) Find the critical points of f.
b) Make use of the derivative test for monotonicity to determine intervals on which f is increasing or decreasing, and the points at which f has a local maximum or minimum.

7.

$$f(x) = x^2 + 4x$$

8.

$$f(x) = -x^2 + 3x - 2$$

9.
$$f(x) = x - \frac{1}{3}x^3$$

10.
$$f(x) = -4x + 8x^3$$

11.
$$f(x) = 1 - 8x^2 + 8x^4$$

12.
$$f(x) = \frac{1}{4}x^4 - \frac{8}{3}x^3 + 10x^2 - 16x + 20$$

Hint: One of the critical points is 4.

13.
$$f(x) = \frac{3}{x^2 - x - 2}$$

14.
$$f(x) = \frac{4x^2 + 4x - 8}{x^2 + x - 12}$$

15.
$$f(x) = \frac{x^2 + 2x + 1}{x + 2}$$

16
$$f(x) = x(x - 2)^{4/5}$$

17.
$$f(x) = x\,(x - 4)^{1/3}$$

18.
$$f(x) = \left(x^2 - 4\right)^{2/3}$$

In problems 19-26, given the function f, make use of the derivative test for monotonicity to determine the absolute minimum and the absolute maximum of f on the given interval, provided that such a value exists. If you claim that such a value does not exist, provide an explanation.

19.
$$f(x) = -2x + \frac{1}{2}x^2 + \frac{1}{3}x^3;\ \text{on}\ [0, +\infty)$$

20.
$$f(x) = 3 - x^2 + x^4\ \text{on}\ (-\infty, 0].$$

21.
$$f(x) = \frac{2x^2 - 8x + 1}{x - 4}\ \text{on}\ (-\infty, 4).$$

22.
$$f(x) = x^2 + \frac{16}{x}\ \text{on}\ (0, +\infty)$$

23.
$$f(x) = \frac{4x^2 + 4x - 8}{x^2 + x - 12}\ \text{on}\ (-4, 3).$$

24.
$$f(x) = \frac{x^2 + 2x + 1}{x + 2}\ \text{on}\ (-2, \infty).$$

25.
$$f(x) = x(x - 2)^{4/5}\ \text{on}\ [0, \infty).$$

26.
$$f(x) = \left(x^2 - 4\right)^{2/3}\ \text{on}\ [0, \infty).$$

3.2   The Mean Value Theorem

We will suggest a practical search procedure for the determination of the absolute extrema of continuous functions on closed and bounded intervals. We will also discuss **the Mean Value Theorem**. This theorem provides a theoretical basis for the derivative test for monotonicity and some other general facts that will be useful in later chapters.

Absolute Extrema on Closed and Bounded Intervals

In Section 3.1 we saw that that the absolute maximum or minimum of a function on a given set may not exist. Here is another example:

Example 1 Let $f(x) = \sin(x)$. Determine the absolute maximum and the absolute minimum of f on the interval $(0, \pi/2)$. Justify your response if such a value does not exist.

Solution

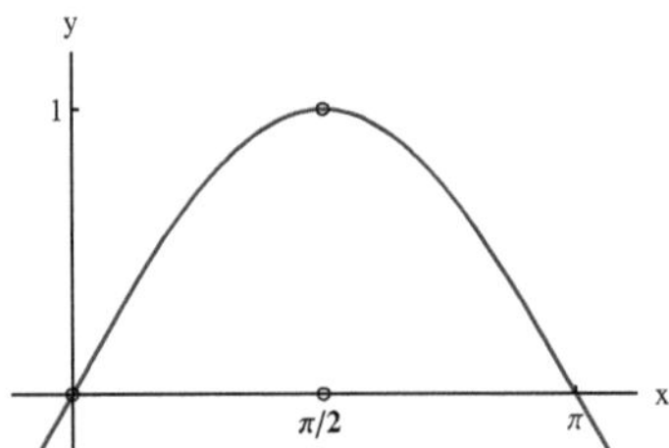

Figure 1: The sine function does not attain absolute extrema on $(0, \pi/2)$

The sine function is continuous on the entire number line. In particular, sine is continuous on the open interval $(0, \pi/2)$. We have

$$\lim_{x \to 0+} \sin(x) = \sin(0) = 0, \quad \lim_{x \to \pi/2-} \sin(x) = \sin(\pi/2) = 1.$$

Therefore, f attains values that are arbitrarily close 0 and 1. Since f is monotone increasing on $(0, \pi/2)$ the only candidate for the absolute maximum of sine on $(0, \pi/2)$ is 1, and the the only candidate for the absolute minimum of sine on $(0, \pi/2)$ is 0. Since sine does not attain either value on the open interval $(0, \pi/2)$, it does not have an absolute maximum or absolute minimum on $(0, \pi/2)$. On the other hand, sine does attain its absolute extrema on the closed interval $[0, \pi/2]$. $\square$

The following theorem guarantees the existence of these values under certain conditions:

Theorem 1 Assume that the function f is continuous on the closed and bounded interval $[a, b]$. Then f attains its absolute maximum and absolute minimum values on $[a, b]$.

Let's stress the fact that the validity of the theorem requires the continuity of f at a from the right and the continuity of f at b from the left, as well as its continuity at each point in the interior of $[a, b]$, i.e., in the open interval (a, b). In Example **??** the intervals that we considered are unbounded. In Example **??** the function is not continuous at each point of the given interval. In Example 1 the given interval is not closed.

We will leave the proof of Theorem 1 to a course in advanced calculus. At this point, we are interested in the practical consequences of the theorem. Assume that f is continuous on $[a, b]$ and differentiable in the interior (a, b), with the exception of a finite number of points at which it is not differentiable. If the absolute maximum or the absolute minimum of f is attained at a point c in the interior (a, b), then f has a local maximum or minimum at c, since c is contained in an open interval that is contained in (a, b). Therefore, c must be a critical point of f. It is also possible that an absolute extremum is attained at one of the endpoints a or b. Therefore, you may follow the following **search procedure** for the absolute maximum and the absolute minimum of f on the closed and bounded interval $[a, b]$, without determining the intervals on which f is increasing or decreasing:

1. **Determine the critical points of f in (a, b).**

2. **Calculate the values of f at its critical points in (a, b) and at the endpoints a and b.**

3. **The maximum of the values that have been calculated is the absolute maximum of f on $[a, b]$, and the minimum of the values is the absolute minimum of f on $[a, b]$.**

Example 2 Let

$$f(x) = x - \frac{1}{6}x^3.$$

Determine the absolute maximum and the absolute minimum of f on the interval $[-1, 2]$.

Solution

Since the polynomial f is differentiable on the entire number line, the only critical points of f are its stationary points, i.e., the points at which f' is 0. Since

$$f'(x) = 1 - \frac{1}{2}x^2 = 0 \Leftrightarrow x = \pm\sqrt{2}$$

the only stationary point of f in the interval $[-1, 2]$ is $\sqrt{2}$. We have

$$f\left(\sqrt{2}\right) = \sqrt{2} - \frac{1}{6}\left(\sqrt{2}\right)^3 = \sqrt{2} - \frac{2^{3/2}}{6} \cong 0.942\,809$$

We need to calculate the values at the endpoints of $[-1, 2]$ as well:

$$f(-1) = -\frac{5}{6} \cong -0.833\,333 \text{ and } f(2) = \frac{2}{3} \cong 0.666\,667$$

When we compare $f\left(\sqrt{2}\right)$, $f(-1)$ and $f(2)$, we see that the maximum is $f\left(\sqrt{2}\right)$ and the minimum is $f(-1)$. Therefore, the absolute maximum of f on $[-1, 2]$ is

$$f\left(\sqrt{2}\right) = \sqrt{2} - \frac{2^{3/2}}{6} \cong 0.942\,809\,,$$

and the absolute minimum of f on $[-1, 2]$ is

$$f(-1) = -\frac{5}{6} \cong -0.833\,333\,.$$

Figure 2 shows the graph of f on the interval $[-1, 2]$. The picture is consistent with the conclusions that we reached using the suggested search procedure for absolute extrema. $\square$

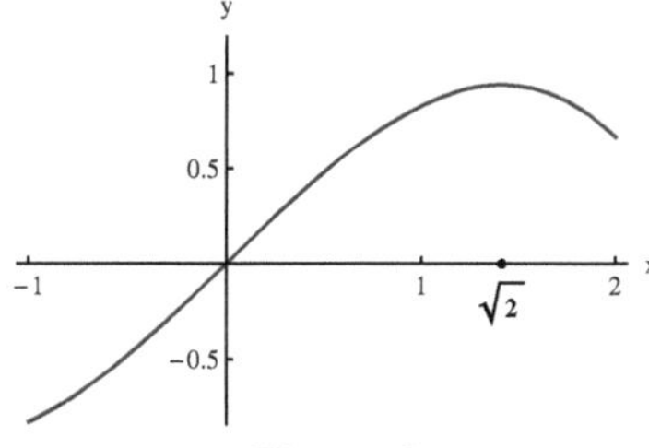

Figure 2

Example 3 Let $f(x) = (x^2 - 4)^{2/5}$. Determine the absolute maximum and the absolute minimum of f on the interval $[-4, 3]$.

Solution

The function is continuous on the entire number line. Indeed, f can be expressed as $F \circ G$, where $F(u) = u^{2/5}$, $G(x) = x^2 - 4$, and both F and G are continuous on $\mathbb{R}$. In particular, f

is continuous on the closed and bounded interval $[-4, 3]$, and we can implement the suggested search procedure to find the absolute extrema of f on $[-4, 3]$.

By the chain rule,

$$f'(x) = \frac{d}{dx}(x^2 - 4)^{2/5} = \left(\frac{2}{5}(x^2 - 4)^{-3/5}\right)(2x) = \frac{4x}{(x^2 - 4)^{3/5}}$$

if $x^2 - 4 \neq 0$, i.e., if $x \neq -2$ and $x \neq 2$. The function is not differentiable at ± 2 (show that the graph of f has cusps at $(\pm 2, 0)$). Thus, -2 and 2 are critical points of f. We have $f'(0) = 0$, so that 0 is also a critical point.

We have

$$f(\pm 2) = 0, \ f(0) = 4^{2/5} \cong 1.741\,1, \ f(-4) = 12^{2/5} \cong 2.701\,92 \text{ and } f(3) = 5^{2/5} \cong 1.903\,65$$

Therefore, the absolute maximum of f on $[-4, 3]$ is $f(-4) = 12^{2/5}$, and the absolute minimum of f on $[-4, 3]$ is $f(\pm 2) = 0$. Figure 3 shows the graph of f. $\square$

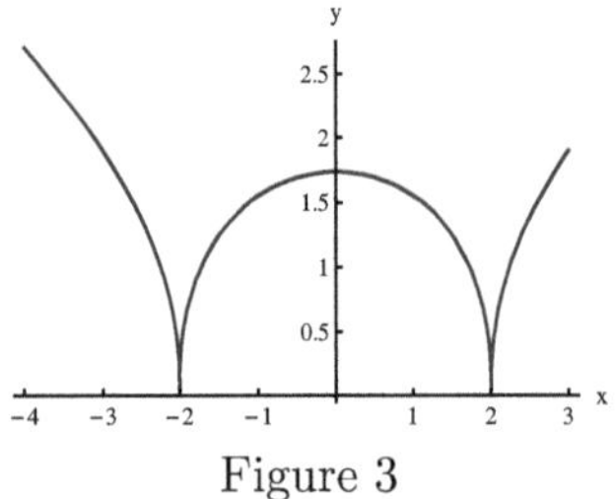

Figure 3

It is not always possible to determine the exact values of local or absolute extrema, as in the following example.

Example 4 Let

$$f(x) = \sin(x) - \frac{1}{2}\sin(2x) + \frac{1}{3}\sin(3x) - \frac{1}{4}\sin(4x).$$

a) Determine the critical points of f in the interval $[1, 3]$ with the help of your calculator.

b) Determine the absolute maximum and the absolute minimum of f on $[1, 3]$ (round to 6 significant digits).

Solution

a) Since the trigonometric polynomial f is differentiable on the entire number line, the only critical points of f are the points at which f' is 0. Figure 4 displays the graphs of f and f' on the interval $[0, 3]$.

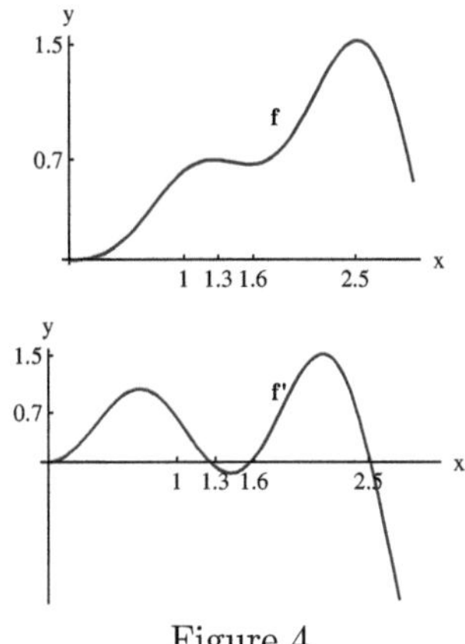

Figure 4

We have

$$f'(x) = \frac{d}{dx}\left(\sin(x) - \frac{1}{2}\sin(2x) + \frac{1}{3}\sin(3x) - \frac{1}{4}\sin(4x)\right)$$

$$= \cos(x) - \frac{1}{2}\left(2\cos(2x)\right) + \frac{1}{3}\left(3\cos(3x)\right) - \frac{1}{4}\left(4\cos(4x)\right)$$

$$= \cos(x) - \cos(2x) + \cos(3x) - \cos(4x).$$

It appears that the graph of f' intersects the x-axis near the points 1.3, 1.6 and 2.5. Thus, we are interested in the solutions of the equation

$$f'(x) = \cos(x) - \cos(2x) + \cos(3x) - \cos(4x) = 0$$

which are near these points. These solutions are

$$x_1 \cong 1.25664, \ x_2 \cong 1.5708, \ \text{and} \ x_3 \cong 2.51327$$

It is practical to use the approximate equation solver of your calculator. You may also approximate some of the solutions with the help of Newton's method, as an exercise in Newton's method. We have

$$f(x_1) \cong 0.699, \ f(x_2) \cong 0.666667, \ f(x_3) \cong 1.52728$$

We need to calculate the values of f at the endpoints of the interval $[1,3]$ as well:

$$f(1) \cong 0.623\,063 \ \text{ and } \ f(3) \cong 0.552\,344$$

We see that the maximum of the above values is $f(x_3)$ and the minimum value is $f(1)$. Therefore, the absolute maximum of f on $[1,3]$ is $f(x_3) \cong 1.52728$ and the absolute minimum of f on $[1,3]$ is $f(1) \cong 0.623\,063$. $\square$

Rolle's Theorem and the Mean Value Theorem

Now we will discuss the Mean Value Theorem that provides the theoretical basis of the derivative test for monotonicity and some other theorems of calculus. Let us begin with a theorem that will lead to the Mean Value theorem:

Theorem 2 (ROLLE'S THEOREM) Assume that f is continuous on $[a, b]$, differentiable at each point in (a, b) and $f(a) = f(b)$. There exists $c \in (a, b)$ such that $f'(c) = 0$.

Graphically, Rolle's Theorem predicts the existence of at least one point c between a and b such that the tangent line to the graph of f at $(c, f(c))$ is horizontal, if $f(a) = f(b)$ (there may be more than one such point). This is illustrated in Figure 5.

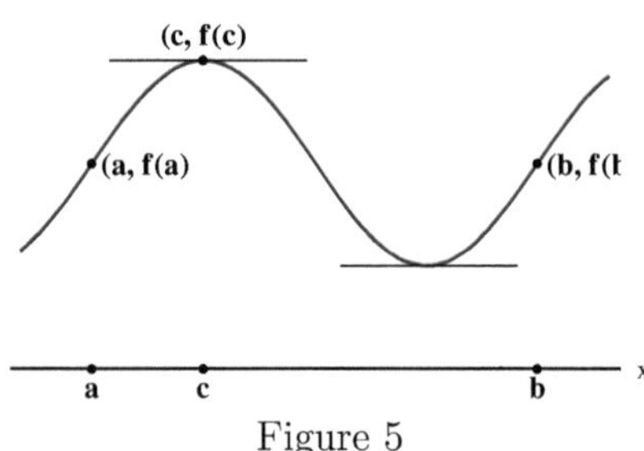

Figure 5

The proof of Rolle's Theorem

If f is a constant on $[a, b]$, we have $f'(x) = 0$ for each $x \in (a, b)$. Therefore, we need to consider the case of a function f that is not constant on $[a, b]$. By Theorem 1 f attains its maximum and minimum values on $[a, b]$. We cannot have both of these values equal to the common value of the function at a and b: This would imply that f is constant on $[a, b]$. Assume that the maximum value of f on $[a, b]$ is different from $f(a)$, and therefore different from $f(b) = f(a)$. Therefore, if $f(c)$ is that maximum value, we must have $c \in (a, b)$. This implies that f has a local maximum at c, so that we must have $f'(c) = 0$ (Theorem 2 of Section 4.1). Similarly, if the minimum value of f on $[a, b]$ is different from $f(a)$ and $f(b)$, and f attains that value at $c \in (a, b)$, we must have $f'(c) = 0$. ■

The Mean Value Theorem follows from Rolle's Theorem:

Theorem 3 (THE MEAN VALUE THEOREM) Assume that f is continuous on $[a, b]$ and differentiable on (a, b). There exists $c \in [a, b]$ such that

$$f(b) - f(a) = f'(c)(b - a).$$

We can express the Mean Value Theorem by writing

$$f'(c) = \frac{f(b) - f(a)}{b - a}.$$

Since the expression on the right is the slope of the secant line that connects the point $(a, f(a))$ to $(b, f(b))$, the Mean Value predicts the existence of at least one point in the open interval (a, b) such that the tangent line at the corresponding point on the graph of f is parallel to that secant line. Figure 6 illustrates the graphical meaning of the Mean Value Theorem.

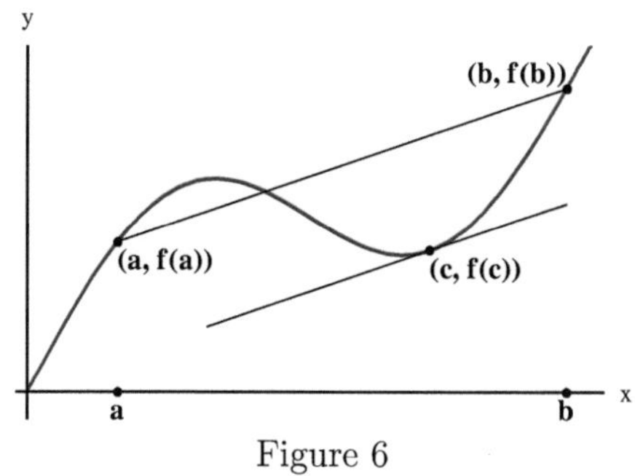

Figure 6

The Proof of the Mean Value Theorem

Let g be the linear function whose graph is the secant line that passes through the points $(a, f(a))$ and $(b, f(b))$, as in Figure 9. The slope of the secant line is

$$\frac{f(b) - f(a)}{b - a}.$$

Therefore,

$$g(x) = f(a) + \left(\frac{f(b) - f(a)}{b - a} \right)(x - a)$$

(the above expression is the point-slope form of the equation of the line that passes through the points $(a, f(a))$ and $(b, f(b))$, with basepoint a). Set

$$h(x) = f(x) - g(x) = f(x) - f(a) - \left(\frac{f(b) - f(a)}{b - a} \right)(x - a).$$

We have $h(a) = h(b) = 0$. By Rolle's Theorem, there exists $c \in (a, b)$ such that $h'(c) = 0$. Since

$$h'(x) = f'(x) - \frac{f(b) - f(a)}{b - a},$$

we have

$$h'(c) = 0 \Leftrightarrow f'(c) = \frac{f(b) - f(a)}{b - a},$$

so that

$$f(b) - f(a) = f'(c)(b - a).$$

∎

We will need a generalization of the Mean Value Theorem in our discussion of L'Hôpital's rule in Appendix D:

Theorem 4 (THE GENERALIZED MEAN VALUE THEOREM) Assume that f and g are continuous on $[a, b]$ and differentiable in (a, b). Then there exists $c \in [a, b]$ such that

$$f'(c)[g(b) - g(a)] = g'(c)[f(b) - f(a)].$$

Note that the Mean Value Theorem follows from the Generalized Mean Value Theorem if we set $g(x) = x$. In this case, $g'(x) = 1$, so that

$$f'(c)\left[g(b) - g(a)\right] = g'(c)\left[f(b) - f(a)\right] \Rightarrow f'(c)(b - a) = f(b) - f(a).$$

The Proof Of Theorem 4

Set

$$h(x) = \left[f(x) - f(a)\right]\left[g(b) - g(a)\right] - \left[f(b) - f(a)\right]\left[g(x) - g(a)\right].$$

Then,

$$h(a) = 0 \text{ and } h(b) = 0.$$

By Rolle's Theorem, there exists $c \in (a, b)$ such that $h'(c) = 0$. We have

$$h'(x) = f'(x)\left[g(b) - g(a)\right] - g'(x)\left[f(b) - f(a)\right].$$

Therefore,

$$h'(c) = 0 \Leftrightarrow f'(c)\left[g(b) - g(a)\right] = g'(c)\left[f(b) - f(a)\right].$$

■

The Mean Value Theorem is an "existence theorem" that is used to prove other theorems such as **the derivative test for monotonicity)**. Let's restate that theorem:

Assume that f is continuous on the interval J and that f is differentiable at each x in the interior of J. If $f'(x) > 0$ for each x in the interior of J, then f is increasing on J. If $f'(x) < 0$ for each x in the interior of J, then f is decreasing on J.

Proof

Let x_1 and x_2 belong to the interval J, and $x_1 < x_2$. Then, the interval $[x_1, x_2]$ is contained in J, and the open interval (x_1, x_2) is in the interior of J. Thus, f is continuous on $[x_1, x_2]$ and differentiable in the interior of $[x_1, x_2]$, so that the Mean Value Theorem is applicable to f on the interval $[x_1, x_2]$: There exists a point $c \in (x_1, x_2)$ such that

$$f(x_2) - f(x_1) = f'(c)(x_2 - x_1).$$

If we assume that $f'(x) > 0$ for each x in the interior of J, we have $f'(c) > 0$. Therefore,

$$f(x_2) - f(x_1) = f'(c)(x_2 - x_1) > 0,$$

so that $f(x_2) > f(x_1)$. This shows that f is increasing on J if $f'(x) > 0$ for each x in the interior of J.

If we assume that $f'(x) < 0$ for each x in the interior of J, we have

$$f(x_2) - f(x_1) = f'(c)(x_2 - x_1) < 0,$$

so that $f(x_2) < f(x_1)$. This shows that f is decreasing on J. ■

We will include two other consequences of the Mean Value Theorem in this section for future reference. Intuitively, a function whose derivative has the constant value 0 should be a constant. This is indeed the case:

Proposition 1 Assume that $f'(x) = 0$ for each x in an interval J. Then f must be a constant on J.

Proof

Let a be a point in J. By the Mean Value Theorem, if x is an arbitrary point in J, and $x > a$, there exists a point c between a and x such that

$$f(x) - f(a) = f'(c)(x - a)$$

Since $f'(c) = 0$, this implies that $f(x) - f(a) = 0$, i.e., $f(x) = f(a)$. Similarly, if $x < a$, there exists a point between x and a such that

$$f(a) - f(x) = f'(c)(a - x),$$

so that $f(a) - f(x) = 0$, i.e., $f(x) = f(a)$. Thus, we have shown that $f(x) = f(a)$ for any x in J, so that f has the constant value $f(a)$ on the interval J. ∎

Two functions that have the same derivative differ at most by a constant:

Corollary **Assume that $f'(x) = g'(x)$ for each x in an interval J. Then there exists a constant C such that $g(x) = f(x) + C$ for each $x \in J$.**

Proof

Set $h(x) = g(x) - f(x)$. Then $h'(x) = g'(x) - f'(x) = 0$ for each $x \in J$. By Proposition 1 there exists a constant C such that $h(x) = C$, i.e., $g(x) - f(x) = C$ for each $x \in J$. Therefore, $g(x) = f(x) + C$ for each $x \in J$. ∎

Problems

In problems 1-8, given the function f and the interval J, determine the absolute extrema of f on J by implementing the search procedure for the extrema of a continuous function on a closed and bounded interval (you need not determine the intervals on which f is increasing/decreasing).

1.
$$f(x) = \frac{1}{3}x^3 + x^2 - 3x, \ J = [-4, 4].$$

2.
$$f(x) = \frac{1}{3}x^3 - \frac{3}{2}x^2 - 4x, \ J = [0, 6].$$

3.
$$f(x) = \frac{1}{4}x^4 - x^3 - 2x^2 + 12x, \ J = [-1, 4]$$

Hint: One of the critical points is 3.

4.
$$f(x) = \frac{x^2 - 4x + 1}{x - 4}, \ J = [1, 3.5].$$

5.
$$f(x) = x^2 \left(4 - x^2\right)^{2/3}, \ [0, 2].$$

6.
$$f(x) = \sin\left(\frac{x}{4}\right), \ J = [0, 5\pi]$$

7.
$$f(x) = \cos^4(x), \ J = \left[\frac{\pi}{3}, \frac{3\pi}{4}\right].$$

8.
$$f(x) = \sin(x) - \cos(x), \ J = [0, \pi].$$

In problems 9 and 10, confirm that the Mean Value Theorem is valid for the function f on the interval J:

9.
$$f(x) = x^3, \ J = [1, 3]$$

10.
$$f(x) = \frac{1}{3}x^3 - \frac{1}{2}x^2 - 2x, \ J = [1, 4]$$

3.3   Concavity and Extrema

In Sections 3.1 and 3.2 we considered the relationship between the sign of the derivative of a function and its increasing/decreasing behavior. In this section we will investigate the increasing/decreasing behavior of the derivative of a function. This will correspond to the determination of the intervals on which the graph of f is "bending up" or "bending down". The computations will involve the second derivative. We will also establish a useful link between the sign of the second derivative and the local or absolute extrema of a function.

Concavity

We will describe the intuitive notion that a graph is "bending up" or "bending down" in precise terms. Let's begin by looking at a specific case.

Example 1 Let $f(x) = x^3$.

The function is increasing on the entire number line. In particular, f is increasing on each of the intervals $(-\infty, 0]$ and $[0, +\infty)$. Nevertheless, Figure 1 indicates that there is a difference between the behavior of f on $(-\infty, 0]$ and the behavior of f on $[0, +\infty)$. The part of the graph of f that corresponds to the interval $(-\infty, 0]$ appears to be "bending down", and the part of the graph corresponding to $[0, +\infty)$ appears to be "bending up".

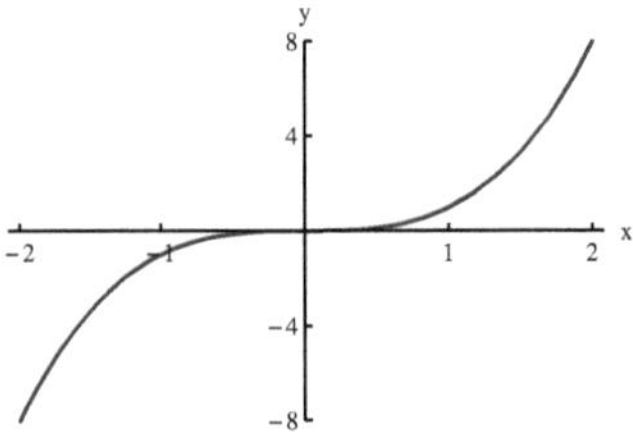

Figure 1: $f(x) = x^3$

These observations correspond to the fact that the slope of the graph of f at $(x, f(x))$ is a decreasing function of x on $(-\infty, 0]$ and an increasing function of x in $[0, +\infty)$. Indeed, the slope of the graph of f at $(x, f(x))$ is $f'(x) = 3x^2$. Figure 2 shows the graph of f'. The derivative function f' is decreasing on $(-\infty, 0]$ and increasing on $[0, +\infty)$. $\square$

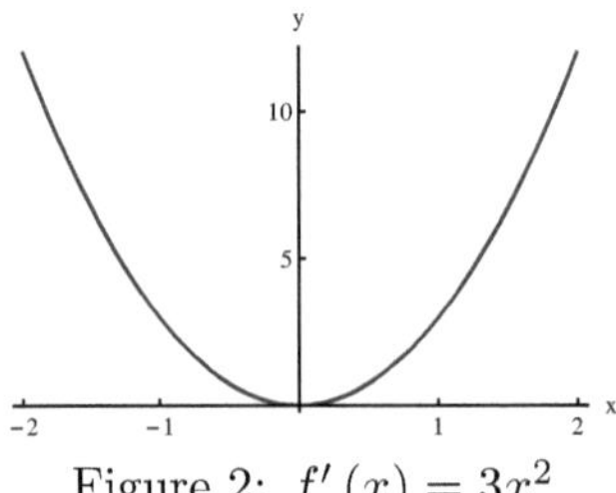

Figure 2: $f'(x) = 3x^2$

The usual terms for "bending up" and "bending down" are "concave up" and "concave down", respectively:

Definition 1 Assume that a function f is continuous on an interval J and differentiable in the interior of J. The graph of f is **concave up** on J if f' is increasing in the interior of J. The graph of f is **concave down** on J if f' is decreasing in the interior of J.

Thus, in Example 1 the graph of f is concave down on the interval $(-\infty, 0]$, and concave up on the interval $[0, +\infty)$.

Note that Definition 1 does not require the differentiability of the given function at the endpoints of the interval in question, even though it requires the appropriate one-sided continuity of the function if such a point belongs to the interval.

Example 2 Let $f(x) = x^{1/3}$.

We have

$$f'(x) = \frac{d}{dx}\left(x^{1/3}\right) = \frac{1}{3}x^{-2/3} = \frac{1}{3x^{2/3}}$$

if $x \neq 0$. The function is not differentiable at 0, but it is continuous for each $x \in \mathbb{R}$, including 0. Therefore, f is continuous on the intervals $(-\infty, 0]$ and $[0, +\infty)$. We have

$$\lim_{x \to 0+} f'(x) = \lim_{x \to 0-} f'(x) = +\infty,$$

so that the graph of f has a vertical tangent at $(0,0)$. Figure 3 shows the graphs of f and f'. The derivative is increasing in the interior of $(-\infty, 0]$, i.e., on the interval $(-\infty, 0)$. Therefore, the graph of f is concave up on $(-\infty, 0]$. The derivative is decreasing on $(0, +\infty)$, i.e., in the interior of $[0, +\infty)$. Therefore, the graph of f is concave down on $[0, +\infty)$. $\square$

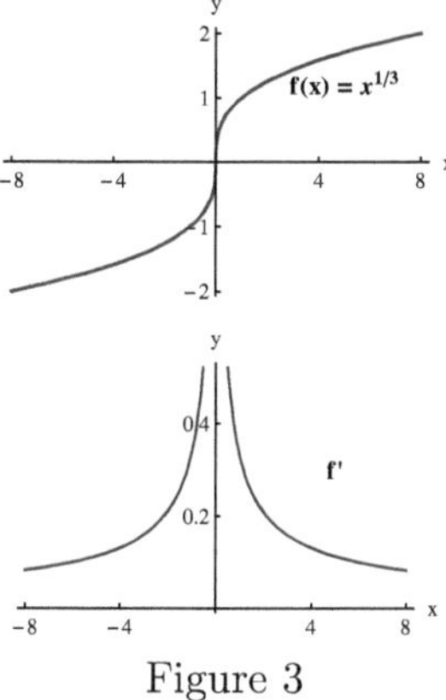

Figure 3

Remark 1 If we interpret $f'(x)$ as the rate of change of f at x, and the graph of f is concave up on the interval J, the rate of change of f is increasing in the interior of J. Similarly, if the graph of f is concave down on an interval, the rate of change of the function is decreasing in the interior of that interval. $\Diamond$

We have defined concavity in terms of the increasing/decreasing behavior of the derivative. It is possible to provide alternative descriptions of concavity: Assume that the graph of f is concave up on the interval J, and that the interval $[a, b]$ is contained in J. Then, the graph of f on (a, b) lies below the secant line that joins $(a, f(a))$ to $(b, f(b))$, as illustrated in Figure 4.

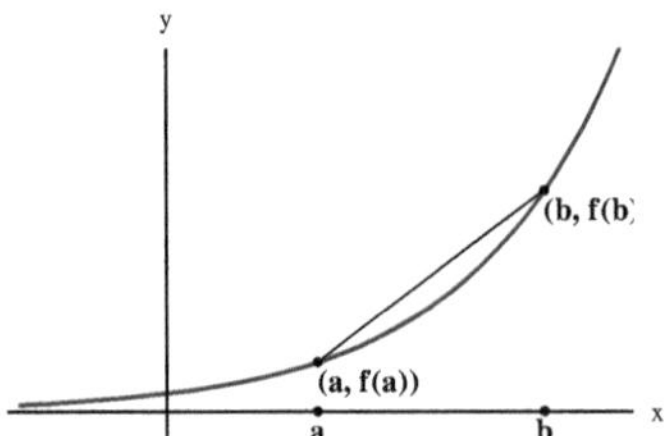

Figure 4: The graph of f is below a secant line if it is concave up

If the graph of f is concave down on J, the graph of f on (a, b) lies above the secant line that joins $(a, f(a))$ to $(b, f(b))$, as illustrated in 5.

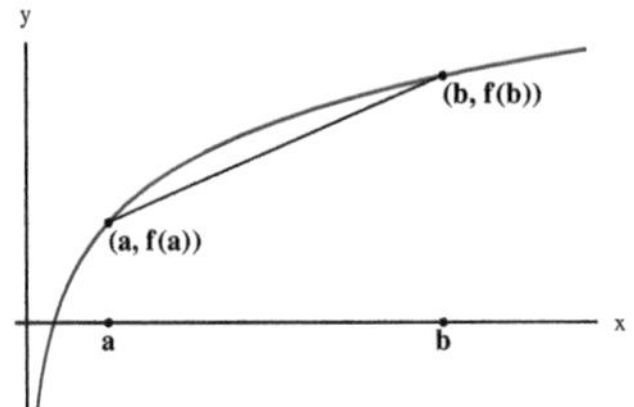

Figure 5: The graph of f is above a secant line if it is concave down

Still another description of concavity involves tangent lines: Assume that the graph of f is concave up on the interval J, and that c is a point in the interior of J. Then, there exists an open interval (a, b) containing c such that the graph of f on (a, b) lies above the tangent line to the graph of f at $(c, f(c))$, i.e., $f(x) > L_c(x) = f(c) + f'(c)(x - c)$ for all $x \in (a, b)$ such that $x \neq c$, as illustrated in Figure 6.

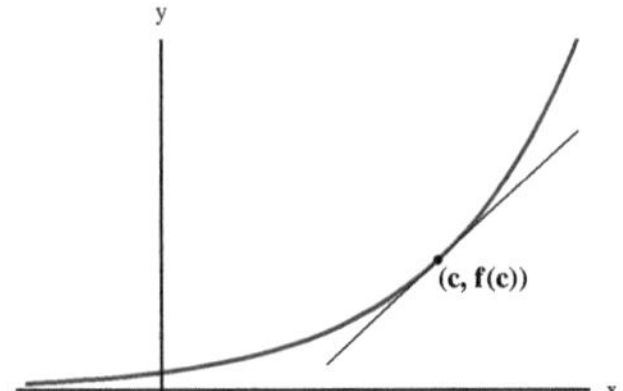

Figure 6: The graph of f is above a tangent line if it is concave up

If the graph of f is concave down on J and c is a point in the interior of J then there exists an open interval (a, b) containing c such that the graph of f on (a, b) lies above the tangent line to the graph of f at $(c, f(c))$as illustrated in Figure 7.

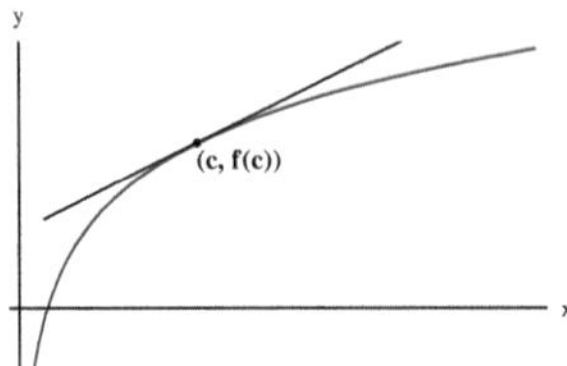

Figure 7: The graph of f is below a tangent line if it is concave down

Example 3 Let $f(x) = -3x + 4x^3$. Determine the intervals on which the graph of f is concave up/concave down.

Solution

We have to determine the intervals on which f' is increasing/decreasing in order to determine the concavity of the graph of f. We can apply the derivative test to f'. We have

$$f'(x) = \frac{d}{dx}\left(-3x + 4x^3\right) = -3 + 12x^2.$$

Thus, $(f')'(x) = f''(x) = 24x$, so that $f''(0) = 0$ and $f''(x) < 0$ if $x < 0$, and $f''(x) > 0$ if $x > 0$. Therefore, f' is decreasing on the interval $(-\infty, 0]$ and increasing on the interval $[0, +\infty)$. Thus, the graph of f is concave down on $(-\infty, 0]$ and concave up on $[0, +\infty)$.
Figure 8 displays the graphs of f and f'. The picture is consistent with our conclusions. The graph of f' indicates that f' is decreasing on $(-\infty, 0]$, and the graph of f appears to be "bending down" on the same interval. Similarly, the pictures indicate that f' is increasing on $[0, +\infty)$, and the graph of f is "bending up" on that interval. $\square$

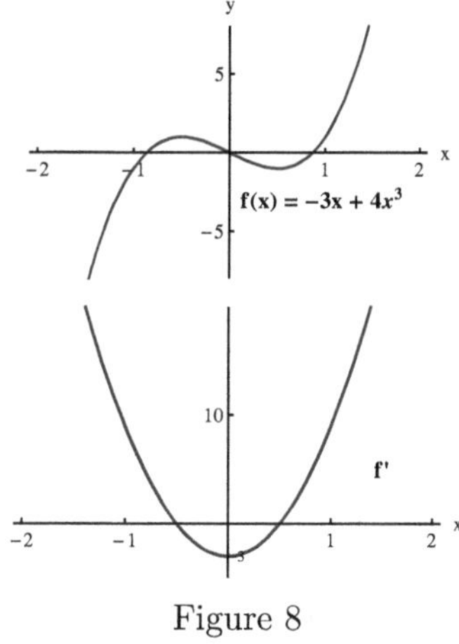

Figure 8

We have a name for the points on the graph of a function at which the concavity of the graph changes:

Definition 2 A point $(a, f(a))$ on the graph of a function is an **inflection point** (or "point of inflection") of the graph of f if f is continuous at a, and a is contained in an open interval (b, c) such that the graph of f is concave up on $(b, a]$ and concave down on $[a, c)$, or vice versa.

Thus, if f is the function of Example 3, $(0, 0)$ is the only inflection point of the graph of f. This is also true for the function f of Example 2, since f' is increasing on the interval $(-\infty, 0)$ and decreasing on the interval $(0, +\infty)$. Note that the function of Example 2 is continuous at 0 but not differentiable at 0.

Remark 2 Assume that f is differentiable at a. Then the graph of f has an inflection point at $(a, f(a))$ if and only if f' has a local maximum or minimum at a. This is an immediate consequence of the definition of concavity in terms of the increasing/decreasing behavior of f'. $\Diamond$

As in Example 3, we can apply the derivative test for monotonicity to f' in order to determine the concavity of the graph of f. This leads to the second derivative test for concavity:

THE SECOND DERIVATIVE TEST FOR CONCAVITY Assume that f is continuous on the interval J and that $f''(x)$ exists for each x in the interior of the interval J.
a) If $f''(x) > 0$ for each x in the interior of J, then the graph of f is concave up on J.
b) If $f''(x) < 0$ for each x in the interior of J, then the graph of f is concave down on J.

Proof

Assume that $f''(x) > 0$ for each x in the interior of J. Since $f'' = (f')'$, we apply the derivative test for monotonicity to f' and conclude that f' is increasing in the interior of J. Therefore, the graph of f is concave up on J. Similarly, if $f''(x) = (f')'(x) < 0$ for each x in the interior of J, f' is decreasing in the interior of J, so that the graph of f is concave down on J. $\blacksquare$

Example 4 Let

$$f(x) = 1 - \frac{1}{2}x^2 + \frac{1}{24}x^4.$$

Determine the intervals on which the graph of f is concave up/concave and the inflection points of the graph of f.

Solution

We will apply the second derivative test for concavity. We have

$$f'(x) = \frac{d}{dx}\left(1 - \frac{1}{2}x^2 + \frac{1}{24}x^4\right) = -x + \frac{1}{6}x^3,$$

and

$$f''(x) = \frac{d}{dx}\left(-x + \frac{1}{6}x^3\right) = -1 + \frac{1}{2}x^2.$$

Therefore,

$$f''(x) = 0 \Leftrightarrow -1 + \frac{1}{2}x^2 = 0 \Leftrightarrow x = -\sqrt{2} \text{ or } x = \sqrt{2}.$$

Table 1 summarizes the relationship between the sign of f'' and the concavity of the graph of f. The graph of f is concave up on the intervals $(-\infty, -\sqrt{2}]$ and $[\sqrt{2}, +\infty)$, and concave down on the interval $[-\sqrt{2}, \sqrt{2}]$. The inflection points of the graph of f are

$$\left(-\sqrt{2}, f\left(-\sqrt{2}\right)\right) = \left(-\sqrt{2}, \frac{1}{6}\right) \text{ and } \left(\sqrt{2}, f\left(\sqrt{2}\right)\right) = \left(\sqrt{2}, \frac{1}{6}\right).$$

x		$-\sqrt{2}$		$\sqrt{2}$	
Sign of f''	$+$	0	$-$	0	$+$
Concavity	concave up	inflection pt.	concave down	inflection pt.	concave up

Table 1

Figure 9 shows the graphs of f, f' and f''. The picture is helpful in visualizing the relationships between the concavity of the graph of f, the increasing/decreasing behavior of f' and the sign of f''. In particular, the picture is consistent with the fact the inflection points on the graph of f correspond to the points at which f' has local extrema. $\square$

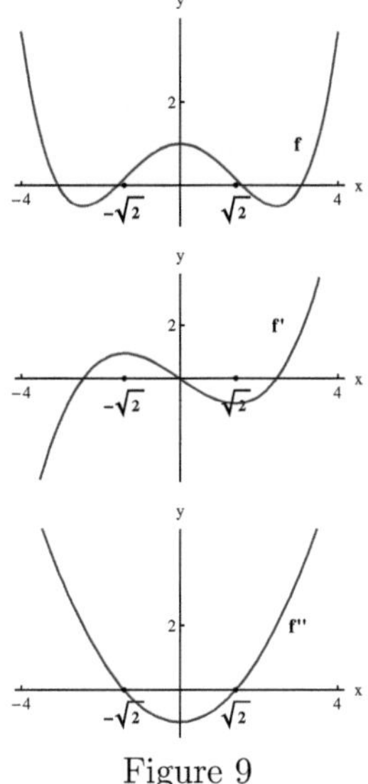

Figure 9

Assume that the second derivative of a function f exists in some open interval containing the point a, $f''(a) = 0$, and the sign of f'' changes at a. Theorem **??** leads to the fact that the graph of f has an inflection point at $(a, f(a))$. For example if f'' is positive on the left of a and negative to the right of a, then the graph of f is concave up on the left of a and concave down the right of a. Let's record this fact:

Corollary 1 Assume that $f''(a) = 0$, and that f'' changes sign at a, i.e., there exists an open interval (b, c) containing a such that $f''(x) > 0$ if $b < x < a$ and $f''(x) < 0$ if $a < x < c$, or vice versa. Then f' has a local extremum at a and the graph of f has an inflection point at $(a, f(a))$.

Example 5 Let

$$f(x) = x^2 + \frac{1}{x}, \quad x \neq 0.$$

Determine the intervals on which the graph of f is concave up/concave down, the local extrema of f', and the inflection points of the graph of f.

Solution

a) We have

$$f'(x) = \frac{d}{dx}\left(x^2 + \frac{1}{x}\right) = 2x - \frac{1}{x^2}, \quad x \neq 0,$$

and

$$f''(x) = \frac{d}{dx}\left(2x - \frac{1}{x^2}\right) = 2 + \frac{2}{x^3}, \quad x \neq 0.$$

The stationary points of f' are the points at which f'' is 0:

$$f''(x) = 0 \Leftrightarrow 2 + \frac{2}{x^3} = 0 \Leftrightarrow 2x^3 + 2 = 0 \Leftrightarrow x^3 = -1 \Leftrightarrow x = -1.$$

We have to take into account the discontinuity of f' and f'' at 0. Thus, we have to examine the sign of f'' in the intervals $(-\infty, -1]$, $[-1, 0)$ and $(0, +\infty)$. Table 2 summarizes the relationship between the sign of f'' (you may sample a point in each interval), the increasing/decreasing behavior of f', and the concavity of the graph of f. The derivative function f' is increasing on $(-\infty, -1]$, decreasing on $[-1, 0)$ and increasing on $(0, +\infty)$ (f' is not defined at 0). Therefore, f' has only one local extremum, i.e., a local maximum at -1. The graph of f is concave up on the intervals $(-\infty, -1]$ and $(0, +\infty)$, and concave down on $[-1, 0)$. The inflection point on the graph of f is $(-1, f(-1)) = (-1, 0)$.

x		-1		0	
Sign of f''	$+$	0	$-$	undefined	$+$
f'	increasing	max	decreasing	undefined	increasing
Concavity	concave up	inflection pt.	concave down	undefined	concave up

Table 2

Figure 10 displays the graphs of f, f' and f''. The picture helps us visualize the relationships between the sign of f'', the increasing/decreasing behavior of f', and the concavity of the graph of f. $\square$

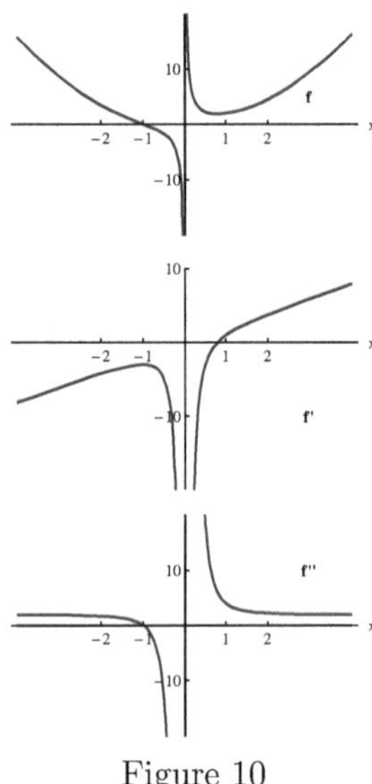

Figure 10

In some cases we may have to resort to approximate calculations, as in the following example:

Example 6 Let $f(x) = \sin^3(x)$.

a) Make use of your calculator in order to plot the graphs of f, f' and f'' on the interval $[0, \pi]$.
b) Determine the subintervals of $[0, \pi]$ on which the graph of f is concave up/concave down, the local extrema of f' in $(0, \pi)$ and the inflection points of the graph of f corresponding to points in $(0, \pi)$. Calculate approximate values with the help of your calculator (round to 6 significant digits).

Solution

a) We have

$$f'(x) = \frac{d}{dx} \sin^3(x) = 3\sin^2(x)\cos(x),$$

and

$$f''(x) = \frac{d}{dx}\left(3\sin^2(x)\cos(x)\right) = 3\left(2\sin(x)\cos(x)\right)\cos(x) + 3\sin^2(x)\left(-\sin(x)\right)$$
$$= 6\sin(x)\cos^2(x) - 3\sin^3(x).$$

Figure 11 displays the graphs of f, f' and f'' on the interval $[0, \pi]$.

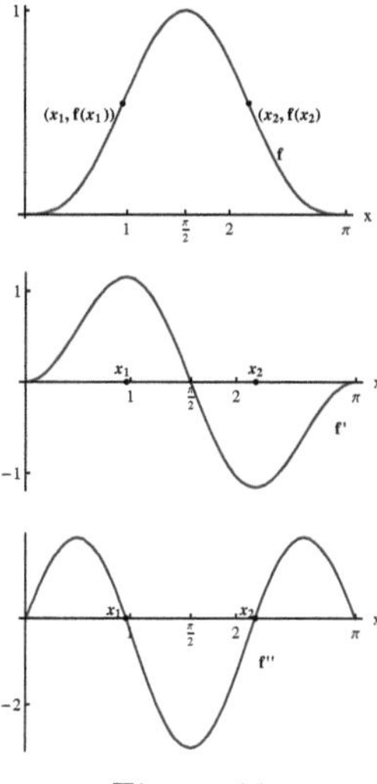

Figure 11

b) We have

$$f''(x) = 0 \Leftrightarrow 3\sin(x)\left(2\cos^2(x) - \sin^2(x)\right) = 0$$
$$\Leftrightarrow \sin(x) = 0 \text{ or } 2\cos^2(x) - \sin^2(x) = 0.$$

The case $\sin(x) = 0$ yields the points 0 and π in the interval $[0, \pi]$. The case

$$2\cos(x) - \sin^2(x) = 0$$

has to be dealt with numerically. Figure 11 indicates that f'' has zeroes near 1 and 2 in the interior of the interval $[0, \pi]$. By means of the approximate equation solver of your calculator, you can confirm that these points are $x_1 = 0.955317$ and $x_2 = 2.18628$, rounded to 6 significant digits. Table 3 summarizes the relationship between the sign of f'' (you may sample points in the relevant intervals), the increasing/decreasing behavior of f', and the concavity of the graph of f. The inflection points of the graph of f that correspond to x_1 and x_2 are

$$(x_1, f(x_1)) \cong (0.955317, 0.544331) \text{ and } (x_2, f(x_2)) \cong (2.18628, 0.544331).$$

$\square$

x	0		x_1		x_2		π
sign of f''		$+$	0	$-$	0	$+$	
f'		increasing	max	decreasing	min	increasing	
concavity		concave up	inflection pt.	concave down	inflection pt.	concave up	

Table 3

The Second Derivative and Extrema

Assume that we are looking for the absolute maximum and the absolute minimum of a function f on an interval J, and that $a \in J$ is a stationary point of f, i.e., $f'(a) = 0$. The function need not attain its absolute maximum or absolute minimum values on J at a, but information about the concavity of the graph of f leads to conclusions about the extrema of f:

THE SECOND DERIVATIVE TEST FOR ABSOLUTE EXTREMA   **Assume that f is continuous on the interval J, differentiable in the interior of J, and $f'(a) = 0$. Then,**
a) f attains its absolute maximum on the interval J at a if $f''(x) < 0$ for each x in the interior of J,
b) f attains its absolute minimum on the interval J at a if $f''(x) > 0$ for each x in the interior of J.

It is helpful to interpret the second derivative test for absolute extrema graphically: If $f''(x) < 0$ for each x in the interior of J, then the graph of f over J is concave down, and the tangent line to the graph of f at $(a, f(a))$ is horizontal since $f'(a) = 0$. Therefore, f must attains its absolute maximum on J at a, as illustrated in Figure 12.

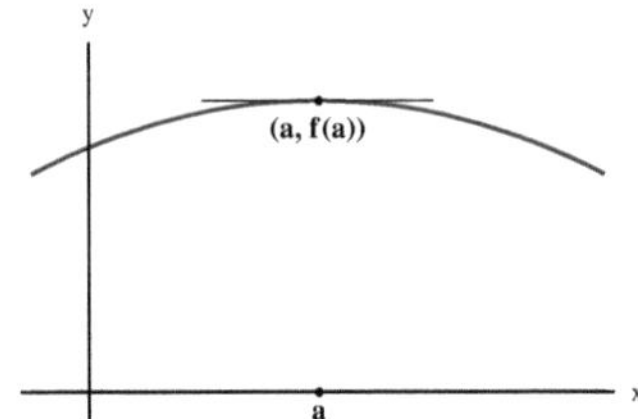

Figure 12: The graph of f is concave down and f has a maximum at a

Similarly, if $f''(x) > 0$ for each x in the interior of J, then the graph of f over J is concave up and the tangent line to the graph of f at $(a, f(a))$ is horizontal since $f'(a) = 0$. Therefore, f must attains its absolute minimum on J at a, as illustrated in Figure 13.

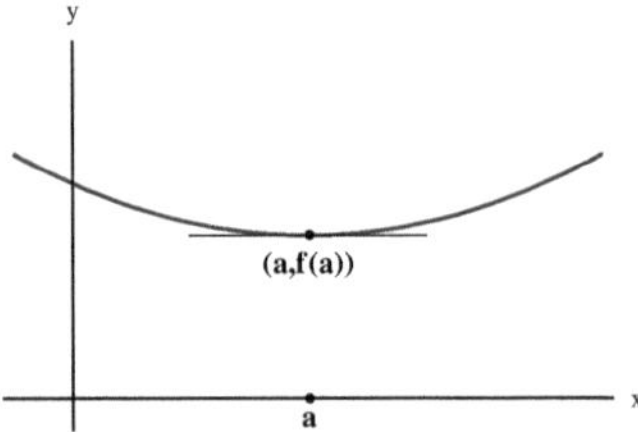

Figure 13: The graph of f is concave up and f has a minimum at a

The Proof of the second derivative test for absolute extrema

Assume that $f'(a) = 0$ and $f''(x) < 0$ for each x in the interior of J. By the application of derivative test for monotonicity to f', f' is decreasing in the interior of J. Since $f'(a) = 0$, we must have $f'(x) > 0$ if x is to the left of a, and $f'(x) < 0$ if x is to the right of a. By the application of the derivative test for monotonicity to f, f is increasing on the part of J to the left of a, and f is decreasing on the part of J to the right of a. Therefore, f has an absolute maximum at a. The proof of part b) is similar. ∎

Example 7 Let

$$f(x) = x + \frac{1}{x-2}.$$

Show that f has an absolute maximum on the interval $(-\infty, 2)$ and an absolute minimum on the interval $(2, +\infty)$ by applying the second derivative test for an absolute extremum.

Solution

We have

$$f'(x) = \frac{d}{dx}\left(x + \frac{1}{x-2}\right) = 1 - \frac{1}{(x-2)^2}.$$

Therefore,

$$f'(x) = 0 \Leftrightarrow 1 - \frac{1}{(x-2)^2} = 0 \Leftrightarrow x = 1 \text{ or } x = 3.$$

Thus, the stationary points of f are 1 and 3.

We have

$$f''(x) = \frac{d}{dx}\left(1 - \frac{1}{(x-2)^2}\right) = \frac{2}{(x-2)^3}.$$

Therefore, $f''(x) < 0$ for each $x < 2$. By the second derivative test for absolute extrema, f attains its absolute maximum on $(-\infty, 2)$ at the stationary point 1 ($f(1) = 0$). We have $f''(x) > 0$ for each $x \in (2, +\infty)$. Therefore, f attains its absolute minimum on $(2, +\infty)$ at 3 ($f(3) = 4$)).

Figure 14 shows the graph of f. The picture is consistent with our analysis. $\square$

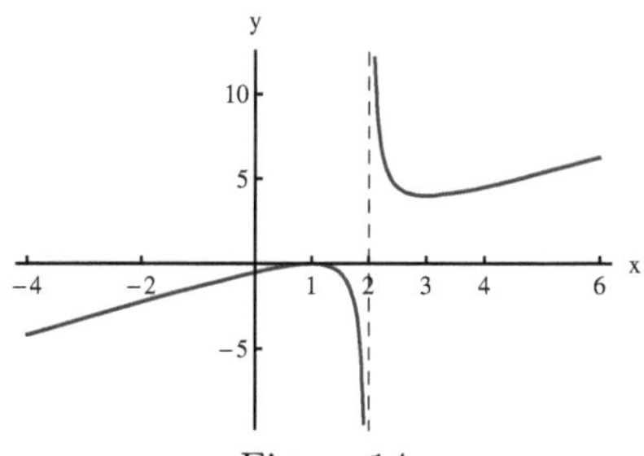

Figure 14

The second derivative test for absolute extrema has a useful local version. We know that a necessary condition for a differentiable function to have a local maximum or a local minimum at a point a is that $f'(a) = 0$, i.e., a must be a stationary point of f. We also know that the condition is not sufficient for f to have a local extremum at a. On the other hand, a function does attain a local maximum or a local minimum value at a stationary point if the second derivative of a function is not zero at that point:

THE SECOND DERIVATIVE TEST FOR LOCAL EXTREMA **Assume that a is a stationary point of f, i.e., $f'(a) = 0$ and that $f''(a)$ exists. Then,**
a) f has a local maximum at a if $f''(a) < 0$,
b) f has a local minimum at a if $f''(a) > 0$.

A Plausibility Argument:

Let's assume that f'' is continuous at a and $f''(a) < 0$. Since $f''(x) \cong f''(a)$ if x is near a, there should be an open interval (b, c) containing a such that $f''(x) < 0$ for each x in (b, c). By the second derivative test for absolute extrema f attains its absolute maximum on (b, c) at

a. Therefore, f has a local maximum at a. Similarly, if $f''(a) > 0$, there should be an open interval (b, c) containing a such that $f''(x) > 0$ for each $x \in (b, c)$. By the second derivative test for absolute extrema f attains its absolute minimum on (b, c) at a. Therefore, f has a local minimum at a. ∎

You can find the proof of Theorem ?? at the end of this section.

The second derivative test for local extrema enables us to determine the nature of a stationary point of a function quickly without having to apply the derivative test for monotonicity:

Example 8 Let

$$f(x) = -2x + \frac{1}{2}x^2 + \frac{1}{3}x^3.$$

Find the points at which f has a local maximum or a local minimum by applying the second derivative test for local extrema.

Solution

We begin by identifying the stationary points of f. We have

$$f'(x) = \frac{d}{dx}\left(-2x + \frac{1}{2}x^2 + \frac{1}{3}x^3\right) = -2 + x + x^2.$$

Therefore,

$$f'(x) = 0 \Leftrightarrow -2 + x + x^2 = 0 \Leftrightarrow x = -2 \text{ or } x = 1.$$

We have

$$f''(x) = \frac{d}{dx}\left(-2 + x + x^2\right) = 1 + 2x.$$

Therefore,

$$f''(-2) = -3 < 0 \text{ and } f''(1) = 3 > 0.$$

By the second derivative test for local extrema, f has a local maximum at -2 and a local minimum at 1.

Figure 15 shows the graph of the function of Example 8. The graph is consistent with our conclusions. □

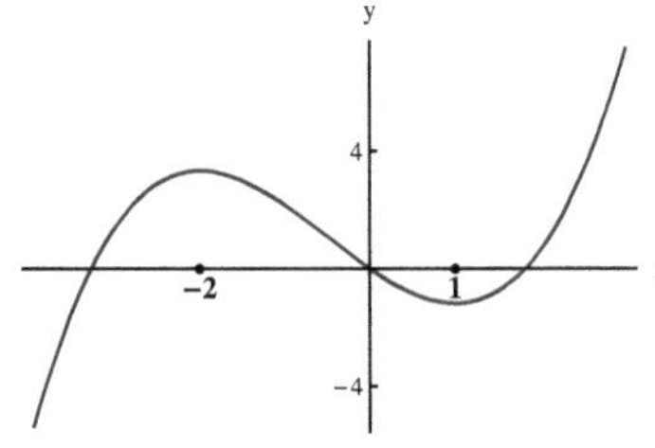

Figure 15

The Proof of the Second Derivative Test for Local Extrema

Assume that $f''(a) > 0$. Since $f''(a) = (f')'(a)$, and $f'(a) = 0$,

$$\lim_{h \to 0} \frac{f'(a+h) - f'(a)}{h} = \lim_{h \to 0} \frac{f'(a+h)}{h} = f''(a) > 0.$$

Therefore, there exists a sufficiently small $\delta > 0$ such that

$$\frac{f'(a+h)}{h} > 0$$

if $h \neq 0$ and $-\delta < h < \delta$. Assume that $\delta > h > 0$. Then, $f'(a+h) > 0$ also. By the derivative test for monotonicity, f is increasing on $[a, a+\delta)$. If $-\delta < h < 0$, the inequality

$$\frac{f'(a+h)}{h} > 0$$

implies that $f'(a+h) < 0$. Thus, f is decreasing on $(a - \delta, a]$. We conclude that f attains a local minimum at a.

The case where $f''(a) < 0$ is handled in a similar manner. ∎

Problems

In problems 1-6 make use of the second derivative test for concavity in order to determine the intervals on which f is concave up/concave down and the x-coordinates of the inflection points of the graph of f.

1.
$$f(x) = x - \frac{1}{3}x^3$$

2.
$$f(x) = \frac{1}{12}x^4 - \frac{1}{6}x^3 - x^2$$

3.
$$f(x) = \frac{x}{x^2 + 1}$$

4.
$$f(x) = \frac{x^2 - x + 1}{x - 1}$$

5.
$$f(x) = \sin^2(x)$$

(consider only the interval $[0, \pi]$).
Hint:

$$\cos^2(x) - \sin^2(x) = \cos(2x).$$

6.
$$f(x) = \tan(x).$$

(consider only the interval $(-\pi/2, \pi/2)$.

In problems 7-9.
a) Make use of the second derivative test for concavity in order to determine the intervals on which f is concave up/concave down and the x-coordinates of the inflection points of the graph of f.
b) Determine the absolute maximum and minimum of f' on D, provided that such values exist.

7.
$$f(x) = \frac{1}{6}x^3 + x^2, \ D = (-\infty, +\infty).$$

8.
$$f(x) = \frac{1}{x^2 + 1}, \ D = (-\infty, +\infty).$$

9.
$$f(x) = \frac{1}{12}x^4 - \frac{1}{6}x^3 - 3x^2 + 20, \ D = (-\infty, 0].$$

10. Let

$$f(x) = x^2 + \frac{1}{x}$$

Make use of the second derivative test for absolute extrema to determine the absolute minimum of f on $(0, +\infty)$.

11. Let

$$f(x) = \frac{1}{3}x^3 - \frac{1}{2}x^2 - 6x.$$

Make use of the second derivative test for absolute extrema to determine the absolute maximum of f on $(-\infty, 0)$.

In problems 12-15, determine the local extrema of f with the help of the second derivative test for local extrema.

12.

$$f(x) = 8x^3 - 4x$$

13.

$$f(x) = 8x^4 - 8x^2 + 1$$

14.

$$f(x) = \frac{x}{x^2 + 4}$$

15.

$$f(x) = \frac{x^2 + 2x + 1}{x + 2}$$

3.4 Sketching the Graph of a Function

We have developed tools that are helpful in analyzing the behavior of a function. We are able to determine the finite or infinite limits of a function at a point $a \in R$ or at $\pm\infty$. Thus, we are able to determine the **vertical, horizontal** or **oblique asymptotes** for the graph of a function f. We are able to determine the intervals on which f is increasing or decreasing, and the local **maxima** or **minima** of f by examining the sign of its derivative. We can determine the intervals on which the graph of f is **concave up** or **concave down**, and **the inflection points** on the graph of f by examining the sign of the second derivative of f. All that information enables us to obtain a rough sketch of the graph of f, without the help of a graphing utility, if the expression $f(x)$ is not too complicated. In this section we will discuss a few typical cases where it is feasible to come up with a reasonable sketch of the graph of a function.

A Strategy for Sketching a Graph

1. Determine **the domain of f, the vertical asymptotes for the graph of f and** $\lim_{x \to \pm\infty} f(x)$, if applicable. Determine **the horizontal, oblique or "curved" asymptotes for the graph of f.** It is helpful to notice whether f is **odd** or **even**, since such properties of f lead to the symmetry of the graph of f with respect to the origin or with respect to the vertical axis (of course, a function need not be odd or even).

2. Make use of **the derivative test for monotonicity:** Compute f', determine **the critical points of f, and the increasing/decreasing behavior of f** on the intervals that are separated from each other by critical points or the points at which f is discontinuous. A by-product is the determination of **the local extrema of f**, and the absolute extrema of f on intervals of interest. You should determine whether the graph of f has a **vertical tangent** or a **cusp** at a critical point where it is not differentiable.

3. Steps 1 and 2 are usually sufficient to give a good idea about the graph of f. If necessary, compute f'' and make use of **the second derivative test for concavity** in order to determine **the intervals on which the graph of f is concave up or concave down, and the inflection points on the graph of f.** Such information enables us to refine the initial graph.

4. Produce a sketch of the graph of f that is consistent with the results of the previous steps.

We will illustrate the implementation of the above strategy in a few cases. We need not implement every step of the strategy in a given case.

The first example involves a polynomial:

Example 1 Let

$$f(x) = \frac{1}{5}x^5 - \frac{20}{3}x^3 + 64x.$$

Implement the suggested strategy to sketch the graph of f.

Solution

The polynomial f is continuous on the entire number line. The graph of f is a continuous curve without any breaks. There are no vertical asymptotes. As for the behavior of f at $\pm\infty$, we can factor the highest power of x, as suggested in Section 1.7:

$$f(x) = x^5 \left(\frac{1}{5} - \frac{20}{3x^2} + \frac{64}{x^4} \right).$$

We have

$$\lim_{x \to +\infty} x^5 = +\infty, \quad \lim_{x \to +\infty} \left(\frac{1}{5} - \frac{20}{3x^2} + \frac{64}{x^4} \right) = \frac{1}{5} > 0.$$

Therefore, $\lim_{x \to +\infty} f(x) = +\infty$. We also have

$$\lim_{x \to -\infty} x^5 = -\infty, \quad \lim_{x \to -\infty} \left(\frac{1}{5} - \frac{20}{3x^2} + \frac{64}{x^4} \right) = \frac{1}{5} > 0,$$

so that $\lim_{x \to -\infty} f(x) = -\infty$.

Note that f is an odd function $(f(-x) = f(x)$ since $(-x)^n = -x^n$ when n is an odd integer). Therefore, the graph of f is symmetric with respect to the origin.

We have

$$f'(x) = \frac{d}{dx} \left(\frac{1}{5}x^5 - \frac{20}{3}x^3 + 64x \right) = x^4 - 20x^2 + 64$$

Since f is differentiable everywhere, the only critical points of f are its stationary points. If we set $u = x^2$,

$$x^4 - 20x^2 + 64 = u^2 - 20u + 64 = 0 \Leftrightarrow u = 4 \text{ or } u = 16$$

(check). Therefore, the stationary points of f are ± 2 and ± 4. Table 1 summarizes the relationship between the sign of f', the increasing/decreasing behavior of f and the local maxima and minima of f.

x		-4		-2		2		4	
sign of $f'(x)$	$+$	0	$-$	0	$+$	0	$-$	0	$+$
f	incr.	loc. max.	decr.	loc. min.	incr.	loc. max.	decr.	loc. min.	incr.

Table 1

Note that the values of f at its stationary points are

$$f(-4) \cong -34, \ f(-2) \cong -81, \ f(2) \cong 81 \text{ and } f(4) \cong 34.$$

Such information is useful in sketching the graph of a function.

We have

$$f''(x) = \frac{d}{dx} \left(x^4 - 20x^2 + 64 \right) = 4x^3 - 40x = 4x \left(x^2 - 10 \right).$$

Therefore,

$$f''(x) = 0 \Leftrightarrow x = 0 \text{ or } x = \pm\sqrt{10} \cong \pm 3.2$$

Table 2 summarizes the relationship between the sign of f'' and the concavity of the graph of f.

x		$-\sqrt{10}$		0		$\sqrt{10}$	
sign of $f''(x)$	$-$	0	$+$	0	$-$	0	$+$
concavity	down	infl. pt.	up	infl. pt.	down	infl. pt	up

Table 2

Thus, the inflection points on the graph of f correspond to $\pm\sqrt{10}$ and 0.

$$f\left(-\sqrt{10}\right) \cong -55, \ f(0) = 0 \text{ and } f\left(\sqrt{10}\right) \cong 55.$$

Figure 1 shows the graph of f. $\square$

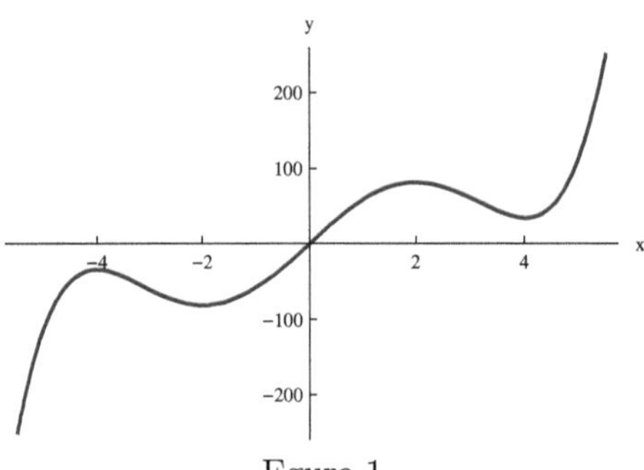

Fgure 1

The next example involves a rational function:

Example 2 Let

$$f(x) = -3x + 10 - \frac{1}{x-3}$$

Implement the suggested strategy to sketch the graph of f. You don't have to discuss concavity.

Solution

The domain of f consists of all x other than 3. We have

$$\lim_{x \to 3}(-3x - 10) = -19, \ \lim_{x \to 3-}\left(-\frac{1}{x-3}\right) = +\infty \text{ and } \lim_{x \to 3+}\left(-\frac{1}{x-3}\right) = -\infty.$$

Therefore,

$$\lim_{x \to 3-} f(x) = +\infty \text{ and } \lim_{x \to 3+} f(x) = -\infty.$$

Thus, the line $x = 3$ is a vertical asymptote for the graph of f.

We also have

$$\lim_{x \to +\infty}(-3x + 10) = -\infty, \ \lim_{x \to -\infty}(-3x + 10) = +\infty \text{ and } \lim_{x \to \pm\infty}\left(-\frac{1}{x-3}\right) = 0.$$

Therefore,

$$\lim_{x \to +\infty} f(x) = -\infty \text{ and } \lim_{x \to -\infty} f(x) = +\infty.$$

We have

$$\lim_{x \to \pm\infty} (f(x) - (-3x - 10)) = \lim_{x \to \pm} \left(-\frac{1}{x-3}\right) = 0.$$

Therefore, the graph of f gets closer and closer to the line $y = -3x - 10$ as $x \to \pm\infty$. The line $y = -3x - 10$ is an oblique asymptote for the graph of f at $\pm\infty$.

We have

$$f'(x) = \frac{d}{dx}\left(-3x + 10 - \frac{1}{x-3}\right) = -3 - \frac{d}{dx}\left(\frac{1}{x-3}\right)$$

$$= -3 - \left(\frac{-1}{(x-3)^2}\right) = \frac{-3x^2 + 18x - 26}{(x-3)^2}.$$

Therefore,

$$f'(x) = 0 \Leftrightarrow -3x^2 + 18x - 26 = 0 \Leftrightarrow x = 3 + \frac{\sqrt{3}}{3} \cong 3.6 \text{ or } x = 3 - \frac{\sqrt{3}}{3} \cong 2.4.$$

These are the only critical points of f since f is differentiable on its entire domain. Table 3 summarizes the relationship between the sign of f', the increasing/decreasing behavior of f and the local extrema of f. We have to make special provision for 3 at which f is not defined.

x		$3 - \frac{\sqrt{3}}{3}$		3		$3 + \frac{\sqrt{3}}{3}$	
sign of $f'(x)$	$-$	0	$+$	undef.	$+$	0	$-$
f	decr,	loc. min.	incr.	vert. asymp.	incr.	loc. max.	decr.

Table 3

We have

$$f\left(3 - \frac{\sqrt{3}}{3}\right) = 1 + 2\sqrt{3} \cong 4.5 \text{ and } f\left(3 + \frac{\sqrt{3}}{3}\right) = 1 - 2\sqrt{3} \cong -2.5$$

Figure 2 shows the graph of f. $\square$

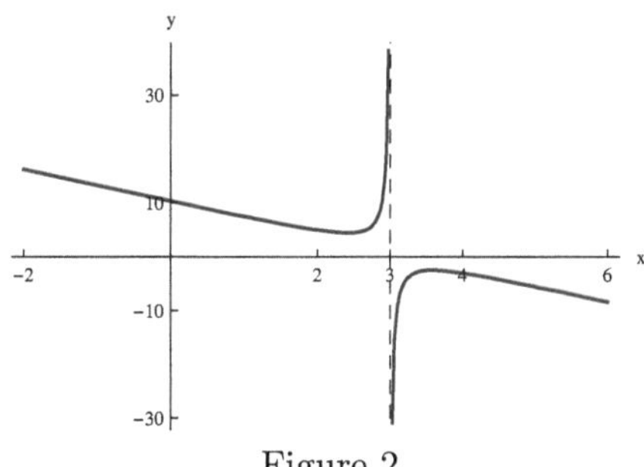

Figure 2

The next example involves a critical point at which the function is not differentiable.

Example 3 Let

$$f(x) = x^2 (x - 4)^{4/5}.$$

Implement the suggested strategy to sketch the graph of f. You don't have to discuss concavity.

Solution

The function is continuous on the entire number line, since both x^2 and

$$(x-4)^{4/5} = \left((x-4)^{1/5}\right)^4$$

define such functions. Since

$$\lim_{x\to\pm\infty} x^2 = +\infty \text{ and } \lim_{x\to\pm\infty} (x-4)^{4/5} = +\infty,$$

we have $\lim_{x\to\pm\infty} f(x) = +\infty$.

We need to calculate f':

$$f'(x) = \frac{d}{dx}\left(x^2 (x-4)^{4/5}\right) = \left(\frac{d}{dx}\left(x^2\right)\right)(x-4)^{4/5} + x^2 \left(\frac{d}{dx}(x-4)^{4/5}\right)$$

$$= 2x(x-4)^{4/5} + x^2 \left(\frac{4}{5}(x-4)^{-1/5}\right)$$

$$= 2x(x-4)^{4/5} + \frac{4x^2}{5(x-4)^{1/5}}$$

$$= \frac{10x(x-4) + 4x^2}{5(x-4)^{1/5}} = \frac{14x^2 - 40x}{5(x-4)^{1/5}}$$

if $x \neq 4$. You can show that f is not differentiable at 4. We have

$$\lim_{x\to 4} \left(14x^2 - 40x\right) = 64 > 0,\ 5(x-4)^{1/5} > 0 \text{ if } x > 4 \text{ and } \lim_{x\to 4}\left(5(x-4)^{1/5}\right) = 0.$$

Therefore,

$$\lim_{x\to 4+} f'(x) = \lim_{x\to 4+}\left(\left(14x^2 - 40x\right)\left(\frac{1}{5(x-4)^{1/5}}\right)\right) = +\infty.$$

Similarly, $\lim_{x\to 4-} f'(x) = -\infty$. Therefore, the graph of f has a cusp at $(4, f(4)) = (4, 0)$.

We determine the stationary points of f:

$$f'(x) = 0 \Leftrightarrow 14x^2 - 40x = 0 \Leftrightarrow x = 0 \text{ or } x = \frac{20}{7} \cong 2.9$$

Table 4 summarizes the relationships between the sign of f', the increasing/decreasing behavior of f and the critical points of f.

x		0		$\frac{20}{7}$		4	
sign of $f'(x)$	$-$	0	$+$	0	$-$	undef.	$+$
f	decr.	loc. min.	incr.	loc. max.	decr.	loc. min.	incr.

Table 4

We have $f(20/7) \cong 9.1$.

Figure 3 shows the graph of f. $\square$

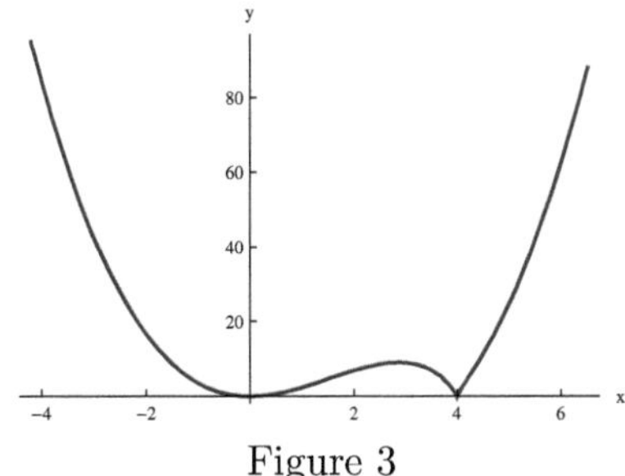

Figure 3

The following example involves a trigonometric function.

Example 4 Let

$$f(x) = \frac{1}{1 - 2\sin(x)}.$$

The function is periodic with period 2π. Implement the suggested strategy to sketch the graph of f on the interval $[0, 2\pi]$. You don't have to discuss concavity.

Solution

The function is continuous at x if $1 - 2\sin(x) \neq 0$. We have

$$1 - 2\sin(x) = 0 \Leftrightarrow \sin(x) = \frac{1}{2} \Leftrightarrow x = \frac{\pi}{6} + 2n\pi \text{ or } x = \frac{5\pi}{6} + 2n\pi,$$

where n is an integer. The relevant points are $\pi/6$ and $5\pi/6$. We have $1 - 2\sin(x) > 0$ if $x < \pi/6$ and x is close to $\pi/6$. Since $\lim_{x \to \pi/6}(1 - 2\sin(x)) = 0$, we have $\lim_{x \to \pi/6-} f(x) = +\infty$. If $x > \pi/6$ and x is close to $\pi/6$, we have $1 - 2\sin(x) < 0$. Therefore, $\lim_{x \to \pi/6+} f(x) = +\infty$. Similarly,

$$\lim_{x \to 5\pi/6-} f(x) = -\infty \text{ and } \lim_{x \to 5\pi/6+} f(x) = +\infty.$$

Thus, the lines $x = \pi/6$ and $x = 5\pi/6$ are vertical asymptotes for the graph of f.

We have

$$f'(x) = \frac{d}{dx}\left(\frac{1}{1 - 2\sin(x)}\right) = \frac{2\cos(x)}{(1 - 2\sin(x))^2}.$$

Therefore, $f'(x) = 0$ if $\cos(x) = 0$. Thus, the stationary points of f in the interval $[0, 2\pi]$ are $\pi/2$ and $3\pi/2$. Note that the sign of $f'(x)$ is determined by the sign of $\cos(x)$. Table 5 summarizes the the relationships between the sign of f', the increasing/decreasing behavior of f and the stationary points of f (the function is differentiable on its domain).

x	0		$\pi/6$		$\pi/2$		$5\pi/6$		$3\pi/2$		2π
sign $f'(x)$	+	+	undef.	+	0	−	undef.	−	0	+	+
f		incr.	undef.	incr.	loc. max.	decr.	undef.	decr.	loc. min.	incr.	

Table 5

It is worth noting that

$$f(0) = f(2\pi) = 1, \ f(\pi/2) = -1 \text{ and } f(3\pi/2) = 1/3.$$

Figure 4 shows the graph of f on $[0, 2\pi]$. $\square$

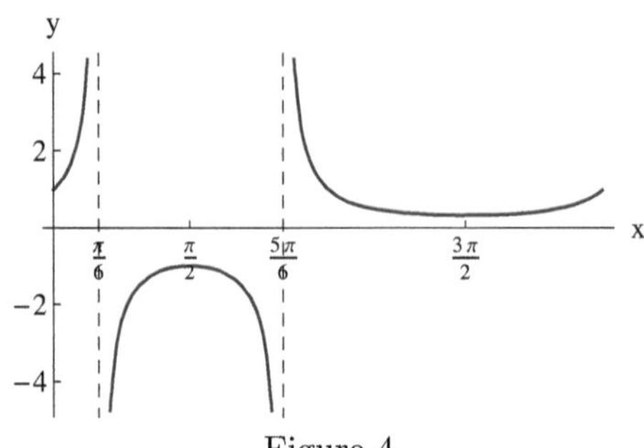

Figure 4

Problems

In problems 1-4,
a) Determine $\lim_{x\to+\infty} f(x)$ and $\lim_{x\to-\infty} f(x)$,
b) Determine the intervals on which f is increasing/decreasing, and the points at which f has a local maximum or minimum,
c) Determine the intervals on which the graph of f is concave up/concave down, and the x-coordinates of the inflection points on the graph of f.
d) Sketch the graph of f. If you expect the graph to be symmetric with respect to the vertical axis or with respect to the origin, your sketch should reflect the relevant symmetry.

1.
$$f(x) = x^3 - 2x$$

2.
$$f(x) = x^4 - x^2 - 2$$

3.
$$f(x) = \frac{1}{3}x^3 - \frac{1}{2}x^2 - 12x$$

4.
$$f(x) = \frac{1}{4}x^4 - \frac{3}{2}x^2 - 2x \quad (\text{Hint:} f'(2) = 0).$$

In problems 5-7,
a) Determine the domain of f, the vertical asymptotes for the graph of f and the relevant infinite limits,
b) Determine $\lim_{x\to+\infty} f(x)$, $\lim_{x\to-\infty} f(x)$, and the horizontal or oblique asymptotes for the graph of f,
c) Determine the intervals on which f is increasing/decreasing, and the points at which f has a local maximum or minimum,
d) Sketch the graph of f. If you expect the graph to be symmetric with respect to the vertical axis or with respect to the origin, your sketch should reflect the relevant symmetry.

5.
$$f(x) = \frac{4x^2 - 4x - 19}{x^2 - x - 6}$$

6.
$$f(x) = \frac{10x^2 - 15x + 4}{2x^2 - 3x + 1}$$

7.

$$f\left(x\right) = 4x + \frac{1}{x-2}$$

In problems 8-10,
a) Determine the domain of f,
b) Determine $\lim_{x \to -\infty} f\left(x\right)$ and $\lim_{x \to +\infty} f\left(x\right)$, if applicable,
c) Determine the domain of f' and the points at which the graph of f has a vertical tangent or a cusp,
d) Determine the intervals on which f is increasing/decreasing, and the points at which f has a local maximum or minimum,
e) Sketch the graph of f. If you expect the graph to be symmetric with respect to the vertical axis or with respect to the origin, your sketch should reflect the relevant symmetry.

8. $f\left(x\right) = x^2(x-3)^{1/5}$

9. $f\left(x\right) = x(x+3)^{3/4}$

10. $f\left(x\right) = x^2\left(x-2\right)^{4/5}$

3.5 Applications of Maxima and Minima

In the previous sections of this chapter we developed powerful tools in order to find the maxima and minima of functions. In this section we will discuss some applications.

Optimization

Let us begin with some geometric applications.

Example 1 Determine the dimensions of the rectangle that has the greatest area among all rectangles which are inscribed in a circle of radius r.

Solution

We will consider the circle of radius r whose center is the origin of the xy-plane. Thus, the circle is the graph of the equation $x^2 + y^2 = r^2$. Due to the symmetries, it is sufficient to consider inscribed rectangles whose sides are parallel to the coordinate axes, and we can assume that the vertices of the rectangles are on the unit circle. After all, we wish to maximize the area, and if the rectangle is strictly within the unit disk, we can enlarge the rectangle to one which has its vertices on the circle.

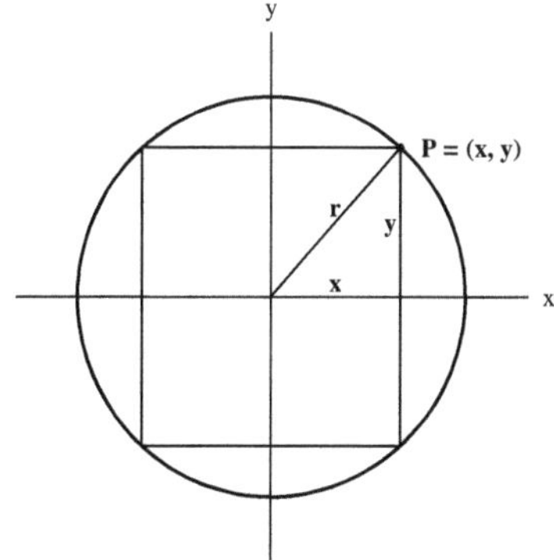

Figure 1

With reference to Figure 1, the inscribed rectangle is completely determined by the point $P = (x, y)$. We have $y = \sqrt{r^2 - x^2}$. Therefore, the area of the rectangle is

$$(2x)(2y) = 4xy = 4x\sqrt{r^2 - x^2}.$$

We will maximize the square of the area in order to maximize the area, since the expressions will be easier to work with, Thus, let's set

$$f(x) = \left(4x\sqrt{r^2 - x^2}\right)^2 = 16x^2\left(r^2 - x^2\right) = 16r^2x^2 - 16x^4.$$

Since $0 < x < r$, we would like to determine the absolute maximum of f on the interval $(0, r)$. We have $f(0) = f(r) = 0$. The cases $x = 0$ and $x = r$ lead to the degenerate cases where the "rectangles" are intervals with 0 area. In any case, the search procedure for the determination of the absolute extrema of a continuous function on a closed and bounded interval is applicable, as we discussed in Section 3.2. Since f is differentiable at any $x \in \mathbb{R}$, the only critical points of f are its stationary points, i.e., points x such that $f'(x) = 0$. We have

$$f'(x) = \frac{d}{dx}\left(16r^2x^2 - 16x^4\right) = 32r^2x - 64x^3 = 32x\left(r^2 - 2x^2\right).$$

Therefore,

$$f'(x) = 0 \Leftrightarrow x = 0 \text{ or } x = \pm\frac{r}{\sqrt{2}}.$$

Thus, the only critical point of f in the interior of the interval $[0, r]$ is $r/\sqrt{2}$.

Figure 2 shows the graph of f that corresponds to $r = 2$.

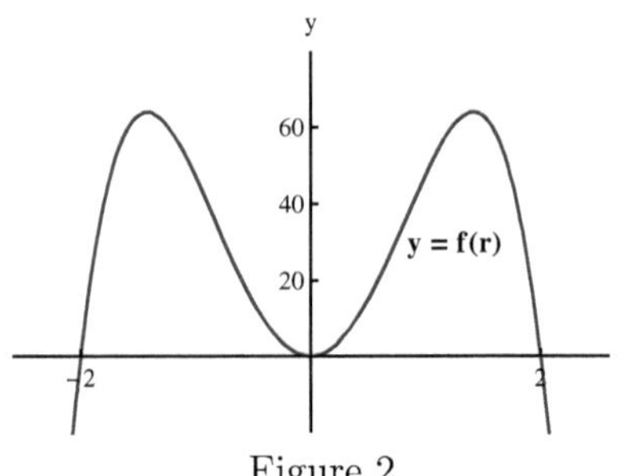

Figure 2

We have

$$f\left(\frac{r}{\sqrt{2}}\right) = 16r^2x^2 - 16x^4\Big|_{x=r/\sqrt{2}} = 16\left(\frac{r}{\sqrt{2}}\right)^2\left(r^2 - \left(\frac{r}{\sqrt{2}}\right)^2\right) = 4r^4.$$

Therefore, $f\left(r/\sqrt{2}\right) > 0 = f(0) = f(r)$. Thus, the absolute maximum of f on $[0, r]$ is $4r^4$, and f attains this value at $x = r/\sqrt{2}$. Therefore, the maximum area of a rectangle that is inscribed in a circle of radius r is

$$\sqrt{f\left(\frac{r}{\sqrt{2}}\right)} = \sqrt{4r^2} = 2r.$$

The dimensions of a rectangle with maximum area are

$$2x = 2\left(\frac{r}{\sqrt{2}}\right) = \sqrt{2}r,$$

and

$$2y = 2\sqrt{r^2 - \left(\frac{r}{\sqrt{2}}\right)^2} = 2\sqrt{r^2 - \frac{1}{2}r^2} = 2\left(\frac{r}{\sqrt{2}}\right) = \sqrt{2}r.$$

Therefore, an inscribed rectangle with maximum area is a square whose sides have length $\sqrt{2}r$. Note that the ratio of the area of such a square and the area of the disc of radius r is

$$\frac{2r^2}{\pi r^2} = \frac{2}{\pi} \cong 0.64.$$

$\square$

Example 2 Find the points on the parabola $y = x^2$ which are closest to the point $(0, 4)$.

Solution

Let $P = (x, y)$ be an arbitrary point on the parabola so that $y = x^2$, as shown in Figure 3.

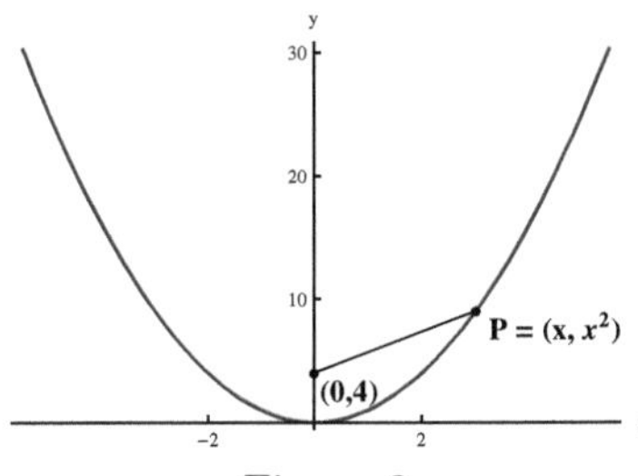

Figure 3

The distance of P from $(0, 4)$ is

$$\sqrt{x^2 + (x^2 - 4)^2} = \sqrt{x^4 - 7x^2 + 16}.$$

We will minimize the distance if we minimize the square of the distance. Thus, let's set $f(x) = x^4 - 7x^2 + 16$, so that f represents the square of the distance of an arbitrary point on the parabola from the point $(0, 4)$. Note that $f(-x) = f(x)$, so that f is even. We would like to minimize $f(x)$ as x varies on the entire number line. Unlike Example 1, we are not able to confine x to a closed and bounded interval that can be determined immediately, and implement the search procedure for the absolute extrema of a continuous function on a closed and bounded interval. We will make use of the derivative test for monotonicity.

We have

$$f'(x) = \frac{d}{dx}\left(x^4 - 7x^2 + 16\right) = 4x^3 - 14x = 2x\left(2x^2 - 7\right).$$

Therefore,

$$f'(x) = 0 \Leftrightarrow 2x\left(2x^2 - 7\right) = 0 \Leftrightarrow x = 0 \text{ or } x = \pm\sqrt{\frac{7}{2}}.$$

Table 1 summarizes the relationship between the sign of f' and the increasing/decreasing behavior of f.

x		$-\sqrt{\frac{7}{2}}$		0		$\sqrt{\frac{7}{2}}$	
sign of f'	$-$	0	$+$		$-$	0	$+$
f	decreasing	local min.	increasing	local max.	decreasing	local min.	increasing

Table 1

Note that

$$\lim_{x \to \pm\infty} f(x) = \lim_{x \to \pm\infty} \left(x^4 - 7x^2 + 16\right) = \lim_{x \to \pm\infty} x^4 \left(1 - \frac{7}{x^2} + \frac{16}{x^4}\right) = +\infty,$$

since

$$\lim_{x \to \pm\infty} x^4 = +\infty \text{ and } \lim_{x \to \pm\infty} \left(1 - \frac{7}{x^2} + \frac{16}{x^4}\right) = 1 > 0.$$

The data displayed in Table 1 shows that f attains its absolute minimum on the entire number line at $\pm\sqrt{7/2}$ (we noted that f is even). Therefore, the points on the parabola at a minimum distance from the point $(0,4)$ are

$$\left(-\sqrt{\frac{7}{2}}, \frac{7}{2}\right) \text{ and } \left(\sqrt{\frac{7}{2}}, \frac{7}{2}\right).$$

These points are at a distance

$$\sqrt{f(\pm\sqrt{\frac{7}{2}})} = \sqrt{\frac{15}{4}} = \frac{\sqrt{15}}{2} \cong 1.936\,49$$

from $(0,4)$.

Figure 4 displays the graph of f, and Figure 5 indicates the points on the parabola $y = x^2$ at a minimum distance from $(0,4)$. $\square$

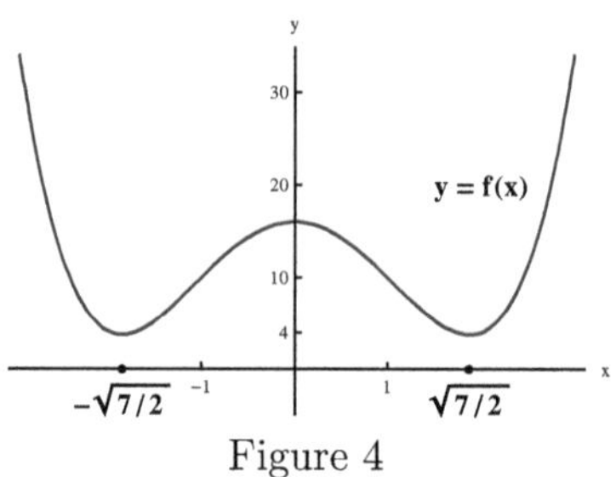

Figure 4

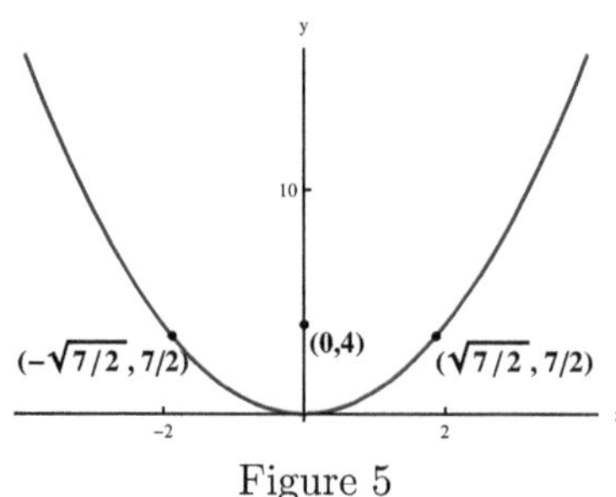

Figure 5

Let us look at an example that involves the minimization of the cost of producing a certain item:

Example 3 Assume that a manufacturer must produce cans in the shape of right circular cylinders. The volume of each can is required to be 250 cubic centimeters. Otherwise, the manufacturer is free to choose the dimensions of the can. The cost per square centimeter of the material for the top and the bottom of the can is twice as much as the cost per square centimeter of the material for the lateral surface. How should the manufacturer determine the dimensions of the can in order to minimize the cost?

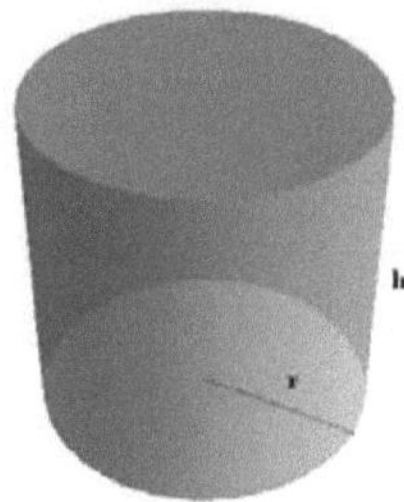

Figure 6

Solution

Let's assume that the material for the lateral surface of the can costs k cents per cm^2, and that the material for the top and the bottom of the can costs $2k$ cents per cm^2. If the cross section of the can is a circle of radius r (centimeters), and the height of the can is h (centimeters), the total area of the top and the bottom is $2\pi r^2$, so that the cost of the material for the top and the bottom is $2\pi r^2 \times 2k$ cents. The area of the lateral surface of the can is $2\pi rh$ (You can imagine that a vertical cut is made, and the lateral surface is laid out on a flat surface: The shape is a rectangle of base length $2\pi r$, the perimeter of the cross section, and height h). Therefore, the contribution of the lateral surface to the cost is $2\pi rh \times k$ cents. Thus, the total cost is

$$2\pi r^2 \times 2k + 2\pi rh \times k = 4\pi r^2 k + 2\pi rhk = \left(4\pi r^2 + 2\pi rh\right)k.$$

cents. The cost is minimized if $4\pi r^2 + 2\pi rh$ is minimized. This involves two variables r and h. We will eliminate one of the variables by making use of the requirement that the volume of the can must be 250 cm^3. Therefore,

$$\text{Volume} = \pi r^2 h = 250,$$

so that

$$h = \frac{250}{\pi r^2}.$$

Now we can express $4\pi r^2 + 2\pi rh$ as a function of the radius r:

$$4\pi r^2 + 2\pi rh = 4\pi r^2 + 2\pi r\left(\frac{250}{\pi r^2}\right) = 4\pi r^2 + \frac{500}{r}.$$

Let us set

$$f(r) = 4\pi r^2 + \frac{500}{r}.$$

We must determine $r_0 > 0$ such that $f(r_0) \leq f(r)$ for each $r > 0$. Thus, we must determine the point r_0 at which f attains its absolute minimum on the interval $(0, +\infty)$. There are no additional restrictions on r (A small r corresponds to a tall and skinny can, and a large r

corresponds to a short and fat can: Certain practical considerations impose restrictions on r and h, but these have not been stipulated). Therefore, we will implement the derivative test for monotonicity on the interval $(0, +\infty)$. We have

$$f'(r) = \frac{d}{dr}\left(4\pi r^2 + \frac{500}{r}\right) = 8\pi r - \frac{500}{r^2} = \frac{8\pi r^3 - 500}{r^2}.$$

Therefore,

$$f'(r) = 0 \iff 8\pi r^3 - 500 = 0 \iff r = \left(\frac{500}{8\pi}\right)^{1/3}.$$

Table 2 summarizes the relationship between the sign of f' and the increasing/decreasing behavior of f.

r	0		$\left(\dfrac{500}{8\pi}\right)^{1/3}$	
sign of f'	undefined	$-$	0	$+$
f	undefined	decreasing	minimum	increasing

Table 2

Note that

$$\lim_{r \to 0+} f(r) = \lim_{r \to 0+}\left(4\pi r^2 + \frac{500}{r}\right) = +\infty,$$

since

$$\lim_{r \to 0+} 4\pi r^2 = 0 \text{ and } \lim_{r \to 0+} \frac{500}{r} = +\infty.$$

We also have $\lim_{r \to +\infty} f(r) = +\infty$, since

$$\lim_{r \to +\infty} 4\pi r^2 = +\infty \text{ and } \lim_{r \to +\infty} \frac{500}{r} = 0.$$

Thus, f attains its absolute minimum on the interval $(0, +\infty)$ at

$$r_0 = \left(\frac{500}{8\pi}\right)^{1/3} \cong 2.709\,63$$

Figure 7 shows the graph of f on the interval $(0, 8)$. The picture is consistent with our analysis.

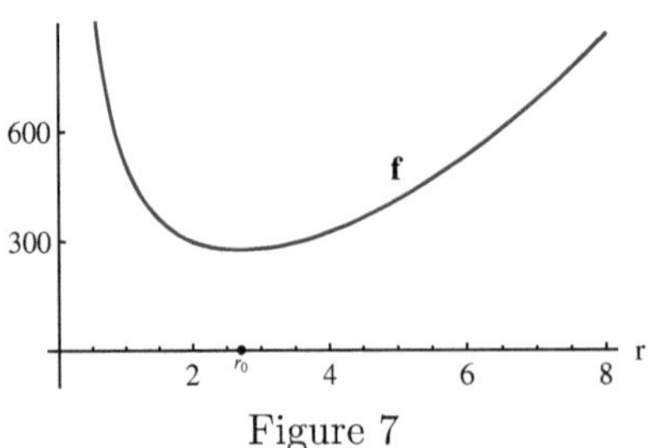

Figure 7

The height of the "optimal can" is

$$h_0 = \frac{250}{\pi r_0^2} = \frac{250}{\pi\left(\dfrac{500}{8\pi}\right)^{2/3}} = 250\left(\frac{8}{500}\right)^{2/3}\frac{1}{\pi^{1/3}} \cong 10.838\,5$$

In practice, it cannot be expected that the manufacturer will manufacture a can of radius 2.70963 and height 10.8385. The precision is up to the negotiations between the manufacturer and the customer with regard to the dimensions of a "suboptimal can". $\square$

Example 4 Assume that two hallways meet at a right angle, as shown in Figure 8. One hallway is 3 meters wide, and the other is 5 meters wide. Determine whether it is possible to carry a ladder which is 10 meters long around the corner horizontally.

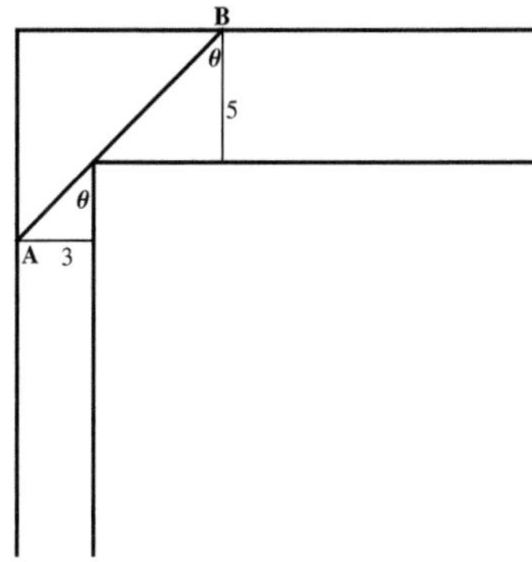

Figure 8

Solution

With reference to Figure 8, the line segment AB must be longer than 10 for any value of the angle θ between 0 and $\pi/2$, so that it is possible to carry the ladder around the corner. The length of AB is

$$f(\theta) = \frac{3}{\sin(\theta)} + \frac{5}{\cos(\theta)}.$$

Thus, we must determine the minimum value of $f(\theta)$, where $0 < \theta < \pi/2$, and see whether that value is greater than 10.

Figure 9 shows the graph of f on the interval $(0, \pi/2)$.

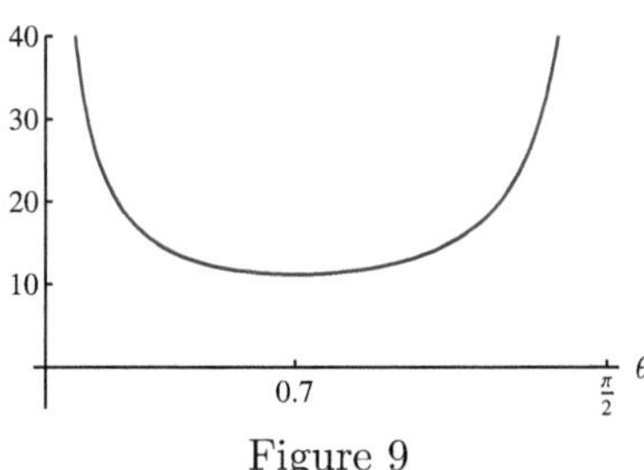

Figure 9

We have

$$\lim_{\theta \to 0+} f(\theta) = \lim_{\theta \to \pi/2-} f(\theta) = +\infty$$

(confirm). Figure 8 indicates that f attains its absolute minimum on the interval $(0, \pi/2)$ at its stationary point that seems to be near 0.7. Let's differentiate f:

$$
\begin{aligned}
f'(\theta) &= \frac{d}{d\theta}\left(\frac{3}{\sin(\theta)} + \frac{5}{\cos(\theta)}\right) \\
&= 3\frac{d}{d\theta}\left(\sin(\theta)\right)^{-1} + 5\frac{d}{d\theta}\left(\cos(\theta)\right)^{-1} \\
&= 3\left(-\left(\sin(\theta)\right)^{-2}\cos(\theta)\right) + 5\left(-\left(\cos(\theta)\right)^{-2}\left(-\sin(\theta)\right)\right) \\
&= -\frac{3\cos(\theta)}{\sin^2(\theta)} + \frac{5\sin(\theta)}{\cos^2(\theta)}.
\end{aligned}
$$

You can check that $f''(\theta) > 0$ for each $\theta \in (0, \pi/2)$, so that the graph of f is concave up on $(0, \pi/2)$ and f attains its absolute minimum on $(0, \pi/2)$ at its stationary point in $(0, \pi/2)$ (Theorem 2 of Section 3.3). You can determine an approximation to θ_0 such that $f'(\theta_0) = 0$ with the help of the approximate equation solver of your calculator, and confirm that $\theta_0 \cong .700\,669$. Therefore, the absolute minimum of f on the interval $(0, \pi/2)$ is $f(\theta_0) \cong 11.194\,1$. Since the ladder in question is 10 meters long, it can be carried around the corner. $\square$

Example 5 (Snell's Law of Refraction) Assume that the speed of light in medium 1 is c_1 kilometers/second, and the speed of light in medium 2 is c_2 kilometers/second.

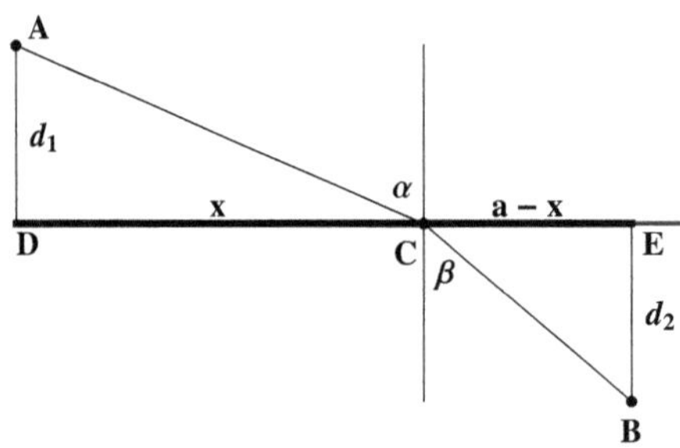

Figure 10

With reference to Figure 10, the point A is in medium 1 and the point B is in medium 2. Let the distance from D to E be a. The horizontal line represents the demarcation between the media. We will assume that light travels from one point to another so that the time of travel is minimized. In particular, light takes the shortest path to travel from A to C and from C to B. Thus, the required path from A to B consists of the line segments AC and CB. **Snell's law of refraction** says that

$$
\frac{\sin(\alpha)}{\sin(\beta)} = \frac{c_1}{c_2}.
$$

The angle α is referred to as **the angle of incidence**, and the angle β is referred to as **the angle of refraction**. Thus, Snell's law says that the ratio of the sine of the angle of incidence to the sine of the angle of refraction is equal to the ratio of the speed of light in the first medium to the speed of light in the second medium.

Let us establish Snell's law. Our task is to determine the point C, i.e., the value of x, so that the time of travel from A to B is minimized.

The time needed for the light to travel from A to C is

$$
\frac{\text{length of } AC}{\text{speed of light in medium 1}} = \frac{\sqrt{d_1^2 + x^2}}{c_1}.
$$

The time needed for the light to travel from C to B is

$$\frac{\text{length of } CB}{\text{speed of light in medium 2}} = \frac{\sqrt{d_2^2 + (a-x)^2}}{c_2}.$$

The only variable is x. Let us set

$$f(x) = \frac{\sqrt{d_1^2 + x^2}}{c_1} + \frac{\sqrt{d_2^2 + (a-x)^2}}{c_2}.$$

Thus, $f(x)$ is the total time that is needed for light to travel from A to C and then from C to B. We need to minimize $f(x)$. We have

$$f'(x) = \frac{d}{dx}\left(\frac{\sqrt{d_1^2 + x^2}}{c_1} + \frac{\sqrt{d_2^2 + (a-x)^2}}{c_2}\right) = \frac{1}{c_1}\frac{x}{\sqrt{d_1^2 + x^2}} - \frac{1}{c_2}\frac{a-x}{\sqrt{d_2^2 + (a-x)^2}}.$$

You can verify that

$$f''(x) = \frac{1}{c_1}\frac{d_1^2}{\sqrt{(d_1^2 + x^2)^3}} + \frac{1}{c_2}\frac{d_2^2}{\sqrt{(d_2^2 + (a-x)^2)^3}} > 0$$

for each $x \in R$. Therefore, the graph of f is concave up on $[0, a]$. We have

$$f'(0) = \frac{1}{c_1}\frac{x}{\sqrt{d_1^2 + x^2}} - \frac{1}{c_2}\frac{a-x}{\sqrt{d_2^2 + (a-x)^2}}\Bigg|_{x=0} = -\frac{a}{c_2\sqrt{d_2^2 + a^2}} < 0,$$

and

$$f'(a) = \frac{1}{c_1}\frac{x}{\sqrt{d_1^2 + x^2}} - \frac{1}{c_2}\frac{a-x}{\sqrt{d_2^2 + (a-x)^2}}\Bigg|_{x=a} = \frac{a}{c_1\sqrt{d_1^2 + a^2}} > 0.$$

Since f' is continuous on $[0, a]$, there exists a point $x_0 \in (0, a)$ such that $f'(x_0) = 0$, by the Intermediate Value Theorem. Since $f''(x) > 0$ for each x, x_0 is the only stationary point of f in the interval $[0, a]$, and f attains its absolute minimum on $[0, a]$ at x_0 (Theorem 2 of Section 3.6). We have

$$f'(x_0) = 0 \Leftrightarrow \frac{1}{c_1}\frac{x}{\sqrt{d_1^2 + x^2}} - \frac{1}{c_2}\frac{a-x}{\sqrt{d_2^2 + (a-x)^2}} = 0$$

$$\Leftrightarrow \frac{1}{c_1}\frac{x_0}{\sqrt{d_1^2 + x_0^2}} = \frac{1}{c_2}\frac{a-x_0}{\sqrt{d_2^2 + (a-x_0)^2}}.$$

With reference to Figure 10, α is the angle of incidence and β is the angle of refraction. We have

$$\sin(\alpha) = \frac{x_0}{\sqrt{d_1^2 + x_0^2}}, \text{ and } \sin(\beta) = \frac{a-x_0}{\sqrt{d_2^2 + (a-x_0)^2}}.$$

Therefore, f attains its minimum value if

$$\frac{1}{c_1}\sin(\alpha) = \frac{1}{c_2}\sin(\beta),$$

i.e.,

$$\frac{\sin(\alpha)}{\sin(\beta)} = \frac{c_1}{c_2}.$$

Thus, we have established Snell's law of refraction. $\square$

Applications to Economics

Let us look at some applications to economics. We will assume that a company produces and sells a single product. We will denote the quantity that is produced by x, and assume that x can be any real number. This is a modeling assumption. After all, in many cases x can only attain positive integer values. For example, a company may produce calculators, in which case x may refer to the number of calculators produced by the company over a period of three months. Therefore, if the result of a calculation is that $x = \sqrt{50000} \cong 223.607$, the decimal can be rounded to 224.

Let us first examine the cost aspects. The **cost function** C is defined so that $C(x)$ denotes the cost of producing quantity x of the product. If there is a fixed cost c per item, the total cost of producing the quantity x is simply cx, so that $C(x) = cx$. In this case C is a simple linear function. More realistic models involve nonlinear cost functions. Cost functions of the form

$$a_0 + a_1 x - a_2 x^2 + a_3 x^3,$$

where a_0, a_1, a_2 and a_3 are certain positive constants are quite common. The term a_0 reflects the cost of maintaining the infrastructure of the company, even if there is no production. Obviously, the reliability of the predictions of a particular model depend on the construction of a realistic cost function. Such an effort involves many practical and theoretical considerations, and some of you may study such issues in other courses. Our discussions will be limited to certain conclusions that may be reached with the help of calculus, given a cost function.

The rate of change of the total cost $C(x)$ with respect to the change in the level of the production may be of interest. Since we identify the rate of change of a function with its derivative, all we have to do is to compute the derivative of the given cost function.

Remark 1 Let $\Delta x > 0$ represent an increase in the production level. Recall our discussion of the differential in Section 3.2. We have

$$C(x + \Delta x) - C(x) \cong C'(x)\Delta x,$$

if Δx is small. But, how small is "small"? Let's assume that $\Delta x = 1$ is small relative to x within the context of the production of many items. Then,

$$C(x + 1) - C(x) \cong C'(x).$$

Thus, $C'(x)$ approximates the change in the total cost corresponding to the increase in the production by a single item. For this reason, economists like to refer to the derivative of the cost function as **marginal cost.** As far as they are concerned, the derivative at x is "the marginal difference" in the cost due to the production of a single extra item. $\Diamond$

Example 6 Let the cost function be

$$C(x) = 100 + 10x - 0.1x^2 + 0.001x^3.$$

You may imagine that a company produces a single type of computer, and that $C(x) \times 100$ is the total cost (in dollars) of producing x computers over a period of three months. The constant term reflects the fact that there is a cost of maintaining the infrastructure even if there is no production.

a) Plot the graph of C on the interval $[0, 100]$ with the help of your graphing utility.
b) Determine level of production at which marginal cost has its minimum value on the interval $[0, 100]$.

Solution

a) Figure 11 shows the graph of the cost function.

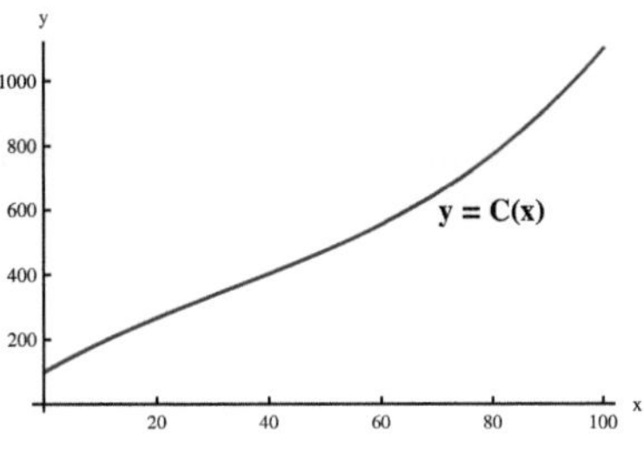

Figure 11: A cost function

b) Notice that the cost function C is an increasing function, as it should be. Since marginal cost is the derivative of the cost function, we have to compute the point at which C' attains its minimum value on the interval $[0, 100]$. We have

$$C'(x) = \frac{d}{dx}\left(100 + 10x - 0.1x^2 + 0.001x^3\right) = 10 - 0.2x + 0.003x^2.$$

Figure 12 shows the graph of C'.

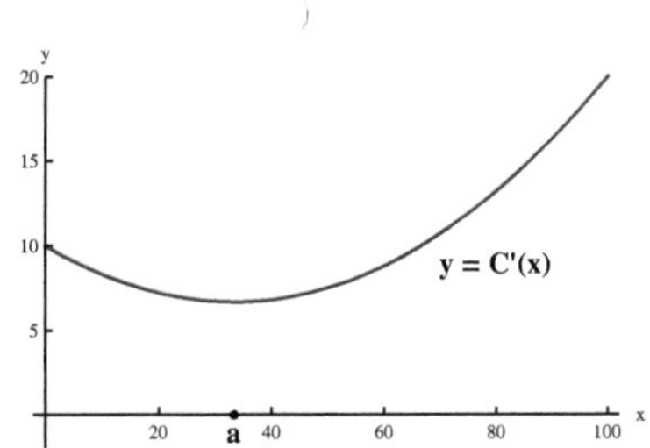

Figure 12: A marginal cost function

Figure 11 indicates that marginal cost decreases up to a certain production level, and then starts to increase. We must find the value of x that minimizes $C'(x)$. We will apply the derivative test for monotonicity to C'. We have

$$(C')'(x) = C''(x) = \frac{d}{dx}\left(10 - 0.2x + 0.003x^2\right) = -0.2 + 0.006x.$$

Therefore,

$$C''(x) = 0 \Leftrightarrow -0.2 + 0.006x = 0 \Leftrightarrow x = a = \frac{1}{3} \times 10^2 \cong 33.3,$$

We have $C''(x) < 0$ if $0 \leq x < a$, and $C''(x) > 0$ if $a < x \leq 100$. Therefore, C' decreases on the interval $[0, a]$ and increases on the interval on the interval $[a, 100]$. Thus, C' attains its minimum value on the interval $[0, 100]$ at a. Note that the graph of the cost function is concave down on $[0, a]$, and concave up on $[a, 100]$. The graph of C has an inflection point at

$$(a, C(a)) = \left(\frac{1}{3} \times 10^2, C\left(\frac{1}{3} \times 10^2\right)\right) \cong (33.3, 359.3).$$

$\square$

Now let us look at the revenue side. If quantity x of the product of the company is sold at a fixed price p per unit, the total revenue is px. But it is not realistic to assume that the product has the same price at all production levels. It is to be expected that the price will be lower if the supply is higher. Thus, the **revenue function** R is of the form

$$R(x) = xp(x),$$

where the function $p(x)$ is not a constant, in general. The function $p(x)$ is referred to as the **price function,** or the **demand function.**

Remark 2 Economists refer to the derivative of the revenue function as **marginal revenue.** As in the case of marginal cost, the approximation

$$R(x+1) - R(x) \cong R'(x)$$

appears to be justified in many cases of practical interest.$\Diamond$

Example 7 Let us consider the specific price function

$$p(x) = 15 - 0.05x.$$

The corresponding revenue function is

$$R(x) = xp(x) = 15x - 0.05x^2.$$

Figure 13 shows the graph of R. $\square$

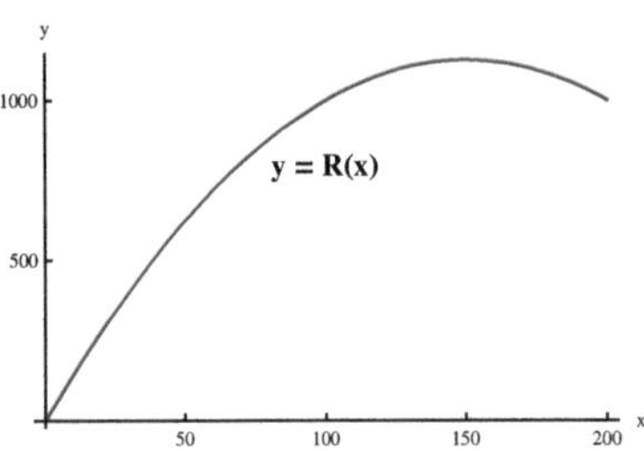

Figure 13: A revenue function

In general, the ultimate goal of a company is to make profit. The **profit function** P is the difference between the revenue function R and the cost function C:

$$P(x) = R(x) - C(x).$$

Proposition 1 **If the profit function P attains a local maximum at a then the marginal revenue at a is the same as the marginal cost at the production level a.**

Proof

If P attains a local maximum at a, we must have $P'(a) = 0$. We have

$$P'(x) = \frac{d}{dx}(R(x) - C(x)) = R'(x) - C'(x),$$

Therefore, $R'(a) = C'(a)$, i.e., the marginal revenue is the same as marginal cost at a.∎

Remark 3 Proposition 1 can be interpreted as follows: The optimal production level is reached when the additional revenue that is obtained by producing one additional item is the same as the additional cost of producing one additional item. Graphically, the line that is tangent to the graph of the revenue function at $(a, R(a))$ is parallel to the line that is tangent to the graph of the cost function at $(a, C(x))$, as illustrated in Figure 14. ◊

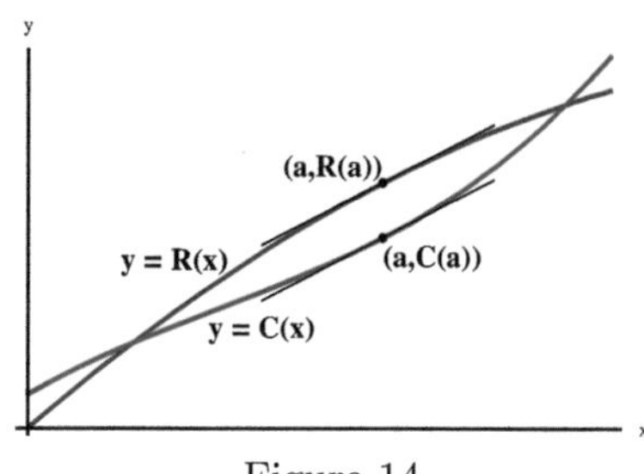

Figure 14

Example 8 Let

$$C(x) = 100 + 10x - 0.1x^2 + 0.001x^3,$$

as in Example 6, and

$$R(x) = 15x - 0.05x^2,$$

as in Example 7. Determine the production level between 0 and 100 so that the profit is maximized.

Solution

We have

$$P(x) = R(x) - C(x)$$
$$= \left(15x - 0.05x^2\right) - \left(100 + 10x - 0.1x^2 + 0.001x^3\right).$$

Figure 15 shows the graph of the profit function. Figure 15 indicates that the profit function attains its absolute maximum on $[0, 100]$ at its stationary point a in that interval. You can confirm this with the help of the derivative test for monotonicity. We have

$$P'(x) = \frac{d}{dx}\left(\left(15x - 0.05x^2\right) - \left(100 + 10x - 0.1x^2 + 0.001x^3\right)\right)$$
$$= 5.0 + .1x - .003x^2.$$

Therefore, the stationary point of P that is in the interval $[0, 100]$ is $60.762\,5$, rounded to 6 significant digits. Thus, the optimal production level a is approximately 61. □

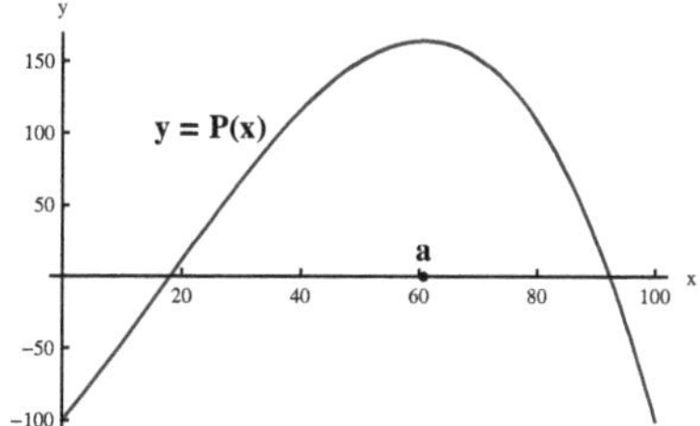

Figure 15: A profit function

Problems

1. Determine two positive numbers whose sum is the minimum among all pairs of positive numbers whose product is 400.

2. Determine the dimensions of a rectangle whose area is the maximum among all rectangles with perimeter 200 meters.

3. Determine the dimensions of a rectangle whose perimeter is the minimum among all rectangles with area 800 m^2.

4. Determine the dimensions of the rectangle that has the largest area among all rectangles that can be inscribed in a semicircle of radius 4.

5. Determine the dimensions of the rectangle of maximum area among all rectangles that can be inscribed in an equilateral triangle of side length 10 such that one side lies on the base of the triangle.

6. Determine the dimensions of the cylinder of largest volume among all right circular cylinders that can be inscribed in a sphere of radius 10.

7. Determine the points on the ellipse

$$\frac{x^2}{4} + y^2 = 1$$

that is closest to the point $(1, 0)$.

8. Assume that a manufacturer must produce cans in the shape of right circular cylinders. The volume of each can is required to be 400 cubic centimeters. Otherwise, the manufacturer is free to choose the dimensions of the can. The cost per square centimeter of the material for the top and the bottom of the can is the same as the cost per square centimeter of the material for the lateral surface. How should the manufacturer determine the dimensions of the can in order to minimize the cost?

9. A box with a square base and open top is required to have a volume of 1000 cubic centimeters. Determine the dimensions of the box so that the amount of the material that is used in the construction of the box is minimized.

10. A water trough has length 10 meters and a cross section in the shape of an isosceles triangle with sides that are 50 centimetres. Use calculus to determine the length of the top of the triangle so that the volume of the trough is maximized.

11. Assume that a racket launcher is fired at an angle θ from the horizotal ground. The range s is the horizontal distance traveled by a projectile fired by the rocket launcher and is given by the expression

$$s = \frac{v_0^2 \sin(2\theta)}{g},$$

where v_0 is the initial speed of the projectile and g is the constant gravitational acceleration (can be assumed to be 9.8 meters/sec/sec.). Use calculus to determine the value of θ that maximizes the range s.

12 [C] Assume that the total cost $C(x)$ of producing x items of a certain product is

$$C(x) = 0.002x^3 - 0.1x^2 + 4x.$$

a) Plot the graph of $C(x)$ and the marginal cost function $C'(x)$.
b) Determine level of production at which marginal cost has its minimum value. Intepret your responce graphically.

13 [C] Assume that the total cost $C(x)$ of producing x items of a certain product is

$$C(x) = 0.002x^3 - 0.1x^2 + 4x$$

and the corresponding revenue is

$$R(x) = 10x - 0.04x^2.$$

a) Show the graphs of $C(x)$ and $R(x)$ in the same picture.
b) Determine production level so that the profit is maximized. Intepret your response in the language of econmics and graphically.

Chapter 4

Special Functions

Functions such as sine and cosine are special, since they occur frequently and in a variety of contexts. In this chapter we will introduce other special functions and study their basic properties. These are **exponential** and **logarithmic functions, inverse trigonometric functions, hyperbolic** and **inverse hyperbolic functions**. You will see that the solutions of mathematical models of population growth, radioactive decay and compound interest can be expressed in terms of exponential functions. In later chapters, you will see that exponential functions and the other special functions that are introduced in this chapter are indispensable in expressing the solutions of many other mathematical models.

4.1 Inverse Functions

Some of the most important special functions of mathematics are constructed as inverses of familiar functions or their restrictions. In this section we will discuss the concept of an inverse function, and the inverses of appropriate restrictions of sine, cosine and tangent.

The Concept of an Inverse Function

Let's begin by reviewing the definition of the square-root. For any $x \geq 0$, we have $y = \sqrt{x}$ if $x = y^2$ and $y \geq 0$. If we set $f(y) = y^2$, then $y = \sqrt{x}$ is the unique nonnegative solution of the equation $x = f(y)$. Figure 1 illustrates the definition of $\sqrt{x}$ graphically. Note that we have reversed the usual roles of the x and y axes, since we denoted the independent variable of f by y and referred to $x = f(y) = y^2$.

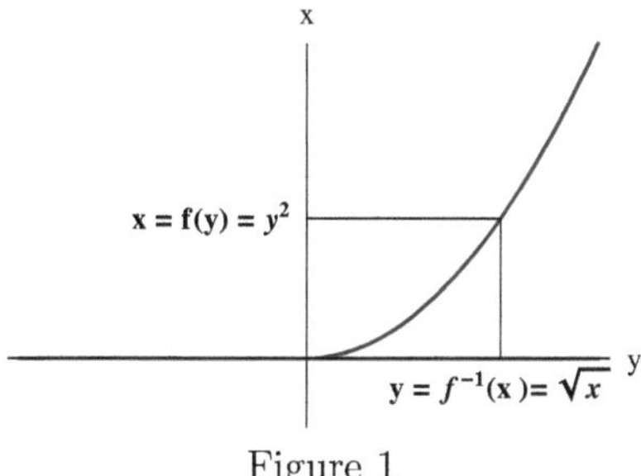

Figure 1

The square-root function is "the inverse of f". Let us consider the general case. Assume that D is the domain of the function f and the range of f is the set J. Let's denote the independent variable of f by y, and the dependent variable by x (note that we have interchanged the usual roles of x and y). Thus, given any $x \in J$, there exists $y \in D$ such that $x = f(y)$. We can define y as a function of x if there exists a unique $y \in D$ such that $x = f(y)$ for each $x \in J$. We will call that function the inverse of f.

Definition 1 Assume that for each x in the range of f there is a **unique** y in the domain of f such that $f(y) = x$. **The inverse** f^{-1} **of** f is defined by the following relationship:

$$y = f^{-1}(x) \Leftrightarrow x = f(y).$$

Thus, the value of f^{-1} at x is the solution of the equation $x = f(y)$, provided that the solution exists and is unique. Figure 2 illustrates the relationship between f and the inverse f^{-1} graphically in the yx-plane (the y-axis is horizontal and the x-axis is vertical).

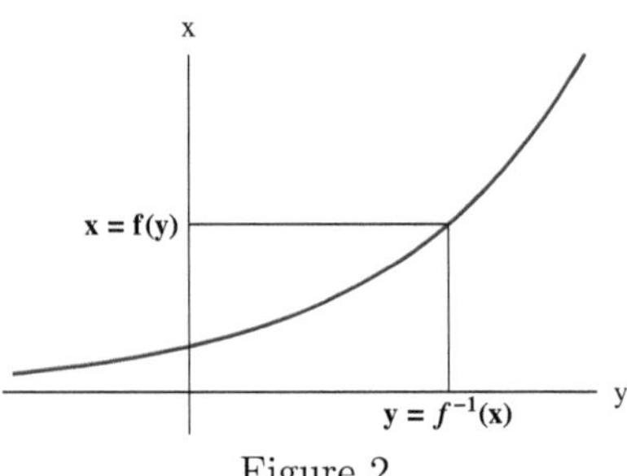

Figure 2

By the relationship between a function f and its inverse f^{-1}, the domain of f^{-1} is the same as the range of f, and the range of f^{-1} is the same as the domain of f. We must emphasize that **the notation** f^{-1} **in the present context should not be confused with the reciprocal** $1/f$ **of the function** f. The meaning of the notation should be clear within a particular context.

Example 1 Let $f(y) = y^2$. We restrict y so that $y \geq 0$. Thus, the domain of f is $[0, +\infty)$. The range of f is also $[0, +\infty)$. Since

$$x = f(y) = y^2 \text{ and } y \geq 0 \Leftrightarrow y = \sqrt{x},$$

we have $f^{-1}(x) = \sqrt{x}$. The domain of f^{-1} consists of all nonnegative numbers. $\square$

If f has an inverse, then f is the inverse of f^{-1}:

$$\left(f^{-1}\right)^{-1} = f.$$

Indeed, we can read the statement

$$y = f^{-1}(x) \Leftrightarrow x = f(y)$$

as

$$x = f(y) \Leftrightarrow y = f^{-1}(x).$$

The square-root function illustrates a function defined by $x^{1/n}$, where n is an even positive integer. We have

$$y = x^{1/n} \Leftrightarrow x = y^n$$

where $x \geq 0$ and $y \geq 0$. Therefore, if we set $f(y) = y^n$, where $y \geq 0$, then $f^{-1}(x) = x^{1/n}$, $x \geq 0$.

Example 2 Let $f(y) = y^3$, where y is an arbitrary real number. The equation $x = f(y) = y^3$ has the unique solution $y = x^{1/3}$ for each $x \in \mathbb{R}$. Thus, $f^{-1}(x) = x^{1/3}$. Figure 3 illustrates the relationship between $x = f(y) = y^3$ and $y = f^{-1}(x) = x^{1/3}$. $\square$

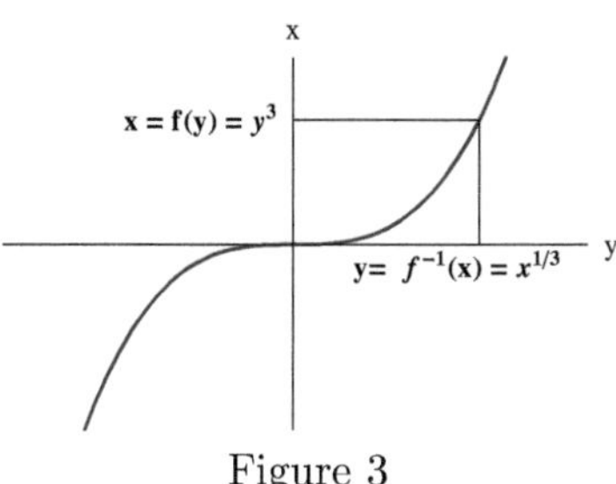

Figure 3

Example illustrates $x^{1/n}$ where n is an odd positive integer. If n is an odd positive integer and $f(y) = y^n$, the equation $x = f(y)$ has the unique solution $y = x^{1/n}$ for each $x \in \mathbb{R}$. Therefore, $f^{-1}(x) = x^{1/n}$ for each $x \in \mathbb{R}$.

We must be clear about the domain of a particular function when we discuss its inverse, and we must not take it for granted that a function has an inverse.

Example 3 As we discussed in Example **??**, the square-root function is the inverse of the function f whose domain is $[0, +\infty)$ and $f(y) = y^2$ for each $y \geq 0$. The range of f is also $[0, +\infty)$. On the other hand, if we set $F(y) = y^2$ for each $y \in \mathbb{R}$, so that the domain of F is the entire number line, the function F is not the same as the function f whose domain is $[0, +\infty)$. Since $f(y) = F(y) = y^2$ for each $y \geq 0$, the function f is the restriction of F to $[0, +\infty)$. Show that F does not have an inverse.

Solution

The definition of an inverse function requires that the equation $x = F(y)$ has a unique solution for each x in the range of F, i.e., for each $x \geq 0$. If $x > 0$, the equation $x = F(y) = y^2$ has two *distinct* solutions, $y = +\sqrt{x}$ or $y = -\sqrt{x}$. Thus, F does not have an inverse. Note that a horizontal line in the upper half of the yx-plane intersects the graph of F at two distinct points, as illustrated in Figure 4. $\square$

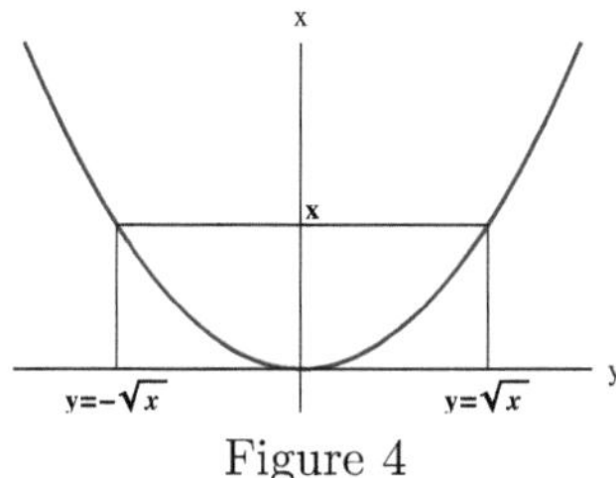

Figure 4

Remark 1 (The Horizontal Line Test) As illustrated in Example 3, **a function F has an inverse if and only if a horizontal line in the yx-plane does not intersect the graph of $x = F(y)$ at more than one point**. This observation is the graphical counterpart of the fact that F has an inverse if and only if the equation has a unique solution y for each x in the range of F. $\Diamond$

Example 4 Let $F(y) = y^2 - 2y + 3$ for each $y \in \mathbb{R}$, and let $f(y) = y^2 - 2y + 3$ for each $y \geq 1$. Thus, f is the restriction of F to the interval $[1, +\infty)$.

a) Sketch the graph of F. Determine the range of F. Show that F does not have an inverse.

b) Sketch the graph of f. Show that f has an inverse and determine f^{-1}. Sketch the graph of f^{-1}.

Solution

a) Since $F(y) = y^2 - 2y + 3 = (y - 1)^2 + 2$, the graph of $x = F(y)$ in the yx-coordinate plane is a parabola whose vertex is at $(1, 2)$, as shown in Figure 5.

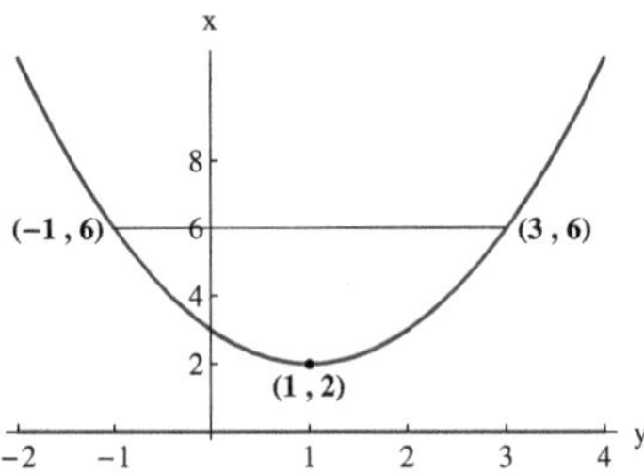

Figure 5: The graph of F fails the horizontal line test

The domain of F is the number line $(-\infty, +\infty)$, and the range of F is the interval $[2, +\infty)$. The function F does not have an inverse. Indeed, Figure 5 indicates that F fails the horizontal line test. For example, the line $x = 6$ intersects the graph of $x = F(y)$ at two distinct points. Algebraically, the equation $6 = F(y)$ has two distinct solutions:

$$6 = (y - 1)^2 + 2 \Leftrightarrow (y - 1)^2 = 4 \Leftrightarrow (y - 1) = \pm 2 \Leftrightarrow y = -1 \text{ or } y = 3.$$

b)

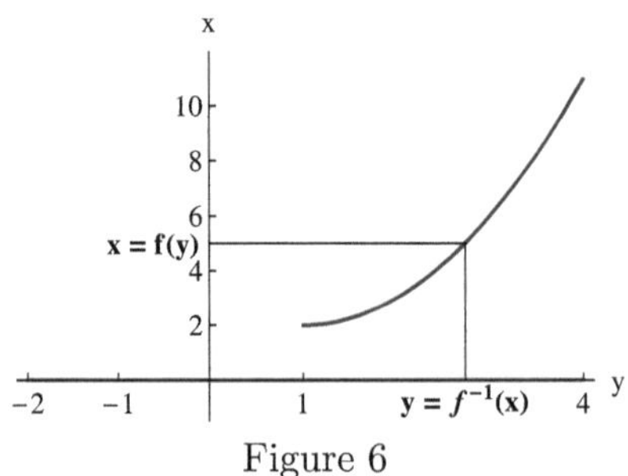

Figure 6

Figure 6 displays the graph of f, the restriction of F to the interval $[1, +\infty)$. The picture suggests that the inverse of f exist and illustrates the definition of $y = f^{-1}(x)$. Let's proceed with the calculations.

Since $f(y) = (y - 1)^2 + 2 \geq 2$ for each $y \geq 1$, the range of f is contained in the interval $[2, +\infty)$. In fact, the range of f is the interval $[2, +\infty)$: Let $x \geq 2$. We must solve the equation $x = f(y)$. We have

$$x = (y - 1)^2 + 2 \Leftrightarrow (y - 1) = \pm\sqrt{x - 2} \Leftrightarrow y = 1 \pm \sqrt{x - 2}.$$

Since the domain of f is $[1, +\infty)$, the relevant solution is uniquely determined as $y = 1 + \sqrt{x - 2}$. Note that $x \geq 2$, so that $y = 1 + \sqrt{x - 2}$ is a real number. Therefore, each $x \in [2, +\infty)$ is in the range of f. Since there exists a unique y in the domain of f such that $x = f(y)$ for each $x \in [2, +\infty)$, i.e., $y = 1 + \sqrt{x - 2}$, the inverse of f exists, and we have

$$f^{-1}(x) = 1 + \sqrt{x - 2}.$$

The domain of f^{-1} is the same as the range of f, i.e., $[2, +\infty)$. The range of f^{-1} is the same as the domain of f, i.e., $[1, +\infty)$.

Figure 7 displays the graph of f^{-1}. $\square$

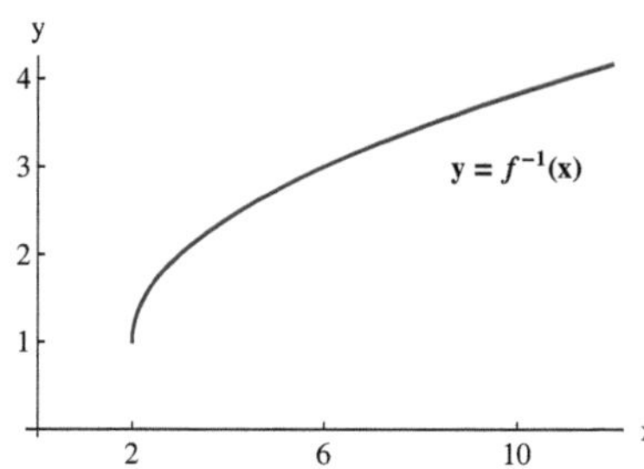

The composition of a function with its inverse results in an identity function:

Assume that the function f has an inverse. Then,

$$(f^{-1} \circ f)(y) = f^{-1}(f(y)) = y \text{ for each } y \text{ in the domain of } f,$$

$$(f \circ f^{-1})(x) = f(f^{-1}(x)) = x \text{ for each } x \text{ in the domain of } f^{-1}.$$

Proof

We have

$$y = f^{-1}(x) \Leftrightarrow x = f(y)$$

where x is in the domain of f^{-1} ($=$ the range of f) and y is in the domain of f. Therefore,

$$\left(f^{-1} \circ f\right)(y) = f^{-1}(f(y)) = f^{-1}(x) = y$$

and

$$\left(f \circ f^{-1}\right)(x) = f\left(f^{-1}(x)\right) = f(y) = x.$$

∎

Example 5 Let $f(y) = (y - 1)^2 + 2$ for each $y \geq 1$, as in Example 4. Confirm that

$$\left(f^{-1} \circ f\right)(y) = f^{-1}(f(y)) = y \text{ for each } y \text{ in the domain of } f,$$

and

$$\left(f \circ f^{-1}\right)(x) = f\left(f^{-1}(x)\right) = x \text{ for each } x \text{ in the domain of } f^{-1}.$$

Solution

As we discussed in Example 4, the domain of f is $[1, +\infty)$ and the range of f is $[2, +\infty)$, so that the domain of f^{-1} is $[2, +\infty)$ and the range of f^{-1} is $[1, +\infty)$. We have $f^{-1}(x) = 1 + \sqrt{x - 2}$ for each $x \geq 2$.

For each $y \in [1, +\infty)$,

$$
\begin{aligned}
\left(f^{-1} \circ f\right)(y) = f^{-1}(f(y)) &= 1 + \sqrt{f(y) - 2} \\
&= 1 + \sqrt{((y-1)^2 + 2) - 2} \\
&= 1 + \sqrt{(y-1)^2} \\
&= 1 + |y - 1| \\
&= 1 + (y - 1) = y,
\end{aligned}
$$

since $y - 1 \geq 0$.

For each $x \in [2, +\infty)$,

$$
\begin{aligned}
\left(f \circ f^{-1}\right)(x) = f\left(f^{-1}(x)\right) &= \left(f^{-1}(x) - 1\right)^2 + 2 \\
&= \left((1 + \sqrt{x - 2}) - 1\right)^2 + 2 \\
&= \left(\sqrt{x - 2}\right)^2 + 2 \\
&= (x - 2) + 2 = x.
\end{aligned}
$$

$\square$

If the scale on the vertical axis is the same as the scale on the horizontal axis, the graph of f^{-1} appears as the reflection of the graph of f with respect to the diagonal $y = x$, as illustrated in Figure 8:

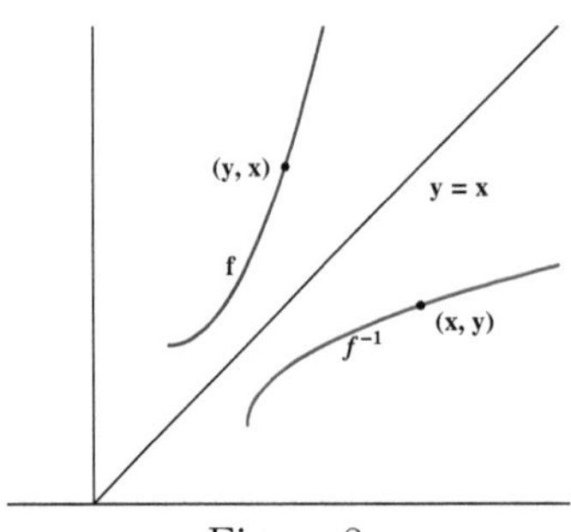

Figure 8

Assume that the inverse of f exists. The graphs of $x = f(y)$ and $y = f(x)$ are symmetric with respect to the diagonal $y = x$, provided that the scale on the vertical axis is the same as the scale on the horizontal axis.

Proof

We will show that a point (x, y) is on the graph of f^{-1} if and only if (y, x) is on the graph of f. Since (x, y) and (y, x) are symmetric with respect to the diagonal $y = x$, the conclusion about the graphs of f and f^{-1} follows. Thus, assume that (x, y) is a point on the graph of f^{-1}. Then, $y = f^{-1}(x)$ so that $x = f(y)$. Therefore, (y, x) is on the graph of f. Conversely, assume that (y, x) is on the graph of f. Then, $x = f(y)$ so that $y = f^{-1}(x)$. Therefore, (x, y) is on the graph of f^{-1}. $\blacksquare$

A word of caution: The graphs of $x = f(y)$ and $y = f^{-1}(x)$ need not appear to be symmetric with respect to the diagonal if the scale on the vertical axis is not the same as the scale on the horizontal axis. For example, Figure 9 displays the graphs of f and f^{-1}, where f is the function of Example 4. The graphs do not appear to be symmetric with respect to the line $y = x$ since the scales on the horizontal and vertical axes are different.

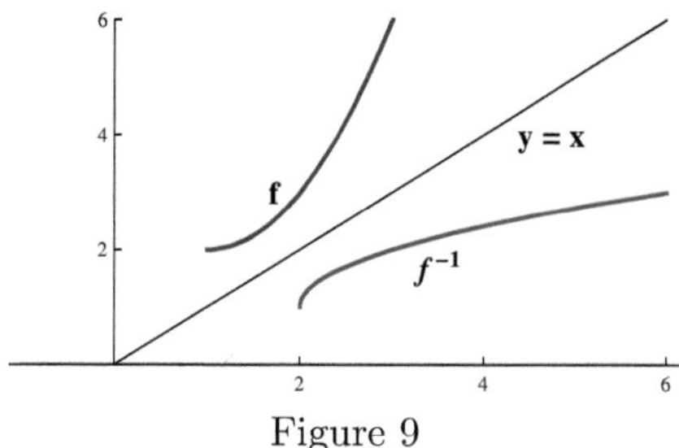

Figure 9

There is a general fact about the existence and continuity of the inverse of a function:

Theorem 1 Assume that f is increasing or decreasing and continuous on the interval I. The range J of f is also an interval. The inverse of f exists and f^{-1} is continuous on J. The function f^{-1} is increasing if f is increasing, and decreasing if f is decreasing.

You can find the proof of Theorem 1 in Appendix C.

Inverse Trigonometric Functions

Appropriate restrictions of sine, cosine and tangent have inverses. These functions are important special functions of mathematics.

Let's begin with the **sine** function. The equation $x = \sin(y)$ has infinitely many solutions for a given $x \in [-1, 1]$. Indeed, if y is a solution of the equation $x = \sin(y)$, then $y + 2n\pi$, $n = \pm 1, \pm 2, \ldots$ are also solutions, since sine is periodic with period 2π:

$$\sin(y + 2n\pi) = \sin(y) = x.$$

The graph of sine fails the horizontal line test for the existence of the inverse function dramatically, as illustrated in Figure 10. Thus, the sine function does not have an inverse.

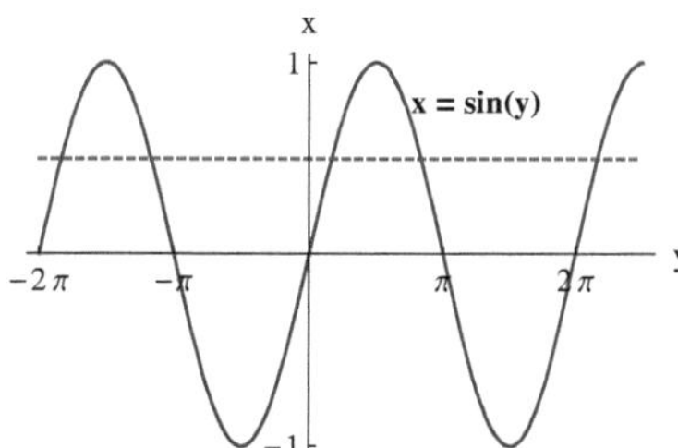

Figure 10: The graph of $x = \sin(y)$ fails the horizontal line test

Let's restrict sine to the interval $[-\pi/2, \pi/2]$ and call the resulting function f. The function f is continuous and increasing on $[-\pi/2, \pi/2]$, and the range of f is the interval $[-1, 1]$. Figure 11 shows the graph of f.

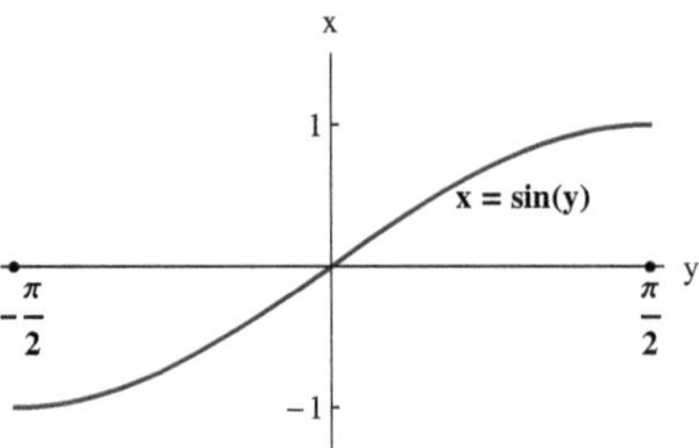

Figure 11: $x = \sin(y)$ on $[-\pi/2, \pi/2]$

By Theorem 1, the inverse of f exists and f^{-1} is continuous on $[-1, 1]$. We have

$$y = f^{-1}(x) \Leftrightarrow x = \sin(y),$$

where $-1 \leq x \leq 1$ and $-\pi/2 \leq y \leq \pi/2$. We will refer to f^{-1} as **arcsine**, and abbreviate it as arcsin. Thus,

$$y = \mathbf{arcsin}(x) \Leftrightarrow x = \sin(y)$$

where $-1 \leq x \leq 1$ and $-\pi/2 \leq y \leq \pi/2$. You can think of $y = \arcsin(x)$ as the unique angle y (in radians) between $-\pi/2$ and $\pi/2$ such that $\sin(y) = x$. Figure 12 illustrates the definition of arcsine.

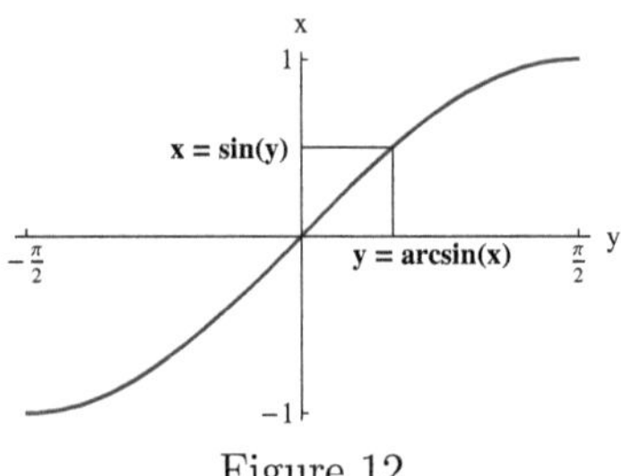

Figure 12

Figure 13 shows the graph of $y = \arcsin(x)$. The graph is symmetric with respect to the origin, and indicates that arcsine is an odd function (exercise).

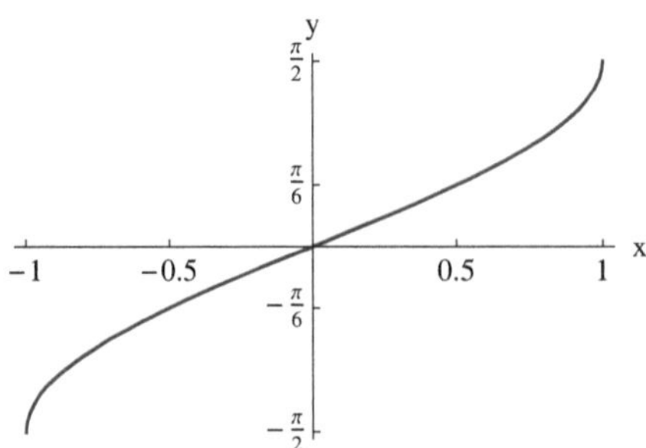

Figure 13: $y = \arcsin(x)$

Another notation for $\arcsin(x)$ is $\sin^{-1}(x)$. We will favor the notation $\arcsin(x)$ in order to avoid any confusion with the reciprocal of sine.

We can determine certain values of arcsine by inspection.

Example 6 Determine

a) $\arcsin(1)$,
b) $\arcsin(-1/2)$.

Solution

a)

$$y = \arcsin(1) \Leftrightarrow \sin(y) = 1 \text{ and } -\frac{\pi}{2} \leq y \leq \frac{\pi}{2}.$$

The only such y is $\pi/2$. Therefore,

$$\arcsin(1) = \frac{\pi}{2}.$$

b)

$$y = \arcsin\left(-\frac{1}{2}\right) \Leftrightarrow \sin(y) = -\frac{1}{2} \text{ and } -\frac{\pi}{2} \leq y \leq \frac{\pi}{2}.$$

The only such y is $-\pi/6$. Thus,

$$\arcsin\left(-\frac{1}{2}\right) = -\frac{\pi}{6}.$$

$\square$

A computer algebra system can supply the exact values of arcsine whenever that is feasible, as in Example 6. In any case, arcsine is one of the built-in functions of any computational utility, and we can always obtain approximate values.

Example 7 Approximate the following values of arcsine with the help of your computational utility (round to 6 significant digits):

a) $\arcsin(1/4)$
b) $\arcsin(-1/3)$.

Solution

a)

$$\arcsin\left(\frac{1}{4}\right) \cong 0.25268$$

b)

$$\arcsin\left(-\frac{1}{3}\right) \cong -0.339837$$

$\square$

As special cases of the relationships between a function and its inverse (Proposition **??**), we have

$$\arcsin(\sin(y)) = y \text{ if } -\frac{\pi}{2} \leq y \leq \frac{\pi}{2},$$

and

$$\sin(\arcsin(x)) = x \text{ if } -1 \leq x \leq 1.$$

We must be careful about the restrictions on x and y, though, since arcsine is not the inverse of sine, but the inverse of the restriction of sine to the interval $[-\pi/2, \pi/2]$. The restriction, $-1 \leq x \leq 1$, is imposed automatically, since the domain of arcsine is the interval $[-1, 1]$. On the other hand, We cannot claim that

$$\arcsin(\sin(y)) = y$$

if y is not in the interval $[-\pi/2, \pi/2]$. For example,

$$\arcsin\left(\sin\left(\pi\right)\right) = \arcsin\left(0\right) = 0 \neq \pi.$$

Figure 14 shows the graph of $x = \arcsin\left(\sin\left(y\right)\right)$ on the interval $[-2\pi, 2\pi]$. The graph is not the line $x = y$, even though it coincides with that line on the interval $[-\pi/2, \pi/2]$.

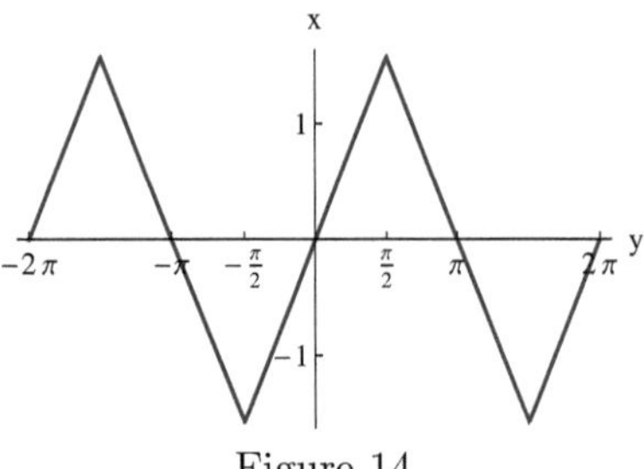

Figure 14

Just as in the case of sine, we cannot define the inverse of the periodic function **cosine**. On the other hand, cosine is continuous and decreasing on $[0, \pi]$, so that the restriction of cosine to the interval $[0, \pi]$ has an inverse. We will refer to that function as **arccosine**, and use the abbreviation arccos:

$$y = \arccos(x) \Leftrightarrow x = \cos(y),$$

where $-1 \leq x \leq 1$ and $0 \leq y \leq \pi$. Thus, the value of arccosine at $x \in [-1, 1]$ is the unique solution y of the equation $\cos(y) = x$ that is in the interval $[0, \pi]$. Figure 15 illustrates the definition of arccosine.

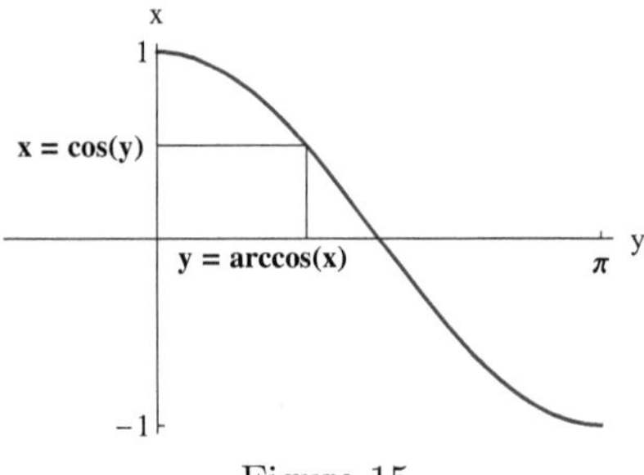

Figure 15

By Theorem 1, arccosine is continuous on $[-1, 1]$. Another notation for $\arccos(x)$ is $\cos^{-1}(x)$. We will favor the notation $\arccos(x)$. Figure 16 shows the graph of arccosine.

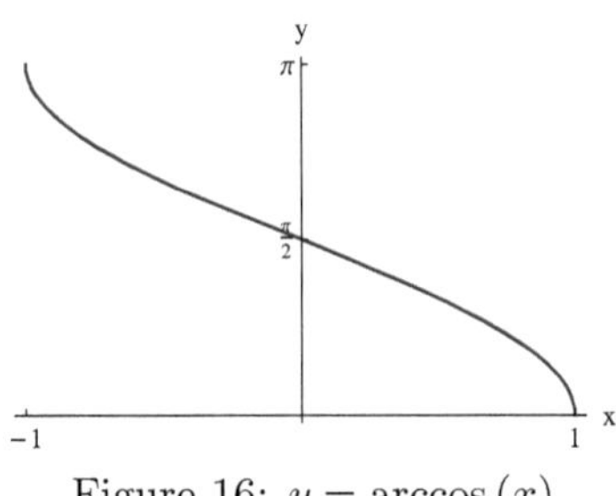

Figure 16: $y = \arccos\left(x\right)$

Some values of arccosine can be determined exactly, by inspection, or with the help of a computer algebra system. On the other hand, arccosine is a built-in function on any computational utility, so that accurate approximations to an arbitrary value of arccosine are readily available.

Example 8
a) Determine the exact value of arccos(−1/2) by inspection.
b) Approximate arccos(1/4) with the help of your computational utility (round to 6 significant digits).

Solution

a)
$$y = \arccos\left(-\frac{1}{2}\right) \Leftrightarrow \cos(y) = -\frac{1}{2} \text{ and } 0 \le y \le \pi.$$

The only such angle is
$$\pi - \frac{\pi}{3} = \frac{2\pi}{3}.$$

Therefore,
$$\arccos\left(-\frac{1}{2}\right) = \frac{2\pi}{3}.$$

b) We have
$$\arccos\left(\frac{1}{4}\right) \cong 1.31812$$

$\square$

Remark 2 The functions arcsine and arccosine are related to each other in a simple manner. In the next section we will show that
$$\arccos(x) + \arcsin(x) = \frac{\pi}{2}, \quad -1 \le x \le 1.$$

Let us set $\alpha = \arccos(x)$ and $\beta = \arcsin(x)$. Thus,
$$\cos(\alpha) = x \text{ and } 0 \le \alpha \le \pi,$$

and
$$\sin(\beta) = x \text{ and } -\frac{\pi}{2} \le \beta \le \frac{\pi}{2}.$$

Therefore, the above relationship between arcsine and arccosine can be expressed as
$$\alpha + \beta = \frac{\pi}{2}.$$

Figure 17 illustrates this relationship if α and β are between 0 and $\pi/2$. $\Diamond$

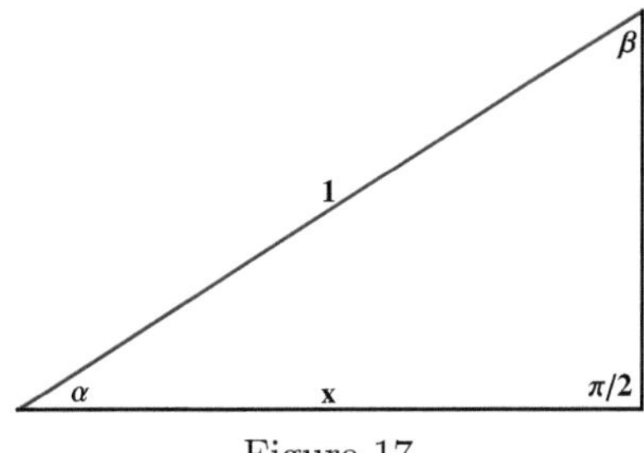

Figure 17

The function **tangent** is periodic with period π, and its range is the entire number line. Therefore, the equation $\tan(y) = x$ has infinitely many solutions for any real number x, so that the inverse of tangent does not exist. On the other hand, **the restriction of tangent to the open interval** $(-\pi/2, \pi/2)$ is continuous, increasing and has range equal to $\mathbb{R}$, so that it has an inverse that is continuous on the entire number line. We will refer to that function as **arctangent**, and use the abbreviation **arctan**. Thus,

$$y = \arctan(x) \Leftrightarrow x = \tan(y),$$

where x is an arbitrary real number and $-\pi/2 < y < \pi/2$. You may think of y as the unique angle between $-\pi/2$ and $\pi/2$ such that $\tan(y) = x$. Figure 18 illustrates the definition of arctangent. Another notation for $\arctan(x)$ is $\tan^{-1}(x)$.

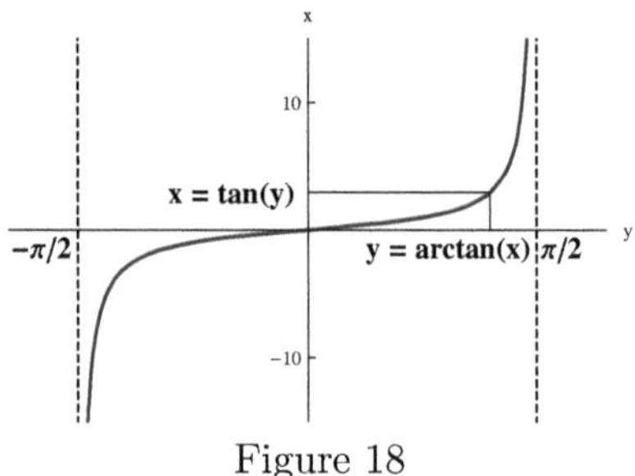

Figure 18

Figure 19 shows the graph of arctangent. The graph of arctangent is symmetric with respect to the origin, and indicates that arctangent is an odd function (exercise).

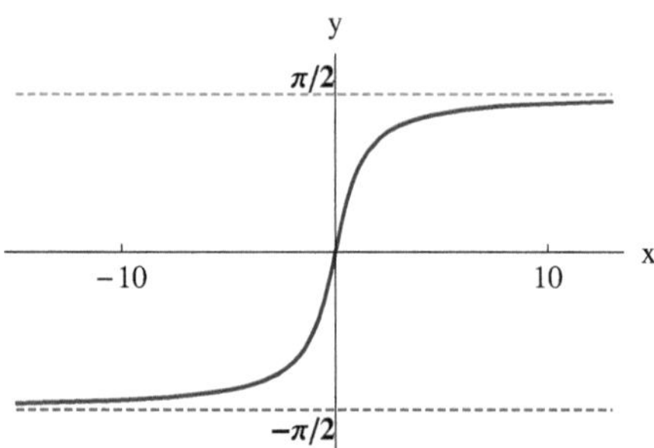

Figure 19: $y = \arctan(x)$

We have

$$\lim_{x \to +\infty} \arctan(x) = \frac{\pi}{2} \quad \text{and} \quad \lim_{x \to -\infty} \arctan(x) = -\frac{\pi}{2}.$$

These facts are parallel to the facts,

$$\lim_{y \to \frac{\pi}{2}-} \tan(y) = +\infty \quad \text{and} \quad \lim_{y \to -\frac{\pi}{2}+} \tan(y) = -\infty.$$

Example 9

a) Determine arctan(-1) by inspection.

b) Approximate arctan$(1/2)$ with the help of your computational utility (round to 6 significant digits).

Solution

a)

$$y = \arctan(-1) \Leftrightarrow \tan(y) = -1 \text{ and } -\frac{\pi}{2} < y < \frac{\pi}{2}.$$

The only such angle y is $-\pi/4$. Therefore,

$$\arctan(-1) = -\frac{\pi}{4}.$$

b) We have

$$\arctan\left(\frac{1}{2}\right) \cong .463\,648.$$

$\square$

In the next section we will calculate the derivatives of inverse functions.

Problems

In problems 1-3, determine f^{-1}, the inverse of f. Identify the domain and range of f and f^{-1}.

1.. $f(y) = y^3 - 8$

2. $f(y) = \sqrt{2y + 6}$

3. $f(y) = \dfrac{y - 2}{y + 2}$, $y > -2$.

4.

a) Let $F(y) = (y - 4)^2$ for each $y \in \mathbb{R}$. Sketch the graph of F. Show that F does not have an inverse.

b) Let $f(y) = (y - 4)^2$, $y \leq 4$, so that f is the restriction of the function F defined in part (a) to the interval $(-\infty, 4]$. Sketch the graph of f. Show that f^{-1} exists and determine f^{-1} (you need to specify the domain of f^{-1})

5.

a) Let $F(y) = y^2 + 6y + 11$ for each $y \in \mathbb{R}$. Sketch the graph of F. Show that F does not have an inverse.

b) Let $f(y) = y^2 + 6y + 11$ for each $y \geq -3$. Thus, f is the restriction of F to the interval $[-3, +\infty)$. Sketch the graph of f. Show that f has an inverse. Determine the domain and the range of f^{-1}, then determine $f^{-1}(x)$ for each x in the domain of f^{-1}. Sketch the graphs of f and f^{-1}.

In problems 6 and 7, confirm that f^{-1} is the inverse of f by showing that $(f^{-1} \circ f)(y) = y$ for each y in the domain of f and $(f \circ f^{-1})(x) = x$ for each x in the domain of f^{-1}.

6..

$$f(y) = y^3 + 8, \ f^{-1}(x) = \sqrt[3]{x - 8}$$

7.

$$f(y) = \frac{3y + 2}{2y + 1}, \ f^{-1}(x) = \frac{2 - x}{2x - 3}$$

In problems 8-19, evaluate the value of the function without the use of a calculator.

8. $\arcsin(-1)$

9. $\arcsin\left(\dfrac{\sqrt{3}}{2}\right)$

10. $\arcsin\left(\dfrac{\sqrt{2}}{2}\right)$

11. $\arcsin\left(-\dfrac{1}{2}\right)$

12. $\arccos\left(\dfrac{1}{2}\right)$

13. $\arccos\left(-\dfrac{\sqrt{3}}{2}\right)$

14. $\arccos(0)$

15. $\arccos\left(-\dfrac{\sqrt{2}}{2}\right)$

16. $\arctan\left(\sqrt{3}\right)$

17. $\arctan\left(\dfrac{1}{\sqrt{3}}\right)$

18. $\arctan(-1)$

19. $\arctan(0)$

4.2   The Derivative of an Inverse Function

In Section 4.1 we discussed the concept of an inverse function and introduced inverse trigonometric functions. In this section we will establish a general formula for the derivative of an inverse function and use the formula to determine the derivatives of inverse trigonometric functions.

The General Expression

Let's assume that the function f and its inverse f^{-1} are differentiable in their respective domains, and discover the relationship between their derivatives. By Proposition 1 of Section 4.1,

$$\left(f \circ f^{-1}\right)(x) = f\left(f^{-1}(x)\right) = x$$

for each x in some interval. Therefore,

$$\frac{d}{dx}\left(f \circ f^{-1}\right)(x) = \frac{d}{dx}(x) = 1.$$

By the chain rule,

$$\frac{d}{dx}\left(f \circ f^{-1}\right)(x) = \left(\frac{d}{dy}f(y)\bigg|_{y=f^{-1}(x)}\right)\left(\frac{d}{dx}f^{-1}(x)\right) = 1.$$

Therefore,

$$\frac{d}{dy}f(y)\bigg|_{y=f^{-1}(x)} \neq 0,$$

and

$$\frac{d}{dx}f^{-1}(x) = \frac{1}{\dfrac{d}{dy}f(y)\bigg|_{y=f^{-1}(x)}}.$$

In other words, **the derivative of the inverse function f^{-1} at x is the reciprocal of the derivative of f at $f^{-1}(x)$.** Under appropriate hypotheses, the above formula is indeed valid:

Theorem 1 (The Derivative of an Inverse Function) Assume that f is increasing or decreasing and differentiable on an open interval I. If $y = f^{-1}(x) \in I$ and $f'(y) \neq 0$,

then f' is differentiable at x and

$$\frac{d}{dx}f^{-1}(x) = \frac{1}{\left.\dfrac{d}{dy}f(y)\right|_{y=f^{-1}(x)}}.$$

In "the prime notation", we can express the above relationship as

$$(f^{-1})'(x) = \frac{1}{f'(f^{-1}(x))}.$$

You can find the proof of Theorem 1 at the end of this section.

Example 1 Let $f(y) = y^3$ so that $f^{-1}(x) = x^{1/3}$ for each $x \in \mathbb{R}$. Derive the expression for $(f^{-1})'(x)$ for $x \neq 0$ as a consequence of Theorem 1.

Solution

Let $x \neq 0$. We have

$$y = f^{-1}(x) = x^{1/3} \neq 0.$$

Therefore,

$$\frac{d}{dy}f(y) = \frac{d}{dy}(y^3) = 3y^2 \neq 0.$$

By Theorem 1,

$$\frac{d}{dx}\left(x^{1/3}\right) = \frac{d}{dx}f^{-1}(x) = \frac{1}{f'(y)|_{y=f^{-1}(x)}} = \frac{1}{3y^2|_{y=x^{1/3}}} = \frac{1}{3\left(x^{1/3}\right)^2} = \frac{1}{3x^{2/3}} = \frac{1}{3}x^{-2/3}.$$

We have

$$\left.\frac{d}{dy}f(y)\right|_{y=f^{-1}(0)} = 3y^2\big|_{y=0} = 0,$$

so that Theorem 1 does not predict the differentiability of f^{-1} at 0. Indeed, the cube-root function f^{-1} is not differentiable at 0 (Example 4 of Section 2.2). $\square$

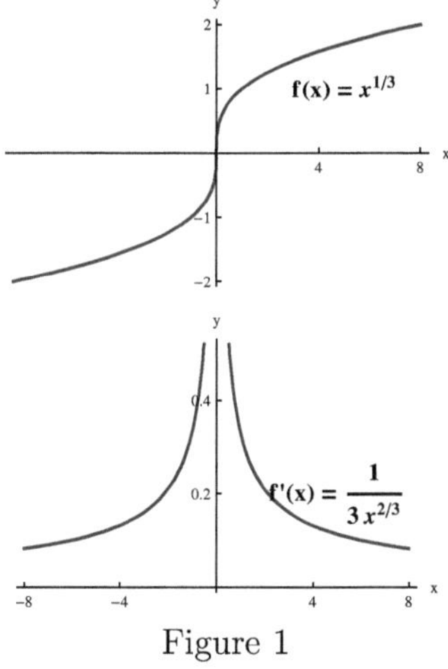

Figure 1

Remark 1 If we refer to f and f^{-1} in terms of their dependent variables, i.e., we set

$$y = y(x) = f^{-1}(x) \Leftrightarrow x = x(y) = f(y),$$

the relationship

$$\frac{d}{dx} f^{-1}(x) = \frac{1}{\dfrac{d}{dy} f(y)\bigg|_{y = f^{-1}(x)}}$$

takes the form

$$\frac{dy}{dx} = \frac{1}{\dfrac{dx}{dy}},$$

where dx/dy is evaluated at $y = y(x)$. Note that the above expression is "formally correct", in the sense that it is correct if we treat the symbolic fractions as if they were genuine fractions. Furthermore, the expression has the appealing feature that it can be considered to be the limiting case of the fraction

$$\frac{\Delta y}{\Delta x} = \frac{1}{\dfrac{\Delta x}{\Delta y}}$$

as $\Delta x \to 0$ and $\Delta y \to 0$. $\Diamond$

The Derivatives of Inverse Trigonometric Functions

Theorem 1 enables us to compute the derivatives of inverse trigonometric functions. Let's begin with the derivative of arcsine:

Proposition 1

$$\frac{d}{dx} \arcsin(x) = \frac{1}{\sqrt{1 - x^2}} \text{ if } -1 < x < 1.$$

Proof

We have

$$y = \arcsin(x) \Leftrightarrow x = \sin(y)$$

where

$$-1 \le x \le 1 \text{ and } -\frac{\pi}{2} \le y \le \frac{\pi}{2}.$$

We will make use of the relationship between the derivative of a function and the derivative of the inverse function, as expressed by Remark 1. Thus,

$$\frac{dy}{dx} = \frac{1}{\dfrac{dx}{dy}} = \frac{1}{\dfrac{d}{dy} \sin(y)} = \frac{1}{\cos(y)}$$

if $\cos(y) \neq 0$. Now,

$$\cos^2(y) = 1 - \sin^2(y) = 1 - x^2,$$

so that

$$\cos(y) = \pm\sqrt{1 - x^2}.$$

Since

$$-\frac{\pi}{2} \le y \le \frac{\pi}{2} \Rightarrow \cos(y) \ge 0,$$

we must disregard the $(-)$ sign. Therefore,

$$\frac{dy}{dx} = \frac{1}{\cos\left(y\right)} = \frac{1}{\sqrt{1-x^2}}$$

if $\cos\left(y\right) \neq 0$, i.e., if $1 - x^2 > 0$. This is the case if $-1 < x < 1$. Therefore,

$$\frac{d}{dx}\arcsin\left(x\right) = \frac{1}{\sqrt{1-x^2}}$$

if $-1 < x < 1$, as claimed. $\blacksquare$

Figure 2 shows the graphs of arcsine and its derivative.

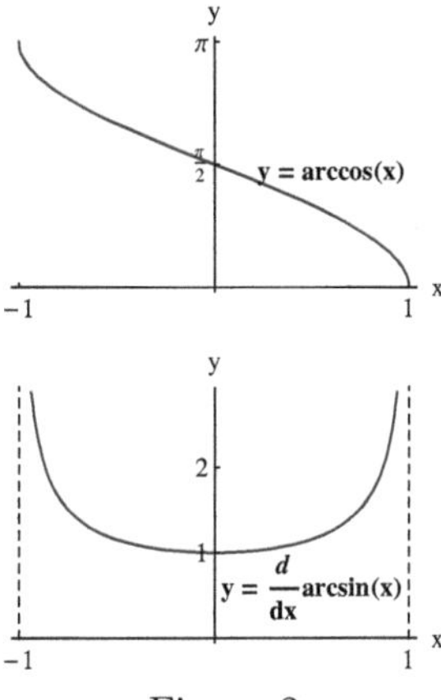

Figure 2

Note that

$$\lim_{x \to -1+} \frac{d}{dx}\arcsin\left(x\right) = \lim_{x \to -1+} \frac{1}{\sqrt{1-x^2}} = +\infty,$$

and

$$\lim_{x \to 1-} \frac{d}{dx}\arcsin\left(x\right) = \lim_{x \to 1-} \frac{1}{\sqrt{1-x^2}} = +\infty.$$

Thus, the tangent line to the graph of arcsine at $(x, \arcsin\left(x\right))$ becomes steeper and steeper as x approaches an endpoint of the domain of definition of arcsine. The graph of sine has vertical tangents at $(-1, -\pi/2)$ and $(1, \pi/2)$.

Example 2

a) Determine $L_{1/2}$, the linear approximation to arcsine based at $1/2$.
b) Use $L_{1/2}$ to approximate $\arcsin\left(.51\right)$. Determine the error in the approximation. Are the numbers consistent with the fact that $L_{1/2}\left(x\right)$ approximates $\arcsin\left(x\right)$ with an error whose magnitude is much smaller than $|x - 1/2|$ if $|x - 1/2|$ is small?

Solution

a) We have

$$\arcsin\left(\frac{1}{2}\right) = \frac{\pi}{6},$$

and

$$\frac{d}{dx}\arcsin(x)\Big|_{x=1/2} = \frac{1}{\sqrt{1-x^2}}\Big|_{x=1/2} = \frac{1}{\sqrt{1-\frac{1}{4}}} = \frac{2}{\sqrt{3}}.$$

Therefore,

$$L_{1/2}(x) = \arcsin\left(\frac{1}{2}\right) + \left(\frac{d}{dx}\arcsin(x)\Big|_{x=1/2}\right)\left(x - \frac{1}{2}\right)$$

$$= \frac{\pi}{6} + \frac{2}{\sqrt{3}}\left(x - \frac{1}{2}\right).$$

Figure 3 shows the graphs of arcsine and $L_{1/2}$.

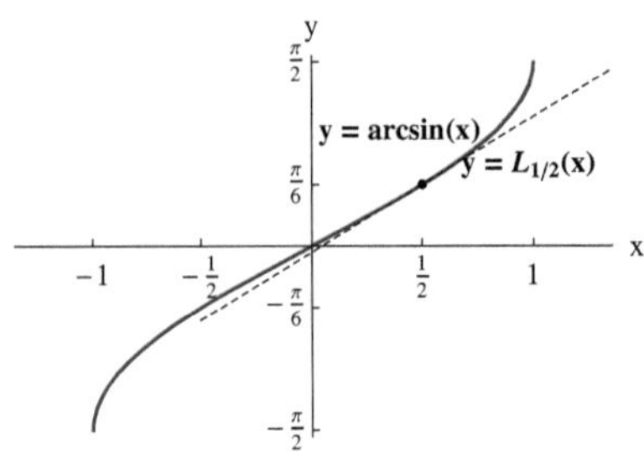

Figure 3

b)

$$\arcsin(.51) \cong L_{1/2}(.51) = \frac{\pi}{6} + \frac{2}{\sqrt{3}}\left(.51 - \frac{1}{2}\right) = \frac{\pi}{6} + \frac{2}{\sqrt{3}}(0.01) \cong .535146$$

We have $\arcsin(.51) \cong .535\,185$, rounded to 6 significant digits. The error in the approximation of $\arcsin(.51)$ by $L_{1/2}(.51)$ is approximately 3.9×10^{-5}. Thus, the error is much smaller than $0.51 - 0.5 = 10^{-2}$. $\square$

We can differentiate arccosine in a similar manner:

Proposition 2

$$\frac{d}{dx}\arccos(x) = -\frac{1}{\sqrt{1-x^2}} \text{ if } -1 < x < 1.$$

Proof

We have

$$y(x) = \arccos(x) \Leftrightarrow x = \cos(y),$$

where

$$-1 \le x \le 1 \text{ and } 0 \le y \le \pi.$$

By the formula for the derivative of an inverse function,

$$\frac{dy}{dx} = \frac{1}{\dfrac{dx}{dy}} = \frac{1}{\dfrac{d}{dy}\cos(y)} = -\frac{1}{\sin(y)}$$

if $\sin(y) \neq 0$. Now,

$$\sin^2(y) = 1 - \cos^2(y) = 1 - x^2,$$

so that

$$\sin (y) = \pm\sqrt{1 - x^2}.$$

Since

$$0 \leq y \leq \pi \Rightarrow \sin (y) \geq 0,$$

we must disregard the $(-)$ sign. Therefore,

$$\frac{dy}{dx} = -\frac{1}{\sin (y)} = -\frac{1}{\sqrt{1 - x^2}}.$$

if $\sin (y) \neq 0$, i.e., $-1 < x < 1$. Thus,

$$\frac{d}{dx} \arccos (x) = -\frac{1}{\sqrt{1 - x^2}}, \quad -1 < x < 1,$$

as claimed. ∎

Figure 4 shows the graphs of arccosine and its derivative.

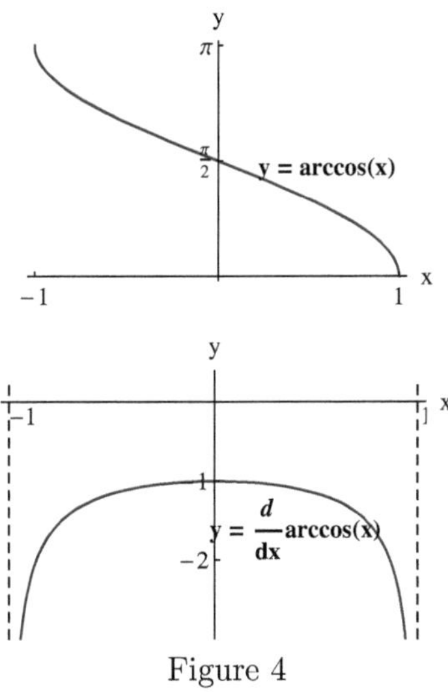

Figure 4

Note that

$$\lim_{x \to -1+} \frac{d}{dx} \arccos (x) = \lim_{x \to -1+} \left(-\frac{1}{\sqrt{1 - x^2}} \right) = -\infty,$$

and

$$\lim_{x \to 1-} \frac{d}{dx} \arccos (x) = \lim_{x \to 1-} \left(-\frac{1}{\sqrt{1 - x^2}} \right) = -\infty.$$

Thus, the graph of arccosine has vertical tangents at $(-1, \pi)$ and $(1, 0)$.

Example 3 In Section 4.1 we claimed that

$$\arccos(x) + \arcsin(x) = \frac{\pi}{2}$$

for each $x \in [-1, 1]$ (Remark 2 of Section 4.1). Prove that this is indeed the case.

Solution

Set

$$f(x) = \arccos(x) + \arcsin(x), \quad -1 \le x \le 1.$$

We would like to show that f is the constant function that has the value $\pi/2$ on the interval $[-1, 1]$. Our strategy will be to show that $f'(x) = 0$ for each $x \in (-1, 1)$. Indeed,

$$\begin{aligned}
f'(x) &= \frac{d}{dx}\left(\arccos(x) + \arcsin(x)\right) \\
&= \frac{d}{dx}\arccos(x) + \frac{d}{dx}\arcsin(x) \\
&= -\frac{1}{\sqrt{1 - x^2}} + \frac{1}{\sqrt{1 - x^2}} = 0
\end{aligned}$$

if $-1 < x < 1$, and f is continuous on $[-1, 1]$. By Theorem 5 of Section 3.2 there exists a constant C such

$$f(x) = \arccos(x) + \arcsin(x) = C$$

for each $x \in [-1, 1]$. If we set $x = 0$,

$$C = \arccos(0) + \arcsin(0) = \frac{\pi}{2} + 0 = \frac{\pi}{2}.$$

Therefore,

$$\arccos(x) + \arcsin(x) = \frac{\pi}{2}$$

for each $x \in [-1, 1]$. $\square$

Finally, we have an elegant and useful formula for the derivative of arctangent:

Proposition 3

$$\frac{d}{dx}\arctan(x) = \frac{1}{1 + x^2} \quad \text{for each } \ x \in \mathbb{R}.$$

Proof

We have

$$y = \arctan(x) \Leftrightarrow x = \tan(y),$$

where x is an arbitrary real number and

$$-\frac{\pi}{2} < y < \frac{\pi}{2}.$$

We use the relationship between the derivative of the inverse function and the derivative of the function:

$$\frac{dy}{dx} = \frac{1}{\dfrac{dx}{dy}} = \frac{1}{\dfrac{d}{dy}\tan(y)} = \frac{1}{\dfrac{1}{\cos^2(y)}} = \cos^2(y).$$

Note that

$$\frac{d}{dy}\tan(y) = \frac{1}{\cos^2(y)} \neq 0$$

for any y in the domain of tangent, so that there is no restriction on x. We must express $\cos^2(y)$ in terms of x. We will make use of the identity

$$\cos^2(y) + \sin^2(y) = 1,$$

and divide both sides by $\cos^2(y)$. Thus,

$$1 + \frac{\sin^2(y)}{\cos^2(y)} = \frac{1}{\cos^2(y)},$$

so that

$$1 + \tan^2(y) = \frac{1}{\cos^2(y)}.$$

Therefore,

$$\cos^2(y) = \frac{1}{1 + \tan^2(y)} = \frac{1}{1 + x^2}.$$

Thus,

$$\frac{d}{dx}\arctan(x) = \cos^2(y) = \frac{1}{1 + x^2}$$

for any $x \in \mathbb{R}$. $\blacksquare$

Figure 5 shows the graphs of arctangent and its derivative.

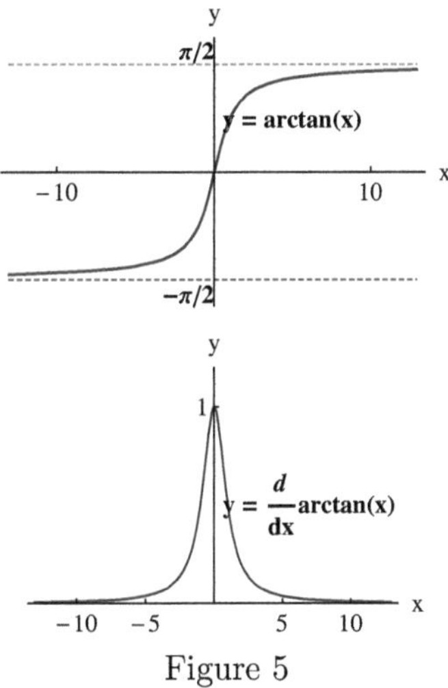

Figure 5

In summary, we have derived the following expressions for the derivatives of arcsine, arccosine and arctangent:

$$\frac{d}{dx}\arcsin(x) = \frac{1}{\sqrt{1 - x^2}} \text{ if } -1 < x < 1,$$

$$\frac{d}{dx}\arccos(x) = -\frac{1}{\sqrt{1 - x^2}} \text{ if } -1 < x < 1,$$

$$\frac{d}{dx}\arctan(x) = \frac{1}{1 + x^2} \text{ for each } x \in \mathbb{R}$$

The Proof of Theorem 1(Optional)

To begin with, it makes sense to speak of f^{-1}. Since f is differentiable on the open interval I, f is continuous on I. Since f is also increasing or decreasing on I, the inverse f^{-1} exists (Theorem 1 of Section 4.1).

Assume that $f'(y) \neq 0$, where $y = f^{-1}(x)$ so that $x = f(y)$. In order to compute the derivative of the inverse function at x, we must form the difference quotient

$$\frac{f^{-1}(x + \Delta x) - f^{-1}(x)}{\Delta x}.$$

We set $y + \Delta y = f^{-1}(x + \Delta x)$ so that $x + \Delta x = f(y + \Delta y)$. Therefore

$$\Delta x = f(y + \Delta y) - x = f(y + \Delta y) - f(y).$$

Figure 6 illustrates the case of an increasing function.

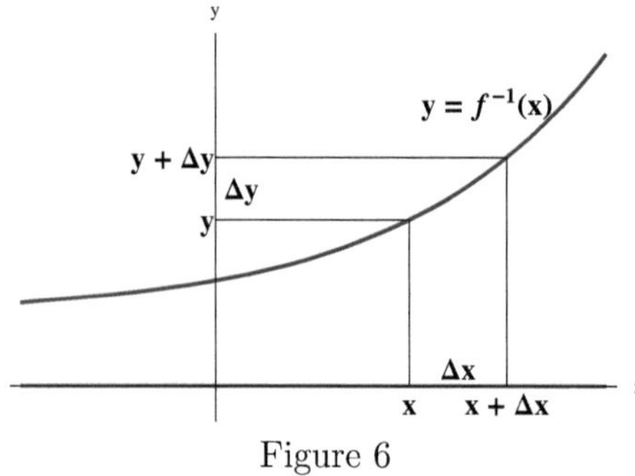

Figure 6

Thus, we can express the difference quotient for f^{-1} as

$$\frac{f^{-1}(x + \Delta x) - f^{-1}(x)}{\Delta x} = \frac{(y + \Delta y) - y}{\Delta x} = \frac{\Delta y}{\Delta x}.$$

We can write

$$\frac{\Delta y}{\Delta x} = \frac{1}{\dfrac{\Delta x}{\Delta y}},$$

if $\Delta x / \Delta y \neq 0$. Since

$$\lim_{\Delta y \to 0} \frac{\Delta x}{\Delta y} = \lim_{\Delta y \to 0} \frac{f(y + \Delta y) - f(y)}{\Delta y} = f'(y) \neq 0,$$

we do have $\Delta x / \Delta y \neq 0$ if $\Delta y \neq 0$ and $|\Delta y|$ is small enough. By Theorem 1 of Section 4.1, both f and f^{-1} are continuous. Therefore, $\Delta x = f(y + \Delta y) - f(y)$ approaches 0 if and only if $\Delta y = f^{-1}(x + \Delta x) - f^{-1}(x)$ approaches 0. Thus,

$$\lim_{\Delta x \to 0} \frac{f^{-1}(x + \Delta x) - f^{-1}(x)}{\Delta x} = \lim_{\Delta x \to 0} \frac{\Delta y}{\Delta x} = \lim_{\Delta x \to 0} \frac{1}{\dfrac{\Delta x}{\Delta y}}$$

$$= \frac{1}{\lim_{\Delta x \to 0} \dfrac{\Delta x}{\Delta y}} = \frac{1}{\lim_{\Delta y \to 0} \dfrac{\Delta x}{\Delta y}} = \frac{1}{f'(y)},$$

as claimed. ∎

Problems

In exercises 1-6, compute $f'(x)$.

1. $f(x) = \arcsin(2x)$

2. $f(x) = \arcsin(x^2)$

3. $f(x) = \arccos(x/9)$

4. $f(x) = \arccos(x^3)$

5. $f(x) = \arctan(x/2)$

6. $f(x) = \arctan(4x^2)$

7. $f(x) = \sqrt{\arctan(x)}$

8. $f(x) = \arcsin(3x) - \sqrt{1 - 9x^2}$

9. $f(x) = \sqrt{1 - x^2}\,\arccos(x)$

10. $f(x) = \dfrac{\arctan(x)}{x^2 + 1}$

11. Let $f(x) = \arcsin(x)$.
a) Determine $L_{1/2}$, the linear approximation to f based at $1/2$.
b) Make use of $L_{1/2}$ to approximate $\arcsin(0.49)$.

12.. Let $f(x) = \arctan(x)$.
a) Determine L_{-1}, the linear approximation to f based at -1.
b) Make use of L_{-1} to approximate $\arctan(-0.9)$.

13. Determine the intervals on which the graph of arcsine is concave up/concave down, and the inflection points of the graph.

14. Determine the intervals on which the graph of arccosine is concave up/concave down, and the inflection points of the graph.

15. Determine the intervals on which the graph of arctangent is concave up/concave down, and the inflection points of the graph.

16. The secant function does not have an inverse since it is periodic with periodic with period 2π. On the other hand, the restriction of secant to the set $[0, \pi/2) \cup (\pi/2, \pi]$, has the following graph. The range of secant is $(-\infty, -1] \cup [1, +\infty)$.

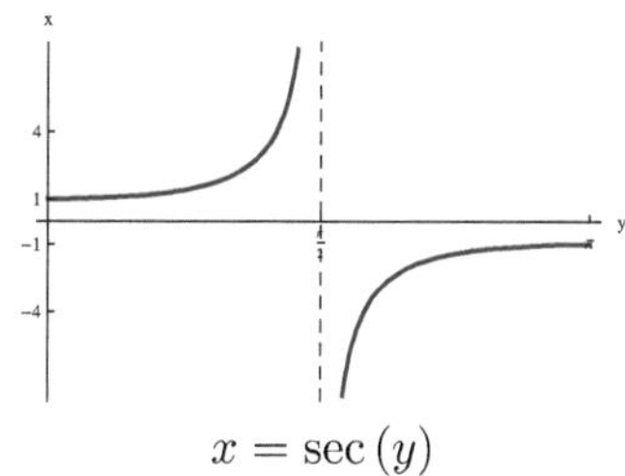

$$x = \sec(y)$$

The inverse exists and is usually defined as arcsecant, abbreviated as arcsec. Thus, arcsec:$(-\infty, -1] \cup [1, +\infty) \to [0, \pi/2) \cup (\pi/2, \pi]$.

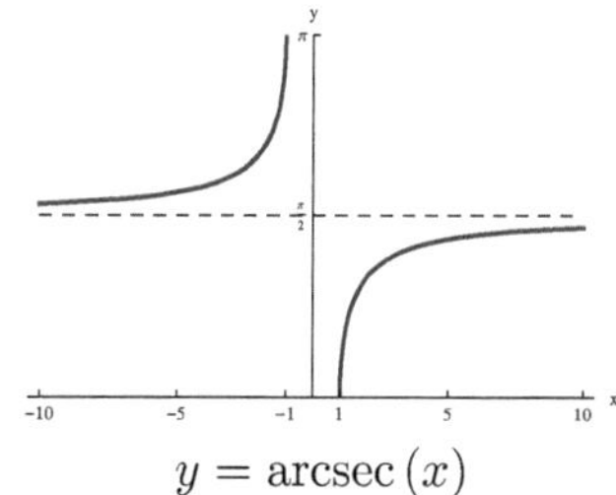

$$y = \operatorname{arcsec}(x)$$

Show that

$$\frac{d}{dx} \operatorname{arcsec}(x) = \frac{1}{|x|\,\sqrt{x^2-1}} \ \text{ if } |x| > 1.$$

4.3 The Natural Exponential Function and the Natural Logarithm

IIn this section we will introduce two important special of functions of mathematics, **the natural exponential function** and **the natural logarithm**. The significance of the adjective "natural" will emerge as you master the material of this section and the following sections.

The Natural Exponential Function

We have studied functions defined by rational powers of x. For example, if $f(x) = x^{3/4}$, we have

$$f(x) = \left(x^{1/4}\right)^3 = x^{1/4} \times x^{1/4} \times x^{1/4},$$

where $x \geq 0$, and $y = x^{1/4}$ means that $x = y^4$ and $y \geq 0$. Here, the exponent $3/4$ is fixed, and the independent variable is x. Now we will consider functions defined by a^x, where a is a given positive number, such as 2 or 10, and the exponent x is the independent variable. We know the meaning of a^x if x is a rational number. How about the meaning of 2^π or $10^{\sqrt{2}}$, where the exponent is an irrational number? We will accept the following fact:

If $a > 0$ there exists a function that is defined on the entire number line such that the value of the function at each rational number x is a^x. We will refer to this function as **the exponential function with base a** and denote it by $\mathbf{exp}_a(x)$. Thus, $\mathbf{exp}_a(x) = a^x$ **for each rational number x. The exponential function with base a is differentiable on the entire number line** and $\mathbf{exp}_a(x) > 0$ for each real number x.

Note that the exponential function is continuous on the entire number line, since differentiability implies continuity.

Let x be an irrational number, and let x_n denote the decimal that is obtained by rounding the decimal expansion of x to n significant digits. Since a finite decimal represents a rational number, we have $\exp_a(x_n) = a^{x_n}$. Since $\lim_{n\to\infty} x_n = x$, we have

$$\exp_a(x) = \lim_{n\to\infty} \exp_a(x_n) = \lim_{n\to\infty} a^{x_n}$$

by the continuity of $\exp_a$. You can imagine that the exponential function with base a extends the definition of a^x from rational exponents to all exponents so that the holes in the graph of $y = a^x$ corresponding to irrational exponents are plugged. Thus, it is natural to denote $\exp_a(x)$ as a^x for each $x \in \mathbb{R}$, rational or irrational. We will refer to the notation a^x for $\exp_a(x)$ as **the exponential notation.**

Example 1 Let's consider $\exp_2$, the exponential function with base 2.

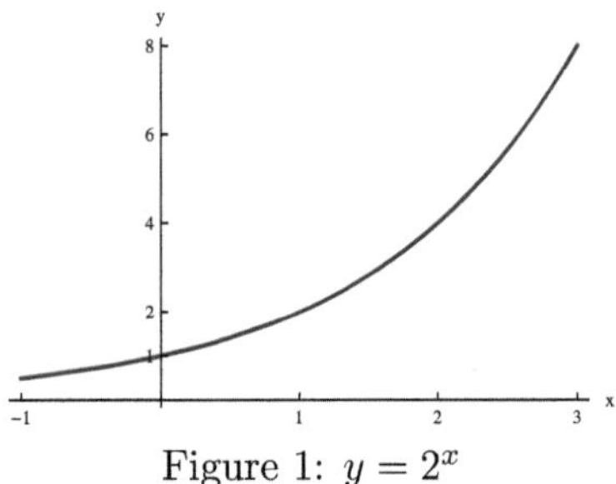

Figure 1: $y = 2^x$

The irrational number π has an infinite decimal expansion:

$$\pi = 3.141\,592\,653\ldots.$$

Let x_n denote the number that is obtained by rounding the decimal expansion of π to n significant digits. Since $\lim_{n\to\infty} x_n = \pi$, we have

$$\lim_{n\to\infty} \exp_2(x_n) = \exp_2(\pi),$$

by the continuity of $\exp_2$ at π. In the exponential notation,

$$\lim_{n\to\infty} 2^{x_n} = 2^\pi.$$

We can obtain the approximation,

$$2^\pi \cong 8.82498,$$

where the decimal has been rounded to 6 significant digits, with the help of our calculator. Table 1 displays x_n, 2^{x_n} and $|2^{x_n} - 2^\pi|$ for $n = 2, 3, 4, 5, 6$. We see that 2^{x_n} approximates 2^π with increasing accuracy as n increases from 2 to 6. This supports the expectation that $\lim_{n\to\infty} 2^{x_n} = 2^\pi$.

n	x_n	2^{x_n}	$|2^{x_n} - 2^\pi|$
2	3.1	8.574\,19	2.5×10^{-1}
3	3.14	8.815\,24	9.7×10^{-3}
4	3.142	8.827\,47	2.5×10^{-3}
5	3.1416	8.825\,02	4.5×10^{-5}
6	3.141\,59	8.824\,96	1.6×10^{-5}

Table 1

□

Remark 1 (Caution) You should not confuse the function defined by 2^x with the power function defined by x^2. The graph of $y = x^2$ is a familiar parabola, and is quite distinct from the graph of $y = 2^x$.◊

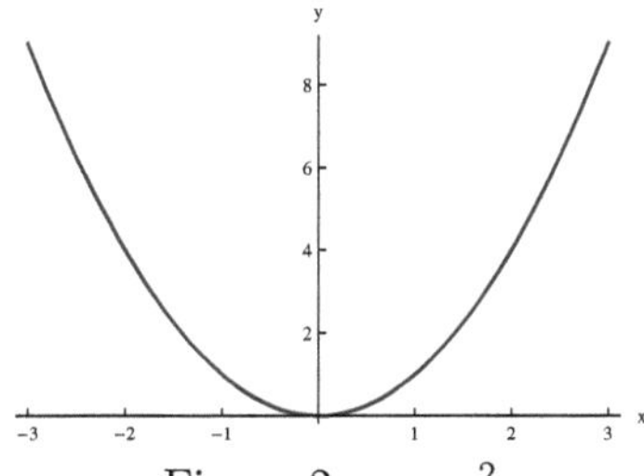

Figure 2: $y = x^2$

Remark 2 (The Rules for Exponents) The exponential notation is convenient, since the rules for exponents that are familiar from precalculus courses are valid for arbitrary exponents: If a is a positive real number, and x, x_1, x_2 are arbitrary real numbers, we have

$$a^{x_1} a^{x_2} = a^{x_1 + x_2},$$

$$a^0 = 1,$$

$$a^{-x} = \frac{1}{a^x},$$

$$\left(a^{x_1}\right)^{x_2} = a^{x_1 x_2}.$$

Thus, we add the exponents in order to multiply a^{x_1} and a^{x_2}. The multiplication of the exponent by (-1) results in the reciprocal of a^x. We raise a^{x_1} to the power x_2 by multiplying x_1 and x_2. $\Diamond$

We stated that the exponential function with base $a > 0$ defines a differentiable function. Let's try to discover the rule for the differentiation of $\exp_a$. The relevant difference quotient is

$$\frac{\exp_a (x + h) - \exp_a (x)}{h} = \frac{a^{x+h} - a^x}{h} = \frac{a^x a^h - a^x}{h} = a^x \left(\frac{a^h - 1}{h}\right).$$

We have

$$\lim_{h \to 0} \left(a^x \left(\frac{a^h - 1}{h}\right)\right) = a^x \lim_{h \to 0} \frac{a^h - 1}{h},$$

by the constant multiple rule for limits, since x is kept fixed. Notice that

$$\lim_{h \to 0} \frac{a^h - 1}{h} = \lim_{h \to 0} \frac{a^h - a^0}{1} = \frac{d}{dx} a^x \bigg|_{x=0}.$$

Therefore,

$$\frac{d}{dx} a^x = \lim_{h \to 0} \frac{a^{x+h} - a^x}{h} = a^x \lim_{h \to 0} \frac{a^h - 1}{h} = a^x \left(\frac{d}{dx} a^x \bigg|_{x=0}\right).$$

If we set

$$C_a = \frac{d}{dx} a^x \bigg|_{x=0},$$

we have

$$\frac{d}{dx} a^x = C_a \, a^x$$

for each real number x. Thus, **the derivative of the exponential function with base a at any $x \in \mathbb{R}$ is a constant multiple of a^x, where the constant is the derivative of the function at 0. We will accept the fact there is a base e such that**

$$C_e = \lim_{h \to 0} \frac{e^h - 1}{h} = 1,$$

so that

$$\frac{d}{dx} e^x = e^x$$

for each real number x. The number e is an irrational number. We have

$$e = 2.718\,281\,828\,459\,\ldots$$

We will refer to the exponential function with base e as the natural exponential function:

Definition 1 The natural exponential function is the exponential function with base e, where e is the number such that

$$\lim_{h \to 0} \frac{e^h - 1}{h} = 1$$

for each $x \in \mathbb{R}$.

By the discussion that led to Definition 1, we have

$$\frac{d}{dx} e^x = e^x \text{ for each } x \in \mathbb{R}.$$

We will abbreviate the natural exponential function as **exp.** Thus,

$$\exp(x) = \exp_e(x) = e^x$$

and

$$\frac{d}{dx} \exp(x) = \exp(x)$$

for each $x \in \mathbb{R}$.

Since e is an irrational number, it may sound strange to label the function defined by e^x as the natural exponential function, rather than the function defined by 2^x or 10^x. The reason for the terminology is that the natural exponential function is indeed "natural" from the point of view of calculus. The discussion that led us to the natural exponential function showed that

$$\frac{d}{dx} a^x = C_a a^x,$$

for an arbitrary base $a > 0$. The constant is 1 if $a = e$. You will see in the next section that C_a does not have such a simple value if $a \neq e$.

The natural exponential function is a built-in function of any computational utility, so that its values can be approximated accurately. Figure 3 shows the graph of the natural exponential function. Figure 3 indicates that the natural exponential function is increasing on $\mathbb{R}$ and its graph is concave up on the entire number line. These observations are consistent with the derivative test for monotonicity and the second derivative test for concavity, since

$$\frac{d^2}{dx^2} e^x = \frac{d}{dx} e^x = e^x > 0$$

for each $x \in \mathbb{R}$.

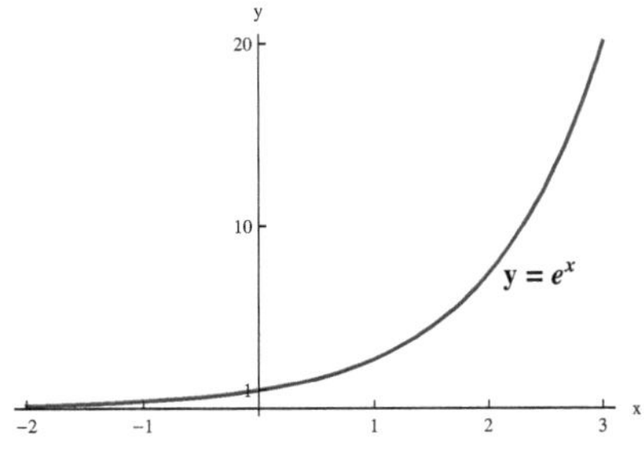

Figure 3 : The graph of **the natural exponential function**

Since $e > 2$, we have

$$e^n > 2^n$$

for each positive integer n. Therefore,

$$\lim_{n \to \infty} e^n = +\infty.$$

Since the natural exponential function is an increasing function, this shows that

$$\lim_{x \to +\infty} e^x = +\infty.$$

In fact, e^x grows very rapidly as x increases: Since

$$\frac{d}{dx} e^x = e^x,$$

and $\lim_{x \to +\infty} e^x = +\infty$, the rate at which e^x increases tends to infinity as x tends to infinity. We will say more about the growth rate of e^x in Section 4.5.

Since

$$e^x = \frac{1}{e^{-x}},$$

we have

$$\lim_{x \to -\infty} e^x = \lim_{x \to -\infty} \frac{1}{e^{-x}} = \lim_{z \to +\infty} \frac{1}{e^z} = 0.$$

In fact, e^x decreases very rapidly as $x \to -\infty$.

Let's list some of the basic properties of the natural exponential function for the record:

1. We have $e^x > 0$ for each $x \in \mathbb{R}$

2. The natural exponential function is increasing and its graph is concave up on the entire number line.

3.

$$\lim_{x \to +\infty} e^x = +\infty \quad \lim_{x \to -\infty} e^x = 0.$$

4. The laws for exponents are valid: If x, x_1 and x_2 are arbitrary real numbers, we have

$$e^{x_1} e^{x_2} = e^{x_1 + x_2},$$

$$e^0 = 1,$$

$$e^{-x} = \frac{1}{e^x},$$

$$\left(e^{x_1}\right)^{x_2} = e^{x_1 x_2}.$$

Example 2 *Determine L_0, the linear approximation the natural exponential function based at 0.*

Solution

Recall that the linear approximation to a function f based at 0 is

$$L_0(x) = f(0) + f'(0)x.$$

If we set $f(x) = e^x$, we have

$$f(0) = e^0 = 1 \text{ and } f'(0) = \frac{d}{dx} e^x \Big|_{x=0} = e^x \big|_{x=0} = e^0 = 1.$$

Therefore,

$$L_0(x) = f(0) + f'(0)x = 1 + x.$$

The graph of the equation L_0 is the tangent line to the graph of $y = e^x$ at $(0, 1)$. Figure 4 shows the graphs of the natural exponential function and L_0. We have $e^x \geq L_0(x)$ for each $x \in \mathbb{R}$, since the graph of the natural exponential function is concave up on $\mathbb{R}$. $\square$

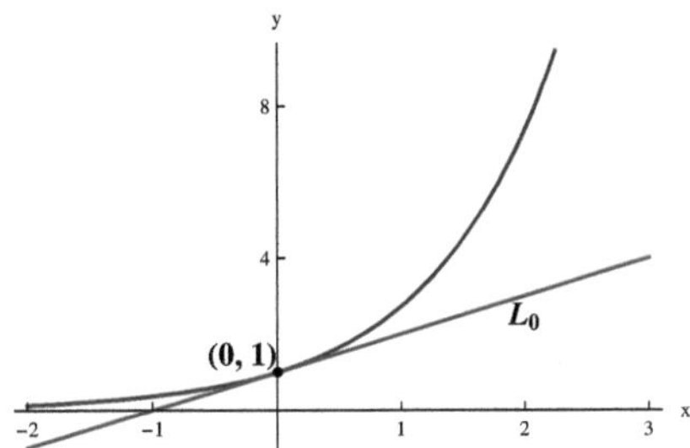

Figure 4: The tangent line to the graph of $y = e^x$ at $(0, 1)$

The reciprocal of the natural exponential function "decays" as $x \to +\infty$:

Example 3 Show that

$$\lim_{x \to +\infty} e^{-x} = 0 \text{ and } \lim_{x \to -\infty} e^{-x} = +\infty.$$

Sketch the graph of $y = e^{-x}$.

Solution

a) We have

$$\lim_{x \to +\infty} e^{-x} = \lim_{x \to +\infty} \frac{1}{e^x} = 0,$$

and

$$\lim_{x \to -\infty} e^{-x} = \lim_{z \to +\infty} e^z = +\infty$$

since $\lim_{x \to +\infty} e^x = +\infty$.
Figure 5 shows the graph of $y = e^{-x}$. $\square$

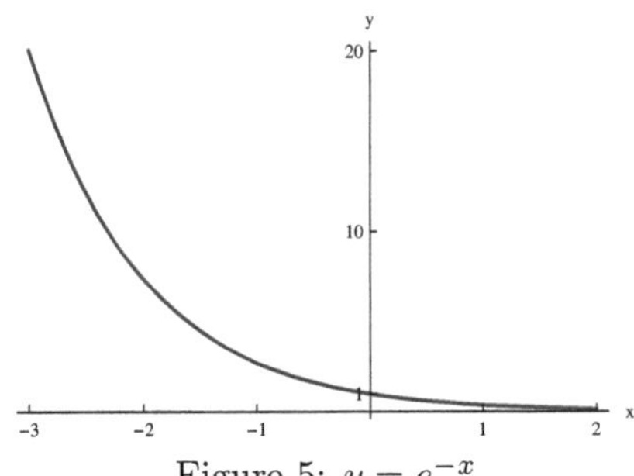

Figure 5: $y = e^{-x}$

Example 4 Let

$$f(x) = e^{-x^2/2}.$$

Determine $f'(x)$.

Solution

We set $u = -x^2/2$. By the chain rule,

$$f'(x) = \left(\frac{d}{du} e^u \Big|_{u=-x^2/2} \right) \left(\frac{du}{dx} \right) = \left(e^u \big|_{u=-x^2/2} \right) \left(\frac{d}{dx} \left(-\frac{x^2}{2} \right) \right)$$

$$= \left(e^{-x^2/2} \right) (-x) = -x e^{-x^2/2}.$$

$\square$

As in Example 4, we will come across many functions of the form $e^{u(x)}$, where the exponent $u(x)$ is differentiable. We have

$$\frac{d}{dx} e^{u(x)} = e^{u(x)} \frac{du}{dx}.$$

Indeed, by the chain rule,

$$\frac{d}{dx} e^{u(x)} = \left(\frac{d}{du} e^u \Big|_{u=u(x)} \right) \left(\frac{du}{dx} \right) = e^{u(x)} \frac{du}{dx}.$$

The Natural Logarithm

The natural logarithm is the inverse of the natural exponential function and will be abbreviated as **ln**:

Definition 2

$$y = \ln(x) \text{ if and only if } x = e^y.$$

Thus, **the natural logarithm is the logarithm with respect to the base e.** Since the values of the natural exponential function are positive, we have $x = e^y > 0$. Therefore, **$\ln(x)$ is defined if and only if $x > 0$, i.e., the domain of the natural logarithm is the open interval $(0, +\infty)$.**
Traditionally, $\ln(x)$ is typed as $\ln x$. We will favor the notation that is consistent with the notation $f(x)$ for the value of a function f at x.

Let's look at the relationship between the natural logarithm and the natural exponential function graphically. Since our starting point is the natural exponential function, and we set $x = e^y$ in order to define $\ln(x)$, we will plot the graph of the natural exponential function in the yx-plane, so that the y-axis is horizontal and the x-axis is vertical, as in Figure 6. Given any $x > 0$, there is a unique y such that $e^y = x$. That y is defined to be $\ln(x)$.

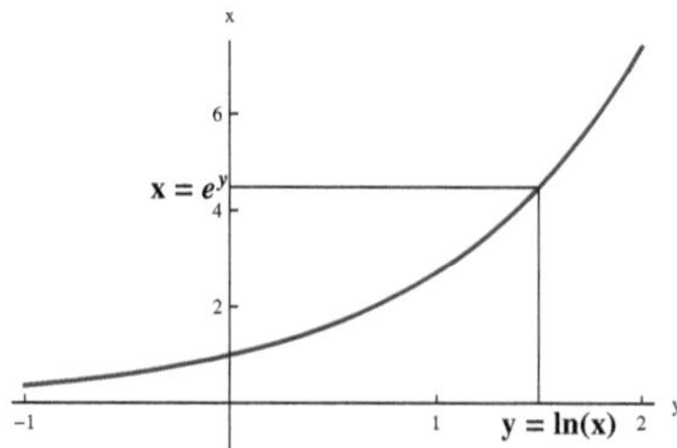

Figure 6: $y = \ln(x) \Leftrightarrow x = e^y$

Figure 7 displays the graph of the natural logarithm. The natural logarithm is a continuous and increasing function on its natural domain, since it is the inverse of the natural exponential function with those properties on $\mathbb{R}$.

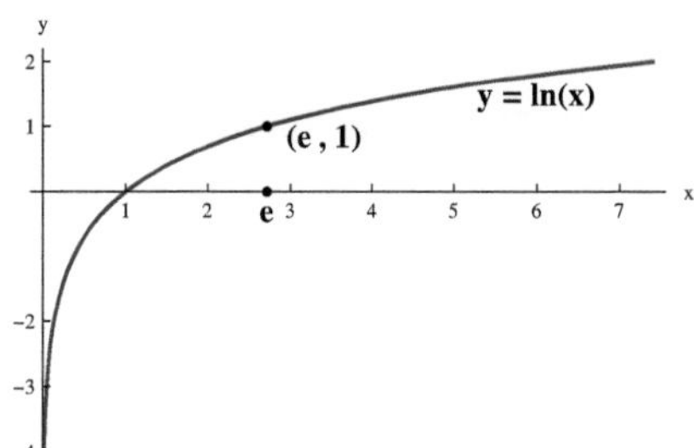

Figure 7: The graph of **the natural logarithm**

The natural logarithm of e is 1, and the natural logarithm of 1 is 0:

$$\ln(e) = 1 \text{ and } \ln(1) = 0.$$

Indeed, $e^1 = e$ so that $\ln(e) = 1$. We have $e^0 = 1$, so that $\ln(1) = 0$.

Since the natural logarithm and the natural exponential function are inverses of each other, their graphs are symmetric with respect to the diagonal $y = x$ if the scale on the vertical axis is the same as the scale on the horizontal axis, as in Figure 8.

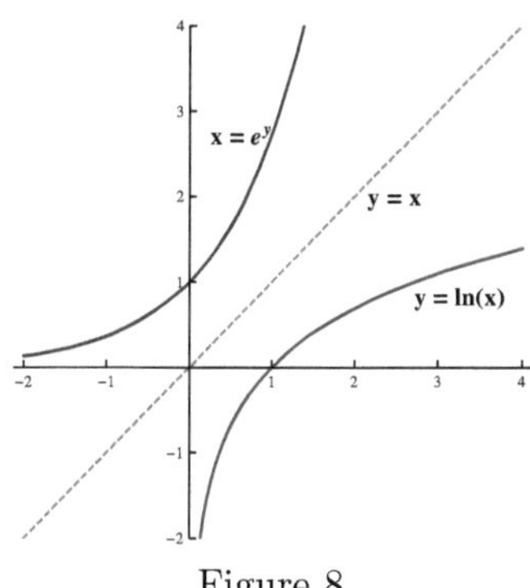

Figure 8

A word of caution: In Figure 9, the scale on the vertical axis is not the same as the scale on the horizontal axis, and the graphs are not symmetric with respect to the diagonal.

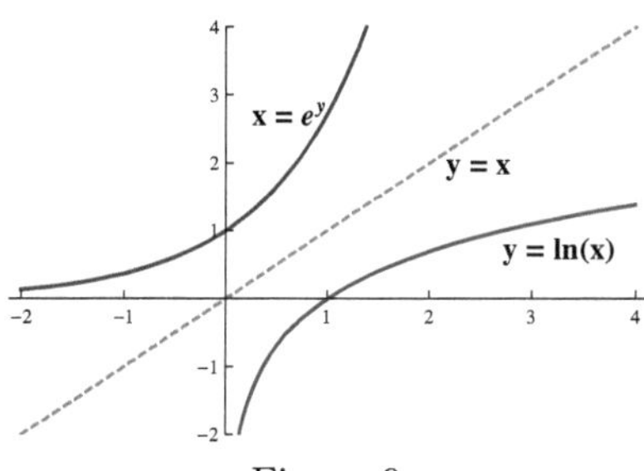

Figure 9

As a special case of the relationship between a function and its inverse, we have the following facts:

$$\ln(e^y) = y \text{ for each } y \in \mathbb{R}$$

and

$$e^{\ln(x)} = x \text{ for each } x > 0.$$

We can obtain the basic algebraic properties of the natural logarithm from the corresponding properties of the natural exponential function:

Assume that r is an arbitrary real number, and x, x_1 and x_2 are positive numbers. Then

$$\ln(x_1 x_2) = \ln(x_1) + \ln(x_2),$$
$$\ln(1/x) = -\ln(x),$$
$$\ln(x^r) = r\ln(x).$$

Proof

Set $y_1 = \ln(x_1)$ and $y_2 = \ln(x_2)$, so that $x_1 = e^{y_1}$ and $x_2 = e^{y_2}$. We have

$$x_1 x_2 = e^{y_1} e^{y_2} = e^{y_1 + y_2}.$$

Therefore,

$$\ln(x_1 x_2) = y_1 + y_2 = \ln(x_1) + \ln(x_2).$$

Thus,

$$0 = \ln(1) = \ln\left(x\left(\frac{1}{x}\right)\right) = \ln(x) + \ln\left(\frac{1}{x}\right),$$

so that

$$\ln\left(\frac{1}{x}\right) = -\ln(x).$$

Finally,

$$x^r = \left(e^{\ln(x)}\right)^r = e^{r\ln(x)},$$

so that

$$\ln(x^r) = \ln\left(e^{r\ln(x)}\right) = r\ln(x).$$

■

Example 5 Simplify the expressions $\ln\left(e^{2/3}\right)$ and $e^{-t\ln(2)}$. Express $\ln(\sqrt{x})$ in terms of $\ln(x)$.

Solution

$$\ln\left(e^{2/3}\right) = \frac{2}{3}\ln(e) = \frac{2}{3},$$

and

$$e^{-t\ln(2)} = \left(e^{-\ln(2)}\right)^t = \left(e^{\ln(1/2)}\right)^t = \left(\frac{1}{2}\right)^t = \frac{1}{2^t}.$$

We have

$$\ln\left(\sqrt{x}\right) = \ln\left(x^{1/2}\right) = \frac{1}{2}\ln(x).$$

□

The expression for the derivative of the natural logarithm is easy to remember:

Proposition 1

$$\frac{d}{dx}\ln(x) = \frac{1}{x}$$

for each $x > 0$.

Proof

We make use of the expression for the derivative of an inverse function that was introduced in Section 4.2: If we set $y = \ln(x)$ so that $x = e^y$, we have

$$\frac{dy}{dx} = \frac{1}{\dfrac{dx}{dy}}.$$

Since

$$\frac{dx}{dy} = \frac{d}{dy}e^y = e^y,$$

we have

$$\frac{d}{dx}\ln(x) = \frac{dy}{dx} = \frac{1}{e^y} = \frac{1}{x}.$$

The expression is valid for each $x > 0$. $\blacksquare$

Since

$$\lim_{x \to +\infty}\frac{d}{dx}\ln(x) = \lim_{x \to +\infty}\frac{1}{x} = 0,$$

the slope of the graph of the natural logarithm at $(x, \ln(x))$ approaches 0 as $x \to +\infty$. Since the derivative of a function at a point is its rate of change at that point, we also conclude that the natural logarithm increases at a rate that approaches 0 as $x \to +\infty$. In Section 4.5 we will say more about the rate at which the natural logarithm increases.

Even though $\ln(x)$ increases at a rate that approaches 0 as $x \to +\infty$, $\ln(x)$ increases beyond all bounds as x increases. If $0 < x < 1$, then $\ln(x) < 0$, since $\ln(1) = 0$ and $\ln$ is an increasing function. As x approaches 0 from the right, $|\ln(x)| = -\ln(x)$ becomes arbitrarily large:

$$\lim_{x \to +\infty}\ln(x) = +\infty \text{ and } \lim_{x \to 0+}\ln(x) = -\infty.$$

Proof

We have

$$\lim_{n \to +\infty}\ln(e^n) = \lim_{n \to \infty}(n\ln(e)) = \lim_{n \to \infty}n = +\infty.$$

This implies that $\lim_{x \to \infty}\ln(x) = +\infty$, since the natural logarithm is an increasing function on $(0, +\infty)$, and the sequence $\{e^n\}_{n=1}^{\infty}$ is an increasing sequence such that $\lim_{n \to \infty}e^n = +\infty$.

We also have

$$\lim_{n \to 0}\ln\left(\frac{1}{e^n}\right) = \lim_{n \to \infty}\ln(e^{-n}) = \lim_{n \to \infty}(-n\ln(e)) = \lim_{n \to \infty}(-n) = -\infty.$$

This implies that $\lim_{x \to 0+}\ln(x) = -\infty$, since the sequence $\{1/e^n\}_{n=1}^{\infty}$ is a decreasing sequence such that $\lim_{n \to \infty}1/e^n = 0$. $\blacksquare$

We can make use of the algebraic properties of the natural logarithm, as stated in Proposition ??, in order to simplify certain expressions. Such simplifications can be helpful in differentiating functions that involve the natural logarithm, as in the following example.

Example 6 Let

$$f(x) = \ln\left((x^2 + 1)^{1/2}(x^2 + 9)^{1/3}\right).$$

Determine $f'(x)$.

Solution

By Proposition **??**,

$$f(x) = \ln\left((x^2 + 1)^{1/2}\right) + \ln\left((x^2 + 9)^{1/3}\right) = \frac{1}{2}\ln(x^2 + 1) + \frac{1}{3}\ln(x^2 + 9).$$

By the linearity of differentiation and the chain rule,

$$\begin{aligned}
f'(x) &= \frac{1}{2}\left(\frac{d}{du}\ln(u)\Big|_{u=x^2+1}\right)\left(\frac{d}{dx}(x^2 + 1)\right) + \frac{1}{3}\left(\frac{d}{du}\ln(u)\Big|_{u=x^2+9}\right)\left(\frac{d}{dx}(x^2 + 9)\right) \\
&= \frac{1}{2}\left(\frac{1}{x^2 + 1}\right)(2x) + \frac{1}{3}\left(\frac{1}{x^2 + 9}\right)(2x) \\
&= \frac{x}{x^2 + 1} + \frac{2x}{3(x^2 + 9)}.
\end{aligned}$$

As an exercise, determine $f'(x)$ without taking advantage of the algebraic properties of the natural logarithm. You will see that the above procedure is much simpler and less prone to errors. $\square$

The following useful fact is a consequence of the chain rule:

Proposition 2 Assume that f is differentiable at x and $f(x) \neq 0$. Then,

$$\frac{d}{dx}\ln(|f(x)|) = \frac{f'(x)}{f(x)}.$$

Proof

Since $f(x) \neq 0$, either $f(x) > 0$ or $f(x) < 0$. Since f is differentiable at x, f is continuous at x. By the continuity of f at x, f does not change sign in some open interval J containing x. Assume that $f(z) > 0$ for each z in J. Then, $|f(z)| = f(z)$ for each $z \in J$. Therefore,

$$\frac{d}{dx}\ln(|f(x)|) = \frac{d}{dx}\ln(f(x)).$$

By the chain rule,

$$\frac{d}{dx}\ln(f(x)) = \left(\frac{d}{du}\ln(u)\Big|_{u=f(x)}\right)\left(\frac{df(x)}{dx}\right) = \frac{1}{f(x)}\frac{df(x)}{dx} = \frac{f'(x)}{f(x)}.$$

Now assume that $f(z) < 0$ for each $z \in J$. Then, $|f(z)| = -f(z)$ for each $z \in J$. Therefore,

$$\frac{d}{dx}\ln(|f(x)|) = \frac{d}{dx}\ln(-f(x)).$$

Again, by the chain rule,

$$\frac{d}{dx}\ln(-f(x)) = \left(\frac{d}{du}\ln(u)\Big|_{u=-f(x)}\right)\left(\frac{d(-f(x))}{dx}\right) = \frac{1}{(-f(x))}\left(-\frac{df(x)}{dx}\right) = \frac{f'(x)}{f(x)}$$

$\blacksquare$

Remark 3 By Proposition 2, we have

$$f'(x) = f(x)\frac{d}{dx}\ln\left(|f(x)|\right).$$

Sometimes it is more convenient to determine $f'(x)$ by differentiating $\ln\left(|f(x)|\right)$ and by making use of the above expression. This procedure is called **logarithmic differentiation.** $\Diamond$

Example 7 Let

$$f(x) = \frac{x^2 - 4}{(x^2 + 1)(x + 3)^2}.$$

Determine $f'(x)$ by logarithmic differentiation.

Solution

We have

$$|f(x)| = \frac{|x^2 - 4|}{(x^2 + 1)(x + 3)^2}$$

if $x \neq -3$. If $x \neq -3$ and $x^2 - 4 \neq 0$, i.e., $x \neq \pm 2$,

$$\ln\left(|f(x)|\right) = \ln\left(|x^2 - 4|\right) - \ln\left(x^2 + 1\right) - \ln\left((x + 3)^2\right)$$
$$= \ln\left(|x^2 - 4|\right) - \ln\left(x^2 + 1\right) - 2\ln\left(x + 3\right).$$

We will make use of the linearity of differentiation and the fact that

$$\frac{d}{dx}\ln\left(|u(x)|\right) = \frac{u'(x)}{u(x)},$$

as in Proposition 2. Thus,

$$\frac{d}{dx}\ln\left(|f(x)|\right) = \frac{d}{dx}\ln\left(|x^2 - 4|\right) - \frac{d}{dx}\ln\left(x^2 + 1\right) - 2\frac{d}{dx}\ln\left(x + 3\right)$$
$$= \frac{2x}{x^2 - 4} - \frac{2x}{x^2 + 1} - \frac{2}{x + 3}.$$

Therefore,

$$f'(x) = f(x)\frac{d}{dx}\ln\left(|f(x)|\right)$$
$$= \left(\frac{x^2 - 4}{(x^2 + 1)(x + 3)^2}\right)\left(\frac{2x}{x^2 - 4} - \frac{2x}{x^2 + 1} - \frac{2}{x + 3}\right)$$
$$= \frac{2x}{(x^2 + 1)(x + 3)^2} - \frac{2x\left(x^2 - 4\right)}{(x^2 + 1)^2(x + 3)^2} - \frac{2\left(x^2 - 4\right)}{(x^2 + 1)(x + 3)^3}.$$

Even though $\ln\left(|f(x)|\right)$ is defined if $x \neq -3$ and $x^2 - 4 \neq 0$, the above expression for $f'(x)$ is valid as long as $x \neq -3$. You can check this by differentiating f directly. $\square$

In the next section we will discuss exponential functions with bases other than e, and the corresponding logarithmic functions. We will also discuss irrational powers of x.

Problems

In problems 1 and 2, determine the domain of the function f.

1. $f(x) = \ln(x - 4)$

2. $f(x) = \ln(5x - 12)$

In problems 3-10, simplify the given expression.

3. $\ln\left(e^4\right)$

4. $\ln\left(e^{-2}\right)$

5. $\ln\left(e^{34/5}\right)$

6. $e^{\ln(6)}$

7. $e^{-\ln(4)}$

8. $e^{\ln(\sqrt{8})}$

9. $\ln\left(x^{1/5}\right)$

10. $\ln\left(\dfrac{1}{x^2}\right)$

In problems 11-14, determine the solutions of the given equation.

11. $3\ln(x) = -2$

12. $2\ln(x) = 5$

13. $e^{-z^2} = 4$

14. $e^{x/2} = e^2$

In problems 15-20, simplify the expression for f by making use of the algebraic properties of the natural logarithm and then determine f'

15. $f(x) = \ln\left(\dfrac{x+2}{x-2}\right)$

16. $f(x) = \ln\left(\dfrac{x}{x^2+1}\right)$

17. $f(x) = \ln\left(\sqrt{x^2 + 16}\right)$

18. $f(x) = \ln\left(\dfrac{1}{\sqrt{x^2 - 9}}\right)$

19. $f(x) = \ln\left((x^2 + 9)^4 (x+1)^2\right)$

20. $f(x) = \ln\left(\dfrac{(x^2 + 16)^{1/3}}{(x-2)^{1/2}}\right)$

In problems 21-23,

a) Determine $\dfrac{d}{dx}\ln(|f(x)|)$,

b) Determine $f'(x)$ by making use of the result of part a) (**"logarithmic differentiation"**).

21. $f(x) = (x-1)(x+1)^3 (x^2 + 4)$

22. $f(x) = \dfrac{(x-1)^4}{(x+4)^2 (x-5)^3}$

23. $f(x) = \dfrac{(x^2 + 16)^4}{(x-4)^3 \sqrt{x^2 - 9}}$

In problems 24-35, determine $f'(x)$.

24. $f(x) = \frac{1}{2}\left(e^x - e^{-x}\right)$

25. $f(x) = \frac{1}{2}\left(e^x + e^{-x}\right)$

26. $f(x) = x^2 e^x$

27. $f(x) = x^3 e^{-x}$

28. $f(x) = \dfrac{e^x - e^{-x}}{e^x + e^{-x}}$

29. $f(x) = \dfrac{e^x}{2e^x + 1}$

30. $f(x) = e^{-x^2}$

31. $f(x) = \exp\left(-\frac{1}{8}(x-1)^2\right)$

32. $f(x) = e^{\sin(x)}$

33. $f(x) = e^{-\cos^2(x)}$

34. $f(x) = e^{-x/4}\sin(2x)$

35. $f(x) = e^{-x/2}\cos\left(\dfrac{x}{4}\right)$

4.4 Arbitrary Bases

In this section we will discuss exponential functions with bases other than e and the corresponding logarithmic functions. We will also discuss functions defined by irrational powers of x.

Exponential Functions with Arbitrary Bases

We can express the exponential function with an arbitrary base $a > 0$ in terms of the natural exponential function:

Assume that $a > 0$. We have

$$\exp_a(x) = a^x = e^{x \ln(a)}$$

for each $x \in \mathbb{R}$.
Indeed, since $a = e^{\ln(a)}$,

$$\exp_a(x) = a^x = \left(e^{\ln(a)}\right)^x = e^{x \ln(a)}$$

for each $x \in \mathbb{R}$.

The exponential function with base 1 is simply the constant function 1. If $a > 1$ then $\ln(a) > 0$. The graph of $y = a^x$ can be obtained by stretching or shrinking the graph of the natural exponential function horizontally. Figure 1 shows the graphs of $y = e^x$, $y = 2^x$ and $y = 10^x$.

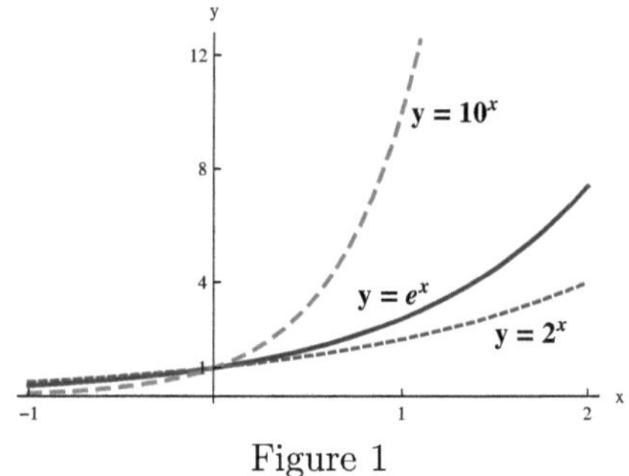

Figure 1

If $0 < a < 1$, then $1/a > 1$ and

$$a^x = \left(a^{-1}\right)^{-x} = \left(\frac{1}{a}\right)^{-x}.$$

Therefore, the graph of $y = a^x$ can be obtained by stretching or shrinking the graph of $y = e^{-x}$ horizontally . Figure 2 shows the graphs of $y = e^{-x}$, $y = 2^{-x}$ and $y = 10^{-x}$.

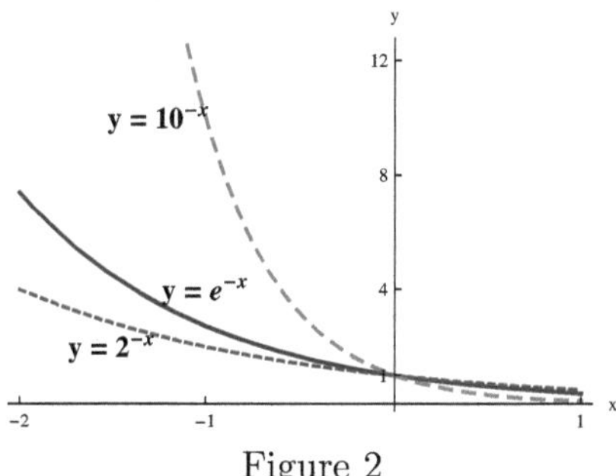

Figure 2

As we remarked in Section 4.3, the **rules for exponents** that you are familiar with from precalculus courses are valid: If a is a positive real number, and x, x_1, x_2 are arbitrary real numbers, we have

$$a^{x_1} a^{x_2} = a^{x_1 + x_2},$$
$$a^0 = 1,$$
$$a^{-x} = \frac{1}{a^x},$$
$$(a^{x_1})^{x_2} = a^{x_1 x_2}.$$

The derivative of an exponential function with an arbitrary base is a constant multiple of that function, as we anticipated in Section 4.3:

For any $a > 0$,

$$\frac{d}{dx} a^x = \ln(a) a^x$$

for each $x \in \mathbb{R}$.
Proof

We express a^x in terms of the natural exponential function and apply the chain rule:

$$\frac{d}{dx} a^x = \frac{d}{dx} e^{x \ln(a)} = \left(\frac{d}{du} e^u \bigg|_{u = x \ln(x)} \right) \frac{d}{dx} (x \ln(a))$$
$$= e^{x \ln(a)} (\ln(a))$$
$$= a^x \ln(a) = \ln(a) a^x.$$

■

Example 1 Determine

$$\frac{d}{dx} 10^x.$$

Solution

We have

$$\frac{d}{dx} 10^x = \ln(10) 10^x.$$

The above expression is not as elegant as the expression

$$\frac{d}{dx} e^x = e^x.$$

We have

$$\ln(10) 10^x \cong (2.302\,59) 10^x.$$

It appears that the natural exponential function is indeed "natural", even though the irrational number e is not a "simple" number such as 10. $\square$

Logarithmic Functions with Arbitrary Bases

Definition 1 Let $a > 0$ and $a \neq 1$. We abbreviate the logarithm of $x > 0$ with respect to the base a by $\log_a(x)$, and set

$$y = \log_a(x) \Leftrightarrow x = a^y.$$

Thus, $\log_a$ **is the inverse of the exponential function with base** a. Figure 3 illustrates the relationship between $x = a^y$ and $y = \log_a(x)$ for a base $a > 1$.

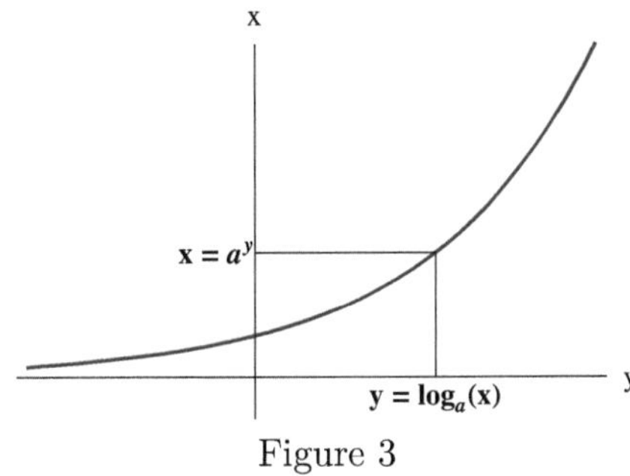

Figure 3

We can express a logarithm with respect to an arbitrary base in terms of the natural logarithm:

Proposition 1 Assume that $a > 0$ and $a \neq 1$. Then

$$\log_a(x) = \frac{\ln(x)}{\ln(a)}$$

for each $x > 0$.

Proof

Let $x > 0$ and $y = \log_a(x)$, so that $x = a^y = e^{y \ln(a)}$. Therefore,

$$y \ln(a) = \ln(x),$$

so that

$$\log_a(x) = y = \frac{\ln(x)}{\ln(a)}.$$

∎

Let $a > 1$. Since

$$\log_a(x) = \frac{\ln(x)}{\ln(a)},$$

and $\ln(a) > 0$ if $a > 1$, the graph of $\log_a$ can be obtained by stretching or shrinking the graph of the natural logarithm vertically. Figure 4 shows the graphs of $y = \ln(x)$, $y = \log_{10}(x)$ and $y = \log_2(x)$.

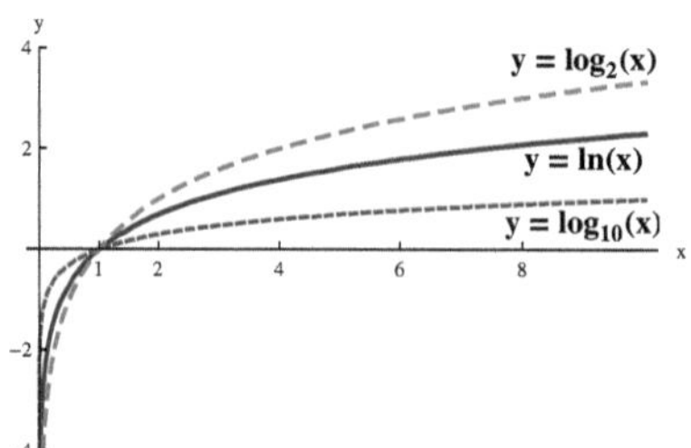

Figure 4

The basic algebraic properties of an arbitrary logarithm can be derived from its representation in terms of the natural logarithm:

Assume that a, x, x_1, x_2 are positive real numbers, $a \neq 1$, and r is an arbitrary real number. Then

$$\log_a(x_1 x_2) = \log_a(x_1) + \log_a(x_2),$$
$$\log_a(1/x) = -\log_a(x),$$
$$\log_a(x^r) = r \log_a(x).$$

The proof is left as an exercise.

Proposition 1 enables us to differentiate any logarithmic function easily:

We have

$$\frac{d}{dx}\log_a(x) = \frac{1}{x \ln(a)}$$

for each $x > 0$.

Proof

By the constant multiple rule for differentiation and Proposition 1,

$$\frac{d}{dx}\log_a(x) = \frac{d}{dx}\left(\frac{\ln(x)}{\ln(a)}\right) = \frac{1}{\ln(a)}\frac{d}{dx}\ln(x) = \frac{1}{\ln(a)}\left(\frac{1}{x}\right) = \frac{1}{x \ln(a)}$$

for each $x > 0$. ■

Example 2 Let $F(x) = \log_2(x)$ and $G(x) = \log_{1/2}(x)$. Determine F' and G'.

Solution

We have

$$F'(x) = \frac{d}{dx}\log_2(x) = \frac{1}{\ln(2)\,x}$$

and

$$G'(x) = \frac{d}{dx}\log_{1/2}(x) = \frac{1}{\ln(1/2)\,x} = -\frac{1}{\ln(2)\,x}.$$

□

Arbitrary Powers of x

For a fixed $a > 0$, a^x defines the exponential function with base a. Let's consider a **power function** defined by x^r, where x is the independent variable, and the exponent r is a fixed real number, rational or irrational. If $x > 0$ we have

$$x^r = \left(e^{\ln(x)}\right)^r = e^{r\ln(x)}.$$

Remark (**Caution**) You should not confuse a power function defined by an expression of the form x^r, where the exponent r is fixed and x is the variable, with an exponential function of the form a^x where the base a is a fixed positive number and the exponent x is the variable. For example, let $f(x) = x^{\sqrt{2}}$. Then

$$f(x) = e^{\sqrt{2}\ln(x)}.$$

Since $1 < \sqrt{2} < 2$, we have

$$\ln(x) < \sqrt{2}\ln(x) < 2\ln(x)$$

if $\ln(x) > 0$, i.e., if $x > 1$. Since the natural exponential function is an increasing function,

$$e^{\ln(x)} < e^{\sqrt{2}\ln(x)} < e^{2\ln(x)} \Rightarrow x < x^{\sqrt{2}} < x^2$$

if $x > 1$. Therefore, the part of the graph of $y = f(x) = x^{\sqrt{2}}$ on $[1, +\infty)$ is between the graphs of $y = x$ and $y = x^2$, as illustrated in Figure 5.

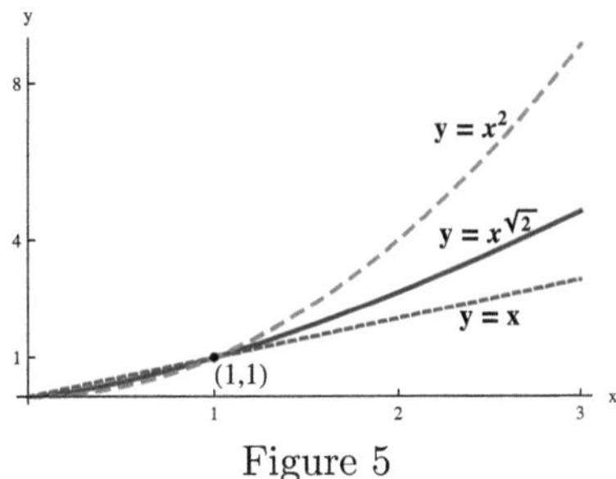

Figure 5

The exponential function

$$g(x) = \left(\sqrt{2}\right)^x = \left(2^{1/2}\right)^x = 2^{x/2} = e^{\ln(2)x/2}$$

is quite different from the power function $f(x) = x^{\sqrt{2}}$. Figure 6 displays the graphs of f and g on the interval $[0, 15]$. $\Diamond$

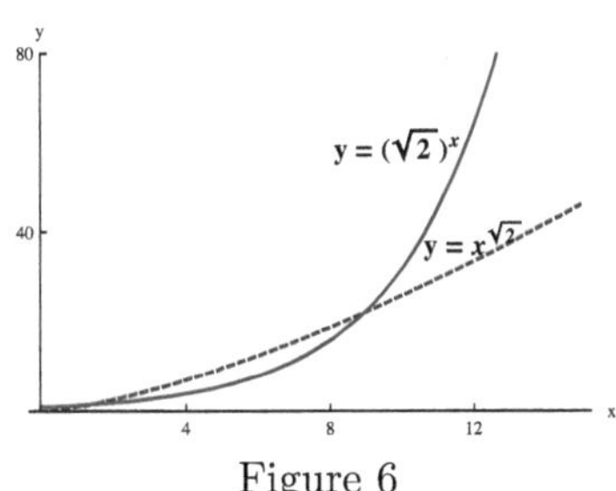

Figure 6

The power rule for the differentiation of x^r, where r is a rational exponent, is valid for any exponent r:

Theorem 1 (The Power Rule for Arbitrary Exponents) Let r be an arbitrary real number. Then

$$\frac{d}{dx}x^r = rx^{r-1}$$

for any $x > 0$.

Proof

We express x^r in terms of the natural exponential function and the natural logarithm, and apply the chain rule: If $x > 0$,

$$\frac{d}{dx}x^r = \frac{d}{dx}e^{r\ln(x)} = \left(\left.\frac{d}{du}e^u\right|_{u=r\ln(x)}\right)\left(\frac{d}{dx}(r\ln(x))\right)$$

$$= e^{r\ln(x)}\left(\frac{r}{x}\right) = x^r\left(\frac{r}{x}\right) = rx^{r-1}.$$

$\blacksquare$

Example 3 Let $f(x) = x^{\sqrt{2}}$ and $g(x) = \left(\sqrt{2}\right)^x$ for each $x > 0$. Determine $f'(x)$ and $g'(x)$.

Solution

By Theorem 1,

$$f'(x) = \frac{d}{dx}x^{\sqrt{2}} = \sqrt{2}x^{\sqrt{2}-1}$$

for each $x > 0$.

On the other hand,

$$g(x) = \left(2^{1/2}\right)^x = 2^{x/2}.$$

By the chain rule and the expression for the derivative of an exponential function,

$$g'(x) = \frac{d}{dx}2^{x/2} = \left(\left.\frac{d}{du}2^u\right|_{u=x/2}\right)\left(\frac{d}{dx}\left(\frac{x}{2}\right)\right) = \left(\ln(2)\,2^u|_{u=x/2}\right)\left(\frac{1}{2}\right) = \frac{1}{2}\ln(2)\,2^{x/2}.$$

$\square$

Some esoteric functions can be defined in terms of the natural exponential function and the natural logarithm, as in the following example.

Example 4 Set

$$f(x) = x^x, \ x > 0.$$

Determine f'.

Solution

We have

$$f(x) = x^x = \left(e^{\ln(x)}\right)^x = e^{x\ln(x)}$$

for each $x > 0$. By the chain rule and the product rule,

$$\frac{d}{dx}x^x = \frac{d}{dx}e^{x\ln(x)} = \left(\left.\frac{d}{du}e^u\right|_{u=x\ln(x)}\right)\left(\frac{d}{dx}(x\ln(x))\right)$$

$$= e^{x\ln(x)}\left(\ln(x) + x\left(\frac{1}{x}\right)\right) = x^x(\ln(x) + 1)$$

for each $x > 0$. $\square$

Problems

In problems 1-6, simplify the given expression.

1. $\log_{10}\left(10^4\right)$

2. $\log_2\left(2^{-3}\right)$

3. $\log_5\left(5^{3/4}\right)$

4. $10^{\log_{10}(7)}$

5. $10^{-\log_{10}(4)}$

6. $2^{\log_2\left(8^{1/4}\right)}$

In problems 7-11, determine the solutions of the given equation.

7. $3\log_{10}(x) = 2$

8. $2\log_{10}(x) = -5$

9. $6\log_2(x) = 4$

10. $10^{2x+1} = 6$

11. $2^{x^2} = 14$

In problems 12-24, determine the derivative:

12. $\dfrac{d}{dx}10^{x^2+1}$

13. $\dfrac{d}{dx}10^{1/x}$

14. $\dfrac{d}{dx}2^{\sqrt{x}}$

15. $\dfrac{d}{dx}3^{-x^2}$

16. $\dfrac{d}{dx}2^{\sin(x)}$

17. $\dfrac{d}{dx}\log_{10}\left(x^2+1\right)$

18. $\dfrac{d}{dx}\log_2\left(\sqrt{x}\right)$

19. $\dfrac{d}{dx}\log_{10}\left(\dfrac{x-1}{x+4}\right)$

20. $\dfrac{d}{dx}\log_2\left(\sin^2(x)\right)$

21. $\dfrac{d}{dx}x^{\sqrt{3}}$

22. $\dfrac{d}{dx}\pi^x$

23. $\dfrac{d}{dx}x^\pi$

24. $\dfrac{d}{dx}x^{1/x}$

25 [C] Plot the graphs of $f(x) = x^\pi$, $g(x) = x^4$ and $h(x) = x^2$ on the interval $[0,2]$ with the help of your graphing utility. What can you say about the relative size of their values?

26. [C] Plot the graphs of $f(x) = x^x$, $g(x) = x$ and $h(x) = e^x$ on the interval $[1,4]$. with the help of your graphing utility. What can you say about the relative size of their values?

4.5   Orders of Magnitude

In this section we will see that an exponential function of the form $a^{\delta x}$, where $a > 1$ and $\delta > 0$, increases much faster than any power of x, and a logarithmic function increases more slowly than any positive power of x as x tends to ∞. Since exponential and logarithmic functions with arbitrary bases can be expressed in terms of the natural exponential function and the natural logarithm, our emphasis will be on the "natural" functions.

Exponentials vs. Powers of x

In Section 4.3 we noted that

$$\lim_{x \to +\infty} e^x = +\infty,$$

since $e > 1$. In fact, e^x grows very rapidly as x increases. Table 1 displays e^n (rounded to 6 significant digits, as usual) for $n = 10, 20, 30, 40$. We see that e^n is much larger than n for the sampled values of n.

n	e^n
10	22026.5
20	4.85165×10^8
30	1.06865×10^{13}
40	2.35385×10^{17}

Table 1

We will see that e^x grows faster than any power of x as $x \to +\infty$. Let's begin by examining a function that involves e^x and x^2.

Example 1 Let

$$f(x) = \frac{e^x}{x^2}.$$

a) Determine $\lim_{x \to 0} f(x)$ and the absolute minimum of f on $(0, +\infty)$.
b) Plot the graph of f on the interval $(0, 8]$ with the help of your graphing utility. Does the picture suggest that $\lim_{x \to +\infty} f(x) = +\infty$?
c) Compute $f(x)$ for $x = 10, 20, 30, 40$. Do the numbers suggest that $\lim_{x \to +\infty} f(x) = +\infty$?

Solution

a) We have

$$\lim_{x \to 0} e^x = e^0 = 1 > 0 \text{ and } \lim_{x \to 0} \frac{1}{x^2} = +\infty.$$

Therefore,

$$\lim_{x \to 0} f(x) = \lim_{x \to 0} \left(e^x \left(\frac{1}{x^2} \right) \right) = +\infty.$$

Thus, f does not have an absolute maximum on $(0, +\infty)$. In order to determine the absolute minimum of f on $(0, +\infty)$ we will apply the derivative test for monotonicity. By the quotient rule,

$$f'(x) = \frac{d}{dx} \left(\frac{e^x}{x^2} \right) = \frac{e^x (x^2) - e^x (2x)}{x^4} = \frac{e^x x (x - 2)}{x^4} = \left(\frac{e^x}{x^3} \right) (x - 2)$$

if $x \neq 0$. We have $f'(x) = 0$ if $x = 2$, so that 2 is the only stationary point of f. Since $e^x > 0$ for each x and $x^3 > 0$ if $x > 0$, the sign of $f'(x)$ is determined by the sign of $x - 2$ if $x > 0$. Thus, $f'(x) < 0$ if $0 < x < 2$ and $f'(x) > 0$ if $x > 2$. Therefore, f is decreasing on the interval $(0, 2]$ and f is increasing on $[0, +\infty)$. Thus, f attains its absolute minimum on $(0, +\infty)$ at 2. We have

$$f(2) = \frac{e^2}{2^2} = \frac{e^2}{4}.$$

b) Figure 1 shows the graph of f. The picture indicates that $\lim_{x \to +\infty} f(x) = +\infty$.

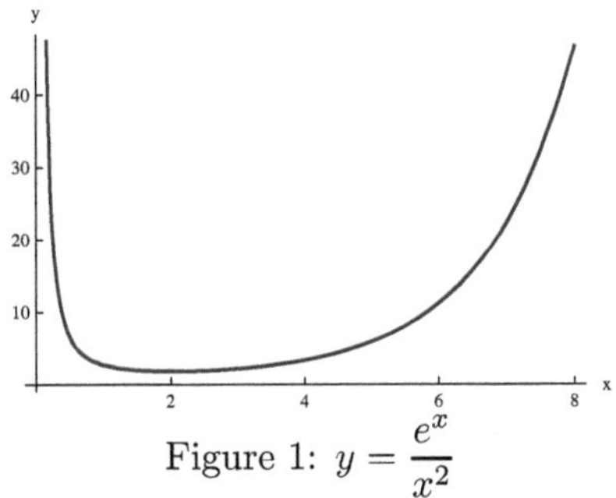

Figure 1: $y = \dfrac{e^x}{x^2}$

c) Table 2 displays $f(x)$ for $x = 10, 20, 30, 40$. The numbers definitely support the expectation that $\lim_{x \to +\infty} f(x) = +\infty$. $\square$

x	$f(x)$
10	2202.65
20	$2.425\,83 \times 10^7$
30	$3.562\,16 \times 10^{11}$
40	$5.884\,63 \times 10^{15}$

Table 2

The function of Example 1 represents the family of functions f_n, where

$$f_n(x) = \frac{e^x}{x^n},$$

and n is a positive integer. Note that $\lim_{x \to +\infty} e^x = +\infty$ and $\lim_{x \to +\infty} x^n = +\infty$. An attempt to evaluate $\lim_{x \to \infty} f(x)$ by setting

$$\lim_{x \to +\infty} \frac{e^x}{x^n} = \frac{\lim_{x \to +\infty} e^x}{\lim_{x \to +\infty} x^n}$$

leads to the indeterminate expression ∞/∞. Actually, e^x increases faster than x^n as $x \to +\infty$, so that the ratio $e^x/x^n \to \infty$:

Proposition 1 We have

$$\lim_{x \to +\infty} \frac{e^x}{x^n} = +\infty$$

for any integer n.

Proof

We have $\lim_{x \to +\infty} e^x = +\infty$, so that statement of the proposition is valid if $n = 0$. If $n < 0$, then $-n > 0$, so that $\lim_{x \to +\infty} x^{-n} = +\infty$. Therefore,

$$\lim_{x \to +\infty} \frac{e^x}{x^n} = \lim_{x \to +\infty} \left(e^x x^{-n}\right) = +\infty,$$

as well.

The statement of the proposition is nontrivial if n is a positive integer. Let's set

$$f_n(x) = \frac{e^x}{x^n}, \quad n = 1, 2, 3, \ldots$$

As in Example 1, we will make use of the derivative test for monotonicity in order to determine the absolute minimum of f_n on the interval $(0, +\infty)$. By the quotient rule,

$$f_n'(x) = \frac{d}{dx}\left(\frac{e^x}{x^n}\right) = \frac{e^x x^n - nx^{n-1}e^x}{x^{2n}} = \frac{e^x x^{n-1}(x-n)}{x^{2n}} = \left(\frac{e^x}{x^{n+1}}\right)(x-n).$$

Since $e^x > 0$ for each x and $x^{n+1} > 0$ if $x > 0$, the sign of $f_n'(x)$ is determined by the sign of $x - n$. We have $f_n'(x) = 0$ if $x = n$. We also have

$$f_n'(x) < 0 \text{ if } 0 < x < n, \text{ and } f_n'(x) > 0 \text{ if } x > n.$$

Therefore, f_n is decreasing on $(0, n]$, and increasing on $[n, +\infty)$. Thus, f_n attains its absolute minimum on $(0, +\infty)$ at $x = n$. We have

$$f_n(n) = \frac{e^n}{n^n} = \left(\frac{e}{n}\right)^n.$$

Therefore,

$$f_n(x) = \frac{e^x}{x^n} \geq \left(\frac{e}{n}\right)^n \text{ for each } x > 0.$$

We can express $f_n(x)$ in terms of $f_{n+1}(x)$:

$$f_n(x) = \frac{e^x}{x^n} = (x)\left(\frac{e^x}{x^{n+1}}\right) = x f_{n+1}(x).$$

Since

$$f_{n+1}(x) \geq \left(\frac{e}{n+1}\right)^{n+1},$$

we have

$$f_n(x) \geq x\left(\frac{e^{n+1}}{(n+1)^{n+1}}\right)$$

for each $x > 0$. Since

$$\lim_{x \to +\infty}\left(\frac{e^{n+1}}{(n+1)^{n+1}}\right)x = +\infty,$$

$\lim_{x \to \infty} f_n(x) = +\infty$, as well. ∎

Example 2 Let

$$f(x) = \frac{2^{x/4}}{x}.$$

Determine $\lim_{x \to +\infty} f(x)$.

Solution

We have

$$f(x) = \frac{2^{x/4}}{x} = \frac{e^{(x/4)\ln(2)}}{x}.$$

Let's set

$$z = \left(\frac{x}{4}\right)\ln(2) \Leftrightarrow x = \left(\frac{4}{\ln(2)}\right)z.$$

Since $\ln(2) > 0$ we have $z \to +\infty$ as $x \to +\infty$. Thus,

$$\lim_{x \to +\infty} f(x) = \lim_{x \to +\infty} \frac{e^{(x/4)\ln(2)}}{x} = \lim_{z \to +\infty} \frac{e^z}{\left(\frac{4}{\ln(2)}\right)z} = \lim_{z \to +\infty}\left(\frac{\ln(2)}{4}\right)\left(\frac{e^z}{z}\right) = +\infty,$$

thanks to Proposition 1. Figure 2 shows the graph of f on the interval $[20, 60]$ (The axes are centered at $(20, 0)$). The picture is consistent with the fact that

$$\lim_{x \to +\infty} \frac{2^{x/4}}{x} = +\infty.$$

$\square$

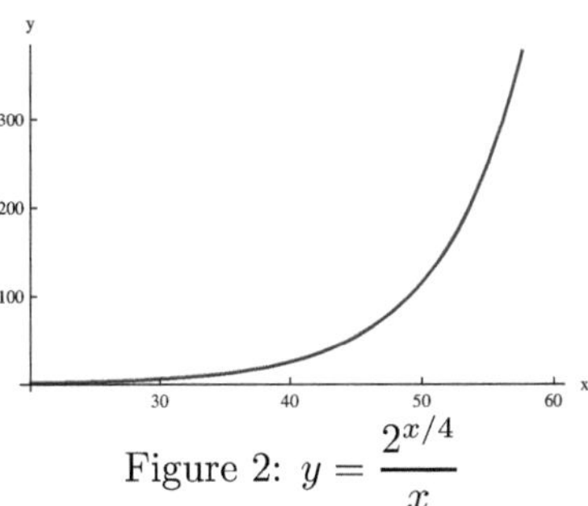

Figure 2: $y = \dfrac{2^{x/4}}{x}$

Example 3 Let $f(x) = x^2 e^{-x}$.

a) Determine $\lim_{x \to +\infty} f(x)$.
b) Determine the absolute maximum of f on $[0, +\infty)$.
c) Plot the graph of f on $[0, 8]$ with the help of your graphing utility. Does the picture support your responses to part a) and part b)?

Solution

a)

$$\lim_{x \to +\infty} f(x) = \lim_{x \to +\infty} x^2 e^{-x} = \lim_{x \to +\infty} \frac{x^2}{e^x} = \lim_{x \to +\infty} \frac{1}{\dfrac{e^x}{x^2}} = 0,$$

since

$$\lim_{x \to +\infty} \frac{e^x}{x^2} = +\infty,$$

by Proposition 1.

b) By the product rule and the chain rule,

$$f'(x) = \frac{d}{dx}\left(x^2 e^{-x}\right) = (2x)\left(e^{-x}\right) + x^2 \left(-e^{-x}\right) = xe^{-x}\left(2 - x\right).$$

Since $e^{-x} > 0$, the sign of $f'(x)$ is determined by the sign of $2 - x$ if $x > 0$. Therefore, $f'(x) = 0$ if $x = 2$ and

$$f'(x) > 0 \text{ if } 0 < x < 2 \text{ and } f'(x) < 0 \text{ if } x > 2.$$

By the derivative test for monotonicity, f is increasing on $[0, 2]$ and decreasing on $[2, +\infty)$ (note that $f(0) = 0$). Therefore, f attains its absolute maximum on the interval $[0, +\infty)$ at 2. We have

$$f(2) = 2^2 e^{-2} = \frac{4}{e^2} \cong 0.541341$$

c) Figure 3 shows the graph of f on the interval $[0, 8]$. The picture is consistent with the fact that $\lim_{x \to +\infty} f(x) = 0$ and our calculation of the absolute maximum of f on $[0, +\infty)$. $\square$

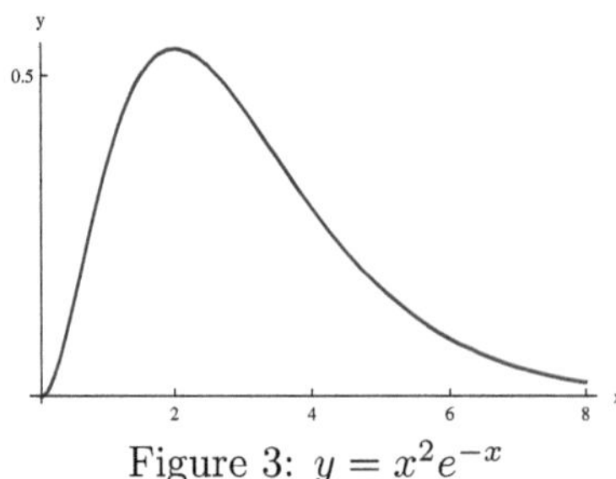

Figure 3: $y = x^2 e^{-x}$

Note that an attempt to evaluate $\lim_{x \to +\infty} x^2 e^{-x}$ as

$$\left(\lim_{x \to +\infty} x^2 \right) \left(\lim_{x \to +\infty} e^{-x} \right) = \left(\lim_{x \to +\infty} x^2 \right) \left(\lim_{x \to +\infty} \frac{1}{e^x} \right)$$

leads to **the indeterminate form** $(+\infty)\,(0)$. $\square$

Logarithmic Growth

We have

$$\lim_{x \to +\infty} \ln (x) = +\infty,$$

and

$$\frac{d}{dx} \ln (x) = \frac{1}{x} > 0,$$

consistent with the fact that the natural logarithm is an increasing function on the interval $(0, +\infty)$.

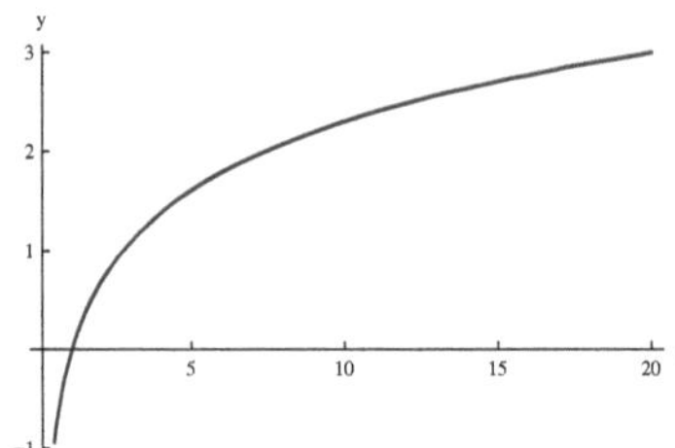

Figure 4: The natural logarithm increases slowly if x is large

But the rate of growth of $\ln(x)$ at x (i.e., $1/x$) tends to 0 as x becomes large. In fact, $\ln (x)$ increases more slowly than any positive power of x as x increases. The following proposition makes this statement more precise:

Proposition 2 Let $a > 0$, $a \neq 1$, and $r > 0$. Then

$$\lim_{x \to +\infty} \frac{\ln (x)}{x^r} = 0.$$

Proof

We set $y = \ln (x)$ so that $x = e^y$, and $y \to +\infty$ as $x \to +\infty$. Therefore,

$$\lim_{x \to +\infty} \frac{\ln (x)}{x^r} = \lim_{y \to +\infty} \frac{y}{(e^y)^r} = \lim_{y \to +\infty} \frac{y}{e^{ry}}.$$

If we set $z = ry$ then $z \to +\infty$ as $y \to +\infty$, since $r > 0$. Thus,

$$\lim_{x \to +\infty} \frac{\ln(x)}{x^r} = \lim_{y \to +\infty} \frac{y}{e^{ry}} = \lim_{z \to +\infty} \frac{z/r}{e^z} = \lim_{z \to +\infty} \frac{1}{r}\left(\frac{1}{\dfrac{e^z}{z}}\right) = 0,$$

since

$$\lim_{z \to +\infty} \frac{e^z}{z} = +\infty$$

by Proposition 1. $\blacksquare$

Note that an attempt to evaluate

$$\lim_{x \to +\infty} \frac{\ln(x)}{x^r}$$

as

$$\frac{\lim_{x \to +\infty} \ln(x)}{\lim_{x \to +\infty} x^r}$$

leads to the **indeterminate form** ∞/∞.

Example 4 Let

$$f(x) = \frac{\ln(x)}{\sqrt{x}}.$$

a) Determine $\lim_{x \to +\infty} f(x)$ and $\lim_{x \to 0+} f(x)$.
b) Determine the absolute maximum of f on the interval $(0, +\infty)$.
c) Plot the graph of on the interval $[1, 100]$ with the help of your graphing utility. Is the picture consistent with your response to part a) and part b)?
d) Compute $f(x)$ for $x = 10^k$, $k = 2, 3, 4, 5$. Do the numbers support your statement concerning $\lim_{x \to +\infty} f(x)$?

Solution

a) We have

$$f(x) = \frac{\ln(x)}{\sqrt{x}} = \frac{\ln(x)}{x^{1/2}}.$$

By Proposition 2 (with $r = 1/2$) $\lim_{x \to \infty} f(x) = 0$.
As for $\lim_{x \to 0+} f(x)$, we have $\lim_{x \to 0+} \ln(x) = -\infty$ and $\lim_{x \to 0+} 1/\sqrt{x} = +\infty$. Therefore,

$$\lim_{x \to 0+} f(x) = \lim_{x \to 0+} \left(\ln(x)\left(\frac{1}{\sqrt{x}}\right)\right) = -\infty.$$

b) By the quotient rule,

$$f'(x) = \frac{d}{dx}\left(\frac{\ln(x)}{x^{1/2}}\right) = \frac{\left(\dfrac{1}{x}\right)x^{1/2} - \ln(x)\left(\dfrac{1}{2}x^{-1/2}\right)}{\left(x^{1/2}\right)^2} = \frac{\dfrac{1}{\sqrt{x}} - \dfrac{\ln(x)}{2\sqrt{x}}}{x} = \frac{2 - \ln(x)}{2x\sqrt{x}}.$$

Since $x\sqrt{x} > 0$ if $x > 0$, the sign of $f'(x)$ is determined by the sign of $2 - \ln(x)$. We have

$$2 - \ln(x) = 0 \Leftrightarrow 2 = \ln(x) \Leftrightarrow e^2 = x.$$

Furthermore,

$$2 - \ln(x) > 0 \text{ if } 0 < x < e^2 \text{ and } 2 - \ln(x) < 0 \text{ if } x > e^2.$$

By the derivative test for monotonicity, f is increasing on $(0, e^2]$, and f is decreasing on $[e^2, +\infty)$. Therefore, f attains its absolute maximum on the interval $(0, +\infty)$ at $e^2 \cong 7.389\,06$. We have

$$f\left(e^2\right) = \left.\frac{\ln(x)}{\sqrt{x}}\right|_{x=e^2} = \frac{\ln\left(e^2\right)}{\sqrt{e^2}} = \frac{2}{e} \cong 0.735759$$

c) Figure 5 shows the graph of f on the interval. The picture is consistent with our response to part a) and part b), but does not provide strong support for the statements about $\lim_{x\to 0+} f(x)$ and $\lim_{x\to +\infty} f(x)$.

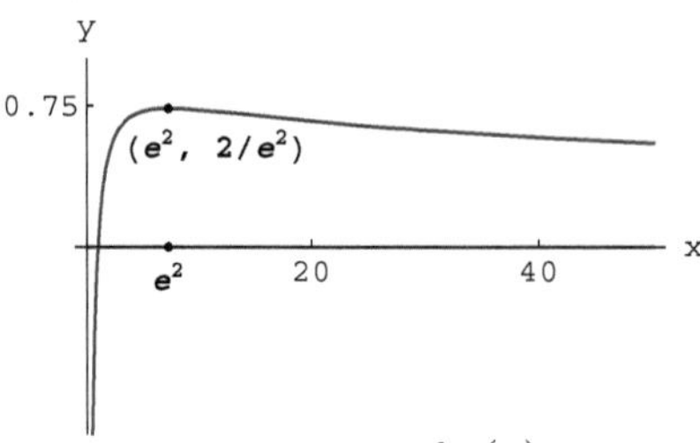

$$\text{Figure 5: } y = \frac{\ln(x)}{\sqrt{x}}$$

d) Table 4 displays $f(x)$ for $x = 10^k$, $k = 2, 3, 4, 5$. The numbers in Table 4 support the statement that $\lim_{x\to +\infty} f(x) = 0$. $\square$

x	$f(x)$
10^2	$.460\,517$
10^3	$.218\,442$
10^4	$9.210\,34 \times 10^{-2}$
10^5	$3.640\,71 \times 10^{-2}$

Table 4

Example 5 Let $f(x) = x\ln(x)$.

a) Determine $\lim_{x\to 0+} f(x)$ and $\lim_{x\to +\infty} f(x)$.
b) Determine the absolute minimum of f on the interval $(0, +\infty)$.
c) Sketch the graph of f.

Solution

a) We set $z = 1/x$ so that $z \to +\infty$ as x approaches 0 from the right. Thus,

$$\lim_{x\to 0+} f(x) = \lim_{x\to 0+} x\ln(x) = \lim_{z\to +\infty} \left(\frac{1}{z}\right)\ln\left(\frac{1}{z}\right) = \lim_{z\to +\infty} \frac{-\ln(z)}{z} = 0$$

by Proposition 2.
Note that an attempt to evaluate $\lim_{x\to 0+} x\ln(x)$ as

$$\left(\lim_{x\to 0+} x\right)\left(\lim_{x\to 0+} \ln(x)\right)$$

leads to **the indeterminate form** $0 \times (-\infty)$.
As for $\lim_{x\to +\infty} f(x)$, $\lim_{x\to +\infty} x\ln(x) = +\infty$ since $\lim_{x\to +\infty} x = +\infty$ and $\lim_{x\to +\infty} \ln(x) = +\infty$.

b) By the product rule,

$$f'(x) = \frac{d}{dx}\left(x\ln(x)\right) = \ln(x) + x\left(\frac{1}{x}\right) = \ln(x) + 1.$$

Therefore,

$$f'(x) = 0 \Leftrightarrow \ln(x) + 1 = 0 \Leftrightarrow \ln(x) = -1 \Leftrightarrow x = e^{-1} = \frac{1}{e} \cong 0.367\,879.$$

Furthermore,

$$f'(x) < 0 \text{ if } 0 < x < \frac{1}{e}, \text{ and } f'(x) > 0 \text{ if } x > \frac{1}{e}.$$

By the derivative test for monotonicity,

$$f \text{ is decreasing on } \left(0, \frac{1}{e}\right] \text{ and increasing on } \left[\frac{1}{e}, +\infty\right).$$

Therefore, f attains its absolute minimum on $(0, +\infty)$ at $1/e$. The corresponding value of f is

$$\frac{1}{e}\ln\left(\frac{1}{e}\right) = -\frac{1}{e}\ln(e) = -\frac{1}{e} \cong -0.367\,879.$$

c) Figure 6 displays the graph of f on the interval $[0, 4]$. Even though f is not defined at 0, the graph is consistent with the fact that $\lim_{x \to 0+} f(x) = 0$. $\square$

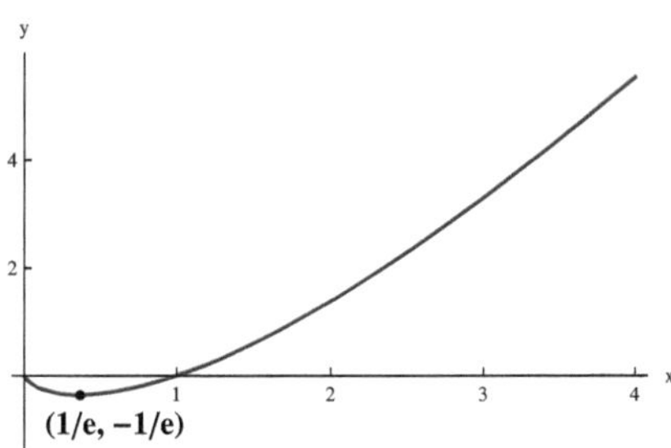

Figure 6: $y = x\ln(x)$

The Natural Exponential Function as a Limit of Polynomials

The natural exponential function can be approximated by polynomials. We will examine such a family of polynomials. In Chapter 9 we will discuss another family of approximating polynomials. Let

$$p_n(x) = \left(1 + \frac{x}{n}\right)^n,$$

where n is a positive integer. Each $p_n(x)$ is a polynomial, and the degree of $p_n(x)$ is n. The first member of this family of polynomials is $p_1(x) = 1 + x$. Note that p_1 is the linear approximation to the natural exponential function based at 0 (Example 2 of Section 5.3). Figure 7 displays the graphs of the natural exponential function and the polynomials

$$p_1(x) = 1 + x, \ p_3(x) = \left(1 + \frac{x}{3}\right)^3 \text{ and } p_{20}(x) = \left(1 + \frac{x}{20}\right)^2.$$

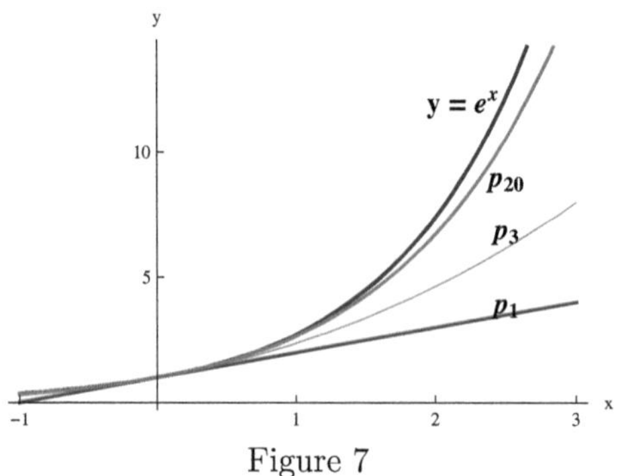

Figure 7

Figure 7 indicates that $p_n(x)$ approximates e^x with increasing accuracy as n increases. This is indeed the case.

Proposition 3 We have

$$\lim_{n \to \infty} \left(1 + \frac{x}{n}\right)^n = e^x \quad \text{for each } x \in \mathbb{R}.$$

In particular,

$$\lim_{n \to \infty} \left(1 + \frac{1}{n}\right)^n = e$$

Since $\lim_{n \to \infty} 1/n = 0$, Proposition 3 follows from the following fact:

Proposition 4

$$\lim_{h \to 0} (1 + xh)^{1/h} = e^x \; .$$

Proof

We have

$$(1 + xh)^{1/h} = \left(e^{\ln(1+xh)}\right)^{1/h} = e^{\ln(1+xh)/h} \; .$$

By the continuity of the natural exponential function,

$$\lim_{h \to 0} e^{\ln(1+xh)/h} = e^{\lim_{h \to 0} \ln(1+xh)/h} \; .$$

Therefore, in order to show that

$$\lim_{h \to 0} (1 + xh)^{1/h} = e^x,$$

it is sufficient to prove that

$$\lim_{n \to \infty} \left(\frac{1}{h} \ln(1 + xh)\right) = x.$$

This is indeed the case: If $x = 0$, the equality is obvious since $\ln(1) = 0$. Let's assume that $x \neq 0$ and set $z = xh$. Then $z \neq 0$ if $h \neq 0$ and $z \to 0$ as $h \to 0$. Therefore,

$$\lim_{h \to 0} \frac{\ln(1 + xh)}{h} = \lim_{h \to 0} \left(\frac{\ln(1 + xh)}{xh}(x) \right) = x \lim_{h \to 0} \frac{\ln(1 + xh)}{xh}$$

$$= x \lim_{z \to 0} \frac{\ln(1 + z)}{z}$$

$$= x \lim_{z \to 0} \frac{\ln(1 + z) - \ln(1)}{z}$$

$$= x \left(\frac{d}{du} \ln(u) \Big|_{u=1} \right) = x \left(\frac{1}{u} \Big|_{u=1} \right) = x.$$

∎

Note that an attempt the evaluate $\lim_{n \to \infty} (1 + 1/n)^n$ as

$$\left(\lim_{n \to \infty} \left(1 + \frac{1}{n} \right) \right)^{\lim_{n \to \infty} n}$$

leads to **the indeterminate form** 1^∞. One may be tempted to say that $1^\infty = 1$, since $1^n = 1$ for each n, but the actual limit need not be 1, as Proposition 3 shows.

Problems

In problems 1-4,
a) Determine the indicated limits and the asymptotes for the graph of f.
b) Use the derivative test to determine the intervals on which f is increasing/decreasing, and the points at which f has a local extremum.
c) Sketch the graph of f.

1.. $f(x) = \dfrac{e^{x/2}}{x}$, $\lim\limits_{x \to 0\pm} f(x)$, $\lim\limits_{x \to \pm\infty} f(x)$

3. $f(x) = \dfrac{\ln(x)}{x^{1/3}}$, $\lim\limits_{x \to 0+} f(x)$, $\lim\limits_{x \to +\infty} f(x)$

2.. $f(x) = x^2 e^{-x}$, $\lim\limits_{x \to \pm\infty} f(x)$

4. $f(x) = \sqrt{x}\ln(x)$, $\lim\limits_{x \to 0+} f(x)$, $\lim\limits_{x \to +\infty} f(x)$

In problems 6-7,
a) Determine the indicated limits and the asymptotes for the graph of f.
b) Use the derivative test to determine the intervals on which f is increasing/decreasing, and the points at which f has a local extremum.
c) Use the second derivative test to determine the intervals on which the graph of f is concave up/concave down, and the x-coordinates of the inflection points of the graph of f.
d) Sketch the graph of f.

5.. $f(x) = xe^{-x^2/4}$, $\lim\limits_{x \to \pm\infty} f(x)$.

7. $f(x) = \dfrac{10e^x}{2 + e^x}$, $\lim\limits_{x \to \pm\infty} f(x)$

6. $f(x) = e^{-x^2/9}$, $\lim\limits_{x \to \pm\infty} f(x)$

In problems 8-11, determine the absolute maximum and the absolute minimum of f on the interval I, provided that such values exist. Justify your response if you claim that such a value does not exist.

8. $f(x) = 10xe^{-4x}$, $I = [0, +\infty)$.   **10.** $f(x) = x^{1/3} \ln(x)$, $I = (0, +\infty)$.

9. $f(x) = \dfrac{e^{2x}}{x^2}$, $I = (0, +\infty)$.   **11.** $f(x) = \dfrac{\ln(x)}{\sqrt{x}}$, $I = (0, +\infty)$

4.6   Exponential Growth and Decay

In this section we will be able to predict the growth of a population or the decay of a radioactive material, based on models that lead to differential equations of the form $y' = ky$, where k is a constant. The solutions can be expressed in terms of the natural exponential function. We will also discuss a difference equation that is related to such a differential equation and compound interest.

Example 1 The Growth of a Population at a Constant Rate

Let $y(t)$ denote the population of a certain country at time t, where t is measured in years. We will assume that the rate of change of the population with respect to t is proportional to $y(t)$, and that the constant of proportionality is a positive number k that is independent of t. Since the rate of change of y with respect to t is $y'(t)$, we have

$$y'(t) = ky(t),$$

so that

$$\frac{y'(t)}{y(t)} = k$$

at each t. The ratio $y'(t)/y(t)$, i.e., the ratio of the rate of change of the population to the population at that time, is referred to as the relative growth rate of the population. Since

$$k = \frac{y'(t)}{y(t)} = \frac{\text{rate of change of population}}{\text{population}} = \frac{\text{number of people/year}}{\text{number of people}},$$

the unit of k is $1/(\text{unit of time}) = 1/\text{year}$. For example, if we are given the information that the population of the country in question grows at the constant rate of 2% per year, what is meant is that the relative growth rate of the population is $0.02/\text{year}$. Thus, $k = 0.02/\text{year}$, and we have

$$\frac{y'(t)}{y(t)} = 0.02,$$

so that

$$y'(t) = 0.02y(t).$$

In order to predict the population at any t, we also need information about the population at a specific time. For example, if the present population is 100 million, we may have $t = 0$ correspond to the present, and state that $y(0) = 100$ (million). $\square$

Example 2 Radioactive Decay at a Constant Rate

A model that involves **a radioactive material** leads to a similar equation. Let $y(t)$ denote the amount of the material at time t. Assume that mass is measured in grams and time is measured in years. We will make the assumption that the rate of change of the material at any time t is proportional to the amount of material at that time. Thus,

$$y'(t) = -ky(t)$$

at any t, where k is a positive constant. The $(-)$ sign corresponds to the fact that amount the material is decreasing with time. The quantity k is referred to as the relative decay rate of the radioactive material, or simply as the decay rate of the material. Thus,

$$\frac{y'(t)}{y(t)} = -k$$

at any t. The unit of k is $1/(\text{unit of time}) = 1/\text{year}$. For example, we may have $k = 0.01/\text{ year}$. In this case,

$$\frac{y'(t)}{y(t)} = -0.01,$$

so that

$$y'(t) = -0.01y(t).$$

In order to predict the amount of the radioactive material at any time t based on this model, we must be given the amount of the material at a certain time. For example, we may set $t = 0$ at the time we have have a sample of 100 grams, so that $y(0) = 100$. $\square$

The Solution of the Differential Equation $y' = ky$

The models for the growth of a population and radioactive decay at a constant rate fit the same mathematical framework: We must determine a function $y(t)$ such that

$$y'(t) = ky(t) \text{ and } y(t_0) = y_0,$$

where k, t_0 and y_0 are given numbers. The expression, $y'(t) = ky(t)$ is a **differential equation**, i.e., an equation that involves a function and its derivative. The condition $y(t_0) = y_0$ is an **initial condition**. The problem of determining the function $y(t)$ so that

$$y'(t) = ky(t) \text{ and } y(t_0) = y_0$$

is referred to as an **initial-value problem.**

We will refer to the ratio

$$\frac{y'(t)}{y(t)} = \frac{\text{rate of change of } y \text{ at } t}{y(t)}$$

as **the relative rate of change of** y at t. Thus, the differential equation $y'(t) = ky(t)$ says that the relative rate of change of y is the constant k.

We can express any solution of the differential equation $y'(t) = ky(t)$ in terms of the natural exponential function:

Theorem 1 Let k be a given constant. The function $y(t)$ is a solution of the differential equation $y'(t) = ky(t)$ if and only if $y(t) = Ce^{kt}$, where C is a constant.

Proof

Let $y(t) = Ce^{kt}$, where C is an arbitrary constant. We must show that $y(t)$ solves the given differential equation. This is a simple exercise in differentiation:

$$y'(t) = \frac{d}{dt}\left(Ce^{kt}\right) = C\frac{d}{dt}e^{kt} = C(ke^{kt}) = k\left(Ce^{kt}\right) = ky(t).$$

Conversely, we must show that any solution of the given differential equation is a constant multiple of e^{kt}. Thus, assume that f solves the differential equation, so that $f'(t) = kf(t)$ for each $t \in R$. Set

$$h(t) = \frac{f(t)}{e^{kt}}.$$

By the quotient rule for differentiation

$$h'(t) = \frac{d}{dt}\left(\frac{f(t)}{e^{kt}}\right) = \frac{f'(t)e^{kt} - f(t)\,ke^{kt}}{\left(e^{kt}\right)^2}.$$

Since $f'(t) = kf(t)$,

$$h'(t) = \frac{f'(t)e^{kt} - f(t)\,ke^{kt}}{\left(e^{kt}\right)^2} = \frac{kf(t)\,e^{kt} - kf(t)\,e^{kt}}{\left(e^{kt}\right)^2} = 0$$

for each $t \in \mathbb{R}$. A function whose derivative is identically 0 must be a constant. Therefore there exists a constant C such that $h(t) \equiv C$. Thus,

$$h(t) = \frac{f(t)}{e^{kt}} = C.$$

Therefore, $f(t) = Ce^{kt}$, as claimed. ∎

By Theorem 1, any constant multiple of e^{kt} solves the differential equation $y' = ky$, and any solution of the differential equation must be in that form. If we had come across the differential equation $y'(t) = ky(t)$ before we knew about the natural exponential function, we would have to discover the natural exponential function! We will say that **the general solution** of the differential equation $y' = ky$ is Ce^{kt}, where C is an arbitrary constant.

Although every function of the form Ce^{kt} is a solution of the differential equation $y' = ky$, there is only one solution that satisfies a given initial condition:

Corollary (Corollary to Theorem 1) Assume that k, t_0 and y_0 are given numbers. The solution of the initial-value problem,

$$y'(t) = ky(t), \ \ y(t_0) = y_0,$$

is unique, and can be expressed as

$$y(t) = y_0\,e^{k(t-t_0)}.$$

In particular, if $y(0) = y_0$,

$$y(t) = y_0\,e^{kt}.$$

Proof

By Theorem 1, the solution $y(t)$ is in the form Ce^{kt}. Therefore, $y(t_0) = y_0$ if and only if

$$Ce^{kt_0} = y_0,$$

so that

$$C = y_0 e^{-kt_0}.$$

Therefore, the solution of the given initial-value problem is unique, and can be expressed as

$$y(t) = Ce^{kt} = (y_0 e^{-kt_0})e^{kt} = y_0 e^{kt-kt_0} = y_0 e^{k(t-t_0)}.$$

If $y(0) = y_0$,

$$y(t) = y_0 e^{kt}.$$

∎

There is no need to memorize the expression that is provided by Corollary . It is sufficient to remember that the general solution of the differential equation $y = ky$ is Ce^{kt}. The constant can be determined for a specific initial condition, as in the following example.

Example 3

a) Determine the general solution of the differential equation,

$$y'(t) = \frac{1}{4}y(t).$$

b) Determine the solution of the initial-value problem,

$$y'(t) = \frac{1}{4}y(t), \; y(1) = 200.$$

Solution

a) By Theorem 1, the general solution of the differential equation $y' = 1/4y$ is

$$y(t) = Ce^{t/4},$$

where C is an arbitrary constant.

b) We have

$$y(1) = 200 \Leftrightarrow Ce^{1/4} = 200 \Leftrightarrow C = 200e^{-1/4}.$$

Therefore, the solution of the given initial-value problem is

$$f(t) = 200e^{-1/4}e^{t/4} = 200e^{(t-1)/4}.$$

Figure 1 shows the graphs of f and two other solutions of the differential equation

$$y'(t) = \frac{1}{4}y(t).$$

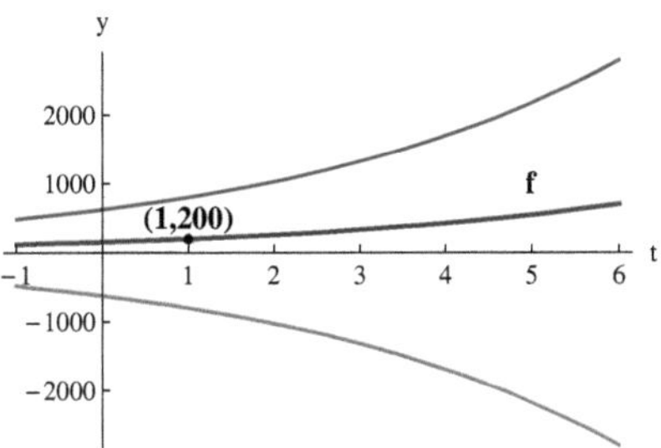

Figure 1: Some solutions of the differential equation $y' = \frac{1}{4}y$

The function f is the only solution whose graph passes through the point $(1, 200)$. Note that

$$\lim_{t \to +\infty} \left| Ce^{t/4} \right| = |C| \lim_{t \to +\infty} e^{t/4} = +\infty$$

if $C \neq 0$. Thus, any solution of the differential equation

$$y'(t) = \frac{1}{4}y(t)$$

grows exponentially, unless it is the solution which is identically 0. This is typical of any solution of a differential equation of the form $y' = ky$ if $k > 0$. $\square$

Example 4

a) Determine the general solution of the differential equation,

$$y'(t) = -\frac{1}{2}y(t).$$

b) Determine the solution of the initial-value problem,

$$y'(t) = -\frac{1}{2}y(t),\ y(0) = 10.$$

Solution

a) The general solution of the differential equation

$$y'(t) = -\frac{1}{2}y(t),$$

is

$$y(t) = Ce^{-t/2},$$

where C is an arbitrary constant.

b) We have

$$y(0) = 10 \Leftrightarrow Ce^{0} = 10 \Leftrightarrow C = 10.$$

Therefore, the solution of the given initial-value problem is

$$g(t) = 10e^{-t/2}.$$

Figure 2 displays the graphs of g and three other solutions of the differential equation

$$y'(t) = -\frac{1}{2}y(t).$$

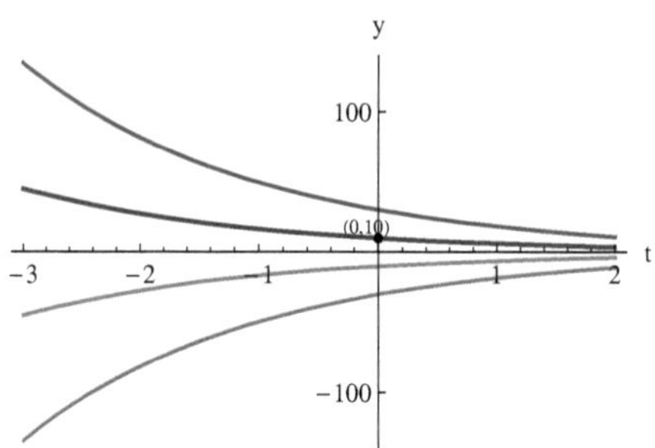

Figure 2: Some solutions of the differential equation $y' = -\frac{1}{2}y$

The function g is the only solution whose graph passes through the point $(0, 10)$. Note that

$$\lim_{x \to +\infty} \left| Ce^{-t/2} \right| = |C| \lim_{t \to +\infty} e^{-t/2} = 0$$

for any constant C. Thus, any solution of the differential equation

$$y'(t) = -\frac{1}{2}y(t)$$

decays exponentially. This is typical of any solution of a differential equation of the form $y' = ky$ if $k < 0$. $\square$

Example 5 Assume that the population of a country grows at the constant rate of 2% per year, and that the present population is 100 million. Let t be time in years, and set $t = 0$ for the present. Denote the population at time t by $y(t)$.

a) Compute $y(t)$.
b) Compute the population of the country 20 years from now.

Solution

a) We are assuming that the relative rate of change of the population has the constant value 0.02/year. Thus,

$$\frac{y'(t)}{y(t)} = 0.02 \Leftrightarrow y'(t) = 0.02y(t).$$

Therefore,

$$y(t) = y(0)e^{0.02t} = 10^8 e^{0.02t}$$

at time t (in years).

b) In particular, the population after 20 years is predicted to be

$$y(20) = 10^8 e^{0.02 \times 20} = 10^8 e^{0.4} \cong 1.49182 \times 10^8 \cong 149,182,000.$$

A growth rate of 2% can be considered to be tolerable in a relatively short span of time (at least in a country where the initial population is not excessive). On the other hand, if we assume that the growth rate stays constant, and predict the population at some distant future, we may be shocked. For example, in the case that we are considering, the population after 100 years is predicted to be

$$y(100) = 10^8 e^{0.02 \times 100} = 10^8 e^2 \cong 7.38906 \times 10^8 \cong 738,906,000.$$

That is probably an overestimate of the population 100 years from now. It is based on the assumption that the population grows at the constant rate of 2% per year. It is more realistic to assume that the growth rate will decrease due to many factors (overcrowding, education, etc.). In Chapter 9 we will discuss a more realistic model for population growth. Figure 3 shows the graph of the predicted population (in millions). $\square$

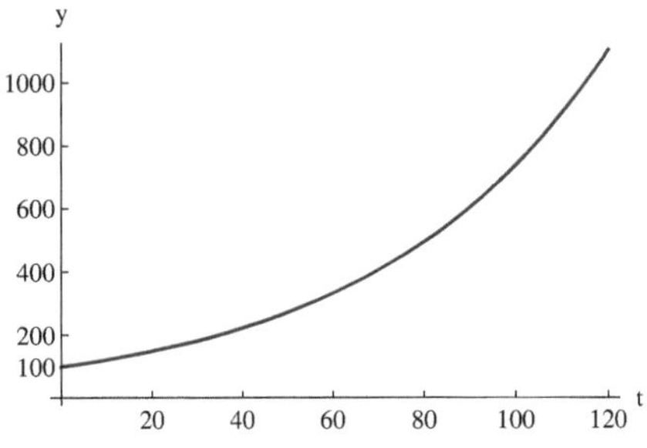

Figure 3: Population Growth

Remark 1 In Example 5 the relative growth rate of the population is 0.02 per year. This does not mean that the population increases by 2% every year. For example,

$$y(1) = 10^8 e^{0.02 \times 1} = 10^8 e^{0.02} \cong 102,020,000 \neq 1.02 \times 10^8$$

$\Diamond$

Definition 1 If $y'(t) = ky(t)$, where $k > 0$, **the doubling time** T is the time such that $y(T) = 2y(0)$.

Proposition 1 If $y' = ky$, where $k > 0$, the doubling time is

$$T = \frac{\ln(2)}{k}.$$

We can express the solution of the initial-value problem

$$y'(t) = ky(t), \ \ y(0) = y_0,$$

as

$$y(t) = y_0 2^{t/T}.$$

Proof

By Corollary ,

$$y(t) = y(0) e^{kt}.$$

Since $y(T) = 2y(0)$,

$$y(T) = y(0) e^{kT} = 2y(0).$$

Therefore,

$$e^{kT} = 2 \Leftrightarrow kT = \ln(2) \Leftrightarrow T = \frac{\ln(2)}{k}.$$

Since

$$k = \frac{\ln(2)}{T},$$

we can express the solution that corresponds to the initial condition $y(0) = y_0$ as

$$y(t) = y_0 e^{kt} = y_0 e^{(\ln(2)/T)t} = y_0 e^{\ln(2)t/T} = y_0 \left(e^{\ln(2)} \right)^{t/T} = y_0 2^{t/T}.$$

■

Example 6 Let $y(t)$ denote the number of bacteria in a culture at time t (in hours). Assume that the number of bacteria in the culture doubles every 5 hours, and that $y(0) = 1000$.

a) Express $y(t)$ in terms of the doubling time.
b) Calculate $y(10)$, $y(20)$ and $y(40)$.

Solution

a) By Proposition 1,

$$y(t) = 1000 \times 2^{t/5}.$$

b) Table 1 displays $y(t)$ for $10, 20$ and 40. Note the rapid rate of growth of the bacteria.$\square$

t	$y(t)$
10	4000
20	16000
40	256000

Table 1

Example 7 Assume that the decay rate of a radioactive material is 10^{-2} per year, and that we have a sample of 10 grams. Let $y(t)$ denote the amount (in grams) that is left at time t (in years), with $t = 0$ corresponding to the present.

a) Determine $y(t)$.
b) Calculate $y(100)$.

Solution

a) Since the decay rate is 0.01,

$$y'(t) = -0.01y(t) \ \text{ and } \ y(0) = 10.$$

Therefore,

$$y(t) = 10e^{-0.01t} \ (\text{grams}).$$

b) The amount that is left after 100 years is

$$y(100) = 10e^{-0.01 \times 100} = 10e^{-1} \cong 3.67879 \text{ grams.}$$

Figure 4 displays the graph of $y(t)$ on the interval $[0, 200]$. $\square$

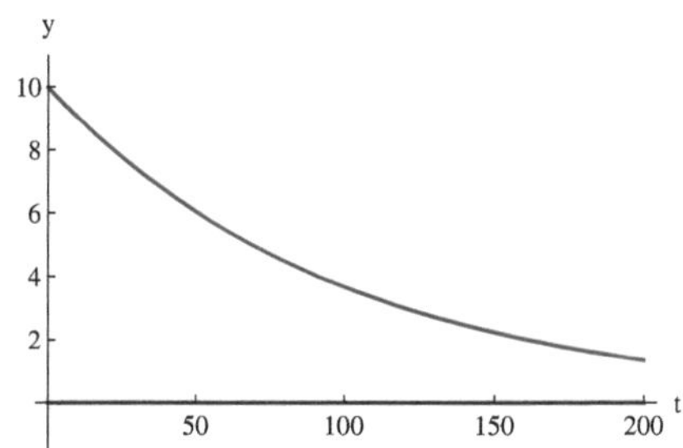

Figure 4: An example of radioactive decay

Definition 2 The **half-life** of a radioactive material is the time it takes for a sample of the material to be reduced to half of the initial amount.

Proposition 2 If the decay rate of a radioactive material is k, the half-life of the material is

$$T = \frac{\ln(2)}{k}.$$

The amount of the material at time t can be expressed as

$$y(t) = \frac{y(0)}{2^{t/T}}.$$

The proof is similar to the proof of Proposition 1 (exercise).

Example 8

a) Determine the half-life T of the radioactive material of Example 7.

b) Calculate $y(kT)$ for $k = 1, 2, 3, 4$.

Solution

a) Since the decay rate is 10^{-2}, the half-life of the material is

$$T = \frac{\ln(2)}{10^{-2}} = 10^2 \times \ln(2) \cong 69.3147 \text{ years.}$$

b) The amount of the material at time t is

$$y(t) = y(0)\,\frac{1}{2^{t/T}} = \frac{10}{2^{t/T}}.$$

Therefore,

$$y(kT) = \frac{10}{2^{kT/T}} = \frac{10}{2^k}.$$

Table 2 displays $y(kT)$ for $k = 1, 2, 3, 4$. $\square$

k	kT	$y(kT)$
1	69.3147	5
2	138.629	2.5
3	207.944	1.25
4	277.259	0.625

Table 2

Compound Interest

Let's consider the meaning of an expression such as **"annual interest rate of 5%, compounded quarterly"**. If you deposit $1000 in your bank account, the amount in your account after 3 months becomes

$$1000 + \frac{0.05}{4} \times 1000 = \left(1 + \frac{0.05}{4}\right)1000.$$

This amount is the principal that is the basis of the calculation for the following 1/4 year. Thus, the amount in your account at the end of 6 months is

$$\left(1 + \frac{0.05}{4}\right)1000 + \frac{0.05}{4} \times \left(1 + \frac{0.05}{4}\right)1000 = \left(1 + \frac{0.05}{4}\right)\left(1 + \frac{0.05}{4}\right)1000$$

$$= \left(1 + \frac{0.05}{4}\right)^2 1000.$$

At the end of 9 months, the amount is

$$\left(1 + \frac{0.05}{4}\right)^2 1000 + \frac{0.05}{4} \times \left(1 + \frac{0.05}{4}\right)^2 1000 = \left(1 + \frac{0.05}{4}\right)^2\left(1 + \frac{0.05}{4}\right)1000$$

$$= \left(1 + \frac{0.05}{4}\right)^3 1000.$$

Now let's consider the general framework for compound interest. Assume that the annual interest rate for deposits in a certain bank is r (in the above example, $r = 0.05$). Let $t = 0$ correspond to the time of the initial deposit. We will assume that there are no withdrawals from the account and that additional deposits are not made. Let us denote the amount of the money in the account at time t (in years) by $Y(t)$. The expression, "**the rate of interest is r per year, compounded at intervals of Δt**" means that

$$Y(t + \Delta t) = Y(t) + r\Delta t Y(t)$$

(we have $\Delta t > 0$, and $\Delta t = 1/4$ in the above example). This is equivalent to the expression

$$\frac{Y(t + \Delta t) - Y(t)}{\Delta t} = rY(t).$$

Thus, **the average change** in Y over the time interval $[t, t + \Delta t]$ is equal to $rY(t)$. We can rewrite the above expression as

$$\frac{\dfrac{Y(t + \Delta t) - Y(t)}{\Delta t}}{Y(t)} = r.$$

We will refer to the ratio

$$\frac{\dfrac{Y(t + \Delta t) - Y(t)}{\Delta t}}{Y(t)}$$

as **the average rate of change of Y on $[t, t + \Delta t]$ relative to $Y(t)$. This quantity is the the interest rate r.**

The equation

$$\frac{Y(t + \Delta t) - Y(t)}{\Delta t} = rY(t)$$

is referred to as a **difference equation**. We can express the solution of such a difference equation in a useful form:

Proposition 3 Assume that r and Δt are constants and $\Delta t > 0$. If

$$\frac{Y(t + \Delta t) - Y(t)}{\Delta t} = rY(t),$$

then

$$Y(k\Delta t) = (1 + r\Delta t)^k Y(0), \quad k = 1, 2, 3, \ldots.$$

Proof
Since

$$\frac{Y(t + \Delta t) - Y(t)}{\Delta t} = rY(t),$$

we have

$$Y(t + \Delta t) - Y(t) = r\Delta t Y(t),$$

so that

$$Y(t + \Delta t) = Y(t) + r\Delta t Y(t) = (1 + r\Delta t)Y(t).$$

Therefore,

$$Y\left(\Delta t\right) = \left(1 + r\Delta t\right) Y\left(0\right),$$

$$Y\left(2\Delta t\right) = \left(1 + r\Delta t\right) Y\left(\Delta t\right) = \left(1 + r\Delta t\right)^2 Y\left(0\right),$$

$$Y\left(3\Delta t\right) = \left(1 + r\Delta t\right) Y\left(2\Delta t\right) = \left(1 + r\Delta t\right)^3 Y\left(0\right),$$

$$\vdots$$

$$Y\left(k\Delta t\right) = \left(1 + r\Delta t\right)^k Y\left(0\right).$$

∎

Remark 2 With reference to Proposition 3 if we set $t = k\Delta t$, so that $k = t/\Delta t$,

$$Y\left(k\Delta t\right) = \left(1 + r\Delta t\right)^k Y\left(0\right) \Rightarrow Y\left(t\right) = \left(1 + r\Delta t\right)^{t/\Delta t} Y\left(0\right).$$

The expression

$$Y\left(t\right) = \left(1 + r\Delta t\right)^{t/\Delta t} Y\left(0\right)$$

defines a function of t and may be used to calculate $Y\left(t\right)$ even if t is not an integer multiple of **"the time step"** Δt. ◊

Example 9 Assume that the annual interest rate is 5%, compounded quarterly. Determine your balance after 10, 20, 30 and 40 years, if your initial deposit is \$1000.

Solution

With the notation of Proposition 3, $r = 0.05$, $\Delta t = 1/4$, and $Y_0 = \$1000$. Therefore

$$Y(k\Delta t) = \left(1 + (0.05)\left(\frac{1}{4}\right)\right)^k 1000.$$

If we set $t = k\Delta t = k/4$, then $k = 4t$. Thus,

$$Y\left(t\right) = \left(1 + \frac{5}{400}\right)^{4t} 1000.$$

Therefore,

$$Y(10) = \left(1 + \frac{5}{400}\right)^{40} 1000 \cong 1643.62,$$

$$Y(20) = \left(1 + \frac{5}{400}\right)^{80} 1000 \cong 2701.48,$$

$$Y(30) = \left(1 + \frac{5}{400}\right)^{120} 1000 \cong 4440.21,$$

$$Y(40) = \left(1 + \frac{5}{400}\right)^{160} 1000 \cong 7298.02.$$

Note the rapid growth of the money in your account. Figure 5 shows the points $(t, Y\left(t\right))$ for $t = 0, 10, 20, \ldots, 60, 80$. ($Y$ in \$ 1000). The growth of the money in the account on a longer time horizon is even more striking.

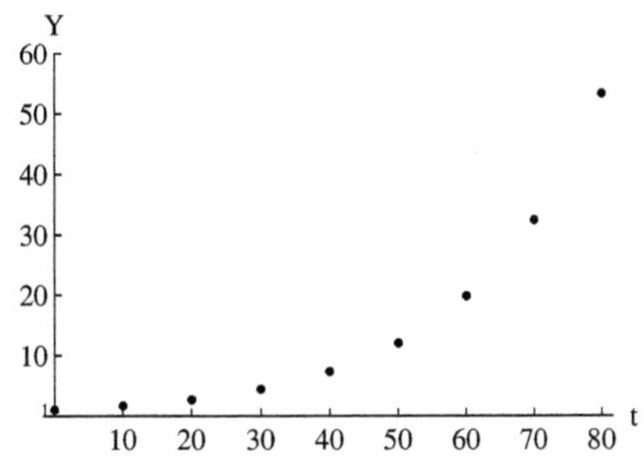

Figure 5: The growth of the money in an account

If we set

$$Y\left(t\right) = \left(1 + \frac{5}{400}\right)^{4t} 1000,$$

we can graph $Y\left(t\right)$ as a continuous function on the interval $[0, 80]$. Figure 6 shows the graph of $Y\left(t\right)$ (in \$ 1000) and the points of Figure 5. $\square$

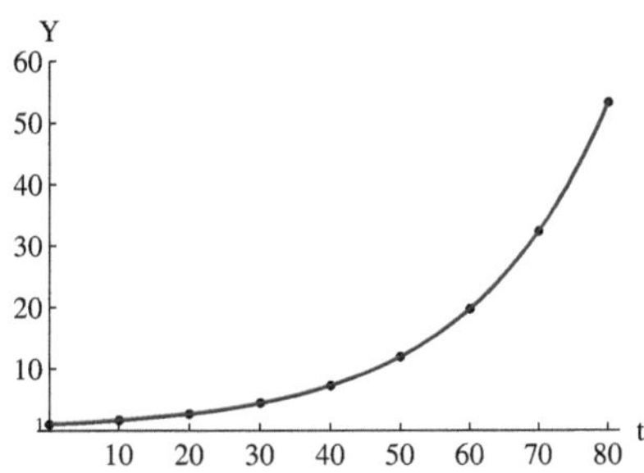

Figure 6: The growth of the money in an account visualized as a continuous curve

Interest may be compounded continuously as well:

Definition 3 Let $y\left(t\right)$ denote the amount of money (say, in dollars) in an account at time t (in years), and assume that there are no deposits or withdrawals after an initial deposit. We say that **the interest rate is r per year** and **the interest is compounded continuously** if the relative rate of growth of y on each the time interval $[t, t + \Delta t]$ is r.

Thus,

$$\frac{y'\left(t\right)}{y\left(t\right)} = r,$$

so that

$$y'\left(t\right) = ry\left(t\right)$$

if the interest rate is r and interest is compounded continuously. By Corollary ,

$$y\left(t\right) = y\left(0\right) e^{rt}.$$

Example 10 Assume that the annual interest rate is 5%, and the initial deposit is \$1000. Determine $y\left(t\right)$, the amount of money in the account at time t (in years) if the interest is compounded continuously.

Solution

We have
$$y(t) = 1000e^{0.05t}.$$

Figure 7 shows the graph of $y(t)$ (in \$ 1000) on the time interval $[0, 80]$. $\square$

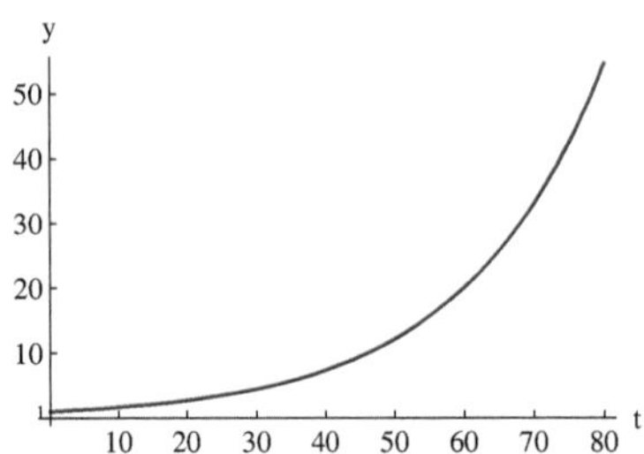

Figure 7: Growth of the money when interest is compounded continuously

Since
$$\frac{y(t + \Delta t) - y(t)}{\Delta t} \cong \frac{dy}{dt},$$

it is reasonable to expect that the solution of the difference equation
$$\frac{Y(t + \Delta t) - Y(t)}{\Delta t} = rY(t),$$

approximates the solution of the initial-value problem
$$\frac{dy}{dt} = ry(t), \ y(0) = y_0$$

if Δt is small and $Y(0) = y_0$. We will denote the solution of the difference equation as $Y_{\Delta t}$ to indicate the dependence on Δt.

Proposition 4 If
$$\frac{Y_{\Delta t}(t + \Delta t) - Y_{\Delta t}(t)}{\Delta t} = rY_{\Delta t}(t), \ Y_{\Delta t}(0) = y_0,$$

and
$$\frac{dy}{dt} = ry(t), \ y(0) = y_0,$$

then
$$\lim_{\Delta t \to 0} Y_{\Delta t}(t) = y(t).$$

Proof

We have
$$y(t) = e^{rt}y_0,$$

By Proposition 3
$$Y_{\Delta t}(k\Delta t) = (1 + r\Delta t)^k y_0.$$

If we set $t = k\Delta t$, so that $k = t/\Delta t$, we have
$$Y_{\Delta t}(t) = (1 + r\Delta t)^{t/\Delta t} y_0$$

(Remark 2). Therefore,

$$Y_{\Delta t}(t) = \left((1 + r\Delta t)^{1/\Delta t} \right)^t y_0.$$

By Proposition 3 of Section 4.5,

$$\lim_{\Delta t \to 0} (1 + r\Delta t)^{1/\Delta t} = e^r.$$

Therefore,

$$\lim_{\Delta t \to 0} Y_{\Delta t}(t) = \lim_{\Delta t \to 0} \left((1 + r\Delta t)^{1/\Delta t} \right)^t y_0 = (e^r)^t y_0 = e^{rt} y_0 = y(t).$$

■

In particular, if the annual interest rate is r, the result of compounding at intervals of length Δt approaches the result of continuous compounding at the same rate r as $\Delta t \to 0$.

Example 11 Assume that the annual interest rate is 5% and the initial deposit is $1000. Calculate $Y_{\Delta t}(10)$ for $\Delta t = 1/4, 1/6, 1/12, 1/32, 1/48$, and compare with $y(10)$.

Solution

As in examples 7 and 8,

$$Y_{\Delta t}(t) = 1000 \left(1 + 0.05\Delta t \right)^{t/\Delta t},$$

and

$$y(t) = 1000 e^{0.05t}.$$

Therefore,

$$Y_{\Delta t}(10) = 1000 \left(1 + 0.05\Delta t \right)^{10/\Delta t}$$

and

$$y(10) = 1000 e^{0.5} \cong 1648.72$$

Table 3 shows $Y_{\Delta t}(10)$ and $|Y_{\Delta t}(10) - y(10)|$ for $\Delta t = 1/4, 1/6, 1/12, 1/24$ and $1/48$. We see that $|Y_{\Delta t}(10) - y(10)|$ decreases as Δt gets smaller. This is consistent with the fact that $\lim_{\Delta t \to 0} Y_{\Delta t}(10) = y(10)$.□

Δt	$Y_{\Delta t}(10)$	$|Y_{\Delta t}(10) - y(10)|$
1/4	1643.62	5.11
1/6	1645.31	3. 41
1/12	1647.01	1. 71
1/24	1647.86	0.86
1/48	1648. 29	0.43

Table 3

The difference equation

$$\frac{Y_{\Delta t}(t + \Delta t) - Y_{\Delta t}(t)}{\Delta t} = rY_{\Delta t}(t)$$

is related to the differential equation

$$\frac{dy}{dt} = ry(t),$$

where r is a constant, without reference to issues of finance. Mathematically, we have replaced the derivative by a difference quotient. The difference equation is referred to as **the Euler difference scheme** for the approximation of the solutions of the differential equation $y'(t) = ry(t)$.

Example 12 Consider the initial-value problem

$$y'(t) = -\frac{1}{2}y(t), \; y(0) = 10,$$

Assume that

$$\frac{Y_{\Delta t}(t + \Delta t) - Y_{\Delta t}(t)}{\Delta t} = -\frac{1}{2}Y_{\Delta t}(t),$$

and $Y_{\Delta t}(0) = 10$. By Proposition 3, $\lim_{\Delta t \to 0} Y_{\Delta t}(t) = y(t)$ for each t. Plot the graphs of $y(t)$ and $Y_{\Delta t}(t)$ for $\Delta t = 10^{-1}$ with the help of your graphing utility. Does the picture support the fact that $Y_{\Delta t}(t)$ approximates $y(t)$ if Δt is small?

Solution

a) By Corollary ,

$$y(t) = 10e^{-t/2},$$

and by Remark 2 ($r = -1/2$ and $\Delta t = 1/10$),

$$Y_{1/10}(t) = 10\left(1 - \frac{1}{2}\left(\frac{1}{10}\right)\right)^{10t}.$$

In Figure 8, the solid curve is the graph of $y(t) = 10e^{-t/2}$ and the dashed curve is the graph of $Y_{1/10}(t)$. We can hardly distinguish between the graphs. $\square$

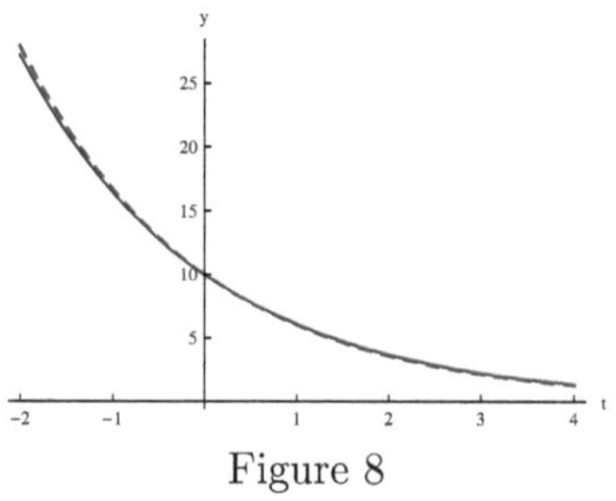

Figure 8

Problems

In problems 1 and 2,
a) Determine the general solution $y(t)$ of the given differential equation.
b) Determine the solutions of the given differential equation corresponding to the initial conditions $y(t_0) = y_0$ where $y_0 = \pm 10$ and ± 20, for the given values of t_0,
c) [C] Make use of your graphing utility to plot the solutions that you calculated in part c).

1.

$$y'(t) = \frac{1}{4}y(t), \; t_0 = 2.$$

2.

$$y'(t) = -\frac{1}{4}y(t), \; t_0 = 1.$$

In problems 3-6,
a) Determine the general solution of the differential equation $y'(t) = ky(t)$.
b) Determine the solution of the differential equation $y'(t) = ky(t)$ that satisfies the given initial condition.

3. $\dfrac{dy}{dt} = -\dfrac{1}{5}y\left(t\right)$, $y\left(1\right) = 10$
5. $\dfrac{dy}{dt} = \dfrac{1}{10}y\left(t\right)$, $y\left(0\right) = 2$

4. $\dfrac{dy}{dt} = \dfrac{1}{5}y\left(t\right)$, $y\left(2\right) = 40$
6. $\dfrac{dy}{dt} = -\dfrac{1}{10}y\left(t\right)$, $y\left(1\right) = 20$

7. Assume that the population of a country grows at the constant rate of 2.5% per year, and that the present population is 100 million. Let t be time in years, and set $t = 0$ for the present. Denote the population at time t by $y\left(t\right)$.
a) Determine $y\left(t\right)$.
b) [C] Determine the population of the country 20, 40 and 80 years from now.
c) [C] Plot the graph of $y\left(t\right)$ on the interval $[0, 80]$.

8. Assume that the population of a country grows at the constant rate of 1.5% per year, and that the present population is 50 million.
a) Let y represent the population after t years. Determine $y\left(t\right)$ (at present $t = 0$).
b) [C] How many years will it take for the population to reach 100 million?

9. Assume that the decay rate of a radioactive material is 0.03 per year, and that we have a sample of 40 grams. Let $y(t)$ denote the amount (in grams) that is left at time t (in years), with $t = 0$ corresponding to the present.
a) Determine $y(t)$.
b) [C] Calculate $y(50)$, $y\left(100\right)$ and $y(150)$
c) [C] Plot the graph of $y(t)$ with the help of your graphing utility.

10. Prove Proposition 2: If the decay rate of a radioactive material is k, the half-life of the material is $T = \dfrac{\ln\left(2\right)}{k}$ and the amount of the material at time t is

$$y\left(t\right) = \frac{y\left(0\right)}{2^{t/T}}.$$

11. Assume that the half life of a radioactive material is 50 years, and the initial amount if the material is 100 grams. Let $y\left(t\right)$ denote the amount of the material (in grams) after t years.
a) Determine $y\left(t\right)$. Express $y\left(t\right)$ in a form that does not involve e explicitly.
b) [C] Calculate $y\left(100\right)$, $y(200)$ and $y\left(400\right)$.

12. Let $y\left(t\right)$ denote the number of bacteria in a culture at time t (in hours). Assume that the number of bacteria in the culture doubles every 6 hours, and that $y\left(0\right) = 1000$.
a) Determine $y\left(t\right)$. Express $y\left(t\right)$ in a simplified form that does not involve e explicitly.
b) [C] Calculate $y\left(12\right)$, $y18)$ and $y\left(24\right)$.

13. Suppose a bacteria culture with constant relative growth rate increases from 1,000 to 8,000 in 12 hours.
a) Determine the doubling time of the culture and $y\left(t\right)$, the number of bacteria in the culture at time t (hours). Express $y\left(t\right)$ in a form that does not involve e explicitly.
b) [C] Calculate $y\left(24\right)$, $y\left(36\right)$, $y\left(48\right)$, $y\left(60\right)$

14 [C] (Carbon Dating). The radioactive substance Carbon-14 has a half-life of 5720 years. If a fossil is found to have 10% of the Carbon-14 that is present in a similar living organism, determine the age of the fossil.

15. [C] An initial sum of $4000 is invested at an annual interest rate of 4%.
a) Compute the balance $Y\left(t\right)$ for $t = 10, 20, 30, 40$ years if interest is compounded semiannually.
b) Determine the function y such that $y\left(t\right)$ is the value of the investment after t years if interest is compounded continuously. Evaluate $y\left(t\right)$ for $t = 10, 20, 30, 40$. Compare with the corresponding values of Y.
c) Plot the points $(t, Y\left(t\right))$ for $t = 10, 20, 30, 40$ and the function y on the same screen.

16. Consider the initial-value problem

$$\frac{dy}{dt} = \frac{1}{3} y\left(t\right), \; y_0 = 2$$

and the solution of the difference equations

$$\frac{Y_{\Delta t}\left(t + \Delta t\right) - Y_{\Delta t}\left(\Delta t\right)}{\Delta t} = \frac{1}{3} Y_{\Delta t}\left(t\right),$$

$$Y_{\Delta t}\left(0\right) = 2.$$

a) Determine $y\left(t\right)$ and $Y\left(t\right)$.

b) [C] Plot the graphs of $y\left(t\right)$ and $Y_{\Delta t}$ for $\Delta t = 10^{-1}$. Does the picture support the fact that $Y_{\Delta t}$ approximates $y\left(t\right)$ if Δt is small?

4.7   Hyperbolic and Inverse Hyperbolic Functions

In this section we will introduce special functions which are combinations of e^x and e^{-x}, and the inverses of these functions. You will see applications of these functions in later chapters.

Hyperbolic Functions

The function **hyperbolic sine** is abbreviated as **sinh** (read as "sinch"):

Definition 1

$$\sinh(x) = \frac{e^x - e^{-x}}{2}$$

for each $x \in \mathbb{R}$.

Figure 1 shows the graph of $y = \sinh\left(x\right)$.

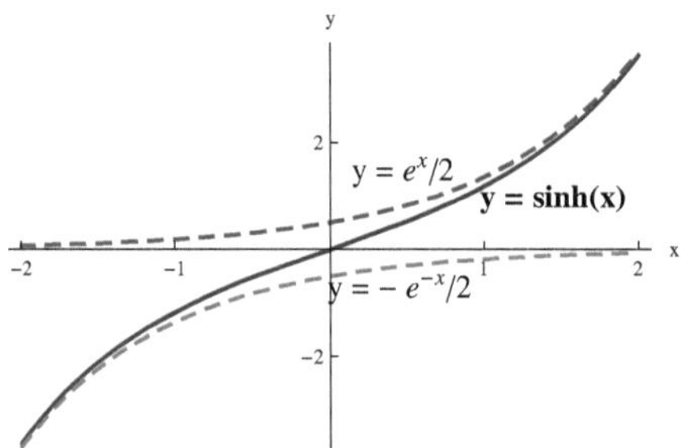

Figure 1: Hyperbolic sine

The function **hyperbolic cosine** is abbreviated as **cosh** (rhymes with "posh"):

Definition 2

$$\cosh(x) = \frac{e^x + e^{-x}}{2}$$

for each $x \in \mathbb{R}$.

Figure 2 shows the graph $y = \cosh(x)$..

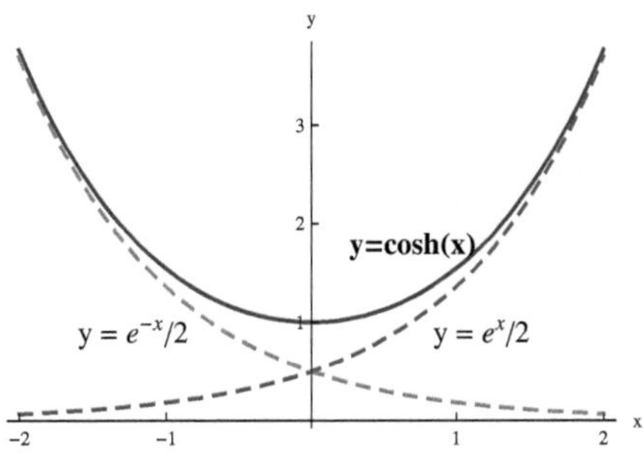

Figure 2: Hyperbolic cosine

Thus, hyperbolic sine and hyperbolic cosine are linear combinations of e^x and e^{-x}. The names of these new functions involve the familiar words "sine" and "cosine", since certain identities and differentiation formulas that involve sinh and cosh resemble identities and formulas that involve sine and cosine. The resemblance is superficial, though. To begin with, sine and cosine are periodic functions whose values lie between -1 and $+1$, whereas sinh and cosh are not periodic, and $|\sinh(x)|$ and $\cosh(x)$ grow exponentially as $|x| \to \infty$. Figure 1 indicates that $\sinh(x) \cong e^x/2$ if x is large, and that $\sinh(x) \cong -e^{-x}/2$ if $x < 0$ and $|x|$ is large. Figure 2 indicates that $\cosh(x) \cong e^x/2$ if x is large, and that $\cosh(x) \cong e^{-x}/2$ if $x < 0$ and $|x|$ is large.The following limits are precise versions of our observations:

We have

$$\lim_{x \to +\infty} \frac{\sinh(x)}{\frac{1}{2}e^x} = 1, \quad \lim_{x \to -\infty} \frac{\sinh(x)}{-\frac{1}{2}e^{-x}} = 1,$$

and

$$\lim_{x \to +\infty} \frac{\cosh(x)}{\frac{1}{2}e^x} = 1, \quad \lim_{x \to -\infty} \frac{\cosh(x)}{\frac{1}{2}e^{-x}} = 1,$$

Proof

We will provide the details for the limits that involve hyperbolic sine. The limits involving hyperbolic cosine are established in a similar manner.

We have

$$\frac{\sinh(x)}{\frac{1}{2}e^x} = \frac{\frac{1}{2}\left(e^x - e^{-x}\right)}{\frac{1}{2}e^x} = \frac{e^x - e^{-x}}{e^x}.$$

We factor e^x that is the dominant term if x is large $(\lim_{x \to +\infty} e^{-x} = 0)$:

$$\frac{\sinh(x)}{\frac{1}{2}e^x} = \frac{e^x - e^{-x}}{e^x} = \frac{e^x\left(1 - e^{-2x}\right)}{e^x} = 1 - e^{-2x}.$$

Therefore,

$$\lim_{x \to +\infty} \frac{\sinh(x)}{\frac{1}{2}e^x} = \lim_{x \to +\infty}\left(1 - e^{-2x}\right) = 1.$$

We have

$$\frac{\sinh(x)}{-\frac{1}{2}e^{-x}} = \frac{e^{-x} - e^x}{e^{-x}}$$

If $x < 0$ and $|x|$ is large, the dominant term is e^{-x} ($\lim_{x \to -\infty} e^x = 0$). We factor e^{-x}:

$$\frac{e^{-x} - e^x}{e^{-x}} = \frac{e^{-x}\left(1 - e^{2x}\right)}{e^{-x}} = 1 - e^{2x}.$$

Therefore,

$$\lim_{x \to -\infty} \frac{\sinh(x)}{-\frac{1}{2}e^{-x}} = \lim_{x \to -\infty} \left(1 - e^{2x}\right) = 1.$$

∎

The following proposition summarizes the basic properties of hyperbolic sine and hyperbolic cosine:

Proposition 1

1. Hyperbolic sine is an odd function, and hyperbolic cosine is an even function.

2. The derivative of $\sinh(x)$ is $\cosh(x)$ and the derivative of $\cosh(x)$ is $\sinh(x)$:

$$\frac{d}{dx}\sinh(x) = \cosh(x) \text{ and } \frac{d}{dx}\cosh(x) = \sinh(x)$$

for each $x \in \mathbb{R}$.

3. Hyperbolic sine is an increasing function on the entire number line.

4. Hyperbolic cosine is decreasing on $(-\infty, 0]$ and increasing on $[0, +\infty)$. Thus, the absolute minimum of hyperbolic cosine on the entire number line is $\cosh(0) = 1$.

Note that the differentiation formulas for sinh and cosh are analogous the differentiation formulas for sine and cosine. Also note the difference: The derivative of cosh does not involve the negative sign.

The Proof of Proposition 1

1. We have

$$\sinh(-x) = \frac{e^{-x} - e^{-(-x)}}{2} = \frac{e^{-x} - e^x}{2} = -\frac{e^x - e^{-x}}{2} = -\sinh(x)$$

for each $x \in \mathbb{R}$. Therefore, sinh is an odd function.

As for hyperbolic cosine,

$$\cosh(-x) = \frac{e^{-x} + e^{-(-x)}}{2} = \frac{e^{-x} + e^x}{2} = \frac{e^x + e^{-x}}{2} = \cosh(x)$$

for each $x \in \mathbb{R}$. Therefore, cosh is an even function.

2. The formulas for the derivatives follow from the fact that

$$\frac{d}{dx}e^x = e^x,$$

with the help of the linearity of differentiation and the chain rule:

$$\frac{d}{dx}\sinh(x) = \frac{d}{dx}\left(\frac{e^x - e^{-x}}{2}\right) = \frac{1}{2}\left(e^x - e^{-x}(-1)\right) = \frac{1}{2}\left(e^x + e^{-x}\right) = \cosh(x).$$

Similarly,

$$\frac{d}{dx}\cosh(x) = \frac{d}{dx}\left(\frac{e^x + e^{-x}}{2}\right) = \frac{1}{2}\left(e^x + e^{-x}(-1)\right) = \frac{1}{2}\left(e^{-x} - e^{-x}\right) = \sinh(x).$$

3. We have

$$\frac{d}{dx}\sinh(x) = \cosh(x) = \frac{e^x + e^{-x}}{2} > 0$$

for each $x \in \mathbb{R}$, since the values of the natural exponential function are positive. By the derivative test for monotonicity, sinh is increasing on R. Note that

$$\sinh(0) = \frac{e^0 - e^0}{2} = \frac{1-1}{2} = 0.$$

Since sinh is increasing, $\sinh(x) < 0$ if $x < 0$ and $\sinh(x) > 0$ if $x > 0$.

4. Since

$$\frac{d}{dx}\cosh(x) = \sinh(x)$$

and $\sinh(x) < 0$ if $x < 0$, cosh is decreasing on $(-\infty, 0]$ by the derivative test for monotonicity. Since

$$\frac{d}{dx}\cosh(x) = \sinh(x) < 0$$

if $x > 0$, cosh is increasing on $[0, +\infty)$, . Therefore, hyperbolic cosine attains its absolute minimum on the number line at 0. We have

$$\cosh(x) \geq \cosh(0) = \frac{e^0 + e^0}{2} = \frac{1+1}{2} = 1$$

for each $x \in \mathbb{R}$. ∎

Example 1 Under certain modeling assumptions, it can be shown that the shape of **a hanging cable** is of the form

$$y = a\left(\cosh\left(\frac{x}{a}\right) - 1\right) + h,$$

where a is a positive constant that depends on the weight of the cable per unit length and the tension of the cable, and h is the height of the lowest point of the cable. Figure 3 illustrates the shape of a hanging cable. □

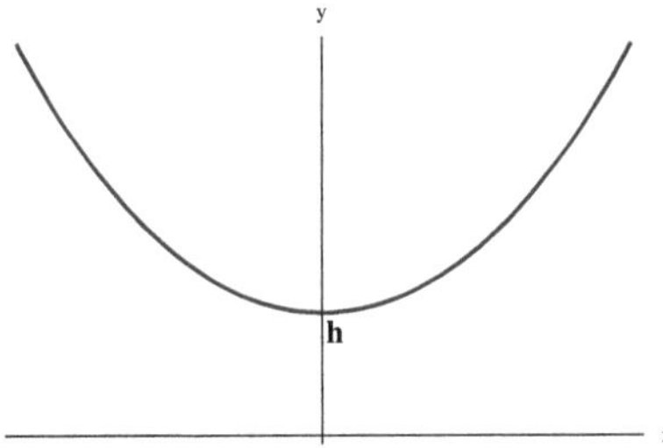

Figure 3: A hanging cable

The trigonometric functions sine and cosine are "circular functions": For any real number x, the point $(\cos(x), \sin(x))$ is on the unit circle, i.e.,

$$\cos^2(x) + \sin^2(x) = 1.$$

There is a similar identity that involves hyperbolic sine and hyperbolic cosine:

Proposition 2 For each real number x

$$\cosh^2(x) - \sinh^2(x) = 1.$$

Proof

The proof follows from the definitions of cosh and sinh in terms of the natural exponential function:

$$\cosh^2(x) - \sinh^2(x) = \left(\frac{e^x + e^{-x}}{2}\right)^2 - \left(\frac{e^x - e^{-x}}{2}\right)^2$$

$$= \frac{1}{4}\left((e^x)^2 + 2(e^x)(e^{-x}) + (e^{-x})^2\right) - \frac{1}{4}\left((e^x)^2 - 2(e^x)(e^{-x}) + (e^{-x})^2\right)$$

$$= \frac{1}{4}e^{2x} + \frac{1}{2} + \frac{1}{4}e^{-2x} - \frac{1}{4}e^{2x} + \frac{1}{2} - \frac{1}{4}e^{-2x} = 1.$$

∎

The above identity leads to an explanation of the term "hyperbolic" that is attached to the names of the new functions. Let's set $u(x) = \cosh(x)$ and $v(x) = \sinh(x)$, and consider the set of points

$$(u(x), v(x)) = (\cosh(x), \sinh(x))$$

in the uv-plane. Since

$$u^2(x) - v^2(x) = \cosh^2(x) - \sinh^2(x) = 1,$$

for each $x \in \mathbb{R}$, the point $(u(x), v(x))$ is on the hyperbola $u^2 - v^2 = 1$ in the uv-plane. Since $u(x) = \cosh(x) \geq 1$ for any $x \in \mathbb{R}$, such a point is on the right branch of the hyperbola. You can imagine that $(u(x), v(x))$ traces the right branch the hyperbola $u^2 - v^2 = 1$, as indicated by the arrows in Figure 4, as x takes on increasing values.

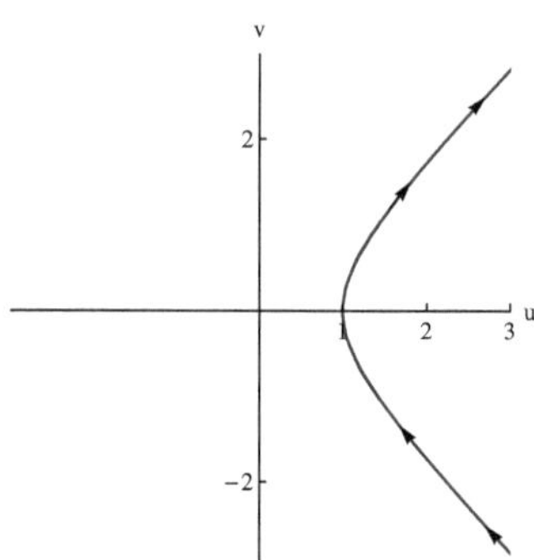

Figure 4: $(\cosh(x), \sinh(x))$ traces part of a hyperbola

You will be asked to confirm the validity of other identities that are analogous to familiar trigonometric identities in the problem set for this section. For example,

$$\sinh(a + b) = \sinh(a)\cosh(b) + \cosh(a)\sinh(b),$$
$$\cosh(a + b) = \cosh(a)\cosh(b) + \sinh(a)\sinh(b).$$

The function **hyperbolic tangent** is the quotient of hyperbolic sine and hyperbolic cosine, and we use the abbreviation **tanh**:

Definition 3

$$\tanh(x) = \frac{\sinh(x)}{\cosh(x)}.$$

We will not suggest that you pronounce tanh in a way that parallels the pronunciation of sinh and cosh. You may read "tanh(x)" as "hyperbolic tangent of x".

Note that **the natural domain of hyperbolic tangent is the entire number line,** since $\cosh(x) \neq 0$ for any $x \in \mathbb{R}$. We can express hyperbolic tangent in terms of the natural exponential function:

$$\tanh(x) = \frac{\sinh(x)}{\cosh(x)} = \frac{e^x - e^{-x}}{e^x + e^{-x}} = \frac{e^x - \dfrac{1}{e^x}}{e^x + \dfrac{1}{e^x}} = \frac{e^{2x} - 1}{e^{2x} + 1}.$$

Figure 5 shows the graph of $y = \tanh(x)$.

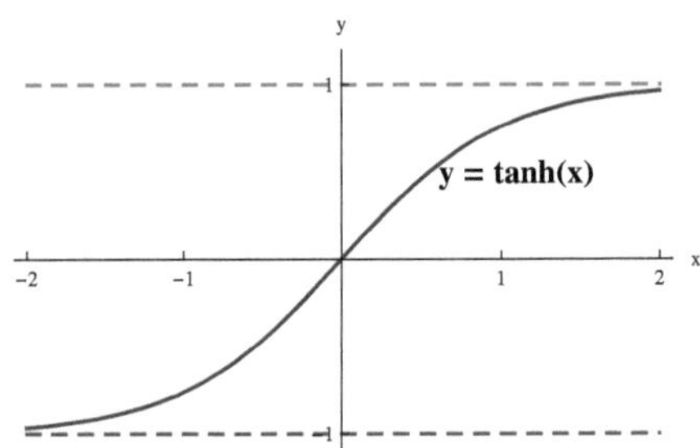

Figure 5: Hyperbolic tangent

The following proposition summarizes the basic properties of hyperbolic tangent:

Proposition 3

1.

$$\lim_{x \to \infty} \tanh(x) = 1 \text{ and } \lim_{x \to -\infty} \tanh(x) = -1.$$

2. Hyperbolic tangent is an odd function.

3.

$$\frac{d}{dx}\tanh(x) = \frac{1}{\cosh^2(x)}$$

for each $x \in \mathbb{R}$.

4. Hyperbolic tangent is an increasing function on the entire number line.

Proof

1. If x is large, the dominant term in the expression

$$\tanh(x) = \frac{e^x - e^{-x}}{e^x + e^{-x}}$$

is e^x, since $\lim_{x \to +\infty} e^{-x} = 0$. Thus,

$$\lim_{x \to +\infty} \tanh(x) = \lim_{x \to +\infty} \frac{e^x - e^{-x}}{e^x + e^{-x}} = \lim_{x \to +\infty} \frac{e^x\left(1 - e^{-2x}\right)}{e^x\left(1 + e^{-2x}\right)} = \lim_{x \to +\infty} \frac{1 - e^{-2x}}{1 + e^{-2x}} = 1.$$

If $x < 0$ and $|x|$ is large, the dominant term in the expression

$$\tanh(x) = \frac{e^x - e^{-x}}{e^x + e^{-x}}$$

is e^{-x}, since $\lim_{x \to -\infty} e^x = 0$. Thus,

$$\lim_{x \to -\infty} \tanh(x) = \lim_{x \to -\infty} \frac{e^x - e^{-x}}{e^x + e^{-x}} = \lim_{x \to -\infty} \frac{e^{-x}(e^{2x} - 1)}{e^{-x}(e^{2x} + 1)} = \lim_{x \to -\infty} \frac{e^{2x} - 1}{e^{2x} + 1} = -1.$$

2. Since sinh is an odd function and cosh is an even function, we have

$$\tanh(-x) = \frac{\sinh(-x)}{\cosh(-x)} = \frac{-\sinh(x)}{\cosh(x)} = -\tanh(x),$$

Therefore, hyperbolic tangent is an odd function.

3. We apply the quotient rule for differentiation, and make use of the identity

$$\cosh^2(x) - \sinh^2(x) = 1.$$

Thus,

$$\frac{d}{dx}\tanh(x) = \frac{d}{dx}\left(\frac{\sinh(x)}{\cosh(x)}\right) = \frac{\cosh(x)\cosh(x) - \sinh(x)\sinh(x)}{\cosh^2(x)}$$

$$= \frac{\cosh^2(x) - \sinh^2(x)}{\cosh^2(x)} = \frac{1}{\cosh^2(x)}.$$

The formula is valid for each $x \in \mathbb{R}$, since $\cosh(x) \neq 0$.

4. Since

$$\frac{d}{dx}\tanh(x) = \frac{1}{\cosh^2(x)} > 0$$

for each $x \in \mathbb{R}$, hyperbolic tangent is increasing on $(-\infty, +\infty)$, by the derivative test for monotonicity. We have

$$\tanh(0) = \frac{\sinh(0)}{\cosh(0)} = \frac{0}{1} = 0,$$

and

$$\lim_{x \to +\infty} \tanh(x) = 1 \text{ and } \lim_{x \to -\infty} \tanh(x) = -1.$$

Therefore, $-1 < \tanh(x) < \tanh(0) = 0$ if $x < 0$, and $0 = \tanh(0) < \tanh(x) < 1$ if $x > 0$. ∎

Example 2 Assume that the force on a falling object due to air resistance is of the form $\delta v^2(t)$, where δ is a positive constant and $v(t)$ is the velocity of the object at time t. We will see in Section 8.4 that

$$v(t) = \sqrt{\frac{mg}{\delta}} \tanh\left(\sqrt{\frac{mg}{\delta}}t\right),$$

where g is the (constant) gravitational acceleration. We have

$$\lim_{t \to +\infty} v(t) = \lim_{t \to +\infty} \sqrt{\frac{mg}{\delta}} \tanh\left(\sqrt{\frac{mg}{\delta}}t\right) = \sqrt{\frac{mg}{\delta}} \lim_{t \to +\infty} \tanh\left(\sqrt{\frac{mg}{\delta}}t\right) = \sqrt{\frac{mg}{\delta}}(1) = \sqrt{\frac{mg}{\delta}}.$$

The limit $\sqrt{(mg)/\delta}$ of $v(t)$ as $t \to \infty$ is referred to as **the terminal velocity** of the object. The graph of the velocity function is as in Figure 6. □

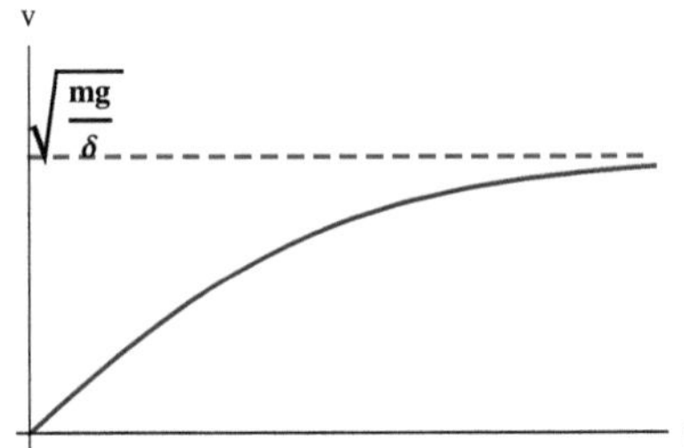

Figure 6: The velocity v of a falling object if air resistance is proportional to v^2

In summary, we have the following expressions for the derivatives of hyperbolic functions:

$$\frac{d}{dx}\sinh(x) = \cosh(x),$$

$$\frac{d}{dx}\cosh(x) = \sinh(x),$$

$$\frac{d}{dx}\tanh(x) = \frac{1}{\cosh^2(x)}.$$

The hyperbolic counterpart of the trigonometric function secant is **hyperbolic secant**, and is abbreviated as **sech**:Exponential Growth and Decay

Definition 4

$$\mathbf{sech}(x) = \frac{1}{\cosh(x)}$$

for each $x \in \mathbb{R}$.

As in the case of hyperbolic tangent, we will not suggest that you pronounce sech in a way that parallels the pronunciation of sinh and cosh. You may read "sech(x)" as "hyperbolic secant of x".

Figure 7 displays the graph of $y = \mathrm{sech}(x)$.

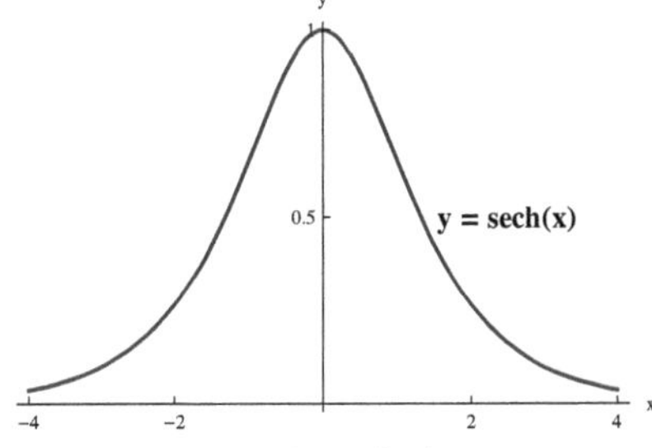

Figure 7: Hyperbolic secant

Example 3

a) Show that hyperbolic secant is an even function.
b) Show that

$$\lim_{x \to \pm\infty} \operatorname{sech}(x) = 0.$$

c) Show that the absolute maximum of hyperbolic secant is $\operatorname{sech}(0) = 1$.

Solution

a) Hyperbolic secant is an even function since cosh is even:

$$\operatorname{sech}(-x) = \frac{1}{\cosh(-x)} = \frac{1}{\cosh(x)} = \operatorname{sech}(x).$$

b)

$$\lim_{x \to \pm\infty} \operatorname{sech}(x) = \lim_{x \to \pm\infty} \frac{1}{\cosh(x)} = 0,$$

since $\lim_{x \to \pm\infty} \cosh(x) = +\infty$.

c) Since

$$\operatorname{sech}(x) = \frac{1}{\cosh(x)},$$

and the absolute minimum of hyperbolic cosine is $\cosh(0) = 1$, the absolute maximum of hyperbolic secant is $\operatorname{sech}(0) = 1$. $\square$

Inverse Hyperbolic Functions

The inverse of hyperbolic sine and the inverse of the restriction of hyperbolic cosine to $[0, +\infty)$ will be useful in the following chapters. We will also discuss the inverse of hyperbolic tangent. All these functions can be expressed in terms of the natural logarithm.

Let's begin with the inverse of hyperbolic sine. Hyperbolic sine is an increasing function on the entire number line. As a linear combination of the continuous functions e^x and e^{-x}, sinh is continuous on R. Since

$$\lim_{x \to -\infty} \sinh(x) = -\infty \text{ and } \lim_{x \to +\infty} \sinh(x) = +\infty,$$

the range of sinh is also $\mathbb{R}$. Therefore, sinh has an inverse whose domain and range are both equal to the entire set of real numbers. The inverse of sinh can be abbreviated as $\sinh^{-1}$ or arcsinh. We will favor arcsinh:

Definition 5

$$y = \textbf{arcsinh}(x) \Leftrightarrow x = \textbf{sinh}(y),$$

where x and y are arbitrary real numbers.

Figure 8 illustrates the relationship between $x = \sinh(y)$ and $y = \operatorname{arcsinh}(x)$ graphically, and Figure 9 displays the graph of $y = \operatorname{arcsinh}(x)$.

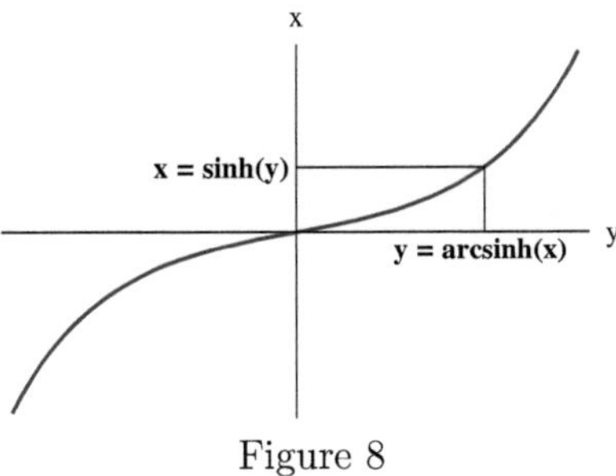

Figure 8

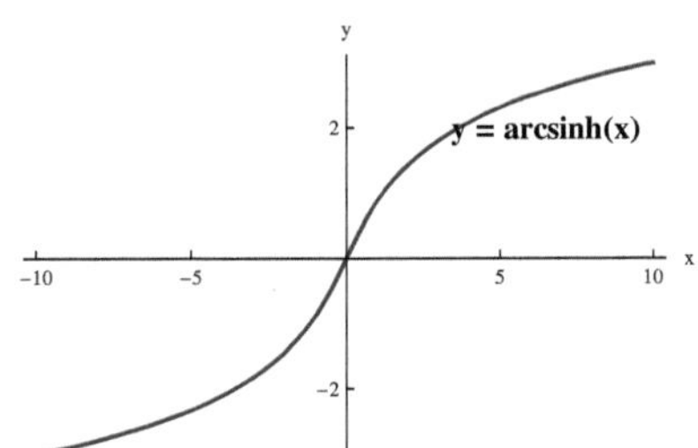

Figure 9: The inverse of hyperbolic sine

We can express the inverse of hyperbolic sine in terms of the natural logarithm:

Proposition 4 We have

$$\operatorname{arcsinh}(x) = \ln(x + \sqrt{x^2 + 1})$$

for each $x \in \mathbb{R}$.

Proof

Given $x \in \mathbb{R}$, the solution of the equation $x = \sinh(y)$, where y is treated as the unknown, leads to $y = \operatorname{arcsinh}(x)$. We have

$$x = \sinh(y) = \frac{e^y - e^{-y}}{2} = \frac{1}{2}e^y - \frac{1}{2e^y}.$$

Let us set $u = e^y$. Then,

$$x = \frac{1}{2}u - \frac{1}{2u} \Leftrightarrow 2xu = u^2 - 1 \Leftrightarrow u^2 - 2xu - 1 = 0.$$

Let us treat x as a parameter, and solve the last equation for u with the help of the quadratic formula:

$$u = \frac{(2x) \pm \sqrt{(2x)^2 + 4}}{2} = \frac{2x \pm 2\sqrt{x^2 + 1}}{2} = x \pm \sqrt{x^2 + 1}.$$

Since $u = e^y > 0$, we must ignore the negative sign. Thus,

$$u = x + \sqrt{x^2 + 1} \Leftrightarrow e^y = x + \sqrt{x^2 + 1} \Leftrightarrow y = \ln(x + \sqrt{x^2 + 1}).$$

Therefore, $\operatorname{arcsinh}(x) = y = \ln(x + \sqrt{x^2 + 1})$, as claimed. ∎

We will need the expression for the derivative of arcsinh:

Proposition 5

$$\frac{d}{dx}\text{arcsinh}\,(x) = \frac{1}{\sqrt{x^2+1}}$$

for each $x \in \mathbb{R}$.

Proof

We will use the expression of arcsinh in terms of the natural logarithm, and the chain rule:

$$
\begin{aligned}
\frac{d}{dx}\text{arcsinh}\,(x) &= \frac{d}{dx}\ln\left(x + \sqrt{x^2 + 1}\right) \\
&= \left(\frac{d}{du}\ln(u)\Big|_{u=x+\sqrt{x^2+1}}\right)\left(\frac{d}{dx}\left(x + \sqrt{x^2 + 1}\right)\right) \\
&= \frac{1}{x + \sqrt{x^2 + 1}}\left(1 + \frac{1}{2\sqrt{x^2 + 1}}(2x)\right) \\
&= \frac{1}{x + \sqrt{x^2 + 1}}\left(1 + \frac{x}{\sqrt{x^2 + 1}}\right) \\
&= \frac{1}{x + \sqrt{x^2 + 1}}\left(\frac{\sqrt{x^2 + 1} + x}{\sqrt{x^2 + 1}}\right) \\
&= \frac{1}{\sqrt{x^2 + 1}}.
\end{aligned}
$$

∎

Hyperbolic cosine does not have an inverse, since the equation $x = \cosh(y)$ has two distinct solutions if $x > 1$. The graph of hyperbolic cosine fails the horizontal line test, as illustrated in Figure 10.

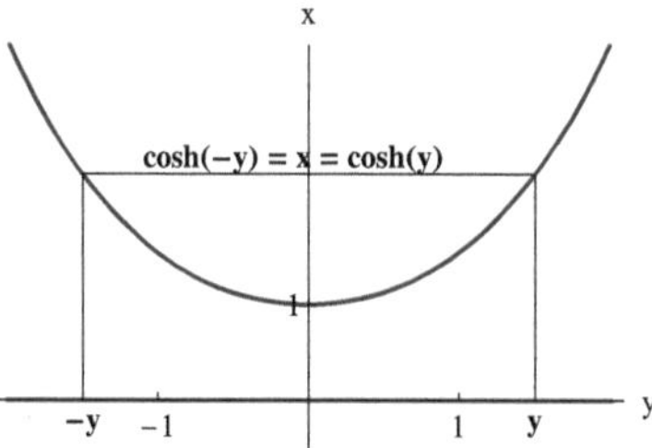

Figure 10: The graph of $x = \cosh(y)$ fails the horizontal line test

On the other hand, if we restrict hyperbolic cosine to the interval $[0, +\infty)$, the new function has an inverse. Indeed, cosh is increasing on $[0, +\infty)$. Since $\cosh(0) = 1$ and $\lim_{y\to+\infty}\cosh(y) = +\infty$, the range of cosh is the interval $[1, +\infty)$. Therefore, the restriction of cosh to $[0, +\infty)$ has an inverse, and the domain of the inverse function is $[1, +\infty)$. We will label the inverse function as arccosh.

Definition 6

$$y = \text{arccosh}(x) \iff x = \cosh(y)$$

where $x \geq 1$ and $y \geq 0$.

Figure 11 illustrates the relationship between $\cosh(y)$ and $\mathrm{arccosh}(x)$ graphically, and Figure 12 displays the graph of $y = \mathrm{arccosh}(x)$.

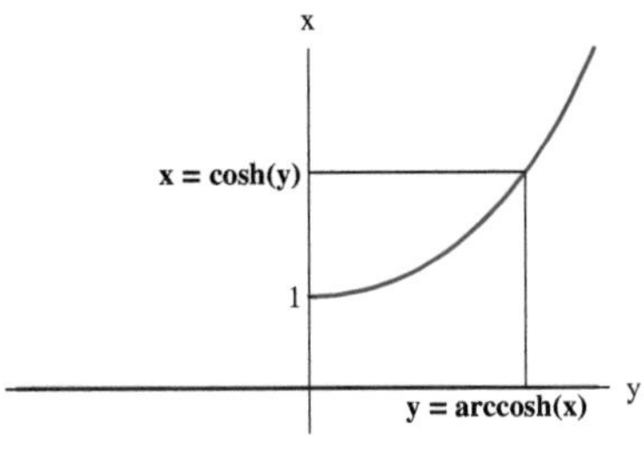

Figure 11

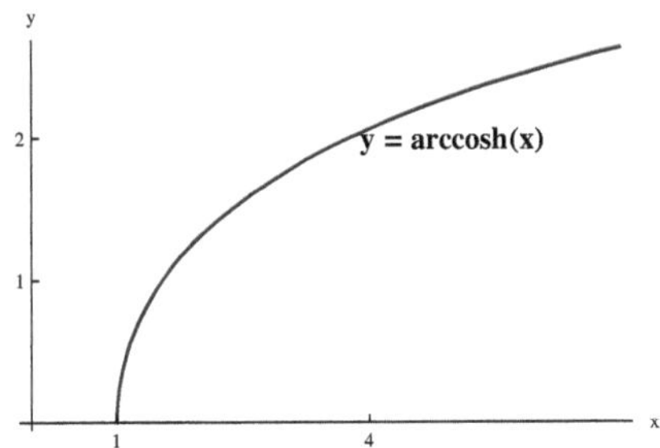

Figure 12: The inverse of hyperbolic cosine

Just as in the case of $\mathrm{arcsinh}(x)$, we can express $\mathrm{arccosh}(x)$ in terms of the natural logarithm:

Proposition 6 **We have**

$$\mathrm{arccosh}\,(x) = \ln(x + \sqrt{x^2 - 1}) \text{ for each } x \geq 1.$$

The proof is similar to the proof of the corresponding fact about arcsinh (Proposition 4), and is left as an exercise.

You can obtain the expression for the derivative of arccosh as in the case of arcsinh:

Proposition 7

$$\frac{d}{dx}\mathrm{arccosh}\,(x) = \frac{1}{\sqrt{x^2 - 1}}$$

if $x > 1$.

Example 4 We have $\mathrm{arccosh}(1) = 0$, since $1 = \cosh(0)$. Therefore, $(1, 0)$ is on the graph of arccosh. Show that the graph has a vertical tangent at $(1, 0)$.

Solution

The function is defined only to the right of 1. We have

$$\frac{d}{dx}\mathrm{arccosh}\,(x) = \frac{1}{\sqrt{x^2 - 1}} = \frac{1}{\sqrt{x + 1}\sqrt{x - 1}}.$$

Since

$$\lim_{x \to 1+} \frac{1}{\sqrt{x+1}} = \frac{1}{\sqrt{2}} > 0 \text{ and } \lim_{x \to 1+} \frac{1}{\sqrt{x-1}} = +\infty,$$

we have

$$\lim_{x \to 1+} \frac{d}{dx}\text{arccosh}\,(x) = \lim_{x \to 1+} \frac{1}{\sqrt{x+1}\sqrt{x-1}} = +\infty.$$

Thus, the graph of arccosh has a vertical tangent at $(1,0)$. $\square$

Hyperbolic tangent is continuous and increasing on $(-\infty, +\infty)$, and

$$\lim_{x \to -\infty} \tanh\,(x) = -1, \quad \lim_{x \to +\infty} \tanh\,(x) = 1.$$

Therefore, the range of hyperbolic tangent is $(-1, 1)$ and its inverse exists. We will denote the inverse of tanh by **arctanh**:

Definition 7

$$\boldsymbol{y = \text{arctanh}(x) \Leftrightarrow x = \tanh(y)}$$

where $-1 < x < 1$ and $y \in \mathbb{R}$.

Figure 13 illustrates the relationship between $y = \text{arctanh}(x)$ and $x = \tanh\,(y)$ graphically, and Figure 14 displays the graph of $y = \text{arctanh}(x)$.

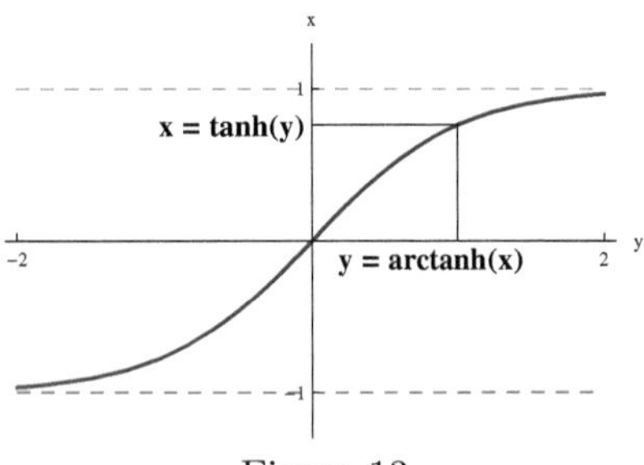

Figure 13

Figure 14: $y = \text{arctanh}(x)$

Just as in the cases of arcsinh and arccosh, we can express arctanh(x) in terms of the natural logarithm, and differentiate arctanh.

Proposition 8

$$\text{arctanh}(x) = \frac{1}{2}\ln\left(\frac{1+x}{1-x}\right), \quad -1 < x < 1.$$

We can differentiate arctanh easily:

Proposition 9

$$\frac{d}{dx}\text{arctanh}\,(x) = \frac{1}{1-x^2}.$$

The proofs of Proposition 8 and Proposition 9 are left as exercises.

In summary, we have the following differentiation formulas for inverse hyperbolic functions:

$$\frac{d}{dx}\text{arcsinh}(x) = \frac{1}{\sqrt{x^2+1}}, \quad x \in \mathbb{R},$$

$$\frac{d}{dx}\text{arccosh}(x) = \frac{1}{\sqrt{x^2-1}}, \quad x > 1,$$

$$\frac{d}{dx}\text{arctanh}(x) = \frac{1}{1-x^2}, \quad -1 < x < 1.$$

Problems

In problems 1-4, evaluate the expression.

1. $\cosh\left(\ln\left(2\right)\right)$

2. $\sinh\left(\ln\left(4\right)\right)$

3. $\tanh\left(\ln\left(3\right)\right)$

4. $\cosh\left(\sinh\left(0\right)\right)$

In problems 5-8, determine $f'\left(x\right).$

5. $f\left(x\right) = \cosh\left(4x\right)$

6. $f\left(x\right) = \sinh\left(3x^2\right)$

7. $f\left(x\right) = \dfrac{1}{\cosh\left(x\right)}$

8.. $f\left(x\right) = \dfrac{1}{\sinh\left(x\right)}$

9. Prove that $\displaystyle\lim_{x\to+\infty} \frac{\cosh\left(x\right)}{\frac{1}{2}e^x} = 1.$

10. Prove that $\displaystyle\lim_{x\to-\infty} \frac{\cosh\left(x\right)}{\frac{1}{2}e^{-x}} = 1.$

In problems 11 and12, establish the given identity.

11. $\sinh\left(a+b\right) = \sinh\left(a\right)\cosh\left(b\right) + \cosh\left(a\right)\sinh\left(b\right)$

12. $\cosh\left(a+b\right) = \cosh\left(a\right)\cosh\left(b\right) + \sinh\left(a\right)\sinh\left(b\right)$

13. Show that

$$\text{arccosh}\left(x\right) = \ln\left(x + \sqrt{x^2-1}\right) \text{ for } x \geq 1.$$

14. Show that

$$\frac{d}{dx}\text{arccosh}\left(x\right) = \frac{1}{\sqrt{x^2-1}}.$$

15. Show that

$$\operatorname{arctanh}(x) = \frac{1}{2} \ln\left(\frac{1+x}{1-x}\right) \text{ for } -1 < x < 1.$$

16. Show that

$$\frac{d}{dx}\operatorname{arctanh}(x) = \frac{1}{1-x^2}.$$

In problem 17-20, determine $f'(x)$:

17. $f(x) = \operatorname{arcsinh}(x/4)$

18. $f(x) = \operatorname{arcsinh}(\sqrt{x})$

19. $f(x) = \operatorname{arccosh}(3x)$

20. $f(x) = \operatorname{arctanh}(e^x)$

4.8 L'Hôpital's Rule

We have come across **indeterminate forms** such as $0/0$, ∞/∞, $0 \times \infty$, $\infty - \infty$ and 1^∞. In this section we will discuss various versions of **L'Hôpital's rule** that lead to the determination of such limits quickly in some cases, even though the implementation of the rule may not be as enlightening as the special techniques that we used previously. L'Hôpital's rule will enable you to evaluate more complicated limits as well.

Samples of Previous Encounters with Indeterminate Forms

We came across the indeterminate form $0/0$ already in chapters 1 and 2, when we introduced the idea of the derivative of a function f at a point a, alias, the slope of the graph of f at $(a, f(a))$. We have

$$f'(a) = \lim_{h \to 0} \frac{f(a+h) - f(a)}{h},$$

so that an attempt to evaluate such a limit by applying the quotient rule for limits leads to the indeterminate form $0/0$, since $\lim_{h \to 0}(f(a+h) - f(a)) = 0$ and $\lim_{h \to 0} h = 0$. For example, $\lim_{h \to 0}\sin(h) = 0$ and $\lim_{h \to 0} h = 0$, so that an attempt to evaluate

$$\lim_{h \to 0} \frac{\sin(h)}{h}$$

as

$$\frac{\lim_{h \to 0}\sin(h)}{\lim_{h \to 0} h},$$

leads the indeterminate form $0/0$. On the other hand, we know that

$$\lim_{h \to 0} \frac{\sin(h)}{h} = \frac{d}{dx}\sin(x)\Big|_{x=0} = 1.$$

You should not get the impression that $0/0$ is 1 from the above example. Since

$$\lim_{h \to 0}(\cos(h) - 1) = \lim_{h \to 0}\cos(h) - 1 = 1 - 1 = 0,$$

an attempt to evaluate

$$\lim_{h \to 0} \frac{\cos(h) - 1}{h}$$

by applying the quotient rule for limits also leads to the indeterminate form $0/0$. We know that

$$\lim_{h \to 0} \frac{\cos(h) - 1}{h} = \frac{d}{dx}\cos(x)\Big|_{x=0} = 0.$$

When we discussed limits at infinity in, we came across the indeterminate form ∞/∞. You should not be tempted to say that ∞/∞ is 1. For example, if

$$f(x) = \frac{2x-1}{x+1},$$

we have $\lim_{x\to+\infty}(x-1) = \infty$ and $\lim_{x\to+\infty}(x+1) = \infty$. We are able to calculate the limit by expressing $f(x)$ as

$$f(x) = \frac{2x\left(1-\dfrac{1}{x}\right)}{x\left(1+\dfrac{1}{x}\right)} = \frac{2\left(1-\dfrac{1}{x}\right)}{1+\dfrac{1}{x}}.$$

Since $\lim_{x\to\infty} 1/x = 0$,

$$\lim_{x\to\infty} f(x) = \lim_{x\to\infty} \frac{2\left(1-\dfrac{1}{x}\right)}{1+\dfrac{1}{x}} = 2.$$

We came across the indeterminate form ∞/∞ when we discussed exponential versus polynomial growth. For example, we showed that

$$\lim_{x\to+\infty} \frac{e^x}{x} = +\infty.$$

We came across the indeterminate form $0 \times (-\infty)$ when we discussed a limit such as

$$\lim_{x\to 0+} x\ln(x),$$

since $\lim_{x\to 0} x = 0$ and $\lim_{x\to 0+} \ln(x) = -\infty$. We were able to show that $\lim_{x\to 0} x\ln(x) = 0$. We came across the indeterminate form 1^∞ when we expressed e as

$$\lim_{x\to+\infty} (1+x)^{1/x}.$$

This example shows that $1^\infty \neq 1$, even though one may be led to that conclusion since $1^n = 1$ for any positive integer n.

The Indeterminate Form $0/0$

Theorem 1 (L'Hôpital's Rule for 0/0) Assume that f and g are differentiable at each x in an open interval J that contains the point a, with the possible exception of a itself, and that $g'(x) \neq 0$ for each $x \in J$. If

$$\lim_{x\to a} f(x) = \lim_{x\to a} g(x) = 0$$

and

$$\lim_{x\to a} \frac{f'(x)}{g'(x)}$$

exists (and is finite) or

$$\lim_{x\to a} \frac{f'(x)}{g'(x)} = \pm\infty.$$

Then

$$\lim_{x\to a} \frac{f(x)}{g(x)} = \lim_{x\to a} \frac{f'(x)}{g'(x)}.$$

The statements are valid if a is replaced by $\pm\infty$, f and g are differentiable in an interval $J = (b, +\infty)$ or $J = (-\infty, b)$, respectively, and $g'(x) \neq 0$ for each $x \in J$.

You can find the proof of Theorem 1 in Appendix D.

A Partial Plausibility Argument for Theorem 1:

We will assume that f' and g' are continuous at a, and that $g'(a) \neq 0$. Then, f and g are also continuous at a, and we have

$$f(a) = \lim_{x \to a} f(x) = 0 \text{ and } g(a) = \lim_{x \to a} g(x) = 0.$$

Let's denote the linear approximations to f and g based at a as $L_a f$ and $L_a g$, respectively. Thus,

$$L_a(x) = f(a) + f'(a)(x - a) = f'(a)(x - a)$$

and

$$L_a g(x) = g(a) + g'(a)(x - a) = g'(a)(x - a).$$

Since $f(x) \cong L_a f(x)$ and $g(x) = L_a g(x)$ if x is close to a, and $g'(a) \neq 0$,

$$\frac{f(x)}{g(x)} \cong \frac{L_a f(x)}{L_a g(x)} = \frac{f'(a)(x - a)}{g'(a)(x - a)} = \frac{f'(a)}{g'(a)}$$

if $x \neq a$ and x is close to a. Therefore, we should have

$$\lim_{x \to a} \frac{f(x)}{g(x)} = \frac{f'(a)}{g'(a)}.$$

By the continuity of $f'(x)$ and $g'(x)$ at a, we also have

$$\lim_{x \to a} \frac{f'(x)}{g'(x)} = \frac{\lim_{x \to a} f'(x)}{\lim_{x \to a} g'(x)} = \frac{f'(a)}{g'(a)},$$

since $g'(a) \neq 0$. Therefore, it plausible that

$$\lim_{x \to a} \frac{f(x)}{g(x)} = \lim_{x \to a} \frac{f'(x)}{g'(x)}.$$

■

Figure 1 illustrates the above plausibility argument.

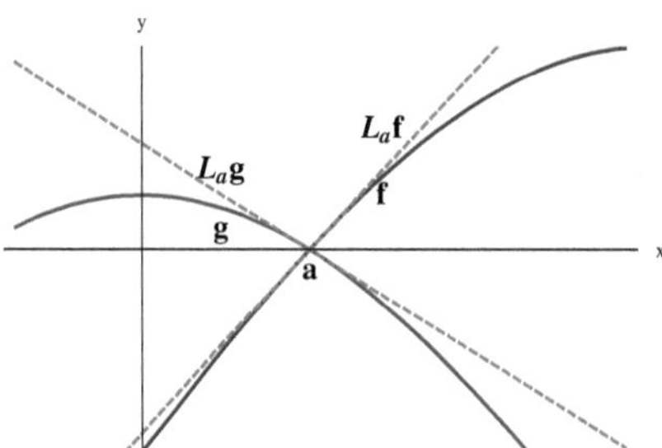

Figure 1:
$$\frac{f(x)}{g(x)} \cong \frac{L_a f(x)}{L_a g(x)} = \frac{f'(a)}{g'(a)}$$

Remark 1 L'Hôpital's rule is valid for one-sided limits as well: You can replace the limits in the statement of Theorem 1 by $\lim_{x \to a+}$ or $\lim_{x \to a-}$. $\Diamond$

Example 1 Determine
$$\lim_{x \to 0} \frac{\ln\left(1 + 2x\right)}{x}.$$

Solution

We have
$$\lim_{x \to 0} \ln\left(1 + 2x\right) = \ln\left(1\right) = 0 \text{ and } \lim_{x \to 0} x = 0,$$
so that we are led to the indeterminate form 0/0. We also have
$$\lim_{x \to 0} \frac{\dfrac{d}{dx} \ln\left(1 + 2x\right)}{\dfrac{d}{dx}\left(x\right)} = \lim_{x \to 0} \frac{\dfrac{1}{1 + 2x}\left(2\right)}{1} = 2 \lim_{x \to 0} \frac{1}{1 + 2x} = 2.$$

Therefore, L'Hôpital's rule is applicable:
$$\lim_{x \to 0} \frac{\ln\left(1 + 2x\right)}{x} = \lim_{x \to 0} \frac{\dfrac{d}{dx} \ln\left(1 + 2x\right)}{\dfrac{d}{dx}\left(x\right)} = 2.$$

Figure 2 shows the graph of
$$f\left(x\right) = \frac{\ln\left(1 + 2x\right)}{x}.$$

The picture is consistent with the fact that $\lim_{x \to 0} f\left(x\right) = 2$. Clearly, f has a removable discontinuity at 0, and the discontinuity can be removed by declaring that $f\left(0\right) = 2$. $\square$

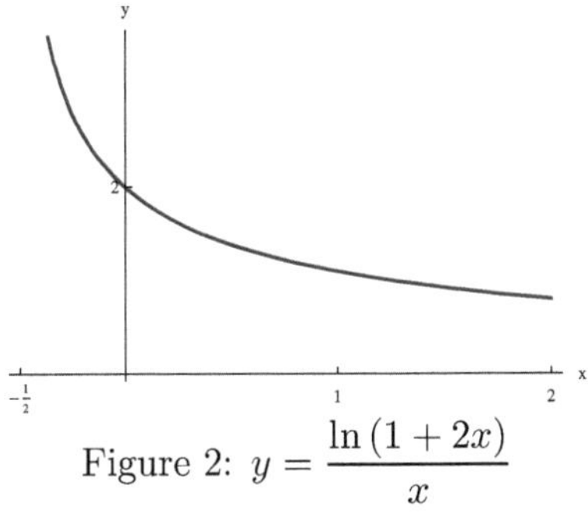

$$\text{Figure 2: } y = \frac{\ln\left(1 + 2x\right)}{x}$$

In some cases, we have to apply L'Hôpital's rule more than once in order to determine a limit:

Example 2 Determine
$$\lim_{x \to 1} \frac{\cos\left(x - 1\right) - 1}{\left(x - 1\right)^2}.$$

Solution

We have
$$\lim_{x \to 1} \left(\cos\left(x - 1\right) - 1\right) = \cos\left(0\right) - 1 = 1 - 1 = 0 \text{ and } \lim_{x \to 1} \left(x - 1\right)^2 = 0,$$
so that we are led to the indeterminate form 0/0. By L'Hôpital's rule,
$$\lim_{x \to 1} \frac{\cos\left(x - 1\right) - 1}{\left(x - 1\right)^2} = \lim_{x \to 1} \frac{\dfrac{d}{dx}\left(\cos\left(x - 1\right) - 1\right)}{\dfrac{d}{dx}\left(x - 1\right)^2} = \lim_{x \to 1} \frac{-\sin\left(x - 1\right)}{2\left(x - 1\right)},$$

provided that the limit on the right-hand side exists (finite or infinite). Since $\lim_{x\to 1} \sin(x-1) = \sin(0) = 0$ and $\lim_{x\to 1} 2(x-1) = 0$, we are still confronted by the indeterminate form $0/0$. If we apply L'Hôpital's rule again,

$$\lim_{x\to 1} \frac{-\sin(x-1)}{2(x-1)} = \lim_{x\to 1} \frac{\dfrac{d}{dx}\left(-\sin(x-1)\right)}{\dfrac{d}{dx}\left(2(x-1)\right)} = \lim_{x\to 1} \frac{-\cos(x-1)}{2} = -\frac{\cos(0)}{2} = -\frac{1}{2}.$$

Therefore,

$$\lim_{x\to 1} \frac{\cos(x-1)-1}{(x-1)^2} = \lim_{x\to 1} \frac{-\sin(x-1)}{2(x-1)} = -\frac{1}{2}.$$

Figure 3 shows the graph of

$$\frac{\cos(x-1)-1}{(x-1)^2}.$$

The picture is consistent with the fact that $\lim_{x\to 1} f(x) = -1/2$. The discontinuity of f at 1 can be removed by declaring that $f(1) = -1/2$. $\square$

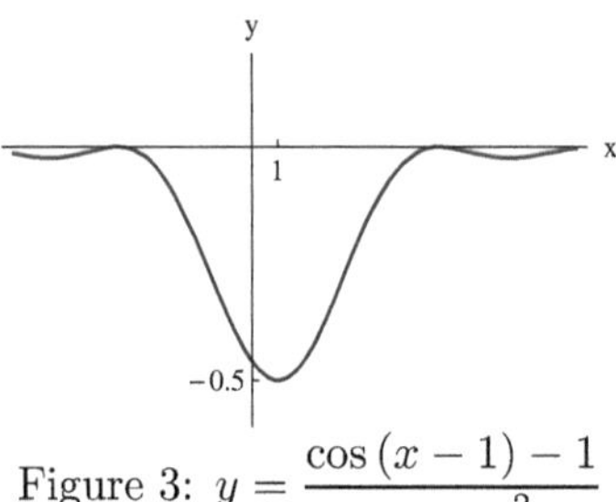

Figure 3: $y = \dfrac{\cos(x-1)-1}{(x-1)^2}$

Caution: A careless application of L'Hôpital's rule can result in an erroneous conclusion, as in the following example.

Example 3 Determine the erroneous step in the following calculation:

$$\lim_{x\to 2} \frac{x^2+x-6}{x^2-2x} = \lim_{x\to 2} \frac{\dfrac{d}{dx}\left(x^2+x-6\right)}{\dfrac{d}{dx}\left(x^2-2x\right)} = \lim_{x\to 2} \frac{2x+1}{2x-2} = \lim_{x\to 2} \frac{2}{2} = 1.$$

Solution

Since $\lim_{x\to 2}\left(x^2+x-6\right) = 0$ and $\lim_{x\to 2}\left(x^2-2x\right) = 0$, L'Hôpital's rule is applicable. We have

$$\lim_{x\to 2} \frac{x^2+x-6}{x^2-2x} = \lim_{x\to 2} \frac{\dfrac{d}{dx}\left(x^2+x-6\right)}{\dfrac{d}{dx}\left(x^2-2x\right)} = \lim_{x\to 2} \frac{2x+1}{2x-2}.$$

Since $\lim_{x\to 2}(2x+1) = 5$ and $\lim_{x\to 2}(2x-2) = 2 \neq 0$, the quotient rule for limits is applicable, and we have

$$\lim_{x\to 2} \frac{2x+1}{2x-2} = \frac{\lim_{x\to 2}(2x+1)}{\lim_{x\to 2}(2x-2)} = \frac{5}{2}.$$

The indicated calculation attempts to apply L'Hôpital's rule erroneously to the calculation of the above limit, even though the application of the quotient rule for limits does not lead to the indeterminate form $0/0$. $\square$

Example 4 Determine

$$\lim_{x \to 0} \frac{x}{\tan(x) - x}.$$

Solution

We have $\lim_{x \to 0} x = 0$, $\lim_{x \to 0} (\tan(x) - x) = 0$. By L'Hôpital's rule,

$$\lim_{x \to 0} \frac{x}{\tan(x) - x} = \lim_{x \to 0} \frac{\dfrac{d}{dx}(x)}{\dfrac{d}{dx}(\tan(x) - x)} = \lim_{x \to 0} \frac{1}{\dfrac{1}{\cos^2(x)} - 1}$$

$$= \lim_{x \to 0} \frac{\cos^2(x)}{1 - \cos^2(x)} = \lim_{x \to 0} \frac{\cos^2(x)}{\sin^2(x)}.$$

We have $\lim_{x \to 0} \cos^2(x) = \cos^2(0) = 1 > 0$. We also have $\sin^2(x) > 0$ if $x \neq 0$ and x is near 0, and $\lim_{x \to 0} \sin^2(x) = 0$. Therefore,

$$\lim_{x \to 0} \frac{x}{\tan(x) - x} = \lim_{x \to 0} \frac{\cos^2(x)}{\sin^2(x)} = +\infty.$$

Figure 4 shows the graph of

$$y = f(x) = \frac{x}{\tan(x) - x}.$$

The picture is consistent with the fact that $\lim_{x \to 0} f(x) = +\infty$. $\square$

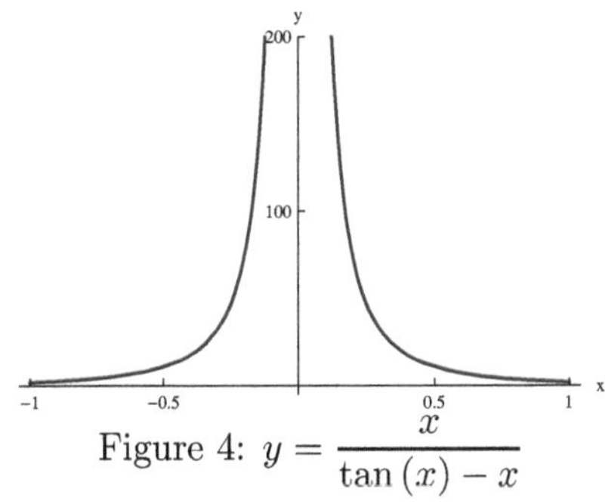

Figure 4: $y = \dfrac{x}{\tan(x) - x}$

The Indeterminate Form ∞/∞

There is a version of L'Hôpital's rule that is applicable to cases which lead to the indeterminate form ∞/∞:

Theorem 2 (L'Hôpital's Rule for ∞/∞) Assume that f and g are differentiable at each x in an open interval J that contains the point a, with the possible exception of a, and that $g'(x) \neq 0$ for each $x \in J$. If

$$\lim_{x \to a} f(x) = \pm\infty, \ \lim_{x \to a} g(x) = \pm\infty,$$

and

$$\lim_{x \to a} \frac{f'(x)}{g'(x)}$$

exists (and is finite) or

$$\lim_{x \to a} \frac{f'(x)}{g'(x)} = \pm\infty,$$

then

$$\lim_{x \to a} \frac{f(x)}{g(x)} = \lim_{x \to a} \frac{f'(x)}{g'(x)}.$$

The statements are valid if a is replaced by $\pm\infty$, f and g are differentiable in an interval of the form $J = (b, +\infty)$ or $J = (-\infty, b)$, respectively, and $g'(x) \neq 0$ for each $x \in J$.

We leave the proof of Theorem 2 to a course in advanced calculus.

Remark 2 The limits in the statement of Theorem 2 can be one-sided. Theorem 2 is also applicable if we have $\lim_{x \to a} f(x) = -\infty$ and $\lim_{x \to a} g(x) = +\infty$, or vice versa. Indeed, if $\lim_{x \to a} f(x) = -\infty$ and $\lim_{x \to a} g(x) = +\infty$, then $\lim_{x \to a} (-f(x)) = +\infty$. Therefore,

$$\lim_{x \to a} \frac{f(x)}{g(x)} = -\lim_{x \to a} \frac{-f(x)}{g(x)} = -\lim_{x \to a} \frac{-f'(x)}{g'(x)} = \lim_{x \to a} \frac{f'(x)}{g'(x)}.$$

$\Diamond$

Example 5 We know that

$$\lim_{x \to +\infty} \frac{x^2}{e^x} = 0.$$

Obtain this fact by using L'Hôpital's rule

$$\lim_{x \to +\infty} \frac{x^2}{e^x}.$$

Solution

We have $\lim_{x \to +\infty} x^2 = \lim_{x \to +\infty} e^x = +\infty$. By L'Hôpital's rule,

$$\lim_{x \to +\infty} \frac{x^2}{e^x} = \lim_{x \to +\infty} \frac{\frac{d}{dx}\left(x^2\right)}{\frac{d}{dx}e^x} = \lim_{x \to +\infty} \frac{2x}{e^x},$$

provided that $\lim_{x \to \infty} 2x/e^x$ exists (finite or infinite).Since $\lim_{x \to +\infty} (2x) = \lim_{x \to +\infty} e^x = +\infty$, we are still confronted with the indeterminate form ∞/∞. We apply L'Hôpital's rule again:

$$\lim_{x \to +\infty} \frac{2x}{e^x} = \lim_{x \to \infty} \frac{2}{e^x} = 0,$$

since $\lim_{x \to \infty} e^x = +\infty$. Therefore,

$$\lim_{x \to +\infty} \frac{x^2}{e^x} = \lim_{x \to +\infty} \frac{2x}{e^x} = \lim_{x \to +\infty} \frac{2}{e^x} = 0.$$

$\square$

Example 6 We know that

$$\lim_{x \to +\infty} \frac{\ln(x)}{\sqrt{x}} = 0.$$

Obtain this fact by using L'Hôpital's rule.

Solution

We have

$$\lim_{x \to +\infty} \ln(x) = +\infty \text{ and } \lim_{x \to +\infty} \sqrt{x} = +\infty,$$

By Theorem 2

$$\lim_{x \to +\infty} \frac{\ln(x)}{\sqrt{x}} = \lim_{x \to +\infty} \frac{\frac{d}{dx}\ln(x)}{\frac{d}{dx}\sqrt{x}} = \lim_{x \to +\infty} \frac{\frac{1}{x}}{\frac{1}{2\sqrt{x}}} = \lim_{x \to +\infty} \frac{2}{\sqrt{x}} = 0.$$

$\square$

Even though L'Hôpital's rule enables us to determine some limits easily as in the above examples, the rule does not solve all limit problems, as the following example demonstrates.

Example 7

a) Show that L'Hôpital's rule will not help you in determining

$$\lim_{x \to 0-} \frac{e^{1/x}}{x}.$$

Determine the above limit.
b) Determine

$$\lim_{x \to 0+} \frac{e^{1/x}}{x}.$$

c) Determine

$$\lim_{x \to \pm\infty} \frac{e^{1/x}}{x}.$$

Solution

a) If we set $u = -1/x$, then $u \to +\infty$ as x approaches 0 through negative values. Therefore,

$$\lim_{x \to 0-} e^{1/x} = \lim_{u \to +\infty} e^{-u} = 0 \text{ and } \lim_{x \to 0-} x = 0,$$

and we are led to the indeterminate form $0/0$. Let's attempt to apply L'Hôpital's rule (Theorem 1). We have

$$\frac{\frac{d}{dx}\left(e^{1/x}\right)}{\frac{d}{dx}(x)} = \frac{-\frac{1}{x^2}e^{1/x}}{1} = -\frac{e^{1/x}}{x^2}.$$

Thus, we still have an indeterminate form $0/0$ if x approaches 0 from the left. Indeed, the indeterminacy involves x^2 instead of x in the denominator, and that is definitely not simpler than the original expression. You can convince yourself that the situation does not improve if you keep differentiating the numerator and denominator. For example,

$$\frac{\frac{d}{dx}\left(-e^{1/x}\right)}{\frac{d}{dx}(x^2)} = \frac{\frac{1}{x^2}e^{1/x}}{2x} = \frac{e^{1/x}}{2x^3}, \quad \frac{\frac{d}{dx}\left(e^{1/x}\right)}{\frac{d}{dx}(2x^3)} = \frac{-\frac{1}{x^2}e^{1/x}}{6x^4} = -\frac{e^{1/x}}{6x^4}.$$

On the other hand, if we set $u = -1/x$ then $u \to +\infty$ as $x \to 0-$. Therefore,

$$\lim_{x \to 0-} \frac{e^{1/x}}{x} = \lim_{u \to +\infty} \frac{e^{-u}}{-u} = -\lim_{u \to +\infty} \frac{1}{\frac{e^u}{u}} = 0.$$

b) Since $\lim_{x\to 0+} 1/x = +\infty$, we have $\lim_{x\to 0+} e^{1/x} = +\infty$. Therefore,

$$\lim_{x\to 0+} \frac{e^{1/x}}{x} = \lim_{x\to 0+} \left(e^{1/x} \left(\frac{1}{x} \right) \right) = +\infty.$$

c) Since $\lim_{x\to \pm\infty} 1/x = 0$,

$$\lim_{x\to \pm\infty} e^{1/x} = e^0 = 1 > 0.$$

Therefore,

$$\lim_{x\to \pm\infty} \frac{e^{1/x}}{x} = \lim_{x\to \pm\infty} \left(e^{1/x} \left(\frac{1}{x} \right) \right) = 1 \times 0 = 0.$$

Figure 5 shows the graph of

$$y = f(x) = \frac{e^{1/x}}{x}$$

The picture is consistent with our determination of $\lim_{x\to 0+} f(x)$, $\lim_{x\to 0-} f(x)$ and $\lim_{x\to \pm\infty} f(x)$.$\square$

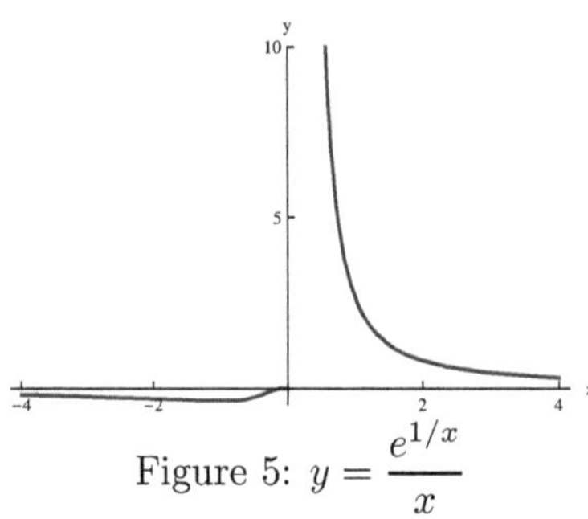

Figure 5: $y = \dfrac{e^{1/x}}{x}$

The Indeterminate Form $0 \times \infty$

If we are confronted with the indeterminate form $0 \times \infty$ (or $0 \times (-\infty)$), we may try to rearrange the relevant expression so that L'Hôpital's rule is applicable, as in the following example:

Example 8 Determine

$$\lim_{x\to 1+} (x-1)^{1/3} \ln(x-1)$$

with the help of L'Hôpital's rule.

Solution

We have $\lim_{x\to 1+} (x-1)^{1/3} = 0$. If we set $u = x - 1$, then u approaches 0 through positive values as x approaches 1 from the right. Therefore, $\lim_{x\to 1+} \ln(x-1) = \lim_{u\to 0+} \ln(u) = -\infty$. An attempt to apply the product rule for limits leads to the indeterminate form $0 \times (-\infty)$. On the other hand we can rearrange the given expression so that L'Hôpital's rule is applicable:

$$(x-1)^{1/3} \ln(x-1) = \frac{\ln(x-1)}{(x-1)^{-1/3}}.$$

We have $\lim_{x\to 1+} \ln(x-1) = -\infty$ and $\lim_{x\to 1+} (x-1)^{-1/3} = +\infty$. By L'Hôpital's rule,

$$\lim_{x\to 1+} \frac{\ln(x-1)}{(x-1)^{-1/3}} = \lim_{x\to 1+} \frac{\frac{d}{dx}\ln(x-1)}{\frac{d}{dx}(x-1)^{-1/3}} = \lim_{x\to 1+} \frac{\frac{1}{x-1}}{-\frac{1}{3}(x-1)^{-4/3}} = -3 \lim_{x\to 1+} (x-1)^{1/3} = 0.$$

Note that we were able to evaluate

$$\lim_{x\to 1+} \frac{\dfrac{1}{x-1}}{-\frac{1}{3}\left(x-1\right)^{-4/3}}$$

easily after simplifying the expression. The implementation of L'Hôpital's rule again would not have simplified matters:

$$\lim_{x\to 1+} \frac{\dfrac{1}{x-1}}{-\frac{1}{3}\left(x-1\right)^{-4/3}} = \lim_{x\to 1+} \frac{-\dfrac{1}{\left(x-1\right)^2}}{\frac{4}{9}\left(x-1\right)^{-7/3}},$$

even though the application of the rule leads to the result. The moral is that you should try to evaluate a limit by means of appropriate simplifications, before you try to implement L'Hôpital's rule repeatedly.

Figure 6 shows the graph of

$$y = f\left(x\right) = \left(x-1\right)^{1/3}\ln\left(x-1\right).$$

The picture is consistent with the fact that $\lim_{x\to 1+} f\left(x\right) = 0$. $\square$

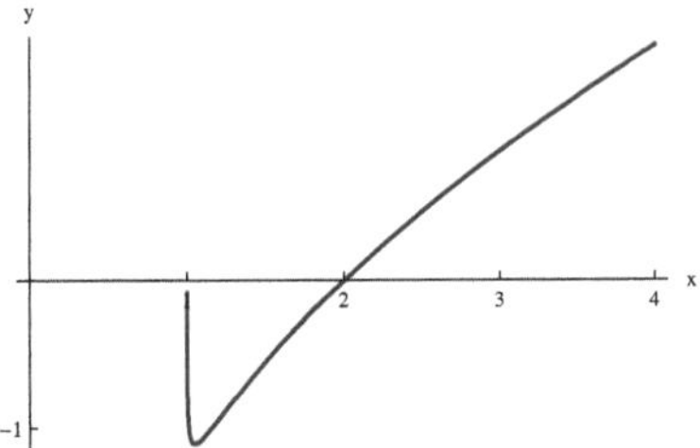

Figure 6: $y = \left(x-1\right)^{1/3}\ln\left(x-1\right)$

The Indeterminate Forms $1^{\infty}, \infty^0$ and 0^0

The recommended strategy in dealing with indeterminate forms such as 1^{∞}, 0^0 or ∞^0 is to express the relevant function in terms of the natural exponential function and the natural logarithm, as we will illustrate by examples.

Example 9 We know that

$$\lim_{n\to\infty}\left(1+\frac{1}{n}\right)^n = e.$$

Determine the above limit with the help of L'Hôpital's rule.

Solution

We will show that

$$\lim_{x\to +\infty}\left(1+\frac{1}{x}\right)^x = e^x,$$

without restricting x to be an integer. Since

$$\lim_{x\to +\infty}\left(1+\frac{1}{x}\right) = 1,$$

we are led to the indeterminate form 1^∞. We will express

$$\left(1 + \frac{1}{x}\right)^x$$

in terms of the natural exponential function and the natural logarithm. Thus,

$$\left(1 + \frac{1}{x}\right)^x = \exp\left(x \ln\left(1 + \frac{1}{x}\right)\right).$$

By the continuity of the natural exponential function,

$$\lim_{x \to +\infty} \left(1 + \frac{1}{x}\right)^x = \lim_{x \to +\infty} \exp\left(x \ln\left(1 + \frac{1}{x}\right)\right) = \exp\left(\lim_{x \to +\infty} x \ln\left(1 + \frac{1}{x}\right)\right).$$

Therefore, it is sufficient to evaluate

$$\lim_{x \to +\infty} x \ln\left(1 + \frac{1}{x}\right).$$

Since

$$\lim_{x \to +\infty} \ln\left(1 + \frac{1}{x}\right) = \ln(1) = 0,$$

we are led to the indeterminate form $0 \times (-\infty)$. We can rewrite the expression in a way that allows us to use L'Hôpital's rule. We have

$$x \ln\left(1 + \frac{1}{x}\right) = \frac{\ln\left(1 + \frac{1}{x}\right)}{\frac{1}{x}},$$

and

$$\lim_{x \to +\infty} \ln\left(1 + \frac{1}{x}\right) = \ln(1) = 0, \quad \text{and} \quad \lim_{x \to +\infty} \frac{1}{x} = 0.$$

Therefore, we are led to the indeterminate form $0/0$. We apply L'Hôpital's rule:

$$\lim_{x \to +\infty} \frac{\ln\left(1 + \frac{1}{x}\right)}{\frac{1}{x}} = \lim_{x \to +\infty} \frac{\frac{d}{dx}\ln\left(1 + \frac{1}{x}\right)}{\frac{d}{dx}\left(\frac{1}{x}\right)} = \lim_{x \to +\infty} \frac{\frac{1}{1 + \frac{1}{x}}\left(-\frac{1}{x^2}\right)}{-\frac{1}{x^2}} = \lim_{x \to +\infty} \frac{1}{1 + \frac{1}{x}} = 1.$$

Thus,

$$\lim_{x \to +\infty} \left(1 + \frac{1}{x}\right)^x = \exp\left(\lim_{x \to +\infty} x \ln\left(1 + \frac{1}{x}\right)\right) = \exp(1) = e.$$

$\square$

The expression ∞^0 is indeterminate. We may be tempted to say that $\infty^0 = 1$, since $a^0 = 1$ for any $a \neq 1$. That is not true. For example,

$$\lim_{x \to +\infty} (e^x)^{1/x} = \lim_{x \to \infty} e = e.$$

If we attempt to evaluate the same limit as

$$\left(\lim_{x \to +\infty} e^x\right)^{\lim_{x \to +\infty} 1/x},$$

we are led to the expression ∞^0.

Example 10 Determine

$$\lim_{x \to +\infty} x^{1/x}.$$

Solution

An attempt to replace x by ∞ and $1/x$ by 0 leads to the indeterminate form ∞^0. On the other hand, we have

$$x^{1/x} = e^{\ln(x)/x},$$

so that

$$\lim_{x \to +\infty} x^{1/x} = \lim_{x \to +\infty} e^{\ln(x)/x} = e^{\lim_{x \to +\infty} \ln(x)/x}.$$

We have

$$\lim_{x \to +\infty} \frac{\ln(x)}{x} = 0$$

(with or without L'Hôpital's rule). Therefore,

$$\lim_{x \to +\infty} x^{1/x} = e^0 = 1.$$

$\square$

The expression 0^0 is indeterminate. You may be tempted to say that $0^0 = 1$ since $a^0 = 1$ for any $a \neq 0$. That is not true. For example,

$$\lim_{x \to 0-} \left(e^{1/x}\right)^x = \lim_{x \to 0-} e = e,$$

even though

$$\lim_{x \to 0-} e^{1/x} = \lim_{u \to -\infty} e^{-u} = 0,$$

as in Example 7, and $\lim_{x \to 0} x = 0$. You may be tempted to say that $0^0 = 0$, since $0^x = 0$ for any $x > 0$, however small x may be. That is not true either, as the above example shows. You should express the given function in terms of the natural exponential function and the natural logarithm when you are confronted with the indeterminate form 0^0.

Example 11 Evaluate

$$\lim_{x \to 0+} x^x.$$

Solution

An attempt to replace x by 0 leads to the indeterminate form 0^0. On the other hand, we have

$$x^x = \exp\left(x \ln\left(x\right)\right),$$

so that

$$\lim_{x \to 0+} x^x = \lim_{x \to 0+} \exp\left(x \ln\left(x\right)\right) = \exp\left(\lim_{x \to 0+} x \ln\left(x\right)\right) = \exp\left(0\right) = 1,$$

since $\lim_{x \to 0+} x \ln\left(x\right) = 0$ (with or without L'Hôpital's rule).

Figure 7 shows the graph of $y = x^x$. The picture is consistent with the fact that $\lim_{x \to 0+} x^x = 1$.

$\square$

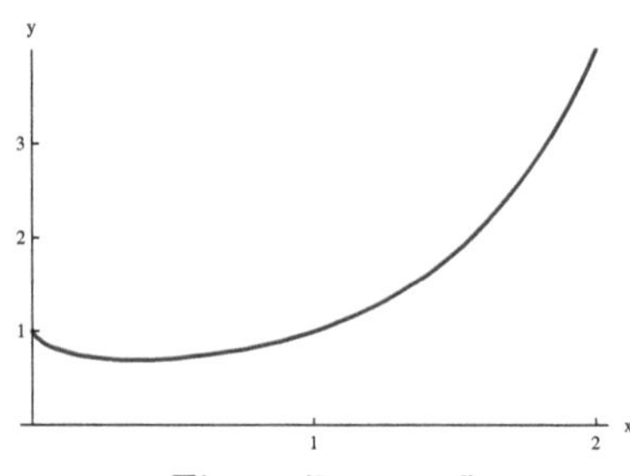

Figure 7: $y = x^x$

The Indeterminate Form $\infty - \infty$

If you come across the indeterminate form $\infty-\infty$, you should obtain an alternative expression for the relevant function so that you can determine the required limit, as in the following example.

Example 12 Determine
$$\lim_{x \to 0+} \left(\frac{3^x}{x} - \frac{2^x}{x} \right).$$

Solution

Since
$$\lim_{x \to 0} 3^x = \lim_{x \to 0} 2^x = 1 > 0$$

and
$$\lim_{x \to 0+} \frac{1}{x} = +\infty,$$

we have
$$\lim_{x \to 0+} \frac{3^x}{x} = \lim_{x \to 0+} \frac{2^x}{x} = +\infty.$$

Thus, we are led to the indeterminate form $\infty - \infty$. We can rearrange the given expression:
$$\frac{3^x}{x} - \frac{2^x}{x} = \frac{3^x - 2^x}{x},$$

We have $\lim_{x \to 0+} (3^x - 2^x) = 1 - 1 = 0$, and $\lim_{x \to 0+} x = 0$. Therefore, L'Hôpital's rule is applicable:

$$\lim_{x \to 0+} \left(\frac{3^x}{x} - \frac{2^x}{x} \right) = \lim_{x \to 0+} \left(\frac{3^x - 2^x}{x} \right) = \lim_{x \to 0+} \frac{\frac{d}{dx}(3^x - 2^x)}{\frac{d}{dx}(x)}$$

$$= \lim_{x \to 0+} \frac{\ln(3)\,3^x - \ln(2)\,2^x}{1} = \ln(3) - \ln(2) = \ln\left(\frac{3}{2}\right) \cong 0.405465$$

Figure 8 shows the graph of
$$y = f(x) = \frac{3^x}{x} - \frac{2^x}{x}.$$

The picture is consistent with the fact that $\lim_{x \to 0+} f(x) = \ln(3/2) \cong 0.4$. $\square$

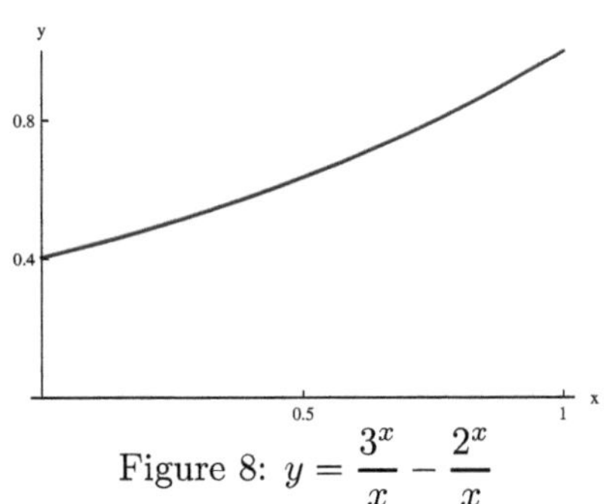

Figure 8: $y = \dfrac{3^x}{x} - \dfrac{2^x}{x}$

Problems

In problems make use of L'Hôpital's rule to determine the limit:

1.
$$\lim_{x \to 0} \frac{\sin(x) - x}{x^3}$$

2.
$$\lim_{x \to 0} \frac{\cos(x) - 1 + \frac{1}{2}x^2}{x^4}$$

3.
$$\lim_{x \to 1} \frac{\ln(x) - x + 1}{(x - 1)^2}$$

4.
$$\lim_{x \to 0} \frac{e^x - 1 - x}{x^2}$$

5.
$$\lim_{x \to 1} \frac{\arctan(x) - \dfrac{\pi}{4} - \frac{1}{2}(x - 1)}{(x - 1)^2}$$

6.
$$\lim_{x \to 0} \frac{\arcsin(x) - x}{x^3}$$

7.
$$\lim_{x \to 1/2} \frac{3\arccos(x) - \pi}{2x - 1}$$

8.
$$\lim_{x \to +\infty} \frac{x^3}{e^{2x}}$$

9.
$$\lim_{x \to +\infty} \frac{e^{x/4}}{x^2}$$

10.
$$\lim_{x \to +\infty} \frac{\ln(x)}{x^{1/3}}$$

11.
$$\lim_{x \to +\infty} \frac{x^{1/4}}{\ln(x)}$$

12.
$$\lim_{x \to 0+} \sqrt{x}\,\ln(x)$$

13.
$$\lim_{x \to +\infty} x^2 e^{-x}$$

14.
$$\lim_{x \to -\infty} x^2 e^x$$

15.
$$\lim_{x \to +\infty} \left(1 + \frac{3}{x}\right)^x$$

16.
$$\lim_{x \to +\infty} \left(1 - \frac{4}{x}\right)^x$$

17.
$$\lim_{x \to +\infty} \left(\frac{x^2}{x + 9}\right)^{1/x}$$

18.
$$\lim_{x \to +\infty} \left(\frac{x^2}{x^2 + 1}\right)^x$$

19.
$$\lim_{x \to 0+} \left(\frac{10^x}{x} - \frac{4^x}{x}\right)$$

Chapter 5

The Integral

In this chapter we will introduce the fundamental concept of **the integral**. The integral of a positive-valued function on an interval corresponds to the **area** of the region between the graph of the function and the interval. The integral of the **velocity function** corresponding to the one-dimensional motion of an object over a time interval yields **the displacement** of the object over that time interval. In later chapters the integral will appear in other contexts such as work, the length of a graph or probability. **The Fundamental Theorem of Calculus** links the integral to the derivative. You will have ample opportunity to appreciate the significance of the Fundamental Theorem throughout the course.

5.1   The Approximation of Area

In this section we will discuss the approximation of the area of a region between the graph of a positive-valued function and an interval.

The Summation Notation

Let's begin by introducing notation that will turn out to be convenient in expressing sums. Given numbers $a_1, a_2, \ldots, a_n$, we can indicate the sum of the numbers as

$$a_1 + a_2 + \cdots + a_n.$$

We can also indicate the sum using **the summation notation**:

$$\sum_{k=1}^{n} a_k$$

(read "sigma a_k as k runs from 1 to n"). The subscript k is **the summation index**, and is a "dummy index", in the sense that it can be replaced by any convenient letter. Thus, both

$$\sum_{j=1}^{n} a_j \text{ and } \sum_{l=1}^{n} a_l.$$

denote the sum $a_1 + a_2 + \cdots + a_n$.

Example 1 The sum of the first n positive integers can be expressed succinctly:

$$\sum_{k=1}^{n} k = 1 + 2 + 3 + \cdots + n = \frac{n(n+1)}{2}.$$

Indeed, if we set $S_n = \sum_{k=1}^{n} k$, we have

$$S_n = 1 + 2 + \cdots + (n-1) + n.$$

Let's add the terms in the opposite order:

$$S_n = n + (n-1) + \cdots + 2 + 1.$$

Therefore,

$$2S_n = (n+1) + ((n-1) + 2) + \cdots + (2 + (n-1)) + (1+n)$$
$$= (n+1) + (n+1) + \ldots + (n+1) + (n+1),$$

where the sum has n terms. Thus,

$$2S_n = n(n+1),$$

so that

$$S_n = \frac{n(n+1)}{2}.$$

□

The following rules are natural and easy to confirm:

$$\sum_{k=1}^{n} (a_k + b_k) = \sum_{k=1}^{n} a_k + \sum_{k=1}^{n} b_k,$$

and

$$\sum_{k=1}^{n} c a_k = c \sum_{k=1}^{n} a_k \quad \text{(the constant rule for sums)}$$

Proof

By the associativity of addition,

$$\sum_{k=1}^{n} (a_k + b_k) = (a_1 + b_1) + (a_2 + b_2) + \cdots + (a_n + b_n)$$
$$= (a_1 + a_2 + \cdots + a_n) + (b_1 + b_2 + \cdots + b_n)$$
$$= \sum_{k=1}^{n} a_k + \sum_{k=1}^{n} b_k.$$

By the distributivity of multiplication with respect to sums,

$$\sum_{k=1}^{n} c a_k = c a_1 + c a_2 + \cdots + c a_n = c(a_1 + a_2 + \cdots + a_n) = c \sum_{k=1}^{n} a_k$$

■

The Area under the Graph of a Function

Assume that f is continuous on the interval $[a, b]$ and $f(x) \geq 0$ for each $x \in [a, b]$. Let G be the region in the xy-plane that is bounded by the graph of f, the interval $[a, b]$ on the x-axis, the line $x = a$ and the line $x = b$. We will refer to G simply as **the region between the graph of f and the interval $[a, b]$**. Our intuitive notion of the area of G is a measure of the size of G. Even though we may not be able to compute the area of G exactly, we should be able to compute approximations. We will devise a strategy that will be based on the approximation of G, in a geometric sense, by unions of rectangles.

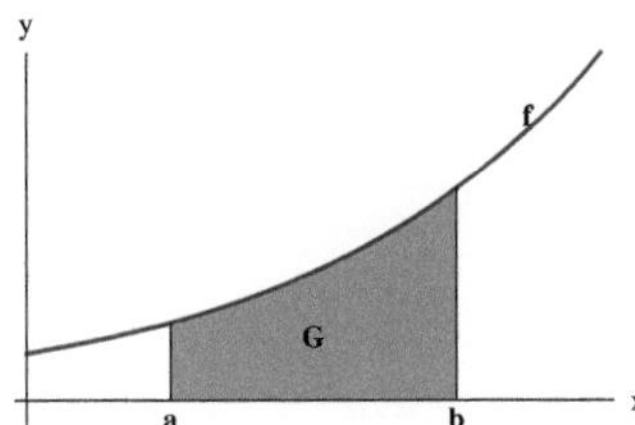

Figure 1: The region between the graph of f and the interval $[a, b]$

Definition 1 The set of points $P = \{x_0, x_1, \ldots, x_{k-1}, x_k, \ldots, x_n\}$ is **a partition of the interval** $[a, b]$ if

$$a = x_0 < x_1 < x_2 < \cdots < x_{k-1} < x_k < \cdots < x_n = b.$$

The interval $[x_{k-1}, x_k]$ is the k**th subinterval** that is determined by the partition P. We will denote the length of the kth subinterval by Δx_k, so that $\Delta x_k = x_k - x_{k-1}$. The maximum of the lengths of the subintervals determined by P is the **norm of the partition** P. We will denote the norm of P by $||P||$, so that $||P||$ is the maximum of $\Delta x_1, \Delta x_2, \ldots, \Delta x_n$. we can abbreviate the expression "maximum of $\Delta x_1, \Delta x_2, \ldots, \Delta x_n$" as $\max_{k=1,\ldots,n} \Delta x_k$ or $\max_k \Delta x_k$. Thus,

$$||P|| = \max_{k=1,\ldots,n} \Delta x_k.$$

Let's sample an arbitrary value of x in the kth subinterval $[x_{k-1}, x_k]$ and denote it by x_k^*. Thus, $x_{k-1} \leq x_k^* \leq x_k$, but there is no other restriction on the choice of x_k^*. Consider the rectangle that has as its base the interval $[x_{k-1}, x_k]$ and has height equal to the value of f at x_k^*. If Δx_k is small, it is reasonable to approximate the area of the slice of G between the lines $x = x_{k-1}$ and $x = x_k$ by the area of such a rectangle.

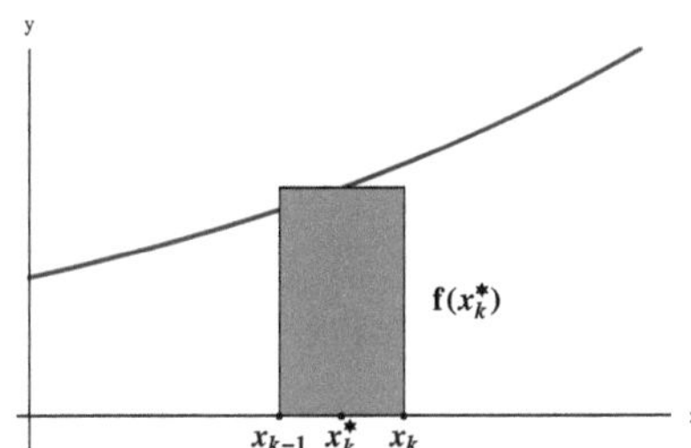

Figure 2: An approximating rectangle

The area of the rectangle is

$$f\left(x_k^*\right)\left(x_k - x_{k-1}\right) = f\left(x_k^*\right) \Delta x_k.$$

The sum of the areas of such rectangles should be a reasonable approximation to the area of G if the maximum of the lengths of the subintervals, i.e., $||P||$ is small:

$$\sum_{k=1}^{n} f\left(x_k^*\right) \Delta x_k \cong \text{Area of } G.$$

We would expect the approximation to be as accurate as desired if $||P|| = \max_k \Delta x_k$ is sufficiently small.

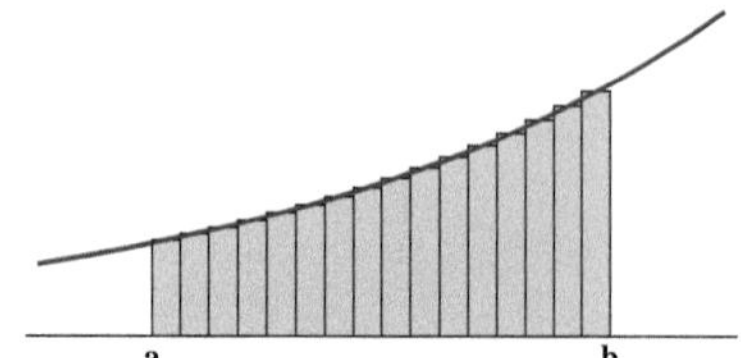

Figure 3: Approximating rectangles

Example 2 Let $f(x) = x^2 + 1$, and let G be the region between the graph of f and the interval $[0, 2]$. Figure 4 shows G.

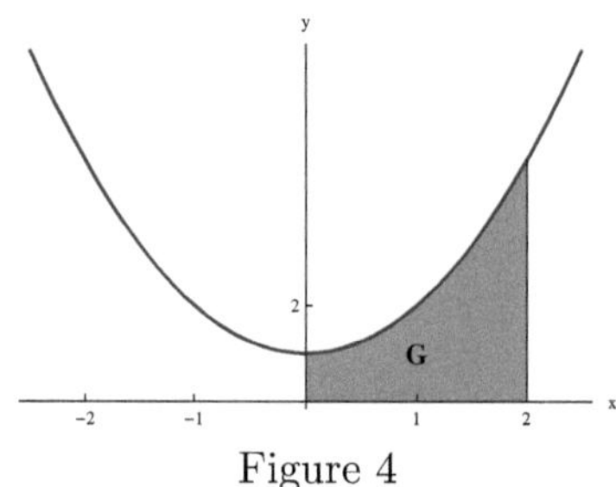

Figure 4

Let
$$P = \{0, 0.5, 1, 1.2, 1.4, 1.6, 1.8, 2\},$$
so that P is a partition of the interval $[0, 2]$. With reference to the notation of Definition 1, we have

$$x_0 = 0, \; x_1 = 0.5, \; x_2 = 1, \; x_3 = 1.2, \; x_4 = 1.4, \; x_5 = 1.6, \; x_6 = 1.8 \text{ and } x_7 = 2.$$

The lengths of the subintervals determined by the partition P are

$$\Delta x_1 = \Delta x_2 = 0.5 \text{ and } \Delta x_3 = \Delta x_4 = \cdots = \Delta x_7 = 0.2.$$

Therefore, the norm of P is 0.5:
$$||P|| = 0.5.$$

Let's form the rectangle of height $f(c_k)$ on the kth subinterval $[x_{k-1}, x_k]$, where c_k is the midpoint of $[x_{k-1}, x_k]$, $k = 1, 2, \ldots, 7$, and approximate the area of the region G by the sum of these rectangles. Figure 5 indicates the rectangles.

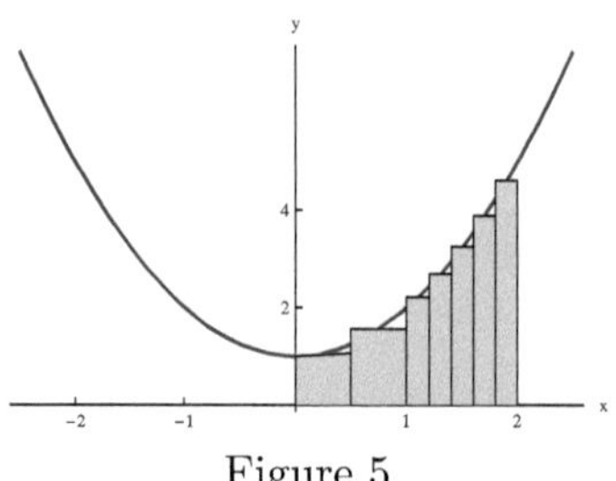

Figure 5

The approximation to the area of G is

$$\sum_{k=1}^{7} f\left(c_k\right) \Delta x_k \cong f\left(0.25\right)\left(0.5\right) + f\left(0.75\right)\left(0.5\right) + f\left(1.1\right)\left(0.2\right)$$
$$+ f\left(1.3\right)\left(0.2\right) + f\left(1.5\right)\left(0.2\right) + f\left(1.7\right)\left(0.2\right) + f\left(1.8\right)\left(0.2\right)$$
$$\cong 4.568\,5.$$

In Section 5.4 we will show that the area of G is

$$\frac{14}{3} \cong 4.666\,67,$$

so that the absolute error of our approximation is about 0.1. For many purposes, the magnitude of the error may be unacceptable. On the other hand, we would expect the error to be as small as desired if the interval $[0, 2]$ is partitioned to subintervals of sufficiently small length. $\square$

In the other examples of this section we will consider the partitioning of an interval $[a, b]$ into n subintervals of equal length, since the corresponding sums can be expressed and computed easily. Thus,

$$\Delta x_k = \Delta x = \frac{b - a}{n} \text{ for } k = 1, 2, \ldots, n,$$

and

$$x_k = a + k\Delta x, \ k = 0, 1, 2, \ldots, n.$$

We will approximate the area of the region between the graph of f and the interval $[a, b]$ by sums of the form

$$\sum_{k=1}^{n} f\left(x_k^*\right) \Delta x_k = \sum_{k=1}^{n} f\left(x_k^*\right) \Delta x = \Delta x \sum_{k=1}^{n} f\left(x_k^*\right).$$

The intermediate points x_k^*, $k = 1, 2, \ldots, n$, can be chosen in many different ways. We will consider the following strategies:

1. A **left-endpoint sum** is obtained by choosing x_k^* to be the left endpoint x_{k-1} of the kth subinterval $[x_{k-1}, x_k]$. We have

$$x_{k-1} = a + (k - 1)\Delta x.$$

We will denote the left-endpoint sum corresponding to the function f and the partitioning of the interval $[a, b]$ to n subintervals of equal length as l_n. Thus,

$$l_n = \sum_{k=1}^{n} f\left(x_{k-1}\right) \Delta x.$$

2. A **right-endpoint sum** is obtained by choosing x_k^* to be the right endpoint x_k of the kth subinterval $[x_{k-1}, x_k]$. We have

$$x_k = a + k\Delta x.$$

We will denote the right-endpoint sum corresponding to the function f and the partitioning of the interval $[a, b]$ to n subintervals of equal length as r_n. Thus,

$$r_n = \sum_{k=1}^{n} f\left(x_k\right) \Delta x.$$

3. A **midpoint sum** is obtained by choosing x_k^* to be the midpoint c_k of the kth subinterval $[x_{k-1}, x_k]$. We have

$$c_k = \frac{x_{k-1} + x_k}{2} = \frac{1}{2}\left(a + (k-1)\,\Delta x + a + k\Delta x\right) = \frac{1}{2}\left(2a + (2k-1)\,\Delta x\right) = a + (k - \frac{1}{2})\Delta x.$$

We will denote the midpoint sum corresponding to the function f and the partitioning of the interval $[a, b]$ to n subintervals of equal length as m_n. Thus,

$$m_n = \sum_{k=1}^{n} f\left(c_k\right) \Delta x.$$

As we will discuss in more detail in the next section, **any of the above sums approximates the area of the region between the graph of f and the interval $[a, b]$ as accurately as desired, provided that f is continuous on $[a, b]$ and Δx is small enough.** Since

$$\Delta x = \frac{b - a}{n},$$

Δx is as small as necessary if n is sufficiently large. Therefore, **the area $A\left(G\right)$ of the region G between the graph of f and the interval $[a, b]$ is the limit of left-endpoint sums, right-endpoint sums or midpoint sums as n tends to infinity:**

$$A\left(G\right) = \lim_{n \to \infty} l_n = \lim_{n \to \infty} r_n = \lim_{n \to \infty} m_n.$$

Example 3 Let $f(x) = x$. The region G between the graph of f and the interval $[0, 1]$ is a triangle whose base has length 1 and whose height is 1. Therefore, the area of G is

$$\frac{1}{2}\left(1\right)\left(1\right) = \frac{1}{2}.$$

Consider the approximation of the area of G by right-endpoint sums r_n. Figure 6 illustrates the rectangles that correspond to $n = 16$. Show that $\lim_{n \to \infty} r_n = $ area of G.

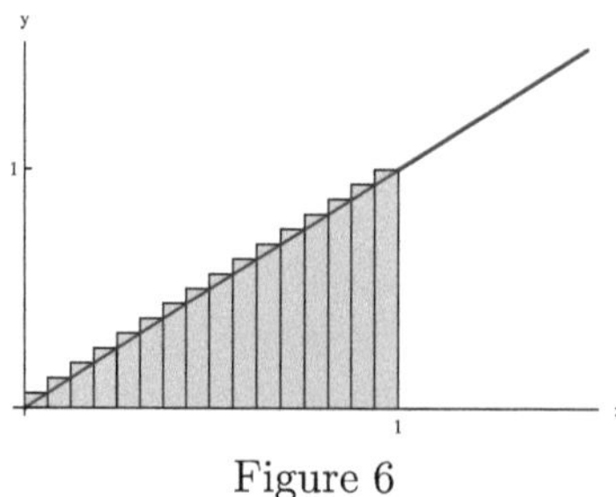

Figure 6

Solution

We have

$$r_n = \sum_{k=1}^{n} f\left(x_k\right) \Delta x = \sum_{k=1}^{n} x_k \Delta x,$$

where

$$\Delta x = \frac{1}{n} \text{ and } x_k = k\Delta x = \frac{k}{n}.$$

Therefore,

$$r_n = \sum_{k=1}^{n} \left(\frac{k}{n}\right)\left(\frac{1}{n}\right) = \sum_{k=1}^{n} \frac{k}{n^2} = \frac{1}{n^2}\sum_{k=1}^{n} k.$$

In Example 1 we showed that

$$\sum_{k=1}^{n} k = \frac{n(n+1)}{2}.$$

Therefore,

$$r_n = \frac{1}{n^2}\sum_{k=1}^{k} k = \frac{1}{n^2}\left(\frac{n(n+1)}{2}\right) = \frac{n(n+1)}{2n^2}.$$

Thus,

$$\lim_{n\to\infty} r_n = \lim_{n\to\infty} \frac{n(n+1)}{2n^2} = \lim_{n\to\infty} \frac{n^2\left(1+\frac{1}{n}\right)}{2n^2} = \lim_{n\to\infty} \frac{1+\frac{1}{n}}{2} = \frac{1}{2}.$$

Therefore, the area of G is $1/2$. $\square$

Example 4 Let $f(x) = x^2$.

a) Sketch the region G between the graph of f and the interval $[1, 3]$.
b) Determine the area of G as the limit of left-endpoint sums. The following expression for the
sum of the squares of the fist n positive integers will be helpful:

$$\sum_{k=1}^{n} k^2 = \frac{1}{6}n(n+1)(2n+1)$$

(as you can confirm by mathematical induction).

Solution

a) Figure 7 shows the graph of f and the region G.

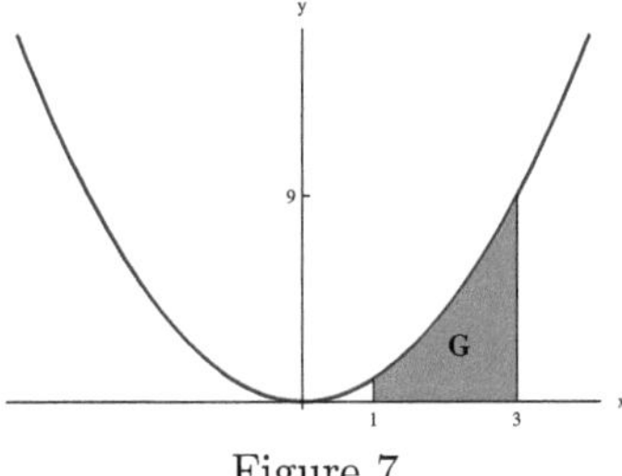

Figure 7

b) The interval $[1, 3]$ is subdivided into n subintervals of length

$$\Delta x = \frac{3-1}{n} = \frac{2}{n}.$$

The corresponding partition consists of the points

$$x_k = 1 + k\Delta x = 1 + k\left(\frac{2}{n}\right), \ k = 0, 1, 2, \ldots, n.$$

Therefore, the corresponding left-endpoint sum for f is

$$l_n = \sum_{k=1}^{n} f\left(x_{k-1}\right) \Delta x = \sum_{k=1}^{n} f\left(1 + (k-1)\left(\frac{2}{n}\right)\right)\frac{2}{n}$$

$$= \frac{2}{n}\sum_{k=1}^{n}\left(1 + \frac{2(k-1)}{n}\right)^2$$

$$= \frac{2}{n}\sum_{k=1}^{n}\left(1 + \frac{4(k-1)}{n} + \frac{4(k-1)^2}{n^2}\right)$$

$$= \frac{2}{n}\left(\sum_{k=1}^{n}1 + \frac{4}{n}\sum_{k=1}^{n}(k-1) + \frac{4}{n^2}\sum_{k=1}^{n}(k-1)^2\right)$$

$$= \frac{2}{n}\sum_{k=1}^{n}1 + \frac{8}{n^2}\sum_{k=1}^{n}(k-1) + \frac{8}{n^3}\sum_{k=1}^{n}(k-1)^2 .$$

We have

$$\sum_{k=1}^{n}1 = 1 + 1 + \cdots + 1 = n,$$

since n terms are added.

We also have

$$\sum_{k=1}^{n}(k-1) = 0 + 1 + 2 + \cdots + (n-1) = \sum_{j=1}^{n-1}j = \frac{(n-1)\,n}{2},$$

as in Example 1 (with n replaced by $n-1$).

Finally,

$$\sum_{k=1}^{n}(k-1)^2 = 0 + 1^2 + 3^2 + \cdots + (n-1)^2 = \sum_{j=1}^{n-1}j^2.$$

We will apply the formula

$$\sum_{k=1}^{n}k^2 = \frac{1}{6}n\,(n+1)\,(2n+1),$$

with n replaced by $n-1$. Thus,

$$\sum_{j=1}^{n-1}j^2 = \frac{1}{6}(n-1)\,(n)\,(2n-1).$$

Therefore,

$$l_n = \frac{2}{n}\sum_{k=1}^{n}1 + \frac{8}{n^2}\sum_{k=1}^{n}(k-1) + \frac{8}{n^3}\sum_{k=1}^{n}(k-1)^2$$

$$= \frac{2}{n}(n) + \frac{8}{n^2}\left(\frac{(n-1)\,n}{2}\right) + \frac{8}{n^3}\left(\frac{1}{6}(n-1)\,(n)\,(2n-1)\right)$$

$$= 2 + \frac{4(n-1)}{n} + \frac{4(n-1)(2n-1)}{3n^2}.$$

Thus,

$$\lim_{n\to\infty} l_n = 2 + 4\lim_{n\to\infty}\frac{n-1}{n} + \frac{4}{3}\lim_{n\to\infty}\frac{(n-1)(2n-1)}{n^2} = 2 + 4 + \frac{4}{3}(2) = \frac{26}{3}.$$

Therefore, the area of the region G between the graph of f and the interval $[1, 3]$ is 26/3.

Figure 8 shows the rectangles corresponding to the partitioning of the interval $[1, 2]$ into 10 subintervals of equal length. $\square$

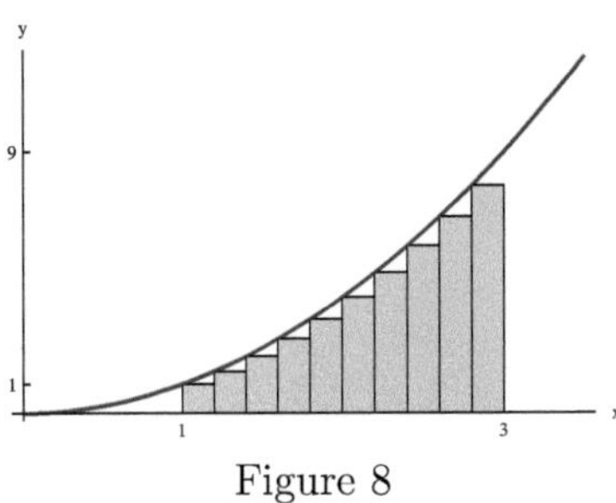

Figure 8

Example 5 Let $f(x) = \sin(x)$. In Section 5.3 we will show that the area of the region G between the graph of f and the interval $[0, \pi]$ is 2.

a) Sketch the region G.
b) Midpoint sums are usually more accurate in approximating the area, compared to left-endpoint sums and right-endpoint sums. Approximate the area of G by midpoint sums that correspond to the partitioning of $[0, \pi]$ to 2^k subintervals of equal length, where $k = 2, \ldots, 7$. Do the numbers support the expectation that it should be possible to approximate the area of G with desired accuracy by a midpoint sum, provided that the length of each subinterval is small enough?

Solution

a) Figure 9 shows the region G.

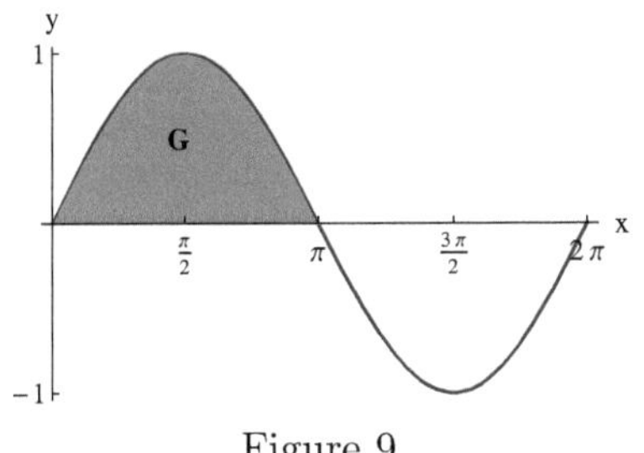

Figure 9

b) We have

$$m_n = \sum_{k=1}^{n} f(c_k)\,\Delta x = \sum_{k=1}^{n} \sin(c_k)\Delta x,$$

where

$$\Delta x = \frac{\pi}{n} \quad \text{and} \quad c_k = \left(k - \frac{1}{2}\right)\Delta x.$$

Figure 10 shows the rectangles corresponding to a partitioning of the interval $[0, \pi]$ to 16 subintervals of equal length.

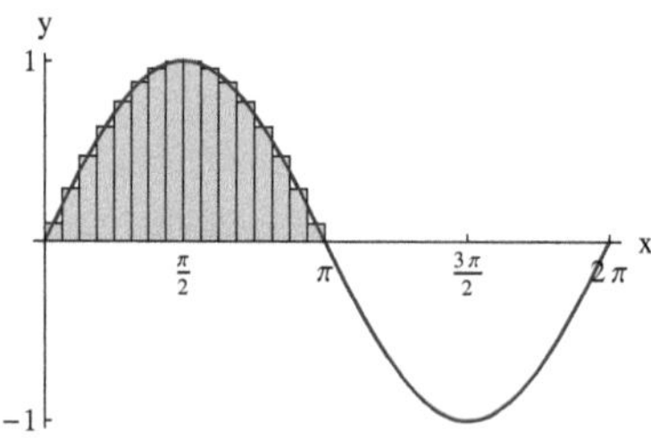

Figure 10

Table 1 displays the relevant data. The numbers support the expectation that it should be possible to approximate the area of G with desired accuracy by a midpoint sum, if the length of each subinterval is small enough. $\square$

n	Δx	m_n	$\lvert m_n - 2 \rvert$
4	.25	2.052 34	5.2×10^{-2}
8	.125	2.012 91	1.3×10^{-2}
16	.0625	2.003 22	3.2×10^{-3}
32	.03125	2.000 8	8.0×10^{-4}
64	.015625	2.000 2	2.0×10^{-4}
128	$7.812\,5 \times 10^{-3}$	2.000 05	5.0×10^{-5}

Table 1

In the next section we will introduce a fundamental concept of calculus, namely **the integral**. You will see that the integral of a positive-valued function can be interpreted as area.

Problems

1.

$$\sum_{k=1}^{5} (4k - 1)$$

2.

$$\sum_{k=1}^{5} \left(2k^3 + k^2 + 1\right)$$

3.

$$\sum_{j=1}^{7} \cos\left(\frac{\pi}{3}j\right)$$

4.

$$\sum_{n=1}^{8} \sin\left(\frac{\pi}{6}n\right)$$

5. Let $S_n = \sum_{k=1}^{n} k^2$.

a) Use mathematical induction to show that

$$S_n = \frac{n\,(n+1)\,(2n+1)}{6}.$$

b) Determine $\lim\limits_{n\to\infty} \dfrac{S_n}{n^3}$.

6. Let $S_n = \sum_{k=1}^{n} k^3$.

a) Use mathematical induction to show that

$$S_n = \left(\frac{n\,(n+1)}{2}\right)^2.$$

b) Determine $\lim\limits_{n\to\infty} \dfrac{S_n}{n^4}$.

[C] In problems 7 and 8, let G be the region between the graph of the function f and the given interval. Sketch the region G. Determine the area of G by making use of known formulas. Determine approximations to the area of G by left- endpoint, right-endpoint and midpoint sums corresponding to subdividing $[a, b]$ into $8, 16, 32$ and 64 subintervals of equal length. Calculate the absolute error of the approximations. Which method provides the most accurate approximations?

7.

$$f(x) = 10 - 2x, \quad [0, 5]$$

8.

$$f(x) = \sqrt{9 - x^2}, \quad [-3, 3]$$

[C] In problems 8 and 9, Let G be the region between the graph of the function f and the given interval. Plot the region G with the help of your graphing utility. The exact value of the area of G is given. Determine approximations to the area of G by midpoint sums corresponding to subdividing $[a, b]$ into $8, 16, 32$ and 64 subintervals of equal length. Calculate the absolute error of the approximations.

9.

$$f(x) = \frac{x}{x^2 + 4}, \quad [1, 2], \quad \text{Area of } G = \frac{3}{2}\ln(2) - \frac{1}{2}\ln(5) \cong 0.235\,002$$

10.

$$f(x) = x^2 \sin(x), \, [0, \pi], \quad \text{Area of } G = \pi^2 - 4 \cong 5.\,869\,6$$

5.2 The Definition of the Integral

In this section we will introduce the fundamental concept of the **integral**. The integral of a positive-valued function on an interval is the area of the region between the graph of the function and the interval. We will be able to interpret the integral of a function that has positive or negative values on an interval as **"the signed area"** of the region between the graph of the function and the interval. In the next section you will see that the **displacement** of an object in one-dimensional motion over a time interval is **the integral of the velocity function** on that interval. In later chapters the integral will appear as the **work** done in moving an object, or as the **probability** that the values of a random variable are in a certain interval.

The Riemann Integral and Signed Area

As in Section 5.1, let $P = \{x_0, x_1, \ldots, x_{k-1}, x_k, \ldots, x_n\}$ be a **partition** of the interval $[a, b]$, so that

$$a = x_0 < x_1 < x_2 < \cdots x_{k-1} < x_k < \cdots < x_{n-1} < x_n = b.$$

Recall that $\|P\|$, **the norm of the partition** P, is the maximum of the lengths of subintervals determined by P:

$$\|P\| = \max_{k=1,\ldots,n} \Delta x_k = \max_{k=1,\ldots,n} (x_k - x_{k-1}).$$

Let f be a function defined on $[a, b]$. As in Section 5.1, we will sample an arbitrary value of x in the kth subinterval $[x_{k-1}, x_k]$ and denote it by x_k^*. Thus, $x_{k-1} \le x_k^* \le x_k$, but there is no other restriction on the choice of x_k^*.

Definition 1 Assume that $P = \{x_0, x_1, \ldots, x_{k-1}, x_k, \ldots, x_n\}$ is a partition of the interval $[a, b]$, and $x_k^* \in [x_{k-1}, x_k]$. A sum of the form

$$\sum_{k=1}^{n} f(x_k^*)\Delta x_k$$

is **a Riemann sum for f on the interval** $[a, b]$.

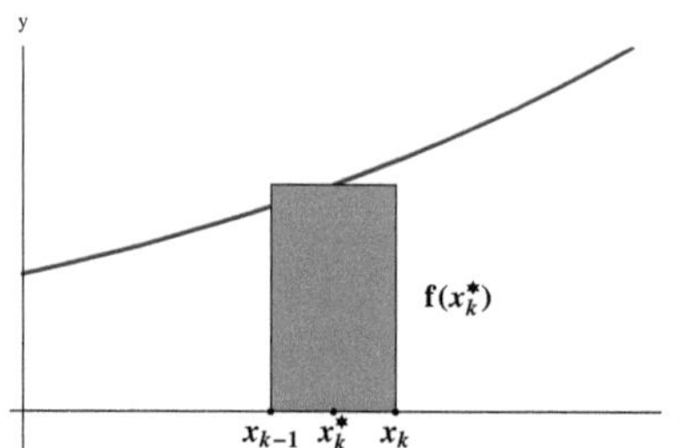

Figure 1: A typical term of a Riemann sum is $f(x_k^*)\Delta x_k$

Thus, a Riemann sum for f on $[a, b]$ approximates the area of the region between the graph of f and the interval $[a, b]$ if $f(x) \geq 0$ for each $x \in [a, b]$ and the norm of the partition is small. Let's lift the restriction on the sign of f, and **assume that any Riemann sum for f on $[a, b]$ approximates a number which depends only on the function f and the interval $[a, b]$ if the norm of the partition is small.** We will denote that number as

$$\int_a^b f(x)\,dx$$

and refer to it as **the Riemann integral of f on $[a, b]$.** You can imagine that we have replaced the summation symbol in the expression

$$\sum_{k=1}^{n} f(x_k^*)\,\Delta x_k$$

by an elongated S, and Δx_k by dx ("dx" within the present context should not be confused with "dx" within the context of the differential, although a connection will arise later). We will also assume that the approximation is as accurate as desired provided that the norm of the partition is small enough. Thus, we can define the Riemann integral of f on $[a, b]$ as follows:

Definition 2 (The informal definition of the integral) We say that a function f is **Riemann integrable on the interval** $[a, b]$ and that **the Riemann integral of f on $[a, b]$** is

$$\int_a^b f(x)\,dx$$

if

$$\left| \sum_{k=1}^{n} f(x_k^*)\,\Delta x_k - \int_a^b f(x)\,dx \right|$$

is as small as desired provided that the norm of the partition $P = \{x_0, x_1, \ldots, x_n\}$ of $[a, b]$ is sufficiently small.

Thus, the Riemann integral of f on $[a, b]$ corresponds to the area of the region between the graph of f and $[a, b]$ if f is positive-valued on $[a, b]$.

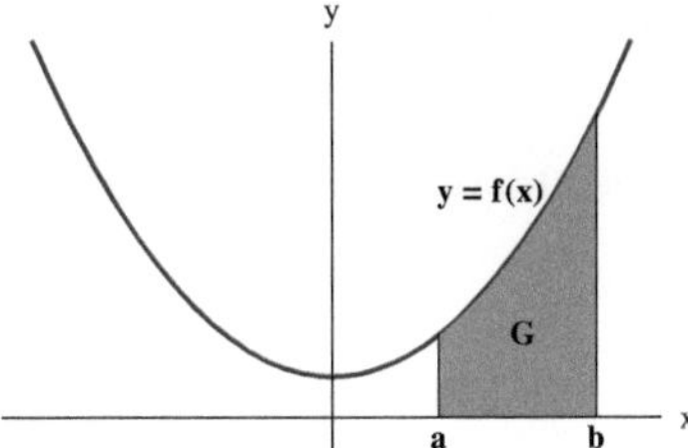

Figure 2: $\int_a^b f(x)\, dx$ is the area of G if f is positive-valued

We may express the relationship between Riemann sums and the Riemann integral by writing

$$\lim_{||P|| \to 0} \sum_{k=1}^{n} f(x_k^*)\, \Delta x_k = \int_a^b f(x)\, dx.$$

You can find the precise definition of the Riemann integral at the end of this section. **Riemann** was a mathematician who made crucial contributions in many areas of mathematics, and played a prominent role in establishing firm foundations for the concept of the integral. Since we will not have occasion to use any other type of integral in this book, we will refer to the Riemann integral simply as "the integral".

In the notation,

$$\int_a^b f(x)\, dx,$$

for the integral of f on $[a, b]$, the number a is referred to as **the lower limit** of the integral, and b as **the upper limit** of the integral. The function f is **the integrand**. The computation of the integral may be described by saying that "f **is integrated from** a **to** b".

We will calculate many integrals in the following sections. Let's determine the integrals of constant functions before we proceed further. If f is constant and has the value $c > 0$, the region between the graph of f and an interval $[a, b]$ is a rectangle with area $c\,(b - a)$. Therefore, we should have

$$\int_a^b f(x)dx = \int_a^b c\,dx = c(b - a).$$

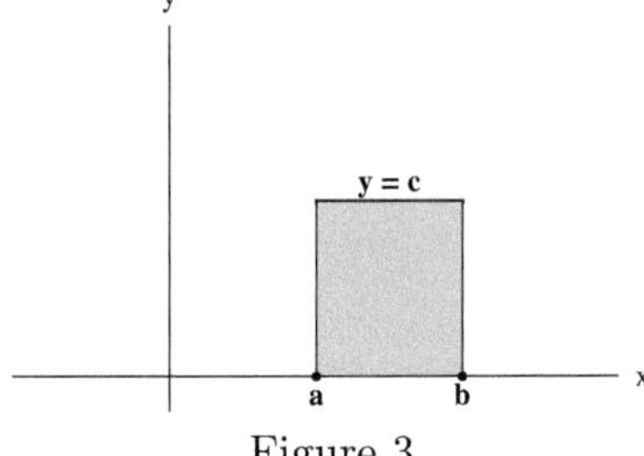

Figure 3

This is the case, irrespective of the sign of c. Indeed, for any partition $\{x_0, x_1, \ldots, x_n\}$ of $[a, b]$ and any choice of the intermediate points x_k^*,

$$\sum_{k=1}^{n} f\left(x_k^*\right) \Delta x_k = \sum_{k=1}^{n} c \Delta x_k = c \sum_{k=1}^{n} \Delta x_k = c\left(b - a\right),$$

since the sum of the lengths of the subintervals is the length of the interval $[a, b]$. Let's record this fact:

Proposition 1 Let f be a constant function, so that $f(x) = c$ for each $x \in \mathbb{R}$, where c is a constant. Then

$$\int_a^b f(x)\, dx = \int_a^b c\, dx = c(b - a).$$

You can find an example of a function that is not Riemann integrable at the end of this section. We have the assurance every continuous function is Riemann integrable:

Theorem 1 Assume that f is continuous on the interval $[a, b]$. Then f is Riemann integrable on $[a, b]$.

The proof of the theorem is left to a course in advanced calculus.

By Theorem 1, a Riemann sum

$$\sum_{k=1}^{n} f\left(x_k^*\right) \Delta x_k$$

for the function f on the interval $[a, b]$ approximates the integral of f on $[a, b]$ as accurately as desired, provided that f is continuous on $[a, b]$ and $\max_k \Delta x_k$ is small enough. In particular, we can approximate an integral by **left-endpoint sums, right-end point sums** or **midpoint sums**, as in Section 5.1 (without the restriction that the functions are positive-valued). If

$$\Delta x = \frac{b - a}{n},$$

Δx is as small as necessary if n is sufficiently large. Therefore,

$$\lim_{n \to \infty} l_n = \lim_{n \to \infty} r_n = \lim_{n \to \infty} m_n = \int_a^b f(x)\, dx,$$

with the notation of Section 5.1.

Example 1 Let $f(x) = x$, as in Example 3 of Section 5.1. In that example, we approximated the area of the region G between the graph of f and the interval $[0, 1]$ by right-endpoint sums. We showed that

$$\lim_{n \to \infty} r_n = \lim_{n \to \infty} \frac{n\left(n + 1\right)}{2n^2} = \frac{1}{2}.$$

Therefore,

$$\int_0^1 f(x)\, dx = \int_0^1 x\, dx = \frac{1}{2}.$$

$\square$

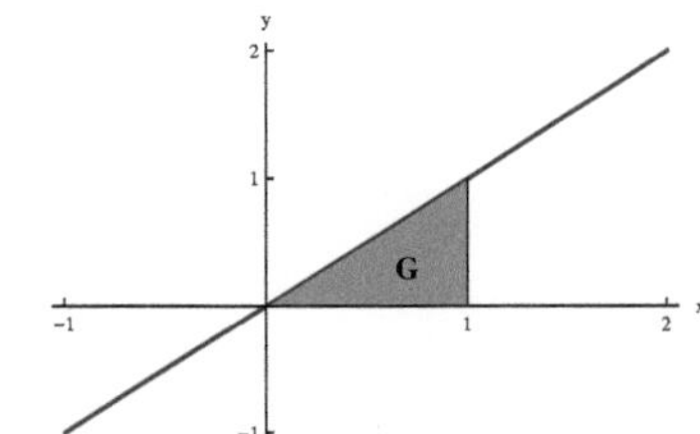

Figure 4: The area of G is $\int_0^1 x\,dx = 0.5$

Let's consider the case of a function f that is continuous on an interval $[a, b]$ and $f(x) \leq 0$ for each $x \in [a, b]$, and interpret the integral of f on $[a, b]$ geometrically. Let G be the region between the graph of f and $[a, b]$, as illustrated in Figure 5.

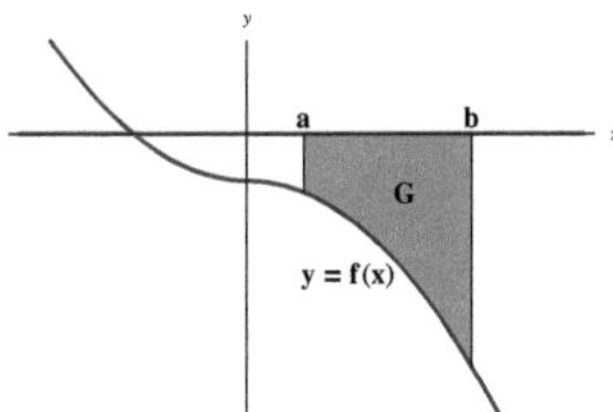

Figure 5: $\int_a^b f(x)\,dx$ is the signed area of G

Let $P = \{x_0, x_1, x_2, \ldots, x_n\}$ be a partition of $[a, b]$, and $x_k^* \in [x_{k-1}, x_k]$, $k = 1, 2, \ldots, n$. If $\|P\|$ is small, the Riemann sum

$$\sum_{k=1}^{n} f(x_k^*)\,\Delta x_k$$

approximates

$$\int_a^b f(x)\,dx.$$

Consider the rectangle R_k that has the vertices $(x_{k-1}, 0)$, $(x_k, 0)$, $(x_{k-1}, f(x_k^*))$ and $(x_k, f(x_k^*))$, as in Figure 6.

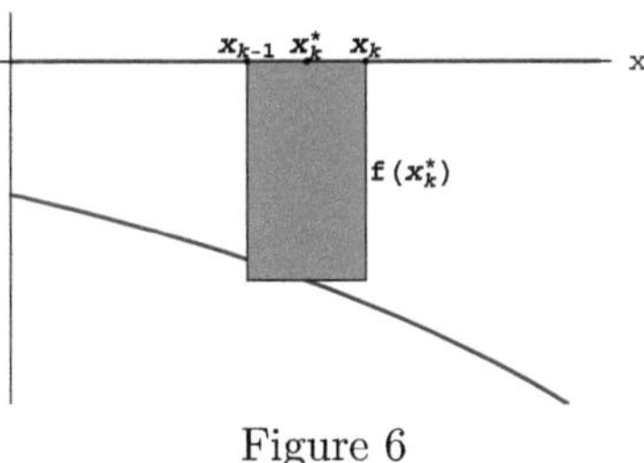

Figure 6

Since $f(x) \leq 0$ for each $x \in [a, b]$, the term $f(x_k^*)\,\Delta x_k$ is $(-1) \times$ (the area of R_k). Thus, the Riemann sum

$$\sum_{k=1}^{n} f(x_k^*)\,\Delta x_k$$

approximates $(-1) \times$ (area of G). We will refer to $(-1) \times$ (area of G) as **the signed area of G. Therefore, we will identify the integral of f on $[a, b]$ with the signed area** of G:

$$\text{The signed area of } G = \int_a^b f(x)\,dx.$$

The area of G is

$$-\int_a^b f(x)\,dx.$$

Example 2 Let $f(x) = \sin(x)$. Figure 7 shows the region G between the graph of f and the interval $[\pi, 4\pi/3]$.

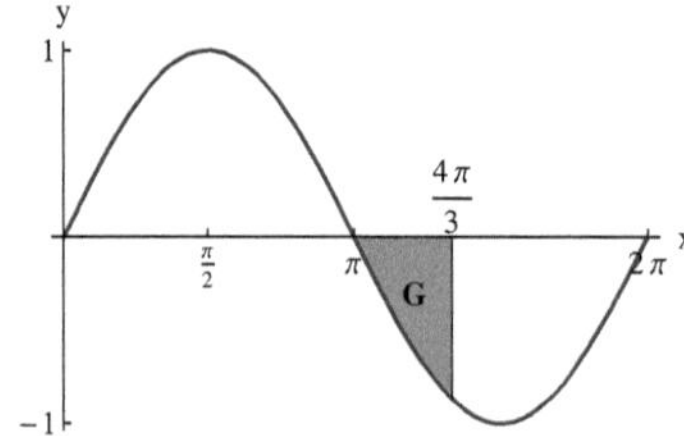

Figure 6: The signed area of G is $-1/2$

We have $\sin(x) \le 0$ if $\pi \le x \le 4\pi/3$. In Section 5.3 we will show that

$$\int_\pi^{4\pi/3} \sin(x)\,dx = -\frac{1}{2}.$$

Therefore, the signed area of G is $-1/2$, and the area of G is

$$-\int_\pi^{4\pi/3} \sin(x)\,dx = -\left(-\frac{1}{2}\right) = \frac{1}{2}.$$

Approximate the integral of sine on $[\pi, 4\pi/3]$ by midpoint sums that correspond to the partitioning of $[0, \pi]$ to 2^k subintervals of equal length, where $k = 2, \ldots, 6$.

Solution

We have

$$m_n = \sum_{k=1}^n \sin(c_k)\,\Delta x,$$

where

$$\Delta x = \frac{\dfrac{4\pi}{3} - \pi}{n} = \frac{\pi}{3n} \quad \text{and} \quad c_k = \left(k - \frac{1}{2}\right)\Delta x, \ k = 1, 2, \ldots . n.$$

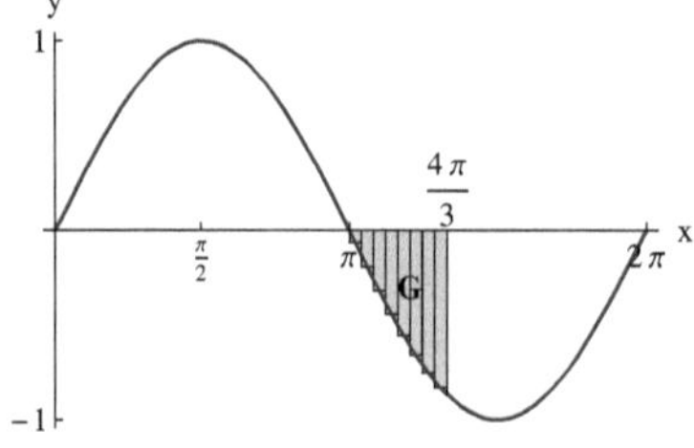

Figure 8

Table 1 displays m_n for $n = 4, 8, 16, 32$ and 64. The numbers in Table 1 are consistent with the fact that

$$\lim_{n \to \infty} m_n = \int_{\pi}^{4\pi/3} \sin(x)\, dx = -\frac{1}{2}.$$

□

n	m_n
4	-0.501431
8	-0.500357
16	-0.500089
32	-0.500022
64	-0.500006

Table 1

With reference to Figure 9, if a function f is positive-valued on $[a, b]$ we should have

$$(\text{area of } G_1) + (\text{area of } G_2) = \text{area of } G_1 \cup G_2.$$

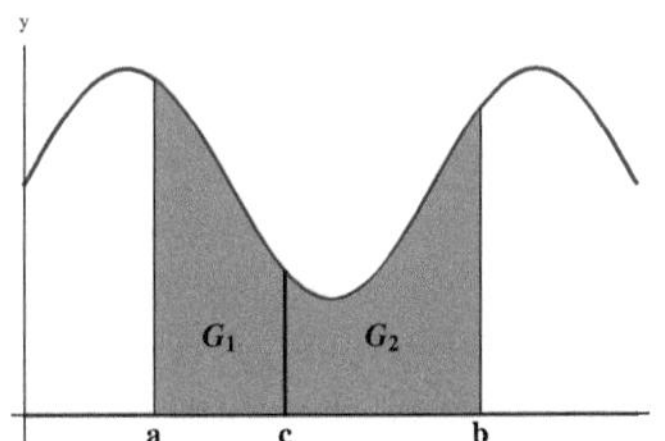

Figure 9: The integral is additive with respect to intervals

Thus, we expect that

$$\int_a^c f(x)dx + \int_c^b f(x)dx = \int_a^b f(x)dx.$$

This is indeed the case, irrespective of the sign of the function. We will refer to this property of the integral as **"the additivity of the integral with respect to intervals"**.

Theorem 2 (The Additivity of the Integral with respect to Intervals) Assume that f is continuous on $[a, b]$ and $a < c < b$. Then

$$\int_a^c f(x)\,dx + \int_c^b f(x)\,dx = \int_a^b f(x)\,dx.$$

We leave the rigorous proof of Theorem 2 to a course in advanced calculus.

Let's assume that f is continuous on the interval $[a, b]$ and that the sign of f changes at a finite number of points in (a, b). In order to be specific, let's assume that $f(c) = 0$, $f(x) > 0$ on (a, c) and $f(x) < 0$ on (c, b), as in Figure 10. With reference to Figure 10, the region G between the graph of f and the interval $[a, b]$ is the union of G_+ and G_-.

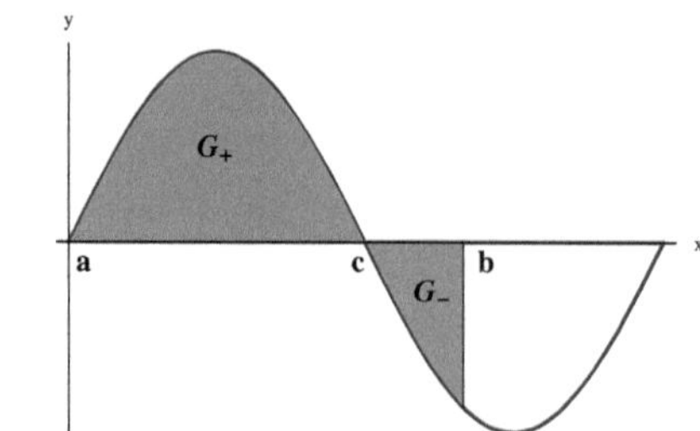

Figure 10: $\int_a^b f(x)dx =$ area of G_+- area of G_-

By the additivity of the integral with respect to intervals,

$$\int_a^c f(x)dx + \int_c^b f(x)dx = \int_a^b f(x)dx.$$

Thus,

$$(\text{area of } G_+) + (\text{signed area of } G_-) = \int_a^b f(x)\,dx.$$

We will identify the signed area of the region $G = G_+ \cup G_-$ with the integral of f on $[a, b]$. The area of G is

$$\int_a^c f(x)dx - \int_c^b f(x)dx.$$

More generally, if a function f is continuous on an interval $[a, b]$, **we will identify the signed area of the region G between the graph of f and $[a, b]$ with the integral of f on $[a, b]$**. If we wish to compute the area of G, we must determine the subintervals of $[a, b]$ on which f has constant sign, and calculate the integral of f on each subinterval. The integral must be multiplied by -1 if the sign of f is negative on the relevant subinterval.

Example 3 Let $f(x) = \sin(x)$.

a) Sketch the region G between the graph of f and the interval $[0, 4\pi/3]$.

b) In Section 5.3 we will show that

$$\int_0^\pi \sin(x) = 2 \quad\text{and}\quad \int_\pi^{4\pi/3} \sin(x)\,dx = -\frac{1}{2}$$

Determine the signed area and the area of G.

c) Approximate

$$\int_0^{4\pi/3} \sin(x)\,dx$$

by midpoint sums corresponding to the partitioning of the interval $[0, 4\pi/3]$ into 2^k subintervals of equal length, where $k = 3, \dots, 7$.

Solution

a) Figure 11 shows the region G.

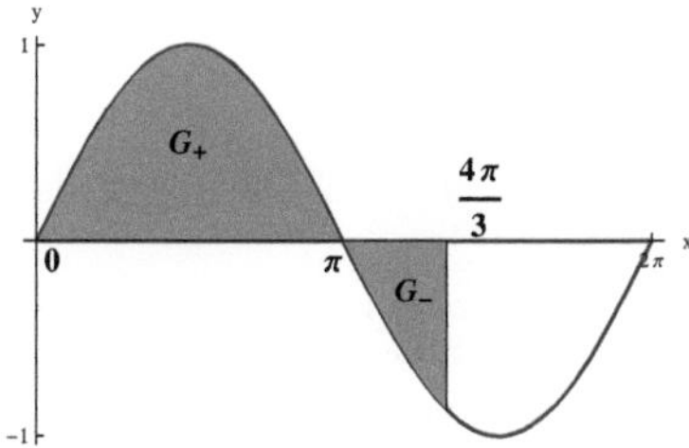

Figure 11

b) With reference to Figure 11,

$$\text{the area of } G_+ = \int_0^{\pi} \sin(x) = 2,$$

and

$$\text{the signed area of } G_- = \int_{\pi}^{4\pi/3} \sin(x)\,dx = -\frac{1}{2}.$$

Since $\sin(x) < 0$ if $\pi < x < 4\pi/3$, the area of G_- is $1/2$.
The signed area of G is

$$\int_0^{4\pi/3} \sin(x)\,dx = \int_0^{\pi} \sin(x)\,dx + \int_{\pi}^{4\pi/3} \sin(x)\,dx = 2 + \left(-\frac{1}{2}\right) = \frac{3}{2},$$

and the area of G is

$$\text{area of } G_+ + \text{ area of } G_- = \int_0^{\pi} \sin(x)\,dx - \int_{\pi}^{4\pi/3} \sin(x)\,dx = 2 - \left(-\frac{1}{2}\right) = \frac{5}{2}.$$

c) The midpoint sum corresponding to the partitioning of the interval $[0, 4\pi/3]$ to n subintervals of equal length is

$$m_n = \sum_{k=1}^{n} f(c_k)\,\Delta x = \sum_{k=1}^{n} \sin(c_k)\Delta x,$$

where

$$\Delta x = \frac{\frac{4\pi}{3}}{n} = \frac{4\pi}{3n} \text{ and } c_k = \left(k - \frac{1}{2}\right)\Delta x.$$

Table 2 displays m_n and

$$\left| m_n - \int_0^{4\pi/3} \sin(x)\,dx \right|$$

for $n = 2^k$, $k = 3, \ldots, 7$. The numbers in Table 2 support the expectation that

$$\lim_{n \to \infty} m_n = \int_0^{4\pi/3} \sin(x)\,dx.$$

□

n	Δx	m_n	$|m_n - 1.5|$
8	$.523\,599$	1.51727	1.7×10^{-2}
16	$.261\,799$	1.50429	4.3×10^{-3}
32	$.130\,9$	1.50107	1.1×10^{-3}
64	$6.544\,98 \times 10^{-2}$	1.50027	2.7×10^{-4}
128	$3.272\,49 \times 10^{-2}$	1.50007	6.7×10^{-5}

Table 2

Remark 1 In the notation

$$\int_a^b f(x)\,dx,$$

the variable x is a **dummy variable**, in the sense that the letter x can be replaced by any other letter. Thus, the expressions

$$\int_a^b f(x)\,dx,\ \int_a^b f(t)\,dt,\ \int_a^b f(\tau)\,d\tau$$

all have the same meaning: The integral of the function f on the interval $[a, b]$. For example,

$$\int_0^\pi \sin(x)\,dx = \int_0^\pi \sin(t)\,dt = \int_0^\pi \sin(u)\,du = 2.$$

This is parallel to the fact that the summation index for a Riemann sum is a dummy index, and we can use any letter to denote the independent variable of the function: The expressions

$$\sum_{k=1}^n f(x_k^*)\,\Delta x_k,\ \sum_{l=1}^n f(x_l^*)\,\Delta x_l,\ \sum_{j=1}^n f(t_j^*)\,\Delta x_j$$

have the same meaning. $\Diamond$

Remark 2 Your computational utility should be able to provide you with an accurate approximation to an integral. The underlying approximation schemes are referred to as **numerical integration schemes**, or **numerical integration rules**. We will see some of these rules in Section 6.5. A computer algebra system such as Maple or Mathematica is able to provide you with the exact value of many integrals. Soon, you will be able to compute the exact values of many integrals yourselves. $\Diamond$

The Integrals of Piecewise Continuous Functions

Theorem 1 states that a function which is continuous on a closed and bounded interval is (Riemann) integrable on that interval. It will be useful to expand the scope of the integral to a wider class of functions.

Assume that f is continuous on the interval (a, b) and

$$\lim_{x \to a+} f(x) \quad \text{and} \quad \lim_{x \to b-} f(x)$$

exist. If we set

$$g(x) = \begin{cases} f(x) & \text{if} \quad x \in (a, b), \\ \lim_{x \to a+} f(x) & \text{if} \quad x = a, \\ \lim_{x \to b-} f(x) & \text{if} \quad x = b, \end{cases}$$

then g is continuous on $[a, b]$. We define the integral of f on $[a, b]$ to be the same as the integral of g on $[a, b]$:

$$\int_a^b f(x)\,dx = \int_a^b g(x)\,dx.$$

This amounts to the fact that

$$\int_a^b f(x)\,dx$$

is approximated by Riemann sums of the form

$$\sum_{k=1}^{n} f\left(x_k^*\right) \Delta x_k,$$

where $f\left(x_0\right)$ should be interpreted as $\lim_{x \to a+} f\left(x\right)$ and $f\left(b\right)$ should be interpreted as $\lim_{x \to b-} f(x)$.

Example 4 Let

$$f(x) = \frac{\sin\left(x\right)}{x}$$

if $x \neq 0$.

a) Discuss the definition of $\int_0^\pi f\left(x\right) dx$.
b) Consider the approximate value of

$$\int_0^\pi \frac{\sin\left(x\right)}{x} dx$$

that you obtain from your computational utility to be the exact value of the integral. Approximate

$$\int_0^\pi \frac{\sin\left(x\right)}{x} dx$$

by midpoint sums corresponding to the partitioning of $[a, b]$ into 10, 20, 40 and 80 subintervals of equal length. Do the numbers support the fact that the integral can be approximated with desired accuracy by Riemann sums, provided that the norm of the partition is small enough?

Solution

a) Since $\sin\left(x\right)$ and x define continuous functions on the number line, the quotient f is continuous on the entire number line, with the exception $x = 0$. We have

$$\lim_{x \to 0} f(x) = \lim_{x \to 0} \frac{\sin\left(x\right)}{x} = 1.$$

If we set

$$g\left(x\right) = \begin{cases} \dfrac{\sin x}{x} & \text{if} \quad x \neq 0, \\ 1 & \text{if} \quad x = 0, \end{cases}$$

then g is continuous on $[0, \pi]$. We set

$$\int_0^\pi \frac{\sin\left(x\right)}{x} dx = \int_0^\pi g\left(x\right) dx.$$

The integral corresponds to the area of the region between the graph of f and the interval $[0, \pi]$, as illustrated in Figure 12.

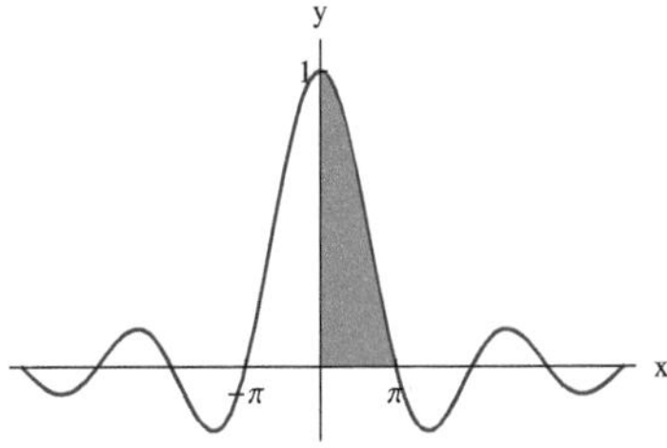

Figure 12

b) We have

$$m_n = \sum_{k=1}^{n} \frac{\sin{(c_k)}}{c_k} \Delta x,$$

where

$$\Delta x = \frac{\pi}{n} \text{ and } c_k = \left(k - \frac{1}{2} \right) \Delta x.$$

Table 3 displays the relevant data. We have

$$\int_0^{\pi} \frac{\sin{(x)}}{x} dx \cong 1.\,851\,94,$$

rounded to 6 significant digits. The numbers in Table 3 support the fact that the integral can be approximated with desired accuracy by Riemann sums, provided that the norm of the partition is small enough.$\square$

n	m_n	$\left\lvert m_n - \int_0^{\pi} f(x)dx \right\rvert$
10	1.85325	1.3×10^{-3}
20	1.85226	3.3×10^{-4}
40	1.85202	8.2×10^{-5}
80	1.85196	2.0×10^{-5}

Table 3

We will say that f is piecewise continuous on the interval $[a, b]$ if f has at most finitely many removable or jump discontinuities in $[a, b]$. Thus, f has (finite) one-sided limits at its discontinuities. In such a case **we will define the integral of f on $[a, b]$ as the sum of its integrals over the subintervals of $[a, b]$ that are separated from each other by the points of discontinuity of f.**

Example 5 Let

$$f(x) = \begin{cases} \sin{(x)} & \text{if} \quad 0 \le x < \pi/2, \\ \cos{(x)} & \text{if} \quad \pi/2 < x < 3\pi/2. \end{cases}$$

Figure 13 shows the graph of f and the region between the graph of f and the interval $[0, 3\pi/2]$.

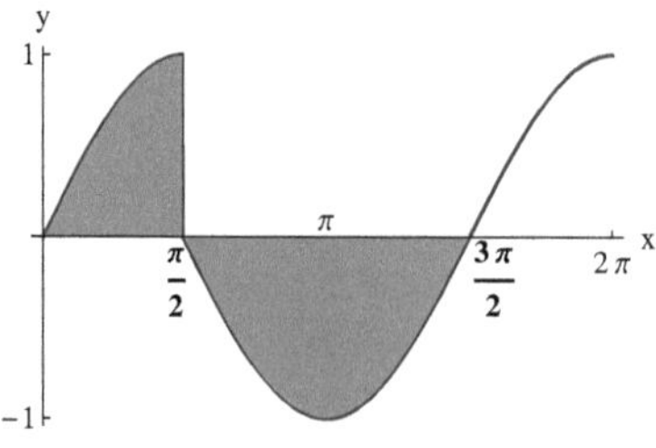

Figure 13

The function f is piecewise continuous on the interval $[0, 3\pi/2]$. Indeed, the only point of discontinuity of f in $[0, 3\pi/2]$ is $\pi/2$. We have

$$\lim_{x \to \pi/2-} f(x) = \lim_{x \to \pi/2} \sin{(x)} = 1,$$

and

$$\lim_{x \to \pi/2+} f(x) = \lim_{x \to \pi/2} \cos(x) = 0.$$

Therefore,

$$\int_0^{3\pi/2} f(x)\, dx = \int_0^{\pi/2} f(x)\, dx + \int_{\pi/2}^{3\pi/2} f(x)\, dx = \int_0^{\pi/2} \sin(x)\, dx + \int_{\pi/2}^{3\pi/2} \cos(x)\, dx.$$

It can be shown that

$$\int_0^{\pi/2} \sin(x)\, dx = 1 \quad \text{and} \quad \int_{\pi/2}^{3\pi/2} \cos(x)\, dx = -2,$$

once we have developed the necessary tools in Section 5.3. Therefore,

$$\int_0^{3\pi/2} f(x)\, dx = 1 - 2 = -1.$$

Thus, the signed area of the region between the graph of f and the interval $[0, 2\pi]$ is -1. $\square$

The Precise Definition of the Integral

We quantify the expressions "with desired accuracy" and "sufficiently small" that appear in the informal definition of the integral (Definition 2):

Definition 3 We say that a function f **is Riemann integrable on the interval** $[a, b]$ and that **the Riemann integral of** f **on** $[a, b]$ is

$$\int_a^b f(x)\, dx$$

if, given any $\varepsilon > 0$ there exists $\delta > 0$ such that

$$\left| \sum_{k=1}^{n} f(x_k^*) \Delta x_k - \int_a^b f(x)\, dx \right| < \varepsilon,$$

where $P = \{x_0, x_1, x_2, \ldots, x_n\}$ is a partition of $[a, b]$, $x_k^* \in [x_{k-1}, x_k]$, $\Delta x_k = x_k - x_{k-1}$ for $k = 1, 2, \ldots, n$, and

$$\|P\| = \max_k \Delta x_k < \delta.$$

You may think of $\varepsilon > 0$ as an arbitrary "error tolerance" that is as small as desired. The positive δ that is referred to in the definition depends on ε, and must be sufficiently small so that the absolute value of the error in the approximation of the integral by *any* Riemann sum

$$\sum_{k=1}^{n} f(x_k^*) \Delta x_k$$

is smaller than ε, provided that $\|P\| < \delta$. We should emphasize that there is complete freedom in the choice of the partition P and the choice of the intermediate points x_k^*, as long as $\|P\| < \delta$.

Remark 3 There are functions that are not Riemann integrable. For example, set

$$f(x) = \begin{cases} 1 & \text{if} \quad x \text{ is rational,} \\ 0 & \text{if} \quad x \text{ is irrational.} \end{cases}$$

We claim that f is not Riemann integrable on $[0, 1]$:

Let's set

$$x_k = \frac{k}{n}, \ k = 0, 1, 2, \ldots, n,$$

so that $\Delta x_k = 1/n$, $k = 1, 2, \ldots, n$, and

$$P_n = \{x_0, x_1, x_2, \ldots, x_n\} = \left\{1, \frac{1}{n}, \frac{2}{n}, \cdots, 1\right\}.$$

If each x_k^* is rational,

$$\sum_{k=1}^{n} f\left(x_k^*\right) \Delta x_k = \sum_{k=1}^{n} (1) \left(\frac{1}{n}\right) = n\left(\frac{1}{n}\right) = 1.$$

If each x_k^* is irrational,

$$\sum_{k=1}^{n} f\left(x_k^*\right) \Delta x_k = \sum_{k=1}^{n} (0) \Delta x_k = 0.$$

It can be shown that there are rational and irrational numbers in any interval, however small it may be. Since $\|P_n\| = 1/n$, and $\lim_{n\to\infty} 1/n = 0$, we can find partitions of arbitrarily small norm and corresponding Riemann sums that are 1 or 0. Therefore, we cannot assert that there is a definite number that is approximated by any Riemann sum with desired accuracy, provided that the norm of the relevant partition is small enough. This rules out the existence of the integral of the function. $\Diamond$

Problems

In problems 1-4 sketch the region G such that $\int_a^b f(x)\, dx$ corresponds to the signed area of G.

1.
$$\int_{-2}^{4} \left(x^2 - 4x\right) dx$$

3.
$$\int_{-\pi/2}^{\pi/2} \sin(2x)\, dx$$

2.
$$\int_{0}^{2} \left(-8 + 12x - 4x^2\right) dx$$

4.
$$\int_{0}^{\pi/2} \cos(2x)\, dx$$

[C] In problems 5-8 the exact value of $\int_a^b f(x)\, dx$ is given. Approximate $\int_a^b f(x)\, dx$ by midpoint sums corresponding to the partitioning of $[a, b]$ to 8, 16, 32 and 64 subintervals of equal length. Compute the absolute. error. Do the numbers indicate that such Riemann sums should approximate the integral with desired accuracy, provided that the norm of the partition is small enough?

5.
$$\int_{1}^{3} \left(x^3 - 2x^2 + 1\right) dx = \frac{14}{3} \cong 4.666\,67$$

7.
$$\int_{1}^{2} x^2 e^{-x} dx = 5e^{-1} - 10e^{-2} \cong 0.486\,044$$

6.
$$\int_{0}^{\pi/3} \sin^2(x)\, dx = \frac{\pi}{6} - \frac{\sqrt{3}}{8} \cong 0.307\,092$$

8.
$$\int_{-2}^{2} \frac{1}{4 + x^2}\, dx = \frac{\pi}{4} \cong 0.785\,398$$

5.3 The Fundamental Theorem of Calculus: Part 1

In sections 5.1 and 5.2 we introduced the concept of the integral. The Fundamental Theorem of Calculus establishes the link between the two fundamental concepts of calculus, namely, the derivative and the integral. We will discuss the first part of the theorem in this section and the second part of the theorem in Section 5.5.

The Fundamental Theorem of Calculus (Part 1)

The first part of the Fundamental Theorem of Calculus states that **the integral of the derivative of a function on an interval is equal to the difference between the values of the function at the endpoints of the interval:**

Theorem 1 (THE FUNDAMENTAL THEOREM OF CALCULUS (Part 1)) Assume that F' is continuous on $[a, b]$ Then

$$\int_a^b F'(x)\,dx = F(b) - F(a).$$

$F'(a)$ and $F'(b)$ can be interpreted as the one sided derivatives $F'_+(a)$ and $F'_-(b)$, respectively.

The Proof of Theorem 1

Let $P = \{x_0, x_1, \ldots, x_{k-1}, x_k, \ldots, x_{n-1}, x_n\}$ be a partition of $[a, b]$, so that $x_0 = a$ and $x_n = b$. We can express the change in the value of F over the interval $[a, b]$ as the sum of the changes in the value of F over the subintervals determined by P:

$$F(b) - F(a) = F(x_n) - F(x_0)$$
$$= [F(x_n) - F(x_{n-1})] + [F(x_{n-1}) - F(x_{n-2})] + \cdots + [F(x_2) - F(x_1)] + [F(x_1) - F(x_0)]$$
$$= \sum_{k=1}^{n} [F(x_k) - F(x_{k-1})].$$

By the Mean Value Theorem (Theorem 3 of Section 3.2), there exists $x_k^* \in (x_{k-1}, x_k)$ such that

$$F(x_k) - F(x_{k-1}) = F'(x_k^*)(x_k - x_{k-1}) = F'(x_k^*)\,\Delta x_k.$$

Therefore,

$$F(b) - F(a) = \sum_{k=1}^{n} F'(x_k^*)\,\Delta x_k.$$

We have

$$\sum_{k=1}^{n} F'(x_k^*)\,\Delta x_k \cong \int_a^b F'(x)\,dx$$

if $\|P\| = \max_k \Delta x_k$ is small, and the approximation is as accurate as desired if $\|P\|$ is small enough. Therefore,

$$F(b) - F(a) \cong \int_a^b F'(x)\,dx,$$

and

$$\left| (F(b) - F(a)) - \int_a^b F'(x)\,dx \right|$$

is as small as desired. This means that the numbers

$$F(b) - F(a) \text{ and } \int_a^b F'(x)\,dx$$

are equal.■

We may refer to the first part of the Fundamental Theorem of Calculus simply as "the Fundamental Theorem of Calculus" or "the Fundamental Theorem", until we introduce the second part of the Fundamental Theorem and a distinction is necessary.

Example 1 Let

$$F(x) = \frac{2}{3}x^{3/2}.$$

By the power rule,

$$F'(x) = \frac{d}{dx}\left(\frac{2}{3}x^{3/2}\right) = \frac{2}{3}\frac{d}{dx}x^{3/2} = \frac{2}{3}\left(\frac{3}{2}x^{1/2}\right) = \sqrt{x}$$

if $x \geq 0$ (we have to interpret $F'(0)$ as $F'_+(0)$). Thus, F' is continuous on $[0,1]$, so that the Fundamental Theorem of Calculus is applicable on $[0,1]$. Therefore,

$$\int_0^1 \sqrt{x}\,dx = \int_0^1 F'(x)\,dx = F(1) - F(0) = \frac{2}{3}.$$

Thus, the area of the region between the graph of $y = \sqrt{x}$ and the interval $[0,1]$ is $2/3$. The region is illustrated in Figure 1. □

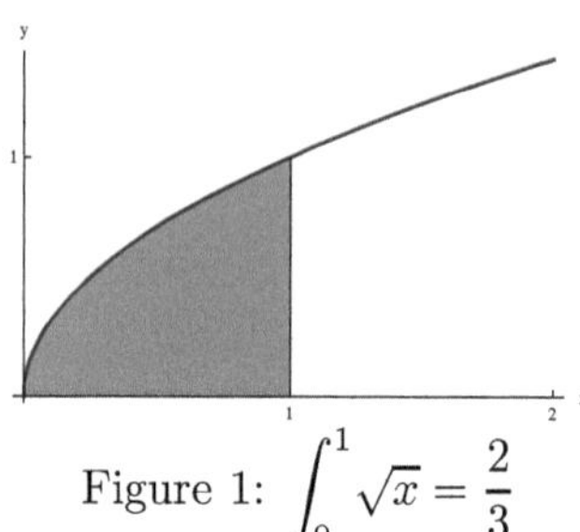

Figure 1: $\displaystyle\int_0^1 \sqrt{x} = \frac{2}{3}$

Example 2 Evaluate

$$\int_0^{\sqrt{\pi/4}} \frac{d}{dx}\cos\left(x^2\right)\,dx$$

Solution

If we set $F(x) = \cos\left(x^2\right)$, we have

$$\int_0^{\sqrt{\pi/4}} \frac{d}{dx}\cos\left(x^2\right)\,dx = \int_0^{\sqrt{\pi/4}} \frac{d}{dx}F(x)\,dx = F\left(\sqrt{\frac{\pi}{4}}\right) - F(0)$$

$$= \cos\left(\frac{\pi}{4}\right) - \cos(0) = \frac{\sqrt{2}}{2} - 1,$$

by Theorem 1.

Note that

$$f\left(x\right) = \frac{d}{dx}\cos\left(x^2\right) = -2x\sin\left(x^2\right),$$

and we have $f\left(x\right) \leq 0$ if $0 \leq x \leq \sqrt{\pi/4}$. Thus, the area of the region G between the graph of f and the interval $\left[0, \sqrt{\pi/4}\right]$ is

$$-\int_0^{\sqrt{\pi/4}} f\left(x\right) dx = -\int_0^{\sqrt{\pi/4}} \frac{d}{dx}\cos\left(x^2\right) dx = 1 - \frac{\sqrt{2}}{2}.$$

Figure 2 shows the region G. $\square$

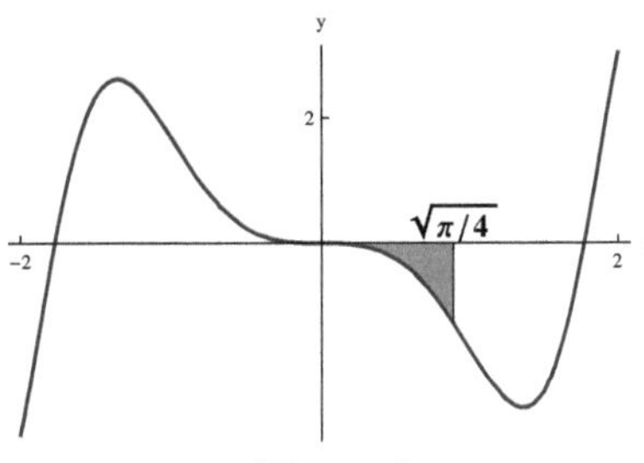

Figure 2

We were able to compute the integrals in the above examples by expressing the integrand as the derivative of a familiar function. We will use this procedure to compute many integrals:

Corollary (Corollary to the Fundamental Theorem of Calculus) Assume that f is continuous on $[a, b]$ and that $F'\left(x\right) = f\left(x\right)$ for each $x \in [a, b]$. Then

$$\int_a^b f\left(x\right) dx = F\left(b\right) - F\left(a\right).$$

Proof

By the Fundamental Theorem of Calculus (Part 1),

$$\int_a^b f\left(x\right) dx = \int_a^b F'\left(x\right) dx = F(b) - F(a)$$

■

As in Theorem 1, $F'\left(a\right)$ and $F'\left(b\right)$ can be interpreted as the one sided derivatives $F'_+\left(a\right)$ and $F'_-\left(b\right)$, respectively.

We may refer to the corollary to the Fundamental Theorem of Calculus simply as "the Fundamental Theorem of Calculus".

Definition 1 A function F is an **antiderivative** of f on an interval J if $F'\left(x\right) = f\left(x\right)$ for each x in J.

The derivative should be interpreted as the appropriate one-sided derivative at an endpoint of the relevant interval.

We will denote $F(b) - F(a)$ as

$$F(x)\big|_a^b.$$

Thus, we can express the Corollary to the Fundamental Theorem of Calculus as follows:

$$\int_a^b f(x)\,dx = F(x)\big|_a^b$$

if F is an antiderivative of f on $[a, b]$.

Example 3 Evaluate

$$\int_4^9 \sqrt{x}\,dx.$$

Solution

With reference to Example 1, if

$$f(x) = \sqrt{x} \text{ and } F(x) = \frac{2}{3}x^{3/2},$$

then F is an antiderivative of f on the interval $[0, +\infty)$, since

$$F'(x) = f(x)$$

for each $x \in (0, +\infty)$, and $F'_+(0) = f(0)$.
Therefore,

$$\int_4^9 \sqrt{x}\,dx = \frac{2}{3}x^{3/2}\bigg|_4^9 = \frac{2}{3}\left(9^{3/2} - 4^{3/2}\right) = \frac{2}{3}(27 - 8) = \frac{38}{3}.$$

$\square$

We have been referring to "an antiderivative of a function". Indeed, a function has infinitely many antiderivatives. On the other hand, any two antiderivatives of the same function can differ at most by an additive constant:

Proposition 1 Let F be an antiderivative of f on the interval J.

a) If C is a constant, then $F + C$ is also an antiderivative of f on J.
b) If G is any antiderivative of f on the interval J, there exists a constant C such that $G(x) = F(x) + C$ for each x in J.

Proof

a) Since F is an antiderivative of f on J, we have

$$\frac{d}{dx}F(x) = f(x) \text{ for each } x \in J.$$

If C is an arbitrary constant,

$$\frac{d}{dx}(F(x) + C) = \frac{d}{dx}F(x) + \frac{d}{dx}(C) = f(x) + 0 = f(x)$$

for each x in J. Therefore, $F + C$ is also an antiderivative of f on the interval J.
b) Since F and G are antiderivatives of f on the interval J, we have

$$\frac{d}{dx}F(x) = f(x) \text{ and } \frac{d}{dx}G(x) = f(x)$$

for each $x \in J$. Therefore, there exists a constant C such that $G(x) = F(x) + C$ for all x in J (Corollary to Theorem 5 of Section 3.2). ∎

By Proposition 1, if F is an antiderivative of f, we can express any antiderivative of f as $F + C$, where C is a constant. We will use the notation

$$\int f(x)dx$$

to denote *any* antiderivative of f and refer to

$$\int f(x)dx$$

as **the indefinite integral of** f. Thus,

$$\int f(x)\,dx = F(x) + C.$$

Example 4 If

$$F(x) = \frac{1}{3}x^3$$

and $f(x) = x^2$, then F is an antiderivative of f (on the entire number line), since

$$\frac{d}{dx}\left(\frac{1}{3}x^3\right) = \frac{1}{3}\left(3x^2\right) = x^2$$

for each $x \in R$. Therefore, we can express the indefinite integral of f as

$$\int x^2 dx = \frac{1}{3}x^3 + C,$$

where C is an arbitrary constant. □

Remark 1 (Caution) We may refer to an integral

$$\int_a^b f(x)\,dx$$

as a **definite integral,** if we feel the need to make a distinction between an integral and an indefinite integral. In spite of the similarities between the terminology and the notation, **the indefinite integral of** f **and the integral of** f **on an interval** $[a, b]$ **are distinct entities.** The (definite) integral

$$\int_a^b f(x)\,dx$$

is a number that can be approximated with arbitrary accuracy by Riemann sums, whereas, the indefinite integral

$$\int f(x)\,dx$$

represents any function whose derivative is equal to the function f. In either case, we will refer to f as **the integrand.** The Fundamental Theorem establishes a link between a definite integral and an indefinite integral:

$$\int_a^b f(x)\,dx = \int f(x)\Big|_{x=a}^{x=b}.$$

◊

Example 5 Let C denote an arbitrary constant. Show that the statements

$$\int 2\sin(x)\cos(x)\,dx = \sin^2(x) + C$$

and

$$\int 2\sin(x)\cos(x)\,dx = -\cos^2(x) + C$$

are both correct.

Solution

We have

$$\frac{d}{dx}\sin^2(x) = 2\sin(x)\cos(x)$$

and

$$\frac{d}{dx}\left(-\cos^2(x)\right) = -2\cos(x)\left(-\sin(x)\right) = 2\sin(x)\cos(x)$$

for each $x \in \mathbb{R}$. Therefore, both $\sin^2(x)$ and $-\cos^2(x)$ are antiderivatives for $2\sin(x)\cos(x)$. Therefore, we can express the indefinite integral of $2\sin(x)\cos(x)$ as

$$\int 2\sin(x)\cos(x)\,dx = \sin^2(x) + C$$

or

$$\int 2\sin(x)\cos(x)\,dx = -\cos^2(x) + C,$$

where C denotes an arbitrary constant.

Since $\sin^2(x)$ and $-\cos^2(x)$ are antiderivatives of the same function, they must differ by a constant. Indeed,

$$\sin^2(x) - \left(-\cos^2(x)\right) = \sin^2(x) + \cos^2(x) = 1$$

for all $x \in \mathbb{R}$. $\square$

Remark 2 We should be able to use any antiderivative of the integrand in order to evaluate an integral. Indeed, if

$$\frac{d}{dx}F(x) = f(x) \text{ and } \frac{d}{dx}G(x) = f(x)$$

for each x in some interval J, there exists a constant C such that $G(x) - F(x) = C$ for each $x \in J$. Therefore,

$$\int_a^b f(x)\,dx = F(x)\big|_{x=a}^{x=b} = F(b) - F(a),$$

and

$$\int_a^b f(x)\,dx = G(x)\big|_{x=a}^{x=b} = G(b) - G(a) = (F(b) + C) - (F(a) + C) = F(b) - F(a).$$

Therefore, we do not have to include an arbitrary constant in the expression for an indefinite integral when we use the indefinite integral to evaluate a definite integral. $\Diamond$

Example 6 With reference to Example 5,

$$\int_{\pi/4}^{\pi/2} 2\sin(x)\cos(x)\,dx = \sin^2(x)\big|_{\pi/2}^{\pi/2} = 1 - \left(\frac{1}{\sqrt{2}}\right)^2 = 1 - \frac{1}{2} = \frac{1}{2}.$$

We also have

$$\int_{\pi/4}^{\pi/2} 2\sin(x)\cos(x)\,dx = -\cos^2(x)\big|_{\pi/4}^{\pi/2} = (0) + \left(\frac{1}{\sqrt{2}}\right)^2 = \frac{1}{2}.$$

$\square$

The Indefinite Integrals of Basic Functions

We will refer to the determination of the antiderivatives of functions as **antidifferentiation**. Traditionally, the term "integration" is also used instead of the term "antidifferentiation", even though we should make a distinction between integrals and antiderivatives. The particular context in which the terms are used should clarify the intended meaning. Antidifferentiation is not as straightforward as differentiation. Computer algebra systems are very helpful in finding the indefinite integrals of many functions. On the other hand, it is convenient to have the indefinite integrals of frequently encountered functions at your fingertips. Let's begin with a short list of indefinite integrals. You will learn about some rules of antidifferentiation in the rest of this chapter and in the next chapter. These rules will enable you to expand the scope of this short list considerably. The letter C denotes an arbitrary constant.

A Short List of Antiderivatives

1. $\displaystyle\int x^r \, dx = \frac{x^{r+1}}{r+1} + C, \ r \neq -1$ (if x^r is defined)

2. $\displaystyle\int \frac{1}{x} dx = \ln(|x|) + C$ (on any interval that that does not contain 0)

3. $\displaystyle\int \sin(\omega x) \, dx = -\frac{1}{\omega}\cos(\omega x) + C$ (ω is a nonzero constant)

4. $\displaystyle\int \cos(\omega x) \, dx = \frac{1}{\omega}\sin(x) + C$ (ω is a nonzero constant)

5. $\displaystyle\int e^x \, dx = e^x + C$

6. $\int a^x \, dx = \dfrac{1}{\ln(a)} a^x + C,$ where $a > 0$

7. $\displaystyle\int \sinh(x) \, dx = \cosh(x) + C$

8. $\displaystyle\int \cosh(x) \, dx = \sinh(x) + C$

9. $\displaystyle\int \frac{1}{x^2+1} dx = \arctan(x) + C$

By the definition of the indefinite integral, each formula is confirmed by differentiation.

1. Let J be an interval that is contained in the natural domain of x^r. By the power rule,

$$\frac{d}{dx}\left(\frac{x^{r+1}}{r+1}\right) = \frac{1}{r+1}\frac{d}{dx}\left(x^{r+1}\right) = \frac{1}{r+1}(r+1)x^r = x^r$$

for each x in J (the derivative may have to be interpreted as a one-sided derivative at 0). Therefore,

$$\int x^r dx = \frac{x^{r+1}}{r+1} + C$$

on the interval J. We will refer to the above antidifferentiation rule as **the reverse power rule** since it is a consequence of the power rule for differentiation.

2. If $x > 0$,

$$\frac{d}{dx}\ln(|x|) = \frac{d}{dx}\ln(x) = \frac{1}{x}.$$

If $x < 0$,

$$\frac{d}{dx}\ln(|x|) = \frac{d}{dx}\ln(-x) = \left(\frac{d}{du}\ln(u)\Big|_{u=-x}\right)\left(\frac{d}{dx}(-x)\right) = \left(\frac{1}{-x}\right)(-1) = \frac{1}{x},$$

with the help of the chain rule.

Therefore,

$$\int \frac{1}{x}dx = \ln(|x|) + C$$

on any interval that does not contain 0.

Figure 3 shows the graph of $y = \ln(|x|)$. Note that $\ln(|x|)$ defines an even function, so that the graph of the function is symmetric with respect to the vertical axis. Also note that

$$\lim_{x \to 0-} \ln(|x|) = \lim_{x \to 0+} \ln(|x|) = -\infty.$$

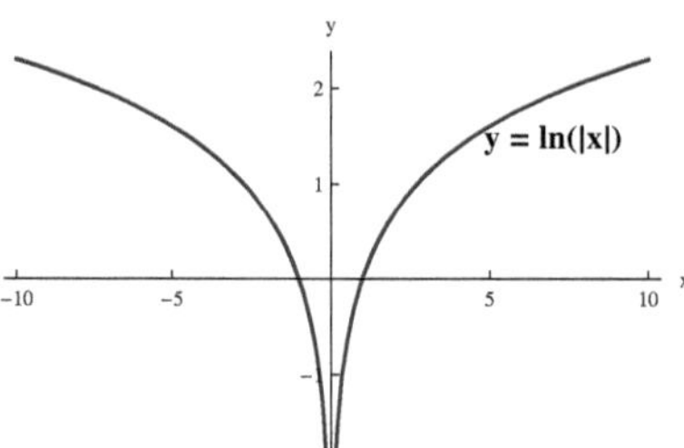

Figure 3: $y = \ln(|x|)$

Formulas 3 - 9 are equivalent to the following differentiation formulas, respectively:

$$\frac{d}{dx}\left(-\frac{1}{\omega}\cos(\omega x)\right) = \sin(\omega x),$$

$$\frac{d}{dx}\left(\frac{1}{\omega}\sin(\omega x)\right) = \cos(\omega x),$$

$$\frac{d}{dx}e^x = e^x,$$

$$\frac{d}{dx}\left(\frac{1}{\ln(a)}a^x\right) = \frac{1}{\ln(a)}\frac{d}{dx}a^x = \frac{1}{\ln(a)}\left(\ln(a)\,a^x\right) = a^x,$$

$$\frac{d}{dx}\cosh(x) = \sinh(x),$$

$$\frac{d}{dx}\sinh(x) = \cosh(x),$$

$$\frac{d}{dx}\arctan(x) = \frac{1}{x^2 + 1},$$

Example 7

a) Determine

$$\int x^{2/3}dx$$

by the reverse power rule. Confirm the result by differentiation.

b) Compute

$$\int_{-8}^{27} x^{2/3}dx.$$

Interpret the integral as signed area.

Solution

a) By the reverse power rule,

$$\int x^{2/3}dx = \frac{x^{2/3+1}}{2/3+1} = \frac{x^{5/3}}{5/3} = \frac{3}{5}x^{5/3} + C,$$

where C is an arbitrary constant.
We have

$$\frac{d}{dx}\left(\frac{3}{5}x^{5/3} + C\right) = \frac{3}{5}\left(\frac{5}{3}x^{2/3}\right) = x^{2/3}$$

for each $x \in \mathbb{R}$. Therefore, the statement

$$\int x^{2/3}dx = \frac{3}{5}x^{5/3} + C$$

is valid on $\mathbb{R}$.

b) By the Fundamental Theorem of Calculus,

$$\int_{-8}^{27} x^{2/3}dx = \frac{3}{5}x^{5/3}\Big| = \frac{3}{5}\left(27^{5/3} - (-8)^{5/3}\right) = \frac{3}{5}(3^5 + 2^5) = 165.$$

Note that f is continuous on $\mathbb{R}$, even though f is not differentiable at 0, so that there is no problem about the existence of an integral of f, or the application of the Fundamental Theorem. Since $x^{2/3} \geq 0$ for each x, the area of the region between the graph of $f(x) = x^{2/3}$ and the interval $[-8, 27]$ is 165. $\square$

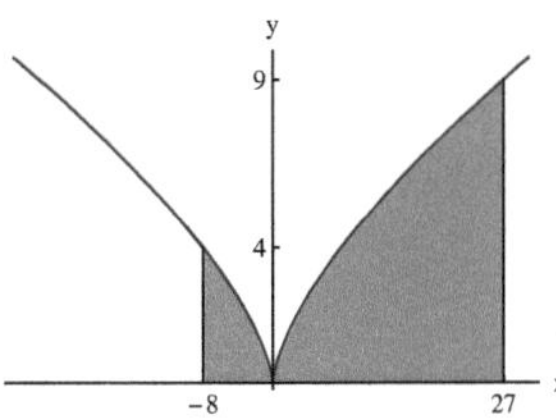

Figure 4: The region between the graph of $y = x^{2/3}$ and $[-8, 27]$

Example 8

a) Determine

$$\int \frac{1}{x^2}dx,$$

and the intervals on which the expression is valid.
b) Compute

$$\int_{-2}^{-1} \frac{1}{x^2}dx.$$

Interpret the integral as signed area.

Solution

a) By the reverse power rule,

$$\int \frac{1}{x^2}dx = \int x^{-2}dx = \frac{x^{-1}}{-1} = -\frac{1}{x} + C,$$

where C is an arbitrary constant. The expression

$$\int \frac{1}{x^2}\,dx = -\frac{1}{x} + C$$

is valid on the interval $(-\infty, 0)$ and on the interval $(0, +\infty)$.

b) By the Corollary to the Fundamental Theorem of Calculus,

$$\int_{-2}^{-1} \frac{1}{x^2}\,dx = -\frac{1}{x}\Big|_{-2}^{-1} = \left(-\frac{1}{(-1)}\right) - \left(-\frac{1}{(-2)}\right) = 1 - \frac{1}{2} = \frac{1}{2}.$$

Since $1/x^2 > 0$, the area of the region between the graph of $f(x) = 1/x^2$ and the interval $[-2, -1]$ is 0.5. Figure 5 illustrates the region. $\square$

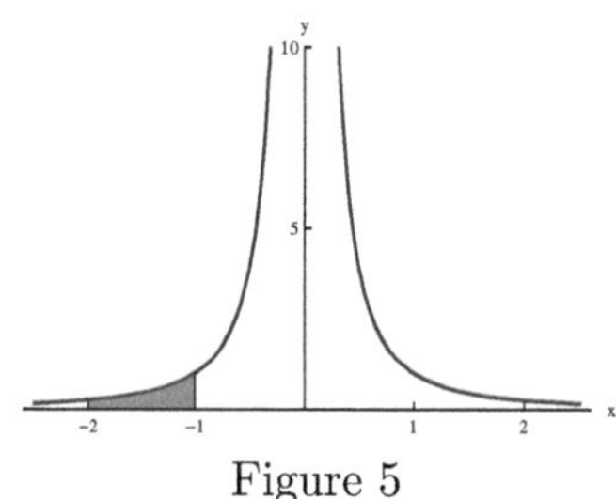

Figure 5

Example 9 Since $x^{-2} > 0$, the following claim cannot be valid:

$$\int_{-1}^{2} \frac{1}{x^2}\,dx = -\frac{1}{x}\Big|_{-1}^{2} = \left(-\frac{1}{2}\right) - (1) = -\frac{3}{2}.$$

Why is the above line incorrect?

Solution

We have

$$\frac{d}{dx}\left(\frac{1}{x^2}\right) = -\frac{1}{x}$$

if and only if $x \neq 0$. But 0 is in the interval $[-1, 2]$, so that the Corollary to the Fundamental Theorem of Calculus (Corollary)cannot be applied as indicated above. $\square$

Example 10 Evaluate

$$\int_{-4}^{-2} \frac{1}{x}\,dx.$$

Solution

Since

$$\int \frac{1}{x}\,dx = \ln\left(|x|\right) + C,$$

on any interval contained in $(-\infty, 0)$ or $(0, +\infty)$, and $[-4, -2]$ is contained in $(-\infty, 0)$, we can use the above indefinite integral to evaluate the given definite integral. By the Fundamental Theorem of Calculus,

$$\int_{-4}^{-2} \frac{1}{x}\,dx = \ln\left(|x|\right)\big|_{-4}^{-2} = \ln\left(|-2|\right) - \ln\left(|-4|\right) = \ln\left(2\right) - \ln(4) \cong -0.693147$$

Thus, the signed area of the region between the graph of the function defined by $1/x$ and the interval $[-4, -2]$ is $\ln(2) - \ln(4)$. The region is illustrated in Figure 6. $\square$

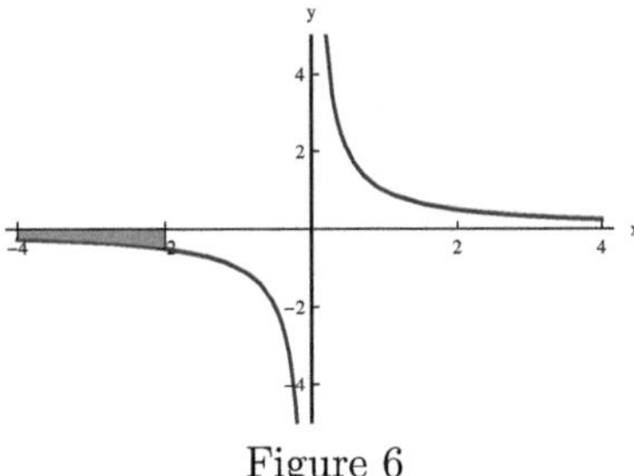

Figure 6

Remark 3 (Caution) We must be careful with the use of the antidifferentiation formula,

$$\int \frac{1}{x}\,dx = \ln(|x|) + C.$$

For example, we might be tempted to write,

$$\int_{-2}^{3} \frac{1}{x}\,dx = \ln(|x|)\big|_{-2}^{3} = \ln(3) - \ln(2).$$

The above statement is not valid since it is not true that

$$\frac{d}{dx}\ln(|x|) = \frac{1}{x}$$

at 0, and $0 \in (-2, 3)$. $\Diamond$

Example 11 Evaluate

$$\int_{0}^{\ln(10)} e^x\,dx.$$

Solution

We have

$$\int e^x\,dx = e^x + C,$$

where C is an arbitrary constant. By the Fundamental Theorem of Calculus,

$$\int_{0}^{\ln(10)} e^x\,dx = e^x\big|_{x=0}^{\ln(10)} = e^{\ln(10)} - e^0 = 10 - 1 = 9.$$

Thus, the area of the region between the graph of the natural exponential function and the interval $[0, \ln(10)]$ is 9. Figure 7 illustrates the region. $\square$

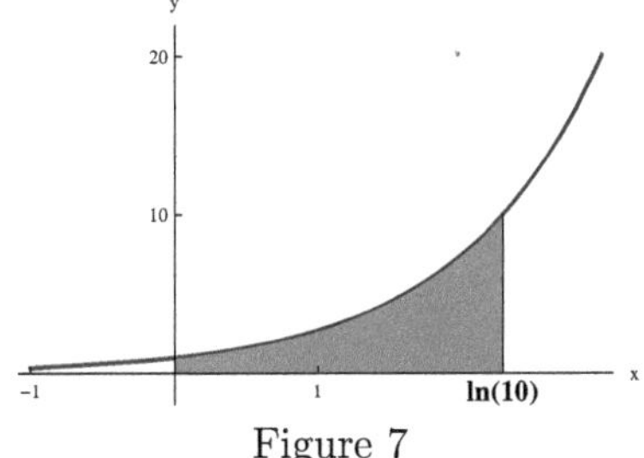

Figure 7

Example 12 Confirm the following claims that were made in Example 4 of Section 5.2:

$$\int_0^\pi \sin(x)dx = 2 \quad \text{and} \quad \int_\pi^{4\pi/3} \sin(x)\,dx = -\frac{1}{2}.$$

Solution

We have

$$\int \sin(x)\,dx = -\cos(x) + C,$$

where C denotes an arbitrary constant, as usual. By the Fundamental Theorem of Calculus ,

$$\int_0^\pi \sin(x)\,dx = -\cos(x)|_0^\pi = -\cos(\pi) - (-\cos(0)) = 1 + 1 = 2,$$

and

$$\int_\pi^{4\pi/3} \sin(x)\,dx = -\cos(x)|_\pi^{4\pi/3} = -\cos\left(\frac{4\pi}{3}\right) - (-\cos(\pi)) = -\left(-\frac{1}{2}\right) - 1 = -\frac{1}{2}.$$

$\square$

The Fundamental Theorem of Calculus and One-Dimensional Motion

Let's interpret the Fundamental Theorem of Calculus within the context of one-dimensional motion. Assume that $f(t)$ is the **position** at time t of an object in one dimensional motion, and let $v(t)$ be its instantaneous **velocity** at time t, so that $v(t) = f'(t)$. Also assume that v is continuous on $[a, b]$. By the Fundamental Theorem of Calculus,

$$\int_a^b v(t)\,dt = \int_a^b f'(t)\,dt = f(b) - f(a).$$

We will refer to the change in the position of the object over the time time interval $[a, b]$ as **the displacement** of the object over that time interval. Thus, **the displacement of the object over the time interval $[a, b]$ is equal to the integral of the velocity function on $[a, b]$**.

Even though the above fact is a direct consequence of the Fundamental Theorem of Calculus, it is helpful to interpret the proof of the theorem within the context of one-dimensional motion. If $P = \{t_0, t_1, t_2, \ldots, t_{n-1}, t_n\}$ is a partition of $[a, b]$, so that $t_0 = a$ and $t_n = b$, we can express the displacement over $[a, b]$ as the sum of the displacements over the subintervals. Thus,

$$
\begin{aligned}
f(b) - f(a) &= f(t_n) - f(t_0)\\
&= [f(t_n) - f(t_{n-1})] + [f(t_{n-1}) - f(t_{n-2}))] + \cdots + [f(t_2) - f(t_1)] + [f(t_1) - f(t_0)]\\
&= \sum_{k=1}^{n} [f(t_k) - f(t_{k-1})].
\end{aligned}
$$

By the Mean Value Theorem, there exists $t_k^* \in (t_{k-1}, t_k)$ such that

$$f(t_k) - f(t_{k-1}) = f'(t_k^*)(t_k - t_{k-1}) = v(t_k^*)\,\Delta t_k.$$

Therefore,

$$\text{Displacement over } [a, b] = \sum_{k=1}^{n} [f(t_k) - f(t_{k-1})] = \sum_{k=1}^{n} v(t_k^*)\,\Delta t_k.$$

Since

$$\sum_{k=1}^{n} v\left(t_k^*\right) \Delta t_k \cong \int_a^b v\left(t\right) dt$$

if $\|P\|$ is small, and the approximation is as accurate as desired provided that $\|P\|$ is small enough,

$$\left| \text{Displacement over } [a, b] - \int_a^b v\left(t\right) dt \right|$$

is arbitrarily small. This is the case if and only if

$$\text{Displacement over } [a, b] = \int_a^b v\left(t\right) dt.$$

In particular the units match. For example, if distance is measured in centimeters and time is measured in seconds, velocity is expressed in terms of centimeters per second. This is consistent with the fact that

$$\text{Displacement over } [a, b] = \int_a^b v\left(t\right) dt \cong \sum_{k=1}^{n} v\left(t_k^*\right) \Delta t_k.$$

Indeed, the unit of $v\left(t_k^*\right) \Delta t_k$ is

$$\frac{\text{centimeter}}{\text{second}} \times \text{ second} = \text{ centimeter.}$$

Graphically, the displacement of the object over the time interval $[a, b]$ is **the signed area** of the region between the graph of the velocity function and the interval $[a, b]$. We must distinguish between the displacement of an object over a time interval and **the distance traveled** by the object over the same time interval. If $v\left(t\right) \leq 0$ for each $t \in [a, b]$, the object is moving in the negative direction over the time interval $[a, b]$. Therefore, the distance traveled is

$$- \int_a^b v\left(t\right) dt.$$

More generally, if we wish to calculate the distance traveled by an object over the time interval $[a, b]$, we need to determine the subintervals of $[a, b]$ on which the velocity has constant sign. If the velocity is negative over a subinterval, the relevant integral must be multiplied by (-1). Graphically, the distance traveled over the time interval $[a, b]$ is **the area** between the graph of the velocity function and the interval $[a, b]$.

Example 13 With the above notation, assume that an object that is attached to a spring has velocity $v\left(t\right) = \cos\left(2t\right).$

a) Sketch the graph of the velocity function on $[0, \pi]$.
b) Determine the displacement of the object over the time interval $[0, 3\pi/4]$.
c) Determine the distance traveled by the object over the time interval $[0, 3\pi/4]$.

Solution

a) Figure 8 shows the graph of the velocity function on $[0, \pi]$.

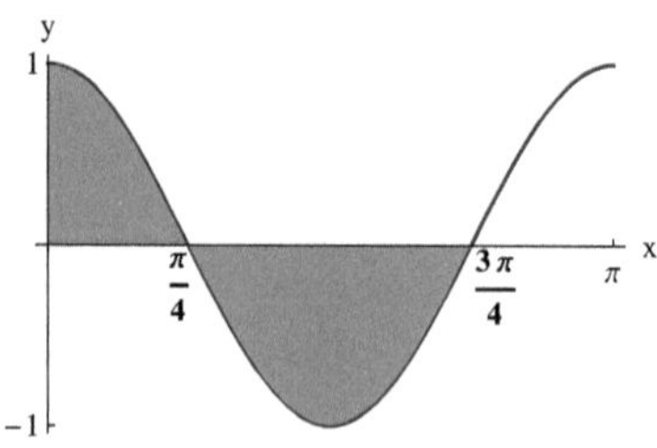

Figure 8

b) Since

$$\int \cos(\omega t)\, dt = \frac{1}{\omega} \sin(\omega t) + C,$$

for any $\omega \neq 0$, we have

$$\int \cos(2t)\, dt = \frac{1}{2} \sin(2t) + C.$$

Therefore, the displacement of the object over the time interval $[0, 3\pi/4]$ is

$$\int_0^{3\pi/4} v(t)\, dt = \int_0^{3\pi/4} \cos(2t)\, dt = \frac{1}{2} \sin(2t)\Big|_0^{3\pi/4}$$

$$= \frac{1}{2} \sin\left(\frac{3\pi}{2}\right) - \frac{1}{2} \sin(0) = -\frac{1}{2}$$

(centimeters).

c) We see that $v(t) > 0$ if $0 < t < \pi/4$ and $v(t) < 0$ if $\pi/4 < t < 3\pi/4$. Thus, the object is moving in the positive direction over the time interval $[0, \pi/4]$ and in the negative direction over the time interval $[\pi/4, 3\pi/4]$. We have

$$\int_0^{\pi/4} v(t)\, dt = \frac{1}{2} \sin(2t)\Big|_0^{\pi/4} = \frac{1}{2} \sin\left(\frac{\pi}{2}\right) - \frac{1}{2} \sin(0) = \frac{1}{2},$$

and

$$\int_{\pi/4}^{3\pi/4} v(t)\, dt = \frac{1}{2} \sin(2t)\Big|_{\pi/4}^{3\pi/4} = \frac{1}{2} \sin\left(\frac{3\pi}{2}\right) - \frac{1}{2} \sin\left(\frac{\pi}{2}\right) = -\frac{1}{2} - \frac{1}{2} = -1.$$

Therefore, total distance traveled is

$$\int_0^{\pi/4} v(t)\, dt - \int_{\pi/4}^{3\pi/4} v(t)\, dt = \frac{1}{2} - (-1) = \frac{1}{2}$$

(centimeters). Graphically, the distance traveled is the area of the region between the velocity function and the interval $[0, 3\pi/4]$. $\square$

Problems

In problems 1-4, evaluate the integral (make use the Fundamental Theorem of Calculus):

1.
$$\int_{\sqrt{\pi/4}}^{\sqrt{\pi/2}} \frac{d}{dx} \sin\left(x^2\right) dx$$

3.
$$\int_{\pi/3}^{\pi/2} \frac{d}{dx} \sqrt{\cos\left(\frac{x}{2}\right)} dx$$

2.
$$\int_{1}^{2} \frac{d}{dx} \sqrt{9 - x^2}\, dx$$

4.
$$\int_{-3}^{2} \frac{d}{dx} \left(\frac{1}{x^2 + 4}\right) dx$$

In problems 5-22, use the Fundamental Theorem of Calculus to evaluate the integral, if applicable, otherwise state why the theorem does not apply.

5.
$$\int_{-3}^{2} x^2\, dx$$

14.
$$\int_{\pi/18}^{\pi/9} \cos\left(9x\right) dx$$

6.
$$\int_{-2}^{3} x^{-2}\, dx$$

15.
$$\int_{-4}^{-2} \frac{1}{x}\, dx$$

7.
$$\int_{4}^{9} \sqrt{x}\, dx$$

16.
$$\int_{-1}^{2} \frac{1}{x}\, dx$$

8.
$$\int_{0}^{27} x^{2/3}\, dx$$

17.
$$\int_{e^{-1}}^{e^2} \frac{1}{x}\, dx$$

9.
$$\int_{1}^{9} \frac{1}{\sqrt{x}}\, dx$$

18.
$$\int_{\ln(3)}^{\ln(10)} e^x\, dx$$

10.
$$\int_{0}^{1} \frac{1}{x}\, dx$$

19.
$$\int_{0}^{\sqrt{\ln(2)}} \frac{d}{dx} e^{-x^2}\, dx$$

11.
$$\int_{\pi/3}^{\pi/2} \sin\left(x\right) dx$$

20.
$$\int_{\log_{10}(2)}^{\log_{10}(4)} 10^x\, dx$$

12.
$$\int_{-\pi}^{-\pi/6} \cos\left(x\right) dx$$

21.
$$\int_{-1}^{0} \frac{d}{dx} \arctan\left(x\right) dx$$

13.
$$\int_{\pi/3}^{\pi/2} \sin\left(\frac{x}{2}\right) dx$$

22.
$$\int_{-\sqrt{3}}^{1/\sqrt{3}} \frac{1}{1 + x^2}\, dx$$

23. Assume that the velocity of an object in one-dimensional motion is t at time t. Calculate the displacement of the object over the time interval $[2, 5]$.

24. Assume that the velocity of an object in one-dimensional motion is $\cos\left(4t\right)$ at time t. Calculate the displacement of the object over the time interval $[\pi/6, \pi/4]$.

5.4   The Fundamental Theorem of Calculus: Part 2

The second part of the Fundamental Theorem of Calculus shows that every continuous function has an antiderivative, even though such an antiderivative may not be expressible in terms of familiar functions. The theorem leads to the definition of new special functions.

Some Properties of the Integral

In preparation for the second part of the Fundamental Theorem of Calculus, we will discuss some general facts about the integral that will be useful in other contexts as well.

Proposition 1 Assume that f and g are continuous on the interval $[a, b]$ and $f(x) \leq g(x)$ for each $x \in [a, b]$. Then

$$\int_a^b f(x)\,dx \leq \int_a^b g(x)\,dx.$$

Figure 1 illustrates the graphical meaning of Proposition 1 if $0 \leq f(x) < g(x)$ for each $x \in [a, b]$: The area of the region between the graph of f and the interval $[a, b]$ is less than the area of the region between the graph of g and $[a, b]$.

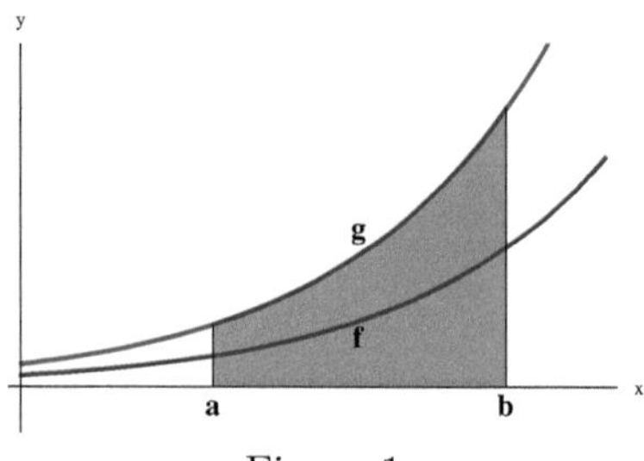

Figure 1

We will leave the rigorous proof of Proposition 1 to a course in advanced calculus. Let's provide a plausibility argument:

Let $P = \{x_0, x_1, \ldots, x_{n-1}, x_n\}$ be a partition of $[a, b]$ and $x_k^* \in [x_{k-1}, x_k]$, $k = 1, 2, \ldots, n$. We have

$$\sum_{k=1}^n f(x_k^*)\Delta x_k \leq \sum_{k=1}^n g\left(x_k^*\right)\Delta x_k,$$

since $f(x) \leq g(x)$ for each $x \in [a, b]$. Since

$$\sum_{k=1}^n f(x_k^*)\Delta x_k \cong \int_a^b f(x)\,dx \quad \text{and} \quad \sum_{k=1}^n g\left(x_k^*\right)\Delta x_k \cong \int_a^b g(x)\,dx,$$

if $\|P\| = \max_k \Delta x_k$ is small, it is plausible that

$$\int_a^b f(x)\,dx \leq \int_a^b g(x)\,dx.$$

∎

Corollary 1 (The Triangle Inequality for Integrals) Assume that f is continuous on $[a, b]$. Then

$$\left| \int_a^b f(x)\,dx \right| \leq \int_a^b |f(x)|\,dx.$$

Proof

It can be shown that $|f|$ is continuous on $[a, b]$ if f is continuous on $[a, b]$. We have

$$-|f(x)| \leq f(x) \leq |f(x)|$$

for each $x \in [a, b]$. By Proposition 1,

$$\int_a^b -|f(x)|\, dx \leq \int_a^b f(x)\, dx \leq \int_a^b |f(x)|\, dx.$$

By the constant multiple rule for integrals,

$$\int_a^b -|f(x)|\, dx = -\int_a^b |f(x)|\, dx.$$

Therefore,

$$-\int_a^b |f(x)|\, dx \leq \int_a^b f(x)\, dx \leq \int_a^b |f(x)|\, dx.$$

The above inequalities imply that

$$\left| \int_a^b f(x)\, dx \right| \leq \int_a^b |f(x)|\, dx.$$

■

We have dubbed the Corollary 1 as "**the triangle inequality for integrals**", since we can
view the inequality

$$\left| \int_a^b f(x)\, dx \right| \leq \int_a^b |f(x)|\, dx$$

as a generalization of the triangle inequality for numbers. Indeed, if $P = \{x_0, x_1, \ldots, x_{n-1}, x_n\}$
is a partition of $[a, b]$ and $x_k^* \in [x_{k-1}, x_k]$ for $k = 1, 2, \ldots, n$, we have

$$\left| \sum_{k=1}^n f(x_k^*)\Delta x_k \right| \leq \sum_{k=1}^n |f(x_k^*)|\, \Delta x_k$$

by the triangle inequality for numbers. If $\|P\| = \max_k \Delta x_k$ is small,

$$\left| \sum_{k=1}^n f(x_k^*)\Delta x_k \right| \cong \left| \int_a^b f(x)dx \right|$$

and

$$\sum_{k=1}^n |f(x_k^*)|\, \Delta x_k \cong \int_a^b |f(x)|\, dx.$$

Therefore, the inequality

$$\left| \int_a^b f(x)dx \right| \leq \int_a^b |f(x)|\, dx$$

is not surprising.

Definition 1 **The mean value** (or **the average value**) of a continuous function f on the
interval $[a, b]$ is

$$\frac{1}{b-a} \int_a^b f(x)\, dx.$$

Thus, the mean value of f on $[a, b]$ is the ratio of the integral of f on $[a, b]$ and the length of the interval $[a, b]$.

The terminology of Definition 1 is reasonable. Indeed, if

$$\Delta x = \frac{b-a}{n}, \ x_k = a + k\Delta x, \ k = 1, 2, \ldots, n,$$

then

$$\sum_{k=1}^{n} f(x_k) \Delta x \cong \int_{a}^{b} f(x)\, dx$$

if Δx is small, i.e., n is large. Therefore,

$$\frac{1}{b-a} \sum_{k=1}^{n} f(x_k) \Delta x \cong \frac{1}{b-a} \int_{a}^{b} f(x)\, dx.$$

We have

$$\frac{1}{b-a} \sum_{k=1}^{n} f(x_k) \Delta x = \frac{1}{b-a} \sum_{k=1}^{n} f(x_k) \left(\frac{b-a}{n}\right) = \frac{1}{n} \sum_{k=1}^{n} f(x_k).$$

Therefore,

$$\frac{1}{n} \sum_{k=1}^{n} f(x_k) \cong \frac{1}{b-a} \int_{a}^{b} f(x)\, dx$$

if n is large. The quantity

$$\frac{1}{n} \sum_{k=1}^{n} f(x_k)$$

is the mean of the values of the function at the points x_k, $k = 1, 2, \ldots, n$.

A Continuous function attains its mean value on an interval:

Theorem 1 (THE MEAN VALUE THEOREM FOR INTEGRALS) **Assume that** f **is continuous on** $[a, b]$**. There exists** $c \in [a, b]$ **such that**

$$f(c) = \frac{1}{b-a} \int_{a}^{b} f(x)\, dx.$$

Proof

Let m and M be the minimum and the maximum value of f on $[a, b]$, respectively. Since

$$m \leq f(x) \leq M$$

for each $x \in [a, b]$, we have

$$\int_{a}^{b} m\, dx \leq \int_{a}^{b} f(x)\, dx \leq \int_{a}^{b} M\, dx,$$

by Proposition 1. Therefore,

$$m(b-a) \leq \int_{a}^{b} f(x)\, dx \leq M(b-a).$$

Thus,

$$m \le \frac{1}{b-a} \int_a^b f(x)dx \le M.$$

By the Intermediate Value Theorem for continuous functions (Theorem 1 of Section 2.9), there exists $c \in [a, b]$ such that

$$f(c) = \frac{1}{b-a} \int_a^b f(x)dx.$$

■

Since

$$f(c) = \frac{1}{b-a} \int_a^b f(x)dx \Rightarrow f(c)(b-a) = \int_a^b f(x)dx,$$

we can interpret the Mean Value Theorem for Integrals in the case of a positive-valued function f graphically: The area of the region between the graph of f and the interval $[a, b]$ is the same as the area of a rectangle that has as its base the interval $[a, b]$ and has height equal to the value of f at some c in $[a, b]$, as illustrated in Figure 2.

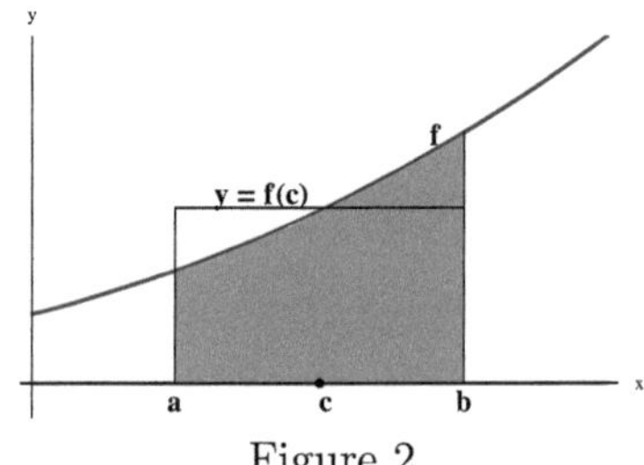

Figure 2

An integral is multiplied by (-1) if the upper and lower limits are interchanged:

Definition 2 Assume that $a < b$. We define

$$\int_b^a f(x)\,dx = - \int_a^b f(x)\,dx.$$

Remark 1 If F is an antiderivative of f, we have

$$\int_b^a f(x)dx = - \int_a^b f(x)dx = -(F(b) - F(a)) = F(a) - F(b).$$

Therefore,

$$\int_b^a f(x)dx = F(x)\big|_b^a,$$

just as

$$\int_a^b f(x)dx = F(x)\big|_a^b.$$

Thus, we need not pay attention to the positions of a and b on the number line relative to each other, when we make use of the Fundamental Theorem to evaluate the integral. ◊

Remark 2 By Definition 2, if $v(t)$ is the **velocity** at time t of an object in **one-dimensional motion**, and f is the corresponding **position function**, we have

$$\int_b^a v(t)dt = -\int_a^b v(t)dt = -\int_a^b f'(t)\,dt = -\left(f(b) - f(a)\right) = f(a) - f(b).$$

Thus, if we imagine that time flows backwards from b to a, the integral of the velocity function from b to a is still the change in the position function. $\Diamond$

Example 1 Determine

$$\int_{\pi/2}^0 \cos(x)\,dx.$$

Solution

By Definition 2,

$$\int_{\pi/2}^0 \cos(x)\,dx = -\int_0^{\pi/2} \cos(x)\,dx = -\left(\int \cos(x)\,dx \Big|_{x=0}^{x=\pi/2}\right)$$

$$= -\left(\sin(x)\big|_0^{\pi/2}\right) = -\left(\sin\left(\frac{\pi}{2}\right) - \sin(0)\right) = -1.$$

We can obtain the same result as follows:

$$\int_{\pi/2}^0 \cos(x)\,dx = \sin(x)\big|_{\pi/2}^0 = \sin(0) - \sin\left(\frac{\pi}{2}\right) = -1.$$

$\square$

We define an integral that has the same lower and upper limits to be 0:

Definition 3

$$\int_a^a f(x)\,dx = 0.$$

The following argument suggests that the above definition is reasonable:
Assume that f is continuous in some open interval that contains the point a and that $|f(x)| \leq M$ for each x in that interval. If the positive integer n is large enough,

$$\left|\int_{a-1/n}^{a+1/n} f(x)\,dx\right| \leq \int_{a-1/n}^{a+1/n} |f(x)|\,dx \leq \int_{a-1/n}^{a+1/n} M\,dx = M\left(\frac{2}{n}\right),$$

with the help of the triangle inequality for integrals. Therefore,

$$\lim_{n\to\infty} \int_{a-1/n}^{a+1/n} f(x)\,dx = 0.$$

Thus, it is natural to set

$$\int_a^a f(x)\,dx = \lim_{n\to\infty} \int_{a-1/n}^{a+1/n} f(x)\,dx = 0.$$

$\blacksquare$

The above definitions enable us to express the generalized version of **the additivity of the integral with respect to intervals:**

Theorem 2 If f is continuous on an interval that contains the points a, b and c, we have

$$\int_a^b f(x)\,dx + \int_b^c f(x)\,dx = \int_a^c f(x)\,dx.$$

Proof

We know that the statement of Thorem 2 is valid if $a < b < c$. Assume that $a < c < b$. Then,

$$\int_a^c f(x)dx + \int_c^b f(x)dx = \int_a^b f(x)dx.$$

Therefore,

$$\int_a^c f(x)dx = \int_a^b f(x)dx - \int_c^b f(x)dx = \int_a^b f(x)dx - \left(-\int_b^c f(x)dx\right)$$

$$= \int_a^b f(x)dx + \int_b^c f(x)dx,$$

as claimed.

Let's consider the case $a = b < c$. Then,

$$\int_a^b f(x)dx + \int_b^c f(x)dx = \int_a^a f(x)dx + \int_a^c f(x)dx = 0 + \int_a^c f(x)dx = \int_a^c f(x)dx.$$

Other cases are handled in a similar fashion. ■

The Second Part of the Fundamental Theorem

We will define functions via integrals. Assume that f is continuous on an interval J that contains the point a. Let us set

$$F(x) = \int_a^x f(t)dt$$

for each $x \in J$. Note that the upper limit of the integral is the variable x, and we used the letter t to denote the "dummy" integration variable (we could have used any letter other than x). If $x > a$, then $F(x)$ is the signed area of the region between the graph of f and the interval $[a, x]$, as illustrated in Figure 3.

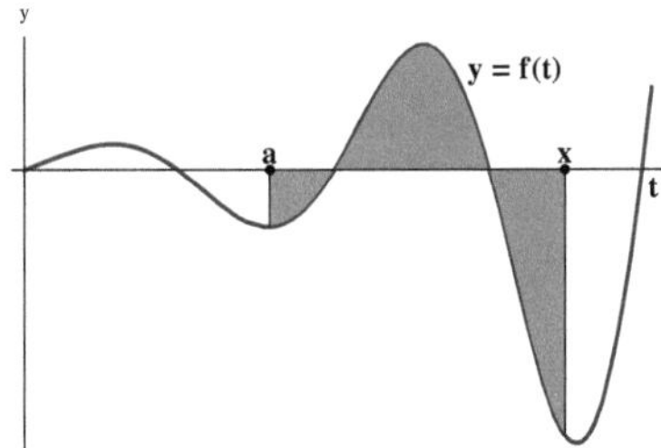

Figure 3: $F(x) = \int_a^x f(t)dt$

If $x < a$, we have

$$F(x) = -\int_x^a f(x)dx,$$

so that $F(x)$ is (-1) times the signed area of the region between the graph of f and the interval $[a, x]$, as illustrated in Figure 4.

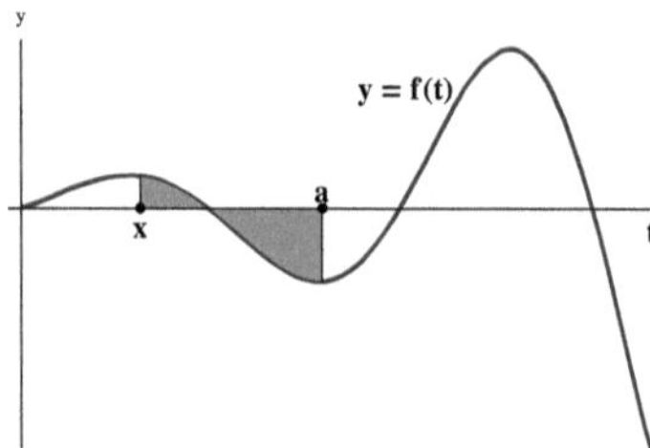

Figure 4: $F(x) = -\int_x^a f(x)dx$

Note that
$$F(a) = \int_a^a f(t)dt = 0.$$

Example 2 Set
$$F(x) = \int_2^x t^2 dt.$$

a) Determine $F(x)$, $F(3)$ and $F(1)$.
b) Interpret the meaning of $F(x)$ graphically. Sketch the graph of F.
c) Determine $F'(x)$.

Solution

a) By the reverse power rule,
$$\int t^2 dt = \frac{t^3}{3} + C,$$

where C is an arbitrary constant. Therefore,
$$F(x) = \int_2^x t^2 dt = \frac{t^3}{3}\Big|_2^x = \frac{x^3}{3} - \frac{2^3}{3} = \frac{1}{3}x^3 - \frac{8}{3}.$$

In particular,
$$F(3) = \int_2^3 t^2 dt = \frac{19}{3} \text{ and } F(1) = \int_2^1 t^2 dt = -\frac{7}{3}.$$

b) We have
$$F(2) = \int_2^2 t^2 dt = 0.$$

If $x > 2$, then $F(x)$ is the area between the graph of $f(t) = t^2$ and the interval $[2, x]$, as illustrated in Figure 5.

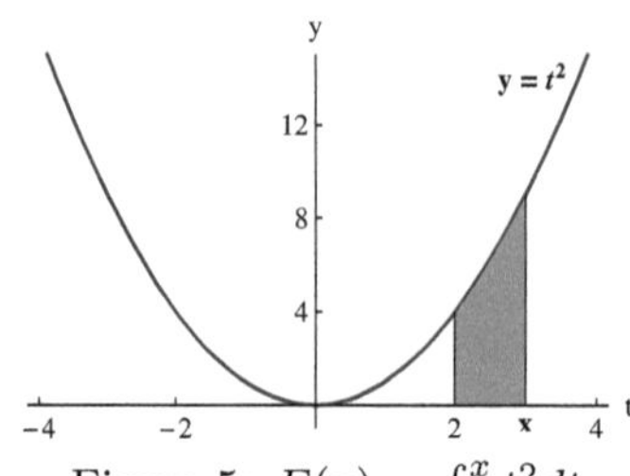

Figure 5: $F(x) = \int_2^x t^2 dt$

If $x < 2$, then

$$F(x) = -\int_x^2 t^2 dt.$$

Therefore, $F(x)$ is (-1) times the area of the region between the graph of $f(t) = t^2$ and the interval $[x, 2]$, as illustrated in Figure 6.

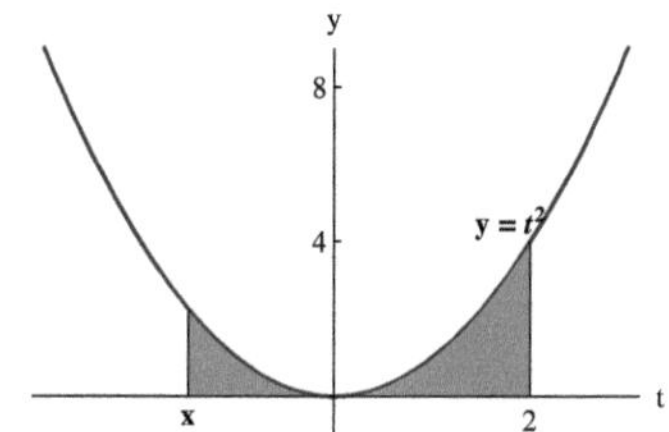

Figure 6: $F(x) = -\int_x^2 t^2 dt$ if $x < 2$

Figure 7 shows the graph of F.

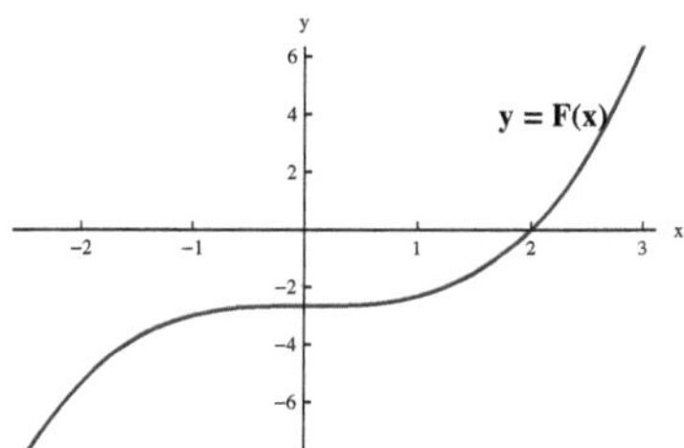

Figure 7: $y = F(x) = \int_2^x t^2 dt$

c) We have

$$F'(x) = \frac{d}{dx} \int_2^x t^2 dt = \frac{d}{dx}\left(\frac{1}{3}x^3 - \frac{8}{3}\right) = \frac{1}{3}\left(3x^2\right) = x^2.$$

Note that x^2 is the value of the integrand t^2 at $t = x$. $\square$

Example 3 Set

$$F(x) = \int_\pi^x \sin(t)\, dt.$$

a) Determine $F(x)$, $F(3\pi/2)$ and $F(\pi/3)$.
b) Interpret the meaning of $F(x)$ graphically. Sketch the graph of F on the interval $[0, 3\pi]$.
c) Determine $F'(x)$.

Solution

a) We have

$$F(x) = \int_\pi^x \sin(t)\, dt = -\cos(t)\Big|_\pi^x = -\cos(x) + \cos(\pi) = -\cos(x) - 1.$$

In particular,

$$F(3\pi/2) = \int_\pi^{3\pi/2} \sin(t)\, dt = -1 \text{ and } F(\pi/3) = \int_\pi^{\pi/3} \sin(t)\, dt = -\frac{3}{2}.$$

b) We have

$$F\left(\pi\right) = \int_{\pi}^{\pi} \sin\left(t\right) dt = 0.$$

If $x > \pi$, $F(x)$ is the signed area of the region between the graph of sine and the interval $[\pi, x]$, as illustrated in Figure 8.

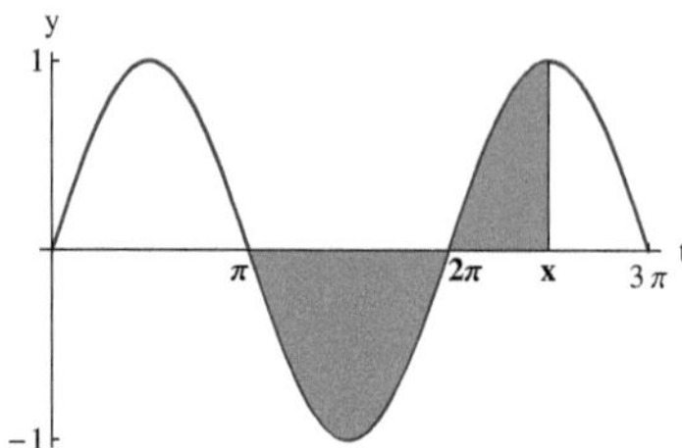

Figure 8: $F(x) = \int_{\pi}^{x} \sin\left(t\right) dt$

If $x < \pi$, we have

$$F\left(x\right) = \int_{\pi}^{x} \sin\left(t\right) dt = - \int_{x}^{\pi} \sin\left(t\right) dt.$$

Therefore, $F(x)$ is (-1) times the signed area of the region between the graph of sine and the interval $[x, \pi]$, as illustrated in Figure 9.

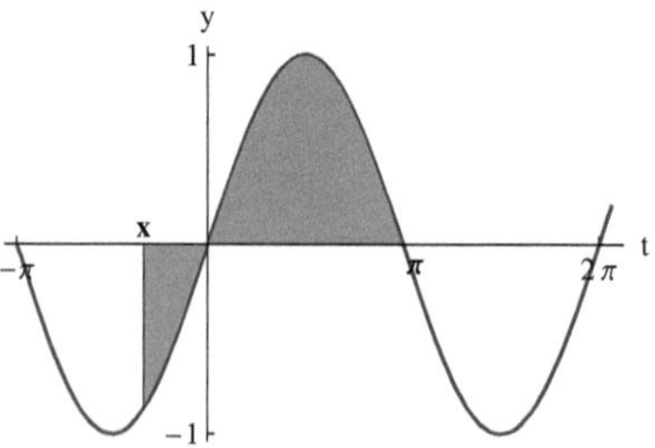

Figure 9: $F(x) = - \int_{x}^{\pi} \sin\left(t\right) dt$

Figure 10 shows the graph of $y = F(x) = -\cos\left(x\right) - 1$ on the interval $[-2\pi, 2\pi]$.

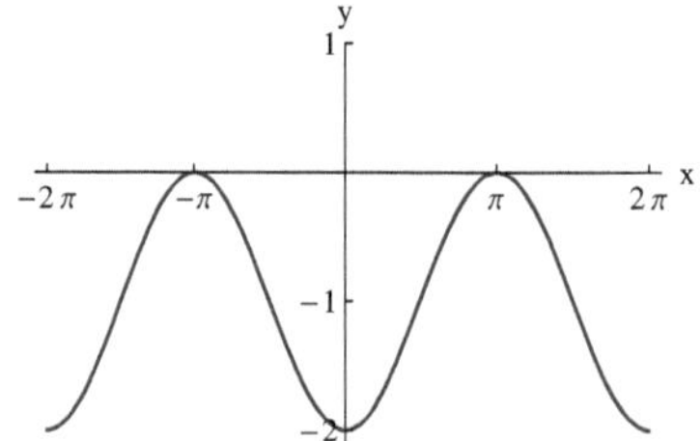

Figure 10: $y = F\left(x\right) = -\cos(x) - 1$

c) We have

$$F'\left(x\right) = \frac{d}{dx} \int_{\pi}^{x} \sin\left(t\right) dt = \frac{d}{dx}\left(-\cos\left(x\right) - 1\right) = \sin\left(x\right).$$

Note that $\sin(x)$ is the value of the integrand $\sin(t)$ at $t = x$. $\square$

In examples 2 and 3, it turned out that

$$\frac{d}{dx}\int_a^x f(t)dt = f(x).$$

That is a general fact:

Theorem 3 (The Fundamental Theorem of Calculus, Part 2) Assume that f is continuous on the interval J, and a is a given point in J. If

$$F(x) = \int_a^x f(t)\,dt,$$

then $F'(x) = f(x)$ for each $x \in J$.

The derivative should be interpreted as the appropriate one-sided derivative at an endpoint of J.

Remark 3 The second part of the Fundamental Theorem of Calculus asserts that

$$\frac{d}{dx}\int_a^x f(t)\,dt = f(x)$$

for each $x \in J$ (provided that f is continuous on J). Therefore,

$$F(x) = \int_a^x f(t)\,dt$$

defines an antiderivative of f on J. $\Diamond$

A Plausibility Argument for Theorem 3:

We have

$$F(x + \Delta x) - F(x) = \int_a^{x+\Delta x} f(t)dt - \int_a^x f(t)dt = \int_x^{x+\Delta x} f(t)dt.$$

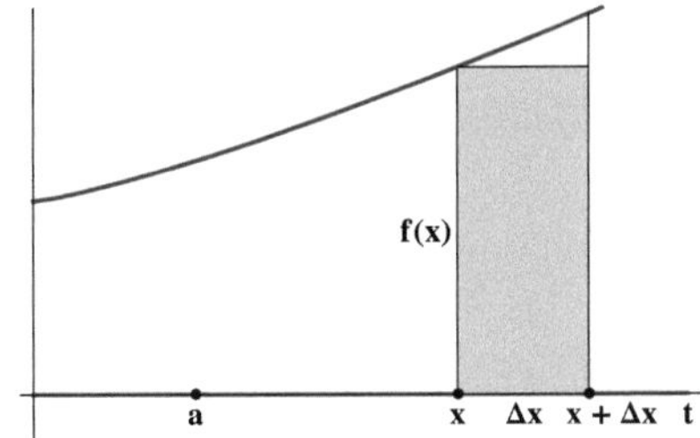

Figure 11: $\int_x^{x+\Delta x} f(t)\,dt \cong f(x)\,\Delta x$

With reference to Figure 11, if $\Delta x > 0$ and small, this quantity is approximately the area of the rectangle that has as its base the interval $[x, x + \Delta x]$ and has height $f(x)$. Therefore

$$F(x + \Delta x) - F(x) \cong f(x)\Delta x,$$

so that

$$\frac{F(x + \Delta x) - F(x)}{\Delta x} \cong f(x)$$

if Δx is small. Thus, it is plausible that

$$F'(x) = \lim_{\Delta x \to 0} \frac{F(x + \Delta x) - F(x)}{\Delta x} = f(x).$$

■

The Proof of Theorem 3

We will show that $F'(x) = f(x)$ at a point x in the interior of J. If x is an endpoint of J, the equality of the appropriate one-sided derivative of F and $f(x)$ is established in a similar manner.

Let $\Delta x > 0$. As in our plausibility argument,

$$F(x + \Delta x) - F(x) = \int_x^{x+\Delta x} f(t)dt.$$

Therefore,

$$\frac{F(x + \Delta x) - F(x)}{\Delta x} = \frac{1}{\Delta x} \int_x^{x+\Delta x} f(t)dt.$$

Thus, the difference quotient is the mean value of f on the interval $[x, x + \Delta x]$. By the Mean Value Theorem for Integrals (Theorem 1), there exists a point $c(x, \Delta x)$ in the interval $[x, x+\Delta x]$ such that

$$\frac{1}{\Delta x} \int_x^{x+\Delta x} f(t)dt = f(c(x, \Delta x))$$

(we have used the notation "$c(x, \Delta x)$" in order to indicate that c depends on x and Δx). Therefore,

$$F'_+(x) = \lim_{\Delta x \to 0+} \frac{F(x + \Delta x) - F(x)}{\Delta x} = \lim_{\Delta x \to 0+} f(c(x, \Delta x)).$$

Since $c(x, \Delta x)$ is between x and $x + \Delta x$, we have $\lim_{\Delta x \to 0} c(x, \Delta x) = x$. Since f is continuous at x,

$$\lim_{\Delta x \to 0+} f(c(x, \Delta x)) = \lim_{\Delta x \to 0} f(c(x, \Delta x)) = f\left(\lim_{\Delta x \to 0} c(x, \Delta x)\right) = f(x).$$

Therefore,

$$F'_+(x) = \lim_{\Delta x \to 0+} f(c(x, \Delta x)) = f(x).$$

If $\Delta x < 0$

$$\frac{F(x + \Delta x) - F(x)}{\Delta x} = \frac{1}{\Delta x} \int_x^{x+\Delta x} f(t)dt = \frac{1}{(-\Delta x)}\left(-\int_x^{x+\Delta x} f(t)dt\right) = \frac{1}{(-\Delta x)} \int_{x+\Delta x}^x f(t)dt.$$

The final expression is the mean value of f on the interval $[x + \Delta x, x]$. By the Mean Value Theorem for Integrals, there exists $c(x, \Delta x) \in [x + \Delta x, x]$ such that

$$\frac{1}{(-\Delta x)} \int_{x+\Delta x}^x f(t)dt = f(c(x, \Delta x)).$$

Therefore,

$$F'_-(x) = \lim_{\Delta x \to 0-} \frac{F(x + \Delta x) - F(x)}{\Delta x} = \lim_{\Delta x \to 0-} f(c(x, \Delta x)) = f\left(\lim_{\Delta x \to 0-} c(x, \Delta x)\right) = f(x),$$

since $c(x, \Delta x)$ is between $x + \Delta x$ and x, and f is continuous at x. Thus,

$$F'(x) = F'_+(x) = F'_-(x) = f(x).$$

■

Example 4 (The function erf) Set

$$F(x) = \int_0^x \frac{2}{\sqrt{\pi}} e^{-t^2}\, dt$$

a) Determine $F'(x)$.
b) Interpret $F(x)$ in terms of area.

Solution

a) By the second part of the Fundamental Theorem of Calculus,

$$F'(x) = \frac{d}{dx} \int_0^x \frac{2}{\sqrt{\pi}} e^{-t^2}\, dt = \frac{2}{\sqrt{\pi}} e^{-x^2}$$

at each $x \in \mathbb{R}$, since

$$f(t) = \frac{2}{\sqrt{\pi}} e^{-t^2}$$

is continuous on $\mathbb{R}$.

b) If $x > 0$, $F(x)$ is the area between the graph of f and the interval $[0, x]$, as illustrated in Figure 12.

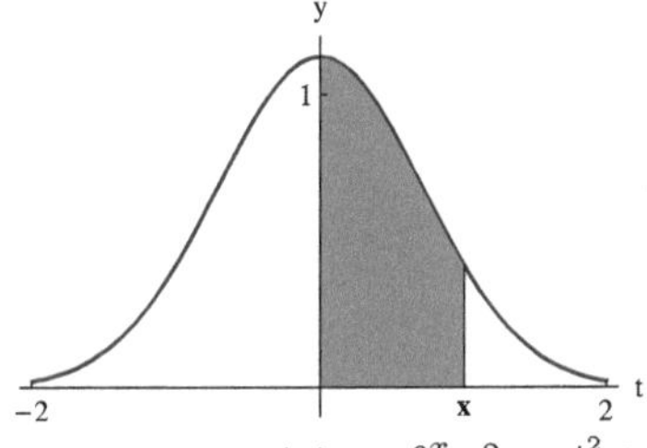

Figure 12: $F(x) = \int_0^x \frac{2}{\sqrt{\pi}} e^{-t^2}\, dt$

If $x < 0$,

$$F(x) = \int_0^x f(t)\, dt = -\int_x^0 f(t)\, dt,$$

so that $F(x)$ is $(-1) \times$ (the area between the graph of f and the interval $[x, 0]$), as illustrated in Figure 13.

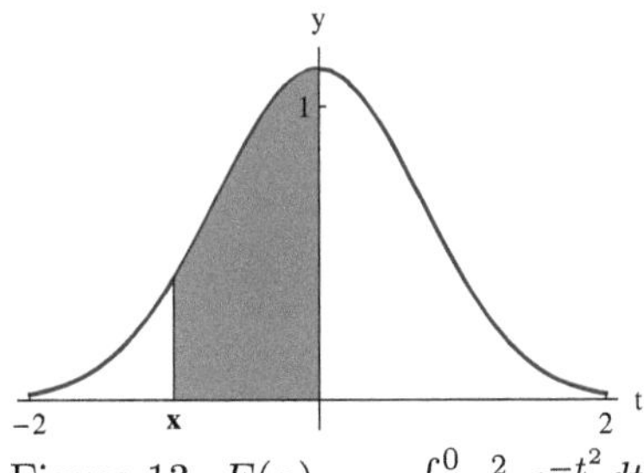

Figure 13: $F(x) = -\int_x^0 \frac{2}{\sqrt{\pi}} e^{-t^2}\, dt$

The function F is a built-in function in computer algebra systems such as Maple or Mathematica, since it occurs in many statistical applications, and is referred to as **the error function erf**. Figure 14 shows the graph of F. $\square$

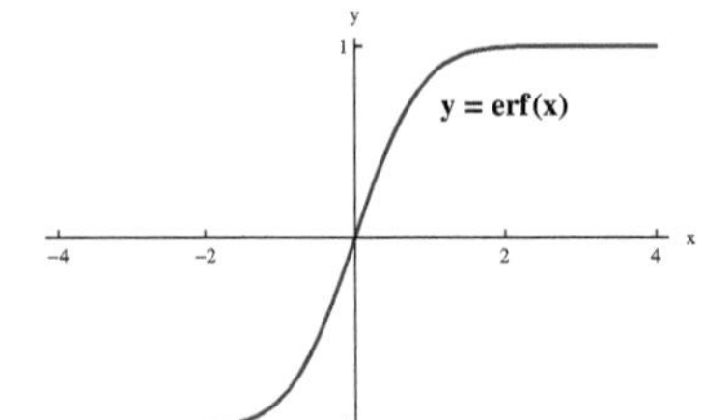

Figure 14: $y = \operatorname{erf}(x) = \int_0^x \frac{2}{\sqrt{\pi}} e^{-t^2} dt$

Example 5 (The natural logarithm defined as an integral)

If $x > 0$, we have

$$\int_1^x \frac{1}{t} dt = \ln(t)\big|_1^x = \ln(x) - \ln(1) = \ln(x),$$

since

$$\frac{d}{dt} \ln(t) = \frac{1}{t}, t > 0.$$

Thus, $\ln(x)$ is the area between the graph of $y = 1/t$ and the interval $[1, x]$ if $x > 1$, as illustrated in Figure 15.

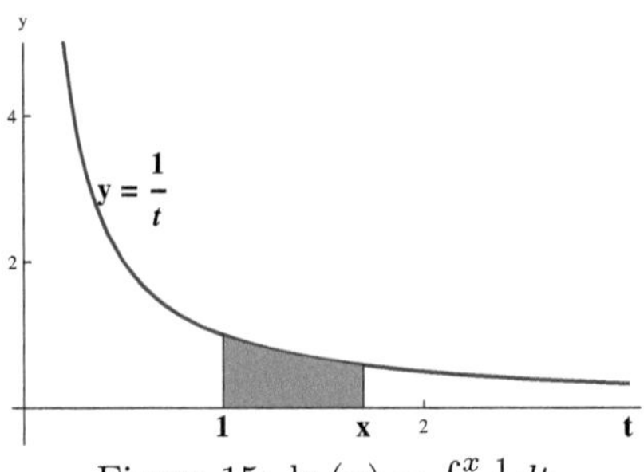

Figure 15: $\ln(x) = \int_1^x \frac{1}{t} dt$

If $0 < x < 1$,

$$\ln(x) = -\int_x^1 \frac{1}{t} dt,$$

so that $\ln(x)$ is $(-1) \times$ (the area between the graph of $y = 1/t$ and the interval $[x, 1]$), as illustrated in Figure 16.

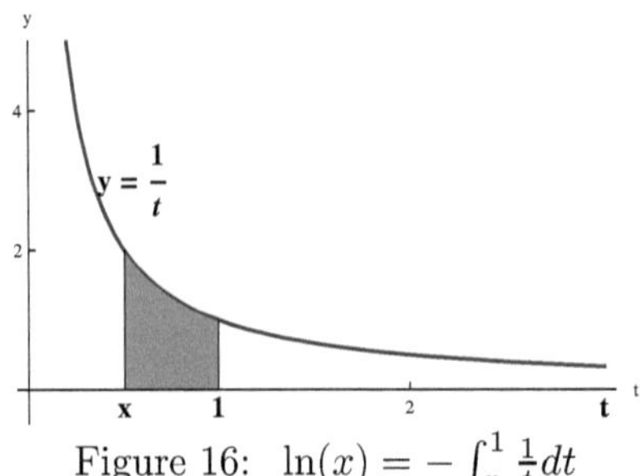

Figure 16: $\ln(x) = -\int_x^1 \frac{1}{t} dt$

If we didn't know about the natural logarithm, and needed an antiderivative if $1/x$, we could have set

$$F(x) = \int_1^x \frac{1}{t} dt, \ x > 0.$$

By the second part of the Fundamental Theorem of Calculus, we have

$$F'(x) = \frac{d}{dx} \int_1^x \frac{1}{t} dt = \frac{1}{x}$$

for each $x > 0$, so that F is an antiderivative of the function defined by $1/x$ on $(0, +\infty)$. Thus, we can introduce the natural logarithm as the function F and define the natural exponential function as its inverse. This approach enables us to derive all the properties of the natural logarithm and the natural exponential function rigorously. This program is carried out in Appendix E. $\square$

Example 6 The sine integral function **Si** is defined by the expression

$$\mathrm{Si}\,(x) = \int_0^x \frac{\sin\,(t)}{t} dt$$

Determine $\mathrm{Si}'\,(x)$.

Solution

Since

$$\lim_{t \to 0} \frac{\sin\,(t)}{t} = 1,$$

if we set

$$f\,(t) = \begin{cases} \dfrac{\sin\,(t)}{t} & \text{if} \ \ t \neq 0, \\[2ex] 1 & \text{if} \ \ t = 0, \end{cases}$$

then f is continuous on the entire number line. We can interpret the integral

$$\int_0^x \frac{\sin\,(t)}{t} dt$$

as

$$\int_0^x f\,(t)\, dt.$$

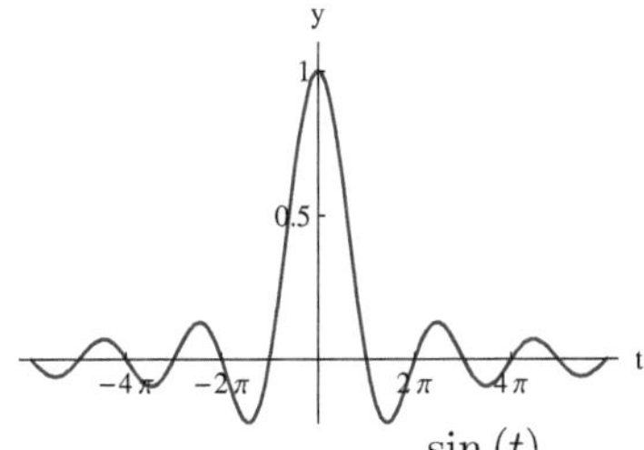

Figure 17: $y = \dfrac{\sin\,(t)}{t}$

Thus, the second part of the Fundamental Theorem of Calculus is applicable:

$$\frac{d}{dx}\,\mathrm{Si}\,(x) = \frac{d}{dx}\int_0^x f\,(t)\,dt = \begin{cases} \dfrac{\sin\,(x)}{x} & \text{if} \quad x = 0, \\[2ex] 1 & \text{if} \quad x = 0. \end{cases}$$

Figure 17 shows the graph of f and Figure 18 shows the graph of the sine integral function Si.
□

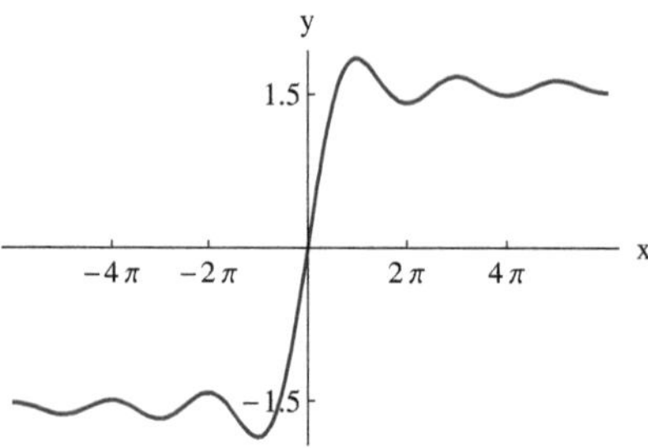

Figure 18: The sine integral function

Example 7 Set

$$F\,(x) = \int_0^x \frac{t^2 - 4}{t^2 + 1}\,dt.$$

a) Determine $F'\,(x)$.
b) Determine the intervals on which F is increasing/decreasing,

Solution

a) The integrand is continuous on the entire number line, since it is a rational function and $t^2 + 1 \neq 0$ for any $t \in \mathbb{R}$. Figure 19 shows the graph of the integrand.

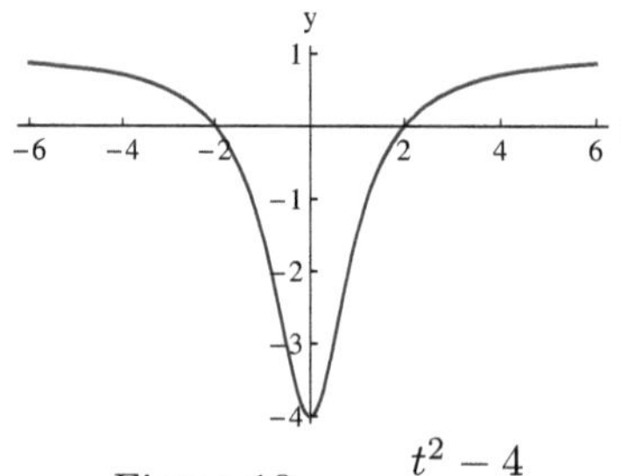

Figure 19: $y = \dfrac{t^2 - 4}{t^2 + 1}$

By the second part of the Fundamental Theorem of Calculus,

$$F'\,(x) = \frac{d}{dx}\int_0^x \frac{t^2 - 4}{t^2 + 1}\,dt = \frac{x^2 - 4}{x^2 + 1} \quad \text{for each } x \in \mathbb{R}.$$

b) We will make of the derivative test for monotonicity. We have

$$F'\,(x) = 0 \Leftrightarrow \frac{x^2 - 4}{x^2 + 1} = 0 \Leftrightarrow x = \pm 2.$$

We also have $F'(x) > 0$ if $x < -2$, $F'(x) < 0$ if $-2 < x < 2$ and $F'(x) > 0$ if $x > 2$. Therefore, F is increasing on $(-\infty, 2]$, decreasing on $[-2, 2]$ and increasing on $[2, +\infty)$.

In the next chapter we will introduce new special functions and we will be able to express $F(x)$ in terms of one of these functions. In the mean time, you can make use of your computational utility to obtain approximates values for F (for example, you can use midpoint sums), and plot a graph of F. Figure 20 shows such a graph.

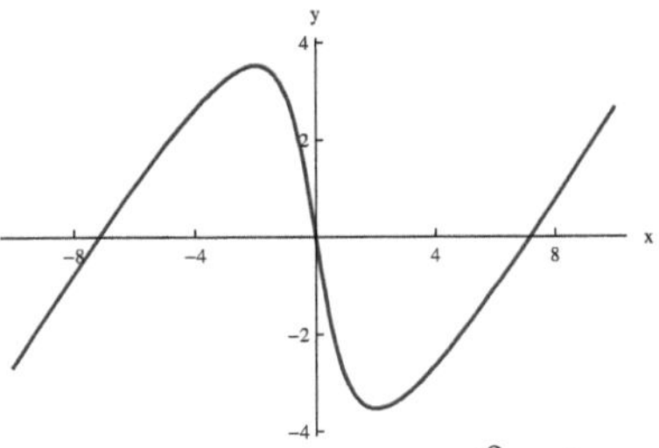

Figure 20: $F(x) = \int_0^x \dfrac{t^2 - 4}{t^2 + 1} dt$

Incidentally,

$$F(2) = \int_0^2 \frac{t^2 - 4}{t^2 + 1} dt \cong -3.535\,74 \text{ and } F(-2) = \int_0^{-2} \frac{t^2 - 4}{t^2 + 1} dt \cong 3.\,535\,74$$

$\square$

Now that we have established the second part of the Fundamental Theorem of Calculus, let us display both parts of the Theorem in a symmetric fashion (the restrictions on the functions have been stated earlier):

THE FUNDAMENTAL THEOREM OF CALCULUS

1.

$$\int_a^x \frac{d f(t)}{dt} dt = f(x) - f(a).$$

2.

$$\frac{d}{dx} \int_a^x f(t) dt = f(x).$$

It is worthwhile to repeat the meaning of the Fundamental Theorem: **The first part of the theorem states that the integral of the derivative of a function on an interval is the difference between the values of the function at the endpoints. The second part of the theorem states that the derivative of the function**

$$\int_a^x f(t) dt$$

is the value of the integrand at the upper limit. We can say that differentiation and integration are reverse operations in the precise sense of the Fundamental Theorem.

We may refer to either part of the Fundamental Theorem of Calculus simply as "the Fundamental Theorem of Calculus".

Problems

In problems 1-8 , compute the indicated derivative.

1. $\dfrac{d}{dx}\displaystyle\int_1^x \dfrac{1}{t^4+1}\,dt$

2. $\dfrac{d}{dx}\displaystyle\int_\pi^x \sqrt{4+\sin^2(t)}\,dt$

3. $\dfrac{d}{dx}\displaystyle\int_x^4 \dfrac{1}{\sqrt{4+u^2}}\,du$

4. $\dfrac{d}{dx}\displaystyle\int_2^{\sin(x)} \sqrt{9-t^2}\,dt$

5. $\dfrac{d}{dx}\displaystyle\int_0^x \cos\left(\dfrac{\pi t^2}{2}\right)\,dt$

6. $\dfrac{d}{dx}\displaystyle\int_0^{\sqrt{x}} \dfrac{1}{\sqrt{(1-t^2)\left(1-\frac{1}{4}t^2\right)}}\,dt$

7. $\dfrac{d}{dx}\displaystyle\int_x^{x^2} t^2\sqrt{t^2+1}\,dt$

8. $\dfrac{d}{dx}\displaystyle\int_0^{\ln(x)} e^{\sqrt{t}}\,dt$

9. Let
$$F(x) = \int_0^x \cos\left(t^2\right)\,dt.$$
Use the second derivative test for local extrema to determine the smallest $a>0$ such that F has a local maximum at a, and the smallest $b>0$ such that F has a local minimum at b.

10. Let
$$F(x) = \int_1^x \dfrac{1}{(t-1)^2+1}\,dt.$$
a) Show that F is increasing on $(-\infty, +\infty)$.
b) Determine the intervals on which the graph of F is concave up/concave down, and the point of inflection

11. Let
$$F(x) = \int_{\pi/2}^x \sqrt{1-\dfrac{1}{2}\sin^2(t)}\,dt.$$

Express
$$\int_{\pi/6}^{5\pi/6} \sqrt{1-\dfrac{1}{2}\sin^2(x)}\,dx$$
in terms of F.

12. Let
$$F(x) = \int_1^x \dfrac{\sin(t)}{t}\,dt.$$

Express
$$\int_{\pi/2}^{\pi} \dfrac{\sin(x)}{x}\,dx.$$
in terms of F.

13. Let
$$F(x) = \int_0^x \dfrac{1}{\sqrt{(1-t^2)\left(1-\frac{1}{4}t^2\right)}}\,dt.$$

Express
$$\int_{1/3}^{1/2} \dfrac{1}{\sqrt{(1-x^2)\left(1-\frac{1}{4}x^2\right)}}\,dx$$
in terms of F.

5.5 Integration is a Linear Operation

In Section 5.3 we introduced the first part of the Fundamental Theorem of Calculus that enabled us to compute the exact value of an integral once we identified an antiderivative of the integrand. We displayed a short of list of the indefinite integrals of some basic functions. In this section we will discuss the indefinite and definite integrals of linear combinations of functions, and expand the scope of our short list considerably. We will also discuss the calculation of the area of a region between the graphs of functions

The Linearity of Indefinite and Definite Integrals

The rules for the indefinite integrals of constant multiples and sums of functions follow from the corresponding rules for differentiation:

Proposition 1

1 (THE CONSTANT MULTIPLE RULE FOR INDEFINITE INTEGRALS) If c is a constant,

$$\int cf(x)\,dx = c\int f(x)\,dx.$$

2 (THE SUM RULE FOR INDEFINITE INTEGRALS)

$$\int (f(x) + g(x))\,dx = \int f(x)\,dx + \int g(x)\,dx.$$

Remark 1 Since the indefinite integral of a function represents any antiderivative of the function, and any two antiderivatives of the same function can differ by a constant, **it should be understood that arbitrary constants can be added to either side of an equality that involves indefinite integrals.** $\Diamond$

The Proof of Proposition 1

1. Assume that F is an antiderivative of f. Thus,

$$\frac{d}{dx}F(x) = f(x)$$

for each x in an interval J, so that

$$\int f(x)\,dx = F(x).$$

By the constant multiple rule for differentiation,

$$\frac{d}{dx}(cF(x)) = c\frac{d}{dx}F(x) = cf(x)$$

for each $x \in J$. Therefore,

$$\int cf(x)\,dx = cF(x) = c\int f(x)\,dx.$$

2. Assume that

$$\frac{d}{dx}F(x) = f(x) \ \text{ and } \ \frac{d}{dx}G(x) = g(x)$$

for each x in an interval J, so that

$$\int f(x)\,dx = F(x) \quad \text{and} \quad \int g(x)\,dx = G(x).$$

By the sum rule for differentiation,

$$\frac{d}{dx}\left(F(x) + G(x)\right) = \frac{d}{dx}F(x) + \frac{d}{dx}G(x) = f(x) + g(x)$$

for each $x \in J$. Therefore,

$$\int \left(f(x) + g(x)\right) dx = F(x) + G(x) = \int f(x)\,dx + \int g(x)\,dx.$$

■

Example 1

a) Determine

$$\int 4\cos(x)\,dx.$$

b) Evaluate

$$\int_{\pi/6}^{\pi/4} 4\cos(x)\,dx.$$

Solution

a) We have

$$\int \cos(x)\,dx = \sin(x)$$

(on the entire number line). By the constant multiple rule for indefinite integrals,

$$\int 4\cos(x)\,dx = 4\int \cos(x)\,dx = 4\sin(x).$$

As in Remark 1, it should be understood that an arbitrary constant C can be added to $4\sin(x)$. If we wish to emphasize this, we may write

$$\int 4\cos(x)\,dx = 4\sin(x) + C.$$

b) The constant in the expression for the indefinite integral is not relevant to the evaluation of the definite integral. Any antiderivative will do for the evaluation of the definite integral with the help of the Fundamental Theorem of Calculus. Thus,

$$\int_{\pi/6}^{\pi/4} 4\cos(x)\,dx = 4\sin(x)\big|_{\pi/6}^{\pi/4} = 4\sin\left(\frac{\pi}{4}\right) - 4\sin\left(\frac{\pi}{6}\right)$$

$$= 4\left(\frac{\sqrt{2}}{2}\right) - 4\left(\frac{1}{2}\right) = 2\sqrt{2} - 2.$$

□

Example 2 Evaluate

$$\int_0^2 \left(x^2 + 1\right) dx.$$

Solution

By the sum rule for indefinite integrals and the reverse power rule,

$$\int \left(x^2 + 1\right) dx = \int x^2 dx + \int 1 dx = \frac{1}{3}x^3 + x + C,$$

where C is an arbitrary constant. The constant can be ignored for the evaluation of the definite integral. By the Fundamental Theorem of Calculus,

$$\int \left(x^2 + 1\right) dx = \frac{1}{3}x^3 + x \Big|_0^2 = \frac{8}{3} + 2 = \frac{14}{3}.$$

$\square$

We can combine parts 1 and 2 of Proposition 1 and obtain the rule for indefinite integrals that corresponds to the linearity of differentiation:

Theorem 1 (THE LINEARITY OF INDEFINITE INTEGRALS) Assume that c_1 and c_2 are constants. Then

$$\int \left(c_1 f(x) + c_2 g(x)\right) dx = c_1 \int f(x) dx + c_2 \int g(x) dx.$$

Proof

By the sum rule for indefinite integrals,

$$\int \left(c_1 f(x) + c_2 g(x)\right) dx = \int c_1 f(x) dx + \int c_2 g(x) dx.$$

By the constant multiple rule for indefinite integrals,

$$\int c_1 f(x) dx + \int c_2 g(x) dx = c_1 \int f(x) dx + c_2 \int g(x) dx.$$

Therefore,

$$\int \left(c_1 f(x) + c_2 g(x)\right) dx = c_1 \int f(x) dx + c_2 \int g(x) dx.$$

∎

Example 3 Let

$$f(x) = (x + 2)(x - 1)(x - 3) = x^3 - 2x^2 - 5x + 6.$$

a) Determine

$$\int f(x) dx.$$

b) Sketch the region G between the graph of f and the interval $[-1, 3]$. Compute the signed area of G and the area of G.

Solution

a) By the linearity of indefinite integrals and the reverse power rule,

$$\int \left(x^3 - 2x^2 - 5x + 6\right) dx = \int x^3 dx - 2\int x^2 dx - 5\int x\, dx + 6\int 1\, dx$$

$$= \frac{x^4}{4} - 2\left(\frac{x^3}{3}\right) - 5\left(\frac{x^2}{2}\right) + 6x + C$$

$$= \frac{1}{4}x^4 - \frac{2}{3}x^3 - \frac{5}{2}x^2 + 6x + C,$$

where C is an arbitrary constant.

b) Figure 1 shows the region G.

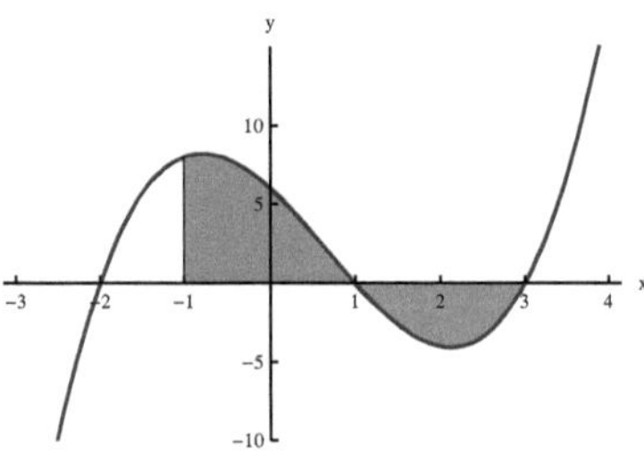

The signed area of G is

$$\int_{-1}^{3} f(x)\, dx = \int f(x)\, dx \Big|_{-1}^{3}$$

$$= \frac{1}{4}x^4 - \frac{2}{3}x^3 - \frac{5}{2}x^2 + 6x \Big|_{-1}^{3}$$

$$= -\frac{9}{4} - \left(-\frac{91}{12}\right) = -\frac{9}{4} + \frac{91}{12} = \frac{16}{3}.$$

Since $f(x) > 0$ if $-1 < x < 1$ and $f(x) < 0$ if $1 < x < 3$, the area of G is

$$\int_{-1}^{1} f(x)\, dx - \int_{1}^{3} f(x)\, dx = \left(\frac{1}{4}x^4 - \frac{2}{3}x^3 - \frac{5}{2}x^2 + 6x \Big|_{-1}^{1}\right) - \left(\frac{1}{4}x^4 - \frac{2}{3}x^3 - \frac{5}{2}x^2 + 6x \Big|_{1}^{3}\right)$$

$$= \frac{32}{3} - \left(-\frac{16}{3}\right) = \frac{32}{3} + \frac{16}{3} = 16.$$

□

A polynomial is a linear combination of a constant function and functions defined by positive-integer powers of x. Therefore we can determine the indefinite integral of any polynomial, as in Example 3.

Example 4

a) Determine

$$\int \left(\sin(x) + \frac{1}{3}\sin(3x)\right) dx$$

b) Compute

$$\int_{\pi/3}^{\pi} \left(\sin(x) + \frac{1}{3}\sin(3x)\right) dx.$$

Solution

a) The formula

$$\int \sin(\omega x)\, dx = -\frac{1}{\omega}\cos(\omega x) + C,$$

where ω is a nonzero constant and C is an arbitrary constant is on the short list of Section 5.3. With the help of the linearity of indefinite integrals,

$$\int \left(\sin(x) + \frac{1}{3}\sin(3x) \right) dx = \int \sin(x)\, dx + \frac{1}{3}\int \sin(3x)\, dx$$

$$= -\cos(x) + \frac{1}{3}\left(-\frac{1}{3}\cos(3x) \right)$$

$$= -\cos(x) - \frac{1}{9}\cos(3x) + C,$$

where C is an arbitrary constant.

b) By the result of part a) and the Fundamental Theorem of Calculus,

$$\int_{\pi/3}^{\pi} \left(\sin(x) + \frac{1}{3}\sin(3x) \right) dx = -\cos(x) - \frac{1}{9}\cos(3x) \Big|_{\pi/3}^{\pi}$$

$$= \left(-\cos(\pi) - \frac{1}{9}\cos(3\pi) \right) - \left(-\cos\left(\frac{\pi}{3}\right) - \frac{1}{9}\cos(\pi) \right)$$

$$= \frac{10}{9} + \frac{7}{18} = \frac{3}{2}.$$

Figure 2 shows the region G between the graph of f and the interval $[\pi/3, \pi]$. Since $f(x) > 0$ if $\pi/3 < x < \pi$, the area of G is

$$\int_{\pi/3}^{\pi} f(x)\, dx = \frac{3}{2}.$$

$\square$

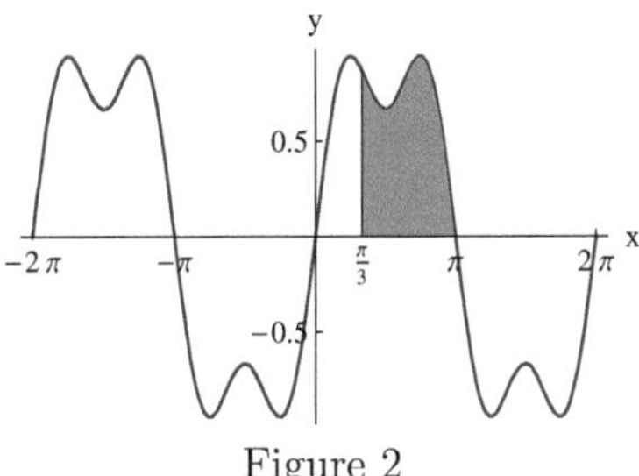

Figure 2

Recall that **a trigonometric polynomial** can be expressed as a linear combination of a constant function and functions defined by $\sin(nx)$ and $\cos(nx)$, where n is a positive integer. We can determine the indefinite integral of a trigonometric polynomial as in Example 4.

As in the above examples, the linearity of indefinite integrals enables us to calculate the definite integrals of linear combinations of functions whose indefinite integrals are known. Nevertheless, we will need to refer to **the linearity of definite integrals** as well.

Proposition 2 Assume that f and g are integrable on the interval $[a, b]$ and c is a constant. Then,

1 (THE CONSTANT MULTIPLE RULE FOR DEFINITE INTEGRALS)

$$\int_a^b cf(x)\,dx = c\int_a^b f(x)\,dx.$$

2 (THE SUM RULE FOR DEFINITE INTEGRALS)

$$\int_a^b (f(x) + g(x))\,dx = \int_a^b f(x)\,dx + \int_a^b g(x)\,dx.$$

As in the case of indefinite integrals, Proposition 2 leads to the linearity of the definite integral:

Theorem 2 (THE LINEARITY OF DEFINITE INTEGRALS) Assume that f and g are integrable on the interval $[a, b]$, and c_1, c_2 are constants. Then

$$\int_a^b (c_1 f(x) + c_2 g(x))\,dx = c_1 \int_a^b f(x)\,dx + c_2 \int_a^b g(x)\,dx.$$

We will leave the rigorous proof of Proposition 2 to a course in advanced calculus. Let's discuss the plausibility of the statements of Proposition 2 under the assumption that f and g are continuous on $[a, b]$:

Let $m_n(f)$ and $m_n(g)$ denote the midpoint sums for f and g, respectively, corresponding to the partitioning of $[a, b]$ to n subintervals of equal length. We have

$$\lim_{n\to\infty} m_n(f) = \int_a^b f(x)\,dx \text{ and } \lim_{n\to\infty} m_n(g) = \int_a^b g(x)\,dx.$$

Therefore,

$$\lim_{n\to\infty} (m_n(f) + m_n(g)) = \lim_{n\to\infty} m_n(f) + \lim_{n\to\infty} m_n(g) = \int_a^b f(x)\,dx + \int_a^b g(x)\,dx.$$

You can confirm that $m_n(f) + m_n(g)$ is a midpoint sum for $f + g$, corresponding to the partitioning of $[a, b]$ to n subintervals of equal length, and $f + g$ is continuous on $[a, b]$. Therefore,

$$\lim_{n\to\infty} (m_n(f) + m_n(g)) = \int_a^b (f(x) + g(x))\,dx.$$

Thus,

$$\int_a^b (f(x) + g(x))\,dx = \int_a^b f(x)\,dx + \int_a^b g(x)\,dx.$$

Similarly, $cm_n(f)$ is a midpoint sum for cf corresponding to the partitioning of $[a, b]$ to n subintervals of equal length, so that

$$\lim_{n\to\infty} cm_n(f) = \int_a^b cf(x)\,dx.$$

Since

$$\lim_{n\to\infty} cm_n(f) = c \lim_{n\to\infty} m_n(f) = c\int_a^b f(x)\,dx,$$

we have

$$\int_a^b cf\,(x)\,dx = c\int_a^b f\,(x)\,dx.$$

∎

Unlike Theorem 1, Theorem 2 is applicable even if we are not able to recognize the antiderivatives of f and g, as in the following example.

Example 5 It is known that

$$\int_0^1 \frac{1}{\sqrt{4-x^2}}\,dx = \frac{\pi}{6} \quad\text{and}\quad \int_0^1 \sqrt{4-x^2}\,dx = \frac{\sqrt{3}}{2} + \frac{\pi}{3}.$$

Determine

$$\int_0^1 \left(\frac{2}{\sqrt{4-x^2}} - 3\sqrt{4-x^2}\right)dx.$$

Solution

By the linearity of the definite integral,

$$\int_0^1 \left(\frac{2}{\sqrt{4-x^2}} - 3\sqrt{4-x^2}\right)dx = 2\int_0^1 \frac{1}{\sqrt{4-x^2}}\,dx - 3\int_0^1 \sqrt{4-x^2}\,dx$$

$$= 2\left(\frac{\pi}{6}\right) - 3\left(\frac{\sqrt{3}}{2} + \frac{\pi}{3}\right)$$

$$= -\frac{2\pi}{3} - \frac{3\sqrt{3}}{2}.$$

□

The Area of a Region between the Graphs of Functions

Thanks to the linearity of integration, we can calculate the area of a region between the graphs of two functions. In order to be specific, let's assume that f and g are continuous on $[a,b]$, $f(c) = g(c)$, where $a < c < b$, $f(x) > g(x)$ if $a \leq x < c$, and $g(x) > f(x)$ if $c < x \leq b$. Figure 3 illustrates such a case.

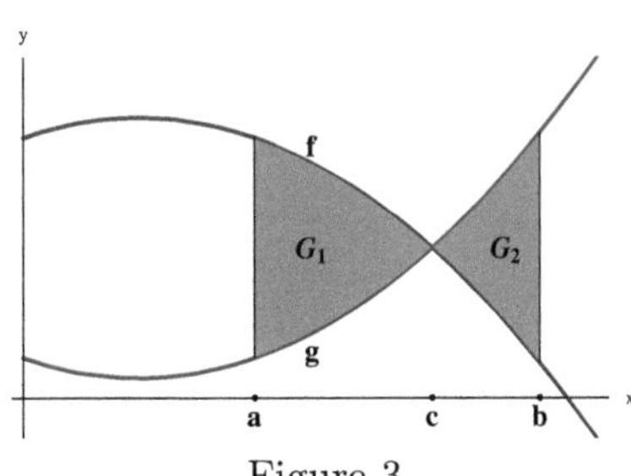

Figure 3

We would like to calculate the area of the region G between the graph of f, the graph of g, the line $x = a$, and the line $x = b$. With reference to Figure 3, the area of G is the sum of the areas of G_1 and G_2. The area of the region G_1 can be obtained by subtracting the area of the region between the graph of g and the interval $[a, c]$ from the area of the region between the graph of f and $[a, c]$. Thus, the area of G_1 is

$$\int_a^c f(x)dx - \int_a^c g(x)dx = \int_a^c \left(f(x) - g(x)\right) dx.$$

Similarly, the area of the region G_2 is

$$\int_c^b \left(g(x) - f(x)\right) dx.$$

Thus, the area of G is

$$\int_a^c \left(f(x) - g(x)\right) dx + \int_c^b \left(g(x) - f(x)\right) dx.$$

Note that $|f(x) - g(x)| = f(x) - g(x)$ if $a \leq x \leq c$, since $f(x) \geq g(x)$ in this case. If $c \leq x \leq b$, then $|f(x) - g(x)| = -\left(f(x) - g(x)\right) = g(x) - f(x)$, since $g(x) \geq f(x)$ for each $x \in [c, b]$. Therefore,

$$\int_a^c \left(f(x) - g(x)\right) dx + \int_c^b \left(g(x) - f(x)\right) dx = \int_a^c |f(x) - g(x)| \, dx + \int_c^b |f(x) - g(x)| \, dx$$

$$= \int_a^b |f(x) - g(x)| \, dx.$$

Thus, the area of G can be expressed as

$$\int_a^b |f(x) - g(x)| \, dx.$$

This fact is true in the general case: **If f and g are continuous on $[a, b]$, the area of the region between the graph of f, the graph of g, the line $x = a$ and $x = b$ is**

$$\int_a^b |f(x) - g(x)| \, dx.$$

As a special case, we can express **the area of the region between the graph of a function f and an interval $[a, b]$ as**

$$\int_a^b |f(x)| \, dx.$$

$(g = 0)$.

Remark 2 We can arrive at the expression for the area of a region between the graphs of functions by going back to the definition of the integral. Thus, assume that the graphs of f and g are as in Figure 4.

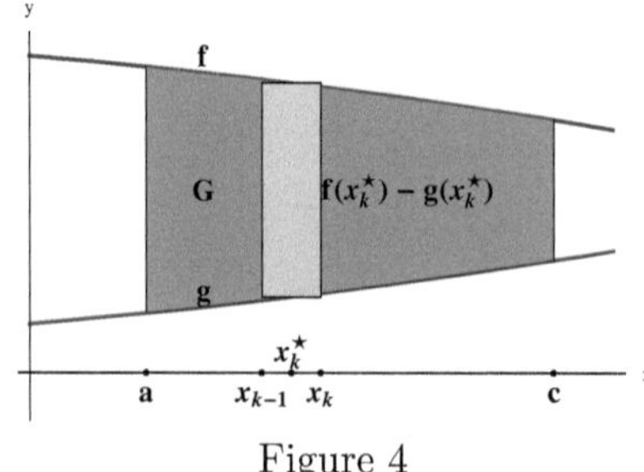

Figure 4

Let $P = \{x_0, x_1, \ldots, x_{n-1}, x_n\}$ be a partition of $[a, c]$. If $\Delta x_k = x_k - x_{k-1}$, and $||P|| = \max_k \Delta x_k$ is small, we can approximate the area of the slice of the region G between the lines $x = x_{k-1}$ and $x = x_k$ by the area of a rectangle whose dimensions are Δx_k and $f(x_k^*) - g(x_k^*)$, where x_k^* is an arbitrary point between x_{k-1} and x_k. The area of such a rectangle is

$$\left(f\left(x_k^*\right) - g\left(x_k^*\right)\right) \Delta x_k,$$

as illustrated in Figure 4. Thus, the sum

$$\sum_{k=1}^{n} \left(f\left(x_k^*\right) - g\left(x_k^*\right)\right) \Delta x_k$$

approximates the area of G if $\max_k \Delta x_k$ is small. But this is a Riemann sum that approximates

$$\int_a^c \left(f\left(x\right) - g\left(x\right)\right) dx.$$

Therefore, we have reached the same expression for the area of G as before.$\Diamond$

Example 6 Let $f(x) = x^2 - 2x - 1$ and $g(x) = -x^2 + 2x + 5$.

a) Sketch the region G between the graph of f, the graph of g, the line $x = 1$ and the line $x = 5$.
b) Calculate the area of G.

Solution

a) Figure 5 shows the region G.

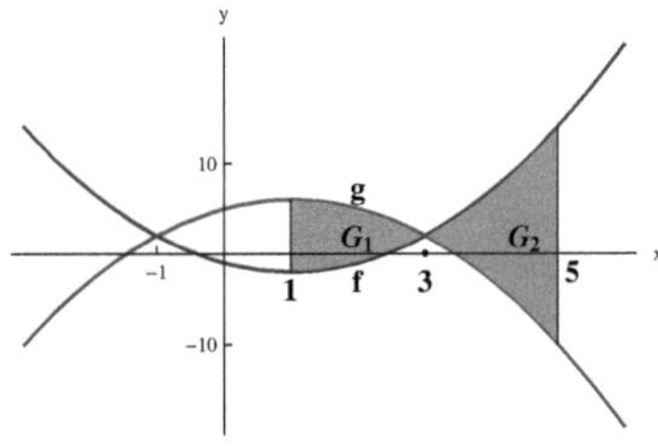

Figure 5

b) In order to determine the x-coordinates of the points at which the graphs of f and g intersect, we need to find the solutions of the equation $f(x) = g(x)$:

$$x^2 - 2x - 1 = -x^2 + 2x + 5 \Leftrightarrow 2x^2 - 4x - 6 = 0$$
$$\Leftrightarrow x - 1 \text{ or } x = 3.$$

With reference to Figure 5, the area of G is the sum of the areas of G_1 and G_2.

$$\text{The area of } G_1 = \int_1^3 \left(g(x) - f(x)\right) dx = \int_1^3 \left(-2x^2 + 4x + 6\right) dx$$
$$= -\frac{2}{3}x^3 + 2x^2 + 6x \Big|_{x=1}^{3} = \frac{32}{3}.$$

$$\text{The area of } G_2 = \int_3^5 (f(x) - g(x))\, dx = \int_3^5 \left(2x^2 - 4x - 6\right) dx$$

$$= \frac{2}{3}x^3 - 2x^2 - 6x \,\Big|_3^5 = \frac{64}{3}.$$

Therefore, the area of G is

$$\frac{32}{3} + \frac{64}{3} = 32.$$

□

Example 7 Let $f(x) = -x^2 + 1$ and $g(x) = \sin(x)$.

a) Plot the graphs of f and g and determine approximations to the points x_1 and x_2 such that $x_1 < 0 < x_2$ and the graphs of f and g intersect at the corresponding points, with the help of your calculator.
b) Express the area of the region G between the graphs of f and g and the lines $x = x_1$ and $x = x_2$ as an integral. Determine an approximation to the integral with the help of your calculator.

Solution

a) Figure 6 shows the region G. The picture indicates that the x-coordinates of the points at which the graphs of f and g intersect are approximately -1.5 and 0.5. We have $x_1 \cong -1.409\,62$ and $x_2 \cong 0.636\,733$, rounded to 6 significant digits.

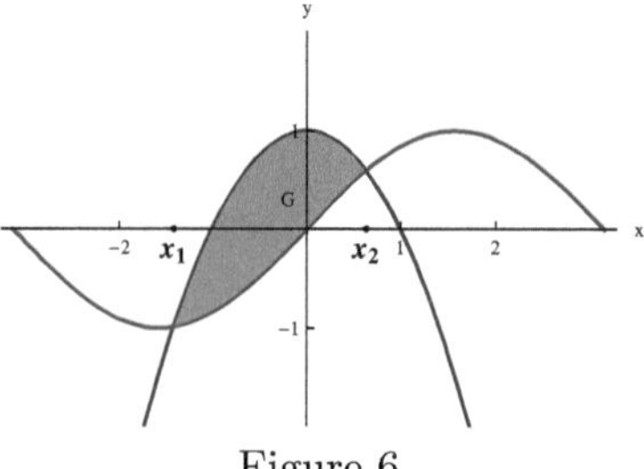

Figure 6

b) Since $f(x) > g(x)$ if $x_1 < x < x_2$, the area of G is

$$\int_{x_1}^{x_2} (f(x) - g(x))\, dx = \int_{x_1}^{x_2} \left(\left(-x^2 + 1\right) - \sin(x)\right) dx$$

$$= -\frac{x^3}{3} + x + \cos(x) \,\Big|_{x_1}^{x_2} \cong 1.67021.$$

□

Remark 3 Assume that $v(t)$ is the **velocity** at time t of an object in one-dimensional motion, and f is the corresponding **position function**. As we saw in Section 4.3, the **displacement of the object over the time interval $[a, b]$** is

$$f(b) - f(a) = \int_a^b v(t)\, dt.$$

The **distance** traveled by the object over the same time interval corresponds to the area of the region between the graph of the velocity function on $[a, b]$ and can be expressed as

$$\int_a^b |v(t)|\, dt.$$

The absolute value of the velocity is the **speed** of the object. Thus, **the distance traveled by the object over a time interval is the integral of the speed of the object over that time interval.** ◊

Example 8 Assume that

$$v(t) = 2\sin\left(\frac{t}{4}\right)$$

is the velocity at time t of an object in one-dimensional motion (in centimeters per second).
a) Sketch the graph of v on the interval $[0, 8\pi]$.
b) Determine the displacement of the object over the time interval $[0, 6\pi]$.
c) Determine the distance traveled by the object over the time interval $[0, 6\pi]$.

Solution

a) Figure 7 shows the graph of v on $[0, 8\pi]$.

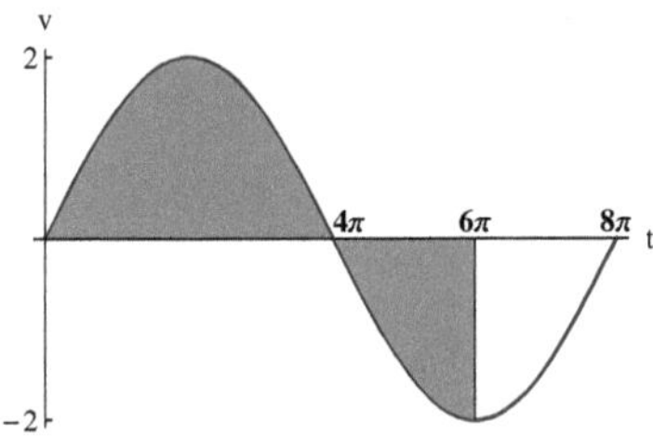

Figure 7

b) The displacement of the object over the time interval $[0, 6\pi]$ is

$$\int_0^{6\pi} v(t)\, dt = \int_0^{6\pi} 2\sin\left(\frac{t}{4}\right) dt = 2\int_0^{6\pi} \sin\left(\frac{t}{4}\right) dt.$$

We have

$$\int \sin\left(\frac{t}{4}\right) dt = -\frac{1}{1/4}\cos\left(\frac{t}{4}\right) + C = -4\cos\left(\frac{t}{4}\right) + C,$$

where C is an arbitrary constant. Therefore,

$$2\int_0^{6\pi} \sin\left(\frac{t}{4}\right) dt = 2\left(-4\cos\left(\frac{t}{4}\right)\Big|_0^{6\pi}\right) = 2\left(-4\cos\left(\frac{3\pi}{2}\right) + 4\cos(0)\right) = 8.$$

Thus, the displacement of the object over the time interval $[0, 6\pi]$ is 8 (centimeters).
c) We have $v(t) > 0$ if $0 < t < 4\pi$ and $v(t) < 0$ if $4\pi < t \le 6\pi$. Therefore the distance traveled by the object over the time interval $[0, 6\pi]$ is

$$\int_0^{6\pi} |v(t)|\, dt = \int_0^{4\pi} v(t)\, dt + \int_{4\pi}^{6\pi} -v(t)\, dt = \int_0^{4\pi} v(t)\, dt - \int_{4\pi}^{6\pi} v(t)\, dt$$

$$= \int_0^{4\pi} 2\sin\left(\frac{t}{4}\right) dt - \int_{4\pi}^{6\pi} 2\sin\left(\frac{t}{4}\right) dt$$

$$= 16 - (-8) = 24$$

(check). Graphically, the area of the region between the graph of the velocity function and the interval $[0, 6\pi]$ is 24 .$\Box$

Problems

In problems 1- 8, determine the indefinite integral.

1.
$$\int \left(1 - \frac{1}{2}x^2 + \frac{1}{24}x^4\right) dx$$

2.
$$\int \left(x - \frac{1}{6}x^3\right) dx$$

3.
$$\int (4x + 3)\left(2x^2 + 1\right) dx$$

4.
$$\int (4\cos(x) - 2\sin(x)) \, dx$$

5.
$$\int (\cos(6t) - 3\sin(2t)) \, dt$$

6.
$$\int \left(\sqrt{x} - \frac{4}{x}\right) dx$$

7.
$$\int \frac{6x^2 + 2x + 1}{x} dx$$

8,
$$\int \left(\sin\left(\frac{t}{6}\right) - 4t^2 + 3\right) dt$$

In problems 9-12 , evaluate the definite integral.

9.
$$\int_{-1}^{3} \left(\frac{1}{2}x + \frac{2}{3}x^2 + 1\right) dx$$

10.
$$\int_{4}^{16} \left(\sqrt{x} + x^{3/2}\right) dx$$

11.
$$\int_{\pi/8}^{\pi/4} (2\cos(4x) - 4\sin(2x)) \, dx$$

12.
$$\int_{\log_2(3)}^{\log_2(4)} (4^x - 2^x) \, dx$$

In problems 13-16, sketch the graphs of the functions f and g. Determine the area of the region between the graphs of f and g, on the interval J if applicable:

13.
$$f(x) = \sqrt{x}, \ g(x) = x/2, \ J = [0, 6]$$

14.
$$f(x) = x^2 - 9, \ g(x) = 9 - x^2, J = [-3, 4]$$

15.
(the bounded region between the graphs of) $f(x) = x^2 - 1, \ g(x) = -x$

16.
$$f(x) = 2x - x^2 + 3, g(x) = x^2 - 2x, J = [0, 6]$$

17. Assume that the velocity of an object in one-dimensional motion is

$$2\sin\left(\frac{t}{4}\right) + \frac{1}{6}\cos(2t)$$

at time t. Calculate the displacement of the object over the time interval $[0, \pi]$

18. Assume that the velocity of an object moving along the x-axis is $\sin(\pi t/4)$ at time t. Determine the distance traveled by the object over the time interval $[0, 6]$.

5.6 The Substitution Rule

The substitution rule for indefinite integrals follows from the chain rule for differentiation. The rule enables us to transform a given antidifferentiation problem to a tractable expression. The definite integral version of the substitution rule is useful in establishing significant general facts.

The Substitution Rule for Indefinite Integrals

Consider the indefinite integral

$$\int \sin\left(x^2\right) 2x\,dx.$$

If we set $u\left(x\right) = x^2$, then

$$\frac{du}{dx} = \frac{d}{dx}\left(x^2\right) = 2x.$$

Therefore

$$\int \sin\left(x^2\right)(2x)\,dx = \int \sin\left(u\right)\frac{du}{dx}\,dx.$$

It is tempting to replace the symbol

$$\frac{du}{dx}\,dx$$

by du, and write

$$\int \sin\left(u\right)\frac{du}{dx}\,dx = \int \sin\left(u\right)\,du.$$

Assume that the above equality is true. Since we know that

$$\int \sin\left(u\right)\,du = -\cos\left(u\right) + C,$$

where C is an arbitrary constant, we are led to claim that

$$\int \sin\left(x^2\right) 2x\,dx = -\cos\left(u\right) + C = -\cos\left(x^2\right) + C.$$

This is indeed the case, as you can check by differentiating the right-hand side.

The procedure that we described above is valid:

Theorem 1 (THE SUBSTITUTION RULE FOR INDEFINITE INTEGRALS) Assume that f is continuous on the interval I, u is a differentiable function on the interval J and $u(x) \in I$ if $x \in J$. Then,

$$\int f\left(u(x)\right)\frac{du}{dx}\,dx = \int f\left(u\right)du$$

where $x \in J$.

The expression

$$\int f(u)du$$

denotes a function of u. It should be understood that the above equality is valid, provided that u is replaced by its expression in terms of x. Since the equality involves indefinite integrals, we are entitled to add arbitrary constants to either side.

The Proof of Theorem 1

By the second part of the Fundamental Theorem of Calculus, the continuous function f has an antiderivative F on the interval I. Thus,

$$\frac{d}{du}F\left(u\right) = f\left(u\right) \Leftrightarrow F\left(u\right) = \int f\left(u\right)du$$

on I. Let's consider the composite function $F \circ u$ on J. By the chain rule,

$$\frac{d}{dx}\left(F \circ u\right)\left(x\right) = \frac{d}{dx}F(u(x)) = \left(\frac{d}{du}F(u)\bigg|_{u=u(x)}\right)\left(\frac{d}{dx}u(x)\right) = f(u(x))\frac{du}{dx}$$

for each $x \in J$. Therefore

$$\int f(u(x))\frac{du}{dx}dx = F\left(u\left(x\right)\right)$$

on J. Since

$$F\left(u\right) = \int f\left(u\right)du,$$

we have

$$\int f(u(x))\frac{du}{dx}dx = F\left(u\left(x\right)\right) = \int f\left(u\right)du\bigg|_{u=u(x)}$$

on J. Therefore,

$$\int f\left(u\left(x\right)\right)\frac{du}{dx}dx = \int f(u)du,$$

with the understanding that the right-hand side is evaluated at $u\left(x\right)$. $\blacksquare$

Example 1 Determine

$$\int \sin^2\left(x\right)\cos\left(x\right)dx.$$

Solution

We set $u(x) = \sin\left(x\right)$. Then,

$$\frac{du}{dx} = \frac{d}{dx}\sin\left(x\right) = \cos\left(x\right).$$

Therefore,

$$\int \sin^2\left(x\right)\cos\left(x\right)dx = \int \left(\sin\left(x\right)\right)^2\cos\left(x\right)dx = \int \left(u\left(x\right)\right)^2\frac{du}{dx}dx$$

By the substitution rule and the reverse power rule,

$$\int \left(u\left(x\right)\right)^2\frac{du}{dx}dx = \int u^2du = \frac{1}{3}u^3 + C,$$

where C is an arbitrary constant. Therefore,

$$\int \sin^2\left(x\right)\cos\left(x\right)dx = \frac{1}{3}u^3 + C = \frac{1}{3}\sin^3\left(x\right) + C.$$

The expression is valid on the entire number line. $\square$

As in the above example, the substitution rule is helpful when it helps us transform the given indefinite integral to a familiar indefinite integral.

Remark 1 It is easy to remember the substitution rule: In the expression

$$\int f(u\,(x))\frac{du}{dx}dx,$$

we can treat

$$\frac{du}{dx}$$

as a "symbolic fraction", carry out "symbolic cancellation" and write

$$\int f(u\,(x))\frac{du}{dx}dx = \int f(u)du.$$

Thus, we can set

$$du = \frac{du}{dx}dx$$

when we implement the substitution rule. There is no need to try to attach a mystical meaning to the symbolic manipulation, though: We are merely describing a practical way to remember the substitution rule. Within the present context, the symbol

$$\frac{du}{dx}dx$$

does not express the value of the differential $du\,(x, dx)$ that we discussed in Section 2.5, even though the notation is the same. $\Diamond$

Example 2 Determine

$$\int x\sqrt{4 - x^2}dx.$$

Solution

Let's set $u = 4 - x^2$. Then

$$\frac{du}{dx} = \frac{d}{dx}\left(4 - x^2\right) = -2x.$$

Therefore,

$$du = \frac{du}{dx}dx = -2xdx.$$

By the substitution rule,

$$\int x\sqrt{4 - x^2}dx = -\frac{1}{2}\int \sqrt{4 - x^2}\,(-2x)\,dx = -\frac{1}{2}\int \sqrt{u}\frac{du}{dx}dx = -\frac{1}{2}\int u^{1/2}du.$$

By the reverse power rule,

$$-\frac{1}{2}\int u^{1/2}du = -\frac{1}{2}\left(\frac{u^{3/2}}{3/2}\right) + C = -\frac{1}{3}u^{3/2} + C,$$

where C is an arbitrary constant. Therefore,

$$\int x\sqrt{4 - x^2}dx = \left.-\frac{1}{3}u^{3/2} + C\right|_{u=4-x^2} = -\frac{1}{3}\left(4 - x^2\right)^{3/2} + C.$$

Note that we can take a shortcut by using the formalism

$$du = \frac{du}{dx}dx,$$

and we won't go wrong. Thus

$$du = \frac{d}{dx}\left(4 - x^2\right) dx = -2x dx,$$

so that we will replace $x dx$ by

$$-\frac{1}{2} du.$$

Therefore,

$$\int x\sqrt{4 - x^2}\,dx = \int \sqrt{4 - x^2}\,x dx = \int \sqrt{u}\left(-\frac{1}{2}\right) du = -\frac{1}{2}\int u^{1/2}\,du,$$

as before. $\square$

Remark 2 Note that the implementation of the substitution rule was successful in Examples 1 and 2, since the rule enabled us to transform the given indefinite integral to a constant multiple of

$$\int u^r\,du,$$

so that we were able to apply the reverse power rule. More generally, if we recognize that the given indefinite integral can be expressed as a constant multiple of

$$\int u^r\left(x\right)\frac{du}{dx}\,dx,$$

the substitution rule leads to $\int u^r\,du$. $\lozenge$

Remark 3 In Section 5.3 we noted that

$$\int \sin\left(\omega x\right) dx = -\frac{1}{\omega}\cos\left(\omega x\right) + C \text{ and } \int \cos\left(\omega x\right) dx = \frac{1}{\omega}\sin\left(\omega x\right) + C,$$

for any constant $\omega \neq 0$, where C denotes an arbitrary constant. The basic formulas are

$$\int \sin\left(x\right) dx = -\cos\left(x\right) + C \text{ and } \int \cos\left(x\right) dx = \sin\left(x\right) + C,$$

since these lead to the more general formulas by the substitution $u = \omega x$. Indeed,

$$u = \omega x \Rightarrow \frac{du}{dx} = \omega.$$

Therefore,

$$\int \sin\left(\omega x\right) dx = \int \sin\left(\omega x\right)\frac{1}{\omega}\omega dx = \frac{1}{\omega}\int \sin\left(u\right)\frac{du}{dx}\,dx = \frac{1}{\omega}\int \sin\left(u\right) du,$$

by the substitution rule. Thus,

$$\int \sin\left(\omega x\right) dx = \frac{1}{\omega}\int \sin\left(u\right) du = -\frac{1}{\omega}\cos\left(u\right) + C = -\frac{1}{\omega}\cos\left(\omega x\right) + C.$$

The formula that involves $\cos\left(\omega x\right)$ can be obtained in a similar manner. $\lozenge$

Example 3

a) Determine

$$\int \tan(x)\,dx.$$

Specify the intervals on which the antidifferentiation formula is valid.

b) Compute

$$\int_{3\pi/4}^{7\pi/6} \tan(x)\,dx.$$

Solution

a) We have

$$\int \tan(x)\,dx = \int \frac{\sin(x)}{\cos(x)}\,dx.$$

Let's set $u = \cos(x)$. Then,

$$\frac{du}{dx} = -\sin(x).$$

Therefore,

$$\int \frac{\sin(x)}{\cos(x)}\,dx = \int \frac{1}{\cos(x)}\,(\sin(x))\,dx$$

$$= \int \frac{1}{u}\left(-\frac{du}{dx}\right)dx$$

$$= -\int \frac{1}{u}\frac{du}{dx}\,dx$$

$$= -\int \frac{1}{u}\,du = -\ln(|u|) + C = -\ln(|\cos(x)|) + C,$$

where C is an arbitrary constant. Thus,

$$\int \tan(x)\,dx = -\ln(|\cos(x)|) + C.$$

The above expression is valid on any interval of the form

$$\left(-\frac{\pi}{2} + n\pi, \frac{\pi}{2} + n\pi\right),\ n = 0, \pm 1, \pm 2, \ldots.$$

b) By part a) and the Fundamental Theorem of Calculus,

$$\int_{3\pi/4}^{7\pi/6} \tan(x)\,dx = -\ln(|\cos(x)|)\Big|_{3\pi/4}^{7\pi/6}$$

$$= -\ln\left(\left|\cos\left(\frac{7\pi}{6}\right)\right|\right) + \ln\left(\left|\cos\left(\frac{3\pi}{4}\right)\right|\right)$$

$$= -\ln\left(\left|-\frac{\sqrt{3}}{2}\right|\right) + \ln\left(\left|-\frac{\sqrt{2}}{2}\right|\right)$$

$$= -\ln\left(\frac{\sqrt{3}}{2}\right) + \ln\left(\frac{\sqrt{2}}{2}\right)$$

$$= -\frac{1}{2}\ln(3) + \ln(2) + \frac{1}{2}\ln(2) - \ln(2)$$

$$= -\frac{1}{2}\ln(3) + \frac{1}{2}\ln(2) \cong -0.202733.$$

The above integral corresponds to the signed area of the region between the graph of tangent and the interval $[3\pi/4, 7\pi/6]$, as indicated in Figure 1. $\square$

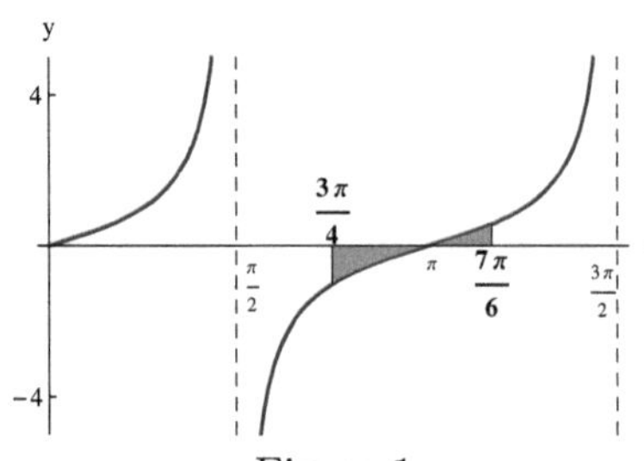

Figure 1

Example 4

a) Determine

$$\int \frac{x}{x^2 + 1}\,dx.$$

b) Evaluate

$$\int_2^4 \frac{x}{x^2 + 1}\,dx$$

Solution

a) If we set $u = x^2 + 1$, we have

$$\frac{du}{dx} = 2x \Rightarrow x = \frac{1}{2}\frac{du}{dx}.$$

Thus,

$$\int \frac{x}{x^2 + 1}\,dx = \int \frac{1}{x^2 + 1}\left(\frac{1}{2}\frac{du}{dx}\right)\,dx = \frac{1}{2}\int \frac{1}{u}\frac{du}{dx}\,dx = \frac{1}{2}\int \frac{1}{u}\,du.$$

Symbolically,

$$du = \frac{du}{dx}\,dx = 2x\,dx \Rightarrow x\,dx = \frac{1}{2}\,du,$$

so that

$$\int \frac{x}{x^2 + 1}\,dx = \int \frac{1}{x^2 + 1}x\,dx = \int \frac{1}{u}\left(\frac{1}{2}\right)\,du = \frac{1}{2}\int \frac{1}{u}\,du.$$

Either way,

$$\int \frac{x}{x^2 + 1}\,dx = \frac{1}{2}\int \frac{1}{u}\,du = \frac{1}{2}\ln\left(|u|\right) + C = \frac{1}{2}\ln\left(\left|x^2 + 1\right|\right) + C = \frac{1}{2}\ln\left(x^2 + 1\right) + C,$$

where C is an arbitrary constant.

b) By part a) and the Fundamental Theorem of Calculus,

$$\int_2^4 \frac{x}{x^2 + 1}\,dx = \frac{1}{2}\ln\left(x^2 + 1\right)\Big|_{x=2}^{4} = \frac{1}{2}\ln\left(17\right) - \frac{1}{2}\ln\left(5\right) \cong 0.611888.$$

The above integral is the area between the graph of

$$y = \frac{x}{x^2 + 1}$$

and the interval $[2, 4]$, as illustrated in Figure 2. $\square$

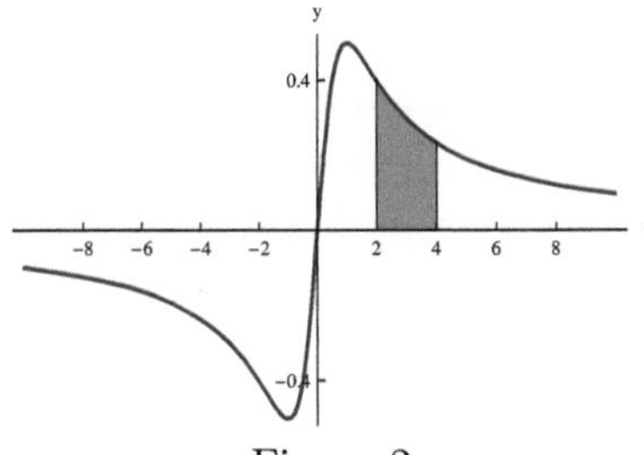

Figure 2

Remark 4 The previous two examples illustrate the appearance of the natural logarithm in many antidifferentiation formulas. Indeed, if we have an indefinite integral that can be expressed as constant multiple of

$$\int \frac{f'(x)}{f(x)}dx,$$

the substitution $u = f(x)$ works:

$$\int \frac{1}{f(x)}f'(x)dx = \int \frac{1}{u}\frac{du}{dx}dx = \int \frac{1}{u}du = \ln(|u|) + C = \ln(|f(x)|) + C,$$

where C is an arbitrary constant. $\Diamond$

The Substitution Rule for Definite Integrals

An indefinite integral that is determined with the help of the substitution rule can be used to evaluate a definite integral, as in the above examples. There is also a version of the substitution rule which applies directly to definite integrals:

Theorem 2 (THE SUBSTITUTION RULE FOR DEFINITE INTEGRALS) As-
sume that f is continuous on the interval determined by $u(a)$ and $u(b)$, and that
du/dx is continuous on the interval $[a, b]$. Then

$$\int_a^b f(u(x))\frac{du}{dx}\,dx = \int_{u(a)}^{u(b)} f(u)\,du.$$

Proof
The substitution rule for definite integrals is derived in a way that is similar to the derivation of the substitution rule for indefinite integrals. Let F be an antiderivative of f in the interval determined by $u(a)$ and $u(b)$. Thus,

$$\frac{d}{du}F(u) = f(u)$$

if u between $u(a)$ and $u(b)$. By the chain rule,

$$\frac{d}{dx}F(u(x)) = \left(\frac{dF}{du}\bigg|_{u=u(x)}\right)\frac{du}{dx} = f(u(x))\frac{du}{dx}$$

if $x \in [a, b]$. The first part of the Fundamental Theorem of Calculus implies that

$$F\left(u(b)\right) - F\left(u\left(a\right)\right) = \int_a^b \frac{d}{dx} F\left(u\left(x\right)\right) dx = \int_a^b f\left(u\left(x\right)\right) \frac{du}{dx} dx.$$

The first part of the Fundamental Theorem of Calculus also implies that

$$F\left(u(b)\right) - F\left(u\left(a\right)\right) = \int_{u(a)}^{u(b)} \frac{dF(u)}{du} du = \int_{u(a)}^{u(b)} f(u) du.$$

Therefore, we must have

$$\int_a^b f\left(u\left(x\right)\right) \frac{du(x)}{dx} dx = \int_{u(a)}^{u(b)} f(u) du,$$

as claimed. ■

Remark 5 (Caution) Even though the substitution rule for definite integrals has an appearance which is similar to the substitution rule for indefinite integrals, Theorem 2 expresses a new rule, since definite and indefinite integrals are different kinds of entities (functions versus numbers). **Also note the change in the limits of integration: The integral on the right-hand side is evaluated from $u(a)$ to $u(b)$, and not from a to b, as in the original integral.** ◊

Example 5 Evaluate

$$\int_0^{\pi/2} \cos^{2/3}\left(x\right) \sin\left(x\right) dx$$

by using the substitution rule for definite integrals.

Solution

We set $u = \cos\left(x\right)$ so that

$$du = \frac{du}{dx} dx = \left(\frac{d}{dx} \cos\left(x\right)\right) dx = -\sin\left(x\right) dx.$$

Therefore,

$$\int_0^{\pi/2} \cos^{2/3}\left(x\right) \sin\left(x\right) dx = \int_{u=\cos(0)}^{u=\cos(\pi/2)} u^{2/3} \left(-\frac{du}{dx}\right) dx$$

$$= -\int_1^0 u^{2/3} du = \int_0^1 u^{2/3} du = \left. \frac{u^{\frac{2}{3}+1}}{\frac{2}{3}+1} \right|_0^1 = \left. \frac{3}{5} u^{5/3} \right|_0^1 = \frac{3}{5}..$$

Figure 3 shows the graph of

$$f\left(x\right) = \cos^{2/3}\left(x\right) \sin\left(x\right).$$

The integral that we calculated is the area of the shaded region. □

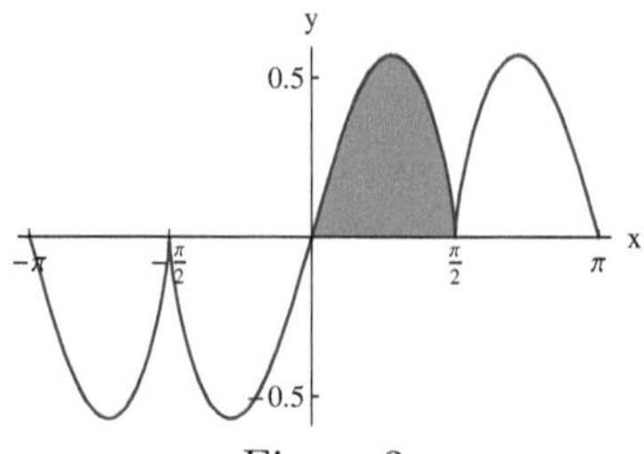

Figure 3

Example 6 Evaluate

$$\int_{\sqrt{\ln(2)}}^{\sqrt{\ln(3)}} e^{-x^2} x\, dx$$

by using the substitution rule for definite integrals.

Solution

We set $u = -x^2$, so that

$$\frac{du}{dx} = -2x, \quad u\left(\sqrt{\ln(2)}\right) = -\ln(2) \ \text{ and } \ u\left(\sqrt{\ln(3)}\right) = -\ln(3).$$

Therefore,

$$\int_{\sqrt{\ln(2)}}^{\sqrt{\ln(3)}} e^{-x^2} x\, dx = \int_{\sqrt{\ln(2)}}^{\sqrt{\ln(3)}} e^{-x^2}\left(-\frac{1}{2}\right)(-2x)\, dx = -\frac{1}{2}\int_{\sqrt{\ln(2)}}^{\sqrt{\ln(3)}} e^{u}\frac{du}{dx}\, dx$$

$$= -\frac{1}{2}\int_{-\ln(2)}^{-\ln(3)} e^{u}\, du$$

$$= -\frac{1}{2}\int_{-\ln(2)}^{-\ln(3)} e^{u}\, du$$

$$= -\frac{1}{2}\left(e^{u}\Big|_{-\ln(2)}^{-\ln(3)}\right)$$

$$= -\frac{1}{2}e^{-\ln(3)} + \frac{1}{2}e^{-\ln(2)}$$

$$= -\frac{1}{2}\left(\frac{1}{e^{\ln(3)}}\right) + \frac{1}{2}\left(\frac{1}{e^{\ln(2)}}\right)$$

$$= -\frac{1}{2}\left(\frac{1}{3}\right) + \frac{1}{2}\left(\frac{1}{2}\right) = \frac{1}{12}.$$

Thus, the area of the region between the graph of $y = e^{-x^2} x$ and the interval $\left[\sqrt{\ln(2)}, \sqrt{\ln(3)}\right]$ that is illustrated in Figure 4 is $1/12$. $\square$

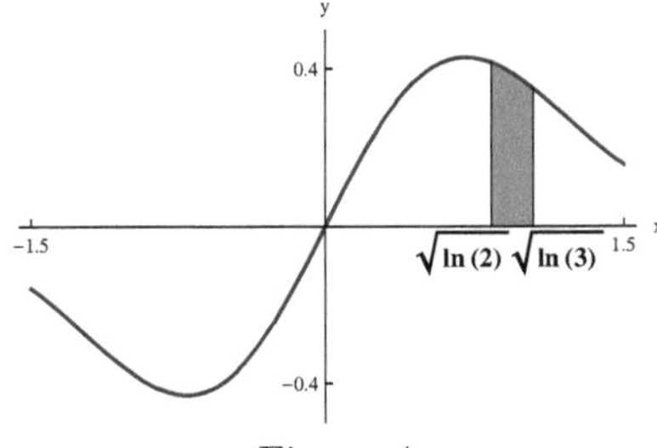

Figure 4

The definite integral version of the substitution rule does not offer an advantage over the indefinite integral version of the rule if

$$\int f\left(u\left(x\right)\right)\frac{du}{dx}\, dx = \int f\left(u\right)\, du$$

and

$$\int f\left(u\right) du$$

can be expressed in terms of familiar functions. On the other hand, the substitution rule for definite integrals leads to useful facts about integrals, as in the following proposition:

Proposition 1

a) If f is even and continuous on $[-a,\, a]$, then

$$\int_{-a}^{a} f\left(x\right) dx = 2 \int_{0}^{a} f\left(x\right) dx.$$

b) If f is odd and continuous on $[-a,\, a]$, then

$$\int_{-a}^{a} f\left(x\right) dx = 0.$$

Both parts of Proposition 1 are plausible. If f is even, the graph of f is symmetric with respect to the vertical axis. With reference to Figure 5, the area of G_L is the same as the area of G_R.

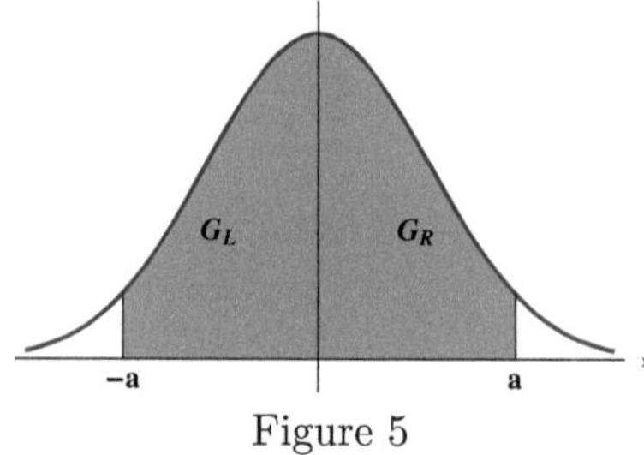

Figure 5

Thus,

$$\int_{-a}^{a} f\left(x\right) dx = \int_{-a}^{0} f\left(x\right) dx + \int_{0}^{a} f\left(x\right) dx = (\text{area of the } G_L) + (\text{area of } G_R)$$

$$= 2 \times (\text{area of } G_R) = 2 \int_{0}^{a} f(x) dx.$$

If f is odd, the graph of f is symmetric with respect to the origin. With reference to Figure 6, the signed area of G_- is $(-1) \times$ the area of G_+.

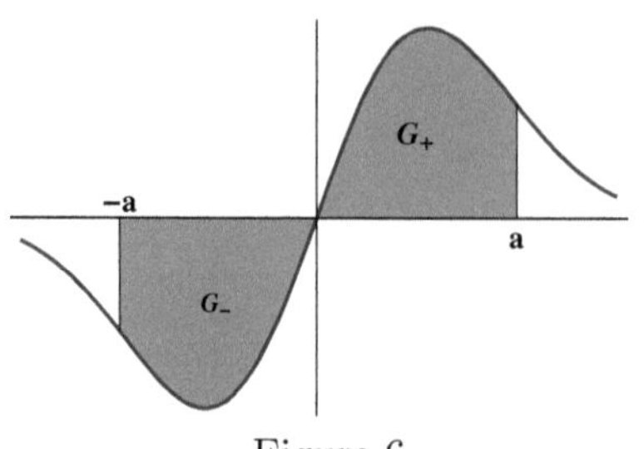

Figure 6

Thus,

$$\int_{-a}^{a} f(x)\,dx = \int_{-a}^{0} f(x)dx + \int_{0}^{a} f(x)\,dx = \text{(the signed area of } G_{-}) + \text{(the area of } G_{+}) = 0.$$

The Proof of Proposition 1

We will prove part a), and leave the similar proof of part b) as an exercise. Thus, assume that f is even. By the additivity of integrals with respect to intervals,

$$\int_{-a}^{a} f(x)dx = \int_{-a}^{0} f(x)dx + \int_{0}^{a} f(x)dx.$$

Since f is even, we have $f(-x) = f(x)$. Therefore,

$$\int_{-a}^{0} f(x)dx = \int_{-a}^{0} f(-x)\,dx.$$

Let us apply the substitution rule to this integral by setting $u = -x$. Then, $du/dx = -1$, so that

$$\int_{-a}^{0} f(-x)dx = -\int_{-a}^{0} f(u)(-1)dx = -\int_{-a}^{0} f(u)\frac{du}{dx}dx$$

$$= -\int_{u(-a)}^{u(0)} f(u)du = -\int_{a}^{0} f(u)du = \int_{0}^{a} f(u)du.$$

Thus,

$$\int_{-a}^{0} f(-x)dx = \int_{0}^{a} f(u)du = \int_{0}^{a} f(x)dx$$

(the variable of integration is a dummy variable). Therefore,

$$\int_{-a}^{a} f(x)dx = \int_{-a}^{0} f(x)dx + \int_{0}^{a} f(x)dx = \int_{0}^{a} f(x)dx + \int_{0}^{a} f(x)dx$$

$$= 2\int_{0}^{a} f(x)dx,$$

as claimed. ∎

Problems

In problems 1-21 , evaluate the indefinite integral.

1. $\int x\left(x^2 + 4\right)^5 dx$

2. $\int x\sqrt{x^2 + 4}dx$

3. $\int x^2\sqrt{x^3 + 1}dx$

4. $\int x^3\left(x^4 + 8\right)^{1/3} dx$

5. $\int \dfrac{x}{\sqrt{4 - x^2}}dx$

6. $\int \dfrac{x^2}{\left(x^3 + 1\right)^4}dx$

7. $\int x\sqrt{x - 16}dx$

8. $\int \sin\left(3x - \dfrac{\pi}{4}\right) dx$

9. $\displaystyle \int \cos\left(\frac{x}{4} + \frac{\pi}{6}\right) dx$

10. $\displaystyle \int \frac{x}{\sqrt{1 - 9x^2}} dx$

11. $\displaystyle \int \sin^4(x)\cos(x)\, dx$

12. $\displaystyle \int \cos^{2/3}(x)\sin(x)\, dx$

13. $\displaystyle \int \frac{1}{x + 4} dx$

14. $\displaystyle \int \frac{4x}{x^2 + 16} dx$

15. $\displaystyle \int \frac{\ln(x)}{x} dx$

16. $\displaystyle \int x e^{-x^2} dx$

17. $\displaystyle \int \frac{e^{\sqrt{x}}}{\sqrt{x}} dx$

18. $\displaystyle \int \left(\frac{1}{4 + x^2} + \frac{1}{\sqrt{9 - x^2}}\right) dx$

19. $\displaystyle \int \frac{1}{16x^2 + 9} dx$

20. $\displaystyle \int \frac{\arctan(x)}{1 + x^2} dx$

21. $\displaystyle \int \frac{1}{\sqrt{1 - 16x^2}} dx$

22. Assume that

$$\int_2^4 f(x)dx = 3 \text{ and } \int_4^{16} f(x)dx = -5.$$

Compute

$$\int_2^4 f(x^2)x\,dx.$$

23. Assume that

$$\int_2^3 f(x)dx = 3 \text{ and } \int_4^9 f(x)dx = 4.$$

Compute

$$\int_4^9 f(\sqrt{x})\frac{1}{\sqrt{x}}dx.$$

In problems 24-29 evaluate the given definite integral.

24. $\displaystyle \int_1^2 \frac{x}{\sqrt{9 - x^2}} dx$

25. $\displaystyle \int_2^3 x^2\left(x^3 + 1\right)^{1/3} dx$

26. $\displaystyle \int_0^{\pi/2} \sin(3t)\, dt$

27. $\displaystyle \int_{\pi/6}^{\pi/2} 3\cos(4x)\, dx$

28. $\displaystyle \int_1^{\sqrt{3}} \frac{1}{\sqrt{4 - x^2}} dx$

29. $\displaystyle \int_4^5 \frac{1}{\sqrt{9 - x^2}} dx$

In problems 30-34, use the substitution rule for definite integrals to evaluate the given integral.

30. $\displaystyle \int_3^4 x\sqrt{x^2 + 1}dx$

31. $\displaystyle \int_0^3 \frac{x}{(x^2 + 4)^{1/3}} dx$

32.
$$\int_{\sqrt{\pi/2}}^{\sqrt{\pi}} x \sin\left(x^2\right) dx$$

34.
$$\int_{-4}^{4} \frac{1}{16 + x^2}\, dx$$

33.
$$\int_{\pi/6}^{\pi} \cos^2\left(x\right) \sin\left(x\right) dx$$

In problems 35and 36, assume that $v\left(t\right)$ is the velocity of an object in one-dimensional motion. Determine the displacement and the distance traveled by the object on the time interval J.

35.
$$v\left(t\right) = \sin\left(4t\right), \quad J = [0, 3\pi/8]$$

36.
$$v\left(t\right) = \cos\left(\frac{t}{2}\right), \quad J = [\pi/2, 3\pi/2]$$

5.7   The Differential Equation $y' = f$

In this section we will take another look at the Fundamental Theorem of Calculus within the framework of differential equations and initial-value problems.

In Section 4.6 we saw that **the general solution** of the differential equation $y'\left(x\right) = ky\left(x\right)$, where k is a constant, is a constant multiple of e^{kx}. Thus, we were able to determine the unique solution of **the initial-value problem**

$$y'\left(x\right) = ky\left(x\right), \ y\left(x_0\right) = y_0,$$

where x_0 and y_0 are given numbers, as

$$y\left(x\right) = y_0 e^{k\left(x - x_0\right)}.$$

Now we will consider differential equations of the form

$$y'\left(x\right) = f\left(x\right),$$

where f is a given function, and initial-value problems of the form

$$y'\left(x\right) = f\left(x\right), \ y\left(x_0\right) = y_0,$$

where x_0 and y_0 are given numbers.

The Differential Equation $y' = f$ and the Fundamental Theorem

We have $y'\left(x\right) = f\left(x\right)$ for each x in an interval J if and only if y is an antiderivative of f on J. Therefore, we can express y as the indefinite integral of f:

$$y\left(x\right) = \int f\left(x\right) dx.$$

We will refer to

$$\int f\left(x\right) dx$$

as .the **general solution** of the differential equation $y'\left(x\right) = f\left(x\right)$. The indefinite integral involves an arbitrary constant. The value of the constant is determined uniquely if **an**

initial condition of the form $y(x_0) = y_0$ is specified, so that the solution of **the initial-value problem**,

$$y'(x) = f(x), \ y(x_0) = y_0$$

is uniquely determined.

Example 1

a) Determine the general solution of the differential equation

$$y'(x) = 2x.$$

b) Determine the solution of the initial-value problem

$$y'(x) = 2x \text{ and } y(2) = 5.$$

Solution

a) We have $y'(x) = 2x$ if and only if

$$y(x) = \int 2x \, dx = x^2 + C,$$

where C is an arbitrary constant. Thus,

$$y(x) = x^2 + C$$

is the general solution of the differential equation $y'(x) = 2x$. Since C is an arbitrary constant, the general solution represents infinitely many functions that differ from x^2 by the addition of a constant. Figure 1 displays the members of this family of functions corresponding to $C = -4, 1, 4$. If (x, y) is on one of the solution curves, the slope of the line that is tangent to that particular solution curve at (x, y) is $2x$. Thus, the tangent lines to the solution curves corresponding to a given x are parallel to each other.

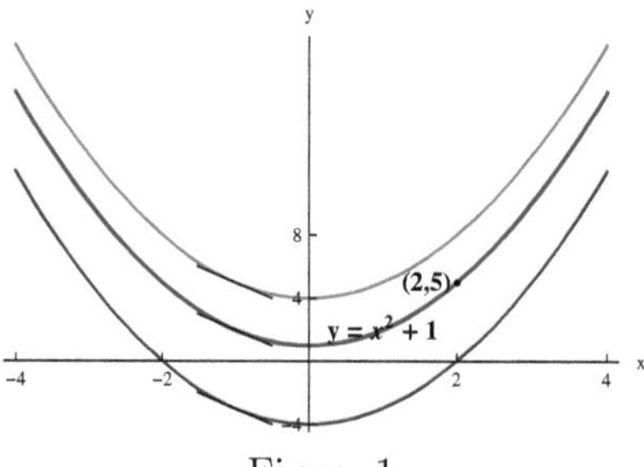

Figure 1

b) Since $y(x) = x^2 + C$ is the general solution of the given differential equation, we have

$$y(2) = 5 \Leftrightarrow 2^2 + C = 5 \Leftrightarrow C = 1.$$

Therefore, the required solution is

$$y(x) = x^2 + 1.$$

The graph of $y = x^2 + 1$ is the only member of the family of curves $y = x^2 + C$ that passes through the point $(2, 5)$. $\square$

Example 2

a) Determine the general solution of the differential equation

$$y'(x) = \sin(4x).$$

b) Determine the solution of the initial-value problems,

$$y'(x) = \sin(4x), \quad y(\pi/4) = 2,$$

and

$$y'(x) = \sin(4x), \quad y(\pi/4) = -2.$$

Sketch the graphs of the solutions.

Solution

a) We have $y'(x) = \sin(4x)$ if and only if

$$y(x) = \int \sin(4x)dx = -\frac{1}{4}\cos(4x) + C,$$

where C is an arbitrary constant. This is the general solution of the differential equation

$$y'(x) = \sin(4x).$$

b) With reference to part a),

$$y\left(\frac{\pi}{4}\right) = -\frac{1}{4}\cos(\pi) + C = \frac{1}{4} + C.$$

Therefore,

$$y(\pi/4) = 2 \Leftrightarrow \frac{1}{4} + C = 2 \Leftrightarrow C = \frac{7}{4}.$$

Thus, the solution of the initial-value problem

$$y'(x) = \sin(4x), \quad y(\pi/4) = 2$$

is

$$F(x) = -\frac{1}{4}\cos(4x) + \frac{7}{4}.$$

Similarly, the solution of the initial-value problem

$$y'(x) = \sin(4x), \quad y(\pi/4) = -2$$

is

$$G(x) = -\frac{1}{4}\cos(4x) - \frac{9}{4}.$$

Figure 2 displays the graphs of F and G. Note that

$$F'(x) = G'(x) = \sin(4x),$$

so that the tangent line to the graph of F at the point $(x, F(x))$ is parallel to the tangent line to the graph of G at $(x, G(x))$. $\square$

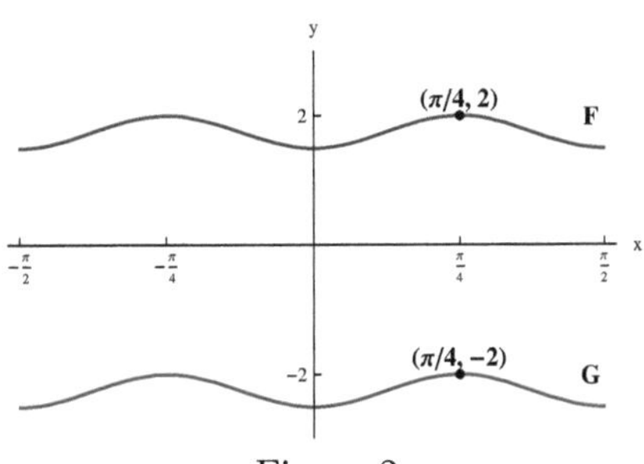

Figure 2

In the above examples, we were able to determine the relevant indefinite integral in terms of familiar functions. This need not be the case. Nevertheless, any continuous function has an antiderivative by the second part of the Fundamental Theorem of Calculus: We have

$$\frac{d}{dx} \int_a^x f(t)\, dt = f(x)$$

for each $x \in J$ if f is continuous on the interval J and a is a fixed point in J. Therefore, we can express the general solution of the differential equation $y' = f$ on the interval J as

$$y(x) = \int_a^x f(t)\, dt + C,$$

where a is some point in J and C is a constant. If we are given an initial condition of the form $y(x_0) = y_0$, it is convenient to set $a = x_0$. In this case,

$$y(x) = C + \int_{x_0}^x f(t)\, dt,$$

so that

$$y_0 = y(x_0) = C + \int_{x_0}^{x_0} f(t)\, dt = C.$$

Therefore $C = y_0$, and **the unique solution of the initial-value problem**

$$y'(x) = f(x), \ \ y(x_0) = y_0$$

can be represented as

$$y(x) = y_0 + \int_{x_0}^x f(t)\, dt.$$

In the above expression, we may or may not be able to express the integral in terms of familiar functions. In any case, the values of the solution can be approximated by approximating the integral.

Example 3

a) Express the solution of the initial-value problem,

$$y'(x) = \sin\left(x^2\right), \ y(2) = 3,$$

in terms of an integral.
b) Compute approximations to $y(3)$ and $y(4)$ with the help of the approximate integration facility of your computational utility.
c) Plot the graph of the solution of part a) on the interval $[0, 4]$ with the help of your computational/graphing utility.

Solution
a) We can express the solution as

$$y(x) = 3 + \int_2^x \sin\left(t^2\right)\, dt.$$

b) We have

$$y(3) = 3 + \int_2^3 \sin\left(t^2\right)\, dt \cong 2.968\,79,$$

and

$$y\,(4) = 3 + \int_2^4 \sin\left(t^2\right)\,dt \cong 2.942\,36.$$

c) Figure 3 shows the graph of the solution on $[0,4]$. $\square$

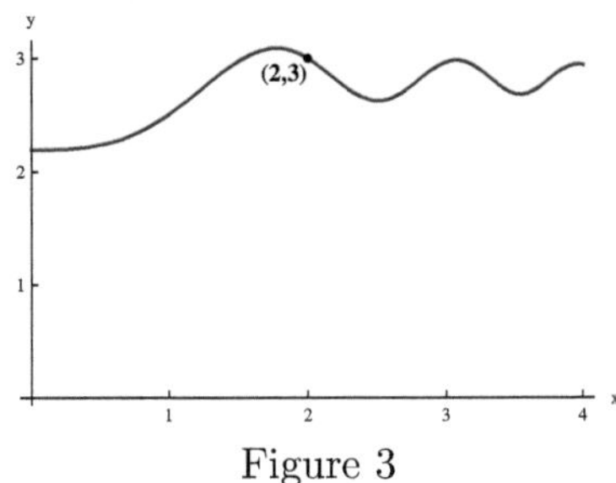

Figure 3

Acceleration, Velocity and Position

Let's consider the relationships between acceleration, velocity and position within the framework of initial-value problems. Assume that $f\,(t)$ is the **position**, $v(t)$ is the **velocity** and $a(t)$ is the **acceleration** at time t of an object in one-dimensional motion. Velocity is the rate of change of position, and acceleration is the rate of change of velocity:

$$v(t) = \frac{df}{dt} \text{ and } a(t) = \frac{dv}{dt}.$$

When we introduced these concepts initially, we assumed that the position was given, and calculated velocity and acceleration by differentiation. Now we are able to begin with a given acceleration function, and determine the velocity and position functions successively. Thus assume that $a(t)$ is given. The velocity function $v(t)$ is the solution of the differential equation

$$\frac{dv}{dt} = a\,(t)\,.$$

We have seen that such a differential equation does not have a unique solution. On the other hand, if an initial condition is specified, the solution is uniquely determined. Thus, assume that the velocity at a certain instant t_0 is v_0, so that $v\,(t_0) = v_0$. We can express the solution of the initial-value problem

$$\frac{dv}{dt} = a\,(t)\,,\; v(t_0) = v_0,$$

as

$$v(t) = v_0 + \int_{t_0}^t a\,(\tau)\,d\tau.$$

The position function is uniquely determined if the position of the object is specified at some instant. If $f(t_0) = f_0$, the position function is the solution of an initial-value problem

$$\frac{df}{dt} = v(t),\; f(t_0) = f_0.$$

The solution can be expressed as

$$f\,(t) = f_0 + \int_{t_0}^t v(\tau)\,d\tau.$$

Example 4 Assume that an object is falling under the influence of gravitational acceleration of 9.8 meters/second/second. The effect of air resistance is neglected. We model the motion as one-dimensional motion so that the number line is vertical, points downward, and the origin coincides with the point at which the object is released. We assume that the object is released from rest. Thus, with the above notation, $a(t) = 9.8$, $v(0) = 0$ and $f(0) = 0$. Determine $v(t)$ and $f(t)$ at any instant t before the object hits the ground.

Solution

We have

$$\frac{dv}{dt} = a(t) = 9.8, \ v(0) = 0.$$

Therefore,

$$v(t) = \int_0^t a\left(\tau\right) d\tau = \int_0^t 9.8 d\tau = 9.8t \text{ (meters/sec.)}.$$

We have

$$\frac{df}{dt} = v(t) = 9.8t \text{ and } f(0) = 0.$$

Therefore,

$$f(t) = \int_0^t v\left(\tau\right) d\tau = \int_0^t 9.8\tau d\tau = \left.\frac{9.8}{2}\tau^2\right|_0^t = 4.9t^2 \text{ (meters)}.$$

$\square$

Example 5 Assume that the acceleration of an object in simple harmonic motion is $20\cos\left(6t\right)$ at the instant t. Determine the velocity and the position of the object at the instant t if $v(\pi/6) = 0$ and $f\left(\pi/6\right) = 2$ (with the notation preceding Example 4).

Solution

We have

$$\frac{dv}{dt} = a(t) = 20\cos\left(6t\right) \text{ and } v\left(\pi/6\right) = 0.$$

Therefore,

$$v(t) = \int_{\pi/6}^t 20\cos\left(6\tau\right) d\tau = \left.\frac{10}{3}\sin\left(6\tau\right)\right|_{\pi/6}^t$$
$$= \frac{10}{3}\sin\left(6t\right) - \frac{10}{3}\sin\left(\pi\right)$$
$$= \frac{10}{3}\sin\left(6t\right).$$

We have

$$\frac{df}{dt} = v(t) = \frac{10}{3}\sin\left(6t\right) \text{ and } f\left(\frac{\pi}{6}\right) = 2.$$

Therefore,

$$f(t) = 2 + \int_{\pi/6}^{t} \frac{10}{3} \sin(6\tau)\, d\tau$$

$$= 2 + \left(-\frac{10}{18} \cos(6\tau) \Big|_{\pi/6}^{t} \right)$$

$$= 2 + \left(-\frac{10}{18} \cos(6t) + \frac{10}{18} \cos(\pi) \right)$$

$$= 2 + \left(-\frac{10}{18} \cos(6t) - \frac{10}{18} \right)$$

$$= \frac{13}{9} - \frac{5}{9} \cos(6t).$$

Figure 4 shows the graph of f. Note that the motion is periodic with period $\pi/3$. $\square$

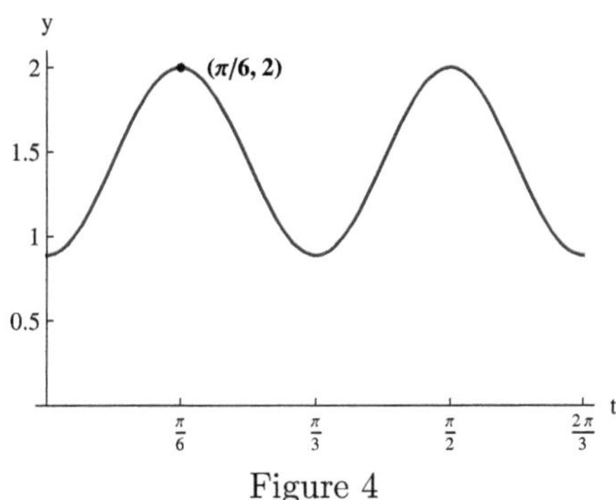

Figure 4

Problems

In problems 1-6, solve the given inital-value problem:

1.
$$y' = \frac{x}{\sqrt{x^2 + 1}}, \quad y(0) = 4.$$

2.
$$y'(x) = x^{1/3}, \quad y(8) = 6$$

3.
$$y'(x) = \sin(x/3), \quad y(0) = -1$$

4.
$$y'(x) = \cos(4x - \pi/3), \quad y(\pi/4) = 5$$

5.
$$y'(x) = e^{2x}, \quad y(0) = 5$$

6.
$$y'(t) = te^{t^2}, \quad y(-2) = 1$$

In problems 7-9, express the solution of the initial-value problem in terms of an integral. Do not evaluate the integral.

7.
$$y'(t) = \sin(t^2), \quad y(2) = 4$$

8.
$$y'(t) = \frac{1}{\sqrt{4 + t^2}}, \quad y(4) = -10.$$

9.
$$y'(t) = e^{-t^2/2}, \quad y(4) = -10$$

10. Assume that $v(t) = \sin(4t)$ is the velocity at time t of an object in one dimensional motion. Let f denote the position function. Assume that $f(\pi/4) = -4$. Determine $f(t)$.

11. Assume that the acceleration of an object in one dimensional motion at the instant t is $3\cos(6t)$. Determine the velocity and the position of the object at the instant t if $v(\pi/6) = 10$ and $f(\pi/6) = 4$.

Appendix A

Precalculus Review

A.1 Solutions of Polynomial Equations

Completion of the Square and the Quadratic Formula

A quadratic equation is of the form

$$ax^2 + bx + c = 0,$$

where a, b and c are given numbers, $a \neq 0$, and x denotes the unknown. The **quadratic formula** enables us to solve such an equation. The formula is based on the **completion of the square:**

$$ax^2 + bx + c = a\left(x^2 + \frac{b}{a}x\right) + c = a\left(x + \frac{b}{2a}\right)^2 - \frac{b^2}{4a} + c.$$

Thus, we have "completed the square". We will use the symbol " $\Leftrightarrow$ " to mean "if and only if". We have

$$ax^2 + bx + c = 0 \Leftrightarrow a\left(x + \frac{b}{2a}\right)^2 - \frac{b^2}{4a} + c = 0$$

$$\Leftrightarrow a\left(x + \frac{b}{2a}\right)^2 = \frac{b^2}{4a} - c$$

$$\Leftrightarrow \left(x + \frac{b}{2a}\right)^2 = \frac{b^2 - 4ac}{4a^2}$$

$$\Leftrightarrow x + \frac{b}{2a} = \pm\frac{\sqrt{b^2 - 4ac}}{2a}$$

$$\Leftrightarrow x = \frac{-b \pm \sqrt{b^2 - 4ac}}{2a}.$$

Thus, we obtained **the quadratic formula** for the solution of the quadratic equation $ax^2 + bx + c$.

$$x = \frac{-b \pm \sqrt{b^2 - 4ac}}{2a}$$

Example 1

a) Solve the equation $x^2 - 2x - 8 = 0$ by completing the square.
b) Solve the equation $x^2 - 2x - 8 = 0$ by using the quadratic formula.

Solution

a) We have

$$x^2 - 2x - 8 = (x-1)^2 - 9$$

Therefore,

$$x^2 - 2x - 8 = 0 \Leftrightarrow (x-1)^2 - 9 = 0 \Leftrightarrow (x-1)^2 = 9$$
$$\Leftrightarrow x - 1 = \pm 3 \Leftrightarrow x = -2 \text{ or } x = 4.$$

b)

$$x = \frac{-(-2) \pm \sqrt{(-2)^2 - 4(1)(-8)}}{2} = \frac{2 \pm \sqrt{4+32}}{2}$$
$$= 1 \pm \frac{\sqrt{36}}{2} = 1 \pm 3.$$

Therefore, the solutions are -2 and 4, as we determined in part a).$\square$
If the solutions of the equation $ax^2 + bx + c = 0$ are x_1 and x_2, we have

$$ax^2 + bx + c = a(x - x_1)(x - x_2).$$

We will refer to factors such as $x - x_1$, and more generally, factors of the form $mx - d$, where m and d are constants, as **linear factors. Thus, a quadratic expression can be expressed as as a product of linear factors, once we solve the corresponding equation.** Conversely, if the quadratic expression $ax^2 + bx + c$ has been factored as $a(x - x_1)(x - x_2)$, then x_1 and x_2 are the solutions of the equation $ax^2 + bx + c = 0$. Even though you may have gone the "factoring route" in your precalculus courses for the determination of the roots of a quadratic equation, it is recommended that you follow the "quadratic formula route" or complete the square. After all, the solutions are not always rational numbers.

Example 2

a) Solve the equation $2x^2 - 12x + 13 = 0$.
b) Express $2x^2 - 12x + 13$ as a product of linear factors.

Solution

a) We will use the quadratic formula. Thus, $2x^2 - 12x + 13 = 0$ if and only if

$$x = \frac{-(-12) \pm \sqrt{(-12)^2 - 4(2)(13)}}{2(2)} = \frac{12 \pm \sqrt{144 - 104}}{4}$$
$$= \frac{12 \pm \sqrt{40}}{4} = 3 \pm \frac{\sqrt{10}}{2}.$$

Therefore the solutions are

$$x_1 = 3 - \frac{\sqrt{10}}{2} \text{ and } x_2 = 3 + \frac{\sqrt{10}}{2}.$$

b) By part a),

$$2x^2 - 12x + 13 = 2(x - x_1)(x - x_2) = 2\left(x - \left(3 - \frac{\sqrt{10}}{2}\right)\right)\left(x - \left(3 + \frac{\sqrt{10}}{2}\right)\right).$$

If we wish, we can express the factorization in a more elegant (if not meaningful) form:

$$2\left(x - \left(3 - \frac{\sqrt{10}}{2}\right)\right)\left(x - \left(3 + \frac{\sqrt{10}}{2}\right)\right) = 2\left(x - 3 + \frac{\sqrt{10}}{2}\right)\left(x - 3 - \frac{\sqrt{10}}{2}\right)$$
$$= \frac{1}{2}\left(2x - 6 + \sqrt{10}\right)\left(2x - 6 - \sqrt{10}\right).$$

$\square$

The expression $b^2 - 4ac$ is called **the discriminant** of the equation $ax^2 + bx + c = 0$. Since the solutions of the equation are

$$\frac{-b \pm \sqrt{b^2 - 4ac}}{2a},$$

these solutions are distinct real numbers if the discriminant is positive, as in Example 1 and Example 2. If the discriminant is 0, there is a single "**repeated root**"

$$x_1 = -\frac{b}{2a},$$

so that

$$ax^2 + bx + c = a\left(x - x_1\right)^2.$$

Example 3

a) Determine the discriminant and the solutions of the equation $3x^2 - 12x + 12 = 0$.
b) Express $3x^2 - 12x + 12$ as a product of linear factors.

Solution

a) The discriminant of the equation $3x^2 - 12x + 12 = 0$ is

$$(-12)^2 - 4\,(3)\,(12) = 144 - 144 = 0.$$

Therefore, the only solution of the equation is

$$x_1 = -\frac{(-12)}{6} = 2.$$

b) By the result of part a),

$$3x^2 - 12x + 12 = 3\left(x - x_1\right)^2 = 3\left(x - 2\right)^2.$$

$\square$

Complex Solutions

If the discriminant of a quadratic equation is negative, the equation has **complex roots.** Even though we will not have to deal with complex numbers for quite a while, let us review a few basic facts about complex numbers, since you may come across equations with complex roots.

The equation $x^2 = -1$ does not have real solutions, since $x^2 \geq 0$ for any real number x. We declare that i is the "**imaginary number**" such that $i^2 = -1$. We may denote i as $\sqrt{-1}$. A **complex number** is an expression of the form $a + ib$, where a and b are real numbers. A complex number of the form $a + (0)\,i$ is identified with the real number a. A number of the form

ib is an **imaginary number**. The operations of addition and subtraction are extended to the set of complex numbers so that the rules of arithmetic remain valid:

$$(a + ib) + (c + id) = (a + c) + i(b + d),$$
$$(a + ib)(c + id) = (ac + i^2 bd) + i(ad + bc) = (ac - bd) + i(ad + bc).$$

Given the complex number $z = a + ib$, the complex conjugate $\bar{z}$ is defined as

$$\bar{z} = a - ib.$$

Note that

$$z\bar{z} = a^2 + b^2.$$

Division extends to complex numbers: If $c^2 + d^2 \neq 0$,

$$\frac{a + ib}{c + id} = \frac{(a + ib)(c - id)}{(c + id)(c - id)} = \frac{(ac + bd) + i(-ad + bc)}{c^2 + d^2} = \frac{ac + bd}{c^2 + d^2} + i\left(\frac{-ad + bc}{c^2 + d^2}\right).$$

Example 4

a) Determine the solutions of the equation $3x^2 - 12x + 16 = 0$.
b) Express $3x^2 - 12x + 16$ as a product of linear factors.

Solution

a) By the quadratic formula, x is a solution of the equation $3x^2 - 12x + 16 = 0$ if and only if

$$x = \frac{12 \pm \sqrt{144 - 192}}{6} = \frac{12 \pm \sqrt{-48}}{6}$$
$$= 2 \pm \frac{\sqrt{48}}{6}i = 2 \pm \frac{4\sqrt{3}}{6}i = 2 \pm \frac{2\sqrt{3}}{3}i.$$

Therefore, the solutions of the given equation are

$$x_1 = 2 - \frac{2\sqrt{3}}{3}i \text{ and } x_2 = 2 + \frac{2\sqrt{3}}{3}i$$

b) By part a),

$$3x^2 - 12x + 16 = 3(x - x_1)(x - x_2) = 3\left(x - \left(2 - \frac{2\sqrt{3}}{3}i\right)\right)\left(x - \left(2 + \frac{2\sqrt{3}}{3}i\right)\right).$$

$\square$

In Example 4 the given quadratic equation has complex roots with nonzero imaginary part. The corresponding quadratic expression has been expressed as the product of linear factors that involve complex numbers, but the linear factors cannot be expressed as $cx + d$ where c and d are real numbers. Such a quadratic expression will be referred to as being irreducible over the real numbers, or simply as **an irreducible quadratic expression**. Thus, the expression $3x^2 - 12x + 16$ of Example 4 is an irreducible quadratic expression.

Higher-Order Polynomials

An expression of the form

$$a_n x^n + a_{n-1} x^{n-1} + \cdots + a_1 x + a_0$$

where $a_0, a_1, \ldots, a_n$ are given numbers, and $a_n \neq 0$ is a **polynomial of degree** n. The number a_k is **the coefficient** of x^k. Thus, a quadratic expression $ax^2 + bx + c$ is a polynomial of degree 2, and an expression of the form $mx + d$ is a polynomial of degree at most 1 (the degree is 0 if $m = 0$, so that the expression is simply a number).

If

$$p(x) = a_n x^n + a_{n-1} x^{n-1} + \cdots + a_1 x + a_0$$

is a polynomial of degree n with real coefficients, then there is a theorem that says that $p(x)$ can be expressed as a product of linear factors and irreducible quadratic factors with real coefficients. Equivalently, it is possible to find all the solutions of the **polynomial equation** $p(x) = 0$, at least in principle (the solutions of the equation $p(x) = 0$ are also called **the roots of the polynomial** $p(x)$). Unfortunately, there is no general formula that is comparable to the quadratic formula for the solutions of the equation $p(x) = 0$ if the degree of $p(x)$ is greater than 4. There are formulas for $n = 3$ and $n = 4.$, but they are not as practical to use as the quadratic formula, and we will not include their expressions. **Computer algebra systems** such as Maple or Mathematica can provide exact solutions based on such formulas. In general, when we have to deal with polynomials of degree greater than 2, either some factors will be known, so that your knowledge of the quadratic case will be sufficient in order to determine the necessary solutions, or you will use the approximate solution capabilities of your computational utility.

Example 5 Let $p(x) = x^3 - x^2 - 2x + 2$.

a) Given that $(x - 1)$ is a linear factor of $p(x)$, determine all the linear and the irreducible factors of $p(x)$.

b) Determine all the solutions of the equation $p(x) = 0$.

Solution

a) Since we are given the information that $(x - 1)$ is a factor of $p(x)$, the remainder should be 0 when we divide $p(x)$ by $(x - 1)$ (long division will do). We have

$$p(x) = (x - 1)\left(x^2 - 2\right)$$

(confirm). Since

$$x^2 - 2 = \left(x - \sqrt{2}\right)\left(x + \sqrt{2}\right),$$

we have

$$p(x) = (x - 1)\left(x - \sqrt{2}\right)\left(x + \sqrt{2}\right).$$

b) By the result of part a), the solutions of the equation $x^3 - x^2 - 2x + 2 = 0$ are 1, $\sqrt{2}$ and $-\sqrt{2}$.$\square$

Example 6 Let $p(x) = x^3 + x + 2x^2 + 2$. Given that -2 is a solution of the equation $p(x) = 0$, determine all the linear and the irreducible factors of $p(x)$ and the solutions of $p(x) = 0$.

Solution

Since -2 is a solution of $p(x)$, $x + 2$ is a linear factor of $p(x)$. We divide $p(x)$ by $x + 2$ and find that

$$p(x) = (x + 2)\left(x^2 + 1\right).$$

The factor $x^2 + 1$ is an irreducible quadratic factor, since the solutions of the equation $x^2 + 1 = 0$ are $\pm i$. Thus, the solutions of $p(x) = 0$ are -2, i and $-i$.$\square$

Problems

In problems 1 - 4, solve the given equation by completing the square.

1.
$$x^2 + x - 6 = 0$$

3.
$$x^2 - 4x - 4 = 0$$

2.
$$3x^2 - 6x - 1 = 0$$

4.
$$2x^2 + 12x + 9 = 0$$

In problems 5 - 8,
a) Solve the given equation by using the quadratic formula,
b) Express the quadratic expression as a product of linear factors
(solutions and the linear factors may involve complex numbers).

5.
$$x^2 - 6x + 7 = 0$$

7.
$$x^2 - 6x + 13 = 0$$

6.
$$3x^2 + 9x - 30 = 0$$

8.
$$x^2 - 4x + 7 = 0$$

In problems 9-12, given a linear factor $x - a$ of $p(x)$,
a) Solve the equation $p(x) = 0$,
b) Express $p(x)$ as a product of linear and irreducible quadratic factors (with real coefficients).

9.
$$p(x) = x^3 - 19x + 30, \ a = 2$$

11.
$$p(x) = x^3 - 7x^2 + 19x - 13, \ a = 1$$

10.
$$p(x) = x^3 + x^2 - 8x - 6, \ a = -3$$

12.
$$p(x) = x^3 + 2x^2 + 2x + 40, \ a = -4$$

A.2   The Binomial Theorem

We will have to expand expressions of the form

$$(a + b)^n,$$

where n is a positive integer. If $n = 2$, we have

$$(a + b)^2 = a^2 + 2ab + b^2.$$

If $n = 3$,
$$(a + b)^3 = (a + b)(a + b)^2 = a^3 + 3a^2b + 3ab^2 + b^3.$$

The Binomial Theorem enables us to write the expansion of $(a + b)^n$ for any positive integer n.

We will need to use the **factorial** notation. We declare that $0! = 1$ and $1! = 1$. If $n = 2, 3, 4, \ldots$ ("$\ldots$" means "and so on"), then $n!$ (read "n **factorial**") is defined as the product of the positive integers from 1 to n. Thus,

$$n! = (1)(2)(3)\ldots(n - 1)(n)$$

For example, $2! = 2$, $3! = (1)(2)(3) = 6$ and $4! = (1)(2)(3)(4) = 24$.
If n and k are nonnegative integers, and k is less than or equal to n, the symbol

$$\binom{n}{k}$$

(read "n **choose** k") stands for

$$\frac{n!}{k!\,(n-k)!}.$$

This is the number of ways you can select k objects from a collection of n objects. For example,

$$\binom{n}{1} = n, \quad \binom{n}{2} = \frac{n\,(n-1)}{2}, \quad \binom{n}{n-1} = n, \quad \binom{n}{n} = 1.$$

Now we can state **the Binomial Theorem:**

Theorem 1 If n is a positive integer,

$$(a+b)^n = a^n + \binom{n}{1}a^{n-1}b + \binom{n}{2}a^{n-2}b^2 + \cdots + \binom{n}{k}a^{n-k}b^k + \cdots + \binom{n}{n-1}ab^{n-1} + \binom{n}{n}b^n$$

$$= a^n + na^{n-1}b + \frac{n\,(n-1)}{2}a^{n-2}b^2 + \cdots + \binom{n}{k}a^{n-k}b^k + \cdots + nab^{n-1} + b^n.$$

The **Pascal triangle** provides a convenient way to remember specific cases of the binomial expansions:

$$
\begin{array}{ccccccccccccc}
&&&&&& 1 &&&&&& \\
&&&&& 1 && 1 &&&&& \\
&&&& 1 && 2 && 1 &&&& \\
&&& 1 && 3 && 3 && 1 &&& \\
&& 1 && 4 && 6 && 4 && 1 && \\
& 1 && 5 && 10 && 10 && 5 && 1 & \\
1 && 6 && 15 && 20 && 15 && 6 && 1 \\
\end{array}
$$

The pattern is straightforward: Consider the top row as Row 0 (consisting of a single 1) and the next row as Row 1 (a pair of 1's). Each subsequent row begins and ends with a 1, and has one more entry than the row directly above it. Each intermediate entry of a given row is the sum of the two numbers in the previous row immediately to the left and right. For example, Row 2 begins and ends with a 1, and has three entries. To obtain the middle entry, add the numbers in Row 1 that are immediately to the left and right of the position for the middle entry of Row 2. Thus, the entry in question is 1+1, or 2. Similarly, the first 15 that appears in Row 6 is obtained by adding the entries from Row 5 to the left and right: 5+10 =15.

The entries in row n of the Pascal triangle provide the binomial coefficients for the expansion of $(a+b)^n$. For example,

$$(a+b)^0 = 1,$$
$$(a+b)^1 = a + b,$$
$$(a+b)^2 = a^2 + 2ab + b^2,$$
$$(a+b)^3 = a^3 + 3a^2b + 3ab^2 + b^3,$$
$$(a+b)^4 = a^4 + 4a^3b + 6a^2b^2 + 4ab^3 + b^4,$$
$$(a+b)^5 = a^5 + 5a^4b + 10a^3b^2 + 10a^2b^3 + 5ab^4 + b^5.$$

Problems

In problems 1 - 4, simplify the given expression.

1.
$$\binom{6}{3}$$

3.
$$\binom{n}{3} \text{ where } n \geq 3$$

2.
$$\binom{5}{2}$$

4.
$$\binom{n}{5} \text{ where } n \geq 5$$

In problems 5 - 8, expand the given expression by using the Binomial Theorem (it will be practical to make use of Pascal's triangle).

5.
$$(x+h)^4$$

7.
$$(4-h)^3$$

6.
$$(a-b)^5$$

8.
$$(x+2y)^3$$

In problems 9 - 12, simplify the given expression.

9.
$$\frac{(2+h)^3 - 8}{h}$$

10.
$$\frac{(x+h)^2 - x^2}{h}$$

11.
$$\frac{(x+h)^3 - x^3}{h}$$

12.
$$\frac{\dfrac{1}{(x+h)^4} - \dfrac{1}{x^4}}{h}$$

(the denominator can involve the term $(x+h)^4$ without being expanded).

A.3   Inequalities, the Number Line and the Absolute Value

The counting numbers, $1, 2, 3, \ldots$ (as in Section A 2, "$\ldots$" means "and so on") are referred to as **positive integers** or **natural numbers.** The set of **integers** consists of **positive integers, negative integers**, i.e., $-1, -2, -3, \ldots$, and 0. **Rational numbers** are numbers which can be expressed as fractions of the form p/q where p and q are integers and $q \neq 0$. Even though the set of rational numbers is adequate to do arithmetic (sums, products and quotients of rational numbers are also rational numbers), it is not adequate for the purposes of calculus. Indeed, even simple geometric problems lead to **irrational numbers,** i.e., numbers which are not fractions of integers, as the ancient Greeks knew. For example, the length of a diagonal of a square whose sides are of unit length is the irrational number $\sqrt{2}$. The circumference of a circle of unit diameter is the irrational number π. It is assumed that you are familiar with the arithmetic operations of addition, subtraction, multiplication and division of real numbers. The approximation of real numbers by decimals is discussed in Section A1.4.. We will denote the set of natural numbers by $\mathbb{N}$, the set of integers by $\mathbb{Z}$, the set of of rational numbers by $\mathbb{Q}$, and the set of all real numbers by $\mathbb{R}$.

A matter of notation: We will use the symbol "$\Rightarrow$" to indicate that the statement on the left of the symbol implies the statement on the right of the symbol. As in Section A 1, we will use the symbol "$\Leftrightarrow$" for the equivalence of the statements on either side of the symbol. In that case, we can also use the abbreviation "iff" for "if and only if'. For example, if a, b and c are real numbers,

$$a = b \Rightarrow a + c = b + c.$$

We have

$$a = b \Leftrightarrow -a = -b.$$

Inequalities

If a and b denote arbitrary real numbers, either $a < b$ (a is less than b) or $a > b$ (a is greater than b) or $a = b$. It is assumed that you are familiar with the basic properties of **inequalities**. For example,

$$a > 0 \text{ and } b > 0 \Rightarrow ab > 0,$$
$$a < 0 \text{ and } b < 0 \Rightarrow ab > 0,$$
$$a > 0 \text{ and } b < 0 \Rightarrow ab < 0$$

(the product of numbers of the same sign is positive, and the product of numbers of opposite sign is negative),

$$a < b \text{ and } b < c \Rightarrow a < c$$

(the **transitive property** of inequality),

$$a < b \Rightarrow a + c < b + c$$

(we can add the same number to both sides of an inequality without changing the direction of the inequality),

$$a < b \text{ and } c > 0 \Rightarrow ac < bc$$

(we can multiply both sides of an inequality by the same positive number without changing the direction of the inequality),

$$a < b \text{ and } c < 0 \Rightarrow ac > bc$$

(the direction of the inequality is reversed if both sides of the inequality are multiplied by the same negative number),

$$0 < a < b \Rightarrow \frac{1}{a} > \frac{1}{b}$$

(the direction of the inequality is reversed when we consider the reciprocals of positive numbers).
The notation, $a \leq b$, means that $a < b$ or $a = b$. Similarly,

$$a \geq b \Leftrightarrow a > b \text{ or } a = b$$

Example 1 Determine the set of all real numbers x such that

$$\frac{1}{x - 4} < \frac{1}{x - 2}.$$

Solution

The expressions on either side of the inequality are defined if $x \neq 4$ and $x \neq 2$. We will consider the cases $x < 2$, $2 < x < 4$, and $x > 4$ separately.

We have $-2 > -4$, so that $x - 2 > x - 4$ for each real number x (we added x to both sides of the inequality $-2 > -4$).

If we consider the case $x > 4$,

$$x - 2 > x - 4 > 0 \Rightarrow \frac{1}{x - 2} < \frac{1}{x - 4}.$$

If we consider the case $x < 2$, we have $x - 2 < 0$ and $x - 4 < 0$, so that

$$x - 2 > x - 4 \Rightarrow \left(\frac{1}{x - 2}\right)(x - 2) < \left(\frac{1}{x - 2}\right)(x - 4) \Rightarrow 1 < \frac{x - 4}{x - 2}$$

(we multiplied both sides of the first inequality by the negative number $x - 2$). We continue:

$$1 < \frac{x-4}{x-2} \Rightarrow \frac{1}{x-4} > \left(\frac{x-4}{x-2}\right)\left(\frac{1}{x-4}\right) \Rightarrow \frac{1}{x-4} > \frac{1}{x-2}$$

(we multiplied both sides of the first inequality by the negative number $x - 4$).

Finally, let us consider the case $2 < x < 4$. In this case $x - 2 > 0$ and $x - 4 < 0$. We have

$$x - 2 > x - 4 \Rightarrow 1 > \frac{x-4}{x-2}$$

(we multiplied both sides of the first inequality by $x - 2 > 0$). Then,

$$1 > \frac{x-4}{x-2} \Rightarrow \frac{1}{x-4} < \frac{1}{x-2}$$

(we multiplied both sides of the first inequality by $1/(x-4) < 0$).
Thus, we conclude that

$$\frac{1}{x-4} < \frac{1}{x-2} \Leftrightarrow 2 < x < 4.$$

$\square$

The Number Line

We will review the correspondence between real numbers and points on a line. This correspondence helps us to visualize and describe subsets of the set of real numbers. For example, we will be able to picture the solution of Example **??**, i.e., the set of all real numbers x such that $2 < x < 4$.

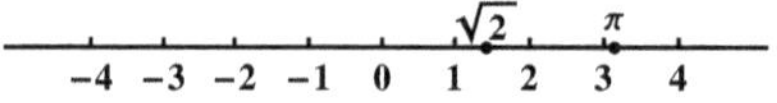

Figure 1: The number line

Points on a line are associated with real numbers as follows: A point on the line is selected as the **origin**. The origin corresponds to 0. A **unit length** is selected and the point corresponding to 1 is placed at unit distance from the origin. The origin and the point corresponding to 1 determine the positive direction along the line, and the opposite direction is the negative direction. Usually, we place the line horizontally, and select the positive direction to the right. If x is a positive number, the point corresponding to x is placed at a distance x from the origin, in the positive direction. If x is a negative number, the corresponding point is at the distance $-x$ from the origin, in the negative direction. Thus, we establish a correspondence between the set of real numbers and a line. We will refer to the line as **the number line**, and identify the number x with the point that corresponds to x. Thus, we may refer to "the point 2" or "the number 2", for example. We have $a < b$ iff a is to the left of b on the number line (assuming that the positive direction of the line is towards the right).

We will use the standard notation for sets. Thus, if A is a set, the fact that x is an element of A will be expressed as "$x \in A$". $A \subset B$ means that the set A is included in the set B, i.e., if $x \in A$ then $x \in B$. We will allow the possibility that $A = B$ when we write $A \subset B$. The union of sets will be denoted by $\cup$. Thus,

$$A \cup B = \{x : x \in A \text{ or } x \in B\}$$

(read the right-hand side as "the set of x such that x is in A or x is in B"). The "or" is an "inclusive or", so that x may belong to both sets. The intersection of A and B consists of all x that belong to both A and B, and will be denoted by $A \cap B$. Thus,

$$A \cap B = \{x : x \in A \text{ and } x \in B\}.$$

Intervals are subsets of the set of real numbers which occur frequently in calculus. If $a < b$, **the open interval** (a, b) with **endpoints** a and b is the set of all points between a and b:

$$(a, b) = \{x \in \mathbb{R} : a < x < b\}.$$

We will usually write

$$(a, b) = \{x : a < x < b\},$$

if it is clear that we are referring to subsets of the set of real numbers $\mathbb{R}$. Note that the open interval (a, b) does not contain the endpoints a and b. We may indicate an open interval as in Figure 2.

Figure 2: An open interval (a, b)

The **closed interval** $[a, b]$ consists of the points which lie between a and b and the endpoints a, b :

$$[a, b] = \{x : a \leq x \leq b\}.$$

We may indicate a closed interval as in Figure 3.

Figure 3: A closed interval $[a, b]$

We may also consider **half-open intervals** of the form

$$[a, b) = \{x : a \le x < b\},$$
$$(a, b] = \{x : a < x \le b\}.$$

Figure 4: An interval $[a, b)$

An **unbounded interval** that consists of all numbers less than a given number b is denoted by $(-\infty, b)$:

$$(-\infty, b) = \{x : x < b\}.$$

There is no need to try to attach a mystical meaning to the symbol $-\infty$. Within the context of intervals, the symbol merely indicates that the interval contains negative numbers whose distance from the origin is arbitrarily large. Similarly,

$$(a, +\infty) = \{x : x > a\},$$
$$(-\infty, b] = \{x : x \le b\},$$
$$[a, +\infty) = \{x : x \ge a\}.$$

Figure 5: An interval of the form $(-\infty, b)$

If J denotes an arbitrary interval, **the interior of** J is the interval which is obtained by deleting those endpoints of J which belong to J. For example, the interior of the open interval (a, b) coincides with itself, and the interior of $[a, b)$ is the open interval (a, b).

Example 2 In Example 2 we determined that

$$\frac{1}{x - 4} < \frac{1}{x - 2} \Leftrightarrow 2 < x < 4.$$

We can describe the set of all such x as the open interval $(2, 4)$, as illustrated in Figure 6. $\square$

Figure 6

Example 3 Let $p(x) = x^2 + x - 6$. Express the set of all real numbers x such that $p(x) \geq 0$ and the set of real numbers x such that $p(x) < 0$ as unions of intervals. Determine the interiors of the intervals.

Solution

We will begin by determining $x \in R$ such that $p(x) = 0$. By **the quadratic formula** (as reviewed in Section A.1), we have $p(x) = 0$ if and only if

$$x = \frac{-1 \pm \sqrt{1 + 24}}{2} = \frac{-1 \pm 5}{2}.$$

Therefore, the solutions of the equation $p(x) = 0$ are -3 and 2. Thus,

$$p(x) = x^2 + x - 6 = (x + 3)(x - 2).$$

We can determine the sign of $p(x)$ from our knowledge about the sign of each factor, since a product $ab > 0$ if a and b have the same sign, and $ab < 0$ if a and b have opposite signs. For example if $x < -3$, we have $x + 3 < 0$ and $x - 2 < 0$, so that $(x + 3)(x - 2) > 0$. Table 1 summarizes the results of these observations ($+$ indicates a positive number and $-$ indicates a negative number)

x		-3		2	
$x + 3$	$-$	0	$+$	$+$	$+$
$x - 2$	$-$	$-$	$-$	0	$+$
$p(x)$	$+$	0	$-$	0	$+$

Table 1

Thus, $p(x) \geq 0$ if and only if $x \leq -3$ or $x \geq 2$, so that

$$\{x : p(x) \geq 0\} = (-\infty, -3] \cup [2, +\infty).$$

The interior of $(-\infty, -3]$ is $(-\infty, -3)$ and the interior of $[2, +\infty)$ is $(2, +\infty)$. We have $p(x) < 0$ iff $-3 < x < 2$. Therefore

$$\{x : p(x) < 0\} = (-3, 2).$$

The interior of the open interval $(-3, 2)$ is equal to itself. $\square$

We will make use of the notion of the absolute value:

Definition 1 If x is an arbitrary real number, **the absolute value of** x is denoted by $|x|$. We have

$$|x| = \begin{cases} x & \text{if } x \geq 0, \\ -x & \text{if } x < 0. \end{cases}$$

Thus, the absolute value of x is the distance of x from the origin. For example,

$$|3| = 3,$$
$$|-3| = -(-3) = 3.$$

Given (real) numbers a and b, we have

$$|a - b| = \begin{cases} a - b & \text{if } a \geq b, \\ b - a & \text{if } a < b. \end{cases}$$

Geometrically, $|a - b|$ **is the distance between the points a and b on the number line.** For example, the distance between the points 2 and 4 is

$$|2 - 4| = |-2| = 2,$$

and the distance between the points 1 and 5 is

$$|1 - 5| = |-4| = 4.$$

Example 4 Express

$$A = \{x : |x - 1| < 2\}$$

as an interval.

Solution

Since A consists of all points whose distance from 1 is less than 2, it is the open interval whose endpoints can be determined by measuring the distance 2 to the left and to the right of 1. Thus,

$$A = \{x : |x - 1| < 2\} = (1 - 2, 1 + 2) = (-1, 3),$$

as illustrated in Figure 7. $\square$

Figure 7: $A = (-1, 3)$

As in Example 4, given $a \in \mathbb{R}$ and $r > 0$, the set $\{x : |x - a| < r\}$ consists of all points whose distance from a is less than r, i.e.,

$$\{x : |x - a| < r\} = (a - r, a + r),$$

and

$$\{x : |x - a| \leq r\} = [a - r, a + r].$$

In particular, if $a = 0$,

$$\{x : |x| \leq r\} = [-r, r].$$

Example 5 Express the set
$$A = \{x : |x - 1| \geq 2\}$$
as a union of intervals.

Solution

The set A consists of all x whose distance from 1 is at least 2. This means that $x \leq -1$ or $x \geq 3$. Therefore, A is the union of the intervals $(-\infty, -1]$ and $[3, +\infty)$, i.e.,
$$A = (-\infty, -1] \cup [3, +\infty).$$
Figure 8 illustrates the set A on the number line. $\square$

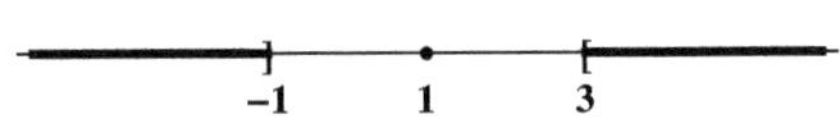

Figure 8

We will make use of the following fact about the absolute value:

Proposition 1 The absolute value of a product is the product of the absolute values:
$$|ab| = |a||b|.$$

Proof

We will consider the following cases:
1. $a \geq 0$ and $b \geq 0$,
2. $a \geq 0$ and $b \leq 0$,
3. $a \leq 0$ and $b \geq 0$
4. $a \leq 0$ and $b \leq 0$.

In the first case, $|a| = a$, $|b| = b$, and $ab \geq 0$ so that
$$|ab| = ab = |a|\,|b|.$$
In the second case, $|a| = a$, $|b| = -b$, and $ab \leq 0$ so that
$$|ab| = -(ab) = a(-b) = |a|\,|b|.$$
In the third case, $|a| = -a$, $|b| = b$, and $ab \leq 0$ so that
$$|ab| = -(ab) = (-a)(b) = |a|\,|b|.$$
In the fourth case, $|a| = -a$, $|b| = -b$, and $ab \geq 0$ so that
$$|ab| = ab = (-a)(-b) = |a|\,|b|.$$

■

We will have occasion to use the triangle inequality:

Theorem 1 (The Triangle Inequality) If a and b are arbitrary real numbers

$$|a + b| \le |a| + |b|.$$

Thus,the absolute value of a sum is less than or equal to the sum of the absolute values.

Proof

Since $a = |a|$ or $a = -|a|$, and $b = |b|$ or $b = -|b|$, we have

$$-|a| \le a \le |a|,$$
$$-|b| \le b \le |b|.$$

Therefore,

$$-|a| - |b| \le a + b \le |a| + |b|,$$

i.e.,

$$-(|a| + |b|) \le a + b \le |a| + |b|.$$

If $a + b \ge 0$,

$$|a + b| = a + b \le |a| + |b|.$$

If $a + b < 0$,

$$-(|a| + |b|) \le a + b \Rightarrow |a| + |b| \ge -(a + b) = |a + b|.$$

Therefore, in all cases,

$$|a + b| \le |a| + |b|.$$

■

Corollary 1 (Corollary to the Triangle Inequality)

If a and b are arbitrary real numbers

$$||a| - |b|| \le |a - b|.$$

Proof

By the triangle inequality,

$$|a| = |a - b + b| \le |a - b| + |b|,$$

so that

$$|a| - |b| \le |a - b|.$$

Similarly,

$$|b| = |b - a + a| \le |b - a| + |a| = |a - b| + |a|,$$

so that

$$|b| - |a| \le |a - b| \Rightarrow |a| - |b| \ge -|a - b|.$$

Thus,

$$-|a - b| \le |a| - |b| \le |a - b|.$$

As in the proof of the triangle inequality, the above inequality leads to the inequality

$$||a| - |b|| \le |a - b|.$$

■

Example 6 Confirm the following inequalities:

a) $|-3 + 4| \leq |-3| + |4|$
b) $||-5| - |3|| \leq |-5 - 3|$.

Solution

a) We have

$$|-3 + 4| = |1| = 1,$$

$$|-3| + |4| = 3 + 4 = 7,$$

and $1 < 7$.

b) We have

$$||-5| - |3|| = |5 - 3| = |2| = 2,$$

$$|-5 - 3| = |-8| = 8,$$

and $2 < 8$. $\square$

Problems

In problems 1 - 12,
a) Determine the set A of points that satisfy the given inequality,
b) Express A as a union of intervals.

1.
$$2x - 6 < 1$$

2.
$$3x + 5 \geq 2$$

3.
$$\frac{1}{x - 4} > 2$$

4.
$$\frac{1}{x + 3} < 1$$

5.
$$x^2 - x - 6 > 0$$

6.
$$x^2 - 5x + 4 \leq 0$$

7.
$$2x^2 - 4x + 6 > 0$$

8.
$$x^2 - 6x + 7 > 0$$

9.
$$|x - 8| < 3$$

10.
$$|x - 4| > 1$$

11.
$$|x - 1| \leq 8$$

12.
$$|x + 3| \leq 5$$

A.4   Decimal Approximations

We make use of decimal approximations to real numbers when we discuss numerical data in connection with the basic concepts of calculus and various applications. The purpose of this section is to fix the relevant terminology and notation.

We can obtain **decimal approximations** to real numbers with the help of a computational utility. For example,

$$\sqrt{2} \cong 1.41421 \text{ and } \pi \cong 3.14159.$$

The irrational number $\sqrt{2}$ is not equal to the decimal 1.41421. On the other hand, $\sqrt{2}$ can be approximated with increasing accuracy by increasing the number of decimal places. For example,

$$\sqrt{2} \cong 1.41421356.$$

We will write

$$\sqrt{2} = 1.41421356\ldots,$$

where "$\ldots$" indicates that an arbitrarily large number of decimal places can be displayed. Similarly,

$$\pi = 3.14159265\ldots$$

Definition 1 A **decimal** is an expression of the form

$$\pm a_0.a_1 a_2 \ldots a_n \ldots,$$

where a_0 is a nonnegative integer and the nth **decimal digit** a_n is an integer between 0 and 9 for $n = 1, 2, 3, \ldots$ .If there exists n such that $a_{n+1} = 0$, $a_{n+2} = 0, a_{n+3} = 0, \ldots$, we identify $\pm a_0.a_1 a_2 \ldots a_n 000 \ldots$ with **the finite decimal** $\pm a_0.a_1 a_2 \ldots a_n$. Otherwise, we will refer to $\pm a_0.a_1 a_2 \ldots a_n \ldots$ as an **infinite decimal**.

Example 1 Consider the decimal 1.41421. The first decimal digit is 4, and the fifth decimal digit is 1. The decimal is a rational number:

$$1.41421 = 1 + \frac{4}{10} + \frac{1}{10^2} + \frac{4}{10^3} + \frac{2}{10^4} + \frac{1}{10^5} = \frac{141421}{100\,000}.$$

$\square$

We can express the decimal 0.000141 as 1.41×10^{-4}. Similarly, we can express the decimal 1410000 as 1.41×10^6. In either case, the alternative expression is easier to read. These are examples of "the scientific notation" that is used to express very small or very large numbers conveniently:

Definition 2 A decimal is expressed in **scientific notation** if it is expressed as

$$\pm a_1.a_2 a_3 \ldots a_n \ldots \times 10^m,$$

where the nth **significant digit** a_n is an integer between 0 and 9 for $n = 1, 2, 3, \ldots$, $a_1 \neq 0$, and the **exponent** m is an integer. If $a_n \neq 0$ and $a_{n+1} = 0$, $a_{n+2} = 0, \ldots$, we say that the decimal has n **significant digits**.

Remark 1 We can count the significant digits of the decimal in its original form by starting with the first nonzero digit and stopping at the last nonzero digit. $\Diamond$

Example 2 We have $0.0001410250 = 1.41025 \times 10^{-4}$ in scientific notation. The decimal has 6 significant digits. The first significant digit is 1 and the sixth significant digit is 5.

We have $1410250 = 1.41025 \times 10^6$ in scientific notation. As in the case of $0.000141025 = 1.41 \times 10^{-4}$, the first significant digit is 1 and the sixth significant digit is 5. $\square$

Definition 3 Assume that x is a real number and that x_{app} is an approximation to x. **The error** in the approximation of x by x_{app} is the difference $x_{app} - x$. **The absolute error** in the approximation of x by x_{app} is the absolute value of the difference $x_{app} - x$, i.e., $|x_{app} - x|$.

Definition 4 We **chop** (or **truncate**) the decimal $\pm a_0.a_1a_2\ldots a_na_{n+1}\ldots$ to n decimal places and obtain $\pm a_0.a_1a_2\ldots a_n$. We **round** $\pm a_0.a_1a_2\ldots a_na_{n+1}\ldots$ to n decimal places as follows: If $a_{n+1} < 5$, we chop the decimal to n decimal places. If $a_{n+1} \geq 5$, we calculate the decimal $\pm(a_0.a_1a_2\ldots a_n + 10^{-n})$, i.e., we discard all the decimal digits past the nth decimal digit and add 1 to a_n.

Example 3

a) Let $x = 0.4599$. Determine x_c and x_r if x_c is obtained by chopping x to 3 decimal places and x_r is obtained by rounding x to 3 decimal places. Compare the absolute error in the approximation of x by x_c and by x_r.
b) Let $x = -3.141593$. Determine x_c and x_r if x_c is obtained by chopping x to 4 decimal places and x_r is obtained by rounding x to 4 decimal places. Compare the absolute error in the approximation of x by x_c and by x_r.

Solution

a) We have $x_c = 0.459$. Therefore,

$$|x_c - x| = x - x_c = 0.4599 - 0.459 = 0.0009 = 9 \times 10^{-4}.$$

Now let's round 0.4599 to 3 decimal places. Since the 4th decimal digit is $9 > 5$,

$$x_r = 0.459 + 10^{-3} = 0.459 + 0.001 = 0.460.$$

Therefore,

$$|x_r - x| = x_r - x = 0.460 - 0.4599 = 0.0001 = 10^{-4}.$$

We see that $|x_r - x| < |x_c - x|$, so that x_r approximates x more accurately than x_c.
b) We have $x_c = -3.1415$. Therefore,

$$|x_c - x| = |-3.1415 - (-3.141593)| = 3.141593 - 3.1415 = 9.3 \times 10^{-5},$$

We have $x_r = -3.1416$. Therefore,

$$|x_r - x| = |-3.1416 - (-3.141593)| = 3.1416 - 3.141593 = 7.0 \times 10^{-6}$$

As in part a), $|x_r - x| < |x_c - x|$, so that x_r approximates x more accurately than x_c. $\square$

Definition 5 We say that **a real number x is represented by the decimal** $a_0.a_1a_2\ldots a_n\ldots$ (a_0 is an arbitrary integer) if the absolute error in the approximation of a by $a_0.a_1a_2\ldots a_n$ is as small as desired if n is sufficiently large. Thus,

$$|a - a_0.a_1a_2\ldots a_n|$$

is as small as we please if n is large enough. In this case we write

$$a = a_0.a_1a_2\ldots a_n\ldots,$$

and say that $a_0.a_1a_2\ldots a_n\ldots$ is a **decimal expansion** of the number a.

We will accept the following fact:

Every real number, rational or irrational, has a decimal expansion.

Since

$$a_0.a_1a_2\ldots a_n = a_0 + \frac{a_1}{10} + \frac{a_2}{10^2} + \cdots + \frac{a_n}{10^n}$$

$$= \frac{a_0 \times 10^n + a_1 \times 10^{n-1} + a_2 \times 10^{n-2} + \cdots + a_n}{10^n},$$

a finite decimal represents a rational number. There are rational numbers that cannot be represented by a finite decimal. For example,

$$\frac{1}{3} = 0.333\ldots,$$

$$\frac{5}{11} = 0.45454545\ldots,$$

as you can confirm by long division. In the infinite decimal expansion of a rational number, a block of numbers keeps repeating, as in the above examples.

The decimal expansion of a rational number need not be unique:

Example 4 We have $1 = 1.000\ldots = 0.999\ldots$.

Indeed, if $0.999\ldots9$ has n decimal places, the absolute error in the approximation of 1 by $0.999\ldots9$ is 10^{-n}, and 10^{-n} is as small as desired if n is sufficiently large. Therefore, $1 = 1.000\ldots = 0.999\ldots$ $\square$

Example 5 The truncation of the decimal expansion of π to 10 significant digits is $3.141\,592\,653$. Let x_n and z_n be the numbers that are obtained by chopping and rounding the decimal expansion of π to n decimal places, respectively. Calculate $|x_n - \pi|$ and $|z_n - \pi|$ for $n = 2, 3, 4, 5$ (express the results in scientific notation and round to 2 significant digits). Do the numbers support the fact that $3.141\,592\,653\ldots$ is the decimal expansion of π? Compare the accuracy of the approximations that are obtained by chopping versus rounding.

Solution

Table 1 displays the required data. We see that $|x_n - \pi|$ gets smaller as n increases from 2 to 5. This supports the fact that $3.141\,592\,653\ldots$ is the decimal expansion of π. We also see that rounding may result in better accuracy than chopping to the same number of decimal places $(n = 4)$. $\square$

n	x_n	z_n	$|x_n - \pi|$	$|z_n - \pi|$
2	3.14	3.14	1.6×10^{-3}	1.6×10^{-3}
3	3.141	3.142	4.1×10^{-4}	4.1×10^{-4}
4	3.1415	3.1416	9.3×10^{-5}	7.3×10^{-6}
5	3.14159	3.14159	2.7×10^{-6}	2.7×10^{-6}

Table 1

In describing the accuracy of an approximation to a very large number or to a very small number, it may be more meaningful to express the absolute error relative to the magnitude of the approximated number:

Definition 6 Assume that $x \neq 0$ and x_{app} is an approximation to x. **The relative error** in the approximation of x by x_{app} is

$$\frac{|x_{app} - x|}{|x|}$$

Since we have the ratio of the absolute error to $|x|$, it may be more appropriate to say "relative *absolute* error", but the more precise language is somewhat clumsy.

Example 6

a) Approximate π by rounding of its decimal expansion to 5 significant digits. Calculate the relative error in the approximation (Express the relative error in scientific notation and round to 2 significant digits).

b) Repeat part a) with π replaced by $\pi \times 10^4$ and $\pi \times 10^{-4}$. Compare the results with the result of part a).

c) Calculate the absolute errors in the approximations of part b). Compare the results with the results of part b).

Solution

a) We have $\pi = 3.41592653\ldots$. If x_r denotes the number that is obtained by rounding the decimal expansion of π to 5 significant digits, then $x_r = 3.1416$. The relative error in the approximation of π by x_r is

$$\frac{|3.1416 - \pi|}{\pi} \cong 2.3 \times 10^{-6}.$$

b) The decimal expansion of $\pi \times 10^4$ is $(3.141592653\ldots) \times 10^4$. If $\tilde{x}_r$ denotes the number that is obtained by rounding the decimal expansion of $\pi \times 10^4$ to 5 significant digits, then $\tilde{x}_r = 3.1416 \times 10^4$. The relative error in the approximation of $\pi \times 10^4$ by $\tilde{x}_r$ is

$$\frac{|\tilde{x}_r - \pi \times 10^4|}{\pi \times 10^4} = \frac{|3.1416 \times 10^4 - \pi \times 10^4|}{\pi \times 10^4} = \frac{|3.1416 - \pi|}{\pi} \cong 2.3 \times 10^{-6}.$$

Similarly, when we consider $\pi \times 10^{-4} = (3.141592653\ldots) \times 10^{-4}$, and denote the number that is obtained by rounding the decimal expansion of $\pi \times 10^{-4}$ to 5 significant digits by $\hat{x}_r$, we have $\hat{x}_r = 3.1416 \times 10^{-4}$. The relative error in the approximation of $\pi \times 10^{-4}$ by $\hat{x}_r$ is

$$\frac{|\hat{x}_r - \pi \times 10^{-4}|}{\pi \times 10^{-4}} = \frac{|3.1416 \times 10^{-4} - \pi \times 10^{-4}|}{\pi \times 10^{-4}} \cong 2.3 \times 10^{-6}.$$

Thus, the relative error in the approximation of $\pi \times 10^4$ or $\pi \times 10^{-4}$ by rounding the relevant decimal to 5 significant digits is the same as the relative error in the approximation of π by rounding the decimal expansion of π to 5 significant digits.

c) The absolute errors are

$$|\tilde{x}_r - \pi \times 10^4| = |3.1416 \times 10^4 - \pi \times 10^4| \cong 7.3 \times 10^{-2},$$

and

$$|\hat{x}_r - \pi \times 10^{-4}| = |3.1416 \times 10^{-4} - \pi \times 10^{-4}| \cong 7.3 \times 10^{-10}.$$

These numbers are quite different from each other and from the numbers of part b).$\square$

As illustrated in Example 6, **the number of significant digits in a decimal approximation to a number is a measure of the relative error of the approximation.**

Remark 2 A computational utility rounds or chops the decimal expansions of a number to a certain number of significant digits at each step of a numerical calculation. **The numerical results that are displayed in this book have been obtained with the help of a calculator that rounds decimals to 14 significant digits. The results are displayed by rounding decimals to 6 significant digits, and errors or relative errors are displayed in scientific notation, rounded to 2 significant digits, unless stated otherwise.** $\Diamond$

We will display a large number of significant digits very rarely, and you should be able to duplicate almost all the results that are displayed in this book. Nevertheless, you should be aware of the limitations of calculations based on the rounding or chopping of decimals to a certain number of significant digits. The following example illustrates what can go wrong when we deal with numbers that are very close to each other:

Example 7 Consider the calculation of

$$\frac{\sqrt{2+h} - \sqrt{2}}{h},$$

where h is a small positive number (soon you will see that such calculations are relevant to a central idea of calculus).

Table 2 shows the numbers that correspond to $h = 10^{-k}$, for $k = 3, 4, 5, 6$, as performed by a calculator that bases the calculations on rounding decimals to 14 significant digits (as in Remark 3, the results are displayed by rounding to 6 significant digits).

h	$\dfrac{\sqrt{2+h} - \sqrt{2}}{h}$
10^{-3}	.353509
10^{-4}	.353549
10^{-5}	.353553
10^{-6}	.353553

Table 2

The numbers in Table 2 indicate that the given expression approximates $.353\,553$ if h is small. Indeed, we can determine that number by rationalizing the numerator:

$$\frac{\sqrt{2+h} - \sqrt{2}}{h} = \left(\frac{\sqrt{2+h} - \sqrt{2}}{h}\right)\left(\frac{\sqrt{2+h} + \sqrt{2}}{\sqrt{2+h} + \sqrt{2}}\right)$$

$$= \frac{(2+h) - 2}{h\left(\sqrt{2+h} + \sqrt{2}\right)}$$

$$= \frac{1}{\sqrt{2+h} + \sqrt{2}}$$

Therefore, if h is small,

$$\frac{\sqrt{2+h} - \sqrt{2}}{h} = \frac{1}{\sqrt{2+h} + \sqrt{2}} \cong \frac{1}{\sqrt{2} + \sqrt{2}} = \frac{1}{2\sqrt{2}} \cong .353\,553.$$

On the other hand, if we ask the same calculator to calculate

$$\frac{\sqrt{2+h} - \sqrt{2}}{h}$$

for $h = 10^{-13}$, the answer is 0. This is due to the fact that both $\sqrt{2 + 10^{-13}}$ and $\sqrt{2}$ have been approximated by the same decimal, namely,

$$1.4142135623731.$$

(If you try to duplicate the numbers with your calculator, you may get slightly different numbers if the calculations of your calculator are based on the rounding of decimals to a number of significant digits other than 14). If we replace the given expression by the algebraically equivalent expression

$$\frac{1}{\sqrt{2+h}+\sqrt{2}},$$

the same calculator will have no problem.

This example illustrates **the danger of subtracting nearly equal numbers**. There is no need to panic, though: We will never demand from a calculator more than it can handle. If the calculations are based on the rounding of decimals to 14 significant digits, it is prudent to restrict h in an expression such as

$$\frac{\sqrt{2+h}-\sqrt{2}}{h}$$

so that $|h|$ is not less than 10^{-10}. $\square$

Problems

[C] In problems 1 - 6,
a) Round the decimal expansion of the given number to 6 decimal places.
b) Calculate the absolute error in the appproximation of the number by that decimal (assuming that your calculator provided the exact value). Express the absolute error in scientific notation and with 2 significant digits.

1. $\sqrt{2}$

2. $\sqrt{3}$

3. $\dfrac{1}{6}$

4. π^2

5. $\dfrac{1}{19}$

6. $\sqrt{101}$

[C] In problems 7 - 10,
a) Round the decimal expansion of the given number to 6 significant digits.
b) Calculate the relative error in the appproximation of the number by that decimal (assuming that your calculator provided the exact value).

7. $\sqrt{504}$

8. $(8.04)^{1/3}$

9. $700^{1/4}$

10. $\dfrac{100}{6}$

A.5 The Coordinate Plane

The Cartesian Coordinate System

Let us place two copies of the number line on the plane so that they intersect at their respective origins at a right angle. We will usually place one copy horizontally, so that the other copy will be vertical. In this case, the horizontal copy will be referred to as **the horizontal axis** and the vertical copy will be referred to as **the vertical axis**. The horizontal and vertical axes meet at **the origin**. Given a point P in the plane, other than the origin, we associate a pair of numbers with P as follows: The first number, say x, corresponds to the point of intersection of the vertical line through P and the horizontal axis. The second number, say y, corresponds to the point of intersection of the horizontal line through P and the vertical axis. We associate with P the pair of numbers, x, y, in that order, and denote this **ordered pair** as (x, y).

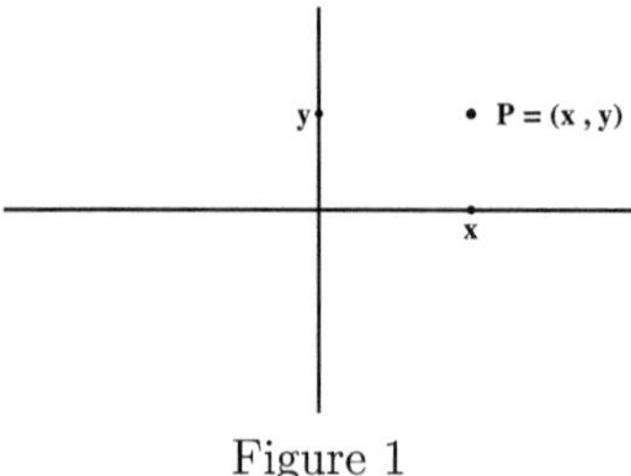

Figure 1

The ordered pairs (x_1, y_1) and (x_2, y_2) are declared to be equal if and only if $x_1 = x_2$ and $y_1 = y_2$. We will identify the point P with the associated ordered pair (x, y) and write $P = (x, y)$. The first entry of the ordered pair is referred to as **the first coordinate** of the point P and the second entry is referred to as **the second coordinate** of P. The ordered pair $(0, 0)$ corresponds to origin. The system that associates ordered pairs of numbers with points as we described is called **the Cartesian coordinate system** after the French philosopher-mathematician **Reneé Descartes**. The system is also referred to as **the rectangular coordinate system**. If we choose to denote the first coordinate of a generic point P as x, we will refer to the horizontal axis as the x-**axis** and if choose to denote the second coordinate generically as y, we will refer to the vertical axis as the y-**axis**. In this case, the plane will be referred to as the xy-**coordinate plane**. We may use letters other than x to denote the first coordinate of a generic point, and letters other than y to denote the second coordinate of a generic point, even though these are the most popular choices.

The first quadrant consists of the points (x, y) such that $x > 0$ and $y > 0$, the second quadrant consists of the points (x, y) such that $x < 0$ and $y > 0$, the third quadrant consists of the points (x, y) such that $x < 0$ and $y < 0$, and the fourth quadrant consists of the points (x, y) such that $x > 0$ and $y < 0$.

Example 1 Let

$$P_1 = (4, 2), \ P_2 = (-4, 2), \ P_3 = (-4, -2) \text{ and } P_4 = (4, -2).$$

Figure 2 shows the points P_1, P_2, P_3 and P_4.

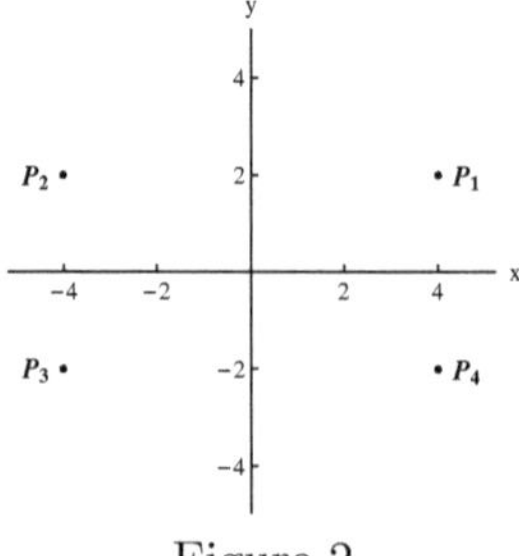

Figure 2

P_1 is in the first quadrant, P_2 is in the second quadrant, P_3 is in the third quadrant, and P_4 is in the fourth quadrant. Notice that P_1 and P_2 are symmetric with respect to the vertical axis: **Points of the from (x, y) and $(-x, y)$ are symmetric with respect to the vertical axis.**

The points P_1 and P_3 are symmetric with respect to the origin: **Points of the form (x,y) and $(-x,-y)$ are symmetric with respect to the origin.** $\square$

Assume that $F(x,y)$ is an expression involving x and y, and that c is a given real number. The set of points (x,y) in the xy-coordinate plane such that $F(x,y) = c$ is called **the graph of the equation $F(x,y) = c$.** Lines, parabolas, circles, ellipses and hyperbolas are such graphs and should be familiar from precalculus courses.

Lines

Lines are the graphs of equations of the form $ax + by = c$, where a, b and c are given numbers. We may refer to the graph simply as "the line $ax + by = c$".
If the equation is of the form $x = c$, the line consists of all points (x,y) such that $x = c$. There is no restriction on the y-coordinate. The line is a vertical line which passes through the point $(c,0)$.

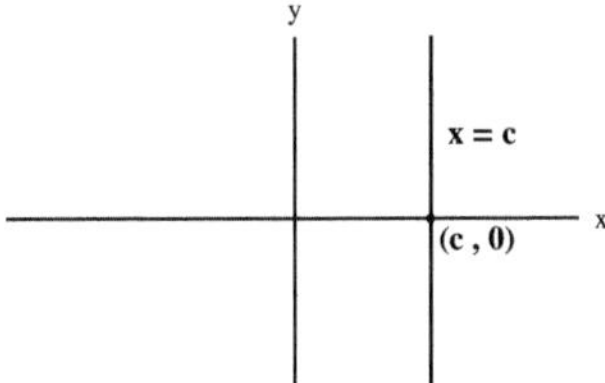

Figure 3: A vertical line

If the equation is of the form $y = c$, the line consists of all points (x,y) such that $y = c$. There is no restriction on the first coordinate and the line is horizontal.

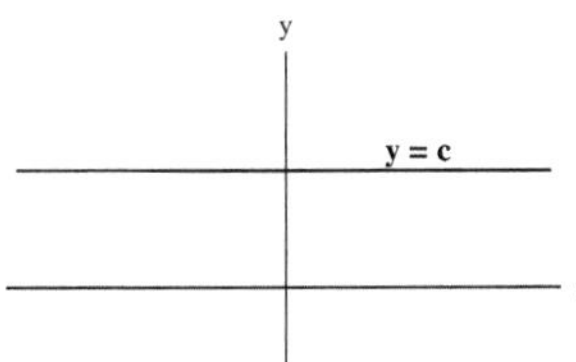

Figure 4: A horizontal line

The line $ax + by = c$ is not vertical if $b \neq 0$. We can express the relationship between x and y in the form,

$$y = mx + d.$$

In this case, m is the **slope of the line**: If (x_1, y_1) and (x_2, y_2) are arbitrary points on the line such that $x_1 \neq x_2$, we have

$$\frac{y_2 - y_1}{x_2 - x_1} = \frac{(mx_2 + d) - (mx_1 + d)}{x_2 - x_1} = \frac{m(x_2 - x_1)}{x_2 - x_1} = m$$

(slope = "rise over run"). If $x = 0$ then $y = d$, so that the line intersects the y-axis at the point $(0, d)$. The point $(0, d)$ is the **y-intercept** of the line. The equation $y = mx + d$ is said to be in **the slope-intercept form.**

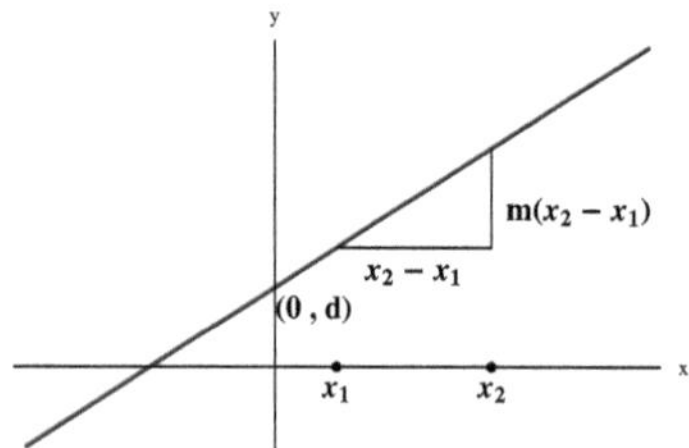

Figure 5: A line with slope m and y-intercept $(0, d)$

In many applications, a specific point on a line will have significance. Assume that $P_0 = (x_0, y_0)$ is that point. If m is the slope of the line, and $P = (x, y)$ is an arbitrary point on the line, we have

$$m = \frac{y - y_0}{x - x_0},$$

so that

$$y - y_0 = m\left(x - x_0\right),$$

i.e.,

$$y = y_0 + m\left(x - x_0\right).$$

We will refer to this form of the equation of a line as **the point-slope form with basepoint** (x_0, y_0).

Example 2

a) Determine the point-slope form of the line that passes through the points $(2, 3)$ and $(4, 7)$ with basepoint $(2, 3)$.
b) Determine the slope-intercept of the line of part a). Sketch the line.

Solution

a) The slope of the line is

$$\frac{7 - 3}{4 - 2} = \frac{4}{2} = 2.$$

If we consider $(2, 3)$ to be the basepoint and (x, y) is an arbitrary point on the line, we have

$$\frac{y - 3}{x - 2} = 2,$$

so that

$$y = 3 + 2(x - 2).$$

This is the point-slope form of the equation of the given line with basepoint $(2, 3)$.

b) We have

$$y = 3 + 2(x - 2) = 2x - 1.$$

This is the slope-intercept of the equation of the line.

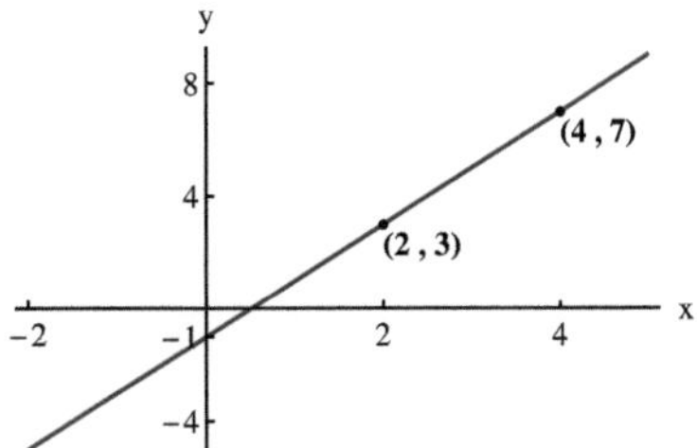

Figure 6: A line with positive slope

Figure 6 shows the line. As in many computer-generated graphs in this book, the scale on the vertical axis is not the same as the scale on the horizontal axis. If (x_1, y_1) and (x_2, y_2) are points on the line, and $x_2 > x_1$, we have $y_2 > y_1$. Indeed,

$$\frac{y_2 - y_1}{x_2 - x_1} = 2,$$

so that

$$y_2 - y_1 = 2(x_2 - x_1) > 0.$$

Just as in this example, if a line has a positive slope, and we imagine that a point moves along the line, the vertical coordinate of the point increases as its horizontal coordinate increases. $\square$

Example 3

a) Determine the point-slope form of the line that passes through the points $(2, 3)$ and $(4, 1)$ with basepoint $(2, 3)$.
b) Determine the slope-intercept of the line of part a). Sketch the line.

Solution

a) The slope of the line is

$$\frac{1 - 3}{4 - 2} = \frac{-2}{2} = -1.$$

Therefore, the point-slope of the equation of the line with basepoint $(2, 3)$ is

$$y = 3 - (x - 2).$$

b) The slope-intercept form of the equation of the line is

$$y = -x + 5.$$

Figure 7 shows the line.

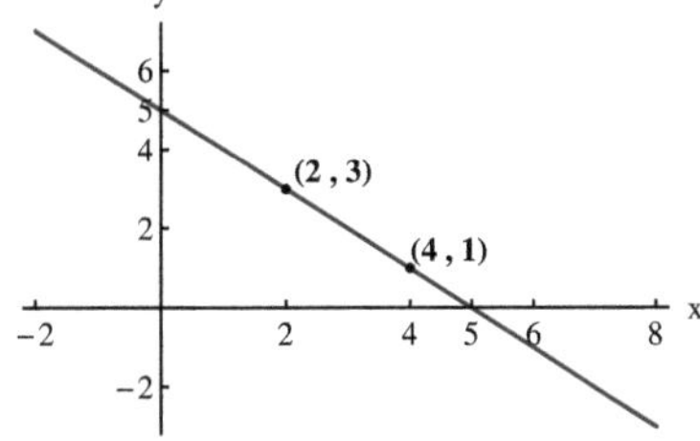

Figure 7: A line with negative slope

If (x_1, y_1) and (x_2, y_2) are points on the line, and $x_2 > x_1$, we have $y_2 < y_1$. Indeed,

$$\frac{y_2 - y_1}{x_2 - x_1} = -1,$$

so that

$$y_2 - y_1 = -(x_2 - x_1) < 0.$$

Just as in this example, if a line has a negative slope, and we imagine that a point moves along the line, the vertical coordinate of the point decreases as its horizontal coordinate increases. $\square$

Any two vertical lines are **parallel** to each other. Non-vertical lines are **parallel** to each other if and only if they have the same slope. Two distinct lines that are parallel to each other do not intersect. Two non-parallel lines intersect at a single point.

Example 4 Determine the point at which the lines $y = 2x - 1$ and $y = -x - 6$ intersect. Sketch the lines, and indicate the point of intersection.

Solution

The point $P = (x, y)$ is the point of intersection of the given lines iff the coordinates of P satisfy the equation of each line. Thus, we need to have

$$y = 2x - 5 \text{ and } y = -x + 4.$$

We substitute $2x - 5$ for y in the second equation:

$$2x - 5 = -x + 4 \Leftrightarrow 3x = 9 \Leftrightarrow x = 3.$$

Therefore,

$$y = 2x - 5 = 2(3) - 5 = 1.$$

Thus, $P = (3, 1)$ is the point at which the given lines intersect. Figure 8 shows the lines and the point P. $\square$

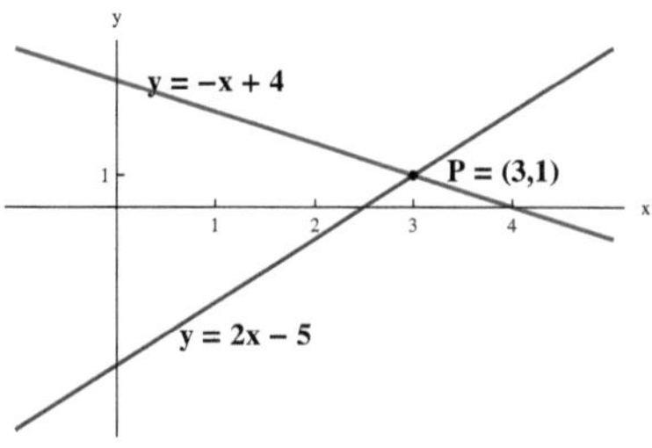

Figure 8

A horizontal line is **perpendicular** to a vertical line. Given lines with slopes m_1 and m_2, the lines are perpendicular to each other if and only if $m_1 m_2 = -1$.

Example 5 Determine the line that is perpendicular to the line $y = -2x + 4$ and passes through the point $(1, 2)$. Sketch the lines.

Solution

The slope of the line $y = -2x + 4$ is -2. If m is the slope of any line that is perpendicular to the given line, we must have

$$-2m = -1 \Rightarrow m = \frac{1}{2}.$$

The point-slope form of the equation of the line that passes through $(1, 2)$ and has slope $1/2$ is

$$y = 2 + \frac{1}{2}(x - 1).$$

Figure 9 shows the lines (Caution: the scale on the vertical axis must be the same as the scale on the horizontal axis in order to confirm the orthogonality of the lines visually). $\square$

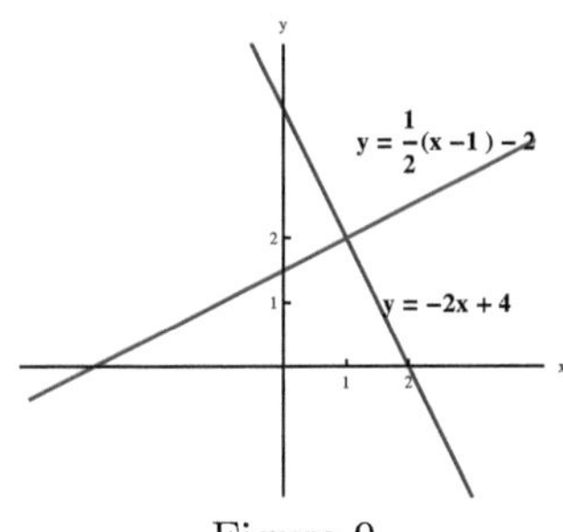

Figure 9

Recall that the formula for the distance between the points $P_1 = (x_1, y_1)$ and $P_2 = (x_2, y_2)$ is consistent with the Pythagorean Theorem, as illustrated in Figure 10: If $|P_1 P_2|$ denotes the distance between P_1 and P_2, we have

$$|P_1 P_2|^2 = (x_2 - x_1)^2 + (y_2 - y_1)^2,$$

so that

$$|P_1 P_2| = \sqrt{(x_2 - x_1)^2 + (y_2 - y_1)^2}.$$

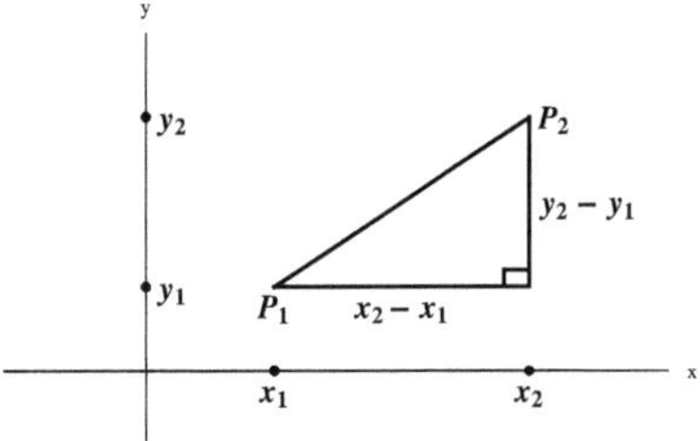

Figure 10

Parabolas

Now let us consider the graphs of equations of the form $y = ax^2 + bx + c$, where $a, b,$ and c are given numbers, and $a \neq 0$. Thus, $ax^2 + bx + c$ is a **quadratic expression**. The graph of such an equation in the xy-plane is a **parabola**. We may interchange the roles of x and y, and consider the graph of an equation of the form $x = ay^2 + by + c$ in the xy-plane. Such a graph is also a parabola if $a \neq 0$.

The graphs of the equations $y = x^2$, $y = -x^2$, $x = y^2$ and $x = -y^2$ are shown in Figures 11. The **completion of the square** in the relevant quadratic expression enables us to show that a parabola resembles one of these cases.

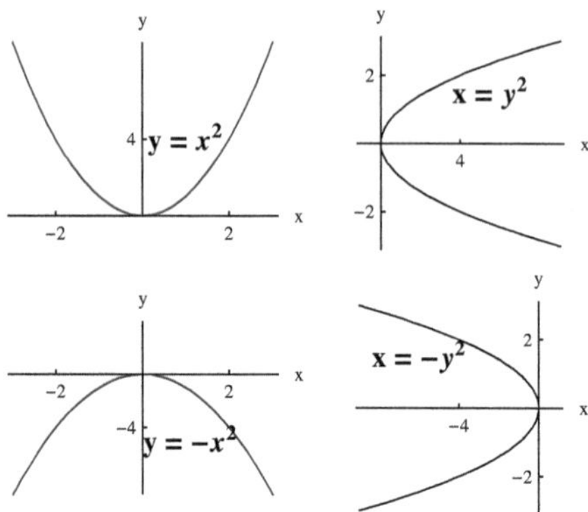

Figure 11: Basic parabolas

Example 6 Consider the parabola that is the graph of the equation

$$y = 2x^2 - 12x + 13$$

in the xy-plane.

a) Determine the lowest point on the parabola.
b) Determine the points at which the parabola intersects the coordinate axes.
c) Sketch the graph of the parabola.

Solution

a) We complete the square:

$$y = 2\left(x^2 - 6x\right) + 13 = 2\left(x - 3\right)^2 - 18 + 13 = 2\left(x - 3\right)^2 - 5.$$

Since $2\left(x - 3\right)^2 \geq 0$, we have

$$y = 2\left(x - 3\right)^2 - 5 \geq -5$$

for each $x \in \mathbb{R}$, and $y = -5$ if $x = 3$. Thus, the lowest point on the parabola is $(3, -5)$. We may refer to the point $(3, -5)$ as the **vertex** of the parabola.

b) The points at which the parabola intersects the x-axis have the form $(x, 0)$. Therefore, we need to solve the equation $2x^2 - 12x + 13 = 0$. We can use the quadratic formula, or make use of part a):

$$2x^2 - 12x + 13 = 0 \Leftrightarrow 2\left(x - 3\right)^2 - 5 = 0 \Leftrightarrow \left(x - 3\right)^2 = \frac{5}{2} \Leftrightarrow x - 3 = \pm\sqrt{\frac{5}{2}}$$

$$\Leftrightarrow x = 3 \pm \sqrt{\frac{5}{2}}.$$

Therefore, the parabola intersects the x-axis at the points

$$\left(3+\sqrt{\frac{5}{2}},0\right) \cong (4.58114,0) \quad \text{and} \quad \left(3-\sqrt{\frac{5}{2}},0\right) \cong (1.41886,0).$$

The point of intersection with the y-axis is obtained by setting $x = 0$ in the equation, $y = 2x^2 - 12x + 13$. Thus, the point in question is $(0, 13)$.

Note that (x, y) is on the parabola if and only if

$$y = 2(x-3)^2 - 5 \Leftrightarrow y + 5 = 2(x-3)^2.$$

Therefore, if we set $Y = y + 5$ and $X = x - 3$, we have $Y = 2X^2$. Thus, the equation of the parabola is simpler in the XY-plane. The origin in the XY-plane coincides with the vertex of the parabola. The X-axis is parallel to the x-axis, and the Y-axis is parallel to the y-axis: We have **translated the origin in the** xy-plane to the vertex of the parabola by introducing the new coordinate system. Figure 12 shows the parabola .$\square$

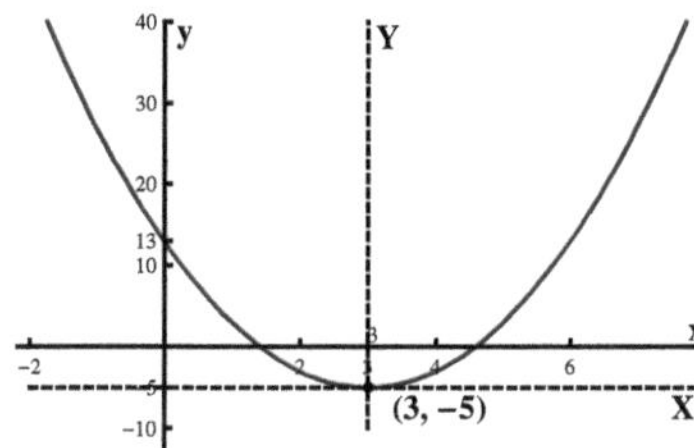

Figure 12

Example 7 Consider the parabola that is the graph of the equation

$$x = 3y^2 - 12y + 13$$

in the xy-plane.

a) Determine the point on the parabola that is farthest to the left.
b) Determine the points at which the parabola intersects the coordinate axes.
c) Sketch the graph of the parabola.

Solution

a) We complete the square:

$$x = 3y^2 - 12y + 13 = 3(y^2 - 4y) + 13 = 3(y-2)^2 - 12 + 13 = 3(y-2)^2 + 1.$$

Therefore,

$$x = 3(y-2)^2 + 1 \geq 1$$

for each $y \in \mathbb{R}$, and $x = 1$ iff $y = 2$. Thus, the point $(1, 2)$ is the point on the parabola that is farthest to the left.

b) Since $x = 13$ if $y = 0$, the parabola intersects the x-axis at the point $(13, 0)$. In order to determine the points of intersection with the y-axis, we need to solve the equation $3y^2 - 12y + 13 = 0$:

$$3y^2 - 12y + 13 = 0 \Leftrightarrow 3(y-2)^2 + 1 = 0$$
$$\Leftrightarrow 3(y-2)^2 = -1.$$

Since $3\left(y-2\right)^2 \geq 0$ for each $y \in \mathbb{R}$, the equation does not have solutions that are real numbers. Therefore, the parabola does not intersect the y-axis. We may refer to the point $(1,2)$ as the vertex of the parabola.

Note that (x,y) is on the parabola if and only if

$$x = 3\left(y-2\right)^2 + 1 \Leftrightarrow x - 1 = 3\left(y-2\right)^2.$$

Therefore, if we set $Y = y - 2$ and $X = x - 1$, we have $X = 3Y^2$. Thus, the equation of the parabola is simpler in the XY-plane. The origin in the XY-plane coincides with the vertex of the parabola: We have translated the origin in the xy-plane to the vertex of the parabola. Figure 13 shows the parabola. $\square$

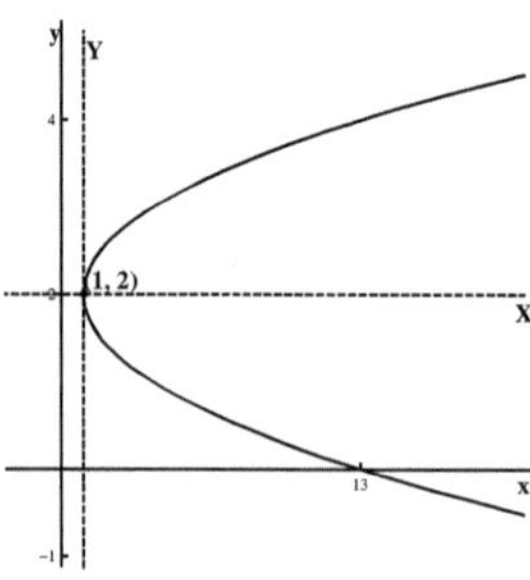

Figure 13

Circles, Ellipses and Hyperbolas

Aside form lines and parabolas, shapes such as **ellipses** and **hyperbolas** should be familiar from precalculus courses. Let us begin with special ellipses, namely **circles**. A circle with center at the point (a,b) and radius $r > 0$ is the set of points (x,y) in the xy-coordinate plane whose distance from (a,b) is r. Since the distance from (x,y) to (a,b) is

$$\sqrt{\left(x-a\right)^2 + \left(y-b\right)^2},$$

the circle in question consists of the points (x,y) such that

$$\sqrt{\left(x-a\right)^2 + \left(y-b\right)^2} = r.$$

If we square both sides of the equation, we obtain the equation,

$$\left(x-a\right)^2 + \left(y-b\right)^2 = r^2.$$

Thus, the graph of the above equation is a circle with center (a,b) and radius r.

Example 8 Show that the graph of the equation

$$x^2 - 6x + 13 + y^2 - 4y = 16$$

in the xy-plane is a circle. Determine the center and the radius of the circle.

Solution

We need to express the equation in the form

$$\left(x-a\right)^2 + \left(y-b\right)^2 = r^2.$$

In order to do this, we will complete the squares:

$$x^2 - 6x + 13 + y^2 - 4y = (x-3)^2 - 9 + 13 + (y-2)^2 - 4.$$

Therefore, the equation is

$$(x-3)^2 + (y-2)^2 = 16.$$

Thus, the graph of the equation is the circle with radius 4 centered at $(3,2)$. If we set $X = x - 3$ and $Y = y - 2$, the circle is the graph of the equation

$$X^2 + Y^2 = 16.$$

The introduction of the new coordinate system corresponds to the translation of the origin in the xy-plane to the center of the circle. Figure 14 shows the circle. $\square$

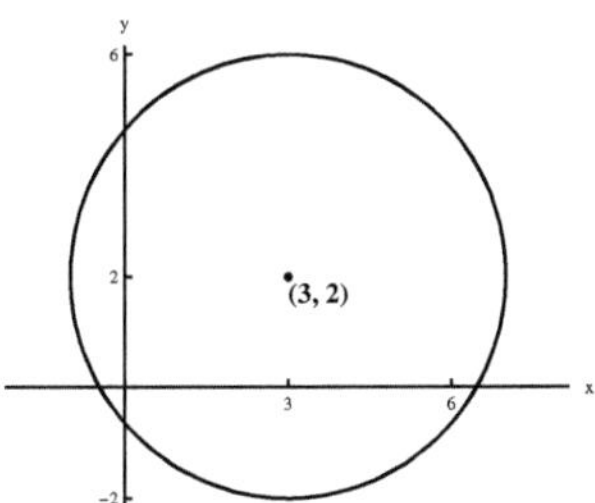

Figure 14

The circle of radius 1 which has as its center the origin plays a distinguished role in the discussion of trigonometric functions. We will refer to this circle as **the unit circle**. Thus, the unit circle is the graph of the equation $x^2 + y^2 = 1$ in the xy-plane.

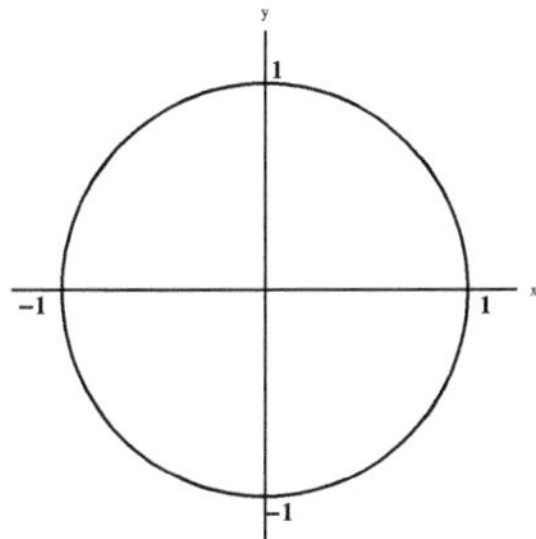

Figure 15: The unit circle

More generally, if A, C, D, E and F are given numbers, and both A and C are positive, the graph of an equation of the form

$$Ax^2 + Cy^2 + Dx + Ey + F = 0$$

is an **ellipse**. We can complete the squares, as in Example 9, and transform the equation to the "**standard form**"

$$\frac{(x-x_0)^2}{a^2} + \frac{(y-y_0)^2}{b^2} = 1,$$

where $a > 0$ and $b > 0$. The ellipse is **centered at** (x_0, y_0), and the lines $x = x_0$ and $y = y_0$ are the **axes** of the ellipse.

If A and C are of opposite signs, the graph of the equation

$$Ax^2 + Cy^2 + Dx + Ey + F = 0$$

is a **hyperbola**. The completion of the squares leads to one of the standard forms,

$$\frac{(x - x_0)^2}{a^2} - \frac{(y - y_0)^2}{b^2} = 1,$$

or

$$\frac{(y - y_0)^2}{b^2} - \frac{(x - x_0)^2}{a^2} = 1.$$

Example 9 Consider the equation

$$9x^2 + 4y^2 - 18x + 16y = 11$$

a) Express the equation in a standard form and identify the graph of the equation as an ellipse or hyperbola.
b) Sketch the graph of the equation.

Solution

a) We complete the squares:

$$
\begin{aligned}
9x^2 + 4y^2 - 18x + 16y &= 9x^2 - 18x + 4y^2 + 16y \\
&= 9\left(x^2 - 2x\right) + 4\left(y^2 + 4y\right) \\
&= 9\left(x - 1\right)^2 - 9 + 4\left(y + 2\right)^2 - 16 \\
&= 9\left(x - 1\right)^2 + 4\left(y + 2\right)^2 - 25.
\end{aligned}
$$

Therefore,

$$
\begin{aligned}
9x^2 + 4y^2 - 18x + 16y = 11 &\Leftrightarrow 9\left(x - 1\right)^2 + 4\left(y + 2\right)^2 - 25 = 11 \\
&\Leftrightarrow 9\left(x - 1\right)^2 + 4\left(y + 2\right)^2 = 36 \\
&\Leftrightarrow \frac{(x - 1)^2}{4} + \frac{(y + 2)^2}{9} = 1.
\end{aligned}
$$

Thus, the graph of the equation is an ellipse centered at $(1, -2)$.

b) The major axis of the ellipse is along the line $y = -2$, and its minor axis is along the line $x = 1$. If we set $X = x - 1$ and $Y = y + 2$, the ellipse is the graph of the equation

$$\frac{X^2}{4} + \frac{Y^2}{9} = 1$$

in the XY-plane. Figure 16 shows the ellipse. The introduction of the new coordinate system corresponds to the translation of the origin on the xy-plane to "the center of symmetries" of the ellipse. $\square$

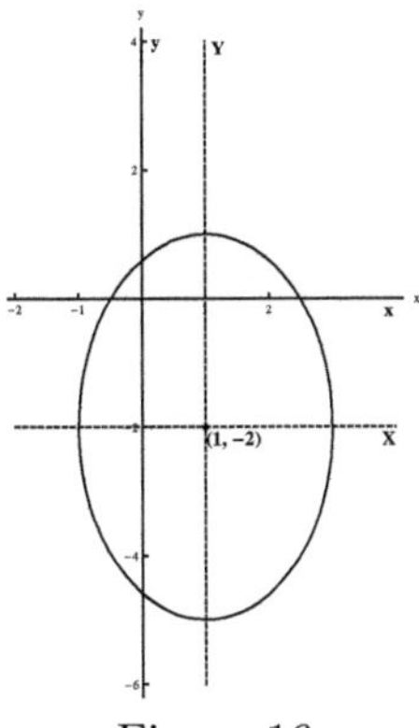

Figure 16

Example 10 Consider the equation

$$x^2 + 4x - 4y^2 + 24y = 48$$

a) Express the equation in a standard form and identify the graph of the equation as an ellipse or hyperbola.
b) Sketch the graph of the equation.
Solution
a) We complete the squares:

$$
\begin{aligned}
x^2 + 4x - 4y^2 + 24y &= x^2 + 4x - 4\left(y^2 - 6y\right) \\
&= (x + 2)^2 - 4 - 4(y - 3)^2 + 36 \\
&= (x + 2)^2 - 4(y - 3)^2 + 32.
\end{aligned}
$$

Therefore,

$$
\begin{aligned}
x^2 + 4x - 4y^2 + 24y = 48 &\Leftrightarrow (x + 2)^2 - 4(y - 3)^2 + 32 = 48 \\
&\Leftrightarrow (x + 2)^2 - 4(y - 3)^2 = 16 \\
&\Leftrightarrow \frac{(x + 2)^2}{16} - \frac{(y - 3)^2}{4} = 1.
\end{aligned}
$$

Thus, the graph of the equation is a hyperbola.

b) If we set $X = x + 2$ and $Y = y - 3$, the hyperbola is the graph of the equation

$$\frac{X^2}{16} - \frac{Y^2}{4} = 1$$

in the XY-plane. The introduction of the new coordinate system corresponds to the translation of the origin on the xy-plane to "the center of symmetries" of the hyperbola. Figure 17 shows the hyperbola.

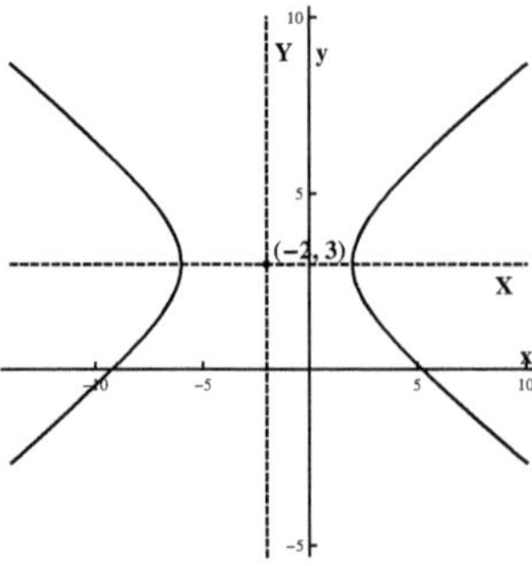

Figure 17

Problems

In problems 1 - 8, sketch the point with the given coordinates in the Cartesian coordinate plane (with the same scale on the horizontal and vertical axes). Identify the quadrant where the point is located.

1. $(3, 2)$

2 $(-3, 2)$

3. $(-3, -2)$

4. $(3, -2)$

5. $(5, 3)$

6. $(3, 5)$

7. $(3, 3)$

8. $(-3, -3)$

In problems 9-14,
a) Determine the point-slope form of the equation of the line that passes through the points P_1 and P_2 with basepoint P_1,
b) Determine the slope-intercept form of the equation of the line of part a).
c) Sketch the line.

9. $P_1 = (2, 1)$, $P_2 = (4, 3)$

10. $P_1 = (2, -3)$, $P_2 = (5, 1)$

11. $P_1 = (-4, -1)$, $P_2 = (-6, 2)$

12. $P_1 = (3, 4)$, $P_2 = (-1, 6)$

13. $P_1 = (-4, 2)$, $P_2 = (2, 6)$

14. $P_1 = (-2, -4)$, $P_2 = (2, 3)$

In problems 15-18, determine the point at which the given lines intersect, unless the lines are parallel.

15. $y = 3x + 4$, $y = 2x - 1$

16. $y = -2x + 5$, $y = 3x - 4$

17. $y = \frac{1}{2}x - 3$, $y = \frac{1}{2}x + 6$

18. $y = 5x - 2$, $y = 3x + 1$

In problems 19 and 20,
a) Show that the given lines are perpendicular,
b) Determine the points at which the lines intersect,
c) Sketch the lines in the coordinate plane with the same scale on the horizontal and vertical axes. Is the picture consistent with the claim that the lines are perpendicular?

19. $y = 4x + 5$, $y = -\dfrac{1}{4}x + 3$ $\qquad$ 20. $y = 3x - 4$, $y = -\dfrac{1}{3}x + 2$

In problems 21 - 24, calculate the distance between the points P_1 and P_2.

21. $P_1 = (3,1)$, $P_2 = (6,5)$ $\qquad$ 23. $P_1 = (-3,-2)$, $P_2 = (-1,4)$

22. $P_1 = (4,1)$, $P_2 = (2,5)$ $\qquad$ 24. $P_1 = (2,-4)$, $P_2 = (-3,5)$

In problems 25-30,
a) Determine the vertex of the parabola with the given equation,
b) Determine the points at which the parabola intersects the coordinate axes,
c) Sketch the graph of the parabola.

25. $y = x^2 + 4x + 7$ $\qquad$ 28. $y = -3x^2 - 6x + 1$

26. $y = 2x^2 - 4x - 3$ $\qquad$ 29. $x = y^2 - 4y - 2$

27. $y = -x^2 + 8x - 10$ $\qquad$ 30. $x = -3y^2 - 6y - 1$

In problems 31-36,
a) Complete the squares in order to identify the graph of the given equation as a circle, ellipse
or hyperbola,
c) Sketch the graph.

31. $x^2 - 6x + y^2 + 2y + 6 = 0$ $\qquad$ 34. $x^2 - 6x + 4y^2 - 16y + 9 = 0$

32. $x^2 + 8x + y^2 - 4y + 11 = 0$ $\qquad$ 35. $9x^2 - 72x - 25y^2 + 150y = 82$

33. $4x^2 - 16x + 9y^2 + 18y = 0$ $\qquad$ 36. $4x^2 + 40x - 9y^2 - 72y = 60$

A.6   Special Angles and Trigonometric Identities

This section of Appendix 1 augments the review of trigonometric functions in Section 1.2. We
review the "triangle picture", and discuss certain trigonometric identities.

In Section 1.2 we reviewed the radian measure: With reference to Figure 1, The length of the
arc AP on the unit circle is the radian measure of the angle AOP.

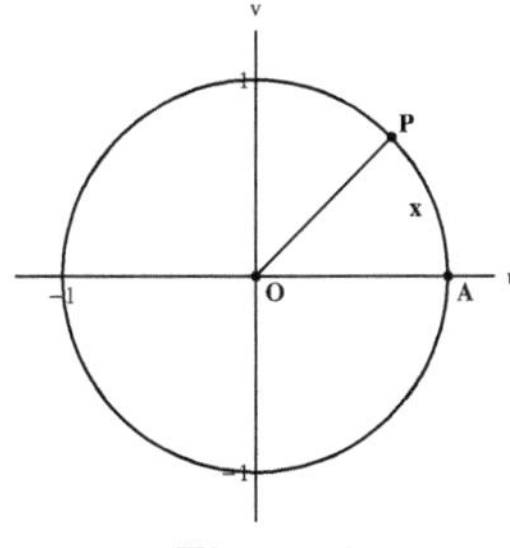

Figure 1

Figure 2 shows a circle of radius $r > 0$. With reference to Figure 2, the length of the arc AP is
$r\theta$ if the radian measure of the angle AOP is θ. The area of the circular sector AOP is

$$\frac{1}{2}r^2\theta.$$

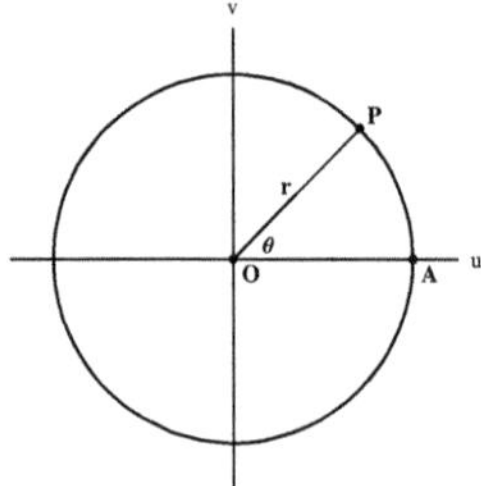

Figure 2

Let's review the connection between the unit circle picture and the triangle picture when we refer to sine, cosine and tangent.

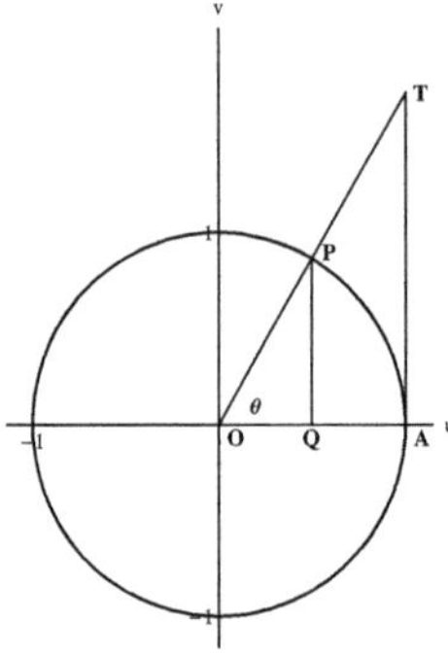

Figure 3

With reference to Figure 3, the angle AOP is θ (radians), so that the horizontal coordinate of P is $\cos(\theta)$ and the vertical coordinate of P is $\sin(\theta)$. We have

$$\frac{QP}{OP} = \frac{AT}{OT} \Rightarrow \sin(\theta) = \frac{AT}{OT} \left(= \frac{\text{opposite}}{\text{hypotenuse}} \right),$$

$$\frac{OQ}{OP} = \frac{OA}{OT} \Rightarrow \cos(\theta) = \frac{OA}{OT} \left(= \frac{\text{adjacent}}{\text{hypotenuse}} \right),$$

and

$$\tan(\theta) = \frac{\sin(\theta)}{\cos(\theta)} = \frac{QP}{OQ} = \frac{AT}{OA} \left(= \frac{\text{opposite}}{\text{adjacent}} \right),$$

by similar triangles.

Angles such as $\pi/3$ and $\pi/6$ are "special", since the exact values of the trigonometric functions corresponding to such angles are readily available. We can recall such values by referring to a couple of triangles. Let us begin with an equilateral triangle such that each side has length 2.

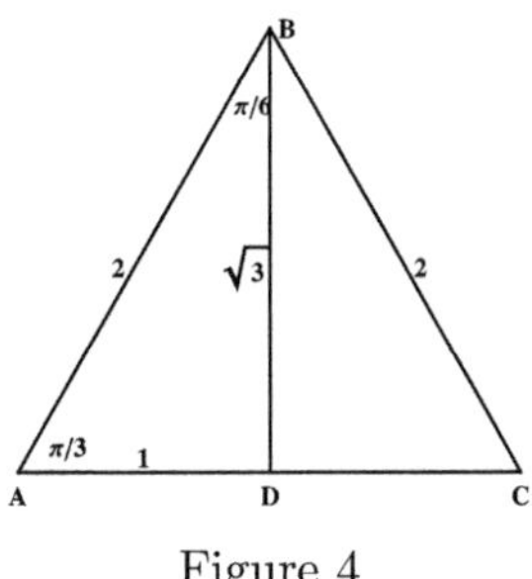

Figure 4

With reference to Figure 4, the line segment BD is the perpendicular bisector of the side AC. By the theorem of Pythagoras, the length of BD is $\sqrt{3}$. The angle CAB is $\pi/3$ (radians), and the angle ABD is $\pi/6$. We have

$$\sin\left(\frac{\pi}{6}\right) = \cos\left(\frac{\pi}{3}\right) = \frac{1}{2},$$

$$\cos\left(\frac{\pi}{6}\right) = \sin\left(\frac{\pi}{3}\right) = \frac{\sqrt{3}}{2},$$

$$\tan\left(\frac{\pi}{6}\right) = \frac{\sin\left(\frac{\pi}{6}\right)}{\cos\left(\frac{\pi}{6}\right)} = \frac{1}{\sqrt{3}} \text{ and } \tan\left(\frac{\pi}{3}\right) = \frac{\sin\left(\frac{\pi}{3}\right)}{\cos\left(\frac{\pi}{2}\right)} = \sqrt{3}.$$

Another special angle is $\pi/4$. Consider an isosceles right triangle as in Figure 5.

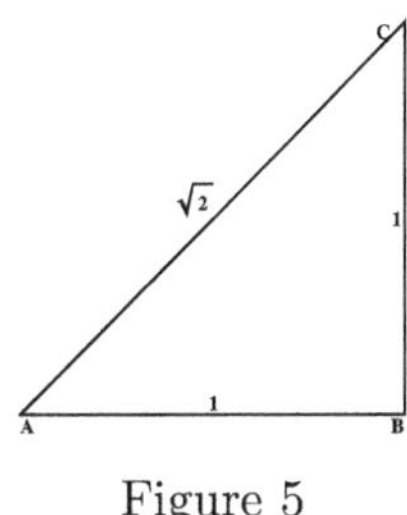

Figure 5

With reference to Figure 5, the length of the hypotenuse AC is $\sqrt{2}$ the theorem of Pythagoras. Therefore,

$$\sin\left(\frac{\pi}{4}\right) = \cos\left(\frac{\pi}{4}\right) = \frac{1}{\sqrt{2}}\left(=\frac{\sqrt{2}}{2}\right),$$

and

$$\tan\left(\frac{\pi}{4}\right) = 1.$$

We can determine the values of trigonometric functions at other angles that are related to $\pi/6$, $\pi/3$ or $\pi/4$ with the help of the "unit circle picture".

Example 1 Indicate the point on the unit circle that corresponds to x (radians) and determine $\sin(x)$, $\cos(x)$ and $\tan(x)$ if

a)

$$x = \frac{2\pi}{3},$$

b)

$$x = \frac{11\pi}{4}.$$

Solution

a) We have

$$\frac{2\pi}{3} = \pi - \frac{\pi}{3}.$$

Figure 6 shows the point P corresponding to $2\pi/3$.

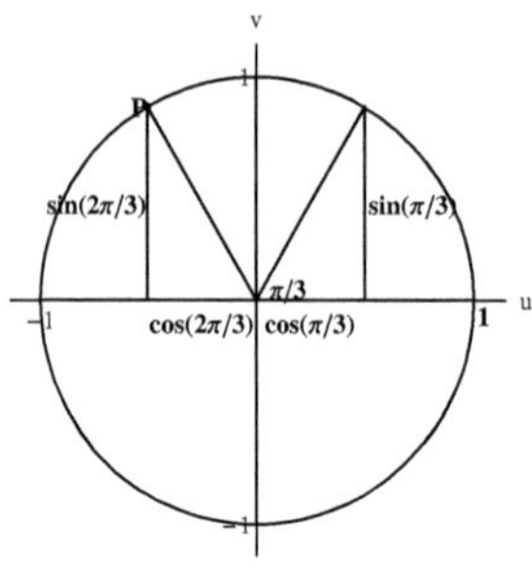

Figure 6

With reference to Figure 6,

$$\sin\left(\frac{2\pi}{3}\right) = \sin\left(\frac{\pi}{3}\right) = \frac{\sqrt{3}}{2}, \quad \cos\left(\frac{2\pi}{3}\right) = -\cos\left(\frac{\pi}{3}\right) = -\frac{1}{2}, \quad \tan\left(\frac{2\pi}{3}\right) = -\sqrt{3}.$$

c) We have

$$x = \frac{11\pi}{4} = \frac{8}{4}\pi + \frac{3}{4}\pi = 2\pi + \frac{3\pi}{4}.$$

Figure 7 shows the point that corresponds to x.

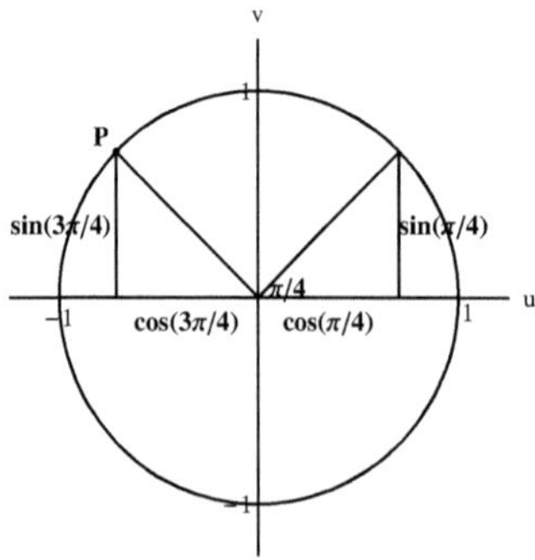

Figure 7

With reference to Figure 7,

$$\sin\left(\frac{11\pi}{4}\right) = \sin\left(\frac{3\pi}{4}\right) = \sin\left(\pi - \frac{\pi}{4}\right) = \sin\left(\frac{\pi}{4}\right) = \frac{\sqrt{2}}{2},$$

$$\cos\left(\frac{11\pi}{4}\right) = \cos\left(\frac{3\pi}{4}\right) = \cos\left(\pi - \frac{\pi}{4}\right) = -\cos\left(\frac{\pi}{4}\right) = -\frac{\sqrt{2}}{2},$$

$$\tan\left(\frac{11\pi}{4}\right) = \frac{\sin\left(\frac{11\pi}{4}\right)}{\cos\left(\frac{11\pi}{4}\right)} = -1.$$

☐

We will make use of some identities that involve sine and cosine. We use traditional notation:

$$\sin^2(x) = (\sin(x))^2 \text{ and } \cos^2(x) = (\cos(x))^2.$$

Proposition 1 We have

$$\cos^2(x) + \sin^2(x) = 1$$

for each real number x.

Proof

Proposition 1 is an immediate consequence of the definition of sine and cosine: The point $P = (\cos(x), \sin(x))$ is on the unit circle, so that the square of its distance from the origin is 1:

$$\cos^2(x) + \sin^2(x) = (\cos(x))^2 + (\sin(x))^2 = 1.$$

■

It is not true that $\sin(x_1 + x_2) = \sin(x_1) + \sin(x_2)$ or $\cos(x_1 + x_2) = \cos(x_1) + \cos(x_2)$ for each pair of real numbers x_1 and x_2. On the other hand, we have **the addition** (and subtraction) **formulas for sine and cosine**, and these formulas will turn out to be very useful:

Proposition 2 Let a and b be arbitrary real numbers. We have

$$\begin{aligned}
\sin(a + b) &= \sin(a)\cos(b) + \cos(a)\sin(b),\\
\sin(a - b) &= \sin(a)\cos(b) - \cos(a)\sin(b),\\
\cos(a + b) &= \cos(a)\cos(b) - \sin(a)\sin(b),\\
\cos(a - b) &= \cos(a)\cos(b) + \sin(a)\sin(b).
\end{aligned}$$

Proof

Let

$$P = (\cos(a), \sin(a)) \text{ and } Q = (\cos(b), \sin(b)).$$

With reference to Figure 8, P goes to P' and Q goes to Q' by a rotation of $-b$, where

$$P' = (\cos(a - b), \sin(a - b)) \text{ and } Q' = (1, 0).$$

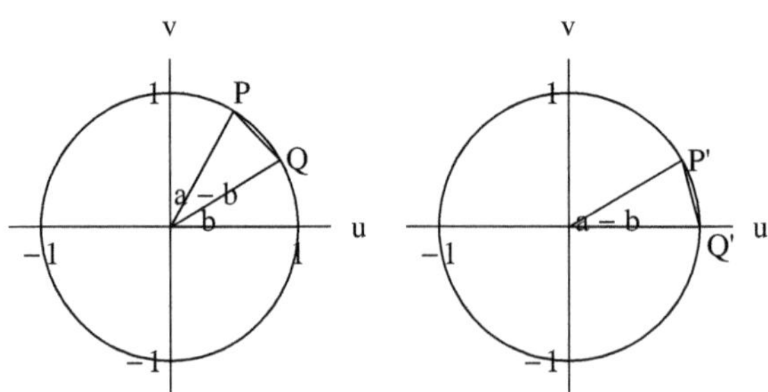

Figure 8

Since rotations preserve distances, we have

$$\text{length of the line segment } PQ = \text{length of the line segment } P'Q'.$$

Now

$$(\text{length of the line segment } PQ)^2 = (\cos(a) - \cos(b))^2 + (\sin(a) - \sin(b))^2,$$

and

$$(\text{length of the line segment } P'Q')^2 = (\cos(a - b) - 1)^2 + \sin^2(a - b).$$

Therefore,

$$(\cos(a) - \cos(b))^2 + (\sin(a) - \sin(b))^2 = (\cos(a - b) - 1)^2 + \sin^2(a - b).$$

We have

$$(\cos(a - b) - 1)^2 + \sin^2(a - b) = \cos^2(a - b) - 2\cos(a - b) + 1 + \sin^2(a - b)$$
$$= 2 - 2\cos(a - b).$$

and

$$(\cos(a) - \cos(b))^2 + (\sin(a) - \sin(b))^2 = \cos^2(a) - 2\cos(a)\cos(b) + \cos^2(b)$$
$$+ \sin^2(a) - 2\sin(a)\sin(b) + \sin^2(b)$$
$$= 2 - 2\cos(a)\cos(b) - 2\sin(a)\sin(b).$$

Therefore,

$$2 - 2\cos(a - b) = 2 - 2\cos(a)\cos(b) - 2\sin(a)\sin(b),$$

so that

$$\cos(a - b) = \cos(a)\cos(b) + \sin(a)\sin(b),$$

as claimed. If we replace b by $-b$, we obtain the addition formula for cosine, since cosine is even and sine is odd:

$$\cos(a + b) = \cos(a)\cos(-b) + \sin(a)\sin(-b) = \cos(a)\cos(b) - \sin(a)\sin(b).$$

As in Section 1.2, the fact that

$$\sin(x) = \cos(x - \frac{\pi}{2})$$

follows from the difference formula for cosine. Thus,

$$\sin\left(\frac{\pi}{2} - x\right) = \cos\left(\frac{\pi}{2} - x - \frac{\pi}{2}\right) = \cos(-x) = \cos(x).$$

Therefore,

$$\sin\left(a+b\right) = \cos\left(a+b-\frac{\pi}{2}\right)$$
$$= \cos\left(a-\frac{\pi}{2}\right)\cos\left(b\right) - \sin\left(a-\frac{\pi}{2}\right)\sin\left(b\right)$$
$$= \sin\left(a\right)\cos\left(b\right) + \sin\left(\frac{\pi}{2}-a\right)\sin\left(b\right)$$
$$= \sin\left(a\right)\cos\left(b\right) + \cos\left(a\right)\sin\left(b\right).$$

This is the addition formula of sine. The difference formula for sine follows:

$$\sin\left(a-b\right) = \sin\left(a\right)\cos\left(-b\right) - \cos\left(a\right)\sin\left(-b\right) = \sin\left(a\right)\cos\left(b\right) + \cos\left(a\right)\sin\left(b\right).$$

■

We will have occasion to use the following consequences of the addition formulas:

$$\mathbf{\sin(2x) = 2\sin(x)\cos(x),}$$

$$\mathbf{\cos(2x) = \cos^2(x) - \sin^2(x),}$$
$$\mathbf{\sin^2(x) = \frac{1 - \cos^2(2x)}{2},}$$
$$\mathbf{\cos^2(x) = \frac{1 + \cos(2x)}{2}.}$$

The first identity follows from the addition formula for sine:

$$\sin\left(2x\right) = \sin\left(x+x\right) = \sin\left(x\right)\cos\left(x\right) + \cos\left(x\right)\sin\left(x\right) = 2\sin\left(x\right)\cos\left(x\right).$$

The second identity follows from the addition formula for cosine:

$$\cos\left(2x\right) = \cos\left(x+x\right) = \cos\left(x\right)\cos\left(x\right) - \sin\left(x\right)\sin\left(x\right) = \cos^2\left(x\right) - \sin^2\left(x\right).$$

If we replace $\cos^2\left(x\right)$ by $1 - \sin^2\left(x\right)$ in the above expression, we obtain

$$\cos\left(2x\right) = \left(1 - \sin^2\left(x\right)\right) - \sin^2\left(x\right) = 1 - 2\sin^2\left(x\right),$$

so that

$$\sin^2\left(x\right) = \frac{1 - \cos\left(2x\right)}{2},$$

as claimed. Similarly, if we replace $\sin^2\left(x\right)$ by $1 - \cos^2\left(x\right)$ in the identity $\cos\left(2x\right) = \cos^2\left(x\right) - \sin^2\left(x\right)$, we obtain

$$\cos\left(2x\right) = \cos^2\left(x\right) - \left(1 - \cos^2\left(x\right)\right) = -1 + 2\cos^2\left(x\right),$$

so that

$$\cos^2\left(x\right) = \frac{1 + \cos\left(2x\right)}{2}.$$

We will have occasion to refer to the **law of cosines**:

Proposition 3 With reference to Figure 9,

$$\mathbf{a^2 = b^2 + c^2 - 2bc\cos(\theta).}$$

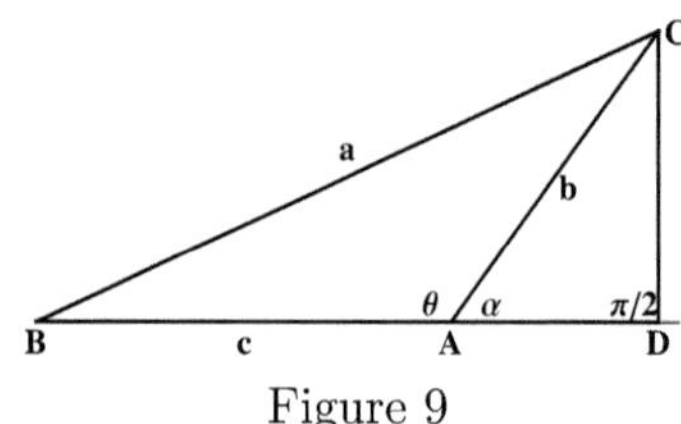

Figure 9

Proof

By the Theorem of Pythagoras,

$$a^2 = \left(b\cos\left(\alpha\right) + c\right)^2 + \left(b\sin\left(\alpha\right)^2\right).$$

Therefore,

$$\begin{aligned}
a^2 &= b^2\cos^2\left(\alpha\right) + 2bc\cos\left(\alpha\right) + c^2 + b\sin^2\left(\alpha\right) \\
&= b^2\left(\cos^2\left(\alpha\right) + \sin^2\left(\alpha\right)\right) + c^2 + 2bc\cos\left(\pi - \theta\right) \\
&= b^2 + c^2 + 2bc\left(\cos\left(\pi\right)\cos\left(\theta\right) - \sin\left(\pi\right)\sin\left(\theta\right)\right) \\
&= b^2 + c^2 - 2bc\cos\left(\theta\right).
\end{aligned}$$

■

Problems

In problems 1 - 6, indicate the point on the unit circle that corresponds to x (radians) and determine $\sin\left(x\right)$ and $\cos\left(x\right)$.

1. $x = \dfrac{3\pi}{2}$

2. $x = -\pi$

3. $x = \dfrac{4\pi}{3}$

4. $x = \dfrac{5\pi}{6}$

5. $x = -\dfrac{\pi}{3}$

6. $x = -\dfrac{\pi}{4}$

In problems 7-14, determine $\sin\left(x\right)$, $\cos\left(x\right)$, $\tan\left(x\right)$ and $\sec\left(x\right)$.

7. $x = \dfrac{3\pi}{4}$

8. $x = \dfrac{5\pi}{2}$

9. $x = \dfrac{7\pi}{2}$

10. $x = \dfrac{5\pi}{4}$

11. $x = \dfrac{7\pi}{6}$

12. $x = \dfrac{7\pi}{3}$

13. $x = -\dfrac{2\pi}{3}$

14. $x = -\dfrac{5\pi}{4}$

In problems 15 - 18, transform the given expression to an expression that is a linear combination of products of powers of $\sin\left(x\right)$ and $\cos\left(x\right)$ by making use of the basic identities.

15. $\cos(4x)$

16. $\sin(4x)$

17. $\cos(3x)$

18. $\sin(3x)$

Appendix B

Some Theorems on Limits and Continuity

You must be familiar with the precise definitions of continuity and the limits, as discussed in Section 1.4, in order to be able to read this section.

Finite Limits and Continuity

Theorem 1 (Limits of arithmetic combinations of functions)

Assume that $\lim_{x \to a} f(x)$ and $\lim_{x \to a} g(x)$ exist. Then,
a)
$$\lim_{x \to a} (f(x) + g(x)) = \lim_{x \to a} f(x) + \lim_{x \to a} g(x),$$

b)
$$\lim_{x \to a} f(x)g(x) = \left(\lim_{x \to a} f(x) \right) \left(\lim_{x \to a} g(x) \right),$$

c)
$$\lim_{x \to a} \frac{f(x)}{g(x)} = \frac{\lim_{x \to a} f(x)}{\lim_{x \to a} g(x)} \quad \text{if } \lim_{x \to a} g(x) \neq 0.$$

Proof

Let $\lim_{x \to a} f(x) = L_1$ and $\lim_{x \to a} g(x) = L_2$.
a) Given $\varepsilon > 0$, there exists $\delta_1 > 0$ and $\delta_2 > 0$ such that

$$|f(x) - L_1| < \frac{\varepsilon}{2} \text{ if } 0 < |x - a| < \delta_1,$$

and

$$|g(x) - L_2| < \frac{\varepsilon}{2} \text{ if } 0 < |x - a| < \delta_2.$$

Set $\delta = \min(\delta_1, \delta_2)$. If $0 < |x - a| < \delta$,

$$
\begin{aligned}
|(f(x) + g(x)) - (L_1 + L_2)| &= |(f(x) - L_1) + (g(x) - L_2)| \\
&\leq |f(x) - L_1| + |g(x) - L_2| \text{ (by the triangle inequality)} \\
&< \frac{\varepsilon}{2} + \frac{\varepsilon}{2} = \varepsilon.
\end{aligned}
$$

Therefore,

$$\lim_{x \to a} (f(x) + g(x)) = L_1 + L_2 = \lim_{x \to a} f(x) + \lim_{x \to a} g(x),$$

as claimed.

b) We have

$$
\begin{aligned}
|f(x)g(x) - L_1 L_2| &= |f(x)g(x) - L_1 g(x) + L_1 g(x) - L_1 L_2| \\
&= |(f(x) - L_1)g(x) + L_1(g(x) - L_2)| \\
&\leq |f(x) - L_1| |g(x)| + |L_1| |g(x) - L_2| \text{ (by the triangle inequality).}
\end{aligned}
$$

Choose $\delta_0 > 0$ so that

$$0 < |x - a| < \delta_0 \Rightarrow |g(x) - L_2| < 1.$$

Then,

$$|g(x)| = |g(x) - L_2 + L_2| \leq |g(x) - L_2| + |L_2| < 1 + |L_2|.$$

Thus, if $0 < |x - a| < \delta_0$, we have

$$
\begin{aligned}
|f(x)g(x) - L_1 L_2| &\leq |f(x) - L_1| |g(x)| + |L_1| |g(x) - L_2| \\
&< (1 + |L_2|) |f(x) - L_1| + |L_1| |g(x) - L_2|.
\end{aligned}
$$

Given $\varepsilon > 0$, pick $\delta > 0$ such that $\delta < \delta_0$ and

$$0 < |x - a| < \delta \Rightarrow |f(x) - L_1| < \frac{\varepsilon}{2(1 + |L_2|)} \text{ and } |g(x) - L_2| < \frac{\varepsilon}{2(1 + |L_1|)}.$$

If $0 < |x - a| < \delta$, we have

$$
\begin{aligned}
|f(x)g(x) - L_1 L_2| &< (1 + |L_2|) |f(x) - L_1| + |L_1| |g(x) - L_2| \\
&< (1 + |L_2|) \left(\frac{\varepsilon}{2(1 + |L_2|)} \right) + |L_1| \left(\frac{\varepsilon}{2(1 + |L_1|)} \right) \\
&< \frac{\varepsilon}{2} + \frac{\varepsilon}{2} = \varepsilon.
\end{aligned}
$$

Therefore, $\lim_{x \to a} f(x) g(x) = L_1 L_2 = (\lim_{x \to a} f(x))(\lim_{x \to a} g(x))$.

c) We assume that $L_2 = \lim_{x \to a} g(x) \neq 0$. It is enough to show that

$$\lim_{x \to a} \frac{1}{g(x)} = \frac{1}{\lim_{x \to a} g(x)} = \frac{1}{L_2},$$

since

$$\lim_{x \to a} \frac{f(x)}{g(x)} = \lim_{x \to a} \left(f(x) \frac{1}{g(x)} \right) = \left(\lim_{x \to a} f(x) \right) \left(\lim_{x \to a} \frac{1}{g(x)} \right),$$

by part b).

Now,

$$\left| \frac{1}{g(x)} - \frac{1}{L_2} \right| = \left| \frac{L_2 - g(x)}{g(x) L_2} \right| = \frac{|g(x) - L_2|}{|g(x)| |L_2|}.$$

Pick δ_0 so that

$$|g(x) - L_2| < \frac{|L_2|}{2}$$

if $0 < |x - a| < \delta_0$. Then,

$$\begin{aligned}
|g(x)| = |(g(x) - L_2) + L_2| &= |L_2 - (L_2 - g(x))| \\
&\geq |L_2| - |L_2 - g(x)| \\
&> |L_2| - \frac{|L_2|}{2} = \frac{|L_2|}{2},
\end{aligned}$$

with the help of the corollary to the triangle inequality. Therefore,

$$\left| \frac{1}{g(x)} - \frac{1}{L_2} \right| = \frac{|g(x) - L_2|}{|g(x)||L_2|} < \frac{|g(x) - L_2|}{\left(\frac{|L_2|}{2}\right)|L_2|} = \left(\frac{2}{L_2^2}\right)|g(x) - L_2|$$

if $0 < |x - a| < \delta_0$.
Given $\varepsilon > 0$, pick $\delta < \delta_0$, so that

$$|g(x) - L_2| < \left(\frac{L_2^2}{2}\right)\varepsilon$$

if $0 < |x - a| < \delta$. Then,

$$\left| \frac{1}{g(x)} - \frac{1}{L_2} \right| < \left(\frac{2}{L_2^2}\right)|g(x) - L_2| < \left(\frac{2}{L_2^2}\right)\left(\frac{L_2^2}{2}\right)\varepsilon = \varepsilon.$$

Therefore,

$$\lim_{x \to a} \frac{1}{g(x)} = \frac{1}{L_2} = \frac{1}{\lim_{x \to a} g(x)},$$

as claimed. ∎

Theorem 2 (The limit of a composite function) Assume that $\lim_{x \to a} g(x) = L$ and f is continuous at L. Then,

$$\lim_{x \to a} (f \circ g)(x) = \lim_{x \to a} f(g(x)) = f(L).$$

Proof

Let $\varepsilon > 0$ be given. Since f is continuous at L, there exists $\delta_0 > 0$ such that $|f(u) - f(L)| < \varepsilon$ if $|u - L| < \delta_0$. Since $\lim_{x \to a} g(x) = L$, there exists $\delta > 0$ such that $|g(x) - L| < \delta_0$ if $0 < |x - a| < \delta$. Therefore,

$$0 < |x - a| < \delta \Rightarrow |g(x) - L| < \delta_0 \Rightarrow |f(g(x)) - f(L)| < \varepsilon.$$

Thus,

$$\lim_{x \to a} f(g(x)) = f(L) = f\left(\lim_{x \to a} g(x)\right),$$

as claimed. ∎

Theorem 3 (The Squeeze Theorem) Assume that $h(x) \leq f(x) \leq g(x)$ for all $x \neq a$ in an open interval containing a, and

$$\lim_{x \to a} h(x) = \lim_{x \to a} g(x) = L.$$

Then $\lim_{x \to a} f(x) = L$ as well.

Proof

Let $\varepsilon > 0$ be given. Since $\lim_{x \to a} h(x) = L$, there exists $\delta_1 > 0$ such that

$$0 < |x - a| < \delta_1 \Rightarrow |h(x) - L| < \varepsilon.$$

Therefore,,

$$0 < |x - a| < \delta_1 \Rightarrow L - \varepsilon < h(x).$$

Since $\lim_{x \to a} g(x) = L$, there exists $\delta_2 > 0$ such that

$$0 < |x - a| < \delta_2 \Rightarrow |g(x) - L| < \varepsilon.$$

Therefore,

$$x \neq a \text{ and } |x - a| < \delta_2 \Rightarrow g(x) < L + \varepsilon.$$

Let us set $\delta = \min(\delta_1, \delta_2)$ (i.e., the minimum of δ_1 and δ_2). If $0 < |x - a| < \delta$,

$$L - \varepsilon < h(x) \le f(x) \le g(x) < L + \varepsilon,$$

so that

$$L - \varepsilon < f(x) < L + \varepsilon.$$

Thus,

$$0 < \ |x - a| < \delta \Rightarrow |f(x) - L| < \varepsilon.$$

Therefore, $\lim_{x \to a} f(x) = L$, as claimed. $\blacksquare$

Theorem 4 (The Continuity of Rational Powers of x) Assume that $f(x) = x^r$, where r is a rational number. Then f is continuous on any interval that is contained in its natural domain.

Proof

Let n be a positive integer and let m be an arbitrary integer.

a) If $f(x) = x^n$, then f is continuous on the entire number line:
Let a be an arbitrary real number. We have

$$f(a + h) = a^n + na^{n-1}h + \frac{n(n-1)}{2}a^{n-2}h^2 + \cdots + h^n.$$

Therefore,

$$\begin{aligned}
\lim_{h \to 0} f(a + h) &= a^n + \left(na^{n-1}\right) \lim_{h \to 0} h + \frac{n(n-1)}{2}a^{n-2} \lim_{h \to 0} h^2 + \cdots + \lim_{h \to 0} h^n \\
&= a^n = f(a).
\end{aligned}$$

b) If $f(x) = x^{-n} = 1/x^n$, then f is continuous on $(-\infty, 0)$ and $(0 + \infty)$, since f is the quotient of the continuous functions defined by 1 and x^n and $x^n \neq 0$ if $x \neq 0$.

c) Let $f(x) = x^{1/n}$. We will show that f is continuous on $[0, +\infty)$ (if n is odd, the proof of the continuity of f on $(-\infty, 0]$ is similar). Let's begin by considering $a > 0$. We will make use of the identity

$$B^n - A^n = (B - A)\left(B^{n-1} + B^{n-2}A + B^{n-3}A^2 + \cdots + A^{n-1}\right).$$

Let's replace B by $x^{1/n}$ and A by $a^{1/n}$:

$$x - a = \left(x^{1/n} - a^{1/n}\right)\left(\left(x^{1/n}\right)^{n-1} + \left(x^{1/n}\right)^{n-2}\left(a^{1/n}\right) + \left(x^{1/n}\right)^{n-3}\left(a^{1/n}\right)^2 + \cdots + \left(a^{1/n}\right)^{n-1}\right)$$

$$= \left(x^{1/n} - a^{1/n}\right)\left(x^{(n-1)/n} + x^{(n-2)/n}a^{1/n} + x^{(n-3)/n}a^{2/n} + \cdots a^{(n-1)/n}\right).$$

Therefore,

$$x^{1/n} - a^{1/n} = \frac{x - a}{x^{(n-1)/n} + x^{(n-2)/n}a^{1/n} + x^{(n-3)/n}a^{2/n} + \cdots a^{(n-1)/n}}.$$

We will restrict x so that

$$|x - a| < \frac{a}{2}.$$

Then $x > a/2 > 0$. Thus,

$$\left|x^{1/n} - a^{1/n}\right| \le \frac{|x - a|}{a^{(n-1)/n}}.$$

Therefore, given $\varepsilon > 0$, in order to ensure that $\left|x^{1/n} - a^{1/n}\right| < \varepsilon$, it is sufficient to have $|x - a| < a/2$ and

$$\frac{|x - a|}{a^{(n-1)/n}} < \varepsilon \Leftrightarrow |x - a| < a^{(n-1)/n}\varepsilon$$

With reference to the precise definition of continuity, we can set

$$\delta = \min\left(a^{(n-1)/n}\varepsilon, \frac{a}{2}\right),$$

so that $|f(x) - f(a)| = \left|x^{1/n} - a^{1/n}\right| < \varepsilon$ if $|x - a| < \delta$. Therefore, f is continuous at a. As for $a = 0$, given $\varepsilon > 0$,

$$|f(x) - f(0)| = \left|x^{1/n} - 0^{1/n}\right| = x^{1/n} < \varepsilon$$

if $0 \le x < \varepsilon^n$. Therefore, f is continuous at 0 from the right (the continuity is two-sided if n is odd).

d) Finally, assume that $f(x) = x^r$, where r is an arbitrary rational number. Set $r = m/n$, where m is an integer and $|m|$ and n do not have any common factor other than 1. Since $f(x) = x^{m/n} = \left(x^{1/n}\right)^m$, we can express f as $F \circ G$, where $G(x) = x^{1/n}$ and $F(u) = u^m$. Since the composite function $F \circ G$ is continuous at a if G is continuous at a and F is continuous at $G(a)$, the continuity of f on its natural domain is obtained from the results of the previous steps. ∎

Theorem 5 (The Continuity of Sine and Cosine)

Proof

The proof depends on the following inequalities:

$$|\sin(a + h) - \sin(a)| < |h| \text{ and } |\cos(a + h) - \cos(a)| < |h|$$

for an arbitrary real number a and $h \neq 0$.

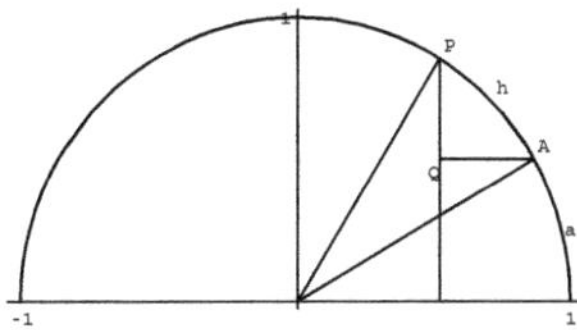

Figure 1

With reference to Figure 1 , the length of the line segment $\overline{QP}$ should be less than the length of the arc AP. But the length of $\overline{QP}$ is

$$\left|\sin\left(a+h\right)-\sin(a)\right| = \sin\left(a+h\right)-\sin(a),$$

and the length of the arc AP is h. Therefore, we have

$$\left|\sin\left(a+h\right)-\sin(a)\right| < |h|\,.$$

Similarly,

$$\left|\cos\left(a+h\right)-\cos\left(a\right)\right| = \cos\left(a\right)-\cos\left(a+h\right) = \text{ length of } \overline{QA},$$

and the length of the line segment $\overline{QA}$ is less than the length of the arc AP. Therefore, we have

$$\left|\cos\left(a+h\right)-\cos\left(a\right)\right| < |h|\,.$$

Thus, given any $\varepsilon > 0$, if $|h| < \varepsilon$ we have

$$\left|\sin\left(a+h\right)-\sin(a)\right| < |h| < \varepsilon,$$

and

$$\left|\cos\left(a+h\right)-\cos\left(a\right)\right| < |h| < \varepsilon$$

$(\delta = \varepsilon)$. Therefore, sine and cosine are continuous at a. $\blacksquare$

Theorem 6 Polynomials, rational functions, sine, cosine, tangent and secant are continuous on their respective natural domains.

Proof

Let $P\left(x\right)$ be a polynomial of order n so that

$$P\left(x\right) = a_0 + a_1 x + a_2 x^2 + \cdots + a_n x^n,$$

where the coefficients $a_0, a_1, a_2, \ldots, a_n$ are given numbers. Since a positive integer power of x and a constant define continuous functions on $\mathbf{R}$, the same is true for each term $a_k x^k$, $k = 0, 1, \ldots, n$, and therefore for $P\left(x\right)$ which is the sum of these terms.

A rational function is a quotient of polynomials. Therefore such a function is continuous at a point where the denominator does not vanish, i.e., at a point in the natural domain of the function.

We have already established the continuity of sine and cosine on the entire number line. Since

$$\tan\left(x\right) = \frac{\sin\left(x\right)}{\cos\left(x\right)} \text{ and } \sec\left(x\right) = \frac{1}{\cos\left(x\right)},$$

these functions are continuous at any x such that $\cos\left(x\right) \neq 0$. $\blacksquare$

Infinite Limits and Limits at Infinity

Proposition 1
a) Assume that $f\left(x\right) > 0$ provided that $x > a$ and x is sufficiently close to a and $\lim_{x\to a+} f\left(x\right) = 0$. Then

$$\lim_{x\to a+} \frac{1}{f\left(x\right)} = +\infty.$$

b) Assume that $f(x) < 0$ provided that $x > a$ and x is sufficiently close to a and $\lim_{x \to a+} f(x) = 0$. Then

$$\lim_{x \to a+} \frac{1}{f(x)} = -\infty.$$

Proof

a) By the given conditions on f, given $M > 0$, there exists $\delta > 0$ such that $0 < f(x) < 1/M$ if $a < x < a + \delta$. Then,

$$\frac{1}{f(x)} > M.$$

Therefore, $\lim_{x \to a+} f(x) = +\infty$.

Part b) follows from part a), since

$$\lim_{x \to a+} \frac{1}{f(x)} = -\infty \Leftrightarrow \lim_{x \to a+} \left(-\frac{1}{f(x)}\right) = +\infty.$$

∎

Proposition 2

a) If $\lim_{x \to a+} f(x) > 0$ or $\lim_{x \to a+} f(x) = +\infty$, and $\lim_{x \to a+} g(x) = +\infty$ then,

$$\lim_{x \to a+} f(x)\, g(x) = +\infty.$$

b) If $\lim_{x \to a+} f(x) > 0$ or $\lim_{x \to a+} f(x) = +\infty$, and $\lim_{x \to a+} g(x) = -\infty$ then,

$$\lim_{x \to a+} f(x)\, g(x) = -\infty.$$

Proof

a) Assume that $\lim_{x \to a+} f(x) = L > 0$ (finite). Given $M > 0$, there exists $\delta > 0$ such that

$$a < x < a + \delta \Rightarrow |f(x) - L| < \frac{L}{2} \text{ and } g(x) > \frac{2M}{L}$$

Then,

$$f(x) > \frac{L}{2},$$

so that

$$f(x)\, g(x) > \frac{L}{2}\left(\frac{2M}{L}\right) = M.$$

Therefore, $\lim_{x \to a+} f(x)\, g(x) = +\infty$.
Now assume that $\lim_{x \to a+} f(x) = +\infty$. Given $M > 0$, there exists $\delta > 0$ such that

$$a < x < a + \delta \Rightarrow f(x) > \sqrt{M} \text{ and } g(x) > \sqrt{M}.$$

Then,

$$f(x)\, g(x) > \left(\sqrt{M}\right)\left(\sqrt{M}\right) = M.$$

Therefore, $\lim_{x \to a+} f(x)\, g(x) = +\infty$.
Part b) follows from part a), since $\lim_{x \to a+} f(x) = -\infty \Leftrightarrow \lim_{x \to a+} (-f(x)) = +\infty.$∎

Proposition 3

a) If $\lim_{x \to a+} f(x) = L$, where L is finite, or $\lim_{x \to a+} f(x) = +\infty$, and $\lim_{x \to a+} g(x) = +\infty$, then

$$\lim_{x \to a+} (f(x) + g(x)) = +\infty.$$

b) If $\lim_{x \to a+} f(x) = L$, **where** L **is finite, or** $\lim_{x \to a+} f(x) = -\infty$, **and** $\lim_{x \to a+} g(x) = -\infty$, **then**

$$\lim_{x \to a+} (f(x) + g(x)) = -\infty.$$

Proof

a) Assume that $\lim_{x \to a+} f(x) = L$, where L is finite. Let $M > 0$ be given. We can assume that $M > |L|$ so that $M - L > 0$. There exists $\delta > 0$ such that

$$a < x < a + \delta \Rightarrow f(x) > L - 1 \text{ and } g(x) > M - L + 1.$$

Then,

$$f(x) + g(x) > (L - 1) + (M - L + 1) = M.$$

Therefore, $\lim_{x \to a+} (f(x) + g(x)) = +\infty$.
Now assume that $\lim_{x \to a+} f(x) = +\infty$. Given $M > 0$, there exists $\delta > 0$ such that

$$a < x < a + \delta \Rightarrow f(x) > \frac{M}{2} \text{ and } g(x) > \frac{M}{2}.$$

Then, $f(x) + g(x) > M$. Therefore, $\lim_{x \to a+} (f(x) + g(x)) = +\infty$.
Part b) follows from part a), since

$$\lim_{x \to a+} f(x) = -\infty \Leftrightarrow \lim_{x \to a+} (-f(x)) = +\infty.$$

$\blacksquare$

Theorem 7 (Sequential Characterization of Continuity) If a function f is continuous at a point a, and $\lim_{n \to \infty} a_n = a$, then

$$\lim_{n \to \infty} f(a_n) = f(a).$$

Proof

Let $\varepsilon > 0$ be given. Since f is continuous at a, there exists $\delta > 0$ such that $|f(x) - f(a)| < \varepsilon$ if $|x - a| < \delta$. Since $\lim_{n \to \infty} a_n = a$, there exists a positive integer N such that $|a_n - a| < \delta$ if $n \geq N$. Thus,

$$n \geq N \Rightarrow |a_n - a| < \delta \Rightarrow |f(a_n) - f(a)| < \varepsilon.$$

Therefore, $\lim_{n \to \infty} f(a_n) = f(a)$, as claimed. $\blacksquare$

Appendix C

The Continuity of an Inverse Function

Theorem Assume that f is strictly increasing or decreasing and continuous on the interval J. The range I of f is also an interval. The inverse of f exists and f^{-1} is continuous on I. The function f^{-1} is increasing if f is increasing, and decreasing if f is decreasing.

Proof

We will assume that f is increasing on its domain, the interval J (the case of a decreasing function is similar).

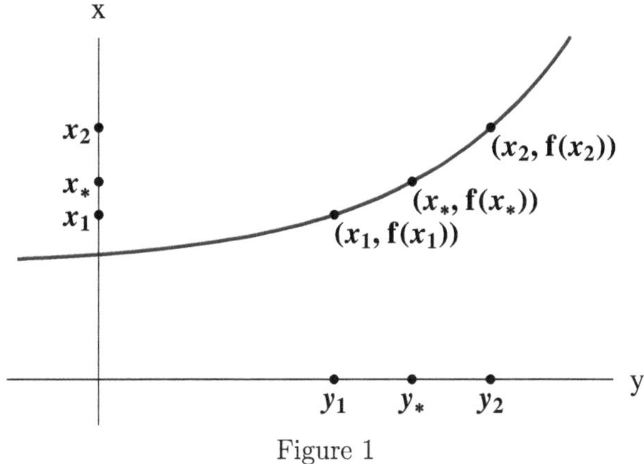

Figure 1

Let's first show that the range of f is an interval. Thus, assume that $x_1 = f(y_1)$ and $x_2 = f(y_2)$ are points in the range of f, and that $x_1 < x_2$. Let $x_* \in (x_1, x_2) = (f(y_1), f(y_2))$. We must show that x_* is in the range of f. Indeed, By the Intermediate Value Theorem, there exists $y_* \in J$ such that $f(y_*) = x_*$

Now we must show that the solution of the equation $x_* = f(y)$ is unique in J for each x_* in the range I of f. Indeed, if $f(y_*) = f(y_{**}) = x_*$, where y_* and y_{**} are in J, we must have $y_* = y_{**}$ since f is increasing on J: If $y_* < y_{**}$, then $f(y_*) < f(y_{**})$, and if $y_{**} < y_*$, then $f(y_{**}) < f(y_*)$.

Now we are entitled to speak of the inverse f^{-1} of f:

$$y = f^{-1}(x) \text{ for any } x \in I \Leftrightarrow x = f(y), \text{ where } y \in J$$

(You can easily show that f^{-1} is increasing).

The proof of the continuity of f^{-1} is somewhat more involved. We will consider the case of a point a in the interior of I (the appropriate one-sided continuity is discussed in a similar manner at an endpoint of I which belongs to I).

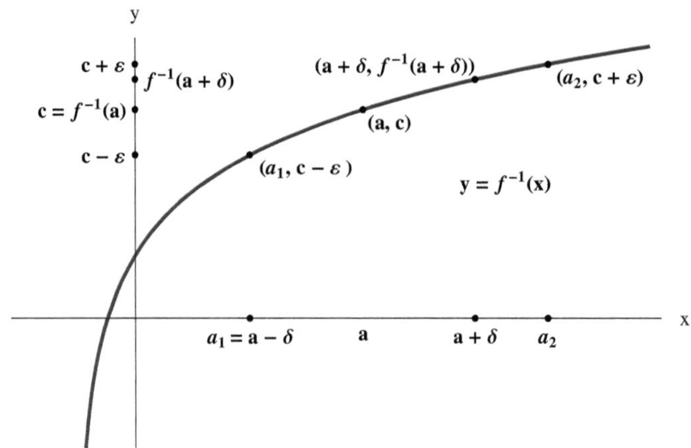

Figure 2

Let $c = f^{-1}(a)$, so that $a = f(c)$. Let $\varepsilon > 0$ be a given error tolerance. With reference to Figure 2, let $a_1 = f(c - \varepsilon)$ and $a_2 = f(c + \varepsilon)$, so that $f^{-1}(a_1) = c - \varepsilon$ and $f^{-1}(a_2) = c + \varepsilon$. Since f is increasing, so is f^{-1}. Therefore, if $a_1 < x < a_2$, then $c - \varepsilon = f^{-1}(a_1) < f^{-1}(x) < f^{-1}(a_2) = c + \varepsilon$. Set δ to be the minimum of $|a - a_1|$ and $|a - a_2|$. Then,

$$|x - a| < \delta \Rightarrow x \in (a_1, a_2) \Rightarrow c - \varepsilon < f^{-1}(x) < c + \varepsilon$$
$$\Rightarrow |f^{-1}(x) - c| = |f^{-1}(x) - f^{-1}(a)| < \varepsilon.$$

This establishes the continuity of f^{-1}. ∎

Appendix D

L'Hôpital's Rule (A Proof)

Theorem 1 (The Generalized Mean Value Theorem) Assume that f and g are continuous on $[a, b]$ and differentiable in (a, b). Then there exists $c \in [a, b]$ such that

$$f'(c)[g(b) - g(a)] = g'(c)[f(b) - f(a)].$$

Proof

Set

$$h(x) = [f(x) - f(a)][g(b) - g(a)] - [f(b) - f(a)][g(x) - g(a)].$$

Then,

$$h(a) = 0 \text{ and } h(b) = 0.$$

By Rolle's Theorem, there exists $c \in (a, b)$ such that $h'(c) = 0$. We have

$$h'(x) = f'(x)[g(b) - g(a)] - g'(x)[f(b) - f(a)].$$

Therefore,

$$h'(c) = 0 \Leftrightarrow f'(c)[g(b) - g(a)] = g'(c)[f(b) - f(a)].$$

■

Note that the Mean Value Theorem follows from the Generalized Mean Value Theorem if we set $g(x) = x$. In this case, $g'(x) = 1$, so that

$$f'(c)[g(b) - g(a)] = g'(c)[f(b) - f(a)] \Rightarrow f'(c)(b - a) = f(b) - f(a).$$

We will prove the version of **L'Hôpital's rule** that is relevant to the indeterminate form 0/0:

Theorem 2 Assume that f and g are differentiable at each x in an open interval J that contains the point a, with the possible exception of a itself, and that $g'(x) \neq 0$ for each $x \in J$. If $\lim_{x \to a} f(x) = \lim_{x \to a} g(x) = 0$ and

$$\lim_{x \to a} \frac{f'(x)}{g'(x)}$$

exists. Then

$$\lim_{x \to a} \frac{f(x)}{g(x)} = \lim_{x \to a} \frac{f'(x)}{g'(x)}.$$

Proof

Since $\lim_{x \to a} f(x) = \lim_{x \to a} g(x) = 0$, the functions f and g are continuous on J if we declare that $f(a) = g(a) = 0$. To begin with, let's show that $g(x) \neq 0$ if $x \in J$ and $x \neq a$. Indeed, if $x \in J$, $x \neq a$ and $g(x) = 0$, we must have $g'(c) = 0$ for some c between a and x, by Rolle's Theorem, since $g(x) = g(a) = 0$. But, $g'(c) \neq 0$ since $c \in J$. This is a contradiction.

Let $x \in J$ and $x \neq a$. By the Generalized Mean Value Theorem there exists c_x between x and a such that

$$f'(c_x)\left[g(x) - g(a)\right] = g'(c_x)\left[f(x) - f(a)\right].$$

Therefore,

$$f'(c_x)g(x) = g'(c_x)f(x).$$

Since $g(x) \neq 0$ and $g'(c_x) \neq 0$, we can divide and obtain the equality,

$$\frac{f'(c_x)}{g'(c_x)} = \frac{f(x)}{g(x)}.$$

Therefore,

$$\lim_{x \to a} \frac{f(x)}{g(x)} = \lim_{x \to a} \frac{f'(c_x)}{g'(c_x)} = \lim_{x \to a} \frac{f'(x)}{g'(x)},$$

since c_x is between x and a. $\blacksquare$

Appendix E

The Natural Logarithm as an Integral

We will derive the properties of the natural logarithm from its definition as an integral. This leads to the rigorous derivation of the basic properties of the natural exponential function as the inverse of the natural logarithm. The properties of exponential functions with arbitrary positive bases and the corresponding logarithms follow easily.

The Natural Logarithm

Definition 1 We set

$$\ln(x) = \int_1^x \frac{1}{t}\, dt, \ x > 0.$$

As an immediate consequence of the second part of the Fundamental Theorem of Calculus, we have

$$\frac{d}{dx}\ln(x) = \frac{d}{dx}\int_1^x \frac{1}{t}\, dt = \frac{1}{x}$$

for each $x > 0$.

Theorem 1 (The basic algebraic properties of the natural logarithm)
a)

$$\ln(1) = 0$$

b)

$$\ln(ab) = \ln(a) + \ln(b) \ \text{ for any } a > 0 \text{ and } b > 0$$

c)

$$\ln\left(\frac{1}{a}\right) = -\ln(a) \ \text{ for each } a > 0$$

d)

$$\ln(a^r) = r\ln(a) \ \text{ for each } a > 0 \text{ and rational number } r.$$

Proof

a)

$$\ln(1) = \int_1^1 \frac{1}{t}\, dt = 0.$$

b) By the chain rule,

$$\frac{d}{dx}\ln\left(ax\right) = \left(\frac{d}{dx}\ln\left(u\right)\Big|_{u=ax}\right)\left(\frac{d}{dx}\left(ax\right)\right) = \left(\frac{1}{ax}\right)(a) = \frac{1}{x}$$

for each $x > 0$. Therefore,

$$\frac{d}{dx}\left(\ln\left(ax\right) - \ln\left(x\right)\right) = \frac{1}{x} - \frac{1}{x} = 0$$

if $x > 0$. A function whose derivative is identically 0 on an interval must be a constant on that interval. Thus, there exists a constant C such that

$$\ln\left(ax\right) - \ln\left(x\right) = C$$

for each $x > 0$. In particular, if we set $x = 1$,

$$\ln\left(a\right) - \ln\left(1\right) = \ln\left(a\right) = C.$$

Therefore,

$$\ln\left(ax\right) - \ln\left(x\right) = \ln\left(a\right),$$

so that

$$\ln\left(ax\right) = \ln\left(a\right) + \ln\left(x\right)$$

for each $x > 0$. If we have the specific value b for x, we obtain

$$\ln\left(ab\right) = \ln\left(a\right) + \ln\left(b\right),$$

as claimed.

c) For any $a > 0$,

$$0 = \ln\left(1\right) = \ln\left(\frac{a}{a}\right) = \ln\left(a\left(\frac{1}{a}\right)\right) = \ln\left(a\right) + \ln\left(\frac{1}{a}\right),$$

by part b). Therefore,

$$\ln\left(\frac{1}{a}\right) = -\ln\left(a\right),$$

as claimed.

d) We will establish the statement for positive integer powers, negative integer powers, fractional powers, and finally for general rational exponents.

If we set $b = a$ in the statement of part b), we obtain

$$\ln\left(a^2\right) = \ln\left(a\right) + \ln\left(a\right) = 2\ln\left(a\right).$$

By induction,

$$\ln\left(a^n\right) = n\ln\left(a\right)$$

for any positive integer n. By c),

$$\ln\left(a^{-n}\right) = \ln\left(\frac{1}{a^n}\right) = -\ln\left(a^n\right) = -n\ln\left(a\right).$$

We have

$$\ln\left(a\right) = \ln\left(\left(a^{1/n}\right)^n\right) = n\ln\left(a^{1/n}\right).$$

Therefore,

$$\ln\left(a^{1/n}\right) = \frac{1}{n}\ln\left(a\right),$$

where n is a positive integer.

If $r = 0$,

$$\ln(a^r) = \ln(a^0) = \ln(1) = 0 = (0)\ln(a) = r\ln(a).$$

If r is a nonzero rational number, we can express r as m/n, where m is a nonzero integer, and n is a positive integer. We have

$$\ln(a^r) = \ln\left(a^{m/n}\right) = \ln\left(\left(a^{1/n}\right)^m\right) = m\ln\left(a^{1/n}\right) = m\left(\frac{1}{n}\ln(a)\right) = \frac{m}{n}\ln(a) = r\ln(a),$$

as claimed. ∎

Theorem 2

a) The natural logarithm is an increasing function on $(0, +\infty)$

b) The graph of the natural logarithm is concave down on $(0, +\infty)$.

c)

$$\lim_{x \to +\infty} \ln(x) = +\infty \quad \text{and} \quad \lim_{x \to 0+} \ln(x) = -\infty.$$

d) The range of the natural logarithm is the entire set of real numbers.

Proof

a) Since

$$\frac{d}{dx}\ln(x) = \frac{1}{x} > 0 \text{ for each } x > 0,$$

the natural logarithm is increasing on its entire domain $(0, +\infty)$, by the derivative test for monotonicity.

b) Since

$$\frac{d^2}{dx^2}\ln(x) = \frac{d}{dx}\left(\frac{1}{x}\right) = -\frac{1}{x^2} < 0 \text{ for each } x > 0,$$

the graph of the natural logarithm is concave down on $(0, +\infty)$, by the second derivative test for concavity.

c) Since the natural logarithm is a continuous increasing function on $(0, +\infty)$,

$$\lim_{x \to +\infty} \ln(x) = \lim_{n \to \infty} \ln(2^n) = \lim_{n \to \infty}(n\ln(2)) = \ln(2)\lim_{n \to \infty}(n) = +\infty$$

(Note that

$$\ln(2) = \int_1^2 \frac{1}{t}\,dt > \int_1^2 \frac{1}{2}\,dt = \frac{1}{2} > 0).$$

Similarly,

$$\lim_{x \to 0+} \ln(x) = \lim_{n \to \infty} \ln(2^{-n}) = \lim_{n \to \infty}(-n\ln(2)) = \ln(2)\lim_{n \to \infty}(-n) = -\infty.$$

d) Since the natural logarithm is a continuous increasing function on $(0, +\infty)$, and

$$\lim_{x \to +\infty} \ln(x) = +\infty \quad \text{and} \quad \lim_{x \to 0+} \ln(x) = -\infty,$$

its range is all of $\mathbb{R}$, by the Intermediate Value Theorem. ∎

The Natural Exponential Function as the Inverse of the Natural Logarithm

Since we established that the natural logarithm is a continuous increasing function on its domain $(0, +\infty)$, and the range of ln is the entire set of real numbers $\mathbb{R}$, the inverse of ln exists, and the domain of the inverse function is $\mathbb{R}$. The inverse of the natural logarithm is the **natural exponential function exp**:

Definition 2

$$y = \exp(x) \Leftrightarrow x = \ln(y),$$

where $x \in \mathbf{R}$ and $y > 0$.

As a result of the function-inverse function relationship between the natural logarithm and the natural exponential function, we have

$$\ln(\exp(x)) = x \text{ for each } x \in \mathbb{R} \text{ and } \exp(\ln(y)) = y \text{ for each } y > 0.$$

Theorem 3   **(The Basic Algebraic Properties of the Natural Exponential Function)**

a)

$$\exp(0) = 1$$

b)

$$\exp(a + b) = \exp(a)\exp(b)$$

c)

$$\exp(-x) = \frac{1}{\exp(x)}$$

d) For each $a \in \mathbb{R}$,

$$\exp(ra) = (\exp(a))^r \text{ if } r \text{ is a rational number.}$$

Proof

a)

$$y = \exp(0) \Leftrightarrow 0 = \ln(y) \Leftrightarrow y = 1.$$

b) We have

$$\exp(a + b) = \exp(a)\exp(b) \Leftrightarrow \ln(\exp(a + b)) = \ln(\exp(a)\exp(b))$$
$$\Leftrightarrow a + b = \ln(\exp(a)\exp(b)).$$

Indeed,

$$\ln(\exp(a)\exp(b)) = \ln(\exp(a)) + \ln(\exp(b)) = a + b.$$

c)

$$1 = \exp(0) = \exp(x - x) = \exp(x)\exp(-x),$$

by part b). Therefore,

$$\exp(-x) = \frac{1}{\exp(x)}, \quad x \in R,$$

as claimed.

d) Let r be a rational number. We have

$$\exp(ra) = (\exp(a))^r \Leftrightarrow \ln(\exp(ra)) = \ln((\exp(a))^r)$$
$$\Leftrightarrow ra = \ln((\exp(a))^r).$$

Now, $\ln\left((\exp(a))^r\right) = r\ln(\exp(a))$, by Theorem **??**, part d), and $r\ln(\exp(a)) = ra$, so that the statement, $\exp(ra) = (\exp(a))^r$, is valid. $\blacksquare$

We define the number e as the number whose natural logarithm is 1. Thus,

$$\ln(e) = 1 \Leftrightarrow \exp(1) = e.$$

Note that $e > 1$. Indeed,

$$e > 1 \Leftrightarrow \ln(e) > \ln(1) \Leftrightarrow 1 > 0.$$

Corollary (**Corollary to Theorem 3**) If r is a rational number, then

$$\exp(r) = e^r.$$

Proof

By part d) of Theorem 3,

$$\exp(r) = (\exp(1))^r = e^r.$$

$\blacksquare$

We denote $\exp(x)$ as e^x for any $x \in \mathbb{R}$. Thus, the statements of Theorem **??** can be expressed as follows:

a)

$$e^0 = 1$$

b)

$$e^a e^b = e^{a+b}.$$

c)

$$e^{-x} = \frac{1}{e^x}.$$

d) For each $a \in \mathbb{R}$,

$$(e^a)^r = e^{ar}$$

if r is a rational number.

Theorem 4

$$\frac{d}{dx}e^x = e^x.$$

Proof

The natural exponential function is the inverse of the natural logarithm:

$$y = e^x \Leftrightarrow x = \ln(y)$$

for each $x \in \mathbf{R}$ and $y > 0$. Therefore,

$$\frac{d}{dx}e^x = \frac{dy}{dx} = \frac{1}{\dfrac{dx}{dy}} = \frac{1}{\dfrac{d}{dy}\ln(y)} = \frac{1}{\dfrac{1}{y}} = y = e^x.$$

$\blacksquare$

Arbitrary Bases

We define the exponential function with base a in terms of the natural exponential function and the natural logarithm:

Definition 3 Let $a > 0$.

$$\exp_a(x) = a^x = e^{x \ln(a)}$$

for each $x \in \mathbb{R}$.

The basic rules for exponents are valid:

Theorem 5

a)
$$a^0 = 1$$

b)
$$a^{x_1} a^{x_2} = a^{x_1 + x_2}$$

for each x_1 and x_2 in $\mathbb{R}$.

c)
$$a^{-x} = \frac{1}{a^x}$$

for each $x \in \mathbb{R}$.

d)
$$(a^{x_1})^{x_2} = a^{x_1} a^{x_2} \text{ for each } x_1 \text{ and } x_2 \text{ in } \mathbb{R}.$$

Proof

a)
$$a^0 = e^{0(\ln(a))} = e^0 = 1.$$

b)
$$a^{x_1} a^{x_2} = e^{x_1 \ln(a)} e^{x_2 \ln(a)} = e^{(x_1 + x_2) \ln(a)} = a^{x_1 + x_2}.$$

c)
$$a^{-x} = e^{\ln(a)(-x)} = e^{-(x \ln(a))} = \frac{1}{e^{x \ln(a)}} = \frac{1}{a^x}.$$

d)
$$(a^{x_1})^{x_2} = e^{x_2 \ln(a^{x_1})} = e^{x_2 \ln\left(e^{x_1 \ln(a)}\right)} = e^{x_2(x_1 \ln(a))} = e^{x_1 x_2 \ln(a)} = a^{x_1 x_2}.$$

∎

Theorem 6
Let $a > 0$.

a)
$$\frac{d}{dx} a^x = \ln(a) a^x.$$

b) If $a > 1$, a^x defines an increasing function on R. If $0 < a < 1$, a^x defines a decreasing function on $\mathbb{R}$.

c) If $a \neq 1$, the graph of the function defined by a^x is concave up on $\mathbb{R}$.

d) If $a > 1$, we have
$$\lim_{x \to +\infty} a^x = +\infty \text{ and } \lim_{x \to -\infty} a^x = 0.$$

If $0 < a < 1$, we have
$$\lim_{x \to +\infty} a^x = 0 \text{ and } \lim_{x \to -\infty} a^x = +\infty.$$

e) The range of an exponential function with base $a \neq 1$ is $(0, +\infty)$.

Proof

a) By the chain rule,

$$\frac{d}{dx}a^x = \frac{d}{dx}e^{x\ln(a)} = \left(\frac{d}{du}e^u\bigg|_{u=x\ln(a)}\right)\left(\frac{d}{dx}(x\ln(a))\right)$$
$$= \left(e^{x\ln(a)}\right)(\ln(a))$$
$$= a^x \ln(a).$$

b) If $a > 1$,

$$\frac{d}{dx}a^x = \ln(a)\,a^x > 0 \text{ for each } x \in \mathbb{R},$$

since $\ln(a) > 0$ and $a^x > 0$. Therefore a^x defines an increasing function on $\mathbb{R}$, by the derivative test for monotonicity.

If $0 < a < 1$,

$$\frac{d}{dx}a^x = \ln(a)\,a^x < 0 \text{ for each } x \in \mathbb{R},$$

since $\ln(a) < 0$ and $a^x > 0$. Therefore, a^x defines a decreasing function on $\mathbb{R}$.

c) If $a \neq 1$,

$$\frac{d^2}{dx^2}a^x = \frac{d}{dx}\left(\ln(a)\,a^x\right) = \ln(a)\frac{d}{dx}a^x = \ln(a)\left(\ln(a)\,a^x\right) = (\ln(a))^2\,a^x > 0, \; x \in \mathbb{R}.$$

Therefore, the function defined by a^x is concave up on $\mathbb{R}$, by the second derivative test for concavity.

d) If $a > 1$ then $\ln(a) > 0$. Therefore,

$$\lim_{x \to +\infty} a^x = \lim_{x \to +\infty} e^{x\ln(a)} = \lim_{u \to +\infty} e^u = +\infty,$$

and

$$\lim_{x \to -\infty} a^x = \lim_{x \to -\infty} e^{x\ln(a)} = \lim_{u \to -\infty} e^u = 0.$$

The proofs of the corresponding facts for $0 < a < 1$ are similar.

e) If $a > 1$, a^x defines a positive-valued, continuous increasing function on $\mathbb{R}$. By d), the range of the function is $(0, +\infty)$. If $0 < a < 1$, a^x defines a positive-valued, continuous decreasing function on $\mathbb{R}$. By d), the range of the function is again $(0, +\infty)$. $\blacksquare$

Let $a > 0$ and $a \neq 1$. The function defined by a^y is continuous and increasing or decreasing on the entire number line. The range of the function is the set of positive numbers. Therefore, it has an inverse defined on $(0, +\infty)$. The inverse function is referred to as the logarithm with respect to the base a, and abbreviated as $\log_a$.

Definition 4

$$y = \log_a(x) \Leftrightarrow x = a^y$$

where $x > 0$ and $y \in \mathbb{R}$.

The derivation of the basic properties of a logarithm with respect to a positive base $a \neq 1$ proceeds as in Section 4.4.

Appendix F

Answers to Some Problems

Answers to Some Problems of Section 1.1

1.

$$
\begin{aligned}
V(r) &= \frac{4}{3}\pi r^3, \\
V(40) &= \frac{2.56 \times 10^5}{3}\pi \ (\text{cm}^3) \\
&\left(\cong 2.680\,83 \times 10^5\right)
\end{aligned}
$$

3.

$$
\begin{aligned}
A(h) &= 100\pi h, \\
A(40) &= 4000\pi \ (\text{cm}^3) \\
&(\cong 12566.4)
\end{aligned}
$$

5. The given recipe does not define a function.

7.

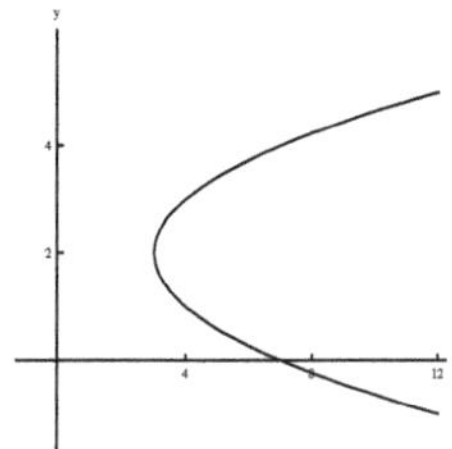

9. D is the entire number line $\mathbb{R}$.

11. $D = (-\infty, -1) \cup (-1, 1) \cup (1, +\infty)$

13. $D = [-2, 2]$.

15. $D = (0, \infty)$.

17. $D = (-\infty, -3] \cup [3, +\infty)$.

19.

21.

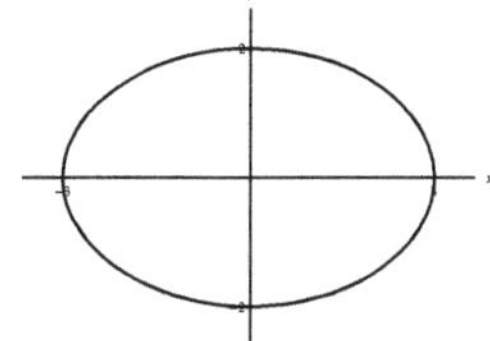

23. Here are two possible computer generated plots for f:

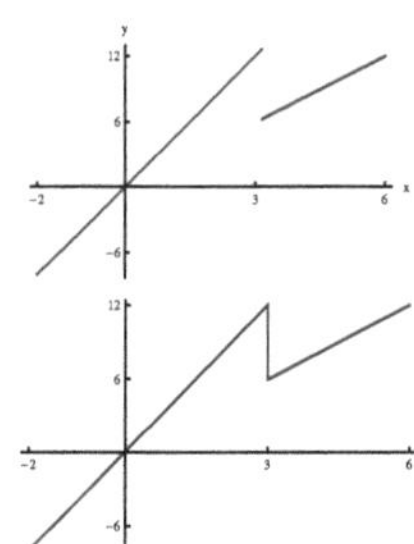

We see that the second plot includes a spurious line segment.

25. a)

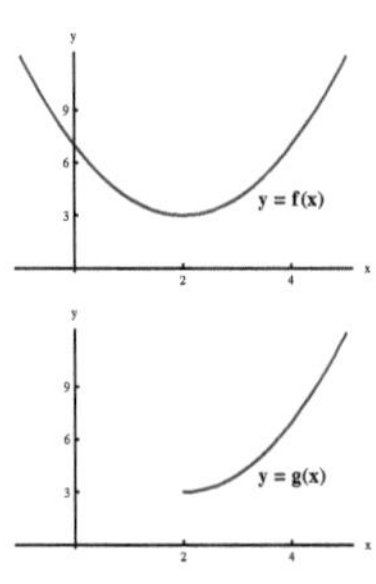

b) g is an odd function.

31.
a)

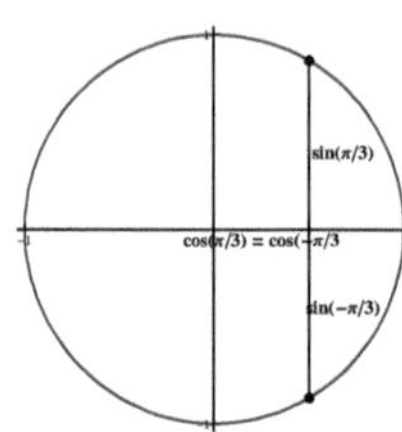

27.
a)

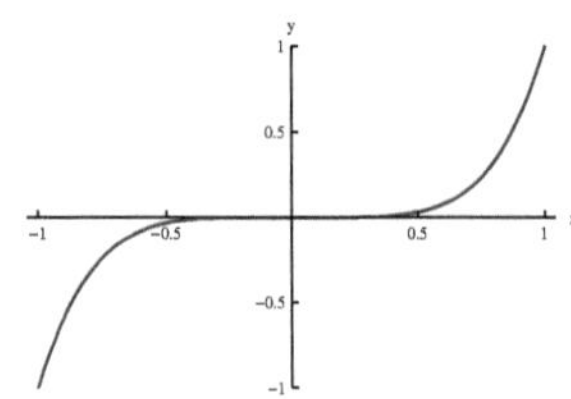

b)

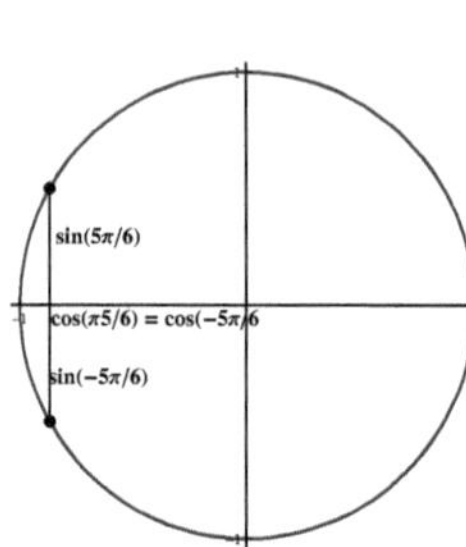

b)

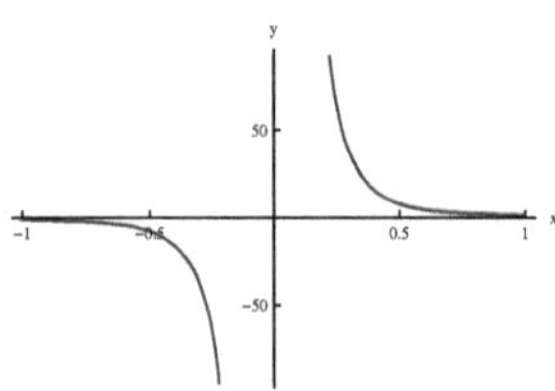

33.
a)

$$\sin\left(1\right) \cong 0.841\,47,$$
$$\sin\left(1+4\pi\right) \cong 0.841\,471,$$
$$\cos\left(1\right) \cong 0.540\,302,$$
$$\cos\left(1+4\pi\right) \cong 0.540\,302.$$

29.
a)

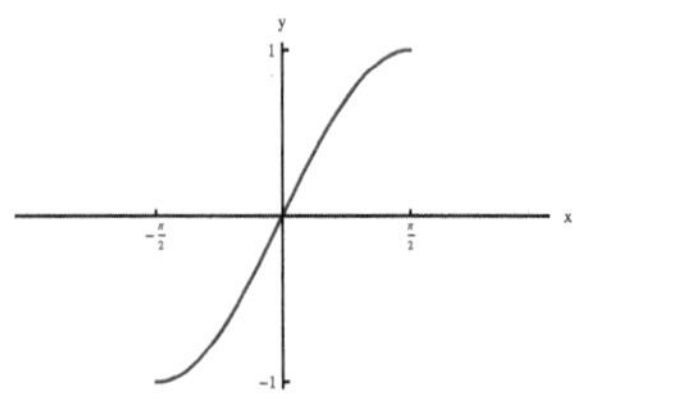

b)

$$\sin\left(2.5\right) \cong 0.598\,472,$$
$$\sin\left(2.5-6\pi\right) \cong 0.598\,472$$
$$\cos\left(2.5\right) \cong -0.801\,144,$$
$$\cos\left(2.5-6\pi\right) \cong -0.801\,144.$$

Answers to Some Problems of Section 1.2

1.

$$(f+g)\left(x\right) = x^2 + x^3 \text{ for each } x \in \mathbb{R}.$$

$$(3f)\left(x\right) = 3x^2 \text{ for each } x \in \mathbb{R}.$$

$$(fg)\left(x\right) = x^5 \text{ for each } x \in \mathbb{R}.$$

$$\left(\frac{1}{f}\right)\left(x\right) = \frac{1}{x^2} \text{ for each } x \neq 0.$$

$$\left(\frac{f}{g}\right)(x) = \frac{1}{x} \text{ for each } x \neq 0.$$

$$(2f - 3g)(x) = 2x^2 - 3x^2 \text{ for each } x \in \mathbb{R}$$

3.

$$(f + g)(x) = \frac{1}{x} + x^2 \text{ for each } x \neq 0.$$

$$(fg)(x) = \frac{x^2}{x} = x \text{ for each } x \neq 0.$$

$$(3f)(x) = \frac{3}{x} \text{ for each } x \neq 0.$$

$$\left(\frac{1}{f}\right)(x) = x \text{ for each } x \neq 0.$$

$$\left(\frac{f}{g}\right)(x) = \frac{1}{x^3} \text{ for each } x \neq 0.$$

5.
a)
$$f(x) = -x + 7.$$

b)

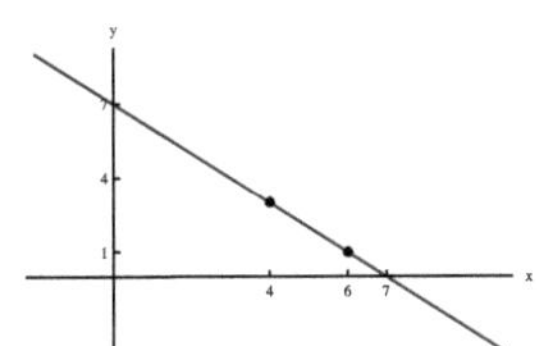

7.
a) The minimum value of f is 5.
The function does not have a maximum value

b) There are no (real) solutions of the equation $f(x) = 0$.

c)

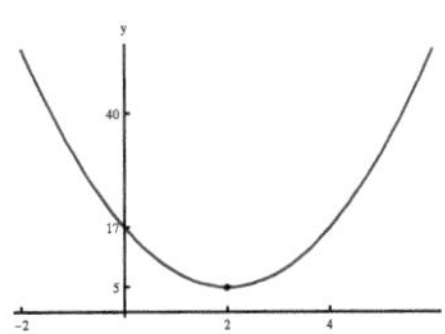

9.
$$(f \circ g)(x) = \frac{1}{x^2} \text{ for each } x \neq 0.$$

$$(g \circ f)(x) = \frac{1}{x^2} \text{ for each } x \neq 0,$$

$$(f \circ f)(x) = x \text{ for each } x \neq 0$$

$$(g \circ g)(x) = x^4 \text{ for each } x \in \mathbb{R}$$

11.

$$(f \circ g)(x) = \sqrt{4 - x^2} \text{ for each } x \in [-2, 2]$$

$$(g \circ f)(x) = \sqrt{4 - x} \text{ for each } x \geq 0$$

$$(f \circ f)(x) = x^{1/4} \text{ for each } x \geq 0$$

$$(g \circ g)(x) = -x^4 + 8x^2 - 12 \text{ for each } x \in \mathbb{R}$$

13.

$$(f \circ g)(x) = \sqrt{|x|} \text{ for each } x \in \mathbb{R}$$

$$(g \circ f)(x) = \sqrt{x} \text{ for each } x \geq 0$$

15.
$$(f \circ g)(x) = \sin^{1/4}(x)$$

for each $x \in [2n\pi, \pi + 2n\pi]$, $n = 0, \pm 1, \pm 2, \pm 3, \ldots$

$$(g \circ f)(x) = \sin\left(x^{1/4}\right) \text{ for each } x \geq 0.$$

17.
a) $[-4, 4]$.
b) $F(u) = \sqrt{u}$ and $G(x) = 16 - x^2$

19.
a) The entire number line $\mathbb{R}$.
b) $F(u) = \sin(u)$ and $G(x) = x^2$.

21.
a) The entire number line $\mathbb{R}$.
b) $F(u) = \sin(u)$ and $G(x) = 4x$.

23.
a) All real numbers that are not odd multiples of $\pm\pi/4$.
b) $F(u) = \tan(u)$ and $G(x) = 2x$.

25.
a) $[0, \pi]$.
b) $F(u) = u^{3/4}$ and $G(x) = \sin(x)$.

27.
a)
$$\frac{\pi}{3}$$

b) $[-\pi/6, \pi/6]$. f is odd.
c)

b) $[0, \pi]$. f is neither even nor odd.
c)

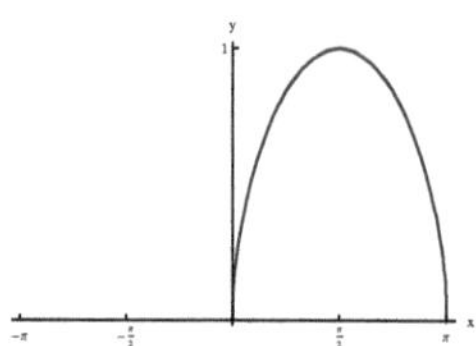

29.
a)

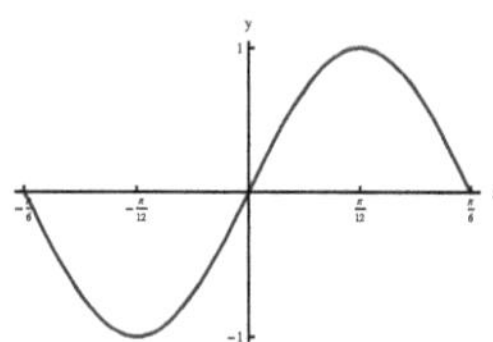

1

b) $(-1/2, 1/2)$. f is odd.
c)

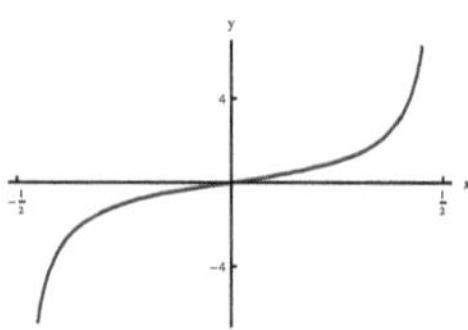

31.
a)

2π

33.
a)

π

b) $[0, \pi/2)$. f is neither even nor odd.
c)

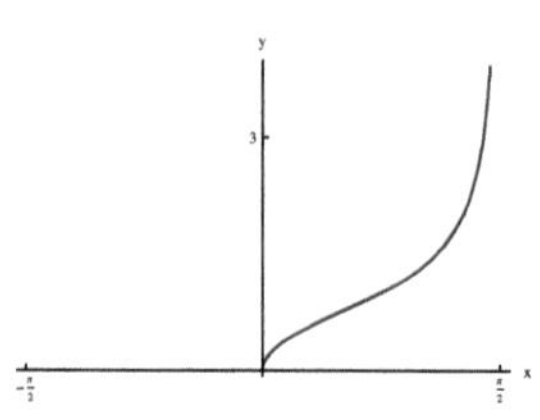

Answers to Some Problems of Section 1.3

1.

$$\lim_{x \to 2+} f(x) = \lim_{x \to 2-} f(x) = 2, \ \lim_{x \to 2} f(x) = 2.$$

f is continuous at 2.

3. f does not have a limit at 2 and is not continuous at 2.

5.

$$\lim_{x \to 4-} f(x) = \lim_{x \to 4+} f(x) = -2, \ \lim_{x \to 4} f(x) = -2.$$

f is continuous at 4.

7. f does not have a limit at -2 and is not continuous at -2.

9.

x	$f(x)$	x	$f(x)$
$4 + 10^{-2}$	5.03	$4 - 10^{-2}$	-4.98
$4 + 10^{-3}$	5.003	$4 - 10^{-3}$	-4.998
$4 + 10^{-4}$	5.0003	$4 - 10^{-4}$	-4.9998
$4 + 10^{-5}$	5.00003	$4 - 10^{-5}$	-4.99998

The function does not have a limit at 4 and is not continuous at 4.

11.

x	$f(x)$	x	$f(x)$
$3 + 10^{-2}$	3.01	$3 - 10^{-2}$	-10^{-2}
$3 + 10^{-3}$	3.001	$3 - 10^{-3}$	-10^{-3}.
$3 + 10^{-4}$	3.0001	$3 - 10^{-4}$	-10^{-4}
$3 + 10^{-5}$	3.00001	$3 - 10^{-5}$	-10^5

The numbers indicate that $\lim_{x \to 3-} f(x)$ does not exist. so that f is not continuous at 3.

Answers to Some Problems of Section 1.4

1. We have
$$\frac{2x^2 - 2}{x - 1} = \frac{2\left(x^2 - 1\right)}{x - 1} = \frac{2\left(x - 1\right)\left(x + 1\right)}{x - 1} = 2\left(x + 1\right) \text{ if } x \neq 1.$$

Therefore, if $x \neq 1$
$$\left|\frac{2x^2 - 2}{x - 1} - 4\right| = |2x + 2 - 4| = |2x - 2| = |2\left(x - 1\right)| = 2|x - 1|.$$

If $x \neq 1$, given $\varepsilon > 0$
$$\left|\frac{2x^2 - 2}{x - 1} - 4\right| < \varepsilon \Leftrightarrow 2|x - 1| < \varepsilon \Leftrightarrow |x - 1| < \frac{\varepsilon}{2}.$$

Thus, we can choose $\delta = \varepsilon/2$. If $|x - 1| < \delta$ and $x \neq 1$ then
$$\left|\frac{2x^2 - 2}{x - 1} - 4\right| < \varepsilon.$$

This proves that
$$\lim_{x \to 1} \frac{2x^2 - 2}{x - 1} = 4$$

5. We have
$$\begin{aligned}
|f\left(1 + h\right) - f\left(1\right)| = \left|3\left(1 + h\right)^2 + \left(1 + h\right) - 1 - 3\right| &= |h\left(3h + 7\right)| \\
&= |h||3h + 7| \leq |h|\left(3|h| + 7\right),
\end{aligned}$$

with the help of triangle inequality. If $|h| < 1$ then
$$|f\left(1 + h\right) - f\left(1\right)| \leq |h|\left(3|h| + 7\right) < 10|h|.$$

Thus, in order to ensure that $|f\left(1 + h\right) - f\left(1\right)| < \varepsilon$ it is sufficient to have
$$|h| < 1 \text{ and } 10|h| < \varepsilon \Leftrightarrow |h| < 1 \text{ and } |h| < \frac{\varepsilon}{10}.$$

Therefore, we can set $\delta = \min\left(1, \varepsilon/10\right)$. By the above calculations, if $|h| < \delta$ then $|f\left(1 + h\right) - f\left(1\right)| < \varepsilon$. This proves that f is continuous at 1.

7. We have
$$|f\left(x\right) - f\left(4\right)| = \left|\frac{1}{x - 2} - \frac{1}{2}\right| = \left|\frac{2 - x + 2}{2\left(x - 2\right)}\right| = \frac{|x - 4|}{2|x - 2|}.$$

If $|x - 4| < 1$ then
$$x - 4 > -1 \Rightarrow x > 3 \Rightarrow x - 2 > 3 - 2 = 1,$$

so that
$$\frac{1}{|x - 2|} = \frac{1}{x - 2} < 1.$$

Thus,
$$|f\left(x\right) - f\left(4\right)| < \frac{1}{2}|x - 4|$$

if $|x - 4| < 1$. Therefore, in order to ensure that $|f\left(x\right) - f\left(4\right)| < \varepsilon$ it is sufficient to have
$$|x - 4| < 1 \text{ and } |x - 4| < 2\varepsilon.$$

Therefore, we can set $\delta = \min(1, 2\varepsilon)$. By the above calculations, if $|h| < \delta$ then $|f(x) - f(4)| < \varepsilon$. This proves that f is continuous at 4.

9. If $x < 3$ then

$$f(x) = \frac{x^2 + x - 12}{x - 3} = \frac{(x+4)(x-3)}{x-3} = x + 4.$$

Therefore,

$$|f(x) - 7| = |(x + 4) - 7| = |x - 3| = 3 - x.$$

Thus, in order to ensure that $|f(x) - 7| < \varepsilon$ it is sufficient to have

$$x < 3 \text{ and } 3 - x < \varepsilon \Leftrightarrow 3 - \varepsilon < x < 3.$$

This proves that $\lim_{x \to 3-} f(x) = 7$.

Answers to Some Problems of Section 1.5

1.b) $2\sqrt{3}$

3. b) 65

5. b) f does not have a limit at 2.

7. b) $\sqrt{7}$

9. b) 0.

11. b) $\lim_{x \to 2} f(x) = 5$.

13. a) $g(x) = \dfrac{x+2}{x^2 + 2x + 4}$, b) $\dfrac{1}{3}$

15. a) $\sin(x)$, b) 1

17. a) $g(x) = x^2 + 3x + 9$, b) 27

19. a) $g(x) = \dfrac{1}{\sqrt{x} + 4}$ b) $\frac{1}{8}$

21. $\dfrac{1}{2}$

23. $\dfrac{7}{4}$

25. 0

27. $f(u) = \sqrt{u}$ and $g(x) = \dfrac{x^2 - 9}{x - 3}$; $\sqrt{6}$

29. $f(u) = \sqrt{u}$ and $g(x) = \sin(x)$; $\dfrac{1}{\sqrt{2}}$

Answers to Some Problems of Section 1.6

1.

$$\lim_{x \to 4+} f(x) = +\infty, \ \lim_{x \to 4-} f(x) = -\infty.$$

3.

$$\lim_{x \to 1+} f(x) = +\infty, \ \lim_{x \to 1-} f(x) = -\infty$$

5.

$$\lim_{x \to -3+} f(x) = -\infty, \ \lim_{x \to -3-} f(x) = +\infty$$

7.

$$\lim_{x \to \pi/2+} \sec(x) = -\infty, \ \lim_{x \to \pi/2-} \sec(x) = +\infty$$

9.

$$\lim_{x \to \pi+} f(x) = -\infty, \ \lim_{x \to \pi-} f(x) = +\infty$$

11.

$$\lim_{x \to 3} f(x) = -\infty.$$

13.

$$\lim_{x \to 2+} f(x) = -\infty, \ \lim_{x \to 2-} f(x) = +\infty$$

Answers to Some Problems of Section 1.7

1.

$$\lim_{x \to \pm\infty} f(x) = 4.$$

The line $y = 4$ is a horizontal asymptote for the graph of f at $\pm\infty$.

3.

$$\lim_{x \to -\infty} f(x) = -\frac{\pi}{2} \text{ and } \lim_{x \to +\infty} f(x) = \frac{\pi}{2}.$$

The line $y = -\pi/2$ is a horizontal asymptote for the graph of f at $-\infty$ and the line $y = \pi/2$ is a horizontal asymptote for the graph of f at $+\infty$.

5.

$$\lim_{x \to \pm} f(x) = +\infty.$$

There no horizontal asymptotes.

7.

a)

$$\lim_{x \to \pm\infty} \frac{3x^2 - 2x - 26}{x^2 - x - 12} = 3.$$

Thus, the line $y = 3$ is the horizontal asymptote for the graph of f at $\pm\infty$.

b)

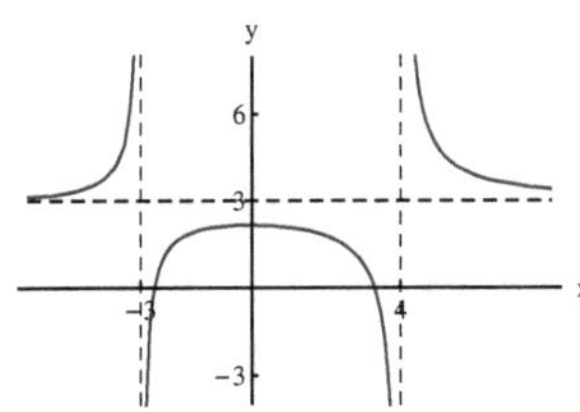

The picture is consistent with the response to part a).

9.

a)

$$\lim_{x \to +\infty} \frac{\sqrt{x^2 + 2x}}{4x + 5} = \frac{1}{4},$$
$$\lim_{x \to -\infty} \frac{\sqrt{x^2 + 2x}}{4x + 5} = -\frac{1}{4}$$

b)

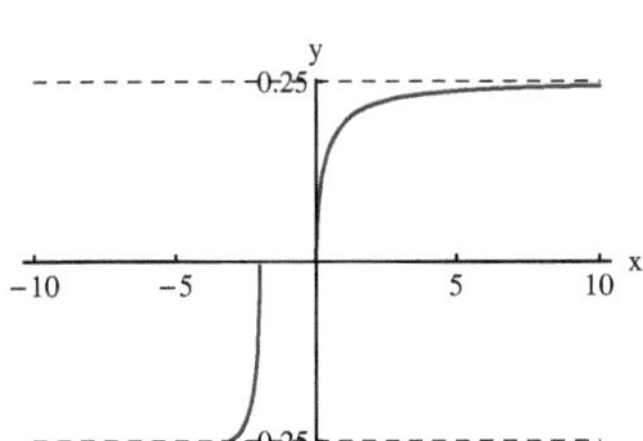

The picture is consistent with the response to part a).

11.

a)

$$\lim_{x \to +\infty} \frac{x^4 + x^2 + 1}{2x^3 + 9} = +\infty,$$
$$\lim_{x \to -\infty} \frac{x^4 + x^2 + 1}{2x^3 + 9} = -\infty$$

b)

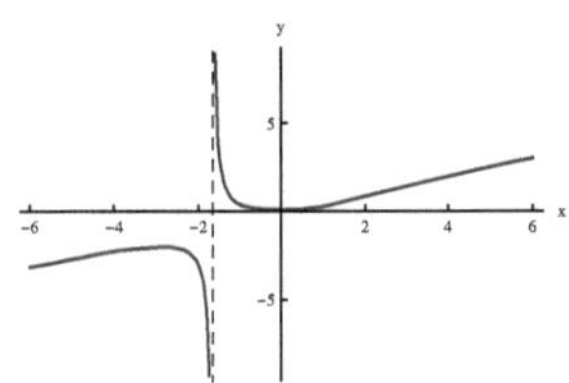

The picture is consistent with the response to part a).

13

a) The line $y = -4x + 3$ is an asymptote for the graph of f at $+\infty$ and $-\infty$.

b)

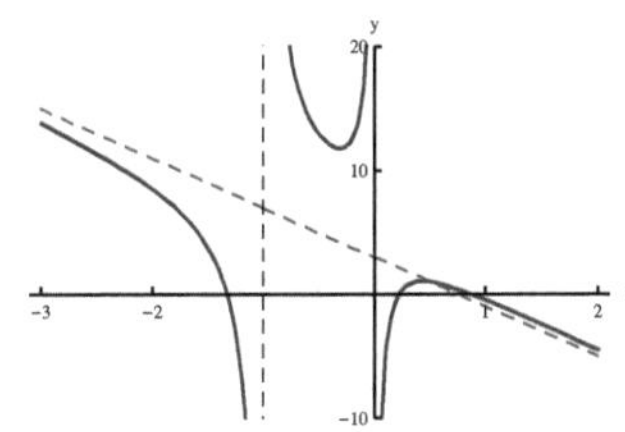

The picture is consistent with the response to part a).

15.

a) The line $y = 2x$ is an asymptote for the graph of f at $+\infty$. The line $y = 0$, i. e., the x-axis is an asymptote for the graph of f at $-\infty$.

b)

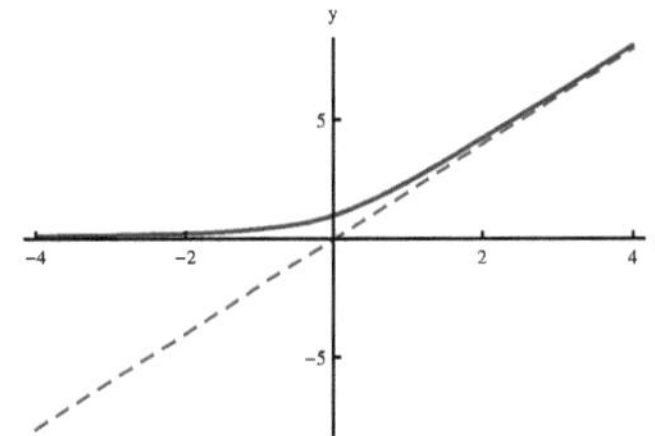

The picture is consistent with the response to part a).

17.

a) The graph of the quadratic function $q(x) = 2x^2 + 3$ is an asymptote to the graph of f at $+\infty$ and $-\infty$.

b)

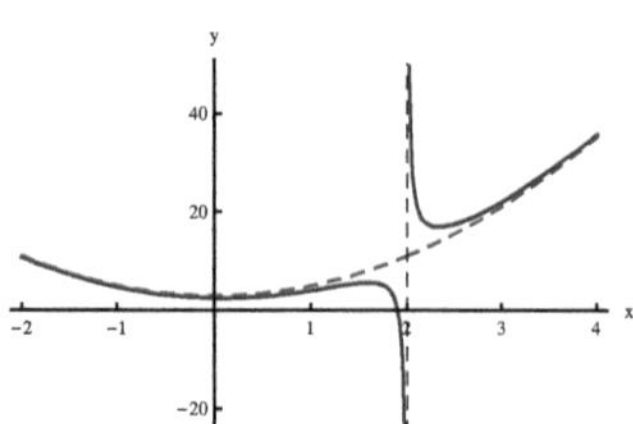

The picture is consistent with the response to part a).

Answers to Some Problems of Section 1.8

1.

$$1,\ 3,\ 5,\ 7$$

3.

$$1,\ -\frac{1}{3},\ \frac{1}{9},\ -\frac{1}{27}$$

7.

$$2,\ 3,\ 5,\ 9$$

9.

$$3,\ 6,\ 33,\ 1086$$

11.

a) 3

b)

13.

a) 0

b)

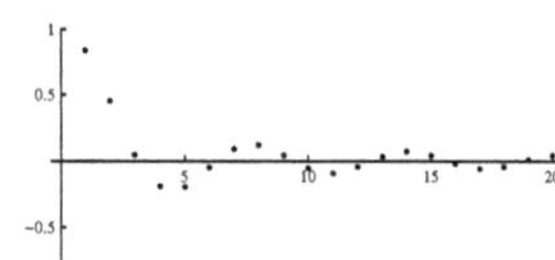

15. 5

17. $+\infty$

19. $\sqrt{\dfrac{1}{3}}$

21. The given sequence does not have limit.

23. The given sequence does not have limit.

25. 0

27. The given sequence does not have a limit (finite or infinite).

Answers to Some Problems of Section 2.1

1.

a)

$$f'(4) = 2;\ y = 3 + 2(x - 4).$$

b)

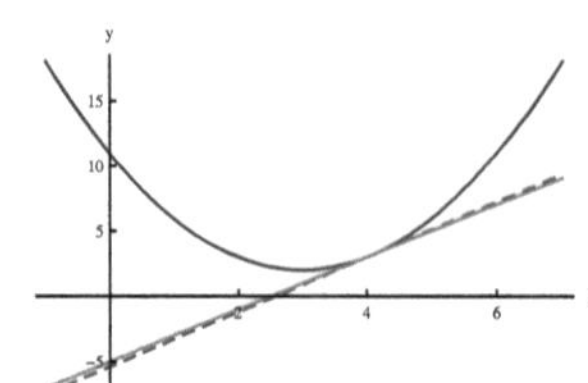

c)

h	$\dfrac{f(4+h) - f(4)}{h}$	$\left\| \dfrac{f(4+h) - f(4)}{h} - f'(4) \right\|$
10^{-1}	2.1	10^{-1}
10^{-2}	2.01	10^{-2}
10^{-3}	2.001	10^{-3}
10^{-4}	2.0001	10^{-4}

3.

a)
$$f'(2) = 8; \; y = 8(x-2)$$

b)

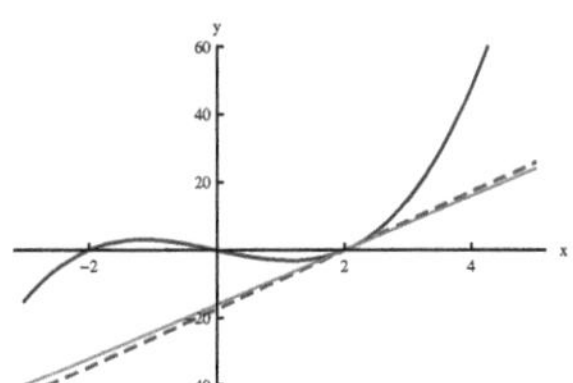

c)

h	$\dfrac{f(2+h) - f(2)}{h}$	$\left\| \dfrac{f(2+h) - f(2)}{h} - f'(2) \right\|$
10^{-1}	8.61	0.61
10^{-2}	8.0601	0.0601
10^{-3}	8.006	6×10^{-3}
10^{-4}	8.0006	6×10^{-4}

5. We have
$$\frac{f(2+h) - f(2)}{h} = 12 + 3h.$$

Therefore,
$$f'(2) = \lim_{h \to 0} (12 + 3h) = 12.$$

7.
$$f'(1) = 4.$$

9. $f(x) = \sqrt{x}$, $a = 16$.

11. $f(x) = x^3$, $a = 2$.

13.

a)
$$-1 + h$$

b)
$$-1.$$

15.

a)
$$f'_{+}(2) = 1; \; f'_{-}(2) = -1$$

b) Since $f'_{+}(2) \neq f'_{-}(2)$, the function f is not differentiable at 2.

17.b) The function is differentiable at 0 and we have $f'(0) = 1$.

19. $f'_{+}(2)$ does not exist.

21. f is differentiable at 3 and $f'(3) = 0$.

23.

a)
$$f'(2) = 36.$$

b)
$$y = 36 + 36(x - 2).$$

25.
$$\frac{f(x+h) - f(x)}{h} = \frac{3(x+h)^2 - 3x^2}{h} = 6x + 3h.$$

Therefore
$$f'(x) = \lim_{h \to 0} \frac{f(x+h) - f(x)}{h} = \lim_{h \to 0}(6x + 3h) = 6x$$

27.

$$f'(x) = 3x^2 + 1.$$

29.

$$f'(x) = -\frac{2}{x^3}. \text{ if } x \neq 0$$

31.

$$\frac{df}{dx} = -\frac{4}{(x-5)^2} \text{ if } x \neq 5$$

33. $f(x) = \dfrac{1}{x^4}$

35. $f(x) = \sin(2x)$

37.
a)

$$f'(x) = \begin{cases} 2x & \text{if} & x < -3, \\ -2x & \text{if} & -3 < x < 3, \\ 2x & \text{if} & x > 3. \end{cases}$$

Answers to Some Problems of Section 2.2

1.

$$\frac{d}{dx}\left(x^5\right) = 5x^4$$

for each $x \in \mathbb{R}$. The domain of the derivative (function) is the entire set of real numbers $\mathbb{R}$.

3.

$$\frac{d}{dx}\left(x^{1/7}\right) = \frac{1}{7}x^{\frac{1}{7}-1} = \frac{1}{7}x^{-6/7} = \frac{1}{7x^{6/7}}.$$

for each $x \neq 0$. The domain of the derivative is $\{x \in \mathbb{R} : x \neq 0\}$.

5.

$$\frac{d}{dx}\left(x^{-3/5}\right) = -\frac{3}{5}x^{-8/5} = -\frac{3}{5x^{8/5}}$$

if $x \neq 0$. The domain of the derivative is $\{x \in \mathbb{R} : x \neq 0\}$.

7.
a)

$$f'(x) = \frac{d}{dx}\left(x^3 - 3\right) = 3x^2$$

for each $x \in \mathbb{R}$. The domain of f' is $\mathbb{R}$.
b) The graph of f does not have any vertical tangents or cusps.
c)

Since f is not differentiable at 3 and -3,

Domain of $f = \{x \in \mathbb{R} : x \neq -3 \text{ and } x \neq 3\}$.

b)

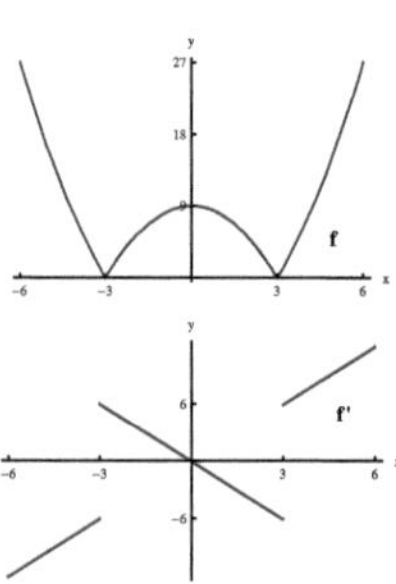

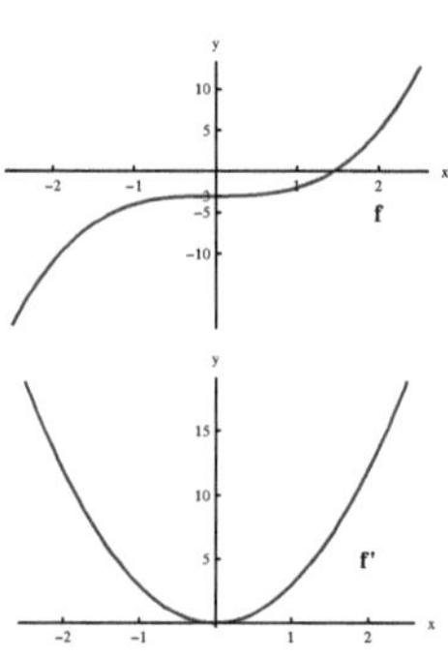

The pictures are consistent with the responses to a) and b).

9.
a)

$$f'(x) = \frac{d}{dx}\left(x^{1/4} + 2\right) = \frac{1}{4}x^{-3/4} = \frac{1}{4x^{3/4}}.$$

The domain of the derivative is the set of all $x > 0$.
b) The graph of f has a vertical tangent at $(0, f(0)) = (0, 2)$.
c)

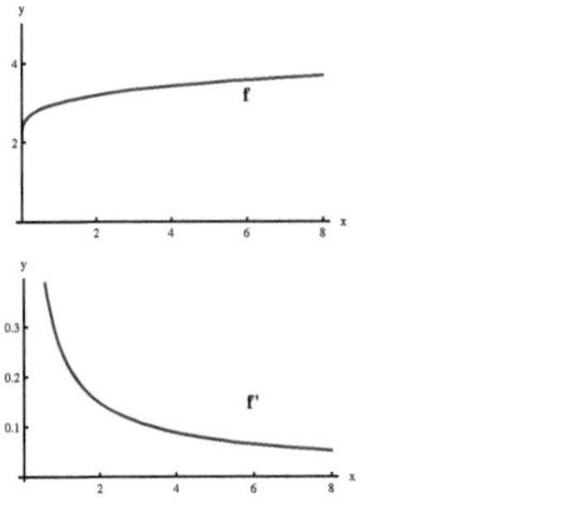

The pictures are consistent with the nonexistence of the derivative at 0, and the existence of a vertical tangent to the graph of f at $(0, 2)$.

11.

a) We have

$$f'(x) = \frac{d}{dx}\left(x^{3/4}\right) = \frac{3}{4}x^{-1/4} = \frac{3}{4x^{1/4}}$$

if $x > 0$. The domain of the derivative is the set of positive numbers $(0, +\infty)$.

b) The graph of f has a vertical tangent at $(0, 0)$.

c)

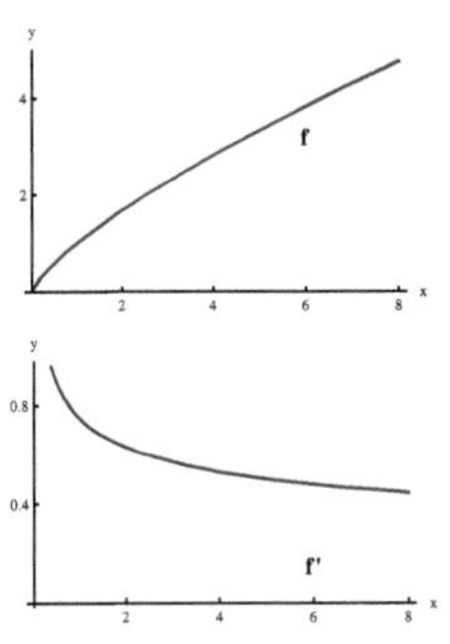

The pictures are consistent with the nonexistence of the derivative at 0, and the existence of a vertical tangent to the graph of f at $(0, 0)$.

13.

a)

$$y = 4 + \frac{1}{3}(x - 8).$$

b)

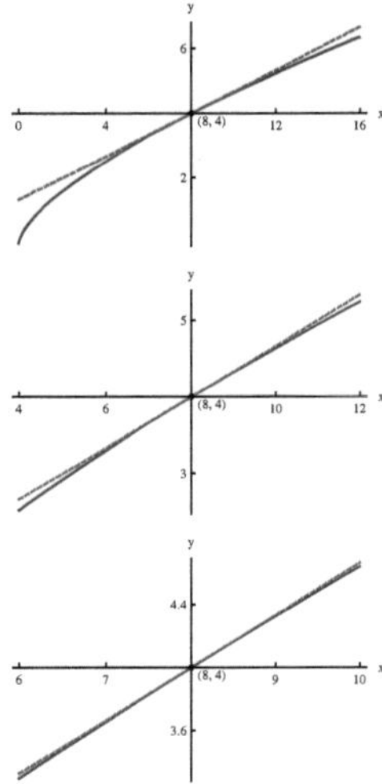

We see that the graph of the function becomes hardly distinguishable from the tangent line as we zoom in towards the point of contact $(8, f(8)) = (8, 4)$ (the axes are centered at $(8, 4)$).

15.

$$f'(x) = -2\sqrt{x}.$$

17.

$$f'(x) = 6x - 4.$$

19.

$$f'(x) = 2 - \frac{3}{2\sqrt{x}} + \frac{6}{x^{1/3}}.$$

Answers to Some Problems of Section 2.3

1. 4

3. 0

5.

a)

$$f'(x) = 2\cos(x).$$

b)

$$f'(\pi/3) = 1.$$

7.

a)

$$f'(x) = 3\cos(x) + 4\sin(x).$$

b)

$$f'(\pi/4) = \frac{7\sqrt{2}}{2}.$$

9.

a)
$$f'(x) = -4\sin(x) + \frac{2}{x^2}.$$

b)
$$f'(\pi) = \frac{2}{\pi^2}$$

11.

a)
$$f'(x) = 2\cos(x) + 3\sin(x).$$

b)
$$f'(\pi/2) = 3.$$

13.

a)
$$y = \frac{1}{2} + \frac{\sqrt{3}}{2}\left(x - \frac{\pi}{6}\right).$$

b)

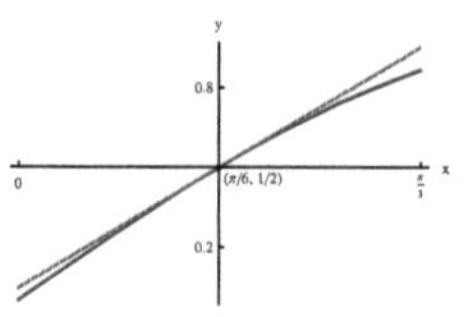
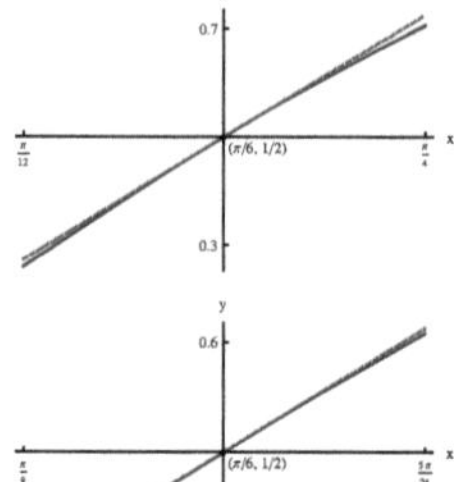

Answers to Some Problems of Section 2.4

1.

a)
$$v(t) = \frac{d}{dt}\left(200t - 5t^2\right) = 200 - 10t,$$

$$a(t) = \frac{d}{dt}\left(200 - 10t\right) = -10.$$

b)
$$v(1) = 200 - 10 = 190, \ a(1) = -10.$$

3.

a)
$$v(t) = \frac{d}{dt}\left(10\sin(t)\right) = 10\cos(t)$$

$$a(t) = \frac{d}{dt}\left(10\cos(t)\right) = 10\left(-\sin(t)\right) = -10\sin(t).$$

b)
$$v\left(\frac{\pi}{6}\right) = 10\cos\left(\frac{\pi}{6}\right) = 10\left(\frac{\sqrt{3}}{2}\right) = 5\sqrt{3},$$

$$a\left(\frac{\pi}{6}\right) = -10\sin\left(\frac{\pi}{6}\right) = -10\left(\frac{1}{2}\right) = -5.$$

Answers to Some Problems of Section 2.5

1.

a)
$$L_3(x) = 12 + 7(x - 3).$$

b)
$$f(3.1) \cong 12.7$$

c)
$$f(3.1) - L_3(3.1) = 0.01.$$

d)

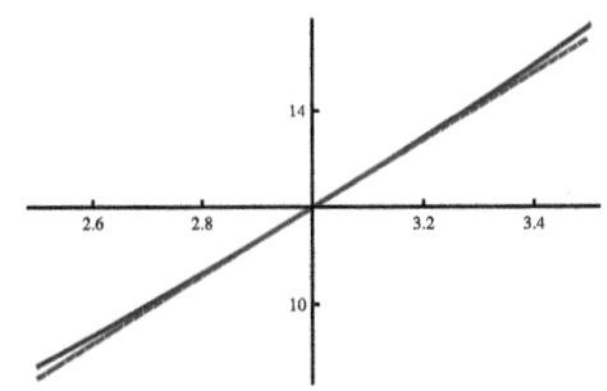

a)
$$L_{16}(x) = 2 + \frac{1}{32}(x - 16).$$

b)
$$f(16.2) \cong 2.0063$$

c)
$$f(16.2) - L_{16}(16.2) \cong -8 \times 10^{-5}.$$

3.

a)
$$L_{1/2}(x) = 4 - 16\left(x - \frac{1}{2}\right).$$

b)
$$f(0.502) \cong 3.968$$

c)
$$f(0.502) - L_{1/2}(0.502) \cong 1.9 \times 10^{-4}$$

d)

5.

7.

a)
$$df(x, \Delta x) = \frac{\Delta x}{2\sqrt{x}}.$$

b)
$$\sqrt{24.9} \cong 4.99.$$

c)
$$f(24.9) - 4.99 \cong -10^{-5}.$$

9. We set $f(x) = x^{1/3}$ and $x = 27$.

$$(26.5)^{1/3} \cong 2.98148$$

11. Set $f(x) = \sin(x)$ and $x = 3\pi/4.$ Therefore, $f'(x) = \cos(x)$, so that

$$df(x, dx) = \cos(x)\, dx.$$

$$\sin\left(\frac{3\pi}{4} + 0.1\right) \cong 0.63640$$

13.

a)
$$A(10.1) - A(10) \cong 2\pi.$$
$$A(10.1) \cong 102\pi.$$

b)
$$\frac{|A(10.1) - 102\pi|}{A(10.1)} \cong 10^{-4}$$

Answers to Some Problems of Section 2.6

1.
$$\left(3x^2 - 4x\right)\left(8x^2 - 7\right) + \left(x^3 - 2x^2 + 9\right)(16x)$$

3.
$$2x\cos(x) - x^2\sin(x).$$

5.
$$\left(24x^2 - 6\right)\cos(x) - \left(8x^3 - 6x + 2\right)\sin(x).$$

7.
$$\begin{aligned}
f'(x) &= -\frac{2}{x^3}\cos(x) - \frac{1}{x^2}\sin(x),\\
f''(x) &= \frac{6}{x^4}\cos(x) + \frac{4}{x^3}\sin(x) - \frac{1}{x^2}\cos(x).
\end{aligned}$$

9.
$$f'(x) = -\frac{8x}{\left(4x^2 + 1\right)^2}$$

for each $x \in \mathbb{R}$.

11.
$$f'(x) = \frac{-2x^2 - 2x - 8}{\left(x^2 - 4\right)^2}$$

if $x \neq 2$ and $x \neq -2$.

13.
$$f'(x) = -\frac{6x}{\left(x^2 - 4\right)^2} + 2$$

if $x \neq 2$ and $x \neq -2$.

15.
$$\frac{-x^2 - 9 + 2x}{\left(x^2 - 9\right)^2}$$

17.
$$2x\tan(x) + \frac{x^2}{\cos^2(x)}$$

19.
$$\frac{\sin(x) + x\cos(x) + x^2\cos(x)}{(1 + x)^2}$$

21.
$$\begin{aligned}
\frac{d}{dx}\cot(x) &= \frac{d}{dx}\left(\frac{\cos(x)}{\sin(x)}\right)\\
&= \frac{-\sin^2(x) - \cos^2(x)}{\sin^2(x)}\\
&= -\frac{1}{\sin^2(x)} = -\csc^2(x).
\end{aligned}$$

23.
a)
$$L_8(x) = \frac{1}{3} + \frac{1}{18}(x - 8).$$

b)
$$f(7.8) \cong \frac{1}{3} + \frac{1}{18}(-0.2) \cong 0.322\,222$$

The absolute error is about 1.9×10^{-4}.

Answers to Some Problems of Section 2.7

1.
$$\frac{x - 1}{\sqrt{x^2 - 2x + 5}}$$

3.
$$\frac{4x}{3\left(x^2 - 16\right)^{1/3}}.$$

5.
$$\frac{8x}{\sqrt{4 + x^2}\left(4 - x^2\right)^{3/2}}.$$

7.
$$10\cos(10x).$$

9.
$$\pi\cos(\pi x)$$

11.
$$\pi\cos(\pi x) - \pi\cos(2\pi x) + \pi\cos(3\pi x)$$

13.
$$\frac{1}{8}\cos\left(\frac{1}{2}x + \frac{\pi}{6}\right)$$

15.
$$-2x\sin\left(x^2\right)$$

17.
$$\frac{1}{2\sqrt{x}}\cos\left(\sqrt{x}\right)$$

19.
$$\frac{\cos(x/2)}{4\sqrt{\sin(x/2)}}$$

21.
$$-\frac{2\sin(1/x)\cos(1/x)}{x^2}$$

23.
$$\frac{x}{\cos^2\left(x^2\right)\sqrt{\tan\left(x^2\right)}}$$

25.
$$\begin{aligned}
f'(x) &= \frac{1}{x^2}\sin\left(\frac{1}{x}\right)\\
f''(x) &= -\frac{2}{x^3}\sin\left(\frac{1}{x}\right) - \frac{1}{x^4}\cos\left(\frac{1}{x}\right)
\end{aligned}$$

27.

a)

$$L_3 \left(x \right) = f \left(3 \right) + f' \left(3 \right) \left(x - 3 \right) = \frac{1}{5} - \frac{3}{125} \left(x - 3 \right).$$

b)

$$f \left(2.8 \right) \cong 0.204\,8$$

29.

a)

$$f' \left(x \right) = -\frac{2 \sin \left(x \right)}{3 \cos^{1/3} \left(x \right)} \frac{d}{dx} \left(\cos \left(x \right) \right)^{2/3}$$

The fundamental period of f is 2π and f is defined on the entire interval $[-\pi.\pi]$. The part of the domain of f' in the interval $[-\pi, \pi]$ consists of all x in $[-\pi, \pi]$ such that $\cos \left(x \right) \neq 0$, i.e., all x in $[-\pi, \pi]$ other than $\pm\pi/2$.

b) The graph of f has a cusps at $(\pi/2, 0)$ and $(-\pi/2, 0)$.

c)

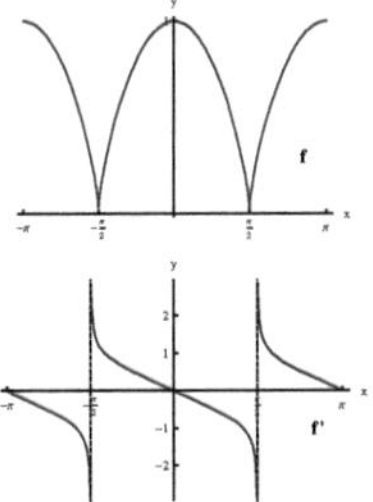

The pictures indicate the existence of cusps in the graph of f at $(\pm\pi/2, 0)$.

31.

a)

$$v \left(t \right) = 24 \cos \left(6t - \frac{\pi}{4} \right),$$
$$a \left(t \right) = -144 \sin \left(6t - \frac{\pi}{4} \right).$$

b)

$$a \left(t \right) = -36 y \left(t \right).$$

c) The fundamental period of the motion is

$$\frac{2\pi}{6} = \frac{\pi}{3},$$

and the amplitude of the motion is 4.

d)

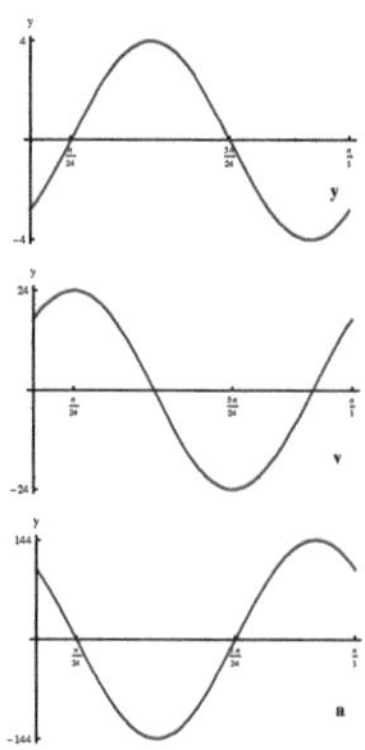

Answers to Some Problems of Section 2.8

1.
$$\left.\frac{dy}{dt}\right|_{x=2,x'(t)=3} = 48.$$

3.
$$\left.\frac{dy}{dt}\right|_{y=\pi/6,x'(t)=4} = \frac{4}{5\sqrt{3}}.$$

5.
$$\left.\frac{dr}{dt}\right|_{A=1000} = \frac{50}{\sqrt{1000\pi}}\ \text{(meters/hour)}.$$

7. The radius is decreasing at the rate of $1/\left(4\pi\right)$ cm/second at the instant $r = 5$ cm.

9.
$$\frac{dh}{dt} = \frac{2}{9}\pi \cong 0.7\ \text{m/sec.}$$

11.
$$\frac{ds}{dt} = \frac{6000}{\sqrt{425}} \cong 291\ \text{miles/hour.}$$

13. Let θ be the angle between the ground level and the ladder:
$$\frac{d\theta}{dt} = -\frac{10}{5\sqrt{84}} \cong -0.22\ \text{radians/sec}$$

Answers to Some Problems of Section 2.9

1.
a)
$$f\left(-1\right) = -\frac{6}{5} < 0 \text{ and } f\left(1\right) = \frac{4}{5} > 0.$$

b) The solution of $f\left(x\right) = 0$ in $[-1,1]$ is 0.246266.

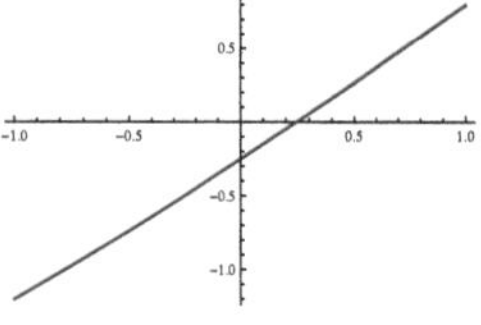

3.
a)
$$f\left(2\right) = 1.\,229\,35 > 0 \text{ and } f\left(4\right) = -0.475\,518 < 0$$

b) The solution of $f\left(x\right) = 0$ in $[2,4]$ is 2.89372.

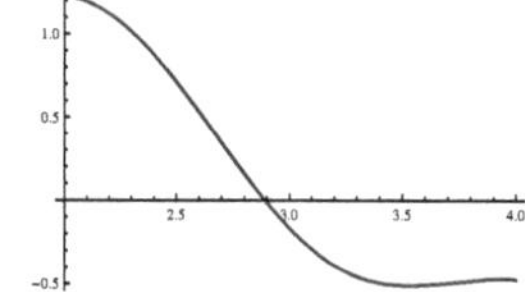

5.
a)

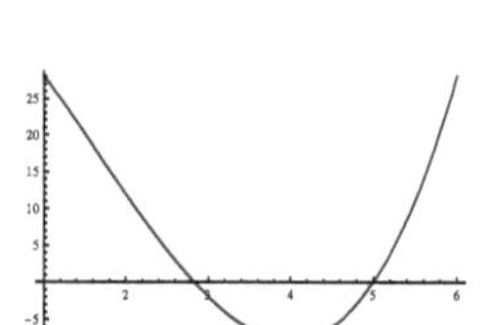

We see that the equation $f(x) = 0$ has solutions near 3 and 5.
b) With $x_0 = 3$,

n	x_{n+1}	$|x_{n+1} - x_n|$
0	2.81818	1.8×10^{-1}
1	2.8284	1.0×10^{-2}
2	2.82843	2.9×10^{-5}

The solution near 3 is $2\sqrt{2} \cong 2.82843$. The absolute error is 2.5×10^{-10}.
With $x_0 = 5$,

n	x_{n+1}	$|x_{n+1} - x_n|$
0	5	0

The exact solution is 5.

7.
a)

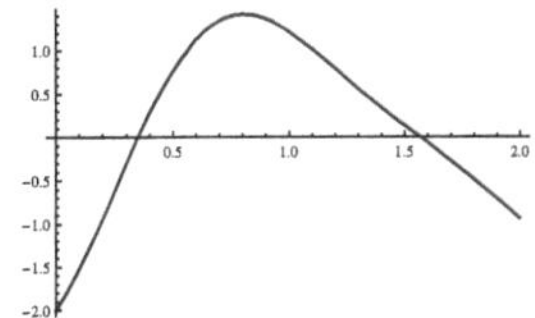

We see that the equation $f(x) = 0$ has solutions near 0.5 and 1.5.
b) With $x_0 = 0.5$,

n	x_{n+1}	$|x_{n+1} - x_n|$
0	0.322279	1.8×10^{-1}
1	0.351239	2.9×10^{-2}
2	0.351649	4.1×10^{-4}
3	0.351649	9.8×10^{-8}

The exact solution is 0.351649 (rounded to 6 significant digits). The absolute error is 5.7×10^{-15}.

With $x_0 = 1.5$,

n	x_{n+1}	$|x_{n+1} - x_n|$
0	1.57023	7×10^{-2}
1	1.5708	5.7×10^{-4}
2	1.5708	3.1×10^{-10}

The exact solution is $\pi/2 \cong 1.5708$ (rounded to 6 significant digits). The absolute error is almost 0.

9.
a)

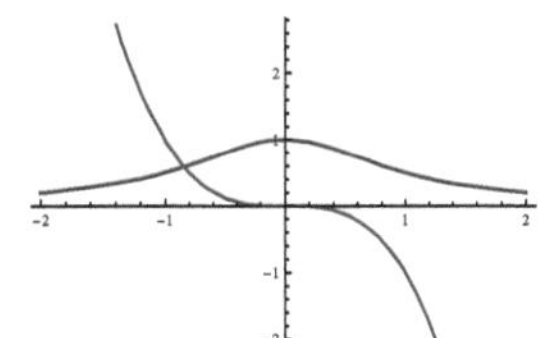

We see that the equation $f(x) = g(x)$ has a solution near -1.
b) We set

$$F(x) = f(x) - g(x) = \frac{1}{x^2 + 1} + x^3,$$

$$G(x) = x - \frac{F(x)}{F'(x)},$$

and $x_{n+1} = G(x_n)$, $n = 0, 1, 2, \ldots$. In order to approximate the solution we set $x_0 = -1$:

n	x_{n+1}	$|x_{n+1} - x_n|$
0	-0.857143	1.4×10^{-1}
1	-0.837939	1.9×10^{-2}
2	-0.83762	3.2×10^{-4}
3	-0.83762	8.7×10^{-8}

The exact solution is -0.83762. The absolute error is 6.4×10^{-15}.

Answers to Some Problems of Section 2.10

1.
a)

$$\left.\frac{dy}{dx}\right|_{x=4, y=-5} = -\frac{4}{5}.$$

b)

$$y = -\sqrt{9 + x^2}$$

$$y^2 - x^2 = 9 \Rightarrow y = \pm\sqrt{9 + x^2},$$

$$\frac{dy}{dx} = -\frac{x}{\sqrt{9 + x^2}}.$$

c)

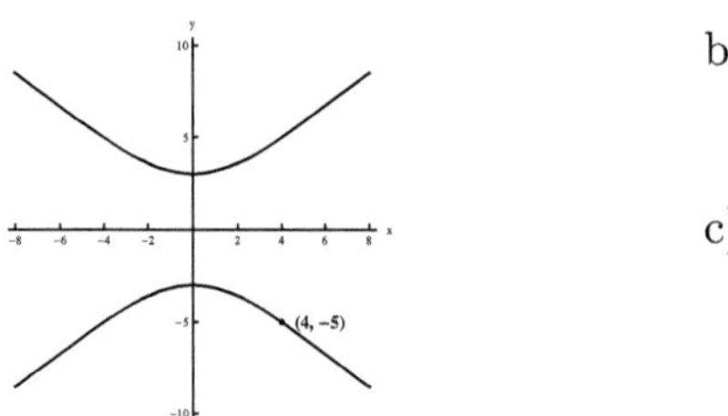

The graph of the equation is a hyperbola. The graph of the function $y(x) = -\sqrt{9 + x^2}$ is the branch of the hyperbola in the lower half-plane.

3.

a)

$$\frac{dy}{dx}(1) = -\frac{\sqrt{5}}{10}.$$

b)

$$y = 2 - \sqrt{x + 4},$$

$$\frac{dy}{dx} = -\frac{1}{2\sqrt{x+4}}.$$

c)

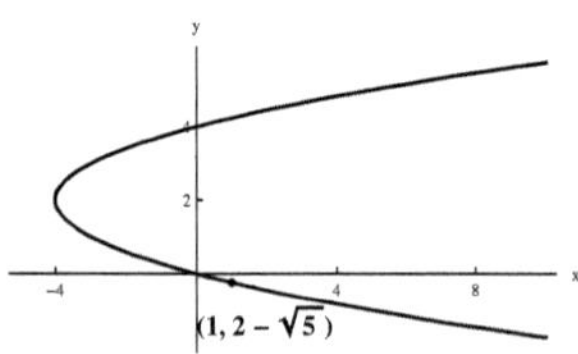

The graph of the equation is a parabola. The graph of the function

$$y(x) = 2 - \sqrt{x + 4}$$

is the part of the parabola that is below the lin $y = 2$.

5.

$$\frac{dy}{dx} = \frac{-2x + y}{-x + 4y}$$

7.

$$\frac{dy}{dx} = \frac{4y}{3y^2 - 4x}$$

9.

$$\frac{dy}{dx} = \frac{\cos(x)}{2\sin(x)}$$

11.

a)

$$\frac{dy}{dx} = \frac{1}{y^2 - 3}$$

b)

$$y = 3 + \frac{1}{6}x.$$

c)

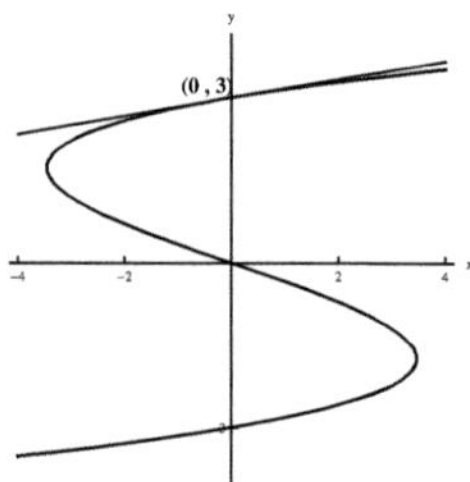

13.

a)

$$\frac{dy}{dx} = \frac{3y^2 - 3x^2}{3y^2 - 6xy}$$

b)

$$y = -1 - \frac{3}{5}(x - 2).$$

c)

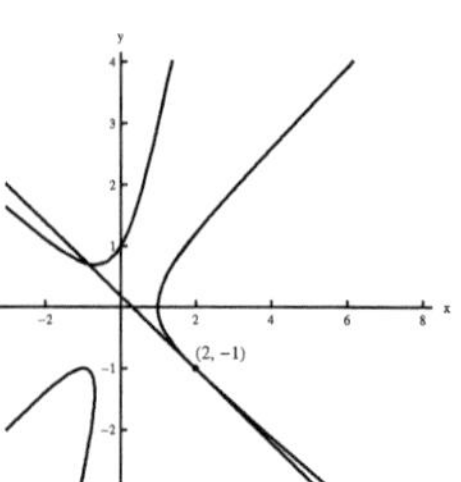

15.

a)

$$\frac{dy}{dx} = \cos^2(y)$$

b)

$$y = \frac{\pi}{4} + \frac{1}{2}(x - 1).$$

c)

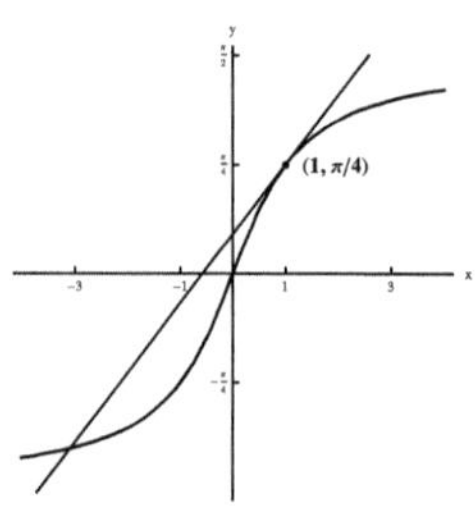

Answers to Some Problems of Section 3.1

1. The only critical point is 2. With the help of the picture, we see that f has a local (and absolute) minimum at $x = 2$.

3. The critical points of f are $-\sqrt{3}$ and $\sqrt{3}$. With the help of the picture, we see that f has a local minimum at $-\sqrt{3}$ and local maximum at $\sqrt{3}$.

5. The critical points of f are -3, $4/7$ and 2. With the help of the picture, we see that f has a local maximum at -3 and a local minimum at $9/7$. The function does not have a local maximum or minimum at the critical point 2.

7.
a) The only critical point of f is -2.
b) f is decreasing on $(-\infty, -2]$, increasing on $[-2, +\infty)$. The function has a local (and absolute) minimum at -2.

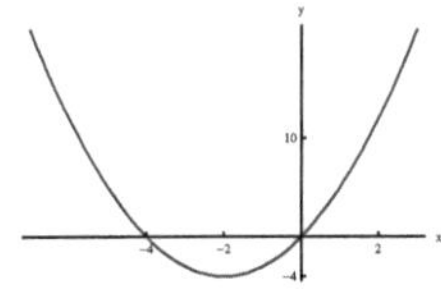

9.
a) The critical points of f are 1 and -1.
b)

f is decreasing on $(-\infty, -1]$, increasing on $[-1, 1]$, decreasing on $[1, +\infty)$.

Thus, f has a local minimum at -1 and a local maximum at 1.

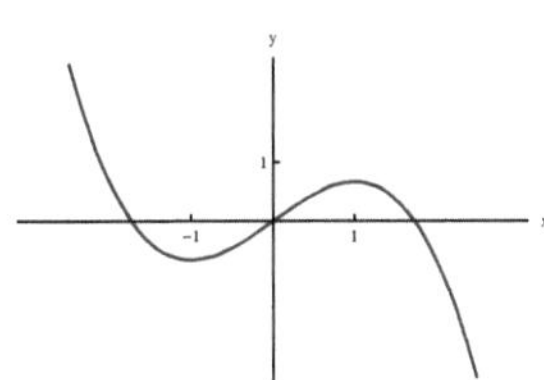

11.
a) The critical points of f are 0, $\pm 1/\sqrt{2}$.
b) f is decreasing on $(-\infty, -1/\sqrt{2}]$, increasing on $[-1/\sqrt{2}, 0]$, decreasing on $[0, 1/\sqrt{2}]$ and increasing on $[1/\sqrt{2}, +\infty)$. Therefore, f has a local minimum at $-1/\sqrt{2}$, a local maximum at 0, and a local minimum at $1/\sqrt{2}$.

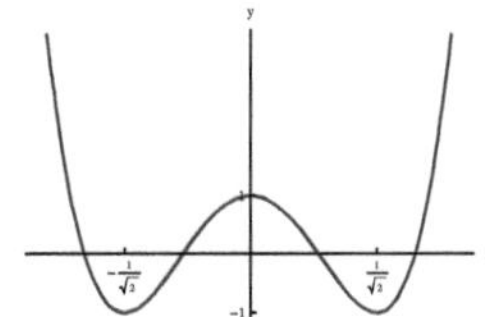

13.
a) The only critical point of f is $1/2$.
b) f is increasing on $(-\infty, -1)$, increasing on $(-1, 1/2]$, decreasing on $[1/2, 2)$, and decreasing on $(2, +\infty)$. Thus, f has a local maximum at $1/2$.

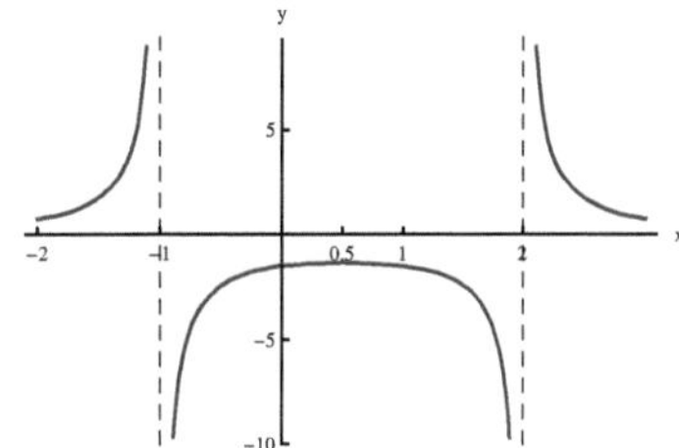

15.
a) The critical points of f are -3 and -1.
b) f is increasing on $(-\infty, -3]$, decreasing on $[-3, -2)$, decreasing on $(-2, -1]$, and increasing on $[-1, +\infty)$. Thus, f has a local maximum at -3 and a local minimum at -1.

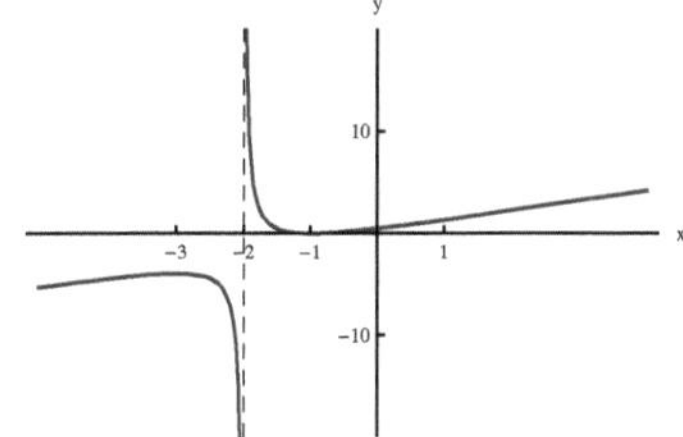

17.
a) The only stationary point of f is 3. The function is defined at 4 but not differentiable at 4, so that 4 is also a critical point.
b) f is decreasing on $(-\infty, 3]$, increasing on $[3, 4]$, and increasing on $[4, +\infty)$. Thus, f has a local minimum at 3 (f does ot have a local maximum or minimum at 4, even though 4 is a critical point of f: the graph of f has a vertical tangent at $(4,0)$).

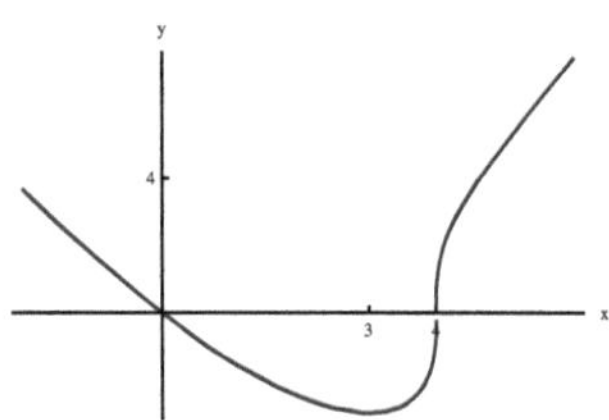

19. The absolute minimum of f on $[0, +\infty)$ is $f(1) = -7/6$. The function does not have an absolute maximum on $[0, +\infty)$ since $\lim_{x \to +\infty} f(x) = +\infty$.

21. The absolute maximum of f on $(-\infty, 4)$ is $f\left(4 - \sqrt{2}/2\right)$. The function does not have an absolute minimum on $(-\infty, 4)$ since $\lim_{x \to -\infty} f(x) = -\infty$.

23. The absolute maximum of f on $(-4, 3)$ is $f(1/2) = 4/9$. The function does not have an absolute minimum on $(-4, 3)$ ($\lim_{x \to 3-} f(x) = -\infty$).

25. f attains its absolute minimum value 0 on $[0, \infty)$ at 0 and 2. On the other hand, f does not have an absolute maximum on $[0, \infty)$ since

$$\lim_{x \to \infty} f(x) = \lim_{x \to \infty} x(x - 2)^{4/5} = \infty.$$

Answers to Some Problems of Section 3.2

1. The absolute maximum of f on $[-4, 4]$ is $f(4) = 76/3$, and the absolute minimum of f on $[-4, 4]$ is $f(1) = -5/3$.

3. The absolute maximum of f on $[-1, 4]$ is $f(4) = 16$, and the absolute minimum of f on $[-1, 4]$ is $f(-1) = -51/4$.

5. The absolute maximum of f on $[0, 2]$ is $f\left(\sqrt{12/5}\right) = 48/5^{5/3}$, and the absolute minimum of f on $[0, 2]$ is $f(0) = f(2) = 0$.

7. The absolute maximum of f on $[\pi/3, 3\pi/4]$ is $f(3\pi/4) = 1/4$, and the absolute minimum of f on $[\pi/3, 3\pi/4]$ is $f(\pi/2) = 0$.

9.

$$f'\left(\sqrt{\frac{13}{2}}\right) = \frac{f(3) - f(1)}{2}.$$

Answers to Some Problems of Section 3.3

1. The graph of f is concave up on $(-\infty, 0]$ and concave down on $[0, +\infty)$. The x-coordinate of the only point of inflection of the graph of f is 0 (the point $(0, 0)$ is the inflection point).

3. the graph of f is concave down on $(-\infty, -\sqrt{3}]$, concave up on $[-\sqrt{3}, 0]$, concave down on $[0, \sqrt{3}]$ and concave up on $[\sqrt{3}, +\infty)$. The x-coordinates of the points of inflection are $-\sqrt{3}$, 0 and $\sqrt{3}$.

5. The graph of f is concave up on $[0, \pi/4]$, concave down on $[\pi/4, 3\pi/4]$ and concave up on $[3\pi/4, \pi]$. The x-coordinates of the points of inflection of the graph of f are $\pi/4$ and $3\pi/4$.

7.
a) The graph of f is concave down on $(-\infty, -2]$ and concave up on $[-2, +\infty)$. Thus, the x-coordinate of the point of inflection on the graph of f is -2.
b) The absolute minimum of f' on $(-\infty, +\infty)$ is $f'(-2) = -2$. The derivative function

f' does not have an absolute maximum on $(-\infty, +\infty)$ since $\lim_{x \to \pm\infty} f'(x) = +\infty$.

9.
a) The graph of f is concave up on $(-\infty, -2]$, concave down on $[-2, 3]$, concave up on $[3, +\infty)$. The x-coordinates of the points of inflection are -2 and 3.
b) The absolute maximum of f' on $(-\infty, 0]$ is $f'(-2) = 22/3$. The derivative does not have an absolute minimum on $(-\infty, 0]$ since $\lim_{x \to -\infty} f'(x) = -\infty$.

11. The absolute maximum of f on $(-\infty, 0)$ is

$$f(-2) = \frac{22}{3}$$

13. f has a local maximum at 0 and local minima at $\pm 1/\sqrt{2}$.

15. f has a local maximum at -3 and a local minimum at -1.

Answers to Some Problems of Section 3.4

1.

a)

$$\lim_{x \to +\infty} f(x) = +\infty, \quad \lim_{x \to -\infty} f(x) = -\infty.$$

b) The function is increasing on $(-\infty, -\sqrt{2/3}]$, decreasing on $\left[-\sqrt{2/3}, \sqrt{2/3}\right]$, and increasing on $[\sqrt{2/3}, +\infty)$. Thus, f has a local maximum at $-\sqrt{2/3}$ and a local minimum at $\sqrt{2/3}$.

c) The graph of f is concave down on $(-\infty, 0]$ and concave up on $[0, +\infty)$. The point $(0, 0)$ is the only inflection point on the graph of f.

d)

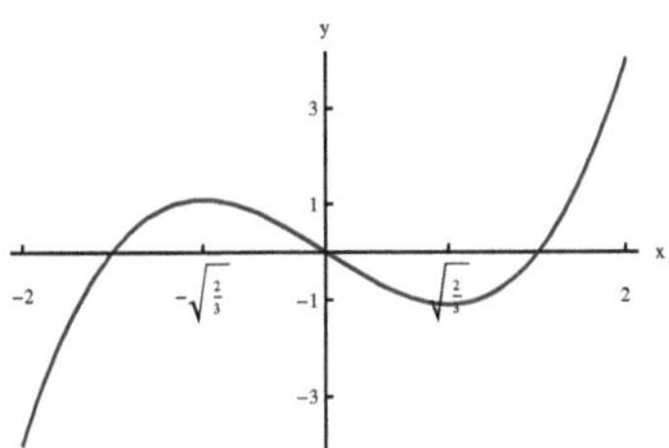

3.

a)

$$\lim_{x \to -\infty} f(x) = -\infty, \quad \lim_{x \to +\infty} f(x) = +\infty.$$

b) The function is increasing on $(-\infty, -3]$, decreasing on $[-3, 4]$ and increasing on $[4, +\infty)$. Therefore, f has a local maximum at -3 and a local minimum at 4.

c) The graph of f is concave down on $(-\infty, -1/2]$ and concave up on $[1/2, +\infty)$. The x-coordinate of the point of inflection is $1/2$.

d)

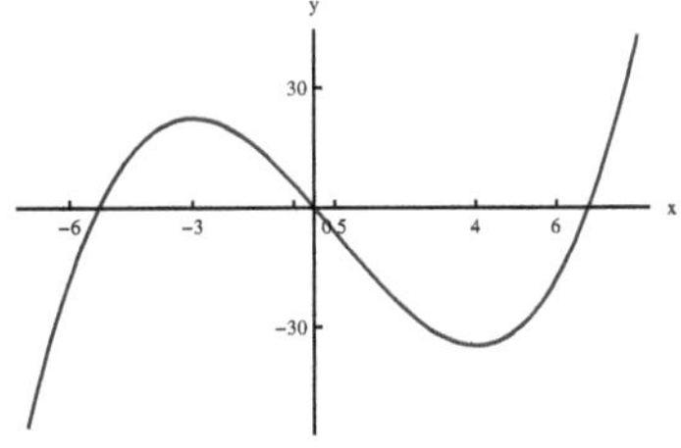

5.

a) The domain of f consists of all x such that $x \neq -2$ and $x \neq 3$. The lines $x = -2$ $x = 3$ are vertical asymptotes for the graph of f.

$$\lim_{x \to -2-} f(x) = +\infty, \quad \lim_{x \to -2+} f(x) = -\infty,$$
$$\lim_{x \to 3-} f(x) = -\infty, \quad \lim_{x \to 3+} f(x) = +\infty.$$

b) The line $y = 4$ is a horizontal asymptote for the graph of f.

c) The function is increasing on $(-\infty, -2)$, increasing on $(-2, 1/2]$, decreasing on $[1/2, 3)$ and decreasing on $(3, +\infty)$. Thus, f has a local maximum at $1/2$.

d)

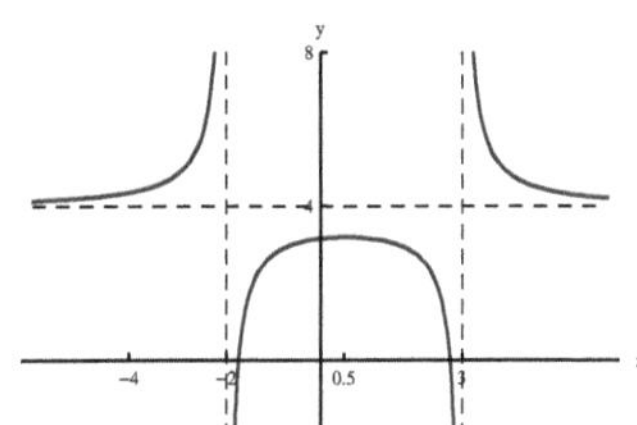

7.

a) The domain of f is the set of all x such that $x = 2$. The line $x = 2$ is a vertical asymptote for the graph of f.

$$\lim_{x \to 2-} f(x) = -\infty, \quad \lim_{x \to 2+} f(x) = +\infty.$$

b)

$$\lim_{x \to -\infty} f(x) = -\infty \text{ and } \lim_{x \to +\infty} f(x) = +\infty.$$

The line $y = 4x$ is an oblique asymptote for the graph of f at $\pm\infty$.

c) The function is increasing on $(-\infty, 3/2]$, decreasing on $[3/2, 2)$, decreasing on $(2, 5/2]$, and increasing on $[5/2, +\infty)$. Thus, f has a local maximum at $3/2$ and a local minimum at $5/2$.

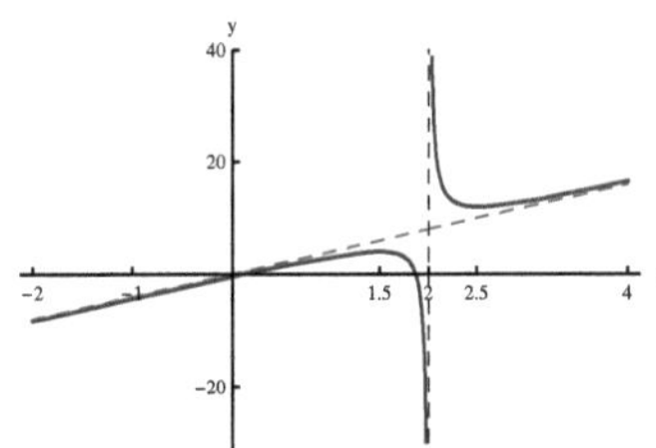

e)

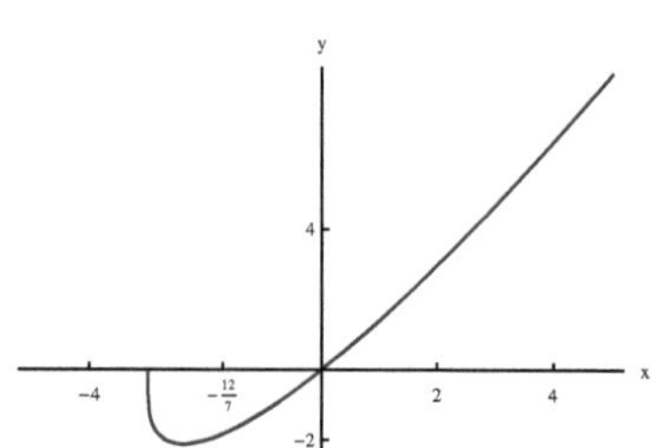

9.

a) The domain of f is $[-3, +\infty)$.

b)
$$\lim_{x \to +\infty} f(x) = +\infty.$$

c) The domain of f' consists of each $x \in \mathbb{R}$ such that $x > -3$. The graph of f has a vertical tangent at $(-3, 0)$.

d) The function is decreasing on $[-3, -12/7]$, and increasing on $[-12/7, +\infty)$. Thus, f has a local minimum at $-12/7$.

Answers to Some Problems of Section 3.5

1.
$$20, 20$$

3.
$$20\sqrt{2}, \ 20\sqrt{2}.$$

5.
$$5, \frac{5}{2}\sqrt{3}.$$

7.
$$\left(\frac{4}{3}, \frac{1}{3}\sqrt{5}\right) \text{ and } \left(\frac{4}{3}, -\frac{1}{3}\sqrt{5}\right).$$

9. The side length is $2000^{1/3}$ and the height is
$$\frac{1000}{2000^{2/3}}.$$

11.
$$\theta = \frac{\pi}{4}.$$

13.

a)

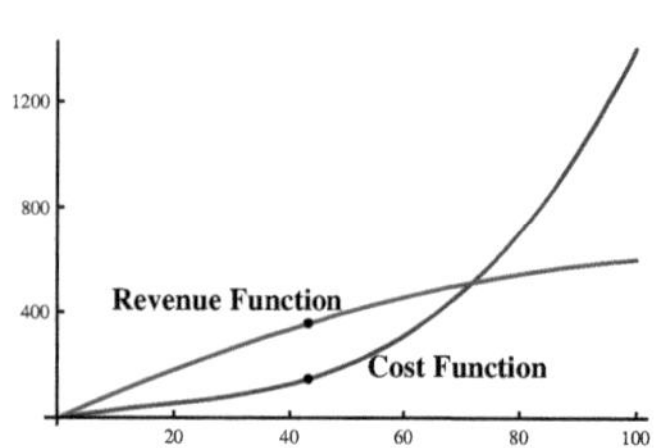

b)
$$10\sqrt{11} + 10$$

Answers to Some Problems of Section 4.1

1.
$$f^{-1}(x) = (x+8)^{1/3} \text{ for each } x \in \mathbb{R}$$

The domain and range of f and f^{-1} is $\mathbb{R}$.

3.
$$f^{-1}(x) = \frac{2(1+x)}{1-x}.$$

The domain f consists of all y other than -2. This is also the range of f^{-1}. The domain of f^{-1} consists of all x other than 1.

5.

a)

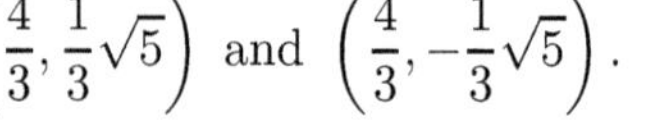

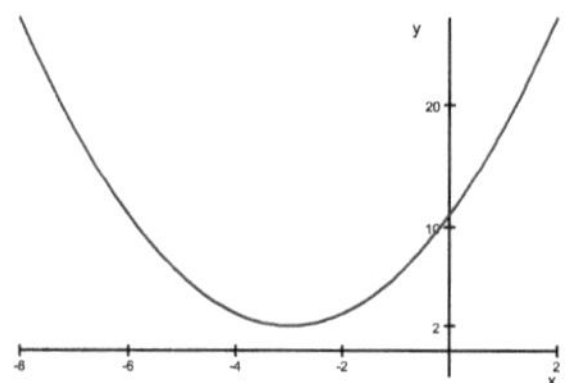

The graph of F fails the horizontal line test. The equation $F(y) = 20$ has two distinct solutions, $:-3 \pm 2\sqrt{7}$. Therefore, F does not have an inverse.

b)

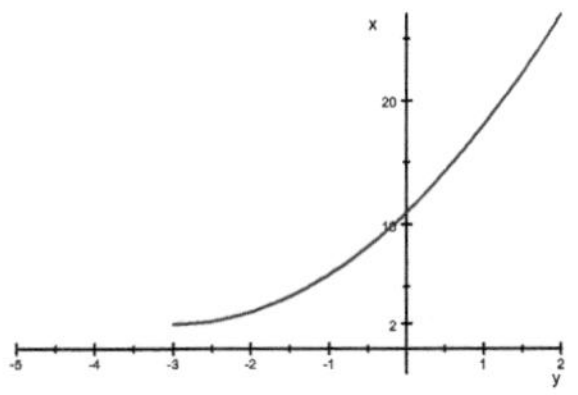

$$y = f^{-1}(x) = -3 + \sqrt{x-2} \text{ for } x \geq 2.$$

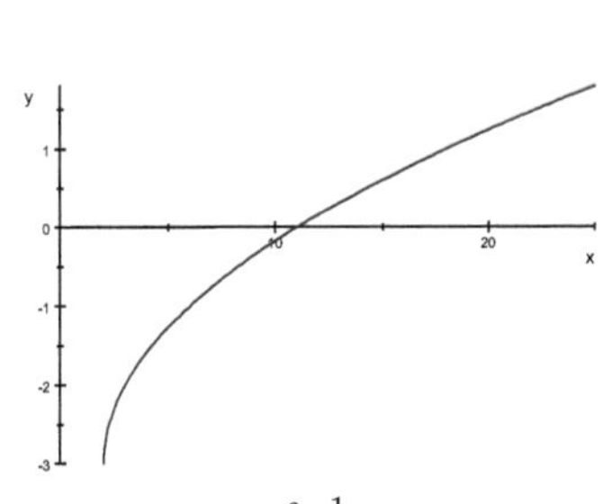

$$f^{-1}$$

7.

$$
\begin{aligned}
\left(f^{-1} \circ f\right)(y) &= f^{-1}\left(f(y)\right) \\
&= f^{-1}\left(\frac{3y+2}{2y+1}\right) \\
&= \frac{2 - \dfrac{3y+2}{2y+1}}{2\left(\dfrac{3y+2}{2y+1}\right) - 3} = y,
\end{aligned}
$$

$$
\begin{aligned}
\left(f \circ f^{-1}\right)(x) &= f\left(f^{-1}(x)\right) \\
&= f\left(\frac{2-x}{2x-3}\right) \\
&= \frac{3\left(\dfrac{2-x}{2x-3}\right) + 2}{2\left(\dfrac{2-x}{2x-3}\right) + 1} = x.
\end{aligned}
$$

We have $y \neq -1/2$ and $x \neq 3/2$.

9.
$$\arcsin\left(\frac{\sqrt{3}}{2}\right) = \frac{\pi}{3}$$

11.
$$\arcsin\left(-\frac{1}{2}\right) = -\frac{\pi}{6}$$

13.
$$\arccos\left(-\frac{\sqrt{3}}{2}\right) = \frac{5\pi}{6}$$

15.
$$\arccos\left(-\frac{\sqrt{2}}{2}\right) = \frac{3\pi}{4}$$

17.
$$\arctan\left(\frac{1}{\sqrt{3}}\right) = \frac{\pi}{6}$$

18.
$$\arctan(-1) = -\frac{\pi}{4}$$

19.
$$\arctan(0) = 0$$

Answers to Some Problems of Section 4.2

1.
$$\frac{d}{dx}\arcsin(2x) = \frac{2}{\sqrt{1-4x^2}}$$

3.
$$\frac{d}{dx}\arccos(x/9) = -\frac{1}{9\sqrt{1-\frac{1}{81}x^2}}$$

5.
$$\frac{d}{dx}\arctan(x/2) = \frac{2}{4+x^2}$$

7.
$$\frac{d}{dx}\sqrt{\arctan(x)} = \frac{1}{2(x^2+1)\sqrt{\arctan(x)}}$$

9.
$$\frac{d}{dx}\left(\sqrt{1-x^2}\arccos(x)\right) = -\frac{x\arccos(x)}{\sqrt{1-x^2}} - 1$$

11.
a)
$$L_{1/2}(x) = \frac{\pi}{6} + \frac{2}{\sqrt{3}}\left(x - \frac{1}{2}\right).$$

b
$$\arcsin(0.49) \cong L_{1/2}(0.49) \cong 0.512\,052$$

13. The graph of arcsine is concave down on $[-1,0]$ and concave up on $[0,1]$. The point of inflection is $(0,0)$.

15. The graph of arctangent is concave up on $(-\infty, 0]$ and concave down on $[0, +\infty)$. The point of inflection is $(0,0)$.

Answers to Some Problems of Section 4.3

1. $(4, +\infty)$

3.
$$\ln\left(e^4\right) = 4$$

5.
$$\ln\left(e^{34/5}\right) = \frac{34}{5}$$

7.
$$e^{-\ln(4)} = \frac{1}{4}$$

9.
$$\ln\left(x^{1/5}\right) = \frac{1}{5}\ln(x)$$

11.
$$e^{-2/3}$$

13. There is no solution.

15.
$$f(x) = \ln(x+2) - \ln(x-2).$$
$$f'(x) = \frac{1}{x+2} - \frac{1}{x-2}.$$

17.
$$\frac{1}{2}\ln\left(x^2 + 16\right).$$
$$f'(x) = \frac{x}{x^2 + 16}.$$

19.
$$4\ln\left(x^2 + 9\right) + 2\ln(x+1).$$
$$f'(x) = \frac{8x}{x^2 + 4} + \frac{2}{x+1}.$$

21.
a)
$$\frac{1}{x-1} + \frac{3}{x+1} + \frac{2x}{x^2+4}$$

b)
$$(x-1)(x+1)^3\left(x^2+4\right)\left(\frac{1}{x-1} + \frac{3}{x+1} + \frac{2x}{x^2+4}\right)$$

23.
a)
$$\frac{8x}{x^2+16} - \frac{3}{x-4} - \frac{x}{x^2-9}$$

b)
$$\frac{\left(x^2+16\right)^4}{(x-4)^3\sqrt{x^2-9}}\left(\frac{8x}{x^2+16} - \frac{3}{x-4} - \frac{x}{x^2-9}\right)$$

25.
$$\frac{1}{2}\left(e^x - e^{-x}\right)$$

27.
$$3x^2 e^{-x} - x^3 e^{-x}$$

29.
$$\frac{e^x}{(2e^x + 1)^2}$$

31.
$$-\frac{1}{4}(x-1)\, e^{-\frac{1}{8}(x-1)^2}$$

33.
$$2\cos(x)\sin(x)\, e^{-\cos^2(x)}$$

35.
$$-\frac{1}{2}e^{-x/2}\cos\left(\frac{x}{4}\right) - \frac{1}{4}e^{-x/2}\sin\left(\frac{x}{4}\right)$$

Answers to Some Problems of Section 4.4

1. 4

3. $\dfrac{3}{4}$

5. $\dfrac{1}{4}$

7. $10^{2/3}$

9. $2^{2/3}$

11. $\pm\sqrt{\log_2(14)}$

13.
$$\frac{d}{dx}10^{1/x} = -\frac{\ln(10)}{x^2}10^{1/x}.$$

15.
$$\frac{d}{dx}3^{-x^2} = -2\ln(3)\,x3^{-x^2}.$$

17.
$$\frac{d}{dx}\log_{10}\left(x^2+1\right) = \frac{2x}{(x^2+1)\ln(10)}$$

19.
$$\frac{d}{dx}\log_{10}\left(\frac{x-1}{x+4}\right)$$
$$= \frac{1}{(x-1)\ln(10)} - \frac{1}{(x+4)\ln(10)}$$

21.

$$\frac{d}{dx}x^{\sqrt{3}} = \sqrt{3}x^{\sqrt{3}-1}.$$

23

$$\frac{d}{dx}x^{\pi} = \pi x^{\pi-1}$$

25.

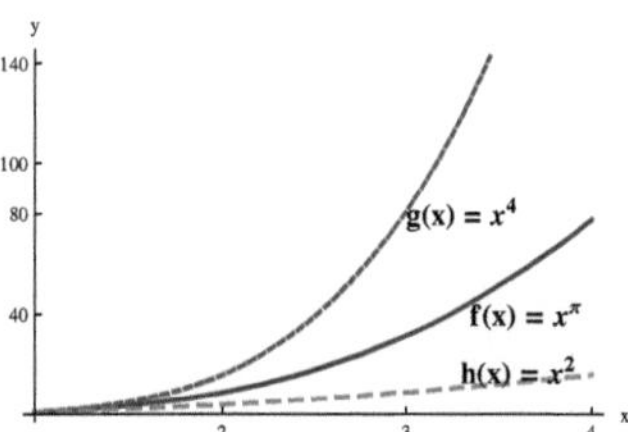

Answers to Some Problems of Section 4.5

1.

a)

$$\lim_{x\to0+}\frac{e^{x/2}}{x} = +\infty, \quad \lim_{x\to0-}\frac{e^{x/2}}{x} = -\infty,$$

$$\lim_{x\to+\infty}\frac{e^{x/2}}{x} = +\infty, \quad \lim_{x\to-\infty}\frac{e^{x/2}}{x} = 0.$$

The vertical axis is a vertical asymptote for the graph of f.

b) The function is decreasing on $(-\infty, 0)$, decreasing on $(0, 2]$, and increasing on $[2, +\infty)$. Therefore, f has a local minimum at 2.

c)

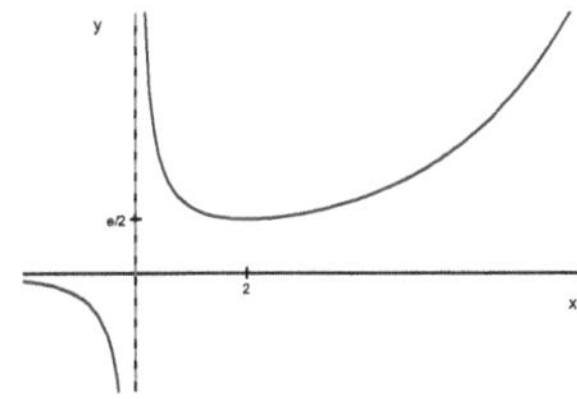

3

a)

$$\lim_{x\to0+}\frac{\ln(x)}{x^{1/3}} = -\infty \text{ and } \lim_{x\to+\infty}\frac{\ln(x)}{x^{1/3}} = 0.$$

The vertical axis is a vertical asymptote for the graph of f, and the horizontal axis is a horizontal asymptote at $+\infty$.

b) The function is increasing on $(0, e^3]$ and decreasing on $[e^3, +\infty)$. Thus, f has a local (and absolute) maximum at e^3.

5.

a)

$$\lim_{x\to\pm\infty} xe^{-x^2/4} = 0.$$

The x-axis is a horizontal asymptote for the graph of f at $\pm\infty$.

b) The function is decreasing on $(-\infty, -\sqrt{2}]$, increasing on $[-\sqrt{2}, \sqrt{2}]$, decreasing on $[\sqrt{2}, +\infty)$. Thus, f has a local (and global) minimum at $-\sqrt{2}$, and a local (and global) maximum at $\sqrt{2}$.

c) The graph of f is concave down on $(-\infty, -\sqrt{6}]$, concave up on $[-\sqrt{6}, 0]$, concave down on $[0, \sqrt{6}]$, concave up on $[\sqrt{6}, +\infty)$. Thus, the x-coordinates of the inflection points are $-\sqrt{6}$, 0, $\sqrt{6}$.

d)

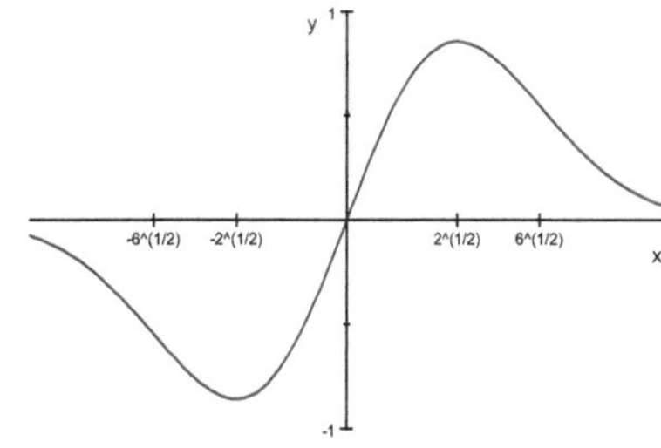

7.

a)

$$\lim_{x \to +\infty} f(x) = 10, \quad \lim_{x \to -\infty} f(x) = 0$$

The line $y = 10$ is a horizontal asymptote for the graph of f at $+\infty$ and the x-axis is a horizontal asymptote for the graph of f at $-\infty$.

b) f is increasing on the entire number line $(-\infty, +\infty)$.

c) The graph of f is concave up on $(-\infty, \ln(2)]$ and concave down on $[\ln(2), +\infty)$. Thus, the x-coordinate of the point of inflection is $\ln(2)$ (the point of inflection is $(\ln(2), 5)$).

d)

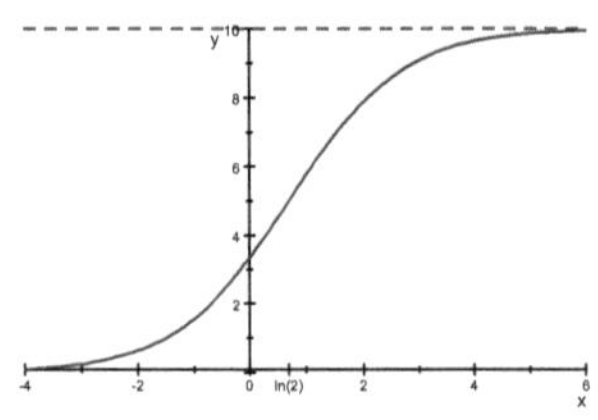

9. f attains its absolute minimum on $(0, +\infty)$

at 1 $(f(1) = e^2)$. The function does not have an absolute maximum on $(0, +\infty)$.

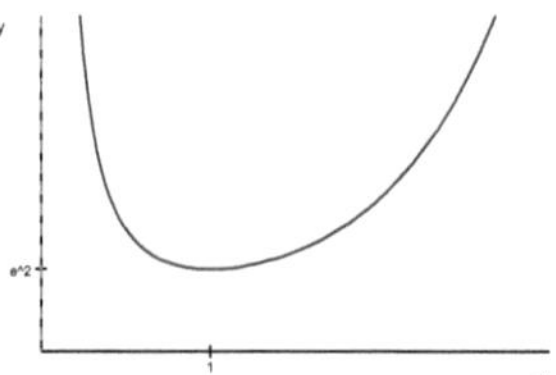

11. f attains its absolute maximum on $(0, +\infty)$ at e^2 $(f(e^2) = 2/e)$. The function does not have an absolute minimum on $(0, +\infty)$.

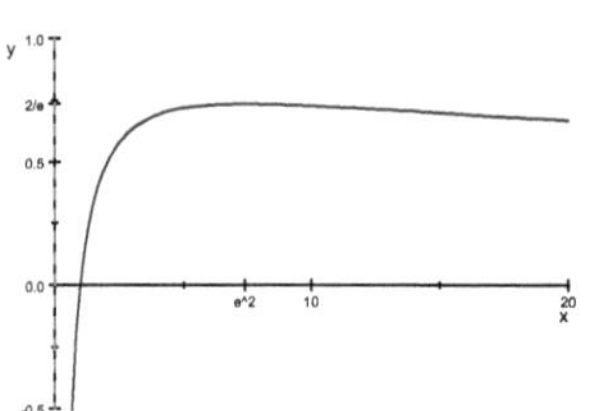

Answers to Some Problems of Section 4.6

1.

a) $Ce^{t/4}$

b) $\pm 10e^{(t-2)/4}$, $\pm 20e^{(t-2)/4}$

c)

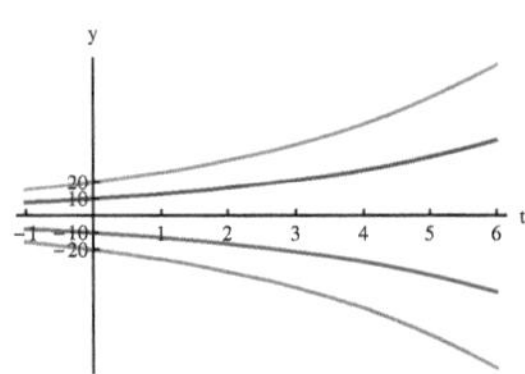

3.

a) $Ce^{-t/5}$

b) $10e^{-(t-1)/5}$

5.

a) $Ce^{t/10}$

b) $2e^{t/10}$

7.

a) $y(t) = 100e^{0.025t}$ million.

b) $y(20) \cong 164.872$ million, $y(40) \cong 271.828$ million, $y(80) \cong 738.906$ million.

c)

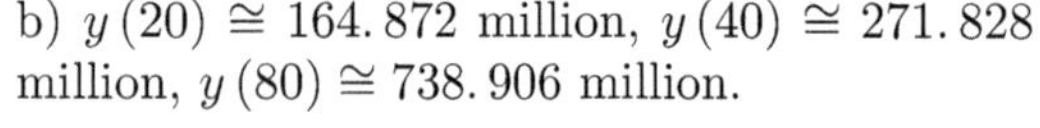

9.

a) $y(t) = 40e^{-0.03t}$ grams.

b) $y(100) \cong 1.99148$ grams, $y(50) \cong 8.92521$ grams, $y(150) \cong 0.444360$ grams

c)

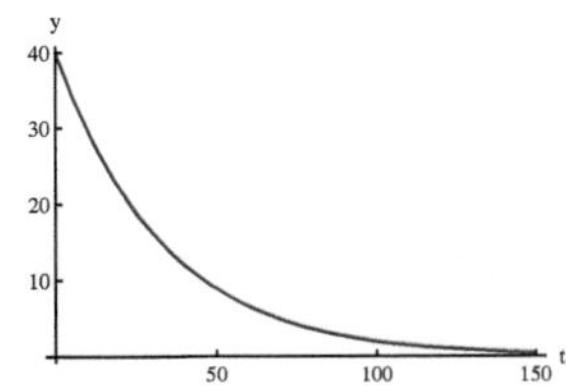

11.

a)
$$y\left(t\right) = \frac{100}{2^{t/50}} \text{ grams.}$$

b)

$$y\left(100\right) = 25 \text{ grams, } y\left(200\right) = 6.25 \text{ grams,}$$
$$y\left(400\right) = 0.390\,625 \text{ grams.}$$

13.

a) The doubling time is 4. $y\left(t\right) = 1000\left(2^{t/4}\right)$.

b)

$$y\left(24\right) = 64\,000, \; y\left(36\right) = 512\,000,$$
$$y\left(48\right) = 4.096 \times 10^{6}, \; y\left(60\right) = 3.276\,8 \times 10^{7}$$

15. Time t is in years.

a) $Y\left(t\right) = 4000\left(1.04^{t}\right)$,

$$Y\left(10\right) \cong 5920.98, \; Y\left(20\right) \cong 8764.49,$$
$$Y\left(30\right) \cong 12973.6, \; Y\left(40\right) \cong 19204.1$$

b) $y\left(t\right) = 4000e^{0.04t}$. Therefore,

$$y\left(10\right) \cong 5967.30, \; y\left(20\right) \cong 8902.16,$$
$$y\left(30\right) \cong 13280.5, \; y\left(40\right) \cong 19812.1$$

We see that continuous compounding has led to somewhat higher numbers.

c)

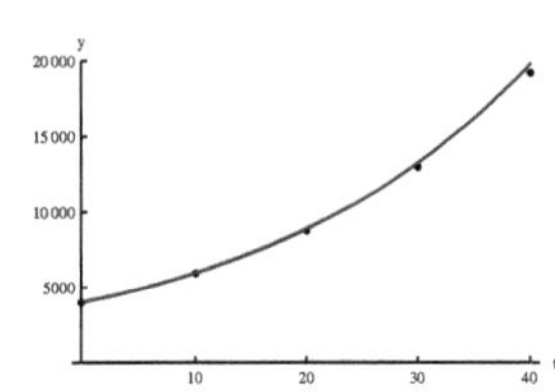

Answers to Some Problems of Section 4.7

1.
$$\cosh\left(\ln\left(2\right)\right) = \frac{5}{4}$$

3.
$$\tanh\left(\ln\left(3\right)\right) = \frac{4}{5}$$

5.
$$\frac{d}{dx}\cosh\left(4x\right) = \frac{1}{4}\sinh\left(4x\right)$$

7.
$$\frac{d}{dx}\left(\frac{1}{\cosh\left(x\right)}\right) = -\frac{\sinh\left(x\right)}{\cosh^{2}\left(x\right)}$$

17.
$$\frac{d}{dx}\operatorname{arcsinh}\left(x/4\right) = \frac{1}{\sqrt{x^{2}+16}}$$

19.
$$\frac{d}{dx}\operatorname{arccosh}\left(3x\right) = \frac{3}{\sqrt{9x^{2}-1}}$$

Answers to Some Problems of Section 4.8

1. $-\dfrac{1}{6}$

3. $-\dfrac{1}{2}$

5. $-\dfrac{1}{4}$

7. $-\sqrt{3}$

9. $+\infty$

11. $+\infty$

13. 0

15. e^{3}

17. 1

19. $\ln\left(5/2\right)$

Answers to Some Problems of Section 5.1

1.

$$\sum_{k=1}^{5}(4k-1)=55$$

3.

$$\sum_{j=1}^{7}\cos\left(\frac{\pi}{3}j\right)=\frac{1}{2}$$

5.

b)

$$\lim_{n\to\infty}\frac{S_n}{n^3}=\frac{1}{3}.$$

7.

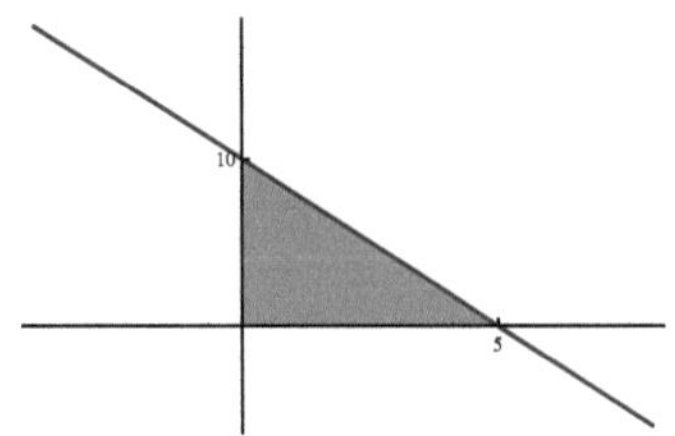

The area of G is

$$\frac{1}{2}(5)(10)=25.$$

$$l_n=\frac{5}{n}\sum_{k=1}^{n}f\left((k-1)\frac{5}{n}\right),$$

where $f(x)=10-2x$.

n	l_n	$|l_n-25|$
8	28.125	3.125
16	26.5625	1.5625
32	25.7813	0.78125
64	25.3906	0.390625

$$r_n=\frac{5}{n}\sum_{k=1}^{n}f\left(k\frac{5}{n}\right)$$

n	r_n	$|r_n-25|$
8	21.875	3.125
16	23.4375	1.5625
32	24.2188	0.78125
64	24.6094	0.390625

$$m_n=\frac{5}{n}\sum_{k=1}^{n}f\left(\left(k-\frac{1}{2}\right)\frac{5}{n}\right).$$

n	m_n	$|m_n-25|$
8	25	0
16	25	0
32	25	0
64	25	0

We see that the left-endpoint sums and right-endpoint sums are not very accurate. Each midpoint sum is the exact value of the area.

9.

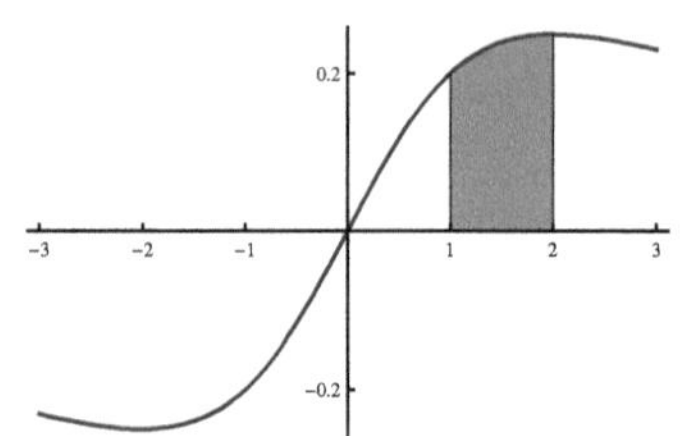

$$m_n=\frac{1}{n}\sum_{k=1}^{n}f\left(1+\left(k-\frac{1}{2}\right)\frac{1}{n}\right)$$

n	m_n	$\left|\frac{3}{2}\ln(2)-\frac{1}{2}\ln(5)-m_n\right|$
8	0.235080	7.81328×10^{-5}
16	0.235021	1.95317×10^{-5}
32	0.235007	4.88284×10^{-6}
64	0.235003	1.22071×10^{-6}

Answers to Some Problems of Section 5.2

1.

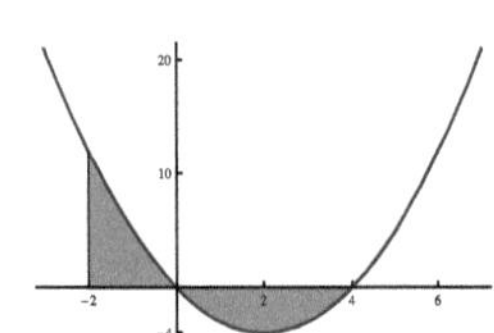

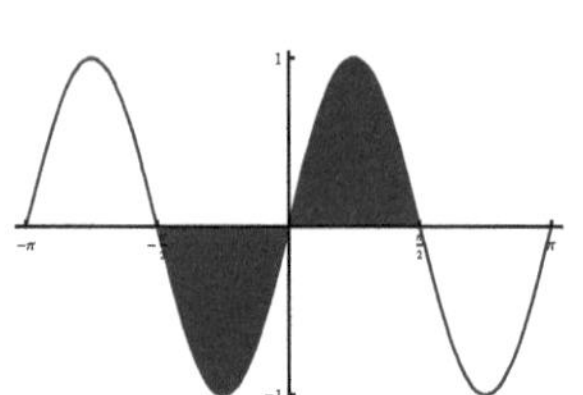

3.

5. Let $f(x) = x^3 - 2x^2 + 1$.

$$m_n = \frac{2}{n} \sum_{k=1}^{n} f\left(1 + \left(k - \frac{1}{2}\right)\frac{2}{n}\right)$$

n	m_n	$\left|\frac{14}{3} - m_n\right|$
8	4.625	4.16667×10^{-2}
16	4.65625	1.04167×10^{-2}
32	4.66406	2.60417×10^{-3}
64	4.66602	6.51042×10^{-4}

7. Let $f(x) = x^2 e^{-x}$.

$$m_n = \frac{1}{n} \sum_{k=1}^{n} f\left(1 + \left(k - \frac{1}{2}\right)\frac{1}{n}\right)$$

n	m_n	$\left|5e^{-1} - 10e^{-2} - m(n)\right|$
8	0.486284	2.39694×10^{-4}
16	0.486104	5.9888×10^{-5}
32	0.486059	1.49698×10^{-5}
64	0.486048	3.74231×10^{-6}

The numbers indicate that such Riemann sums should approximate the integral with desired accuracy, provided that the norm of the partition is small enough.

The numbers indicate that such Riemann sums should approximate the integral with desired accuracy, provided that the norm of the partition is small enough.

Answers to Some Problems of Section 5.3

1. $1 - \dfrac{\sqrt{2}}{2}$

3. $\dfrac{1}{2^{1/4}} - \dfrac{3^{1/4}}{\sqrt{2}}$

5. $\dfrac{35}{3}$

7. $\dfrac{38}{3}$

9. 4

11. $\dfrac{1}{2}$

13. $-\sqrt{2} + \sqrt{3}$

15. $-\ln(2)$

17. 3

19. $-\dfrac{1}{2}$

21. $\dfrac{\pi}{4}$

23. $\dfrac{21}{2}$

Answers to Some Problems of Section 5.4

1.
$$\frac{1}{x^4 + 1}$$

3.
$$-\frac{1}{\sqrt{4 + x^2}}$$

5.
$$\cos\left(\frac{\pi x^2}{2}\right)$$

7.
$$-x^2 \sqrt{x^2 + 1} + 2x^5 \sqrt{x^4 + 1}$$

9. F has a local maximum at $\sqrt{\pi/2}$ and a local minimum at $\sqrt{3\pi/2}$.

11.
$$F\left(\frac{5\pi}{6}\right) - F\left(\frac{\pi}{6}\right)$$

13.
$$F\left(\frac{1}{2}\right) - F\left(\frac{1}{3}\right).$$

Answers to Some Problems of Section 5.5

1.
$$x - \frac{1}{6}x^3 + \frac{1}{120}x^5$$

3.
$$2x^4 + 2x^2 + 2x^3 + 3x$$

5.
$$\frac{1}{6}\sin(6t) + \frac{3}{2}\cos(2t)$$

7..
$$3x^2 + 2x + \ln(|x|)$$

9.
$$\frac{110}{9}$$

11.
$$-\frac{1}{2} - \sqrt{2}$$

13.

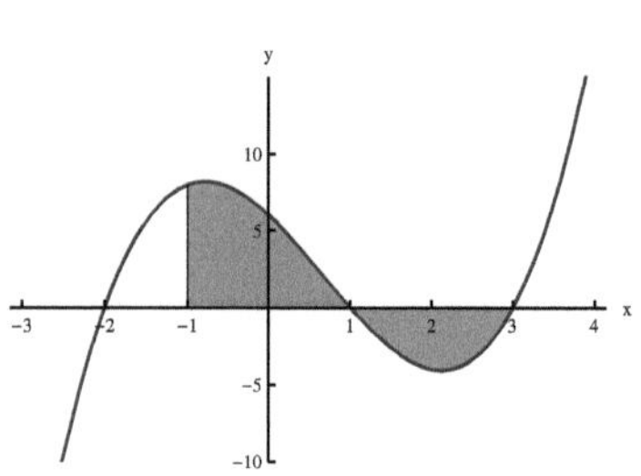

$$\frac{35}{3} - 4\sqrt{6}$$

15.

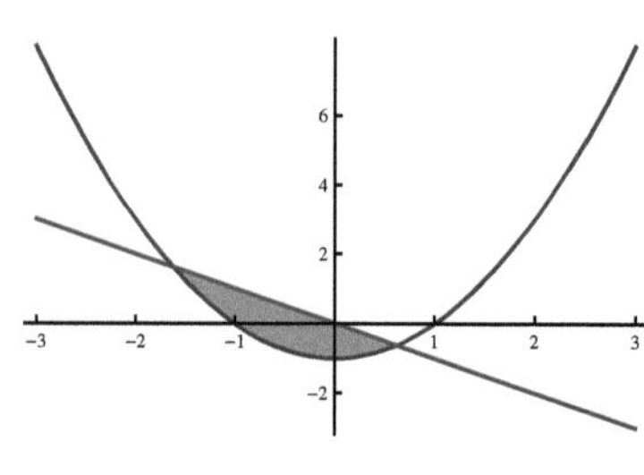

$$\frac{5}{6}\sqrt{5}$$

17.
$$-4\sqrt{2} + 8$$

Answers to Some Problems of Section 5.6

1.
$$\frac{1}{12}\left(x^2 + 4\right)^6$$

3.
$$\frac{2}{9}\left(x^3 + 1\right)^{3/2}$$

5.
$$-\sqrt{4 - x^2}$$

7.
$$\frac{2}{5}\left(x - 16\right)^{5/2} + \frac{32}{3}\left(x - 16\right)^{3/2}$$

9.
$$4\sin\left(\frac{x}{4} + \frac{\pi}{6}\right)$$

11.
$$\frac{1}{5}u^5 = \frac{1}{5}\sin^5\left(x\right)$$

13.
$$\ln\left(|x + 4|\right)$$

15.
$$\frac{\ln^2\left(x\right)}{2}$$

17.
$$2e^{\sqrt{x}}$$

19.
$$\frac{1}{12}\arctan\left(\frac{4}{3}x\right)$$

21.
$$\frac{1}{4}\arcsin\left(4x\right)$$

23.
$$6$$

25.
$$\frac{1}{4}\left(28^{4/3}\right) - \frac{1}{4}\left(9^{4/3}\right)$$

27.
$$-\frac{3\sqrt{3}}{8}$$

29. The integral does not exist.

31.
$$\frac{3}{4}\left(13^{2/3} - 4^{2/3}\right)$$

33.
$$\frac{1}{3} + \frac{1}{3}\left(\frac{3\sqrt{3}}{8}\right)$$

35. The picture shows the graph of velocity:

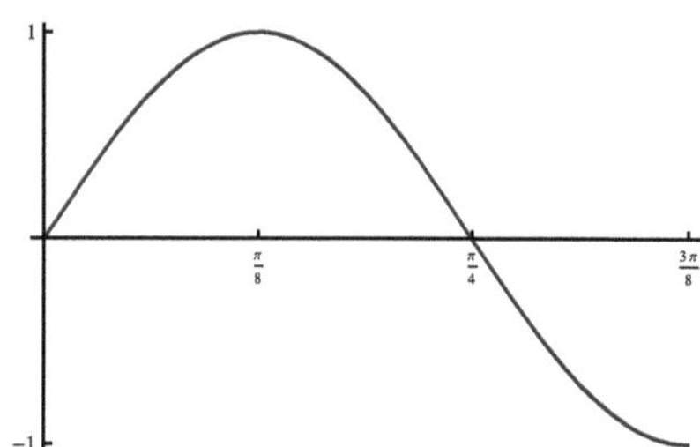

The displacement is

$$\frac{1}{4}$$

The distance traveled is

$$\frac{3}{4}$$

Answers to Some Problems of Section 5.7

1.
$$\sqrt{x^2 + 1} + C$$

3.
$$-3\cos\left(\frac{x}{3}\right) +$$

5.
$$\frac{9}{2} + \frac{1}{2}e^{2x}$$

7.
$$y(t) = 4 + \int_2^t \sin\left(u^2\right) du$$

9.
$$y(t) = -10 + \int_4^t e^{-u^2/2} du$$

11. Velocity is
$$v(t) = \frac{1}{2}\sin(6t) + 10$$

Position is
$$f(t) = -\frac{1}{12}\cos(6t) + 10t + \frac{47}{12} - \frac{5}{3}\pi$$

Answers to Some Problems of A1

1. $x = 2$ or $x = -3$

3. $x = 2 \pm 2\sqrt{2}$

5.
a) $x = 3 \pm \sqrt{2}$
b) $\left(x - 3 - \sqrt{2}\right)\left(x - 3 + \sqrt{2}\right)$

7.
a) $x = 3 \pm 2i$

b) $(x - 3 - 2i)(x - 3 + 2i)$

9.
a) $x = 2$ or 3 or -5
b) $(x - 2)(x - 3)(x + 5)$

11.
a) $x = 1$ or $3 + 2i$ or $3 - 2i$
b) $(x - 2)\left(x^2 - 6x + 13\right)$

Answers to Some Problems of A2

1.
$$\binom{6}{3} = 20.$$

3.
$$\binom{n}{3} = \frac{(n-2)(n-1)(n)}{6}$$

5.
$$(x + h)^4 = x^4 + 4x^3h + 6x^2h^2 + 4xh^3 + h^4$$

7.
$$(4 - h)^3 = 64 - 48h + 12h^2 - h^3$$

9.
$$\frac{(2 + h)^3 - 8}{h} = 12 + 6h + h^2$$

11.
$$\frac{(x + h)^3 - x^3}{h} = 3x^2 + 3xh + h^2$$

Answers to Some Problems of A3

1.
a) $\left\{x \in \mathbb{R} : x < \frac{7}{2}\right\}$
b) $\left(-\infty, \frac{7}{2}\right)$

3.
a) $\left\{x \in \mathbb{R} : 4 < x < 4 + \frac{1}{2}\right\}$
b) $\left(4, 4 + \frac{1}{2}\right)$

5.
a) $\{x \in \mathbb{R} : x > 3 \text{ or } x > -2\}$
b) $(3, +\infty) \cup (-\infty, -2)$

7.
a) $\mathbb{R}$
b) $(-\infty, +\infty)$

9.
a) $\{x \in \mathbb{R} : 5 < x < 11\}$
b) $(5, 11)$

11.
a) $\{x \in \mathbb{R} : -7 \le x \le 9\}$
b) $[-7, 9]$

Answers to Some Problems of A4

1.
a) $1.414\,214$
b) 4.4×10^{-7}

3.
a) $0.166\,667$
b) 3.3×10^{-7}

5.
a) 0.052632

b) 4.2×10^{-7}

7.
a) $22.449\,9$
b) 2×10^{-6}

9.
a) $5.143\,69$
b) 6.4×10^{-7}

Answers to Some Problems of A5

1.

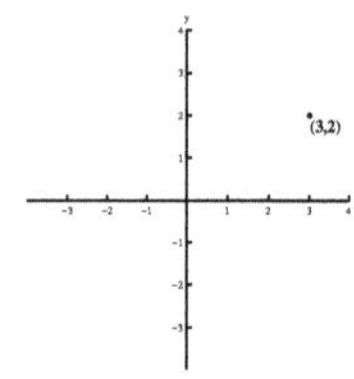

The point is in the first quadrant.

3.

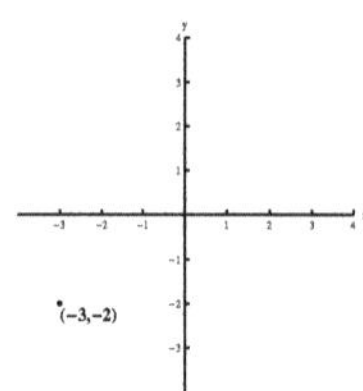

The point is in the third quadrant.

5.

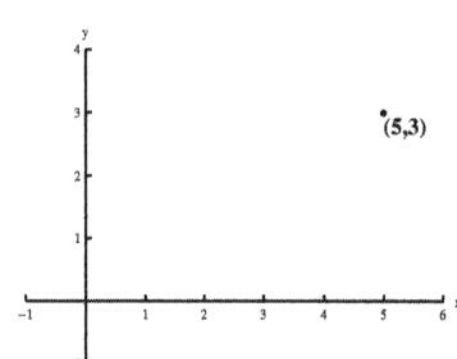

The point is in the first quadrant.

7.

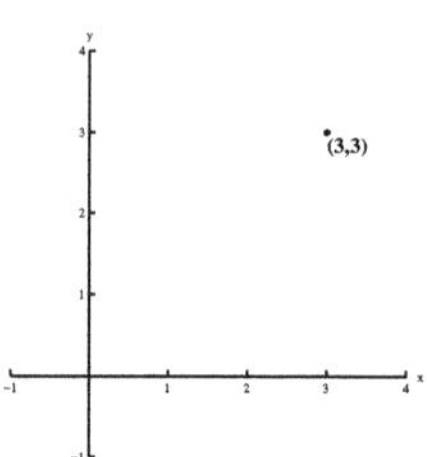

The point is in the first quadrant.

9.
a) 1, $y = 1 + (x - 2)$.
b) $y = x - 1$
c)

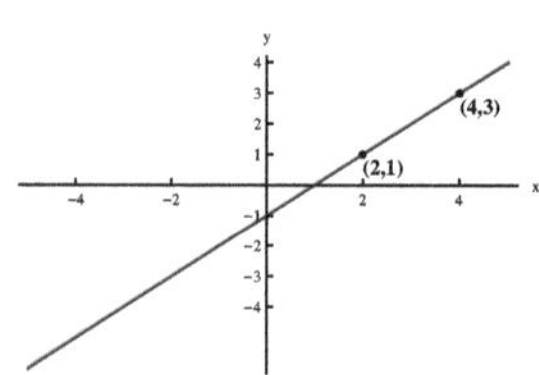

11.
a) $-\dfrac{3}{2}$, $y = -1 - \dfrac{3}{2}(x + 4)$
b) $y = -\dfrac{3}{2}x - 7$
c)

c)

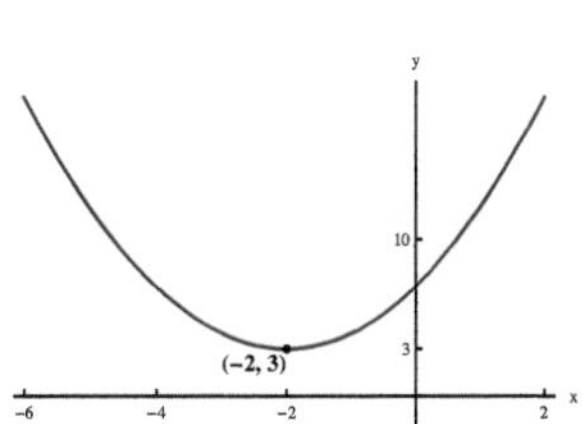

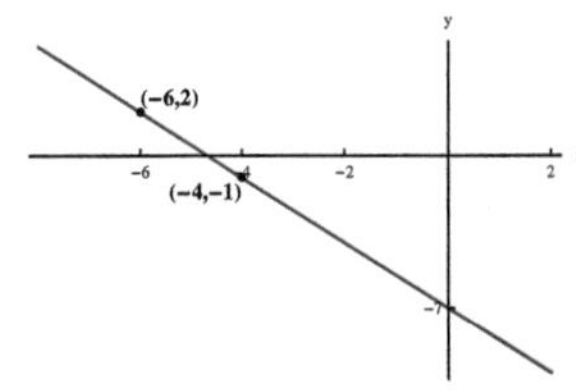

13.

a) $\frac{2}{3}$, $y = 2 + \frac{2}{3}(x + 4)$

b) $y = \frac{2}{3}x + \frac{14}{3}$

c)

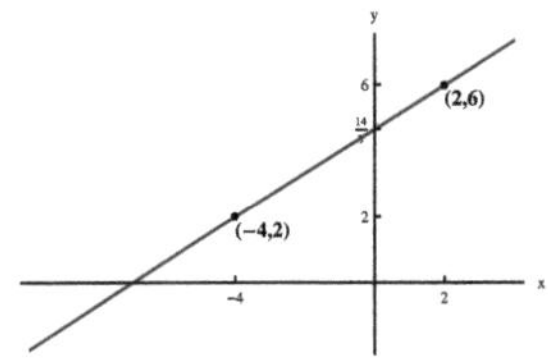

27.

a) $(4, 6)$

b) the parabola intersects the x-axis at $4 + \sqrt{6}$ and $4 - \sqrt{6}$.. It ntersects the y-axis at $(0, -10)$.

c)

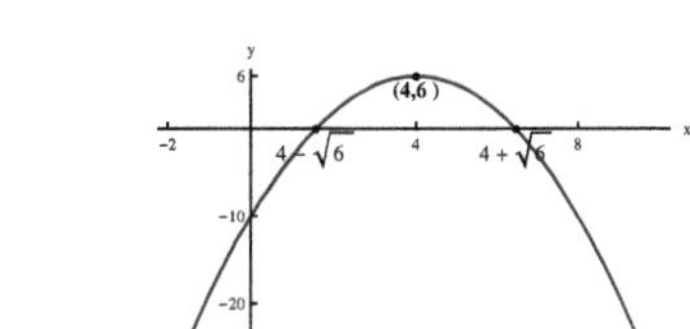

15. $(-5, -11)$

17. Both lines have slope $1/2$, so that they are parallel. Since they do not coincide, they do not intersect.

19.

b) $(-8/17, 53/17)$

c)

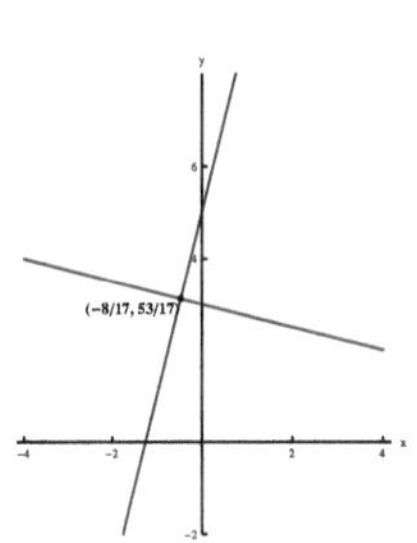

The picture is consistent with the claim that the lines are perpendicular.

21. 5

23. $2\sqrt{10}$

25.

a) $(-2, 3)$

b) The parabola does not intersect the x-axis. It intersects the y-axis at $(0, 7)$.

29.

a) $(-6, 2)$

b) The parabola intersects the x axis at $(-2, 0)$. It intersects the y-axis at $2 + \sqrt{6}$ and $2 - \sqrt{6}$.

c)

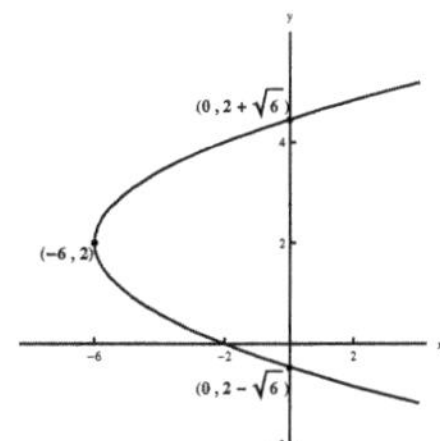

31.

a) $(x - 3)^2 + (y + 1)^2 = 4$. The graph of the equation is a circle of radius 2 centered at $(3, -1)$.

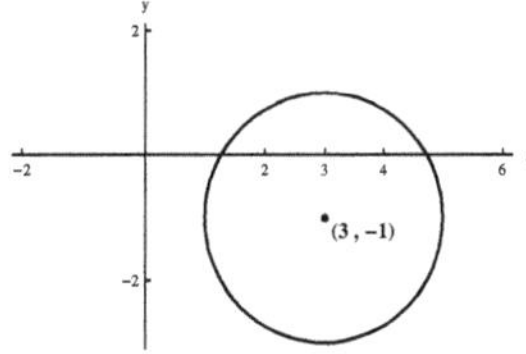

33.

a) $\dfrac{(x-2)^2}{\left(\dfrac{5}{2}\right)^2} + \dfrac{(y+1)^2}{\left(\dfrac{5}{3}\right)^2} = 1.$

The graph of the equation is an ellipse that is centered at $(2, -1)$.

b)

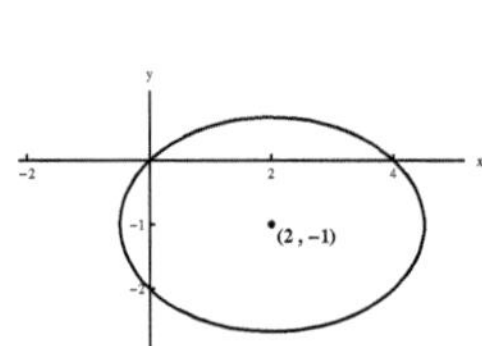

35.

a) $\dfrac{(x-4)^2}{3^2} - \dfrac{(y-3)^2}{5^2} = 1.$

The graph of the equation is a hyperbola that is centered at $(4, 3)$.

b)

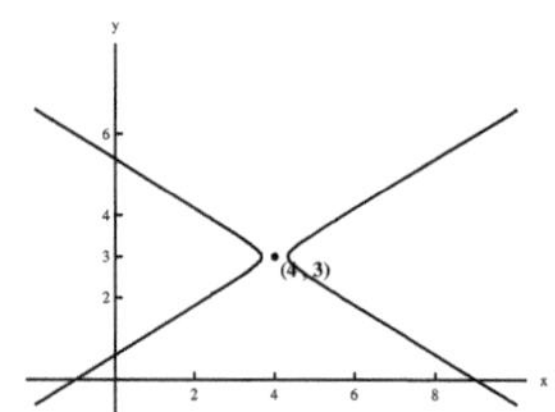

Answers to Some Problems of A6

1.

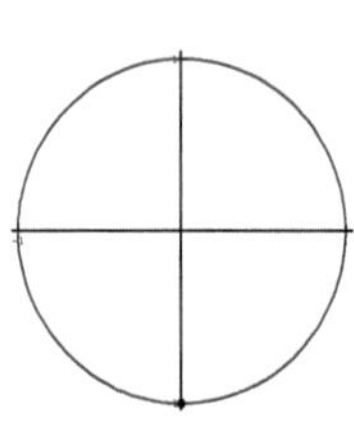

$$\sin\left(\frac{3\pi}{2}\right) = -1, \quad \cos\left(\frac{3\pi}{2}\right) = 0$$

3.

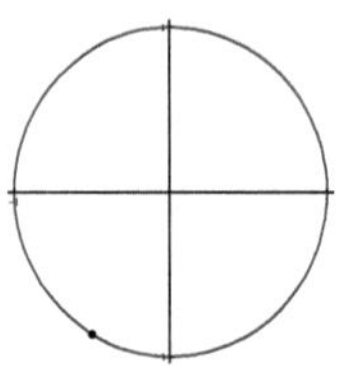

$$\sin\left(\frac{4\pi}{3}\right) = -\frac{\sqrt{3}}{2},$$
$$\cos\left(\frac{4\pi}{3}\right) = -\frac{1}{2}$$

5.

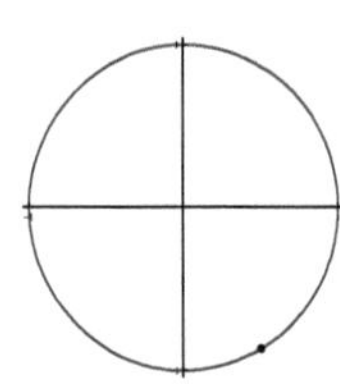

$$\sin\left(-\frac{\pi}{3}\right) = -\frac{\sqrt{3}}{2}, \quad \cos\left(-\frac{\pi}{3}\right) = \frac{1}{2}$$

7.

$$\sin\left(\frac{3\pi}{4}\right) = \frac{\sqrt{2}}{2}, \quad \cos\left(\frac{3\pi}{4}\right) = -\frac{\sqrt{2}}{2},$$
$$\tan\left(\frac{3\pi}{4}\right) = -1, \quad \sec\left(\frac{3\pi}{4}\right) = -\sqrt{2}$$

9.

$$\sin\left(\frac{7\pi}{2}\right) = -1, \quad \cos\left(\frac{7\pi}{2}\right) = 0,$$

$\tan\left(\dfrac{7\pi}{2}\right)$ and $\sec\left(\dfrac{7\pi}{2}\right)$ are undefined.

11.

$$\sin\left(\frac{7\pi}{6}\right) = -\frac{1}{2}, \quad \cos\left(\frac{7\pi}{6}\right) = -\frac{\sqrt{3}}{2},$$
$$\tan\left(\frac{7\pi}{6}\right) = \frac{\sqrt{3}}{3}, \quad \sec\left(\frac{7\pi}{6}\right) = -\frac{2\sqrt{3}}{3}$$

13.

$$\sin\left(-\frac{2\pi}{3}\right) = -\frac{\sqrt{3}}{2}, \quad \cos\left(-\frac{2\pi}{3}\right) = -\frac{1}{2},$$
$$\tan\left(-\frac{2\pi}{3}\right) = \sqrt{3}, \quad \sec\left(-\frac{2\pi}{3}\right) = -2.$$

15.

$$\cos(4x) = \cos^4(x) + \sin^4(x) - 6\cos^2(x)\sin^2(x)$$

17.

$$\cos(3x) = \cos^3(x) - 3\sin^2(x)\cos(x)$$

Appendix G

Basic Derivatives and Integrals

Basic Differentiation Formulas

1. $\dfrac{d}{dx}x^r = rx^{r-1}$

2. $\dfrac{d}{dx}\sin(x) = \cos(x)$

3. $\dfrac{d}{dx}\cos(x) = -\sin(x)$

4. $\dfrac{d}{dx}\sinh(x) = \cosh(x)$

5. $\dfrac{d}{dx}\cosh(x) = \sinh(x)$

6. $\dfrac{d}{dx}\tan(x) = \dfrac{1}{\cos^2(x)}$

7. $\dfrac{d}{dx}a^x = \ln(a)\,a^x$

8. $\dfrac{d}{dx}\log_a(x) = \dfrac{1}{x\ln(a)}$

9. $\dfrac{d}{dx}\arcsin(x) = \dfrac{1}{\sqrt{1-x^2}}$

10. $\dfrac{d}{dx}\arccos(x) = -\dfrac{1}{\sqrt{1-x^2}}$

11. $\dfrac{d}{dx}\arctan(x) = \dfrac{1}{1+x^2}$

Basic Antidifferentiation Formulas

C denotes an arbitrary constant.

1. $\displaystyle\int x^r\,dx = \dfrac{1}{r+1}x^{r+1} + C \;\;(r \neq -1)$

2. $\displaystyle\int \dfrac{1}{x}\,dx = \ln(|x|) + C$

3. $\displaystyle\int \sin(x)\,dx = -\cos(x) + C$

4. $\displaystyle\int \cos(x)\,dx = \sin(x) + C$

5. $\int \sinh(x)\,dx = \cosh(x) + C$

6. $\int \cosh(x)\,dx = \sinh(x) + C$

7. $\displaystyle\int e^x\,dx = e^x + C$

8. $\displaystyle\int a^x\,dx = \dfrac{1}{\ln(a)}a^x + C \;\;(a > 0)$

9. $\displaystyle\int \dfrac{1}{1+x^2}\,dx = \arctan(x) + C$

10. $\int \dfrac{1}{\sqrt{1-x^2}}\,dx = \arcsin(x) + C$

535

Index

Printed by Libri Plureos GmbH in Hamburg,
Germany